IB	IIB	IIIA	IVA	VA	VIA	VIIA	0
							2 **He** 4.00
		5 **B** 10.81	6 **C** 12.01	7 **N** 14.01	8 **O** 16.00	9 **F** 19.00	10 **Ne** 20.18
		13 **Al** 26.98	14 **Si** 28.09	15 **P** 30.97	16 **S** 32.06	17 **Cl** 35.45	18 **Ar** 39.95
29 **Cu** 63.55	30 **Zn** 65.37	31 **Ga** 69.72	32 **Ge** 72.59	33 **As** 74.92	34 **Se** 78.96	35 **Br** 79.90	36 **Kr** 83.80
47 **Ag** 107.87	48 **Cd** 112.40	49 **In** 114.82	50 **Sn** 118.69	51 **Sb** 121.75	52 **Te** 127.60	53 **I** 126.90	54 **Xe** 131.30
79 **Au** 196.97	80 **Hg** 200.59	81 **Tl** 204.37	82 **Pb** 207.19	83 **Bi** 208.98	84 **Po** 210	85 **At** 210	86 **Rn** 222

64 **Gd** 157.25	65 **Tb** 158.92	66 **Dy** 162.50	67 **Ho** 164.93	68 **Er** 167.26	69 **Tm** 168.93	70 **Yb** 173.04	71 **Lu** 174.97

96 **Cm** 247	97 **Bk** 247	98 **Cf** 251	99 **Es** 254	100 **Fm** 253	101 **Md** 256	102 **No** 255	103 **Lw** 257

chemistry and biochemistry: a comprehensive introduction

McGraw-Hill Book Company

New York
St. Louis
San Francisco
Düsseldorf
London
Mexico
Panama
Sydney
Toronto

chemistry and biochemistry: a comprehensive introduction

A. Leslie Neal

Associate Professor of Biochemistry
and Molecular Biology
Cornell University

chemistry and biochemistry: a comprehensive introduction

Library of Congress Catalog Card Number 70-98053

ISBN 07-046135-X

567890 HDBP 754

This book was set in News Gothic by Graphic Services, Inc., printed on permanent paper by Halliday Lithograph Corporation, and bound by The Book Press, Inc. The designer was Marsha Cohen; the drawings were done by BMA Associates, Inc. The editors were James L. Smith, Nancy L. Marcus, and Janet Wagner. Stuart Levine supervised production.

preface

In the past two decades biochemistry has come to be appreciated as a basic science, like chemistry or physics. It is no longer of interest only to the specialist: a basic knowledge of biochemistry can enrich anyone's appreciation of his environment and ability to cope with it.

A knowledge of biochemistry does, however, require some knowledge of basic principles of general and organic chemistry. Previously, only students who were science majors and had several preliminary chemistry courses could be expected to handle a course in biochemistry. This need not be the case. In this text, selected principles of general and organic chemistry presented in the first two sections enable the student to understand the biochemistry presented in the third section. This text represents a one-year course in introductory chemistry and biochemistry for the nonscience major as well as for the major in sciences.

Part One of the text serves as an introduction to general chemistry. It includes a discussion of matter and energy, then defines an atom, emphasizing topics necessary for an understanding of biochemical compounds and reactions. A discussion of chemical bonds and reactions is highly selective so that the student is *not* required to memorize material that does not apply to later discussions. Finally, a discussion of solutions and equilibrium provides an introduction to the practical principles of concentrations and acid-base titrations.

Part Two outlines those aspects of organic chemistry which are necessary for an understanding of the biochemistry in Part Three. Carbon is discussed in detail, with emphasis on electron configuration and structural characteristics which are important in biochemical compounds. The functional classes of organic compounds are discussed next, each in a separate chapter. For each class, reactions, preparations, and structure are illustrated with examples which the student again encounters in Part Three. Part Two serves as an introduction to material which the student uses again but which, when presented in Part Three, is considered as a part of a biochemical reaction or series of reactions. The science becomes more meaningful and more familiar with use. Early and frequent study of the individual reactions makes the more complex study of biochemical pathways a review or, at least, a familiar concept for the student. It has been the author's experience over a number of years that an early introduction to polyfunctional compounds is necessary.

A discussion of intermolecular forces completes Part Two and forms a natural transition for the discussion of biochemistry which follows.

Part Three, the largest section of the book, applies the principles presented in Parts One and Two. Sufficient material is presented to offer a challenge to any student, but the units are arranged so that details can be deleted by an instructor without sacrificing the basic principles. Thus the contents of the course can be tailored to meet the needs of a variety of students.

Color and screening have been used throughout the book as a teaching device. For example, in the Krebs cycle schematic, those carbons which enter the cycle are in a second color so that the student readily focuses on their fate. Since the important pathways are presented both as schematics and as individual reactions, the student has overall perspective as well as detail. Review questions are formulated at the beginning of many chapters to help focus attention on the most important principles. At the end of each chapter a series of diagnostic problems and questions enables the student and the instructor to assess proficiency.

This text represents the summation of a number of years' experimentation with an introductory biochemistry course taught at Cornell University. In a sense, the students helped to design the text; certainly they tested it.

The author wishes to express his appreciation to the many students who inspired the text. The assistance of Dr. John Stezowski and Dr. Joan Griffith is gratefully acknowledged. The patience and encouragement of my wife and family during the writing of this book are deeply appreciated.

A. Leslie Neal

contents

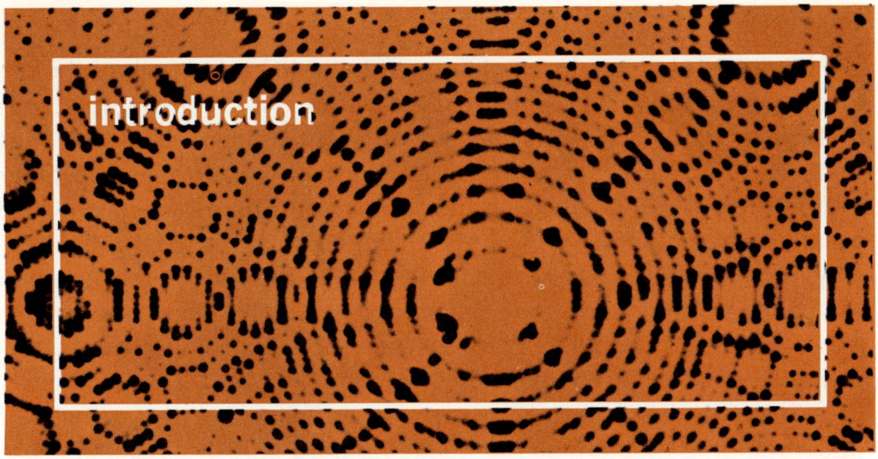

introduction

Although it is a relatively new discipline, biochemistry now constitutes a foundation of the biological sciences. Biochemistry is the specific field of chemistry that deals with the transformations of matter brought about by plants, animals, or microorganisms.

Until recently biochemistry dealt only with the ultimate effects of molecules upon intact animals or organisms, e.g., the effect of vitamins and other dietary substances on the animal's growth rate. Today, besides observing the whole organism, biochemists delve into the innermost secrets of individual cells to study the processes whereby the cell makes its cellular constituents and thus is able to stay alive, increase its size, and multiply. Some cells (bacteria) are able to duplicate themselves exactly within a matter of minutes, whereas others require a longer time. If a living cell in the proper environment is given the necessary starting materials, it is able to make all the thousands of cellular constituents necessary to produce a new cell that is an exact duplicate.

Biochemistry is the study of the dynamic state of matter that is always associated with life. Life is never a stationary process; when the dynamic processes of the cell stop, death results. One might imagine that when a cell component has been formed, it is in place for the rest of the life of the cell, like a brick in a wall. On the contrary, the components of every cell are always being broken down and replaced by new components. Even the constituents of bone undergo constant change. Old material is continuously replaced by new material.

Building new cells and repairing old ones is *not* a haphazard operation. A controlling mechanism must be in force during the rebuilding or duplicating stages if the cells are not to grow out of all proportion to their normal size or to multiply too rapidly (which is what cancer cells do). Also, a master plan must be in force for each of the different types of cell to direct its respective functions. For example, cells that make up the fingernail have a different function from the ones that make up hair; the red blood cells, which carry oxygen to other cells, differ in composition and function from the nerve cells, which transfer electric impulses; the plant cells that can transform light energy into chemical energy are not identical to the cells in the rods and the cones of the eye that are capable of converting light energy into nerve impulses during the process of vision. Explaining these phenomena is part of biochemistry.

Since biochemistry incorporates other fields of study, some basic understanding of chemistry is needed before studying biochemistry, and for this reason the text is divided into three main sections: Part One, General Chemistry; Part Two, Organic Chemistry; and Part Three, Biochemistry.

Because it would not be possible to treat each phase of chemistry extensively in a book of this size, selected principles and reactions are discussed. Many details have been omitted to allow more emphasis on general concepts. The aim is not to make chemists out of students overnight but to provide a basic understanding of the principles of chemistry as they apply to biological systems. The chemistry in this course has immediate and direct application to other fields of study, e.g., botany, zoology, or nutrition.

part one… general chemistry

Biochemistry is the study of chemical reactions which occur in living organisms. Most people are aware of many of the benefits of chemistry in such industries as wire manufacturing, housing, heating, transportation, and food processing. However, the use of chemicals is not limited to external application: all life depends on the incorporation of particular chemicals into the organic system. The necessary chemicals range in complexity from proteins to simple mineral elements like iron and copper. An adult should ingest approximately 1/100,000 oz of copper and 1/10,000 oz of iron per day to keep his body functioning properly.

Your body is essentially a chemical factory taking in raw material in the form of food and changing it into the chemicals necessary for proper growth, maintenance, and reproduction. Your body comprises a group of organized, interrelated chemical reactions, all of which are essential for normal life. An understanding of biochemistry will help you understand your own personal manufacturing plant and that of all other organisms. Biochemistry is your chemistry.

All branches of chemistry are relevant to biochemistry. They are:

Inorganic: the chemistry of compounds in nonliving substances such as minerals
Analytical: the quantitative and qualitative evaluation of a substance
Organic: the chemistry of compounds of carbon
Physical: the measurement of properties and behavior of substances

In Part One, you will become familiar with some basic principles of chemistry which will be your tools for understanding biochemistry.

A field-ion micrograph illustrating the atomic structure of a metal surface. Each single-image dot is an iridium atom. (Courtesy of Erwin W. Mueller, Pennsylvania State University.)

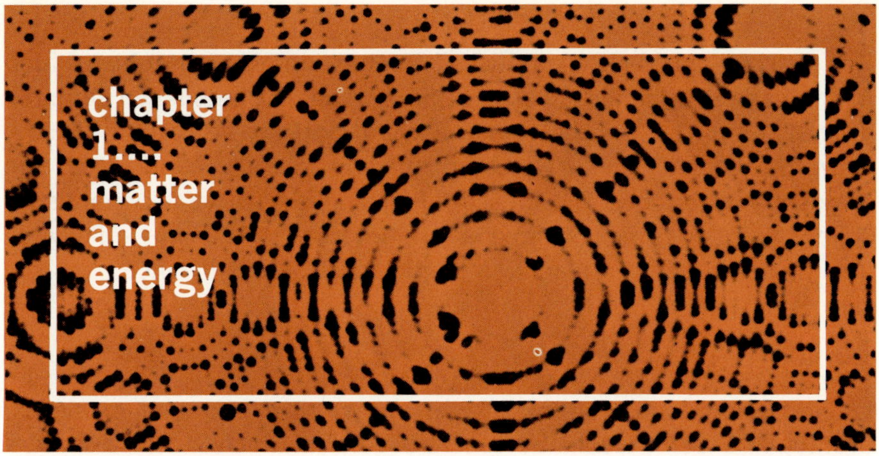

chapter 1.... matter and energy

The study of the molecules and reactions in chemical systems involves the study of matter. What is matter? Is it something that can be seen or felt? Must matter always have weight? What properties does it have? What is energy? Suitable answers must be found to these questions before attempting to use the concept of matter in our study of biochemistry.

1.1 MATTER

All objects which exist in our world, visible or not, are matter. It is not always possible to visualize the concepts involved in chemical reactions. The scientist frequently postulates a model which can be visualized in order to test his theories. Air is an example of matter which is not visible. Although air cannot be seen, we know it is there.

One way to make the presence of air known is to cause it to move; the movement will then be visible *through a reaction* such as a windmill turning. If we examine the components of air by chemical methods, we find that there is 0.03% (by volume) carbon dioxide and 21% (by volume) oxygen. As shown in Table 1.1, there are also other components in air.

TABLE 1.1 APPROXIMATE COMPOSITION OF DRY AIR

Constituent	Percent
Nitrogen	78
Oxygen	21
Argon	0.9
Carbon dioxide	0.03
Hydrogen	0.01
Neon, helium, krypton	0.002

Carbon dioxide is a product of combustion or the burning of fuel. In the case of animals, this fuel can be sugar. Combustion requires oxygen (Chap. 8). Animals inhale oxygen and exhale carbon dioxide. How the ratio of these gases remains reasonably constant in the atmosphere will become clearer as you learn more biochemistry. Briefly, plants use carbon dioxide for photosynthesis, giving off oxygen as a by-product of the reaction. Life depends on this balance.

1.2 COMPONENTS OF MATTER

If one could divide a chunk of iron into smaller and smaller pieces, eventually a point would be reached beyond which it would not be possible to obtain a smaller piece that would still be iron. If one started with approximately $\frac{1}{2}$ in.^{3} of iron and divided it into 6.02×10^{23} equal pieces, the ultimate division of iron would be reached. This unit of iron is an *atom.* The smallest unit that any element could be divided into without loss of chemical activity is an *atom.*

All matter is composed of atoms, but all atoms are not alike. (Details of atomic structure will be discussed in Chap. 2.) Matter is composed of agglomerates of atoms held together in particular ways so that a variety of physical states is possible. Matter can be liquid, gaseous, or solid. Matter cannot be created or destroyed: this basic law is known as the *law of conservation of matter.* Although matter undergoes transformations in state (see Sec. 1.5) and composition, it is *never* totally destroyed or newly created. The law of conservation of matter holds for every chemical reaction. The validity of this law is illustrated throughout our discussion of biochemistry.

1.3 UNITS OF MATTER

The smallest unit of matter having a unique identity is the atom. A substance which has only one kind of atom is an *element.* Examples of elements are iron, copper, and oxygen. Scientists abbreviate the names of elements using symbols which have been adopted by convention. It is impossible to learn chemistry without first becoming familiar with some chemical symbols. A partial list containing the most frequently encountered symbols appears in the inside of the back cover of this book. The periodic table in the front of this book lists all the symbols. Familiarize yourself with the symbols and begin to use them.

Matter which contains more than one kind of atom *chemically* united in definite proportions is a *compound*. The smallest unit which retains all the properties of the compound is called a *molecule*. Each compound has a *fixed ratio* of the component atoms which are bonded to one another. The atoms cannot be separated without a chemical reaction to break these bonds. Bonds are discussed in Chap. 3.

Matter which contains more than one kind of nonbonded atom or molecule and no definite proportion of components is a *mixture*. If a group of people was asked to bring in a mixture of salt and sugar, each person would probably have a sample with a different amount of salt per unit of sugar. Each would have a mixture, although one might be 1 gram (g) of salt to 10 g of sugar and another 10 g of salt to 0.05 g of sugar. A mixture can be resolved into its components by physical means, *not involving a chemical reaction*. This means that if one were patient enough, one could manually separate the grains of salt from the grains of sugar. A compound can *never* be separated into its component atoms by purely physical means.

1.4 PHYSICAL STATES OF MATTER

As we suggested earlier, matter commonly occurs in three states, solid, liquid, and gas (Fig. 1.1).

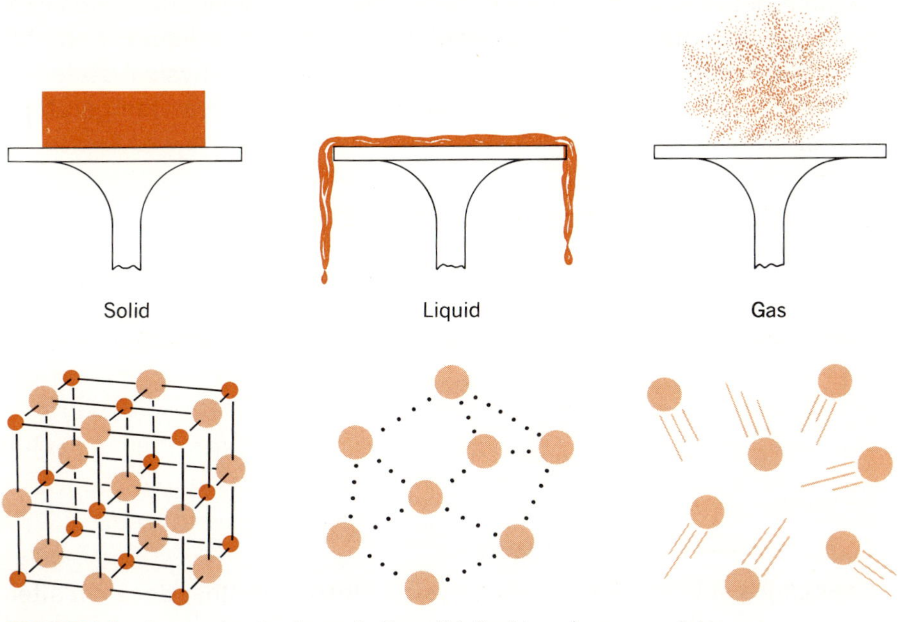

FIGURE 1.1 Intermolecular forces in the solid, liquid, and gaseous states.

Solids have a definite shape and volume. The shape can change, but it will always have the same volume. For example, when 1 in.^{3} of aluminum is rolled into a thin sheet, the shape is different but the volume is the same.

Liquids have a definite volume but take the shape of the container which holds them. Eight ounces of milk (about 240 cm^3) can be contained in a square or round jar, in a glass, or in a bucket; although the shape of the container differs, the amount of liquid stays the same.

Gases take both the shape and the volume of the container which holds them. A gas spreads out into any volume which encloses it.

The particles of gas in a confined volume move rapidly within their confinement regardless of the presence of other gases. This particle movement is called *diffusion.* If a rotten egg is broken at the kitchen sink, it is only a matter of minutes before the odor is noticeable in the farthest corner of the room. The heavier the particles of gas the slower the rate of its diffusion. Liquids diffuse much slower than gases because of the greater attraction between particles, and solids do not diffuse appreciably.

The behavior of the states of matter is explained by the forces existing between atoms in each state. The interatomic attractions in a solid are obviously much stronger than those in a liquid, and the forces in a liquid are stronger than those in a gas. Physical states of matter change according to external conditions, such as temperature and pressure.

1.5 INTERCONVERSION OF PHYSICAL STATES

The commonest example of the change of physical states of matter is

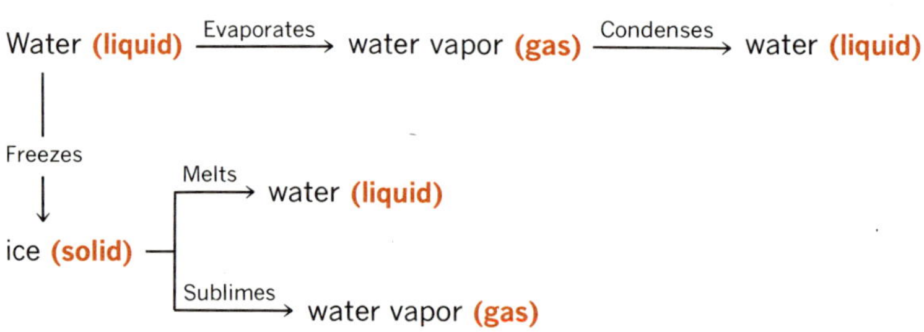

(1.1)

Changing a substance from one physical state to another does *not* alter its chemical composition. This is purely a physical change. Some common terms and their meanings follow:

Evaporation: The conversion of a liquid to a gas usually occurs as the temperature is increased or the pressure is decreased.

Condensation: Since this is the reverse of evaporation, it requires that the temperature be decreased or that the pressure be increased, or both, so that the gas becomes a liquid.

Freezing: Solidification of a liquid is usually observed as the temperature is decreased.

Melting: The reverse of solidification occurs when the temperature of the solid is increased and it becomes a liquid.

Sublimation: A solid can vaporize *directly* to a gas under special conditions without passing through a liquid stage.

Sublimation is a term less familiar than freezing or melting but many of you have witnessed the sublimation of ice, perhaps without realizing it. When laundry is hung outside to dry in temperatures below the freezing point of water, the water in the clothes freezes. Allowed to hang on the line long enough, the clothes eventually dry even though the temperature of the air remains below the freezing point of water. The process is relatively slow, but it does occur. Sublimation of ice is used to dry biological materials which would be inactivated by heating and therefore cannot be dried by boiling off the water. The temperature can be kept at, or slightly below, the freezing point during the entire drying operation if the material is dried by sublimation. The apparatus to do this subjects the frozen sample to a very high vacuum, and the water vapor is removed from the system as the ice sublimes. The term *lyophilization* or *freeze drying* has been applied to this process. This process is extremely important as a biochemical technique for storing biological samples which are unstable.

Carbon dioxide gas can be converted into a solid by subjecting it to a high pressure and low temperature. The Dry Ice thus formed gets its name from the fact that it sublimes instead of melting and forming a liquid phase. Another substance that sublimes readily is iodine. When iodine crystals are gently heated, they sublime, and if the vapor is allowed to come into contact with a cool surface, it changes directly to solid iodine crystals.

All these transformations are summarized in Eq. (1.1). Each transformation takes place under specific external conditions which vary with each substance considered. A fourth state of matter recently elucidated is the *liquid crystal,* an apparently highly ordered state that certain compounds can form between the solid state and melting. These compounds are useful in cancer research and analysis because they are particularly sensitive to minute changes in temperature.

1.6 ENERGY

Energy may be defined as the *ability to do work.* Different kinds of energy have the ability to do different kinds of work. A law similar to that for the conservation of matter applies to energy. The *law of conservation of energy* states that energy may neither be created nor destroyed. It can, however, be transformed from one type to another.

1.7 TYPES OF ENERGY

Kinetic energy is the ability to do work as a result of motion. For example, driving a nail into a piece of wood requires kinetic energy on the part of the hammer. The energy of the moving hammer is transferred to the nail, which is driven into the wood. The faster the hammer is moving, the more kinetic energy it has and the further the nail is driven into the wood. The kinetic energy of the hammer is not lost; it is transferred to the nail and dissipated in overcoming the friction between the wood and the surface of the nail.

An object's kinetic energy depends upon its mass and velocity. Two objects of different weight moving at the same speed have different kinetic energy; the heavier the object the more kinetic energy it has. It is possible to stop a hard-hit baseball with relative ease, but to stop a 100-lb cannonball traveling at the same rate of speed (velocity) as the baseball would require something more than a baseball glove. Kinetic energy depends upon the mass and velocity

$$k_e = \tfrac{1}{2}mv^2 \tag{1.2}$$

where m is mass and v is velocity. The units of energy depend upon the units used to express m and v. If m is in pounds and v in feet per second, k_e is expressed in foot-pounds. If the mass is in grams and v in centimeters per second, k_e is expressed in ergs.

EXAMPLE 1

Given that the mass of hydrogen gas, H_2, is 2 and that of helium gas, He, is 4, determine the ratio of the kinetic energy of these gases when their velocities are identical.

To solve, substitute the values into Eq. (1.2).

$$k_{e,\mathrm{He}} = \tfrac{1}{2} \times 4 \times v_{\mathrm{He}}{}^2$$
$$k_{e,\mathrm{H_2}} = \tfrac{1}{2} \times 2 \times v_{\mathrm{H_2}}{}^2$$

The ratio then is

$$\frac{k_{e,\mathrm{H_2}}}{k_{e,\mathrm{He}}} = \frac{\frac{1}{2} \times 2 \times v_{\mathrm{H_2}}{}^2}{\frac{1}{2} \times 4 \times v_{\mathrm{He}}{}^2} = \frac{2}{4} = \frac{1}{2}$$

since $v_{\mathrm{H_2}}{}^2 = v_{\mathrm{He}}{}^2$.

The helium gas has twice as much energy as hydrogen gas under these conditions.

EXAMPLE 2

If hydrogen, H_2, and hydrogen sulfide, H_2S, have the same kinetic energy, which has the greater velocity? How much faster will it diffuse than the slower one?

To solve the first part one must know that the mass of hydrogen is 2 and of hydrogen sulfide is 34. Substituting these values in Eq. (1.2) and letting subscript 1 refer to hydrogen and subscript 2 refer to hydrogen sulfide, we have

$$\frac{k_{e_1}}{k_{e_2}} = \frac{\frac{1}{2} \times 2 \times v_1{}^2}{\frac{1}{2} \times 34 \times v_2{}^2} = \frac{1 \times v_1{}^2}{17 \times v_2{}^2}$$

Since $k_{e_1} = k_{e_2}$,

$$1 \times v_1{}^2 = 17 \times v_2{}^2$$

or

$$\frac{v_1{}^2}{v_2{}^2} = \frac{17}{1} \qquad \frac{v_1}{v_2} = \frac{\sqrt{17}}{\sqrt{1}} = \frac{4.16}{1} = 4.16$$

The velocity of hydrogen is 4.16 times greater than the velocity of hydrogen sulfide. Hydrogen diffuses 4.16 times faster than hydrogen sulfide since the rate of diffusion is directly proportional to velocity.

Potential energy is the ability to do work by virtue of relative position. A piece of chalk on a table has potential energy because it is pulled by the force of gravity. If the table were not holding it up, the chalk would be set in motion and would fall further toward the center of the earth. Once the chalk starts to fall, its *potential* energy becomes *kinetic* energy. If it hits the floor, the kinetic energy will break the chalk. The pieces of chalk on the floor have potential energy in that they will respond to the force of gravity if the floor gives way.

Water in a reservoir has potential energy. When the water is allowed to flow through a pipe in the bottom of the dam and over a turbine, the potential energy of the water is converted into kinetic energy, which in turn can be used to make a turbine rotate. The water turbine may be connected to a machine to do mechanical work or to a generator to produce electric energy.

Electric energy is the result of the kinetic energy, or movement, of electrons. In a certain sense, electric energy may be likened to energy derived from water flowing through a pipe; the amount of work which can be done depends upon the pressure (force per unit area) of the water and the amount of water flowing through the pipe. The amount of electric work depends upon the pressure of electrons, or voltage, and the amount of electrons passing a given point per unit of time, amperes. The watt is a unit of electric power:

$$\text{Watts} = \text{volts} \times \text{amperes} \tag{1.3}$$

The energy inherent in a substance because of its chemical structure and composition is *chemical* energy. When plants use carbon dioxide, water, and light energy for photosynthesis and give off oxygen, the light energy is stored as chemical energy. When wood is burned or starch or glucose utilized by the body, the chemical energy stored in these plant substances is released and transformed into another kind of energy.

Energy may appear in the form of *heat* or *light,* e.g., conversion of electric energy into heat and light in an electric light bulb. When wood or coal is burned, the chemical energy is converted into heat and light energy and the heat can be used to generate steam for doing mechanical work. Our body converts some of the chemical energy from food to heat to maintain body temperature, but unlike the steam engine it cannot use heat as a source of energy for work. In living organisms, any energy converted into heat is lost as far as useful work is concerned. If this were not so, instead of eating we would merely have to go into a warm room to get recharged with energy.

1.8 INTERCONVERSION OF ENERGY

Living organisms illustrate the conversion of one form of energy into another. Every cell requires energy to maintain the organization within the cell. A continued supply of the right kind of energy is therefore a prerequisite to the well-being of the cell. Living organisms cannot utilize heat energy, and nonphotosynthesizing organisms, i.e., nonplants, must obtain chemical energy from food. The conversion of chemical energy from food to high-energy-containing compounds that can be used subsequently for body functions involves a series of chemical reactions in the cells. The ability of different cells to transport electrons or to build an electric potential is quite varied. For example, nerve impulses are propagated by a slight difference in potential along nerve fibers. Compare these cells with those of the electric eel which can store electrons and release them in quantities large enough to produce a sizable shock.

Cold light produced by the firefly is a good example of conversion of electron flow, or electric energy, into light energy unaccompanied by a large evolution of heat. Sight and hearing involve the conversion of light and mechanical energy into chemical and electric energy.

1.9 SUMMARY

All living organisms are made up of aggregates of matter, which may be in one of several physical states. The relationship between the components of a living organism requires knowledge of the properties and chemical reactions which living matter undergoes. Transformations of matter from one state to another involve a change of energy, which is also found in a variety of forms. Neither matter nor energy can be created or destroyed. Life can therefore be viewed as a dynamic system of transformations and reactions.

QUESTIONS

1. Define:

Element	Energy
Compound	Potential energy

2. Explain the difference between a compound and a mixture.

3. Does the fineness of division of a substance affect its physical state of matter? Give an example.

4. What is meant by sublimation? Diffusion?

5. Which object has the greater kinetic energy?

(*a*) Mass = 150 g; velocity = 2×10^2 cm/sec

(*b*) Mass = 500 g; velocity = 1×10^2 cm/sec

6. Discuss the importance of the following conversions, including as many biological examples as possible:

(*a*) Chemical energy into electric energy

(*b*) Chemical energy into mechanical energy

(*c*) Mechanical energy into electric energy

(*d*) Light energy into electric energy

7. What kinds of energy are involved in each of the following examples?

(*a*) A flower blooming (*b*) A match burning

(*c*) Reading (*d*) Hearing

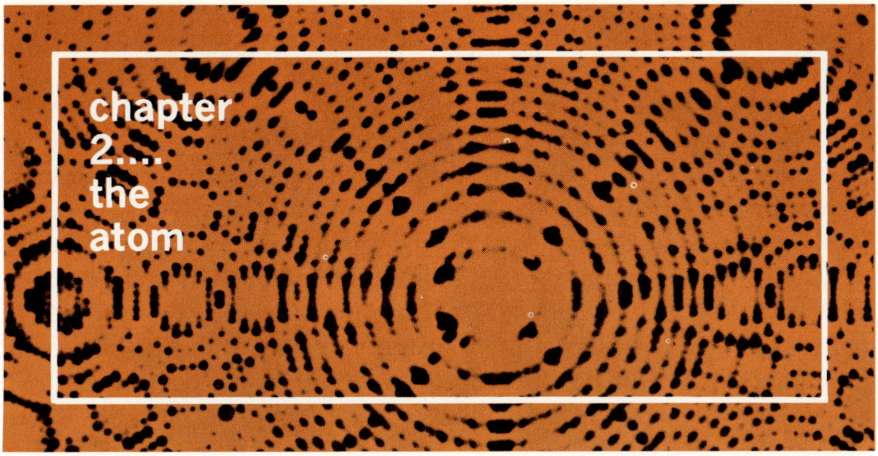

Atoms and the laws which govern the atomic weight of atoms or molecules are discussed in this chapter. Since an atom is too small to be weighed on the most sensitive balance available, how are atomic weights obtained? Review the definition of element and atom given in Chap. 1. What symbols are used to indicate the names of atoms?

2.1 THE ATOM: THE SYMBOL

Only the first letter of the symbol is capitalized. The symbol is an abbreviated way of writing the element. As you become more familiar with the elements, the full significance of the symbols will become clear. It is essential to be able to identify the element by its symbol. See the table in the back cover of this book.

2.2 WEIGHT OF AN ATOM: ATOMIC WEIGHT

If you had a balance that could weigh only in grams, how could you find the average weight of a grain of sand? Certainly a grain of sand weighs less than a gram, but you could weigh 1 g of sand and then count the number of grains of sand in the weighed sample. The average weight per grain would be calculated by dividing the weight of the sample by the number of grains of sand.

How would you determine the relative weights of two objects, M and G? Using the same balance, compare the weight of x number of M's with x number of G's. Let M's weight be 1 unit (regardless of the numerical scale on the balance). If one G then weighs 16 times as much as M, then relative to M's 1 unit of weight, G has 16 units. The relative weight of M to G is expressed as

$$\frac{M}{G} = \frac{1}{16}$$

Now imagine that each G has two hooks on it and each M has only one such hook. If we allow only unlike objects to link, what units of weight are represented by the combination of one G and two M's (see Fig. 2.1)? The answer, as shown in the figure, is a total of 18 units of weight. What percent of the total weight is due to M and what percent is due to G? To calculate percent of M, divide the number of units of M (2) by the total units in the sample (18 in this case) and multiply by 100. Repeat the procedure for G:

$$\tfrac{2}{18} \times 100 = 11.1\%\ M \text{ in sample}$$

$$\tfrac{16}{18} \times 100 = 88.9\%\ G \text{ in sample}$$

Regardless of the units of weight (grams, tons, pounds, etc.), the relative weights of M and G are the same. The percentage composition of M—G—M is also the same.

It will be useful to recognize this problem when it is reversed. Given that the ratio of M/G is equal to 2 and that the percentage of M is 11.1, what is the relative weight of M to G?

$$\%\ G = 100 - 11.1 = 88.9$$

M is present in the smallest percent, hence the smallest amount. Express the relative weight of G in terms of M. If you assume that M is 1 unit, to find units of G divide the percentage of G by the percentage of M

$$\tfrac{88.9}{11.1} = 8 \text{ units of } G \text{ for } \textit{each} \text{ unit of } M$$

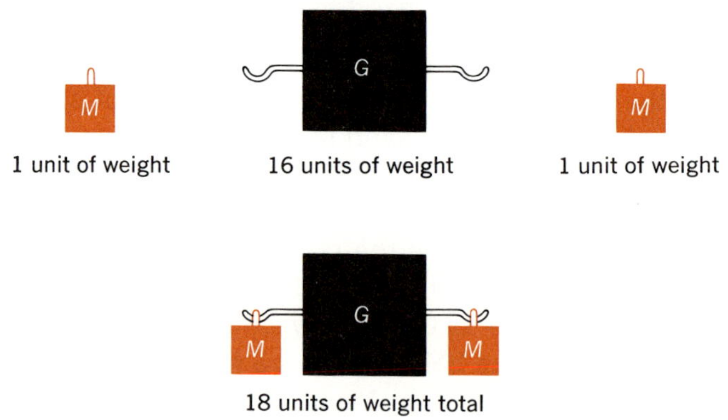

FIGURE 2.1 The combining ratio (weight) of two objects.

Since there are 2 M's, there are 2 × 8, or 16, units of weight for 1 G.

These two procedures are used to determine the relative weights of various atoms.

2.3 THE STANDARD FOR ATOMIC WEIGHTS

To compare the weights of the various elements, the weight of one element must be chosen as a standard. Formerly the standard was oxygen, O, which had arbitrarily been given an atomic weight of 16 amu (atomic mass units), and the weights of other elements were compared with this value. However, at a meeting of the International Union of Pure and Applied Chemistry (IUPAC) in 1961, it was decided to use carbon, C, as the standard of 12.000 amu for comparison. Although this changed the value of oxygen from 16.000 to 15.9994, it is usually rounded to 16.00 amu.

It has been found experimentally that nitrogen, N, is $\frac{14}{12}$ as heavy as carbon, C; hydrogen, H, is $\frac{10}{12}$ as heavy as carbon; and chlorine, Cl, is $\frac{35}{12}$ as heavy as carbon. Thus, we can speak of the atomic weight of these elements as C = 12, N = 14, H = 1.0, Cl = 35 units of weight *without having had to weigh a single isolated atom.* A list of the atomic weights of the elements is given inside the back cover.

2.4 UNITS OF WEIGHT

Since the atomic weights of the elements are relative, any unit may be used. If the unit of weight is the *gram,* we talk about the *gram atomic weight.* If the weight is expressed in pounds or tons, we talk about the pound or ton atomic weight.

EXAMPLE

Chemists have determined that oxygen, O, combines with calcium, Ca, to form calcium oxide, which contains 28.6% O and 71.4% Ca by weight. Because no other product of these two elements contains a different percentage of oxygen and calcium, it can be assumed that oxygen and calcium unite in the simplest ratio of 1 : 1, that is, one atom of calcium and one atom of oxygen per molecule of product. What is the simplest ratio by weight of oxygen to calcium?

Since the weight percentage of oxygen is 28.6% and the number of oxygen present is assumed to be 1, 28.6/28.6 = 1, applying this to the known weight percentage of calcium,

$$\frac{71.4}{28.6} = 2.5 \text{ units of Ca}$$

Since we know, by definition, that 1 unit of oxygen weighs 16 amu, 1 unit of calcium weighs 2.5 times that number, or 40 amu

$$2.5 \times 16 = 40, \text{ or 1 unit of Ca}$$

Furthermore, since calcium and oxygen combine in a ratio of 1 : 1 and their relative weights in amu are 40 and 16, respectively, it follows that there are the same number of calcium atoms in 40 units of weight of calcium as there are oxygen atoms in 16 units of weight of oxygen. *There are the same number of atoms of any element in a given atomic weight of the element.* For example, 1.0 g of H contains the same number of atoms as 40 g of Ca or as 16 g of O. Similarly, 16 tons of O contains the same number of atoms as 1.0 ton of H.

2.5 AVOGADRO'S NUMBER

How many atoms are there in one atomic weight of an element? To answer this question the unit of weight must first be defined. It has been determined experimentally that if the unit of weight is the gram, *the atomic weight expressed in grams of any element contains Avogadro's number, or* 6.02×10^{23}, *atoms.* This is an experimental fact which can be used in theoretical calculations.

A *gram atom* is the weight of that number of atoms which equals the atomic weight expressed in grams, or Avogadro's number of atoms.

To find the number of gram atoms in any given weight of an element, divide the weight of the element (expressed in grams) by the gram atomic weight of the element

$$\text{No. gram atoms} = \frac{\text{wt. in grams}}{\text{g at. wt.}} \qquad (2.1)$$

EXAMPLE 1

If one dozen eggs weighs 565 g, calculate the number of dozens of eggs in (*a*) 50 g and (*b*) 600 g of eggs.

(*a*) No. of dozens = $\frac{50}{565}$ g = 0.09 dozen eggs

(*b*) No. of dozens = $\frac{600}{565}$ g = 1.06 dozen eggs

EXAMPLE 2

If a dozen aspirin tablets weighs 1.524 g, how many dozens are there in 10 lb of aspirin? To calculate this answer the units of weight must be the same. Change the 10 lb to grams (see the conversion tables in Appendix B). Since 1 lb contains 454 g, 10 lb = 454 × 10, or 4,540 g. The number of dozens of aspirin is 4,540/1.524 g, or 2,979 dozen.

The unit number in each of the samples is 12 although the weight of one such unit is different in each example. One egg does not weigh the same amount as one aspirin. In chemistry the unit number is usually Avogadro's number of atoms or, when expressed in grams, 1 g atom. You would not expect the weight of 1 g atom of the various elements to be the same any more than you would expect one dozen eggs to weigh the same as one dozen aspirins.

Fractional parts of a gram atom frequently cause difficulty because it is easily forgotten that the gram atom represents 6.02×10^{23} atoms. The value of 0.01 g atom *does not* mean one one-hundredth of one single atom. In one-millionth (1×10^{-6}) g atom of any element there are 6.02×10^{17} atoms. You can take a teaspoon of sand from a carload of sand without having to split a grain of sand.

PROBLEM 1

You analyze a compound and find it contains only Fe and O in the atomic ratio of 2:3 and has the following composition: Fe = 69.9%, O = 30.1%. What is the gram atomic weight of Fe?

Atomic ratio of Fe/O = $\frac{2}{3}$

Weight ratio Fe/O = 69.9/30.1

Therefore, the weight in grams of Fe that is in combination with 3 g atoms of O (1 g atom of O equals 16 g, see Table 2.1) is

$$\frac{69.9}{30.1} = \frac{x}{3 \times 16}$$

Solving for x,

$$x = 111.36 \text{ g-wt. of 2 g atoms of Fe}$$

$$\frac{111.36}{2} = 55.68 \text{ g-wt. of 1 g atom of Fe}$$

Check on the correct value for the gram atomic weight of Fe and determine your percentage error.

PROBLEM 2

How many gram atoms are there in 45 g of potassium, K? The gram atomic weight of K is 39.1; therefore, in 45 g there would be 45/39.1, or 1.15 g atoms of K.

PROBLEM 3

How many gram atoms are there in 20 lb of O? Since the answer calls for gram atoms, all weights must be expressed in grams.

$$\frac{454 \times 20}{16} = 567.5 \text{ g atoms of O}$$

PROBLEM 4

Calculate (*a*) the number of gram atoms of nitrogen, N, and (*b*) the number of atoms of N contained in 2.17 mg of the element.

(*a*) 1 mg = 0.001 g

2.17 mg = 0.00217 g

1 g atom of N = 14 g

$$\text{No. of g atoms in 0.00217 g} = \frac{0.00217}{14}$$

$$= 0.000155, \text{ or } 1.55 \times 10^{-4}$$

(*b*) No. of atoms in 1.0 g atom = 6.02×10^{23}

No. of atoms in 1.55×10^{-4} g atoms

$$= (6.02 \times 10^{23})(1.55 \times 10^{-4}) = 9.331 \times 10^{19} \text{ atoms}$$

2.6 COMPONENTS OF THE ATOM

Since no two elements are the same, the atoms of one element must be different from the atoms of any other element and the difference must be inherent in the structure of the atoms themselves. An atom has two major parts, the nucleus and the electrons. All atoms are electrically neutral; i.e., they have neither a positive nor a negative charge.

The nucleus

The nucleus is the dense central core of the atom which contains the neutrons and the protons. *Neutrons* are neutral particles, whereas *protons* are positive particles.

Protons and neutrons are almost of equal relative weight, and for practical purposes each is considered to have a value of 1 amu. An electron weighs only 1/1,848 as much as a proton; therefore, we usually consider the weight of an atom to be due to the number of protons and neutrons in its structure. The number of protons in an element is indicated by a subscript as ${}_1H$, ${}_2He$, and ${}_{18}O$.

The electron

Electrons are negative particles which travel around the nucleus in particular patterns at the speed of light. The mass of an electron is 0.06% that of a proton (see Table 2.1). Electrons are negatively charged particles which are indicated by a minus sign.

The atomic number Z of an element is the number of protons in the atom. It also represents the number of electrons in an atom since all atoms are electrically neutral.

$$\text{At. wt. (amu)} = \text{no. of protons} + \text{no. of neutrons} \tag{2.2}$$

TABLE 2.1 RELATIVE WEIGHT OF ATOMIC PARTICLES

Particle	Weight, amu
Neutron	1.00866
Proton	1.00732
Electron	0.00055

From the periodic table (inside cover), the number of neutrons in an atom of any element can easily be calculated

$$\text{No. of neutrons} = \text{at. wt.} - Z \tag{2.3}$$

Several general rules have emerged from the study of nuclear stability in relation to the number of protons and neutrons:

1. For elements of low atomic number, 2 to 20, a ratio of 1 of neutrons to protons indicates stability. Of the 18 nuclei of the atoms in this group, 8 have neutrons equaling protons, and 9 have a difference of 1 between the number of neutrons and the number of protons.
2. When there are more than 20 protons in the nucleus, a ratio of neutrons to protons greater than 1 is necessary for a stable nucleus. This ratio increases with increasing atomic number to a maximum of about 1.8.
3. Nuclei in which neutrons equal protons and both are even numbers are the most stable and the most common. About 80% of the earth's crust is composed of six elements including $^{16}_{8}O$, $^{24}_{12}Mg$, $^{28}_{14}Si$, and $^{40}_{20}Ca$. The stability of these nuclei may be attributed to a complete nuclear subshell analogous to the complete outer shell of electrons (see Sec. 2.10).
4. Atoms containing an odd number of neutrons and an even number of protons or vice versa are fairly stable and occur to about the same extent as those in which the number of neutrons equals the number of protons.
5. Only four stable nuclei having an odd number of both neutrons and protons are known. All the stable ones have an atomic mass of 14 or less. Those of higher atomic mass are very unstable and radioactive.

Table 2.2 summarizes these general rules. Figure 2.2 is a graph of the neutron/proton ratio for stable elements.

TABLE 2.2 THE NUMBER OF NEUTRONS AND PROTONS IN STABLE NUCLEI

Number of		Stable nuclei
Neutrons	**Protons**	
Even	Even	164
Odd	Odd	4
Odd	Even	55
Even	Odd	50

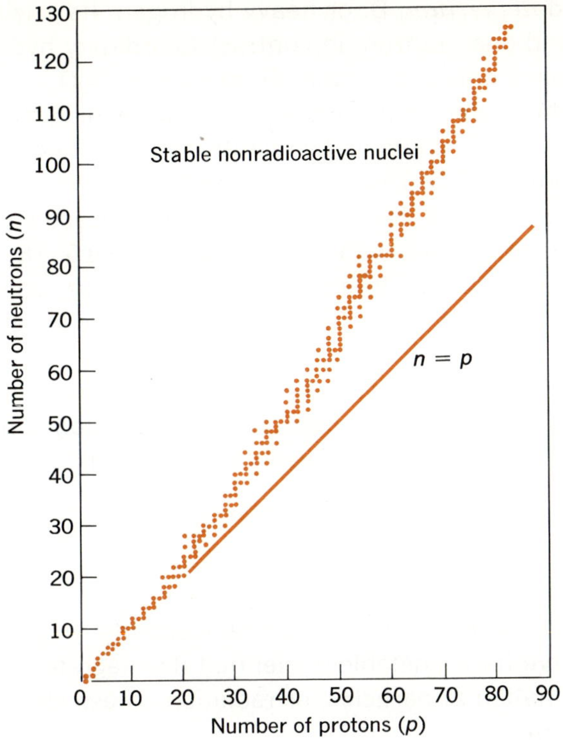

FIGURE 2.2 The ratio of neutrons n to protons p for stable nuclei. *(From Michell J. Sienko and Robert A. Plane, "Chemistry," © 1957 by McGraw-Hill, Inc. Used with permission of McGraw-Hill Book Company.)*

2.7 ISOTOPES

All atoms of any one element contain the same number of protons and electrons, but there may be differences in weight due to a variation in the number of neutrons. Atoms of an element which have the same atomic number but different atomic weight because they contain a different number of neutrons are *isotopes.*

From the relative atomic weights of the elements, it can be seen that a large number of elements have fractional amu values. Chlorine, for example, is 35.5. There are two isotopes of chlorine, one with an amu of 35 and the other 37. Since in a sample of ordinary chlorine gas the two isotopes appear in the ratio of 3:1 by weight, respectively, the average weight of chlorine is 35.5 amu.

Hydrogen gas has an isotope that occurs naturally to a very slight extent, 0.02%, and has a relative atomic weight of 2.0. This isotope is called *deuterium,* D, or heavy hydrogen. Its nucleus contains one proton and one neutron, in contrast to ordinary hydrogen, whose nucleus contains only one proton. *Heavy water,* D_2O, is so named because it contains deuterium instead of hydrogen. A third isotope of hydrogen, tritium, T, contains two neutrons in the nucleus and therefore has an amu of 3.0. This isotope is radioactive whereas deuterium is stable (see Sec. 2.8 for a discussion of radioactivity). There are three isotopes of oxygen, with atomic weights of 16, 18, and 19. The one with 16 amu occurs to an extent of 99.7% in ordinary oxygen. There are 10 isotopes of tin, Sn, ranging in atomic weights from 112 to 124.

In general, elements with an *odd* atomic number have relatively few stable isotopes, whereas those with an *even* atomic number often have several stable isotopes of even atomic weights.

2.8 RADIOACTIVITY

Elements with stable nuclei occur only within a rather narrow range of neutron-proton ratio (see Fig. 2.2). The elements lying to the right or left of the curve have unstable nuclei that disintegrate spontaneously, with the elimination of particles, or radiation. These elements are said to be *radioactive.*

The particles, or radiation, expelled from the nucleus of radioactive elements are of three kinds, namely, alpha (α) particles, beta (β) particles, and gamma (γ) radiation. An α particle is a *helium nucleus* with an atomic mass of 4 and a charge of plus 2, ${}^4_2He^{++}$. A β particle is negatively charged, an electron, ejected at high velocity. The β particle is produced by the disintegration of a neutron into an electron and a proton, the electron being ejected from the nucleus. γ radiation is shortwave, high-energy rays similar to x-rays. Although a neutron splits into a proton and an electron during radioactive decay, *the neutron is not merely a combination of an electron and a proton.*

α and β particles and γ rays can be separated by passing them through a magnetic or electric field (see Fig. 2.3). The path of the α particle bends toward the negative electrode because the particle is positively charged; the path of the negatively charged β particle bends toward the positive electrode; whereas the path of the neutral γ rays is not affected by an electric or magnetic field.

Since the ejection of an α particle is equivalent to a loss of two protons and two neutrons from the nucleus, the atomic number de-

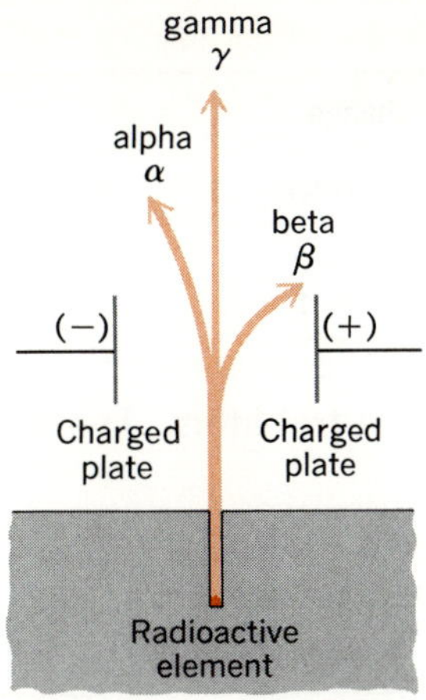

FIGURE 2.3 Separation of α and β particles and γ rays in an electric field.

creases by a value of 2 and the atomic weight decreases by a value of 4. A loss of two protons from the nucleus produces a species with a 2− charge. To balance the electrons and protons, consider that this charged species, called an *ion*, gives two orbital electrons to the He^{++} ion (α particle) to produce a neutral helium atom, ${}^{4}_{2}He$. This is illustrated as follows in the conversion of uranium to thorium:

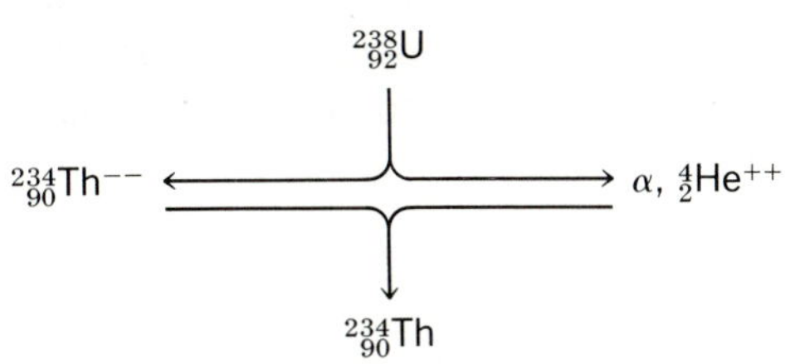

Thorium is also radioactive and disintegrates spontaneously. The ultimate stable end product resulting from uranium decay is ${}^{206}_{82}Pb$. The overall reaction is

$${}^{238}_{92}U \longrightarrow {}^{206}_{82}Pb + 8\alpha + 6\beta$$

(see Appendix C for the intermediary steps).

Emission of a β particle from the nucleus is equivalent to changing a neutron into a proton; this increases the atomic number by 1 but does not change the atomic weight. The initial product is a positive ion, which for the present we can consider as trapping the ejected nu-

TABLE 2.3

	Nuclear change		Change in	
	Neutrons	**Protons**	**Atomic weight**	**Atomic number**
α emission	−2	−2	−4	−2
β emission	−1	+1	0	+1

clear electron to form a neutral atom. The disintegration of $^{14}_{6}C$ to $^{14}_{7}N$ may be depicted as

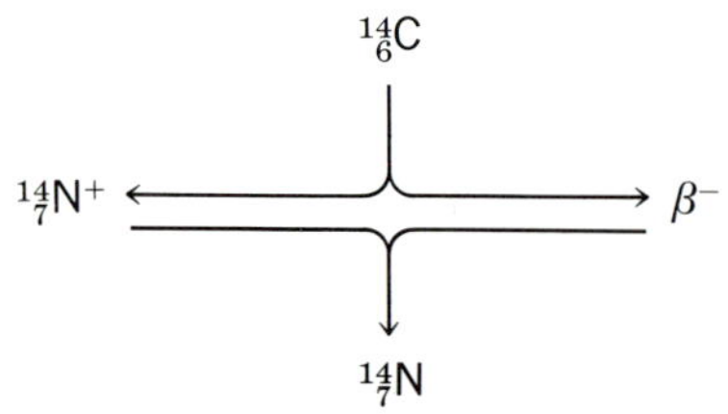

Under actual conditions the ions produced by nuclear disintegration are not always neutralized by their own electrons, as shown above, but by an exchange of electrons with those from the surrounding medium or from other radioactive ions or atoms. For example, the α particle is ejected from the nucleus at a speed of perhaps 30 million mph and travels through air at atmospheric pressure for about 8 cm. During this short distance the α particle collides with millions of air molecules and ionizes them by knocking off many of their orbital electrons. These free electrons can combine with the 2+ α particle to form a neutral He atom.

Nuclear fission occurs when a radioactive element decomposes into two other elements

$$^{235}_{92}U + n \longrightarrow \underset{\text{Unstable}}{[^{236}_{92}U]} \longrightarrow ^{91}_{36}Kr + ^{142}_{56}Ba + 3n$$

$^{235}_{92}U$ requires bombardment with neutrons to start the reaction which becomes a self-sustaining chain reaction because more neutrons are liberated than are required to keep the reaction going. Fission reactions occur almost spontaneously (10^{-9} sec) and unless controlled a mole of uranium would disintegrate in about 0.000001 sec. This rapid a reaction would result in a tremendous explosion with the liberation of energy equal to 2×10^{2}/MeV, or 2×10^{12} cal. Under controlled conditions this energy can be put to useful work. To control such a reaction a moderator must be present to slow down or trap some of the neutrons so that the chain reaction does not go too fast (see Fig. 2.4).

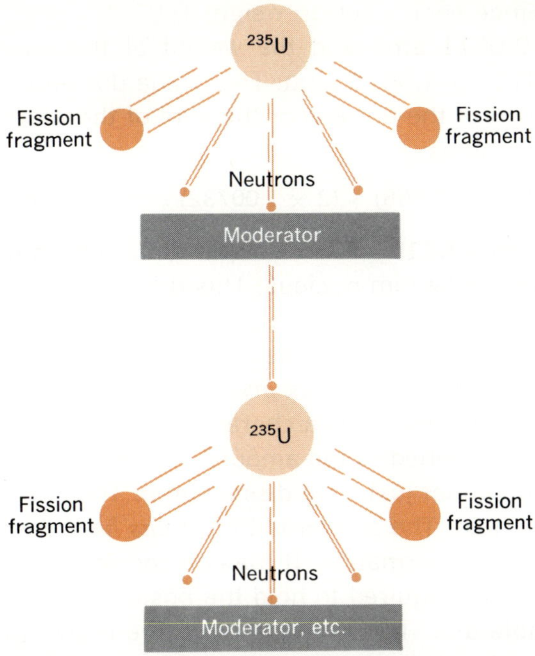

FIGURE 2.4 The controlled nuclear reaction of ^{235}U.

Nuclear power has been used to generate electricity and to propel ships and submarines. Attempts are now being made to devise a small 16-watt nuclear power plant to be inserted into a man's chest to power an artificial heart for about 10 years before recharging with more radioactive element.

The use of ^{14}C in biologically active compounds has made possible many advances of biochemistry in the past decade. Instruments that take advantage of the ionizing power of radioactive emissions measure the amount of radioactive element in a compound. For example, radioactive carbon dioxide, $^{14}CO_2$, was an important tool in determining the chemical steps of photosynthesis. Actively photosynthesizing plants (usually algae) were fed $^{14}CO_2$, the plants were killed after a short time, and the organic compounds containing ^{14}C were isolated and identified. Radioactive elements also play an important role in medicine, especially in the treatment of cancer.

2.9 MASS DEFECT

The helium atom, He, contains two protons, two neutrons, and two electrons. The observed weight of the He nucleus is found by subtracting the weight of the two electrons from the atomic weight of He,

4.003 amu. Since each electron weighs 0.00055 amu, two electrons would weigh 0.0011 amu and the weight of the nucleus would be 4.003 − 0.0011, or 4.0019 amu. This value does not agree with the calculated weight of the nucleus or the sum of the weights of two neutrons and two protons

$$(2 \times 1.00866) + (2 \times 1.00732) = 4.03196 \text{ amu}$$

being greater by 4.03196 − 4.0019, or 0.03006 amu, than the observed mass of the helium nucleus. This difference is called the *mass defect.* When the total mass of the protons and neutrons corresponding to the atomic mass of the nucleus for the atoms of the various elements is calculated, it is *always greater than the observed* nuclear-mass value. Each element has a characteristic mass-defect value.

What has happened to the amount of mass corresponding to the mass defect? It cannot just have disappeared. The law of conservation of matter must hold. The answer is that *it has been converted into energy* during nucleus formation. It may be considered as representing the *binding energy* required to hold the positive nucleons together to produce a stable unit. The same amount of energy would have to be put into the nucleus to separate the nucleons into isolated particles. This means that the nucleus contains less energy than the free neutrons and protons from which the nucleus was made and illustrates a general rule of reactions, namely, that *reactions tend to go in the direction which forms more stable products (of lower energy content) than the reactants.*

The conversion of only 0.03 amu to energy, as in the example of helium, may seem too trivial to mention, but this is far from true, as it represents an enormous amount of energy. Einstein's equation for the equivalence of mass and energy is

$$E_{\text{ergs}} = mc^2 \tag{2.4}$$

where m is mass in grams and c is the velocity of light in centimeters per second.

$$E_{\text{ergs}} = m(3 \times 10^{10})^2$$

Therefore, using the conversion table in Appendix B,

$$E_{\text{cal}} = m(2.15 \times 10^{13})$$

Nuclear rearrangements associated with a change in the nuclear binding energy involve *millions* of times more energy than is represented by a chemical change associated with the binding of electrons. The energy released during the change of only 0.03 g of mass into energy is approximately 645×10^9 cal. This is about the amount of energy the body would derive from the oxidation of some 640×10^4 lb of sugar. You would have to eat about 8.75 tons of sugar every day for a year to get around this pile of sweets!

The conversion of four hydrogen atoms to a helium atom with a concomitant conversion of mass into energy is believed to be the source of the sun's energy ($4\ H \longrightarrow He + \text{energy}$). It has been estimated that the hydrogen in the sun's atmosphere is sufficient to permit the present rate of conversion of mass into energy to continue for at least a billion more years.

It is often more convenient to express large amounts of energy in terms of million electron volts (MeV) instead of calories. An electron volt is the energy acquired by one electron when it is accelerated through a potential difference of one volt. Appropriate substitutions in Einstein's equation (2.4) give for the conversion of mass into energy

$$1\ g = 931\ \text{MeV} \qquad 1\ \text{MeV} = 2.31 \times 10^{10}\ \text{cal} \tag{2.5}$$

Expressed in these units, the mass defect of 0.03 g represents 0.03×931, or 28 MeV.

To compare the binding energy of the nuclei of the various elements, reduce the energy to a common denominator, namely, the binding energy per nuclear particle. For example, there are four particles in the helium nucleus; hence $\frac{28}{4}$, or 7 MeV, would be the binding energy per particle. Oxygen, which has eight neutrons and eight protons in the nucleus, has a mass defect of 0.13648 g per 16 particles. This is equivalent to 0.00853 g per particle, or about 8 MeV per particle. The elements with the highest value for the binding energy per nuclear particle are the most stable, since the amount of energy to separate the nuclear particles is the same as the binding energy.

In nuclear reactions like the conversion of uranium to lead plus energy, the element resulting from the decomposition has a higher binding energy than the element being decomposed.

2.10 ELECTRON ORBITALS

Electrons traveling at the speed of light move in particular patterns around the nucleus. They are also spinning. We may visualize them as *traveling in definite volumes, or orbitals,* not just wandering around

TABLE 2.4 ELECTRON DISTRIBUTION IN AN ATOM

Principal quantum number	Electron shell	Electron suborbital	Number of electrons
$n = 1$	*K*	*s*	2
$n = 2$	*L*	*s*	2
		p	6
$n = 3$	*M*	*s*	2
		p	6
		d	10
$n = 4$	*N*	*s*	2
		p	6
		d	10
		f	14

the nucleus like lost sheep. But perhaps they are "lost"; for if they are traveling at the speed of light (186,000 miles/sec), one cannot say they are at a given place at any given instant. An electron may or may not be paired with another one that is spinning in the opposite direction. Although atomic structure will not be studied in great detail in this book, certain general rules of atomic structure will be discussed to clarify the biochemistry which follows.

Starting closest to the nucleus, the orbitals are defined by the *principal quantum number n* and are indicated as $n = 1$, $n = 2$, $n = 3$, etc. Originally these orbitals were referred to as *K*, *L*, *M*, *N*, etc., shells. After the first orbital, the subsequent ones are divided into suborbitals (see Table 2.4). Since the electron orbitals become larger the further away they are from the nucleus, the maximum number of electrons theoretically possible in any particular n orbital becomes greater as the distance from the nucleus increases.

The *rule of eight,* also called the *octet theory,* states that any n orbital beyond the first one, regardless of its size, if it is the *outermost orbital* of the atom, *contains a maximum of eight electrons.* This is a general rule; there are some exceptions to it.

2.11 DISTRIBUTION OF ELECTRONS

The chemical properties of the elements are directly associated with the electronic configuration of their atoms, and the distribution of electrons about the nucleus of atoms is therefore of great importance. Only a few of the simpler atoms will be considered in detail, but it is *essential* to follow the text by frequently referring to Table 2.4 and Fig. 2.5.

The $n = 1$ *orbital* is the first orbital. It is completely filled when it contains two electrons. Thus this orbital can contain either one or two electrons but no more. The simplest atom, H, has one electron; the next simplest, He, has two electrons in the $n = 1$ orbital. The two electrons of helium are paired; i.e., they are spinning in opposite directions. The electrons of this orbital are called s electrons because they are traveling in a spherical volume equidistant from the nucleus. The $n = 1$ orbital electrons are denoted $1s$ electrons, where the 1 refers to the principal quantum number. The respective electrons in this orbital are indicated as H, $1s^1$ and He, $1s^2$. The superscript number refers to the number of electrons in the orbital and should not be confused with the superscript numbers used in mathematics to indicate exponential functions.

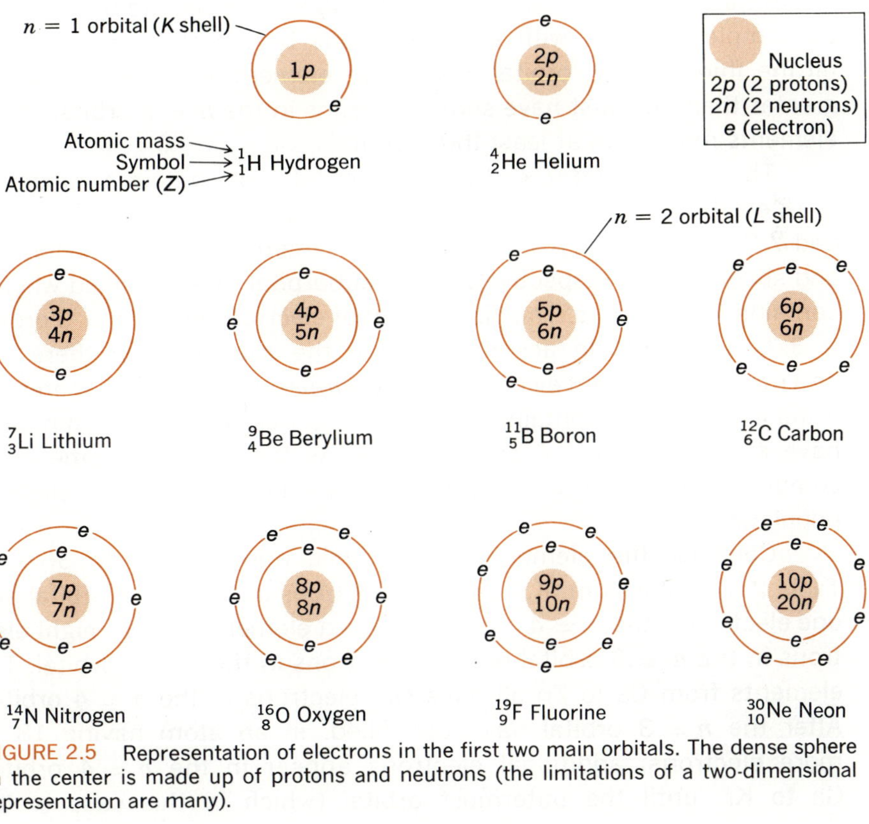

FIGURE 2.5 Representation of electrons in the first two main orbitals. The dense sphere in the center is made up of protons and neutrons (the limitations of a two-dimensional representation are many).

The $n = 2$ *orbital* is the second orbital. It may contain up to a maximum of eight electrons and is divided into two suborbitals, s and p. There can be a total of two electrons in the $2s$ suborbital and six in the $2p$ suborbital, making a maximum of eight electrons total in the $n = 2$ orbital. The $n = 2$ orbital electrons are indicated as $2s^2$, $2p^6$. The $2s$ suborbital must be filled before electrons can enter the $2p$ suborbital.

Atoms with an electron in the $n = 2$ orbital must also have two electrons in the $n = 1$ orbital; the first orbital must be completely filled before the second orbital can accept electrons. Thus, an atom with three electrons, Li, has them distributed as follows: two in the $n = 1$ orbital and one in the $n = 2$ orbital ($1s^2$, $2s^1$). The atom F, atomic number 9, would have two electrons in the $n = 1$ orbital and seven in the $n = 2$ orbital ($1s^2$, $2s^2$, $2p^5$). Since the $n = 2$ orbital can contain a maximum of eight electrons, an atom which has this number of electrons in the $n = 2$ orbital, such as Ne, would have a total of ten electrons around the nucleus ($1s^2$, $2s^2$, $2p^6$). The second orbital must be completely filled with eight electrons before electrons begin to fill the third, $n = 3$, orbital. Hence, an element with atomic number greater than 10 must have some electrons in the $n = 3$ orbital; these elements must have at least three orbitals.

The $n = 3$ *orbital,* or the third orbital, can contain a maximum of 18 electrons. There are three suborbitals in this orbital, namely, s, p, and d. Since the s and p suborbitals can contain a maximum of only two and six electrons, respectively, the d suborbital must be filled when it contains 10 electrons. The distribution of electrons in the $n = 3$ orbital is $3s^2$, $3p^6$, $3d^{10}$. In accordance with the *rule of eight* there can be a maximum of eight electrons in any *major outside orbital;* therefore, if the $n = 3$ orbital contains more than eight electrons, the atom must have at least an $n = 4$ orbital. Thus, as the atoms become more complex, both the $n = 3$ and the $n = 4$ orbital increase in electron population.

Consider the elements with atomic numbers 19 to 36 (see Table 2.5). Potassium, K, has eight electrons in the $n = 3$ orbital and one electron in the $n = 4$ orbital. The next element, Ca, has eight electrons in the $n = 3$ orbital but two electrons in the $n = 4$ orbital. The elements from Ca to Zn all have two electrons in the $n = 4$ orbital. After the $n = 3$ orbital has been filled, in an atom having 18 or more electrons, additional electrons appear in the $n = 4$ orbital, Ga to Kr, until the outermost orbital (which in this case is the

TABLE 2.5 ELECTRON DISTRIBUTION OF SOME ELEMENTS

Atomic number	Element		Electron distribution, n							Periodic group
	Name	Symbol	1	2	3	4	5	6	7	
1	Hydrogen	H	1							I
2	Helium	He	2							0
3	Lithium	Li	2	1						I
4	Berylium	Be	2	2						II
5	Boron	B	2	3						III
6	Carbon	C	2	4						IV
7	Nitrogen	N	2	5						V
8	Oxygen	O	2	6						VI
9	Fluorine	F	2	7						VII
10	Neon	N	2	8						0
11	Sodium	Na	2	8	1					I
12	Magnesium	Mg	2	8	2					II
13	Aluminum	Al	2	8	3					III
14	Silicon	Si	2	8	4					IV
15	Phosphorus	P	2	8	5					V
16	Sulfur	S	2	8	6					VI
17	Chlorine	Cl	2	8	7					VII
18	Argon	Ar	2	8	8					0
19	Potassium	K	2	8	8	1				I
20	Calcium	Ca	2	8	8	2				II
21	Scandium	Sc	2	8	9	2				II
22	Titanium	Ti	2	8	10	2				II
23	Vanadium	V	2	8	11	2				II
24	Chromium	Cr	2	8	12	2				II
25	Manganese	Mn	2	8	13	2				II
26	Iron	Fe	2	8	14	2				II
27	Cobalt	Co	2	8	15	2				II
28	Nickel	Ni	2	8	16	2				II
29	Copper	Cu	2	8	17	2				II
30	Zinc	Zn	2	8	18	2				II
31	Gallium	Ga	2	8	18	3				III
32	Germanium	Ge	2	8	18	4				IV
33	Arsenic	As	2	8	18	5				V
34	Selenium	Se	2	8	18	6				VI
35	Bromine	Br	2	8	18	7				VII
36	Krypton	Kr	2	8	18	8				0
37	Rubidium	Rb	2	8	18	8	1			I
...	...	...	...	...	...	...	...			
54	Xenon	Xe	2	8	18	18	8			0
55	Cesium	Cs	2	8	18	18	8	1		I
...	...	...	...	...	...	...	...			
86	Radon	Rn	2	8	18	32	18	8		0
...	...	...	...	...	...	...	...			
88	Radium	Ra	2	8	18	32	18	8	2	II

$n = 4$ orbital) has eight electrons. Rubidium, Rb, contains one more electron than Kr. It has one electron in the $n = 5$ orbital. This schematic picture of electron orbitals is far from being adequate to explain the real situation since it is impossible to pinpoint the electron at any one spot.

The separation between the boundaries of the various orbitals is not equally sharp for all. The distance is greater between the $n = 1$ and the $n = 2$ orbitals than between the $n = 2$ and the $n = 3$ orbitals. Some of the electrons in the $n = 3$ orbital overlap into the $n = 4$ orbital and vice versa. Thus we visualize the nucleus of the atom as being surrounded by layers of an electron "fog" created by the spinning electrons as they travel at the speed of light. This fog is denser in some areas than in others because the electrons spend more of their time at some average distance from the nucleus. Because the radius of the atom is about 10^5 times the radius of the nucleus, most of the volume

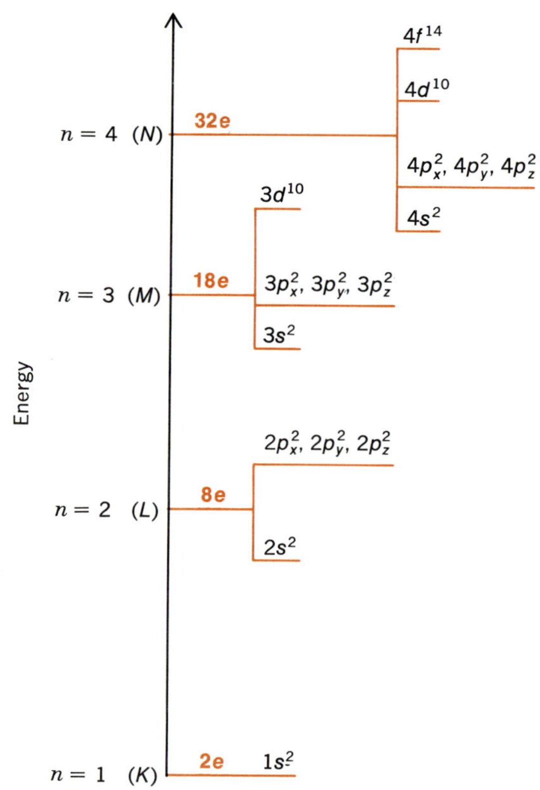

FIGURE 2.6 Electron energy levels schematically drawn. Note the overlap of energy levels in the $n = 3$ and $n = 4$ orbitals.

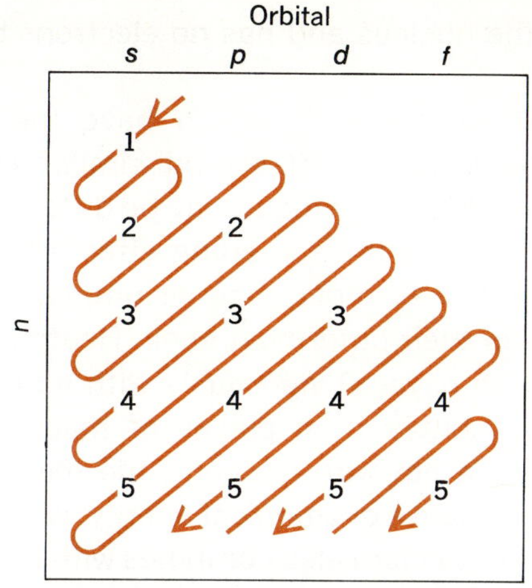

FIGURE 2.7 Order of filling electron orbitals, showing the overlap of energies (compare with Fig. 2.6).

of an atom is due to the relatively large distance the electrons are away from the nucleus.

The further away from the nucleus, the more energy the electrons have. The energy level of the electrons in the various orbitals is illustrated in Fig. 2.6. Electrons enter the lower-energy orbitals first (see Fig. 2.7). It is important to note that the two lowest-energy electrons of the $n = 4$ orbital, the $4s^2$ electrons, have a slightly lower energy level than the $3d$ electrons. Under certain conditions one (or more) of these $3d$ electrons reacts as though it were in the $4s$ orbital. The distribution of the electrons in the $n = 5$ orbital, which has four suborbitals, is even more complex with respect to energy levels.

2.12 IMPORTANCE OF ELECTRON STRUCTURE

Electrons in the outermost shell are called the *valence electrons*. The chemical properties of an element are directly related to these electrons, and the chemical activity of the element is modified by the number of electrons in lower orbitals. The larger the atomic number of the element, the greater the positive charge of the nucleus. It is easy to see that the behavior of a single electron of the outermost shell of a complex atom with a broad electron cloud between it and the nucleus would be different from that of the electron of the hydrogen atom, which is closer to the nucleus and has no electrons between it and the positive nucleus.

In general, those elements which have the *same number of electrons in the outermost orbit* exhibit similar chemical behavior, although the rate at which they react may be different. Such variations in reactivity are caused by the screening effect of the inner electrons and the atomic size. For example, magnesium, calcium, and iron each have two electrons in their outermost shell. Heated in air, magnesium burns (combines with oxygen) vigorously with a brilliant light. Photographic flash bulbs contain thin pieces of magnesium and oxygen. When it is heated by a hot wire filament, the magnesium ignites and oxidizes with a large release of chemical energy, which is converted into light and heat energy. Calcium also combines with oxygen but at a much slower rate than magnesium, and iron combines with oxygen at a slower rate still. The electron structure of the atoms also plays an important role in their biological function. Magnesium, calcium, and iron are all biologically active; each enters into specific reactions and cannot be substituted for another.

QUESTIONS

1. What are the three main components of an atom? What charges are associated with each of these components?

2. Are all atoms of a particular element identical? Explain.

3. What does the term "mass defect" mean?

4. (*a*) Calculate the mass defect for sodium.
(*b*) How many calories and MeV are associated with the mass defect per nucleon in sodium?

5. (*a*) What is the maximum number of electrons that can be contained in each principal-quantum-number orbital?
(*b*) What is the maximum number of electrons that can be contained in an outermost orbital, excluding the first one?

6. Indicate the atomic structure of the following elements. For example, 9 protons and 10 neutrons. *Ans.:* $^{19}_{9}[\ \ ]\ 1s^2, 2s^2, 2p^5$.
(*a*) 4 protons, 5 neutrons (*b*) 14 protons, 12 neutrons
(*c*) 14 protons, 14 neutrons (*d*) 3 electrons, atomic weight 7
(*e*) 8 electrons, atomic weight 16 (*f*) 20 electrons, atomic weight 40

7. If a stable element has an atomic number of 14, what would you estimate its atomic weight to be?

8. If a stable element has an atomic number of 60, would you expect its atomic weight to be (*a*) approximately 60, (*b*) approximately 120, or (*c*) greater than 120?

9. In all probability would the element in Prob. 8 have an even- or an odd-numbered atomic weight?

10. Match the atoms represented by the following electron arrangements with the appropriate periodic group (I to VII or 0):

(*a*) $1s^2$ (*b*) $1s^2, 2s^2$ (*c*) $1s^2, 2s^2, 2p^3$

(*d*) $1s^2, 2s^2, 2p^6$ (*e*) $1s^2, 2s^2, 2p^6, 3s^2$

11. Give the atomic number of the element with the electron distribution $1s^2, 2s^2, 2p^6, 3s^2, 3p^6, 3d^{10}, 4s^2, 4p^6, 4d^{10}, 5s^2, 5p^2$.

12. Calculate the binding energy per nuclear particle (MeV per particle) for uranium and lead.

13. Calculate the number of neutrons per atom in several of the elements listed in Table 2.5. What is the calculated value for chlorine? How is it possible to have a fractional number when neutrons exist only as whole units of matter?

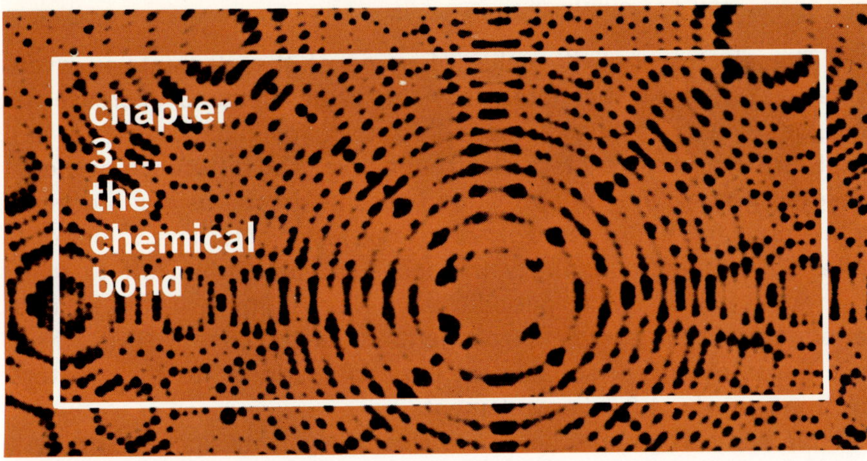

chapter 3.... the chemical bond

Why do some elements enter into certain reactions while others do not? How do atoms unite chemically with one another to form compounds? Why do compounds have a definite composition? The answers to these questions are given in this chapter.

The chemical bond between atoms is very important since the chemical properties of a compound depend on the type of bond. Bonds are formed by an interaction of the outer-shell electrons of atoms. Atoms combine to obtain their full complement of eight electrons in the outermost shell; i.e., they combine to achieve the *most stable* configuration possible.

3.1 ELECTRONEGATIVITY

The extent to which atoms attract electrons of other atoms, i.e., their *electronegativity,* determines the type of chemical bond which the atom forms. The nuclear charge, the screening effect of the electrons surrounding the nucleus, and the overall size of the atoms are factors which influence the electronegativity of atoms. If an atom has a strong attraction for electrons, it tends to pull electrons from atoms which have a lower electron attraction. Atoms which can pull electrons from neighboring atoms are strongly electronegative.

A very strongly electronegative atom may pull some of the electrons completely away from the influence of the nucleus of a very weakly electronegative atom. On the other hand, two atoms that have about the same electronegativity will be able to combine only by sharing electrons since neither atom is electronegative enough to pull electrons completely away from the influence of the nucleus of the other atom.

The electronegativity value of some of the elements is given in Table 3.1. It will be noted that the values tend to decrease as one goes down a particular periodic group; for instance, H, Li, Na, and K are progressively less electronegative. The difference between successive atoms in a group tends to become greater as one proceeds from group I to VII. Elements of group I, exclusive of H, have values of 1 or less. The electronegativity increases in value from left to right in any one period of the periodic table. For example, in the second period, Li to F, the successive elements differ by 0.5 unit. This difference becomes less in subsequent periods. Fluorine has the greatest electronegativity, 4.0, of any element.

Elements in group 0 have zero electronegativity since their outermost shell contains eight electrons. They have slight attraction for electrons of other atoms and are almost inert chemically. Crystalline compounds such as XeF_4 have been prepared in which Xe has 12 electrons in its outer orbit.

3.2 ELECTROVALENT, OR IONIC, BOND

Electrovalent bonds are formed when atoms with a high electronegativity combine with atoms of low electronegativity, e.g., atoms of the elements in groups I, II, and some in III with atoms of the elements in group VII. In general, *a difference of about* 1.7 *units in electronegativity is needed to form an electrovalent bond.* In this type of bond there is a complete shift, or transfer, of one or more electrons from

TABLE 3.1 ELECTRONEGATIVITY VALUES

Period	Group							
	I	**II**	**III**	**IV**	**V**	**VI**	**VII**	**0**
1	H 2.0							He 0
2	Li 1.0	Be 1.5	B 2.0	C 2.5	N 3.0	O 3.5	F 4.0	Ne 0
3	Na 0.9	Mg 1.2	Al 1.5	Si 1.8	P 2.1	S 2.5	Cl 3.0	Ar 0
4	K 0.8	Ca 1.0	Ga 1.6	Ge 1.8	As 2.0	Se 2.4	Br 2.8	Kr 0

Cr	Mn	Fe	Co	Ni	Cu	Zn
1.6	1.5	1.8	1.8	1.8	1.9	1.6

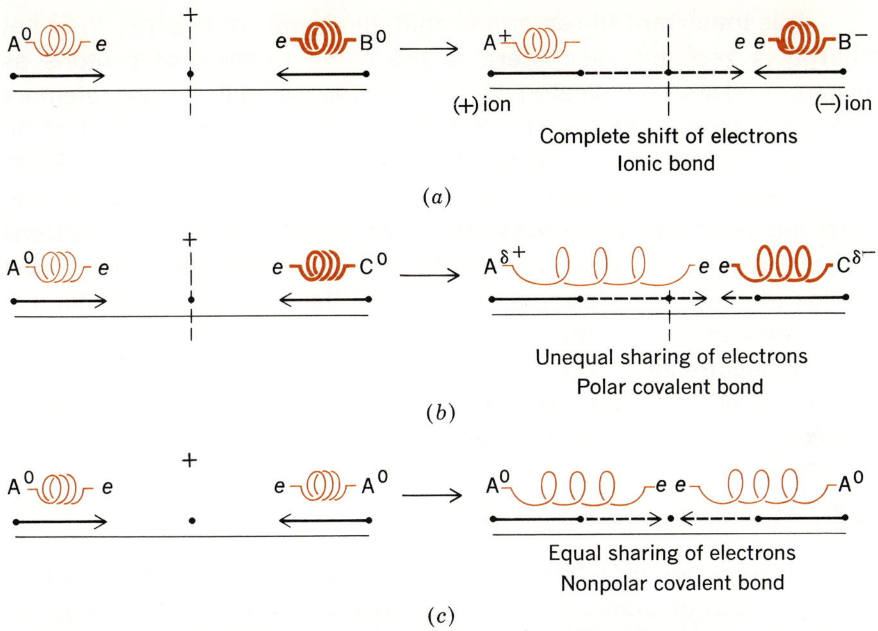

FIGURE 3.1 In this analogy atoms are represented by A, B, and C. Their electronegativity is quantitatively shown by the size of the spring. A heavy-weight spring represents a strongly electronegative atom, a medium-weight spring represents a moderately electronegative atom, and a light-weight spring represents a weakly electronegative atom. The electron is shown as *e*. When atoms combine, it is as if the springs were stretched and hooked together. (*a*) Weakly electronegative atom and a strongly electronegative atom with a complete shift of electrons (difference ≥ 1.7); (*b*) weakly electronegative atom and a strongly electronegative atom with unequal sharing of electrons (difference < 1.7); (*c*) two atoms of the same electronegativity with equal sharing of electrons.

the weakly electronegative atom to the strongly electronegative atom to form ions (Fig. 3.1*a*).

If an atom which donates one electron combines with an atom capable of accepting only one electron, it is logical to reason that only one atom of each of these elements would be needed to make a compound.

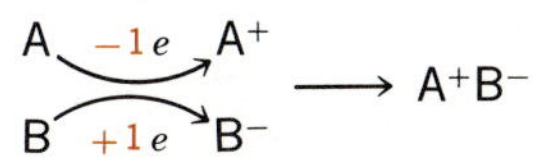

or (3.1)

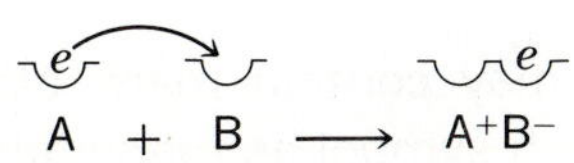

It is important to remember that all atoms are neutral: their net charge is zero because there is the same number of protons as electrons. This is indicated by writing the symbol for the element without a plus or minus superscript. However, after the shift of an electron, the net charge of an atom (A or B) changes [see Eq. 3.1)]. These oppositely charged particles are attracted to each other and form the associated molecule A^+B^-. Of course, this transfer of electrons cannot take place when the atoms are separated; the atoms must collide so that the outermost electrons of the electron donor momentarily come under the influence of the nucleus of the electron acceptor.

An electrically charged atom or group of atoms is an *ion.* Ions are formed when electrons are completely transferred from one atom to another. *Ionic bonds* exist between ions.

3.3 COMBINING RATIO OF ATOMS

All electrons donated by one or more atoms must be accepted by another atom or atoms. The law of conservation of matter applies to electrons. In Eq. (3.2) an atom C donates two electrons to D, which is capable of accepting only two electrons. In this case C becomes a C^{++} ion and D becomes a D^{--} ion. The resulting compound is $C^{++}D^{--}$.

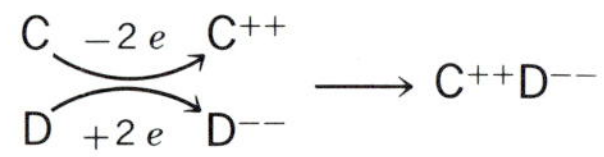

or (3.2)

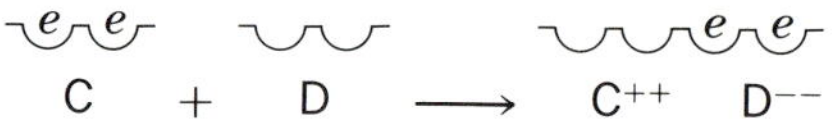

How would an element whose atoms donate two electrons each combine with an element whose atoms can accept only one electron each? In this instance [Eq. (3.3)] it takes two atoms of the electron acceptor to accept the two electrons donated by one atom of the electron donor. The formula for this compound is $C^{++}B^{-}{}_2$.

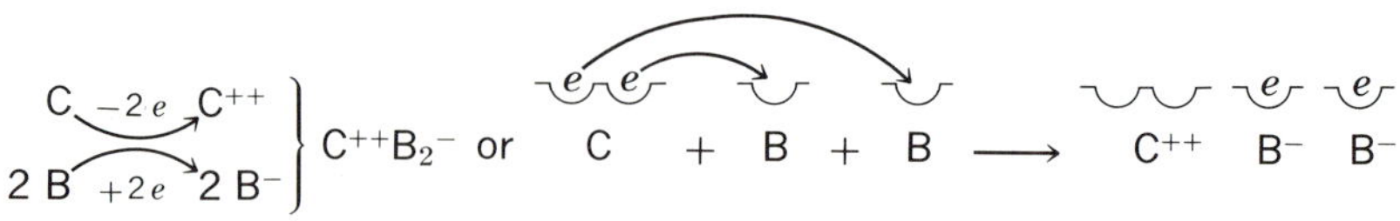

(3.3)

Since compounds may contain many atoms, it would be quite cumbersome to write each symbol as many times as it occurs in a formula. Instead a subscript number is used to indicate how many times a particular atom occurs in a compound. When no subscript is written, the number 1 is understood. For example, the formula $C_{12}H_{22}O_{11}$ means there are 12 atoms of carbon, 22 atoms of hydrogen, and 11 atoms of oxygen in one molecule of the compound. Note particularly that each atom of B [Eq. (3.3)] receives only one electron and becomes B^-. Regardless of the number of B^-'s that occur in a compound each one would have a -1 charge; this is the value placed above the atom. (A car has four wheels: does this mean that if you had two cars, one car would have eight wheels and the other none? Of course not!) The total number of *negative* charges in $C^{++}B_2^-$ is 2. This exactly balances the two *positive* charges of C^{++}. *Every compound must have a net charge of zero* or it cannot exist in the free state.

Another way of stating this is that the sum of the products of electron-donating atoms times the electrons donated must equal the sum of the electron-accepting atoms times electrons accepted.

$$\text{Donating atoms} \times ne^- = \text{accepting atoms} \times ne^-$$

Study the examples in Table 3.2. Note that electron-donating atoms combine with electron-accepting atoms in simple ratios of whole numbers. Since electrons are transferred in their entirety, this ratio depends upon the number of electrons donated or accepted by the respective atoms. There must be a sufficient number of electron-accepting atoms to account for all the electrons donated and vice versa.

TABLE 3.2 COMBINING RATIO OF ATOMS

Number of electrons per atom		Simplest combining ratio of atoms, donor/acceptor	Number of atoms × number of electrons	
Donated	Accepted		Donor	Acceptor
1	1	1:1	$1 \times 1 = 1$	$1 \times 1 = 1$
2	1	1:2	$1 \times 2 = 2$	$2 \times 1 = 2$
3	1	1:3	$1 \times 3 = 3$	$3 \times 1 = 3$
3	2	2:3	$2 \times 3 = 6$	$3 \times 2 = 6$
1	2	2:1	$2 \times 1 = 2$	$1 \times 2 = 2$
2	2	1:1	$1 \times 1 = 1$	$1 \times 1 = 1$
1	3	3:1	$3 \times 1 = 3$	$1 \times 3 = 3$
2	3	3:2	$3 \times 2 = 6$	$2 \times 3 = 6$
3	3	1:1	$1 \times 3 = 3$	$3 \times 1 = 3$

3.4 ELECTRONS PER GRAM ATOM

How many electrons are donated by 1 g atom of A in Eq. (3.1)? Since 1 g atom of any element contains 6.02×10^{23} atoms and in this reaction each atom of A donates one electron, there would be a total of 6.02×10^{23} electrons (Avogadro's number). How many Avogadro's number of electrons are transferred when 1 g atom of C reacts with B in Eq. (3.3)? Each atom of C donates two electrons; therefore, 1 g atom, or 6.02×10^{23} atoms, would donate $2 \times 6.02 \times 10^{23}$, or 12.04×10^{23}, electrons. *There is one Avogadro's number of electrons involved per gram atom for each electrovalent bond formed.*

Valence or combining power

The valence, or combining power, of an element indicates the number of Avogadro's number of electrons shifted when 1 g atom of the element enters into a reaction. Thus, the valence of an element is *neither positive nor negative;* it is an integer. In Eqs. (3.1) to (3.3), A and B have a valence of 1, and C and D have a valence of 2. See Table 3.2 for a summary of the combining ratios of atoms.

3.5 COVALENT BOND

For a covalent bond to be formed between two atoms the orbitals of the outermost electrons of the atoms must overlap. The electrons involved in this type of bond must be shared by the respective atoms because there is not a great enough difference in the electronegativity of the atoms, less than 1.7 units, to form an ionic bond. The bonding electrons become nonspecific with respect to a single atom; they are not associated with any one atom, as in an ionic bond. Consequently, the electron cloud, i.e., the volume in which the electron is traveling, around the atoms is distorted. It becomes increasingly difficult to represent the electronic configuration associated with covalent bonds because the *s*, *p*, *d*, or *f* electrons have different positions of maximum electron density about the nucleus. For an analogy for the formation of a covalent bond see Fig. 3.1*b* and *c*.

A *sigma* (σ) bond exists when the electron cloud of the bond electrons is symmetrical about a line drawn through the nuclei of the atoms (Fig. 3.2). A σ bond may be formed by the interaction of *two s electrons, an s and a p electron,* or *two p electrons.* Although the distribution of electrons in these three cases is not the same, the electron

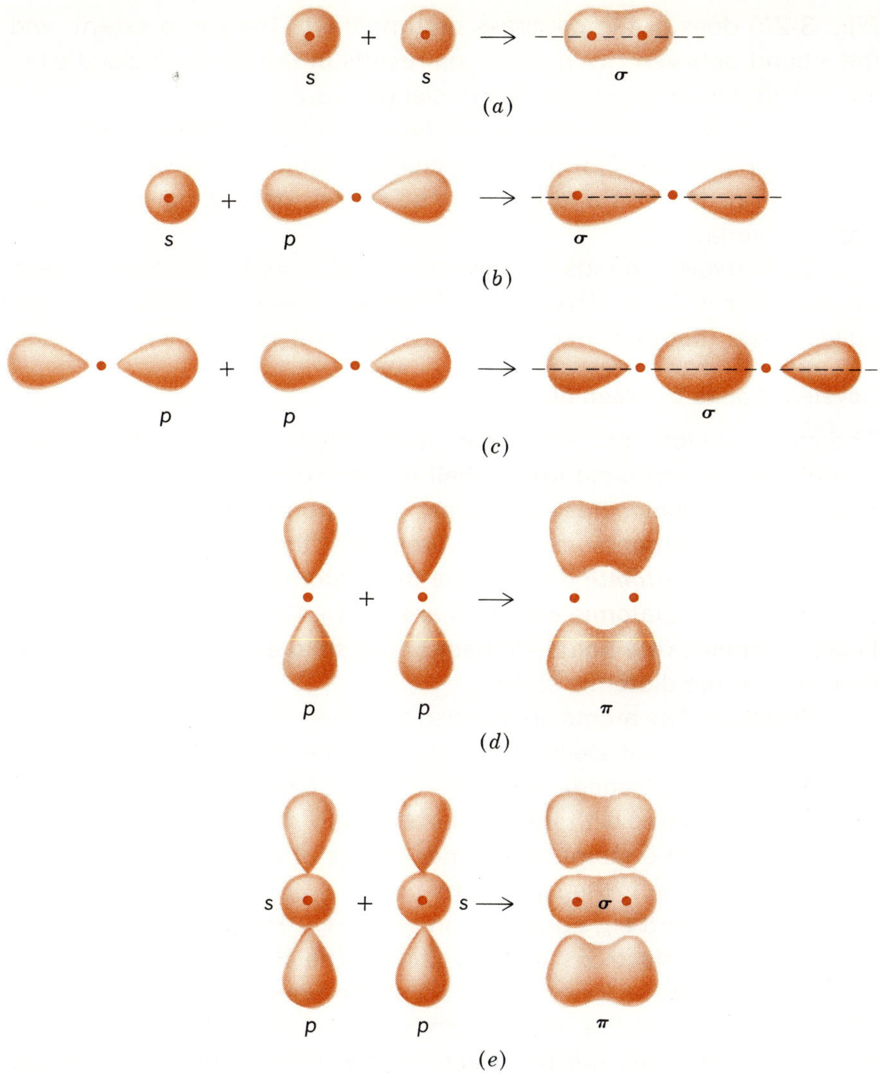

FIGURE 3.2 Types of σ and π bonds: (*a*) σ bond between two *s* electrons; (*b*) σ bond between an *s* and a *p* electron; (*c*) σ bond between two *p* electrons; (*d*) π bond between two *p* electrons; (*e*) double bond involving a σ bond between two *s* electrons and a π bond between two *p* electrons.

cloud is symmetrical about a line drawn between the nuclei of the bonded atoms. The σ bond formed by the interaction of two *s* electrons results in an electron cloud that encompasses both nuclei to the same extent (Fig. 3.2*a*). The σ bond formed between an *s* and a *p* electron

(Fig. 3.2*b*) does not encompass both nuclei to the same extent, and the σ bond between two p electrons results in the electron cloud's being distributed between the two nuclei (Fig. 3.2*c*).

A *pi* (π) *bond* is formed when two p-electron orbitals overlap, as shown in Fig. 3.2*d*. The resulting electron cloud is above and below the nuclei or in front and back of the nuclei, depending upon which orbitals overlap.

Two covalent bonds, a σ and a π bond, may be formed between two nuclei (Fig. 3.2*e*). The nuclei of the two atoms in combination are pulled closer together than if only a single covalent bond were present.

Covalent bonds between like atoms

The gaseous elements which are chemically active, i.e., those which do not have a complete outer shell of electrons, usually exist in the free state as molecules containing two atoms joined by covalent bonds. These elements are called *diatomic* if they have two atoms per molecule and *triatomic* if they have three atoms per molecule. Examples of gaseous diatomic elements are H_2, F_2, Cl_2, N_2, and O_2. Br_2 and I_2 are examples of a liquid element and a solid element, respectively, that exist in the diatomic state.

When two like atoms are bonded together, neither atom is capable of causing a shift of electrons away from the other since each atom has the same electronegativity. Therefore, it can be assumed that the electrons associated with the bond are equally distributed between the atoms and that the atoms have a net charge of zero (Fig. 3.1*c*).

For example, if the two atoms in Fig. 3.2*a* represent two hydrogen atoms, H_a and H_b, combination gives $H_a—H_b$ joined by a σ bond. In $H_a—H_b$ both electrons may be around the H_a nucleus part of the time and around the H_b nucleus part of the time. Part of the time the electrons may be at the opposite ends of the two nuclei or someplace between them. Thus the two electrons forming the bond are equally associated with each nucleus; this is equivalent to stating that each hydrogen atom has a full complement of electrons in its $n = 1$ orbital (see Table 2.4). Electrons can be represented by dots, as shown in Table 3.3. The dividing line between the bond electrons indicates that each hydrogen atom has been assigned one electron, thus giving it the same status as an atom of hydrogen before combination, namely, a zero net charge.

The combination of two Cl atoms to form a molecule of Cl_2 illustrates the formation of bonds between atoms with more than one elec-

TABLE 3.3 ELECTRONIC FORMULAS FOR SOME DIATOMIC MOLECULES[a]

Monatomic species				Diatomic species				
H·	+	H·	⟶	H	:	H	H—H	H_2
:C̤̈l·	+	:C̤̈l·	⟶	:C̤̈l	:	C̤̈l:	Cl—Cl	Cl_2
:Ṇ:	+	:Ṇ:	⟶	:N	⋮	N:	N≡N	N_2
:O̤:	+	:O̤:	⟶	:Ȯ̤	:	Ȯ̤:	O—O	O_2

[a] The ⌐ indicates that the bonding electrons are equally shared by the two atoms.

tron in the outermost shell (Table 3.3). Only the electrons in the last shell are shown. Since the two atoms have the same electronegativity, the electrons making the bond are equally divided between the two atoms. This would give each atom a net charge of zero. The rule of eight has been satisfied in forming these diatomic molecules because each atom has reacted in a manner that gives it a complete outer shell of electrons.

More than one covalent bond may exist between like atoms in molecules. For example, in the nitrogen molecule, N_2, there are *six shared electrons,* three from each atom of nitrogen, thus making a total of three covalent bonds (Table 3.3).

Oxygen exists in the free state as a diatomic molecule, O_2, or as the triatomic molecule, O_3, ozone. Whereas O_2 is stable, O_3 is very unstable and decomposes into O_2. Diatomic oxygen, O_2, is apparently an exception to the rule of eight. Careful studies of its molecular structure indicate that there is only one covalent bond between the two atoms and that the other electron of each atom is associated with the entire molecule rather than taking part in a localized bond. Thus we may write O_2 as O—O.

Covalent bonds between unlike atoms

When two atoms with different electronegativities combine, an unequal distribution of electron density about the two atoms results (see Fig. 3.1*b*). The more electronegative atom has a greater electron density around it than the atom with the lower electronegativity. This is illustrated with HCl as the example in Fig. 3.3. The small arrow indicates the direction of the electron shift. As a result of this unequal distribution of electrons, the molecule is *polar;* i.e., it has a positive and a negative end although true ions do not exist. *Polarity* is the unequal

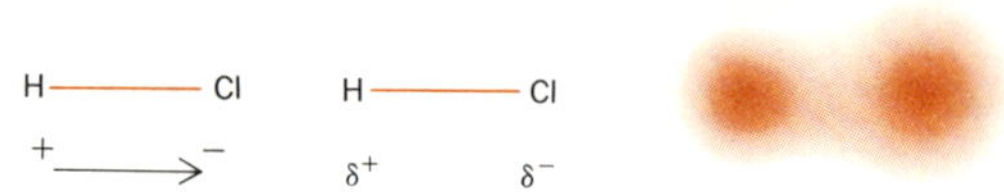

FIGURE 3.3 Electron distribution in a molecule of hydrogen chloride, HCl.

distribution of electrons between atoms. Most compounds are polar to some extent. Polarity is important in determining the direction of reaction as well as some of the physical properties such as solubility of a given compound.

3.6 OXIDATION NUMBER

The oxidation number of an element is the *charge* or *assigned charge* that an atom has when it enters into a chemical reaction. *Free* elements have a *zero oxidation number* since they are electrically neutral until they give up or accept electrons in a reaction.

The oxidation number in Eq. (3.1) of the A^+ ion is $+1$ and that of the B^- ion is -1. An ion represented by C^{++} has an oxidation number of $+2$, whereas ion D^{--} has an oxidation number of -2. In these equations it was indicated that ions are formed first and then the ions are attracted to each other to form the respective molecules. Since ions are charged atoms or groups of atoms resulting from a complete shift of electrons, they have a real oxidation number. However, atoms may have an oxidation number without being ions; the fact that an atom has an oxidation number *does not mean* that it is an ion.

Oxidation numbers are integers. In calculating the oxidation number of the atoms in a compound, it is conventional to assign the electrons of a covalent bond to the *more electronegative* atom even though there has not been a complete shift of electrons. According to this method,

$$\underset{+1}{\mathrm{H}} \quad \underset{-1}{:\ddot{\underset{..}{\mathrm{Cl}}}:} \qquad \text{ox. no.}$$

In almost all common compounds, hydrogen has an oxidation number of $+1$ and oxygen of -2. Knowing these two oxidation numbers, it is a rather simple matter to calculate the oxidation number of other elements in a compound that contains either or both of these elements. A single covalent bond is formed by sharing two electrons, one from each of the atoms concerned. The formation of a double covalent bond involves sharing of four electrons between two atoms.

A triple covalent bond has a total of six electrons shared between the respective atoms. A covalent bond is represented by —, a double covalent bond by =, and a triple covalent bond by ≡.

In the following examples the shift of the electron toward the atom of higher electronegativity will be indicated by a small arrow to aid in calculating the oxidation number of the atoms.

1. Indicate the oxidation number of each atom in water, H_2O. Since oxygen has six electrons in its outer orbit, it would require two hydrogen atoms to furnish the two electrons needed to complete its octet:

H_2O 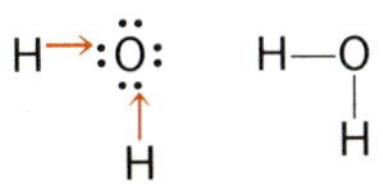

H_2	O	
+1	−2	ox. no.

Water is a nonlinear, polar molecule; the atoms do not lie in a straight line, and each hydrogen atom has its electron shifted away from it toward the more electronegative oxygen. The oxidation number of hydrogen is +1 and of oxygen −2. The entire molecule must have a net charge of zero (total positive = 2 × 1, total negative = 1 × 2).

2. Carbon can have one of several oxidation numbers from −4 to +4. Several compounds of carbon are used in the following examples to illustrate this variation in oxidation number. The valence electrons of carbon are tinted to keep them distinct from the electrons of the elements to which it is bound.

(*a*)

Methane:

CH_4 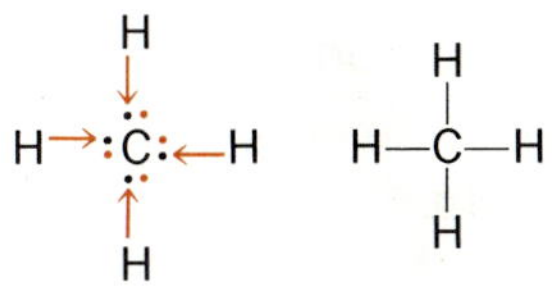

C	H_4	
−4	+1	ox. no.

Electronegativity: H = 2.1, C = 2.5 (see Table 3.1)

Carbon is more electronegative than hydrogen, and so the electron of each of the four hydrogen atoms is shifted toward the carbon atom giving each hydrogen an apparent oxidation number of +1 and carbon a −4 value.

Methyl chloride:

CH_3Cl 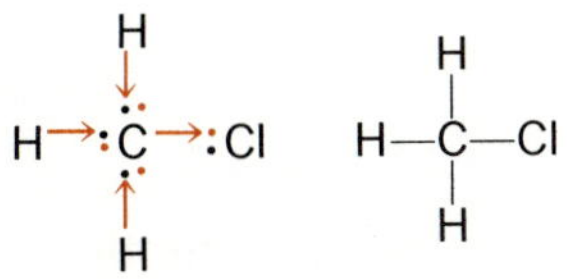

C	H_3	Cl	
−2	+1	−1	ox. no.

Electronegativity: H = 2.1, C = 2.5, Cl = 3.0

In methyl chloride each electron of the three hydrogen atoms is shifted toward the carbon atom, and one electron of the carbon atom has been shifted toward the nucleus of the chlorine atom. The net change is an apparent gain of two electrons for the carbon atom; therefore, carbon is said to have an oxidation number of -2. Each hydrogen atom has an oxidation number of $+1$, and the chlorine atom has a -1 oxidation number.

(*b*)

Methyl alcohol:

CH_3OH

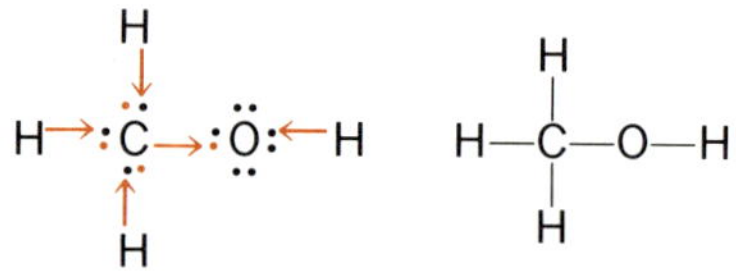

C H_4 O

-2 $+1$ -2 ox. no.

Electronegativity: H = 2.1, C = 2.5, O = 3.5

Another example in which carbon has an oxidation number of -2 is methyl alcohol, which contains four hydrogen atoms, one carbon atom, and one oxygen atom.

(*c*)

Dichloromethane:

CH_2Cl_2

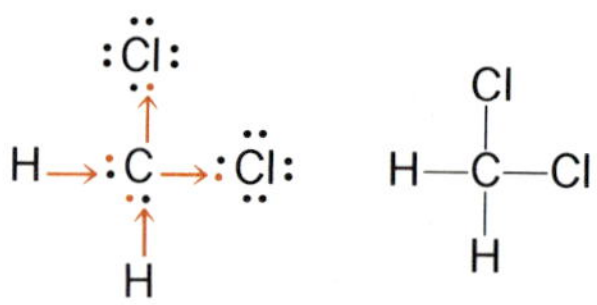

C H_2 Cl_2

0 $+1$ -1 ox. no.

Electronegativity: H = 2.1, C = 2.5, Cl = 3.0

In dichloromethane the carbon atom has apparently gained two electrons, one from each of the two hydrogen atoms; however, it has had two electrons shifted away from its nucleus, one toward each chlorine atom. Therefore, the net change in electron shift of the carbon atom is zero. The oxidation number of carbon is zero in this example. Chlorine has an oxidation number of -1.

(*d*)

Chloroform:

$CHCl_3$

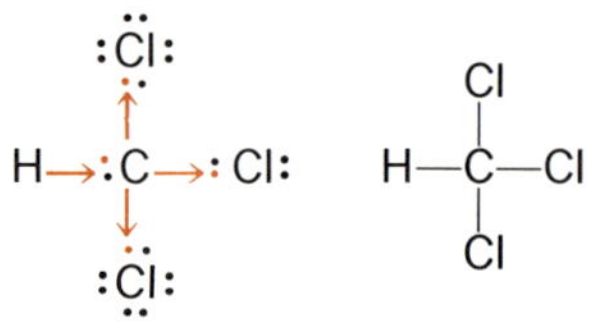

C H Cl_3

$+2$ $+1$ -1 ox. no.

Electronegativity: H = 2.1, C = 2.5, Cl = 3.0

In chloroform the carbon atom has apparently lost three electrons and gained one; thus the net effect is the loss of two electrons, giving carbon an oxidation number of +2. Chlorine has an oxidation number of −1.

3.

Hypochlorous acid:

HOCl H→:Ö:←C̈l: H—O—Cl H O Cl

+1 −2 +1 ox. no.

Electronegativity: H = 2.1, O = 3.5, Cl = 3.0

Chlorine in hypochlorous acid has an oxidation number of +1 because its electron that forms the covalent bond with oxygen has been shifted toward the more electronegative oxygen atom.

3.7 COORDINATE COVALENT BOND

A coordinate covalent bond is formed when one atom furnishes both electrons instead of each atom furnishing one electron. Many compounds contain a coordinate covalent bond between oxygen and chlorine, sulfur, or phosphorus. These three elements furnish two electrons to the more electronegative oxygen to form the coordinate covalent bond:

Chloric acid:

$HClO_3$

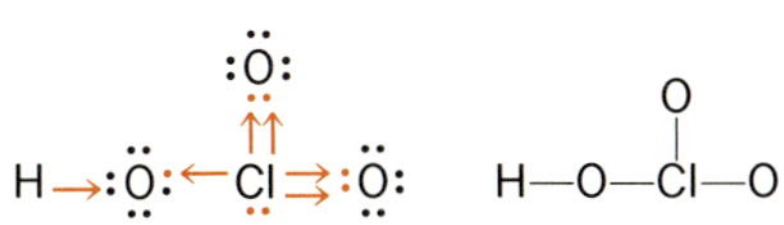

H Cl O

+1 +5 −2 ox. no.

Perchloric acid:

$HClO_4$

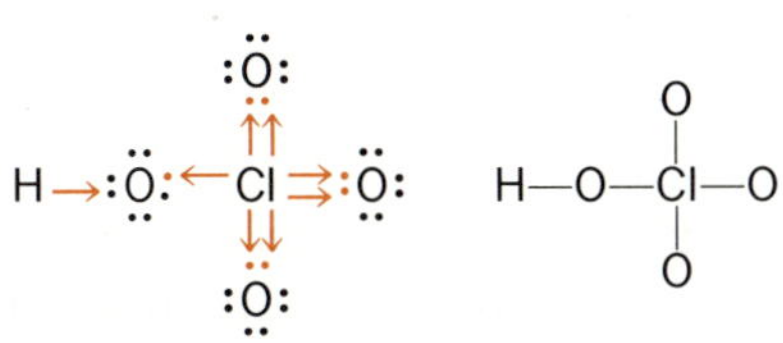

H Cl O

+1 +7 −2 ox. no.

Electronegativity: H = 2.1, O = 3.5, Cl = 3.0

The ⟹ in the formulas indicates a coordinate covalent bond involving two electrons from the atom from which the arrow points. Frequently the bonds between the chlorine atom and oxygen that are coordinate covalent in the above formulas are written as double bonds, in which case the electron-donating atom would have more than eight electrons in its outer shell. For example, perchloric acid would be written:

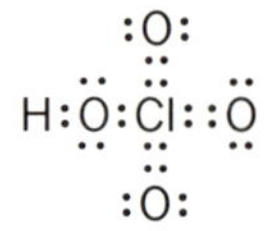

Since electrons are in continual motion and the bond electrons are really *associated with the molecule as a whole,* this manner of writing the formula is only for convenience. The chlorine in chloric acid has had five electrons shifted toward the oxygen atoms; therefore, chlorine has an oxidation number of +5. In perchloric acid there has been a total of seven electrons shifted from the single chlorine atom toward the four oxygen atoms; hence this chlorine has an oxidation number of +7. Is the net oxidation number for these molecules zero? Check this as follows:

Atom Kind	Atom No.	Ox. no.	Total ox. no.
H	1	+1	$1 \times +1 = +1$
Cl	1	+5	$1 \times +5 = +5$
O	3	−2	$3 \times -2 = -6$
			Net ox. no. = 0

Atom Kind	Atom No.	Ox. no.	Total ox. no.
H	1	+1	$1 \times +1 = +1$
Cl	1	+7	$1 \times +7 = +7$
O	4	−2	$4 \times -2 = -8$
			Net ox. no. = 0

The formulas are correct with respect to the rule of eight and the oxidation number of the elements. As illustrated in the chlorine compounds, chlorine may have an oxidation number of −1, +1, +5, or +7.

The ability of sulfur and phosphorus to form coordinate covalent bonds is illustrated in the following examples:

Sulfuric acid:

H_2SO_4

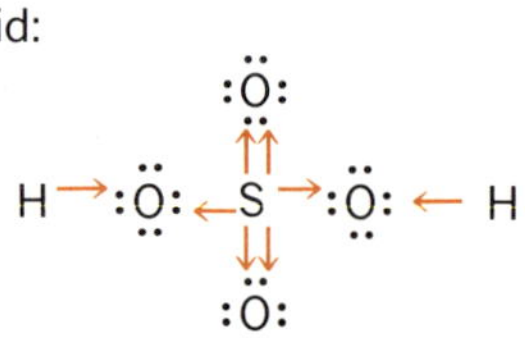

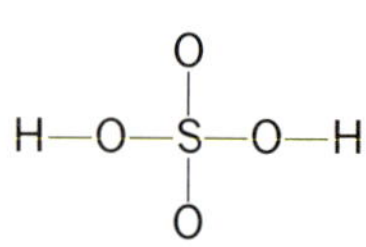

H_2	S	O_4	
+1	+6	−2	ox. no.

Electronegativity: H = 2.1, S = 2.5, O = 3.5

Phosphoric acid:

H_3PO_4 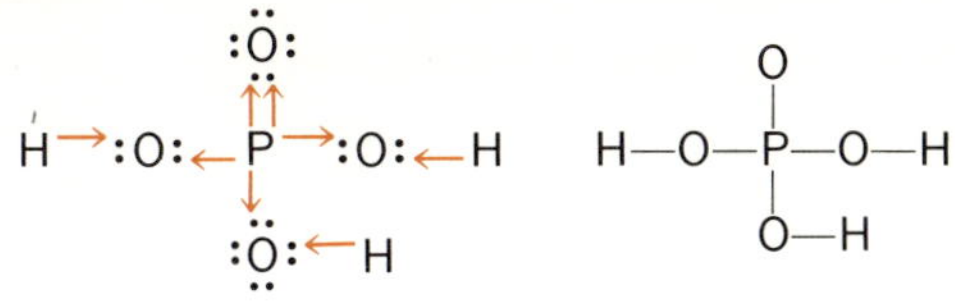

H_3	P	O_4	
+1	+5	−2	ox. no.

Electronegativity: H = 2.1, P = 2.1, O = 3.5

Chromium, Cr, and manganese, Mn, are two mineral elements that in addition to forming ionic bonds ($CrCl_2$, $MnCl_2$) can also form coordinate covalent bonds with oxygen in certain compounds. From the position of these two elements in the periodic table, would you expect them to have a smaller or a larger electronegativity value than oxygen? Are the 3d and the 4s electrons of about equal energy?

The electron distribution in these two elements is

Cr:	$1s^2, 2s^2, 2p^6, 3s^2, 3p^6, \mathbf{3d^4, 4s^2}$	(6 electrons)
Mn:	$1s^2, 2s^2, 2p^6, 3s^2, 3p^6, \mathbf{3d^5, 4s^2}$	(7 electrons)

In $KMnO_4$ the $3d^5$ and $4s^2$ electrons (seven electrons total) of manganese enter into bond formation. Thus, in essence, manganese may react in a manner similar to chlorine with its seven electrons in the outer shell.

Potassium permanganate:

$KMnO_4$ 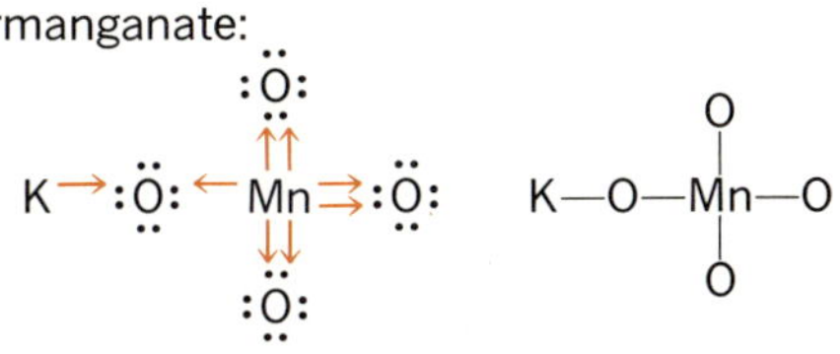

K	Mn	O_4	
+1	+7	−2	ox. no.

Electronegativity: K = 0.8, O = 3.5, Mn = 1.5

As noted in the previous examples of coordinate covalent bonds, oxidation numbers of a given element may vary considerably from one compound to another. The oxidation number of these elements can easily be calculated by determining the difference between the sum of the oxidation number and the unknown element; remember the net oxidation number of a molecule must be zero. In common compounds oxygen may be assumed to have an oxidation number of −2 and elements in the first and second rows of the periodic table may be assumed to have an oxidation number of +1 and +2, respectively. Bearing this in mind, the oxidation number of manganese can be calculated as follows:

$KMnO_4$:	Kind	Atom no.	Ox. no.	Total ox. no.
	K	1	+1	$1 \times +1 = +1$
	O	4	−2	$4 \times -2 = -8$
				Net ox. no. = −7

To have a zero net oxidation number for the molecule of potassium permanganate, the manganese atom must have an oxidation number of +7.

Consider the following manner of arriving at the electronic formula for $KMnO_4$, which is based upon the oxidation numbers of the elements and the rule of eight. If the oxidation number of potassium, K, is considered to be +1 in all common compounds, it must give up one electron to an atom which has a negative oxidation number in the final compound. The only negative-oxidation-numbered element in $KMnO_4$ is oxygen; therefore, K has to be bonded to O, as K:O. Similar reasoning indicates that the O of KO is bonded to an element with a positive oxidation number from which it receives an eighth electron. This positive element is manganese, and the partial formula can be written K:O:Mn. Potassium and oxygen now have their complete complement of outer-shell electrons, and manganese has a +1 oxidation number. For manganese to have an oxidation number of +7 it must have six more electrons shifted away from its nucleus. For this to happen, these six electrons must have been around the manganese nucleus to begin with. When these six electrons are put about the manganese atom, it gives the partial formula $\mathrm{K{:}\underset{\cdot\cdot}{\overset{\cdot\cdot}{O}}{:}\underset{\cdot\cdot}{\overset{\cdot\cdot}{Mn}}{:}}$, in which manganese now has its full complement of electrons in the outer shell. This means that the remaining three atoms of oxygen must be arranged about the manganese in such a manner as to attract two electrons each away from the manganese but not to furnish any electrons for the bond. Otherwise, there would be more than eight electrons around the manganese atom. The oxygen can be added under these conditions by forming a coordinate covalent bond with each of the three oxygen atoms.

Sodium tetrathionate has the molecular formula $Na_2S_4O_6$. What is the oxidation number of sulfur in this compound? The elements with known oxidation numbers are Na = +1 and O = −2. The total positive oxidation number is 2 and the total negative oxidation number is 12. This leaves +10 for the four sulfur atoms, or a value of +2.5 for each sulfur. How can a fractional value for an oxidation number be justified? An electron cannot be split! Is there any reason why each of the sulfur atoms in this compound must have the same oxidation number? The answer to this question is the key to the problem. Not every sulfur atom has the same oxidation number. The total of $+10$ is distributed among four atoms; thus, the value $+2.5$ is an average value for each atom. Unless the distribution of electrons is indicated, the exact oxidation number of each sulfur atom is not known.

3.8 FORMULAS

From the previous discussion, two generalizations can be made: (1) atoms generally combine in accordance with the rule of eight; and (2) the net oxidation number of a molecule must always equal zero. Since the oxidation number is a reflection of the electron shift when atoms combine, the oxidation number for the common elements and radicals must be memorized.

A *radical* is a group of atoms that reacts chemically as a single ion. Radicals have an oxidation number. They do not exist alone in the free state but must be in combination with another radical or element. A list of some common radicals is found in Table 3.4.

The *molecular formula* indicates the kind and number of atoms in the molecule. It does not give any indication of the distribution of electrons or type of bond between the respective atoms. For example, H_2O, H_2SO_4, H_3PO_4, $KClO_3$, $Ca_3(PO_4)_2$, $K_2Cr_2O_7$, and Cl_2 are molecular formulas.

The *structural formula* tells not only the kind and number of atoms in a molecule but also the arrangement of the atoms in the molecule:

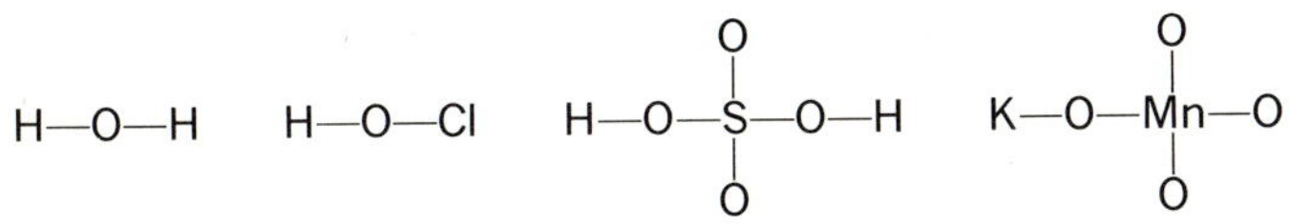

To write correct molecular formulas it is essential to have a zero net oxidation number for the molecule. In inorganic compounds generally the positive-oxidation-number moiety is written first, NH_3 being an exception to this rule.

The procedure for writing formulas is really quite simple. First, write the atoms or radicals with their oxidation numbers. Then determine the least common denominator (LCD) of the oxidation numbers. Divide the oxidation numbers of the atoms or radicals into the LCD.

This quotient is the number of atoms of the respective elements in the molecule. In simple compounds one can take the absolute numerical value of the positive-oxidation-numbered atom and use it as a subscript to the negative-oxidation-numbered unit and vice versa (see Question 1).

By learning to recognize *classes* of compounds, you can easily learn how to name a large number of molecules which you may not previously have encountered. The naming of compounds is part of the language of chemistry; without such aids to communication it would be impossible to talk intelligently about the food we eat, how it is digested, or how our bodies are able to obtain energy from the food.

3.9 NOMENCLATURE

The suffix *-ide*

Classes of compounds containing only two different atoms are named by adding the suffix *-ide* to the stem name, e.g., chloride, sulfide, and oxide from chlorine, sulfur, and oxygen, respectively. To name a specific member of this class the name of the positive-oxidation-numbered element precedes the class name, and the suffix is added to the member having the negative oxidation number.

Formula	*Name*
NaCl	Sodium chloride
$AlCl_3$	Aluminum chloride
K_2S	Potassium sulfide
CaS	Calcium sulfide
KBr	Potassium bromide
Li_2O	Lithium oxide
SO_2	Sulfur dioxide
SO_3	Sulfur trioxide

In the last two examples, the prefixes *di-* and *tri-* indicate the number of oxygens per sulfur atom.

The suffixes *-ous* and *-ic*

Some elements have more than one positive oxidation number. The suffix *-ous* is used to indicate the one with the lower oxidation number and the suffix *-ic* the one with the higher oxidation number. Roman numerals are used in the formula to indicate this difference. Iron, copper, and manganese are examples of elements that exist in two different oxidation states.

Formula	*Name*
$Fe^{++}O$	Ferrous oxide or Fe(II) oxide
$Fe_2^{3+}O_3$	Ferric oxide or Fe(III) oxide
$Fe^{++}Cl_2$	Ferrous chloride or Fe(II) chloride
$Fe^{3+}Cl_3$	Ferric chloride or Fe(III) chloride
$Cu^{+}Br$	Cuprous bromide or Cu(I) bromide
$Cu^{++}Br_2$	Cupric bromide or Cu(II) bromide

Radicals

When the portion of the molecule having the negative oxidation number is a radical, the name of the unit having the *positive* oxidation number is given first.

Formula	*Name*	*Radical*
NaOH	Sodium **hydroxide**	**—OH**
KNO_3	Potassium **nitrate**	**—NO_3**
$Fe_2(SO_4)_3$	Ferric **sulfate**	**—SO_4**
$AlPO_4$	Aluminum **phosphate**	**—PO_4**

When the negative radical has an oxidation number of greater than 1, there is the possibility that more than one kind of positive unit may be in combination with the negative unit, such as $NaKSO_4$, sodium potassium sulfate. $NaHCO_3$ may be called sodium hydrogen carbonate, sodium bicarbonate, or sodium acid carbonate because the presence of the hydrogen makes the compound an acid (see Sec. 4.5). The *bi-* in bicarbonate is used to indicate that hydrogen is one of the positive-oxidation-numbered elements; another example is sodium bisulfite, $NaHSO_3$.

Some radicals contain a common positive element that may have two different oxidation numbers, such as

$$\underset{+4}{S}\,\underset{-2}{O_3^{--}} \quad \text{and} \quad \underset{+6}{S}\,\underset{-2}{O_4^{--}} \quad \text{ox. no.}$$

Compounds in which the positive-oxidation-numbered element is *not hydrogen* and which also contain a radical having the common element with the *higher oxidation* number have the suffix *-ate*. Those in which the common element has the *lower oxidation* number have the suffix *-ite*. This has already been illustrated above with sodium potassium sulfate and potassium bisulfite. The radical NO_3^- is the *nitrate* radical, whereas NO_2^- is the *nitrite* radical. What is the oxidation number of the nitrogen in each of these radicals?

Acids

Certain compounds are classified as acids, and for the present discussion on nomenclature it is sufficient to say that they contain hydrogen as a positive-oxidation-number element in the formula. A more sophisticated discussion of acids is reserved for Chap. 11.

Acids that do not contain oxygen are called *hydro · · · ic* acids.

Formula	*Name*
HCl	Hydrochloric acid
HBr	Hydrobromic acid

Acids which contain oxygen in a radical in which the common element is in the lower oxidation state are called *-ous* acids. Acids that have the common element in its higher oxidation state are called *-ic* acids.

Formula	*Name*	*Radical*
H_2SO_4	Sulfuric acid	—SO_4
H_2SO_3	Sulfurous acid	—SO_3
HNO_3	Nitric acid	—NO_3
HNO_2	Nitrous acid	—NO_2

An examination of the diagrams in Fig. 3.4 and the list of common compounds with their formula and name in Table 3.4 will aid in clarifying the nomenclature of some simple inorganic compounds.

TABLE 3.4 FORMULA AND NAME OF SOME COMMON RADICALS AND INORGANIC COMPOUNDS

Radical		**Compound**	
Name	**Formula**	**Name**	**Formula**
Hydroxide	OH^-	Aluminum hydroxide	$Al(OH)_3$
Nitrate	NO_3^-	Silver nitrate	$AgNO_3$
Nitrite	NO_2^-	Sodium nitrite	$NaNO_2$
Sulfate	SO_4^{--}	Potassium sulfate	K_2SO_4
Sulfite	SO_3^{--}	Calcium sulfite	$CaSO_3$
Bicarbonate	HCO_3^-	Magnesium bicarbonate	$Mg(HCO_3)_2$
Carbonate	CO_3^{--}	Calcium carbonate	$CaCO_3$
Phosphate	PO_4^{3-}	Ferric phosphate	$FePO_4$
Chlorate	ClO_3^-	Potassium chlorate	$KClO_3$
Perchlorate	ClO_4^-	Magnesium perchlorate	$Mg(ClO_4)_2$
Arsenate	AsO_4^{3-}	Sodium arsenate	Na_3AsO_4
Arsenite	AsO_3^{3-}	Cupric arsenite	$Cu_3(AsO_3)_2$
Permanganate	MnO_4^-	Potassium permanganate	$KMnO_4$
Chromate	CrO_4^{--}	Sodium chromate	Na_2CrO_4
Dichromate	$Cr_2O_7^{--}$	Potassium dichromate	$K_2Cr_2O_7$
Ammonium	NH_4^+	Ammonium chloride	NH_4Cl

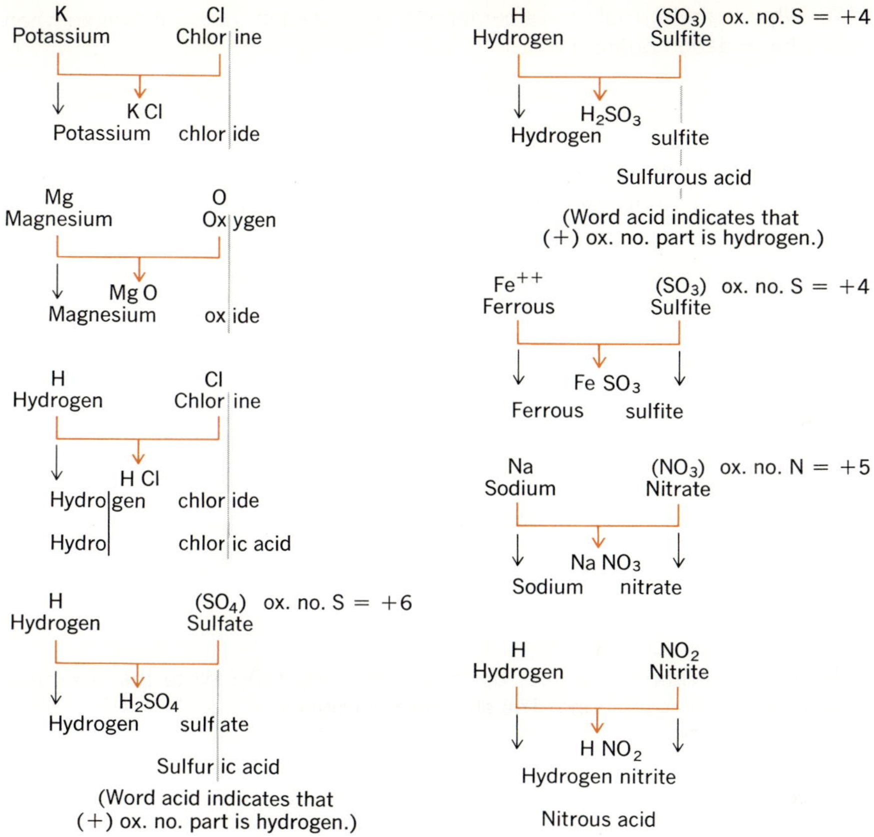

FIGURE 3.4 Scheme for applying the rules of nomenclature of inorganic molecules.

QUESTIONS

1. (*a*) What is the correct formula for the compound composed of magnesium and chlorine?

$$\begin{array}{ll} & \text{Mg} \quad \text{Cl} \\ \text{Ox. no.:} & +2 \quad -1, \quad \text{LCD} = 2 \end{array}$$

The correct formula is $MgCl_2$.

(*b*) Write a correct formula for a compound composed of only nitrogen and hydrogen.

$$\begin{array}{ll} & \text{N} \quad \text{H} \\ \text{Ox. no.:} & -3 \quad +1, \quad \text{LCD} = 3 \end{array}$$

The correct formula is NH_3.

(*c*) What is the formula for a compound composed of iron and oxygen when the oxidation number of iron is +3?

Fe O

Ox. no.: +3 −2, LCD = 6

The correct formula is Fe_2O_3.

(*d*) Write the formula for calcium nitrate (nitrate is NO_3).

Ca NO_3

Ox. no.: +2 −1, LCD = 2

The correct formula is $Ca(NO_3)_2$.

(*e*) What is the formula for ammonium phosphate?

NH_4 PO_4

Ox. no.: +1 −3, LCD = 3

The correct formula is $(NH_4)_3PO_4$.

(*f*) What is the formula for aluminum sulfate?

Al SO_4

Ox. no.: +3 −2, LCD = 6

The correct formula is $Al_2(SO_4)_3$.

2. What is the oxidation number of chromium in $K_2Cr_2O_7$? Write the electronic formula for this compound. The structural formula is

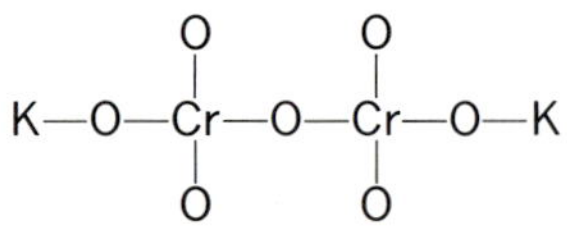

3. Explain how the oxidation number of the respective sulfur atoms was obtained:

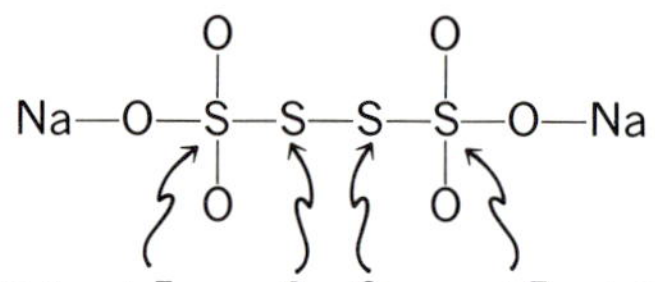

Ox. no.: +5 0 0 +5 ; av. ox. no. = 2.5

4. Define (*a*) valence and (*b*) oxidation number. Note the differences between these terms.

5. Give the name and oxidation number for each of the following radicals:

NO_3	NO_2	MnO_4
SO_3	NH_4	CO_3

6. Write the formula for:

Ammonium phosphate	Cuprous arsenite
Magnesium sulfite	Calcium nitrate

Ferric bicarbonate	Aluminum oxide
Potassium chromate	Copper(I) oxide

7. Write the formula for the following acids:

Hydroiodic	Phosphoric
Hydrosulfuric	Sulfurous
Nitric	Carbonic

8. Write the name for the following compounds:

$Al(OH)_3$	$Mg_3(PO_4)_2$	$(NH_4)_2SO_4$
FeO	$BaSO_3$	$Al(NO_2)_3$

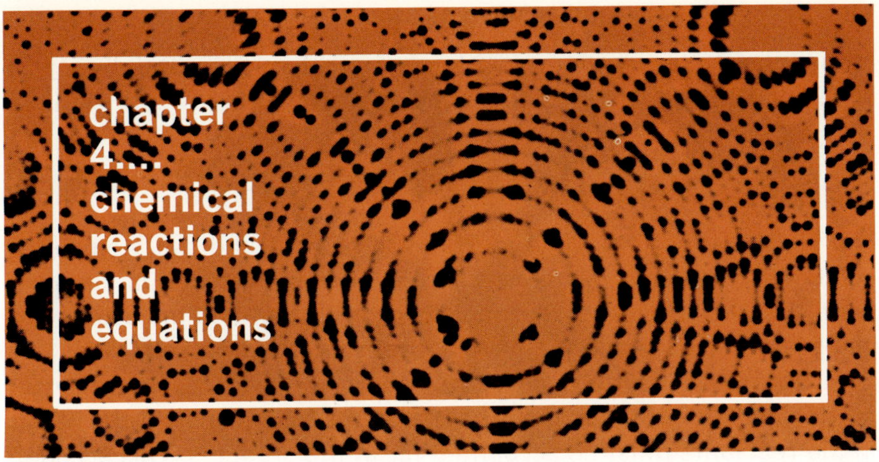

chapter 4 chemical reactions and equations

We live with chemical reactions every second of our lives. In fact, life itself may be considered to be an orderly arrangement of controlled chemical reactions each having a specific role in the overall process.

Which elements can be expected to form ionic bonds? Covalent bonds? What variations are possible in the shape of the electron cloud enveloping a bond? What is a polar covalent bond? How does an atom become charged?

4.1 CHEMICAL EQUATIONS

A chemical equation is a shorthand way of writing a chemical reaction, the compounds reacted, and the compounds formed. How do you determine whether or not a chemical reaction has taken place? At present, we shall use only the criterion that the products are different from the starting materials, or reactants.

To write a chemical equation it is necessary first to write the correct formulas for the reactants and the products of the reaction. For example, the reaction of sodium chloride and silver nitrate to form silver chloride and sodium nitrate is written

$$\underbrace{NaCl + AgNO_3}_{\text{Reactants}} \longrightarrow \underbrace{AgCl + NaNO_3}_{\text{Products}}$$

By convention the reactants are written to the left of the arrow, and the products are written to the right. It is *essential* that these formulas be *correctly written.* The plus sign indicates that the compounds have been *mixed* but no chemical reaction has yet taken place.

If more than one product is formed in a reaction, a mixture

results; hence, each compound formed must be separated by a plus sign. For example, consider the interpretation of

H_2O_2	or	$H_2 + O_2$
Hydrogen peroxide		Hydrogen and Oxygen

The H_2O_2 means that there are two atoms of H and two atoms of O combined chemically in a molecule of hydrogen peroxide. On the other hand, $H_2 + O_2$ indicates that molecular hydrogen and molecular oxygen are mixed together and have not yet reacted chemically.

Regardless of the type of reaction, the law of conservation of matter *must* hold. When chemical reactions take place, all the atoms in the reactants must be accounted for in the products. This accounting procedure is called *balancing* the equation.

4.2 TYPES OF REACTIONS

Numerous types of chemical reaction can occur:

1. *Combination* or *addition:* Two or more compounds react to form only one product

$$A + B \longrightarrow C$$

2. *Decomposition:* One compound reacts to yield two or more products

$$C \longrightarrow A + B$$

3. *Double decomposition* or *double displacement:* Two compounds react to yield two different products

$$AB + CD \longrightarrow AD + CB$$

4. *Substitution:* One element replaces another in a compound

$$AB + C \longrightarrow AC + B$$

5. *Oxidation-reduction:* A change occurs in the oxidation state of at least two of the elements involved in the reaction

$$A^+B^- + C^0 \longrightarrow A^+C^- + B^0$$

Inorganic reactions are usually carried out in an aqueous (water) solution; under these conditions the molecules dissociate into ions. However, since the purpose of this section is to balance equations, discussion of ionic reactions is postponed (see Chap. 6).

4.3 COMBINATION OR ADDITION REACTIONS

A reaction in which two or more compounds combine to produce only one new compound is an *addition reaction.*

EXAMPLE 1

	H_2 +	O_2 ⟶	H_2O	unbalanced
Ox. no.	0	0	+1, −2	

As it now stands this equation is not balanced; i.e., there is one less oxygen atom on the right than on the left. Since the *formulas cannot be changed* without changing the chemical reaction, to balance the equation one must chance the number of whole molecules involved. To get two atoms of O on the right side, we need two whole molecules of H_2O. This is indicated by putting a 2 in front of the formulas: 2 H_2O, which means that there are two of everything in the molecule that follows the number. Thus, in 2 H_2O there are a total of four atoms of hydrogen and two of oxygen. When a 2 is placed in front of H_2O, the oxygen becomes balanced, but now the hydrogen is not.

$$H_2 + O_2 \longrightarrow 2\ H_2O$$

We need four atoms of hydrogen on the left, and we must therefore take two molecules of hydrogen, 2 H_2. The equation is now balanced:

$$2\ H_2 + O_2 \longrightarrow 2\ H_2O$$

EXAMPLE 2

Write a balanced equation for the reaction of hydrogen with chlorine. Since chlorine is a chemically active gas, it exists as a diatomic molecule, Cl_2. The reactants are H_2 and Cl_2. The product of the reaction must be the combination of the two elements, namely, HCl. Is this the correct formula for the product? Yes: since hydrogen has an oxidation number of +1 and chlorine of −1, these elements combine in a 1 : 1 ratio. The equation at this point becomes

	H_2 +	Cl_2 ⟶	HCl	unbalanced
Ox. no.	0	0	+1, −1	

If the same reasoning is followed as in the previous example, it is readily seen that 2 HCl must be formed. The final equation is therefore

$$H_2 + Cl_2 \longrightarrow 2\ HCl$$

EXAMPLE 3

Write a balanced equation for the reaction of iron and sulfur to give ferric sulfide [iron(III) sulfide]:

$$\begin{array}{lccc} & Fe + & S \longrightarrow & Fe_2S_3 \\ \text{Ox. no.:} & 0 & 0 & +3, -2 \end{array}$$

This equation is easily balanced by sight; 2 Fe and 3 S are needed on the left of the equation to balance the reaction. Write the balanced equation for this reaction.

EXAMPLE 4

$$\text{Aluminum} + \text{oxygen} \longrightarrow \text{aluminum oxide}$$

$$\begin{array}{lccccc} & Al & + & O_2 & \longrightarrow & Al_2O_3 \\ \text{Ox. no.:} & 0 & & 0 & & +3, -2 \end{array}$$

Balance the oxygen first since aluminum is a monatomic element and any number placed before it to balance the equation will not necessitate changing the amount of Al_2O_3 formed. The LCD is 6; therefore, to get 6 atoms of oxygen on each side of the equation requires 3 O_2 and 2 Al_2O_3:

$$Al + 3\ O_2 \longrightarrow 2\ Al_2O_3$$

The aluminum is now easily balanced by taking 4 Al, and the balanced equation becomes

$$4\ Al + 3\ O_2 \longrightarrow 2\ Al_2O_3$$

4.4 DECOMPOSITION REACTIONS

The conversion of one compound into two or more simpler substances is called *decomposition*. Although the products of the reaction cannot always be predicted, balanced equations can be written quite easily when the products are known. Often the conditions under which the reaction occurs govern the manner of decomposition and therefore the products produced.

EXAMPLE 1

Calcium carbonate (limestone) when heated to a high temperature decomposes into calcium oxide (quicklime) and carbon dioxide. The following reaction is utilized to make quicklime commercially for use as fertilizer, in manufacturing mortar, and in obtaining iron from iron ore:

$$CaCO_3 \xrightarrow{\Delta} CaO + CO_2$$

The equation is balanced as it stands; the triangle means heat.

EXAMPLE 2

When a solution of magnesium bicarbonate is boiled, the following reactions take place:

$$\begin{array}{lllr} & Mg(HCO_3)_2 & \longrightarrow MgCO_3 + H_2CO_3 & (1) \\ & H_2CO_3 & \longrightarrow H_2O + CO_2 & (2) \\ \hline \text{Overall:} & Mg(HCO_3)_2 & \longrightarrow MgCO_3 + H_2O + CO_2 & (3) \end{array}$$

The final equation (3) is balanced. One of the products formed in reaction (1), H_2CO_3, is unstable in a heated solution and decomposes into CO_2 and H_2O as shown in Eq. (2). When a series of reactions takes place, it is possible to add them as algebraic equations are added.

The overall equation obtained by adding the individual equations, (1) and (2), indicates only the starting materials and the end products; it tells nothing of the intermediate steps involved.

EXAMPLE 3

A solution of $Ca(HCO_3)_2$ when heated decomposes in a manner similar to $Mg(HCO_3)_2$. The final equation is

$$Ca(HCO_3)_2 \xrightarrow{\Delta} CaCO_3 + H_2O + CO_2$$

Temporary hard water contains $Ca(HCO_3)_2$ and $Mg(HCO_3)_2$. When heated, reactions occur to produce $MgCO_3$ and $CaCO_3$. These carbonates are insoluble and precipitate out of solution to form boiler scale in the bottom of tea kettles or in hot-water lines. Thousands of dollars are spent annually to soften water and prevent these reactions from occurring.

EXAMPLE 4

$$\text{Potassium perchlorate} \longrightarrow \text{potassium chloride} + \text{oxygen}$$
$$KClO_4 \longrightarrow KCl + O_2$$

All that is needed to balance this equation is to have 2 O_2 produced.

Example 4 represents a convenient method of preparing oxygen in the laboratory. To facilitate the decomposition of $KClO_4$, manganese dioxide, MnO_2, is added to the reactant; MnO_2 does not enter into the final products but merely catalyzes the decomposition of $KClO_4$. A *catalyst* is defined as a substance that changes the rate of a chemical reaction but does not appear in the products of the reaction.

EXAMPLE 5

When heated to about 300°C, H_2SO_4 decomposes into H_2O, SO_2, and O_2:

$$H_2SO_4 \longrightarrow H_2O + SO_2 + O_2 \quad (1)$$
$$H_2SO_4 \longrightarrow H_2O + SO_2 + \tfrac{1}{2}\,O_2 \quad (2)$$
$$2\,H_2SO_4 \longrightarrow 2\,H_2O + 2\,SO_2 + O_2 \quad (3)$$

Equation (1) can be balanced by placing $\frac{1}{2}$ in front of the O_2 to give Eq. (2). This is a correctly balanced equation, but since it is customary to use whole numbers in an equation, we multiply the entire equation by 2 to give Eq. (3).

4.5 DOUBLE DECOMPOSITION REACTIONS

When two compounds react to form two new compounds, it is called a *double decomposition reaction.* In this type of reaction the positive and negative oxidation moieties exchange partners.

EXAMPLE 1

$$\text{Silver nitrate} + \text{hydrochloric acid} \longrightarrow \text{silver chloride} + \text{nitric acid}$$
$$AgNO_3 + HCl \longrightarrow AgCl\downarrow + HNO_3$$

Ox. no.:	$+1, -1$	$+1, -1$	$+1, -1$	$+1, -1$

The equation is balanced as it stands. The arrow pointing down indicates that the compound is insoluble and precipitates out of solution. Solubility is relative; about 0.001 g of AgCl dissolves in 100 g of H_2O, but this amount is so small that AgCl is considered to be insoluble in water.

EXAMPLE 2

Potassium hydroxide + sulfuric acid ⟶ potassium sulfate + water

	KOH	+	H_2SO_4	⟶	K_2SO_4	+	H_2O
Ox. no.:	+1, −1		+1, −2		+1, −2		+1, −2

The formulas are correct, but the equation is not balanced since more potassium atoms are formed than are used and the same is true of the OH radical. To balance this equation we need 2 KOH and 2 H_2O (2 HOH):

$$2\ KOH + H_2SO_4 \longrightarrow K_2SO_4 + 2\ H_2O$$

4.6 OXIDATION-REDUCTION REACTIONS

Oxidation occurs when the oxidation number of an element becomes more positive (or less negative). *Reduction* is just the opposite of oxidation; i.e., the oxidation number becomes less positive (or more negative). In other words, oxidation is a shift of electrons away from an atom, and reduction is a shift of electrons toward an atom. Oxidation cannot occur without a concomitant reduction because if an atom gives up electrons, another atom must take them on.

EXAMPLE

$$2\ Fe + 3\ S \longrightarrow Fe_2S_3$$

In this equation it is readily seen that Fe^0 unites with S^0 and they become Fe^{3+} and S^{--}, respectively. The oxidation number of iron has changed from 0 to +3 and that of sulfur from 0 to −2; the iron has been oxidized and the sulfur reduced. Three electrons have been shifted away from the influence of the iron atom toward the sulfur atom, which has the greater electronegativity.

An atom does not have to go from a zero oxidation number to a positive value to be oxidized. For example, the Fe(II) atom (oxidation number = +2) in a compound may be converted into iron in the +3 oxidation state. This is an oxidation of Fe(II) to Fe(III). Conversely, a compound containing Fe(III) may enter into a chemical reaction which produces a compound containing Fe(II); the Fe(III) in this reaction has been reduced to Fe(II).

Reactions involving the oxidation-reduction of iron are essential for the biological release of energy from food and also for the conversion of light energy into chemical energy during photosynthesis. In the oxidation of sugar to energy, carbon dioxide, and water, the carbon in sugar is oxidized from 0 to +4 oxidation number, and molecular oxygen is reduced from 0 to −2 oxidation number.

4.7 BALANCING EQUATIONS BY OXIDATION NUMBERS

There are several reasons for balancing equations from the viewpoint of oxidation numbers in addition to the method already described. One aim in balancing reactions is to have the same number of positive and negative oxidation numbers on each side of the equation; otherwise the law of conservation of matter with respect to the electron would be violated. Since each compound must have a net oxidation number of zero, *the net oxidation number of the whole equation must also equal zero.*

It is convenient to talk in terms of *mole* quantities. A mole is the amount of a substance represented by one formula weight. It is the molecular counterpart of the gram atom; a gram molecular weight is a mole expressed in gram units.

We shall first be concerned with the number of electrons shifted from the positive part of the molecule to the negative part, i.e., the positive or negative oxidation number per mole.

	$AlCl_3$	$Ca_3(\mathbf{PO_4})_2$
Ox. no./atom or radical:	+3, −1	+2, −3
+ ox. no./mole:	$+3 \times 1 = +3$	$+2 \times 3 = +6$
− ox. no./mole:	$-1 \times 3 = -3$	$-3 \times 2 = -6$
	Net = 0	Net = 0

EXAMPLE 1

	Aluminum hydroxide	+ hydrochloric acid	⟶ aluminum chloride	+ water
	$Al(\mathbf{OH})_3$	+ HCl	⟶ $AlCl_3$	+ H**OH**
Ox. no.:	+3, −1	+1, −1	+3, −1	+1, −1
+ ox. no./mole:	+3	+1	+3	+1

The LCD is 3; therefore, the number of moles needed of each species to make the oxidation number the same is

$$Al(OH)_3 + 3\ HCl \longrightarrow AlCl_3 + 3\ HOH$$

This is the balanced equation.

EXAMPLE 2

	$Ca(\mathbf{OH})_2$	+ $H_3\mathbf{PO}_4$	⟶ $Ca_3(\mathbf{PO}_4)_2$	+ H**OH**
Ox. no.:	+2, −1	+1, −3	+2, −3	+1, −1
+ ox. no./mole:	+2 × 1 = **+2**	+1 × 3 = **+3**	+2 × 3 = **+6**	+1 × 1 = **+1**
LCD = 6; ∴ no. moles to obtain same ox. no.:	3	2	1	6
Balanced equation:	$3\ Ca(OH)_2$	+ $2\ H_3PO_4$	⟶ $Ca_3(PO_4)_2$	+ 6 HOH

EXAMPLE 3

The next two equations are balanced by considering each side of the equation separately and then equating the half-equations. This method is useful when the reactants or products are fairly complicated.

$$\text{Lead nitrate} + \text{phosphoric acid} \longrightarrow \text{lead phosphate} + \text{nitric acid}$$
$$Pb(NO_3)_2 + H_3PO_4 \longrightarrow Pb_3(PO_4)_2 + HNO_3$$

Reactants:	$Pb(\mathbf{NO}_3)_2$ +	$H_3\mathbf{PO}_4$
Ox. no.:	+2, −1	+1, −3
+ ox. no./mole:	+2	+3
LCD = 6; ∴ no. moles to obtain same ox. no.:	3	2
Balanced half-equation:	$3\ Pb(NO_3)_2$ +	$2\ H_3PO_4$

Products:	$Pb_3(\mathbf{PO}_4)_2$ +	$H\mathbf{NO}_3$
Ox. no.:	+2, −3	+1, −1
+ ox. no./mole:	+6	+1
LCD = 6; ∴ no. moles to obtain same ox. no.:	1	6
Balanced half-equation:	$Pb_3(PO_4)_2$ +	6 HNO_3

Since there is a total of 6(+), or 6(−), oxidation number for each side of the equation, we may equate the two half-reactions, giving

Balanced equation: $3\ Pb(NO_3)_2 + 2\ H_3PO_4 \longrightarrow Pb_3(PO_4)_2 + 6\ HNO_3$

EXAMPLE 4

The following equation will also be balanced by half-equations:

$$Al_2(CO_3)_3 + H_3PO_4 \longrightarrow AlPO_4 + H_2CO_3$$

Reactants:	$Al_2(CO_3)_3$ +	H_3PO_4
Ox. no.:	+3, −2	+1, −3
+ ox. no./mole:	+6	+3
LCD = 6; ∴ no. moles to obtain same ox. no.:	1	2
Balanced half-equation:	$Al(CO_3)_3$ +	2 H_3PO_4

Products:	$AlPO_4$ +	H_2CO_3
Ox. no.:	+3, −3	+1, −2
+ ox. no./mole:	+3	+2
LCD = 6; ∴ no. moles to obtain same ox. no.:	2	3
Balanced half-equation:	2 $AlPO_4$ +	3 H_2CO_3

Balanced equation:

$$Al_2(CO_3)_3 + 2\ H_3PO_4 \longrightarrow 2\ AlPO_4 + 3\ H_2CO_3$$

4.8 BALANCING O/R REACTIONS

Since oxidation-reduction (O/R) reactions involve a complete transfer of electrons from one element to another, these equations can be balanced most easily by following this electron transfer. With a little experience in setting up the format some of the most difficult equations are easily balanced. The procedure may be outlined briefly as follows:

1. Determine which element is oxidized and which is reduced.
2. Connect the element in the reactants that is oxidized to the same element in the products with a line. Do the same with the element reduced.
3. Write the oxidation numbers of the elements oxidized and reduced under the respective elements.
4. Note the change in the oxidation number per atom of the respective elements.
5. Calculate the change in the oxidation number per mole of the respective compounds.
6. Find the LCD of the oxidation number per mole.
7. Determine the number of moles needed to obtain the same change of oxidation number per mole.

In the examples which follow note the similarity of this method to that of balancing double decomposition equations by oxidation numbers. Delta (Δ) indicates the difference between any two given oxidation numbers.

EXAMPLE 1

	Zn	$+\ HCl \longrightarrow$	$H_2\ +$	$ZnCl_2$
Ox. no.:	0	+1	0	+2
Δ ox. no./atom:	2	1	1	2
Δ ox. no./mole:	2	1	2	2
LCD = 2; ∴ no. moles to have same Δ ox. no.:	1	2	1	1
Balanced equation:	$Zn + 2\,HCl \longrightarrow H_2 + ZnCl_2$			

(Lines connect Zn (0) to $ZnCl_2$ (+2), and HCl (+1) to H_2 (0).)

EXAMPLE 2

	HNO_3	$+\ SO_2 \longrightarrow$	$H_2SO_4\ +$	NO_2
Ox. no.:	+5	+4	+6	+4
Δ ox. no./atom:	1	2	2	1
Δ ox. no./mole:	1	2	2	1
LCD = 2; ∴ no. moles to have same Δ ox. no.:	2	1	1	2
Balanced equation:	$2\,HNO_3 + SO_2 \longrightarrow H_2SO_4 + 2\,NO_2$			

(Lines connect HNO_3 (+5) to NO_2 (+4), and SO_2 (+4) to H_2SO_4 (+6).)

EXAMPLE 3

$$MnO_2 + HCl \longrightarrow Cl_2 + MnCl_2 + H_2O$$

	MnO_2	HCl	Cl_2	$MnCl_2$
Ox. no.:	+4	−1	0	+2
Δ ox. no./atom:	2	1	1	2
Δ ox. no./mole:	2	1	2	2
LCD = 2; ∴ no. moles to have same Δ ox. no.:	1	2	1	1

Hence: $MnO_2 + 2\,HCl \longrightarrow Cl_2 + MnCl_2 + (?)\,H_2O$

The equation is balanced from the standpoint of oxidation-reduction but not with respect to the elements not oxidized or reduced. How many moles of H_2O are formed? More Cl^- is needed on the left of the equation; not all the Cl^- of HCl is oxidized to Cl_2, some being needed to unite with Mn^{++}. The elements not oxidized or reduced are balanced by sight, but the molar ratio of substances balanced for O/R *cannot be changed: they are already balanced.* Balance the Cl^- next; there are 4 Cl^- on the right and only 2 Cl^- on the left; thus 2 HCl must be *added* to the left of the equation. This in addition to the 2 HCl needed for balancing the O/R reaction gives a total of 4 H^+ on the left. These 4 H^+ and the 2 O^{--} of MnO_2 form 2 H_2O on the right. The completely balanced equation is

$$MnO_2 + 4\,HCl \longrightarrow Cl_2 + MnCl_2 + 2\,H_2O$$

EXAMPLE 4

$$P_2O_3 \longrightarrow P_2O_4 + P$$

In this reaction the P^{3+} in some molecules of P_2O_3 is oxidized to P^{4+} in P_2O_4 while the P^{3+} in other molecules of P_2O_3 is reduced to P^0. To keep the P that is oxidized separate from that reduced, the equation is written:

	P_2O_3	+ P_2O_3	⟶ P_2O_4	+ P
Ox. no.:	+3	+3	+4	0
Δ ox. no./atom:	3	1	1	3
Δ ox. no./mole:	6	2	2	3
LCD = 6; ∴ no. moles to have same Δ ox. no.:	1	3	3	2
	4			

Hence: $$4\ P_2O_3 \longrightarrow 3\ P_2O_4 + 2\ P$$

EXAMPLE 5

$$SO_2 + H_2S \longrightarrow S + H_2O$$

Rewrite the equation as follows since the sulfur in both SO_2 and H_2S forms S:

	SO_2	+ H_2S	⟶ S	+ S	+ H_2O
Ox. no.:	+4	−2	0	0	
Δ ox. no./atom:	4	2	2	4	
Δ ox. no./mole:	4	2	2	4	
LCD = 4; ∴ no. moles to have same Δ ox. no.:	1	2	2	1	
			3		

Hence: $$SO_2 + 2\ H_2S \longrightarrow 3\ S + (?)\ H_2O$$

The H and O are balanced by sight. Since the final balanced equation would call for 2 H_2O,

$$SO_2 + 2\ H_2S \longrightarrow 3\ S + 2\ H_2O$$

EXAMPLE 6

	I_2	+ $Na_2S_2O_3$	⟶ $Na_2S_4O_6$	+ NaI
Ox. no.:	0	+2	+2.5	−1
Δ ox. no./atom:	1	0.5	0.5	1
Δ ox. no./mole:	2	1	2	1
LCD = 2; ∴ no. moles to have same Δ ox. no.:	1	2	1	2

Hence: $$I_2 + 2\ Na_2S_2O_3 \longrightarrow Na_2S_4O_6 + 2\ NaI$$

EXAMPLE 7

$$Na_2C_2O_4 + KMnO_4 + H_2SO_4 \longrightarrow MnSO_4 + CO_2 + Na_2SO_4 + K_2SO_4 + H_2O$$

	$Na_2C_2O_4$	$KMnO_4$	$MnSO_4$	CO_2
Ox. no.:	+3	+7	+2	+4
Δ ox. no./atom:	1	5	5	1
Δ ox. no./mole:	2	5	5	1
LCD = 10; ∴ no. moles to have same Δ ox. no:	5	2	2	10

Hence: $5\ Na_2C_2O_4 + 2\ KMnO_4 + (?)\ H_2SO_4 \longrightarrow 2\ MnSO_4 + 10\ CO_2 + (?)\ Na_2SO_4 + (?)\ K_2SO_4 + (?)\ H_2O$

To balance K^+ from 2 $KMnO_4$ requires 1 K_2SO_4. To balance Na^+ from 5 $Na_2C_2O_4$ requires 5 Na_2SO_4; 8 H_2SO_4 is required to balance the 8 SO_4^{--} (K_2SO_4, 5 Na_2SO_4, 2 $MnSO_4$). To balance 16 H^+ from 8 H_2SO_4 requires 8 H_2O. The completely balanced equation is

$$5\ Na_2C_2O_4 + 2\ KMnO_4 + 8\ H_2SO_4 \longrightarrow 2\ MnSO_4 + 10\ CO_2 + 5\ Na_2SO_4 + K_2SO_4 + 8\ H_2O$$

4.9 MOLECULAR-WEIGHT CALCULATIONS

The gram molecular weight, or *mole,* represents the formula weight of a compound expressed in grams. It is the weight in grams of one Avogadro's number of molecules. It is the sum of the respective atomic weights of the elements times the number of times they occur in the compound. For monatomic elements, such as Mg, Na, K, or Ca, the gram atomic weight is the same as the molecular weight of the element, whereas with diatomic elements like Cl_2, O_2, and N_2 the molecular weight is twice the atomic weight.

The term mole automatically means that the unit of weight is in grams, but since the weights of the elements are relative, it is possible to express the molecular weight in units other than the gram.

EXAMPLE 1

What is the weight of 1 mole of K_2CO_3?

No. atoms of K/mole = 2	at. wt. K = 39	2 × 39 = 78 g K/mole
No. atoms of C/mole = 1	at. wt. C = 12	1 × 12 = 12 g C/mole
No. atoms of O/mole = 3	at. wt. O = 16	3 × 16 = 48 g O/mole
		Total = 138 g mol. wt. K_2CO_3

EXAMPLE 2

Calculate the gram molecular weight of $K_2C_2O_4$.

No. atoms of K/mole = 2	at. wt. K = 39	2 × 39 = 78 g K/mole
No. atoms of C/mole = 2	at. wt. C = 12	2 × 12 = 24 g C/mole
No. atoms of O/mole = 4	at. wt. O = 16	4 × 16 = 64 g O/mole
		Total = 166 g mol. wt. $K_2C_2O_4$

Consider the following problem before beginning the discussion of equivalent weights. Calculate the weight per wheel for several makes and types of vehicles assuming that the weight is equally distributed per wheel.

The unicycle, a one-wheeled vehicle, is the simplest vehicle from the standpoint that it has only one wheel. The weight per wheel is the total weight of the vehicle divided by 1. The weight per wheel is the same, theoretically, for each unicycle of a particular model and make but would not necessarily be the same for other models and makes. There can be many different models of unicycles, each having a different weight per wheel but each of a given model having the same weight per wheel.

Next consider the bicycle, which has two wheels. The distribution of weight per wheel would be the total weight of the vehicle divided by 2. As above, each model of a bicycle made by a particular manufacturer would weigh the same. However, other manufacturers' bicycles would not weigh the same and hence would have a different value for the weight per wheel. The motorcycle is also a two-wheeled vehicle but would certainly have a greater value for weight per wheel than the bicycle.

The same calculations could be made for vehicles of three, four, six, etc., wheels. Here again the weight per wheel is the gross weight of the toy wagon, car, or the heaviest machine made divided by the number of wheels on the vehicle.

In each of the above examples we have calculated the weight of various vehicles on an equivalent basis, namely, the weight per wheel. If we call this *one equivalent weight,* we can say that a unicycle represents one equivalent of weight, the bicycle two equivalents of weight, the tricycle three equivalents of weight, and so on. Values for one equivalent weight for the different models or types of vehicles are not the same. Equivalent weight depends upon the gross weight and the number of wheels.

Apply this discussion to the study of chemistry. By analogy

Vehicle = atom
Type of vehicle = element
Model of vehicle = atoms of a particular element
No. of wheels = number of electrons shifted when element enters into a chemical reaction
Gross weight = atomic weight of the element

One equivalent weight or $\dfrac{\text{weight}}{\text{wheels}}$

$$= \frac{\text{atomic weight}}{\text{no. of electrons shifted when element enters into a chemical reaction}}$$

The simplest chemical bond is formed when one electron is shifted from the influence of one atom to the environment of another atom which has a greater electronegativity. Therefore, the equivalent weight of an element is equal to the atomic weight of the element divided by the number of electrons shifted away from or toward it during a chemical reaction.

The gram atomic weight expresses the weight in grams of 6.02×10^{23} atoms. The gram equivalent weight of an element is the weight represented by the transfer or shift of one Avogadro's number of electrons.

The number of equivalents per mole is actually the *valence* or *combining power* of the element. Therefore, to calculate the equivalent weight of an element it is necessary to know the valence or the oxidation number of the element. This is easy to determine for those elements in groups I, II, III, and VII of the periodic table. Elements in group I donate one electron when entering into a chemical reaction; therefore, their equivalent weight is the same as their atomic weight. This is also true of elements in group VII: they accept one electron upon entering into a chemical reaction. Elements in groups II and III have equivalent weights which are one-half and one-third of their atomic weights, respectively.

The equivalent weight of *compounds* is calculated in a similar manner. The gram molecular weight is divided by the number of electrons shifted from the positive to the negative part of the molecule. In the compound NaCl, for example, one electron is donated by the sodium, and therefore the equivalent weight would be the same as the molecular weight. In $CaCl_2$ two electrons have been donated by the calcium; therefore the equivalent weight of $CaCl_2$ is one-half the molecular weight. The equivalent weight of $AlCl_3$ is one-third of the gram molecular weight.

The calculation of equivalent weight can be set up as follows:

	$NaCl$	$CaCl_2$	$AlCl_3$
+ ox. no./mole:	+1	+2	+3
Electrons shifted/mole:	1	2	3
G mol. wt.:	58.5	111	127.48
G equiv wt.:	$\frac{58.5}{1} = 58.5$	$\frac{111}{2} = 55.5$	$\frac{127.48}{3} = 42.49$

	$Ca(OH)_2$	$(NH_4)_3PO_4$	$Ca_3(PO_4)_2$
+ ox. no./mole:	+2	+3	+6
Electrons shifted/mole:	2	3	6
G mol. wt.:	74.1	149.09	310.2
G equiv wt.:	$\frac{74.1}{2} = 37.05$	$\frac{149.09}{3} = 49.69$	$\frac{310.2}{6} = 51.7$

A word should be said about calculating the equivalent weights of compounds containing radicals. The radical is a unit and therefore constitutes the negative or positive part of a molecule. This has been illustrated in several of the examples that contain radicals.

In Sec. 4.7 equations were balanced to have the same total oxidation numbers for each constituent of the reaction. In essence, this is exactly what we are doing in calculating the equivalent weights. We are calculating the number of moles of each constituent of a reaction that would be equivalent to a shift of one electron.

EXAMPLE 3

	$Al(OH)_3$	+	HCl	$\longrightarrow$	$AlCl_3$	+	HOH
+ ox. no./mole:	+3		+1		+3		+1
Electrons shifted/mole:	3		1		3		1
Moles/electron shifted:	$\frac{1}{3}$		1		$\frac{1}{3}$		1
Balanced equation:	$\frac{1}{3}\,Al(OH)_3$	+	HCl	$\longrightarrow$	$\frac{1}{3}\,AlCl_3$	+	HOH
No. equiv	1	+	1	$\longrightarrow$	1	+	1

One equivalent of reactants reacts to form one equivalent of products. This is true for all reactions because 1 equiv represents the amount of material that is involved in the shift of one electron in each case. Therefore, 1 equiv of reactants must react to give 1 equiv of each product; otherwise the electrons shifted would not balance.

Using equivalent weights is frequently very convenient because calculations can often be made without bothering to write a complete equation; all that is needed is to use equivalent weights of the compounds under consideration.

EXAMPLE 4

How many grams of H_3PO_4 are required to exactly react with 3.5 g of $Ca(HCO_3)_2$? Although the equation is not necessary for solving this problem, it is written for the sake of emphasis.

$$2\ H_3PO_4 + 3\ Ca(HCO_3)_2 \longrightarrow Ca_3(PO_4)_2 + 6\ H_2O + 6\ CO_2$$

Since 1 equiv wt. of H_3PO_4 reacts with 1 equiv wt. of $Ca(HCO_3)_2$, the problem can be set up as follows:

	H_3PO_4	+	$Ca(HCO_3)_2$
Electrons shifted/mole:	3		2
G mol. wt.:	98		162
G equiv wt.	$\frac{98}{3}$		$\frac{162}{2}$
Reacting ratio:	1 equiv		1 equiv

$$\therefore \quad \frac{\frac{98}{3}}{x} = \frac{\frac{162}{2}}{3.5}$$

$$x = \frac{98 \times 3.5 \times 2}{3 \times 162} = 1.4 \text{ g}$$

EXAMPLE 5

The nitrogen of $Ca(CN)_2$ can be converted into NH_4OH by appropriate reactions. How many milligrams of NH_4OH can be produced from 0.01 g of $Ca(CN)_2$? The only equation required is

	$Ca(CN)_2$	$\longrightarrow$	NH_4OH
Electrons shifted/mole:	2		1
G mol. wt.:	92.12		35
G equiv wt.	$\frac{92.12}{2}$		$\frac{35}{1}$
Reacting ratio:	1 equiv		1 equiv

$$\therefore \quad \frac{\frac{92.12}{2}}{0.01} = \frac{35}{x}$$

$$x = \frac{0.01 \times 35 \times 2}{92.12} = 7.6 \times 10^{-3} \text{ g} = 7.6 \text{ mg}$$

QUESTIONS

1. If 7.6×10^{23} atoms of a monatomic element weigh 10 g, what is the atomic weight of the element?
2. Calculate the gram molecular weight of the following compounds and compare your answers with the correct ones given.

(*a*) $ZnCl_2$ (*b*) H_2SO_4 (*c*) $KMnO_4$
(*d*) $(NH_4)_3PO_4$ (*e*) $Ca(H_2PO_4)_2$ (*f*) $Al(NO_3)_3$
Ans: (*a*) 136.29, (*b*) 98.08, (*c*) 158.03, (*d*) 149.09, (*e*) 234.09, (*f*) 213.

3. Repeat Prob. 2 for the gram equivalent weight.
Ans: (*a*) 68.15, (*b*) 49.04, (*c*) 158.03, (*d*) 49.69, (*e*) 39.01, (*f*) 71.

4. Complete and balance the following equations:
(*a*) $HNO_3 + Ca(OH)_2 \longrightarrow$ (*b*) $(NH_4)_2SO_4 + Ba(OH)_2 \longrightarrow$
(*c*) $NH_4OH + H_3PO_4 \longrightarrow$ (*d*) $Ca(HCO_3)_2 + HCl \longrightarrow$
(*e*) $H_2SO_4 + Ca(OH)_2 \longrightarrow$ (*f*) $Pb(NO_3)_2 + H_3PO_4 \longrightarrow$

5. Which of the following compounds contains carbon in the *highest* oxidation state: CH_4, $CHCl_3$, CCl_4?

6. Column *A* of the following table is a list of reactions. Write in column *B* the answer *yes* if the reaction is an oxidation-reduction reaction and *no* if it is not. In the respective blanks in column *C* write the names of the elements oxidized and reduced as shown for (*a*). Remember that elements in the uncombined state have an oxidation number of zero.

A	*B*	*C*	
		Element	
Reaction	*O/R*	*Ox.*	*Red.*
(*a*) $Fe + S \longrightarrow Fe_2S_3$	Yes	Fe	S
(*b*) $N_2 + H_2 \longrightarrow NH_3$			
(*c*) $H_2O + SO_2 \longrightarrow H_2SO_3$			
(*d*) $H_2 + Cl_2 \longrightarrow HCl$			
(*e*) $KOH + H_2SO_4 \longrightarrow K_2SO_4 + H_2O$			

7. Balance the following O/R reactions:
(*a*) $Sb_2O_3 + I_2 + H_2O \longrightarrow Sb_2O_5 + HI$
(*b*) $KClO_3 + FeSO_4 + H_2SO_4 \longrightarrow KCl + Fe_2(SO_4)_3 + H_2O$
(*c*) $MnSO_4 + PbO_2 + HNO_3 \longrightarrow HMnO_4 + PbSO_4 + Pb(NO_3)_2 + H_2O$
(*d*) $V_2O_2 + KMnO_4 + H_2SO_4 \longrightarrow V_2O_5 + MnSO_4 + K_2SO_4 + H_2O$
(*e*) $Hg(CN)_2 + I_2 \longrightarrow HgI_2 + I(CN)$
Ans: (*a*) $1 + 2 + 2 = 1 + 4$; (*b*) $1 + 6 + 3 = 1 + 3 + 3$; (*c*) $2 + 5 + 6 = 2 + 2 + 3 + 2$; (*d*) $5 + 6 + 9 = 5 + 6 + 3 + 9$; (*e*) $1 + 2 = 1 + 2$.

8. How many milligrams of $(NH_4)_3PO_4$ can be formed from 24 mg of $(NH_4)_2CO_3$?

9. If a solution containing 60 mg of $MgCl_2$ and a solution containing 150 mg of $AgNO_3$ are mixed, how many milligrams of AgCl will be formed? *Hint:* Which reactant limits the reaction?

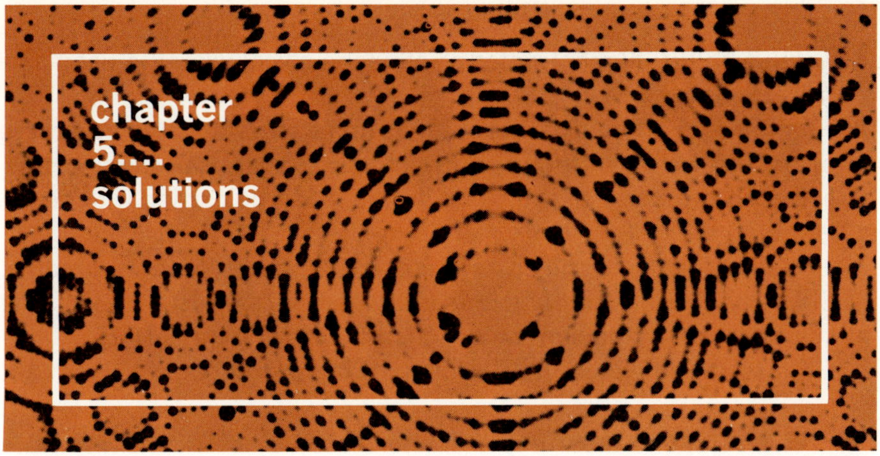

chapter 5.... solutions

Solutions are basic to life. Soil nutrients must be dissolved in water before plants can absorb them. The food we eat is only partly soluble in water; most of the insoluble material must be converted by the process of digestion into a soluble form before it can be absorbed by the body and dissolved in the blood, which is 81 to 86% water. The chemical reactions in the cells of living organisms occur in an aqueous environment; each cell must be bathed in an aqueous solution of nutrients. A person may live for many days without food if sufficient water is available, but without water, he can survive only a few days. Biologically, water is a very important substance; it enters into many chemical reactions.

In previous chapters reactions between elements or compounds have been written as though they reacted in the solid or gaseous state. Although this is possible in some instances, dry chemicals usually do not react. Sometimes only a trace of water is needed to permit a reaction to proceed. For example, under strict anhydrous conditions, $AgNO_3$ and NaCl do not react appreciably at ordinary temperatures, but if a water solution of each of these chemicals is mixed, reaction occurs instantaneously, as evidenced by the immediate formation of the insoluble white precipitate silver chloride, AgCl.

What types of solutions are there? Why do reactants react faster in solution than in their dry state? Why does sugar dissolve in water but not in gasoline, whereas fats are soluble in gasoline but not in water? These are but a few of the questions to be considered in this chapter.

5.1 TYPES OF SOLUTIONS

There are a number of types of solutions, each with an important role to play in the chemistry of our everyday life:

1. *Liquid-liquid:* Alcohol-water or ethylene glycol–water is used as a radiator antifreeze.
2. *Gas-liquid:* Carbon dioxide–water as found in carbonated beverages or oxygen-water as found in the blood. Oxygen must first be dissolved in the water of the blood before it can be carried with hemoglobin to the cells of the body.
3. *Gas-gas:* Nitrogen-oxygen as found in the atmosphere we breathe.
4. *Solid-solid:* Iron and carbon as found in steel (alloys are solid-solid solutions) or germanium and gallium for making transistors.
5. *Gas-solid:* Smog is partly a solution of finely divided particles in air.

The solutions most important to the study of biochemistry are solid-liquid, liquid-liquid, gas-gas, and gas-liquid. Many factors influence the solubility of a substance, e.g., temperature, pressure, polarity, and fineness of the solid. The extent to which a substance dissolves in another substance is also markedly influenced by their interaction, which is affected by the nature of the bond between the atoms of the respective molecules.

The term soluble is relative. If a substance such as sugar dissolves appreciably in water, it is said to be *soluble* in water. A compound such as AgCl is only 0.00013% soluble by weight in water and is therefore said to be insoluble. The term *insoluble* is not really correct since everything dissolves to a minute extent, even glass.

Solutions are useful tools in chemistry. It is inconvenient (and requires special balances) to weigh out very small amounts, such as 0.0002 g of a substance especially when it is a liquid. To overcome this difficulty a convenient weight of the substance is dissolved in a solvent and a portion of the solution used to obtain the desired amount of the solute. For example, with a solution containing 2.0 g of NaCl per 1,000 ml of solution it would be simple to measure out 0.0002 g (0.2 mg) of NaCl by taking 0.1 ml of this solution.

5.2 POLARITY OF SOLIDS IN SOLUTION

If sugar is added to water, the molecules of sugar start to break away and diffuse throughout the water. The sugar dissolves. The water molecules cluster about the sugar molecules on the surface of the crystal and overcome or neutralize the attractive force between the sugar molecules in the crystal. The hydrated sugar molecules are then free to diffuse throughout the solution as individual molecules (see

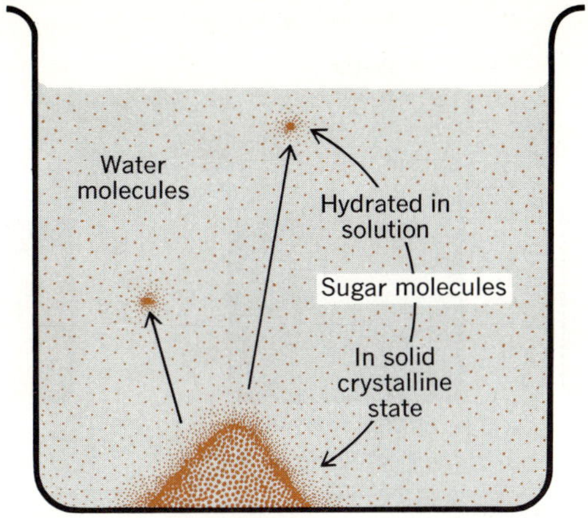

FIGURE 5.1 The mechanism of solution of a solid in a liquid.

Fig. 5.1). Sugar is a polar compound containing a number of —OH groups which can become associated with the polar water molecules.

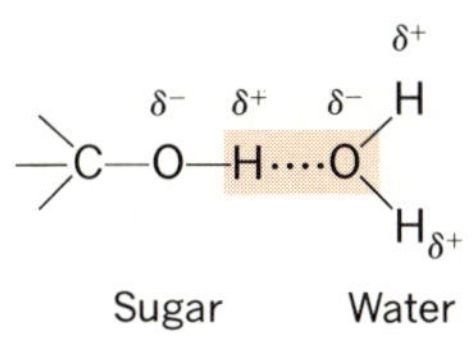

In general, polar compounds tend to be soluble in polar substances.

5.3 SOLVENT AND SOLUTE

The *solvent* is the substance in which another substance, the *solute,* is dissolved. In a solid-liquid system the liquid is the solvent and the solid is the solute, by convention. In a liquid-liquid system the solvent is the component present in the greatest amount. For example, alcohol and water are mutually soluble in all proportions; i.e., they are *miscible* liquids. When 95 ml of alcohol is added to 100 ml of water, the alcohol is the solute and the water the solvent. However, if 25 ml of water is added to 30 ml of alcohol, the alcohol is the solvent and the water the solute. The term *immiscible* applies to two liquids that are not soluble in each other, e.g., oil and water (Fig. 5.2).

In solution the solute exists as (1) individual molecules, (2) relatively small aggregates of molecules clumped together because of their

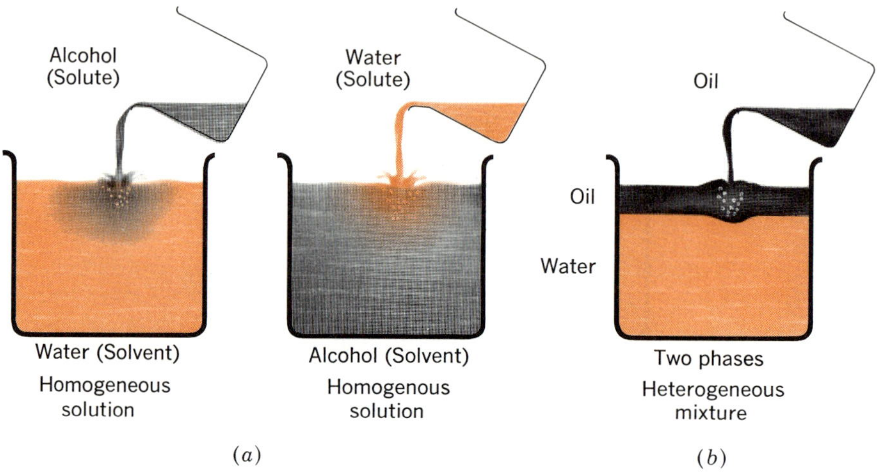

FIGURE 5.2 Mixing of (*a*) two soluble (miscible) liquids and (*b*) two insoluble (immiscible) liquids.

polar nature, or (3) ions if the solute ionized. (Ionization is discussed in Chap. 7.) In a *true* solution the molecules are free to diffuse throughout the entire solution and do not settle out if the solution is allowed to stand. If a polar substance is dissolved in a polar solvent, the dissolved molecules are solvated; if the solvent is water, we speak of the molecules as being hydrated.

In Fig. 5.1, the solid sugar molecules are shown going *into* solution, but there is a limit to which solids can dissolve in liquids. As one adds more solute, the concentration of dissolved solute becomes greater and greater (Fig. 5.3) and the collisions between the dissolved solute molecules become more frequent. Eventually a condition is reached where the added solute no longer dissolves. When dissolved

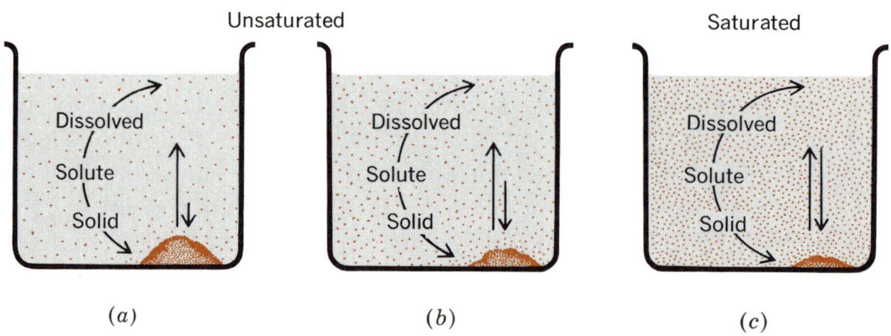

FIGURE 5.3 Diagram of saturated (*c*) and unsaturated (*a*) and (*b*) solutions.

solute molecules are hitting the crystalline lattice and remaining attached to it at the same rate as other molecules are leaving the crystalline lattice, a state of *equilibrium* exists between the dissolved and the undissolved solute. At equilibrium the concentration of the dissolved solute remains constant regardless of the amount of added solid solute provided the external conditions (temperature, pressure) remain the same.

A *saturated solution* is one in which the dissolved solute would be in equilibrium with the undissolved solute if it were present. An *unsaturated solution* is one in which equilibrium between the dissolved solute and the undissolved solute has not yet been attained.

The simplest way to determine whether a solution is unsaturated or saturated is to add a small amount of solute. The added solute will dissolve in an unsaturated solution but not in a saturated one.

A solution which contains more dissolved solute than would be possible if solid solute were present is called a *supersaturated solution.* A supersaturated and an unsaturated solution have in common that no equilibrium exists between the undissolved and the dissolved solute. How a supersaturated solution is prepared and how to determine whether a solution is supersaturated will become apparent after the discussion of solubility which follows.

5.4 CONCENTRATIONS IN SOLUTIONS

Concentration as well as solubility can be expressed in several different terms. The solubility of a substance refers only to the concentration of a saturated solution. There are several ways of expressing concentration.

GRAMS OF SOLUTE PER GRAM OF SOLVENT. If 5 g of solute A is dissolved in 100 g of H_2O, the concentration of A is simply expressed as 5 g per 100 g H_2O.

PERCENTAGE BY WEIGHT. The grams of solute per 100 g of solution are expressed as a percentage. Thus, if 10 g of A is added to 90 g of H_2O, the solution is said to contain 10% by weight.

PERCENTAGE BY WEIGHT-VOLUME. This is the grams of solute per 100 ml of solution; e.g., an 8% (weight per volume) solution contains 8 g of solute in 100 ml of *solution.* (This includes the volume of the 8 g.)

MOLALITY. A 1 molal (m) solution contains 1 mole of solute in 1,000 g of solvent. Fractional moles are, of course, possible.

PROBLEM 1

12 g of CH_3OH is in 50 g of H_2O, what is the molality? Since 1 m = 32 g (1 mol. wt.) CH_3OH in 1,000 g H_2O, 12 g CH_3OH in 50 ml H_2O is how many grams of CH_3OH in 1,000 g H_2O?

$$\frac{12 \text{ g } CH_3OH}{50 \text{ g } H_2O} = \frac{x \text{ g } CH_3OH}{1{,}000 \text{ g } H_2O}$$

Solving for x, x = 240 g CH_3OH

$$\frac{240}{32 \text{ (mol. wt. } CH_3OH)} = 7.5$$

therefore

$$\text{Molality} = 7.5$$

PROBLEM 2

How would you prepare 75 g of a 0.01 m solution of NaCl? The molecular weight of NaCl is 58.5. A 1 m solution would contain 58.5 g NaCl + 1,000 g H_2O; hence, a 0.01 m solution would contain 0.585 g NaCl + 1,000 g H_2O. There would be 0.585 g NaCl in 1,000.585 g of solution. Therefore, 75 g of solution would contain

$$\frac{0.585}{1{,}000.585} = \frac{x}{75} \qquad x = \frac{0.585 \times 75}{1{,}000.585} = 0.044 \text{ g NaCl}$$

To make the solution add 0.044 g NaCl to (75 − 0.044 g) H_2O or 74.96 g H_2O.

MOLE FRACTION. Mole fraction is a numerical expression of the ratio of the number of moles of solute or solvent to the *total number* of moles in the solution.

PROBLEM 3

What is the mole fraction (N_a or N_b) of a solution of A and B? If n_a = moles of A and n_b = moles of B, then

$$\text{Mole fraction of A } (N_a) = \frac{n_a}{n_a + n_b}$$

$$\text{Mole fraction of B } (N_b) = \frac{n_b}{n_a + n_b}$$

PROBLEM 4

What is the mole fraction of NaOH in a 1 m solution of NaOH? Let n_a = moles of NaOH and n_b = moles of water; then

$$n_a = 1.0 \qquad n_b = \frac{1{,}000 \text{ g } H_2O}{18 \text{ (mol. wt. } H_2O)} = 55.6$$

$$N_a = \frac{1}{1 + 55.6} = 0.00589$$

PROBLEM 5

What is the mole fraction of CH_3OH in a solution containing 240 g CH_3OH in 1,000 g H_2O? Let n_a = moles of CH_3OH (mol. wt. = 32) and n_b = moles of water (mol. wt. = 18). Then

$$n_a = \tfrac{240}{32}$$

$$n_b = \tfrac{1{,}000}{18}$$

$$N_a = \frac{\frac{240}{32}}{\frac{1{,}000}{18} + \frac{240}{32}} = 0.77 \text{ mole fraction}$$

NORMALITY. Normality (N) is the *number of equivalents of solute per liter of solution.* To make a 1 N solution the equivalent weight of the substance is determined, and this amount is then dissolved in sufficient solvent to make 1 *liter of final solution.* Note that the material is *not* added to 1 liter of solvent. From the definition of normality, the following equation can be derived:

$$\frac{\text{No. equiv}}{\text{Vol (liters)}} = N \qquad \text{or} \qquad \text{no. equiv} = NV_{\text{liter}}$$

If a 1 N solution contains 1 g equiv wt./liter, then 1.0 ml of this solution contains $\frac{1}{1{,}000}$ equiv wt., or 1 milliequivalent (meq):

$$\frac{\text{No. meq}}{\text{Vol (ml)}} = N \qquad \text{or} \qquad \text{no. meq} = NV_{\text{ml}}$$

PROBLEM 6

How many equivalents are in 500 ml of a 3 N solution of H_3PO_4? V = 0.5 liter; therefore, 0.5 × 3 = 1.5 equiv. There would be 1.5 equiv in 500 ml of a 3 N solution of *any* compound.

PROBLEM 7

How many milliequivalents of $(NH_4)_2SO_4$ would be in 30 ml of 1×10^{-3} N solution of this compound? Substituting in the equation $V_{\text{ml}} \times N = \text{meq}$, we have $30 \times 1 \times 10^{-3}$ meq, or 3×10^{-2} meq.

PROBLEM 8

What is the normality of a solution that contains 0.48 g NaOH per 50 ml? Since the concentration sought is to be expressed as normality, we must find the number of equivalents of NaOH in 1 liter of the solution. Therefore, the number of grams of NaOH must be changed to equivalents. This is accomplished by dividing 0.48 by the equivalent weight of NaOH, which is 40 g; hence, there are $\frac{0.48}{40}$ equiv of NaOH in 50 ml of solution. We must now find the number of equivalents in 1 liter, or 1,000 ml, of solution. This is done by setting up the proportion

$$\frac{0.48/40}{50\text{ ml}} = \frac{x\text{ equiv}}{1{,}000\text{ ml}} \qquad 50x\text{ equiv} = \frac{1{,}000 \times 0.48}{40}$$

$$x\text{ equiv} = \frac{1{,}000 \times 0.48}{50 \times 40} \qquad x\text{ equiv/liter or } N = 0.24$$

In working this problem, we have actually substituted in the equation

$$\text{No. equiv} = NV_{\text{liter}} \qquad \text{or} \qquad N = \frac{\text{no. equiv}}{V_{\text{liter}}}$$

where

$$\text{No. equiv} = \frac{0.48}{40} \qquad \text{and} \qquad V_{\text{liter}} = \frac{50}{1{,}000}$$

therefore

$$N = \frac{0.48/40}{50/1{,}000} = \frac{0.48}{40} \times \frac{1{,}000}{50}$$

If two substances 1 and 2 react, 1 equiv wt. of 1 will exactly react with 1 equiv wt. of 2; therefore, the following equation may be derived (the subscripts 1 and 2 refer to the respective reactants):

$$\text{equiv}_1 = N_1V_1 \qquad \text{equiv}_2 = N_2V_2$$

but

$$\text{equiv}_1 = \text{equiv}_2$$

Hence,

$$N_1V_1 = N_2V_2$$

V_1 and V_2 may be expressed in any units provided they are the same units.

PROBLEM 9

How many equivalents are there in 0.25 liter of a 3 N solution of H_3PO_4? There are $0.25 \times 3 = 0.75$ equiv. There would be 0.75 equiv of any compound in 250 ml of a 3 N solution of the substance. In other words, the *same number of equivalents* are contained in the *same volume of solutions having the same normality.* (In the previous equation, if $N_1 = N_2$, then $V_1 = V_2$.)

PROBLEM 10

If 25 ml of 0.1 N A is mixed with 6 ml of 0.5 N B and if the reaction is A + B $\longrightarrow$ C, (*a*) how many equivalents of C would be formed? (*b*) If the equivalent weight of C is 53.3, what is the normality of C? The number of equivalents of each reactant must be calculated to determine whether one of them limits the reaction. The problem may be set up as follows:

(*a*)
$$\frac{\text{A}}{25\text{ ml } 0.1\ N} + \frac{\text{B}}{6\text{ ml } 0.5\ N} = \frac{\text{C}}{31\text{ ml}}$$

$$(25 \times 0.1 = 2.5\text{ meq}) + (6 \times 0.5 = 3.0\text{ meq}) = 2.5\text{ meq}$$

Substance A limits the reaction (B is in excess by 3.0 − 2.5, or 0.5 meq); therefore, there can be *no more* than 2.5 meq of C produced.

(*b*) The total volume of solution C is 25 + 6 = 31 ml; therefore, the normality of C would be N = meq/ml.

$$N = \frac{2.5}{31} = 0.081$$

PROBLEM 11

How many milliliters of a 0.25 N solution of NaOH are required to exactly neutralize 50 ml of 0.18 N H_2SO_4? We substitute in the equation $V_1N_1 = V_2N_2$. Let subscript 1 refer to NaOH and subscript 2 to H_2SO_4. Then

$$V_1 \times 0.25 = 50 \times 0.18 \qquad V_1 = \frac{50 \times 0.18}{0.25} \qquad V_1 = 36 \text{ ml}$$

PROBLEM 12

How many grams of NaOH are contained in 55 ml of 0.968 N solution? The mention of volume and normality brings to mind the fact that by using these two values one can easily calculate the number of equivalents in the solution. Once this has been done, equivalents can be converted into grams:

$$55 \times 0.968 = 53.24 \text{ meq}$$

Since 1 meq NaOH $=$ 40 mg,

$$53.24 \times 40 = 2{,}129.6 \text{ mg} = 2.1296 \text{ g NaOH}$$

PROBLEM 13

What is the normality of a solution that contains 0.056 mg H_3PO_4 per milliliter? To substitute in the equation $V_{\text{ml}}N =$ meq, one must calculate first the number of milliequivalents of H_3PO_4 in 0.056 mg. There are 3 equiv/mole of this acid and the amu of H_3PO_4 is 98; therefore, 1 mmole $=$ 98 mg and 1 meq $= \frac{98}{3}$ mg. The number of milliequivalents of H_3PO_4 in 0.056 mg is $0.056/\frac{98}{3}$. The normality of the solution is

$$1 \text{ ml} \times N = \frac{0.056}{\frac{98}{3}} \text{ meq} \qquad N = 1.715 \times 10^{-3}$$

MOLARITY. Molarity (M) refers to the number of moles of a compound contained in 1 liter of solution.

$$M = \frac{\text{g solute in 1,000 ml solution}}{\text{g mol. wt. solute}}$$

A 1 molar (1 M) solution is made by dissolving 1 g mol. wt. of a compound in sufficient water to make 1 liter of final solution. Mathematically

$V_l M$ = moles. One does not take a liter of water and add a mole of compound to it.

PROBLEM 14

A saturated aqueous solution of NaOH is about 40%, weight per volume. What is the molarity of this solution? In 1,000 ml of solution there would be 400 g NaOH. Since 1 amu is 40, there would be $\frac{400}{40}$, or 10, moles of NaOH per 1,000 ml. The molarity is 10.

PROBLEM 15

What is the molarity of water? Assume 1 ml = 1 g; then there would be 1,000 g H_2O per liter. The number of moles of water in 1,000 g is $\frac{1{,}000}{18}$, or 55.6. Pure water is therefore 55.6 M.

PROBLEM 16

How many liters of 0.1 M solution can be made from 150 g of H_2SO_4? 150 g = $\frac{150}{98}$ moles. Substituting in the equation $V_l M$ = moles, we have

$$V_l \times 0.1 = \frac{150}{98} \qquad V_l = \frac{150}{98 \times 0.1} = 15.32 \text{ liters}$$

Fractions of a mole of a compound can be obtained by taking the desired amount of a solution of known molarity. For example, suppose one wanted to add 1.32×10^{-3} mmole of an aqueous solution of $AgNO_3$ to a reaction tube.

$$1 \text{ mmole } AgNO_3 = 169.89 \text{ mg} \simeq 170 \text{ mg}$$

therefore

$$1.32 \times 10^{-3} \text{ mmole} = 0.2244 \text{ mg}$$

It would be a waste of time to try to weigh this amount on a balance. Hence we make a more concentrated solution, using an amount that can be weighed more conveniently, and then take the volume of the solution that contains the desired amount of $AgNO_3$. Such a solution might be 224.4 mg per 100 ml. Thus by measuring out 0.1 ml of this solution, one would be taking 1.32×10^{-3} mmole of $AgNO_3$.

5.5 TEMPERATURE VERSUS SOLUBILITY

Temperature affects the solubility of substances in one of three ways: (1) an increase in temperature causes an increase in the solubility (directly proportional); (2) no great effect is noticed with a change in temperature; (3) an increase in temperature causes a decrease in the solubility (inversely proportional).

In general, the solubility of gases in liquids decreases as the temperature increases. The effect of temperature on the solubility of a solid in a liquid depends upon the type of solid and the characteristic of the solvent. The solubility of a solid in a liquid is associated with the energy needed to break the molecules or ions away from the crystal lattice, the nature of the solvent in overcoming these lattice forces (the dielectric constant of the solvent), and the energy released during solvation of the dissolved molecules or ions. Although no exact prediction can be made of which compounds will be more soluble in water as the temperature increases, most inorganic compounds that are soluble in water are more soluble with higher temperature.

5.6 INTERPRETATION OF SOLUBILITY CURVES

Solids and liquids

The solubility curve for KNO_3 is shown in Fig. 5.4. Any point on the curve represents a saturated solution. For example, D, A, and E correspond to saturated solutions at 0, 40, and 60°C, respectively. Thus, at A a saturated solution of KNO_3 in water is 60 g per 100 g water. At a higher or lower temperature the solubility varies according to the curve. Any point under the curve corresponds to an unsaturated solution.

At point B 60 g KNO_3 per 100 g H_2O at 60°C is an unsaturated solution. If the temperature of this solution is lowered until the concentration value is on the solubility curve, point A, the solution is saturated. If, however, the solution is further cooled, say to 20°C, without any KNO_3 crystallizing from the solution, a supersaturated solution would result (point C).

If a crystal of the solute is added to a supersaturated solution, the solute will crystallize out until equilibrium is reached, i.e., until the solution becomes saturated. If a crystal of KNO_3 is added to a solution represented by concentration C (at 20°), KNO_3 will crystallize out until its concentration is the same as its solubility (at 20°); about 35 g will come out of solution.

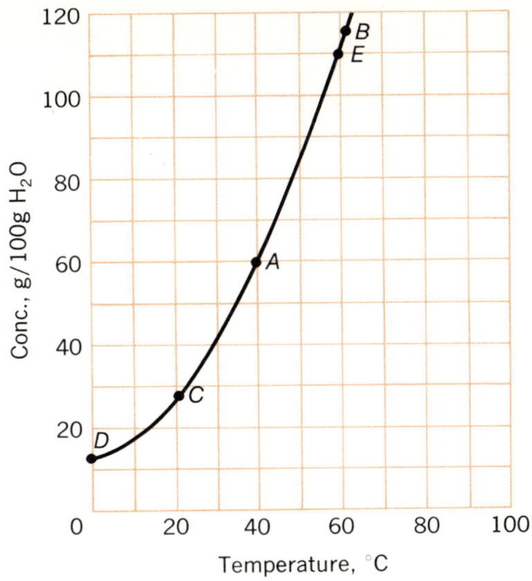

FIGURE 5.4 Solubility curve for potassium nitrate, KNO_3. The concentration is plotted versus the temperature.

How many grams of KNO_3 must be added to a solution whose concentration is represented by point B (Fig. 5.4) to make a saturated solution if the temperature remains constant? One would have to add enough KNO_3 to bring the concentration up to point E on the solubility curve, or about 55 g.

An inspection of the solubility curves for three compounds (Fig. 5.5) shows quite a variation. NaCl, for example, is only about 1.3 g more soluble at 50° than at 0°, whereas the solubility of $Pb(NO_3)_2$ is greatly increased with increasing temperature. The solubility of $NaNO_3$, for example, is 73 g at 0° and 180 g at 100°. Chromium sulfate is an example of a compound whose solubility decreases with increasing temperature.

Solubility curves are useful in determining the approximate solubilities of compounds at any given temperature.

Solubility of a mixture

In general the presence of a second solute has little if any effect on the respective solubilities unless the solutes react with each other or have a common ion. A comparison of the solubility curve of a single component with the curve for a mixture of components can be used to determine the purity of protein solutions and other compounds.

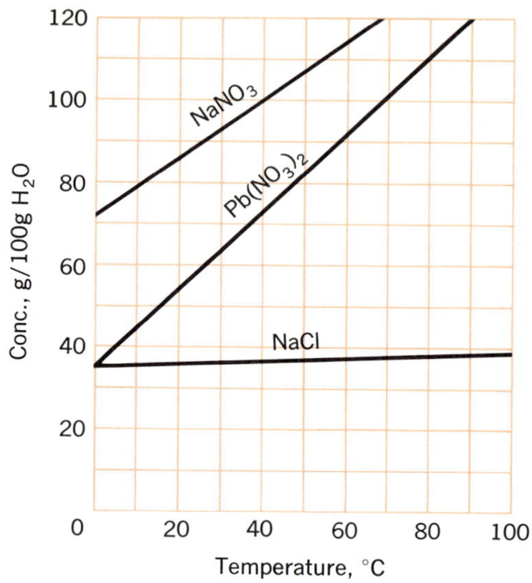

FIGURE 5.5 Solubility curve for three salts, NaCl, $Pb(NO_3)_2$, and $NaNO_3$.

For example, if increasing amounts of a pure compound X are added to a given amount of water and the amount of dissolved X is determined, a plot of the amount dissolved versus the amount added is as shown in Fig. 5.6*a*. Between A and B the solution is unsaturated and all the added solute has dissolved; therefore, the same amount of X is in solution as was added. However, after reaching point B the solution is saturated, and if more solute is added, no more will go into solution and the amount of X in solution remains constant (from B to C). This situation causes a sharp change in the slope of the curve, which remains flat after point B.

If, on the other hand, a mixture of two components of different solubility is used instead of a single pure substance, the curve will be like that in Fig. 5.6*b*. Between A and B the solution is unsaturated with respect to both components, and therefore the amount in solution is equal to the amount of the mixture added. At B, however, the solution is saturated only with respect to the least soluble component; hence, there will be a change in the slope of the curve upon the addition of more of the mixture. Between B and C the solution is becoming more concentrated in the more soluble component (it is still saturated with respect to the least soluble component). At C the solution is saturated with respect to both components and the addition of more of the mixture does not change the concentration of the solution. Hence the curve remains flat.

If two liquids are mutually soluble (miscible) or are insoluble (immiscible), there can be no solubility curve, and so the discussion that follows applies only to partially soluble (partially miscible) liquids. The solubility curve for this type of solution differs from the solid-liquid curves in that a point is eventually reached where the solvent becomes the solute and vice versa. In general as the temperature is increased, the solubility of two partially miscible liquids also increases, one exception being ether-water; ether is more soluble in water (or vice versa) as the temperature *decreases.*

This general solubility statement implies that if two partially soluble liquids are present in two phases (the solution is heterogeneous),

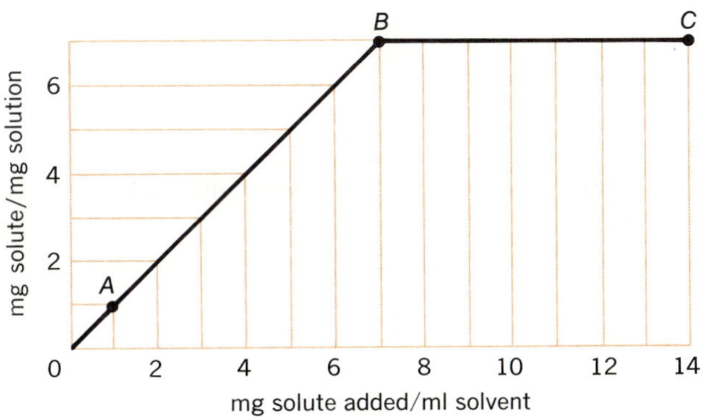

(*a*)

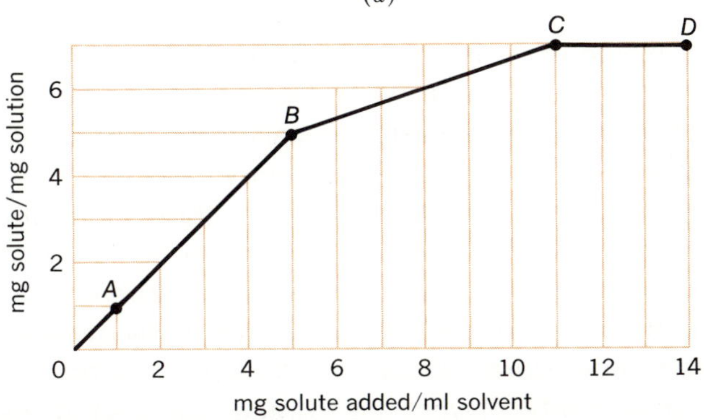

(*b*)

FIGURE 5.6 (*a*) Amount of pure solute added to a solvent versus the amount of solute in solution. (*b*) Same as (*a*) but with a mixture of two solutes.

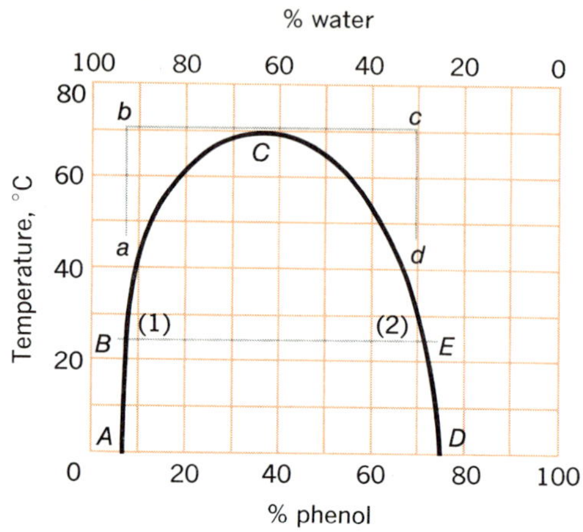

FIGURE 5.7 Solubility curve for phenol-water. All solutions with a concentration outside the curve ACD are homogeneous; those within the curve are heterogeneous and exist in two phases: (1) a solution of water saturated with phenol and (2) a solution of phenol saturated with water. The ratio of the volumes of the two phases at any given temperature and concentration is the same as the ratio of the distance between the given concentration to 100% water and the given concentration to 100% phenol. For example, at 20°C:

Distance from (1) to (2) = (1)•——•——•——•——•(2)
$\qquad \frac{1}{4} \qquad \frac{1}{2} \qquad \frac{3}{4}$

$$\frac{\text{Vol (phenol phase)}}{\text{Vol (water phase)}} = \begin{cases} \dfrac{1/4}{3/4} & \dfrac{1/2}{1/2} & \dfrac{3/4}{1/4} \\ \dfrac{1}{3} & \dfrac{1}{1} & \dfrac{3}{1} \end{cases}$$

the solution would exist as a single phase (become homogeneous) if the temperature could be raised sufficiently high. The temperature above which only one phase exists is called the *critical temperature.*

Phenol-water is an example of a binary mixture, i.e., a mixture of two partially miscible liquids, that illustrates these concepts. The solubility curve is shown in Fig. 5.7. The percentage composition of the solution is indicated on the x axis and the temperature on the y axis. For example, if one starts with 100% water and slowly adds phenol while keeping the temperature constant at 0°, a point is reached at about 6% phenol at which two layers are formed (point A). If the temperature is raised to 25° and kept constant while more phenol is added, a heterogeneous system will form again after the addition of a very little more phenol (point B). Repeating this procedure at different temperatures and plotting a curve as in Fig. 5.7 shows that the solubility of phenol in water does not increase very rapidly up to a temperature of about 50°. At 45° the percentage composition of the mixture is 90% water and 10% phenol. However, as the temperature is increased still further, the solubility of phenol in water increases rapidly to point C, 68.4°. Above this temperature, the critical solution temperature, the two compounds are completely miscible.

If the same procedure is repeated by adding water to 100% phenol, the curve DC (Fig. 5.7) is obtained. At any point inside the curve ACD two liquid phases exist; the phenol layer is saturated with water, and the water layer is saturated with phenol. Since at a given temperature the mutual solubility of the pair is constant, any point along the line BE represents the same percentage of water in phenol or vice versa, but different amounts of the respective phases. It is possible to go from a to d without two phases forming. This is done by increasing the temperature of the solution represented by a to above the critical solution temperature (68.4°), for example, to a temperature represented by point b, and then adding phenol until the concentration represented by c is obtained. When the temperature is lowered to the initial temperature, point d is reached without the appearance of two phases.

5.7 FRACTIONAL CRYSTALLIZATION

The difference in the solubility of compounds frequently affords a convenient means of separating them from each other. In Fig. 5.8 the solubility curves for compounds A and B are given in terms of percentage by weight. The concentration of a solution of these compounds is represented by C_A and C_B at temperature t_3. When the temperature

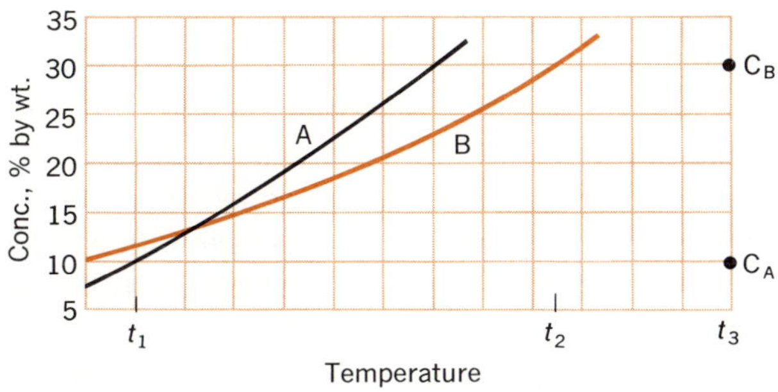

FIGURE 5.8 Solubility curve for two compounds, A and B.

is lowered to t_2, the solution is saturated with respect to component B but is still unsaturated with respect to component A. The temperature can be lowered to t_1 before the solution becomes saturated with respect to A. Between temperatures t_2 and t_1 compound B crystallizes out of solution. At temperatures lower than t_1 both A and B separate out in crystalline form. In practice recrystallization is performed several times to obtain a pure product.

PROBLEM 1

In the previous illustration, how many grams of B could be obtained from 50 g of solution? The initial concentration of B is 30%, and the final concentration of B in the solution is 10% at temperature t_1. Thus, 20 g of B could be obtained per 100 g of solution, or 10 g from 50 g of solution.

The procedure of fractional crystallization can be used when the compounds have different solubility in the same solvent and neither reacts chemically with the other or with the solvent.

If one component of a mixture of two substances is insoluble in a particular solvent whereas the other one is soluble, the two components can be separated quite easily. For example, sugar and charcoal (carbon) can be separated by dissolving the sugar in water and filtering off the insoluble carbon. The sugar solution is then concentrated by evaporating off the water, and the sugar can be crystallized from the concentrated solution.

5.8 PARTITION OR DISTRIBUTION COEFFICIENT α

The partition coefficient of a solute is the ratio of the concentration of the substance in equal volumes of the two phases of liquids that are partially miscible or immiscible.

In a two-phase system of phenol and water in which there are equal volumes of each phase, if substance X, which is equally soluble in both the phenol-water and the water-phenol phase, is added to the mixture, after equilibrium has been reached X will be equally distributed between the two layers of liquids (see Fig. 5.9). There is the same amount of X in the phenol-water phase as in the water-phenol phase. The partition coefficient, water/phenol, would equal 0.5/0.5, or 1.0. Generally the partition coefficient of a substance is independent of the presence of other compounds provided, of course, there is no chemical reaction between them.

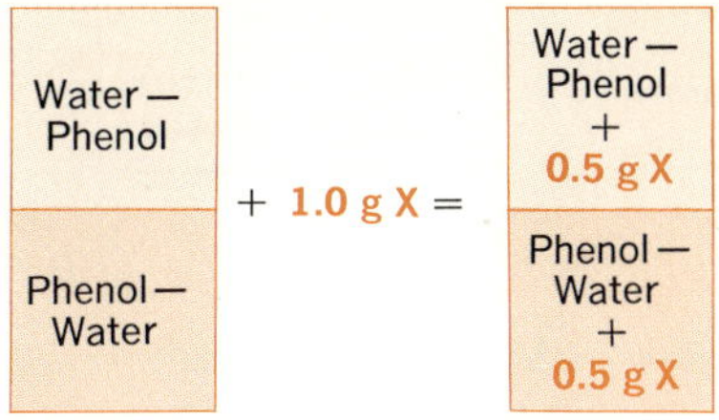

FIGURE 5.9 Distribution of X in water-phenol and phenol-water phases, $\alpha = 1$.

5.9 TECHNIQUES FOR SEPARATION OF BIOCHEMICAL MIXTURES

A procedure for separating extremely small amounts of compounds, called *countercurrent distribution,* is based upon the difference in the distribution coefficient α of the respective compounds. A substance is added to equal volumes of two phases and shaken, and after the solute has come to equilibrium, the top phase is removed by appropriate means. Fresh upper phase (the same volume as removed) is added to the original bottom phase. Fresh lower phase (again, the same volume as originally present) is added to the remaining upper phase (see Fig. 5.10). When equilibrium has again been attained after shaking, the phases are separated and fresh solvents are added as indicated in the diagram.

The amount of solute is greater in the center tube. As the number of separations increases, more and more solute will be distributed in the center tubes and less and less will appear in the end tubes. In Fig. 5.11 the amount of solute per tube is plotted against the number of the tube.

From the definition of α, it follows that if a solute has a distribution coefficient greater than 1, the tube containing the greatest amount of the solute will be to the right of the center tube. The greater the distribution coefficient, the further the peak of the curve is shifted to the right (Fig. 5.11). Conversely, as α becomes less than 1, the peak of the curve is shifted to the left of center. The larger the number of tubes, the better the separation of compounds whose α's are rather close together.

Countercurrent distribution has been extensively used in separating small amounts of biologically active compounds. A decided advantage of this procedure is that the compounds being separated are not subjected to harsh treatment, such as heat, which might affect their

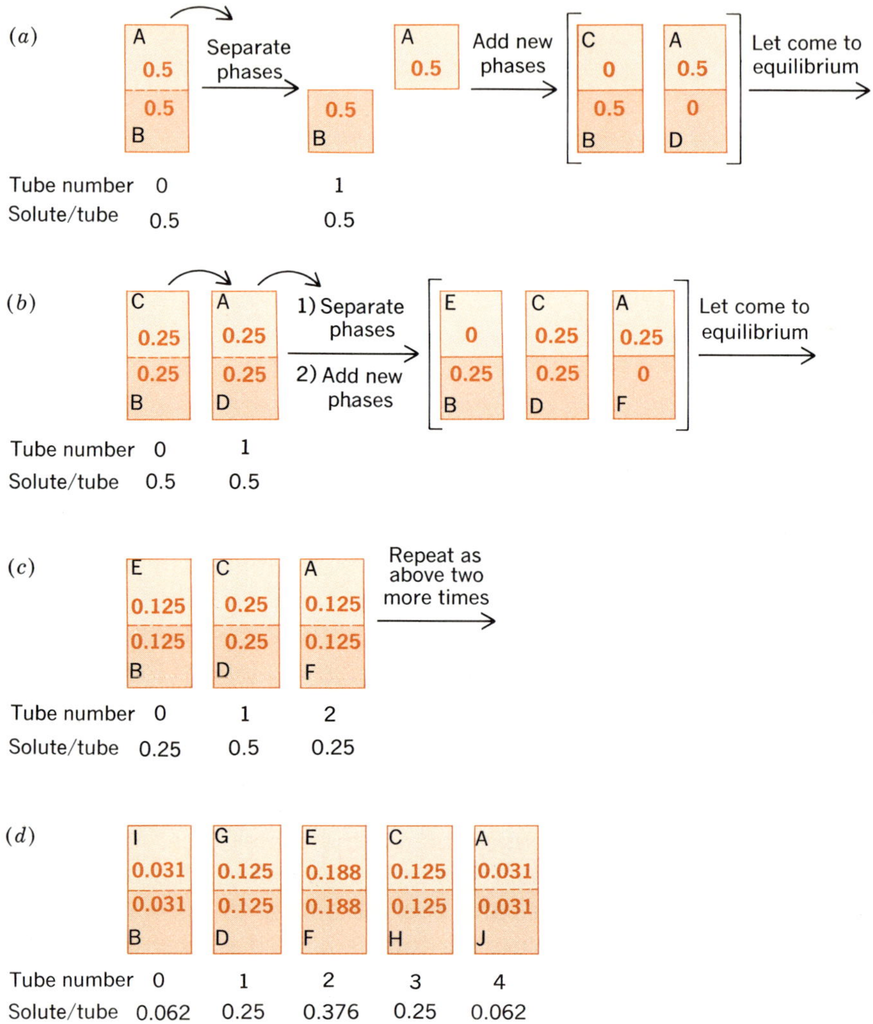

FIGURE 5.10 Countercurrent distribution: distribution of solute, $\alpha = 1$, in solvent after four transfers.

biological activity. One disadvantage is that special equipment is needed for a large number of tubes.

Also based on the distribution coefficient α and involving the same principal as countercurrent distribution is the technique of *paper chromatography*. It is simpler than countercurrent distribution and can be carried out quite conveniently in the laboratory. Instead of using two liquids that must be separated at each stage, paper chromatography uses a solid medium as a support, namely, filter paper, over which a mixture of solvents is allowed to flow by capillary action. The materials to be separated are placed in a small area on the paper near the end that is inserted in the solvent.

The solvent with a greater attraction for the polar molecules of cellulose in the paper moves more slowly than other less polar solvents. This liquid may be considered to be a stationary phase while the less polar solvent is free to move over the paper. If one imagines that the flow of solvents is stopped a number of times, each of an infinitely short duration, each stop would be equivalent to a tube transfer in countercurrent distribution. Theoretically, paper chromatography is

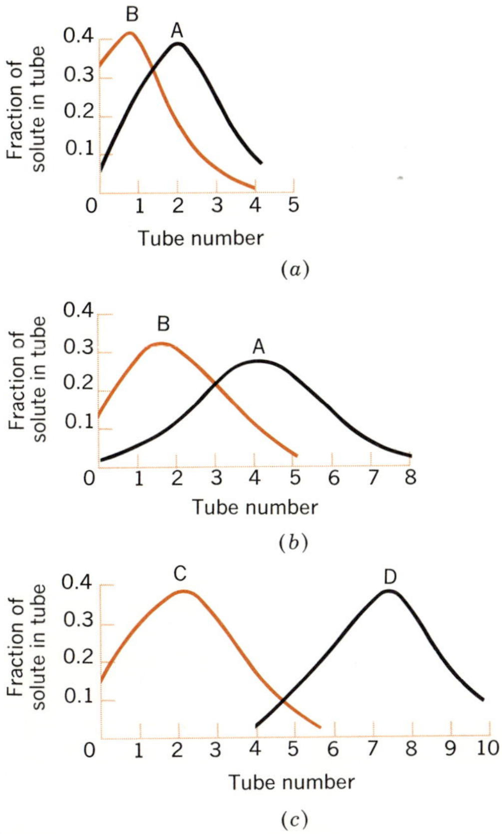

FIGURE 5.11 Countercurrent distribution: (*a*) separation of solutes A ($\alpha = 1$) and B ($\alpha = \frac{1}{3}$) after four transfers; (*b*) separation of solutes A and B after eight transfers; (*c*) separation of solutes C ($\alpha = \frac{1}{3}$) and D ($\alpha = \frac{3}{1}$) after nine transfers.

countercurrent distribution involving infinite numbers of transfers. As a result, separation can be accomplished of compounds that have quite similar distribution coefficients and chemical properties. Paper chromatography can be used to separate both inorganic and organic compounds.

Assume that four different-colored compounds (A, B, C, and D) and a mixture of the four are placed on the filter paper, which is then put in a chamber with a layer of solvent at the bottom (Fig. 5.12*a*). The bottom edge of the paper is immersed in the solvent. After from 2 to 18 hr, depending upon the length of the paper, the paper is removed from the chamber and the solvent front is marked. Obviously, the solvent front cannot be allowed to reach the upper end of the paper since the most soluble component would be lost. The paper is dried, and then the distance each component and the solvent has moved from the point of application is measured. R_f, the position of each spot relative to the flow of the solvent (the solvent front), is calculated according to the equation

$$R_f = \frac{\text{distance compound moves}}{\text{distance solvent moves}}$$

In this example A did not move in this particular solvent (Fig. 5.12*b*); therefore, it has an R_f value of zero. The R_f values of B, C, and D are

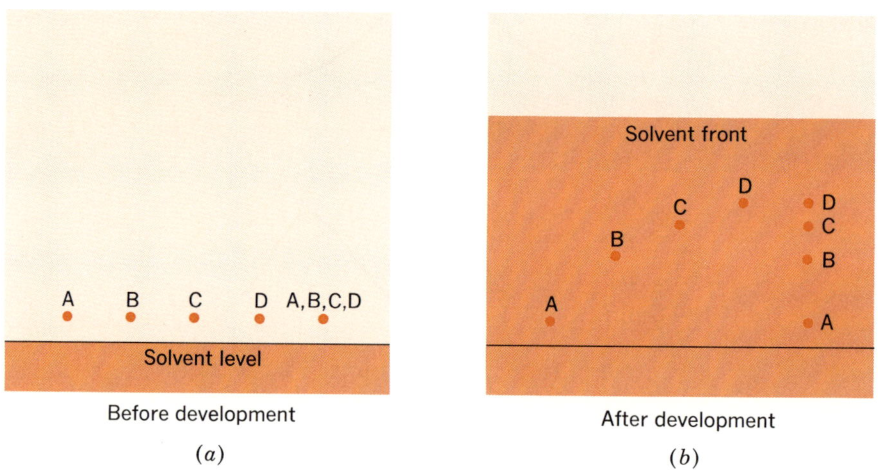

FIGURE 5.12 (*a*) Compounds A, B, C, D and a mixture of all four spotted on paper before chromatography. (*b*) Distribution of compounds A, B, C, and D after development of the chromatogram.

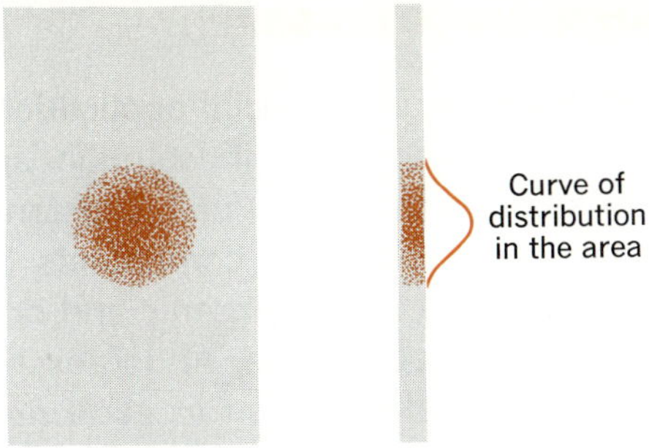

(*a*) (*b*)

FIGURE 5.13 Distribution of solute in an area after development of the chromatogram: (*a*) front view; (*b*) side view through the paper. The distribution is also indicated by the curve drawn on the side of the cross section of the paper.

$$B = \tfrac{8}{20} = 0.4 \qquad C = \tfrac{12}{20} = 0.6 \qquad D = \tfrac{15}{20} = 0.75$$

It is frequently necessary to treat the paper with a suitable reagent in order to locate the area to which a substance has moved. For example, hydrogen sulfide, H_2S, can be used to indicate the presence of a number of ions, e.g., silver, lead, mercury, copper, cadmium, bismuth, arsenic, antimony, tin, iron, cobalt, nickel, and manganese. These ions react with H_2S to form a grey to dark brown or black sulfide. A number of inorganic ions with their respective R_f values in different solvents are listed in Appendix E. Frequently it is necessary to run more than one chromatogram in different solvents to identify the components of a mixture. For example, consider a mixture of Hg^+, Hg^{++}, Sn^{++}, and Sn^{4+}. Neither solvent 1 nor solvent 3 (Appendix E) would perfect a separation of these four components. Solvent 1 would separate Hg^+ and Hg^{++} but not Sn^{++} and Sn^{4+}. Solvent 3, on the other hand, would separate the two forms of Sn but not the two forms of Hg.

If one could determine the concentration of material in the spot of a chromatogram, the distribution of the compound would be more concentrated at the center of the spot (see Fig. 5.13). The cross-sectional distribution curve is similar to that obtained for countercurrent distribution.

5.10 ION-EXCHANGE CHROMATOGRAPHY

Ion-exchange chromatography is a special application of displacement or exchange reactions in which ions in solution are exchanged for similarly charged ions bound to an *insoluble* substance (usually a synthetic resin, although some inorganic compounds can be used). By ion-exchange chromatography both inorganic and organic ions can be separated and obtained in a pure state. An ion-exchange resin which exchanges positive ions is called a *cation exchange* resin, and one which exchanges negative ions is called an *anion exchange* resin. In the following reactions R stands for the insoluble ion.

Cation exchange reaction:

$$\underset{\text{Insoluble}}{R^-\text{—}H^+} + Ca^{++} \rightleftharpoons \underset{\text{Insoluble}}{R^-\text{—}Ca^{++}} + H^+$$

Anion exchange reaction:

$$\underset{\text{Insoluble}}{R^+\text{—}OH^-} + Cl^- \rightleftharpoons \underset{\text{Insoluble}}{R^+\text{—}Cl^-} + OH^-$$

The cation exchange resin is shown in the hydrogen state, $R^-\text{—}H^+$, and the anion resin in the hydroxide state, $R^+\text{—}OH^-$. The resin has a large number of exchangeable ions per gram of material. The H^+ of $R^-\text{—}H^+$ is exchanged for such cations as Ca^{++}, Mg^{++}, Na^+, etc., or for organic positive ions. The positive ions in solution become bound to the insoluble resin and are thereby removed from solution. This is also true for negative ions of salts when passed over an anion exchanger; the negative ions replace the OH^- of $R^+\text{—}OH^-$ and become bonded to the insoluble resin, being replaced in the solution by OH^-.

In addition to being insoluble in order to remove ions from solution, ion-exchange materials are designed to have specific properties. For example, some resins bind cations strongly whereas other resins bind cations weakly. In choosing anion-exchange resins, this property must be considered. The type of resin determines what ions can be separated. Also, the nature of the ions in solution influences the order in which ions bind to the resin as well as the order in which the ions are eluted (removed) from the resin. One factor, for example, is the oxidation number. Cations with a higher oxidation number are usually more firmly bound to the ion-exchange compound. When several inorganic ions with the same oxidation number are exposed to a resin, the ions with the greater atomic number are usually bound most firmly to the ion-exchange compound. The ion that is bound the most firmly will displace the more weakly bound ions that are already on the resin. This property permits separation of ions.

Using ion-exchange chromatography, it is possible to effect a separation of ions that cannot be easily achieved by other means. In Fig. 5.14 the ionic state of the resin is indicated by R^-—H^+. Upon addition of a solution of NaCl the Na^+ in solution displaces the H^+ of R^-—H^+ and forms insoluble R^-—Na^+ and the H^+ goes into solution (Fig. 5.14*a*). If a solution, say $CaCl_2$, containing Ca^{++} is added to the column, the Ca^{++} in solution replaces the Na^+ from R^-—Na^+ to form insoluble R^-—Ca^{++} and the Na^+ will return to the solution. The Na^+ moves down the column until it comes into contact with more R^-—H^+, where it becomes bound again as insoluble R^-—Na^+ (Fig. 5.14*b*). The sodium will be layered *below* the calcium on the resin. A third solution (such as $FeCl_3$) containing Fe^{3+} is now added to the column (Fig. 5.14*c*). Since Fe^{3+} is more strongly bound to the resin than Ca^{++}, the soluble Fe^{3+} displaces the Ca^{++} from the resin and forms insoluble R^-—Fe^{3+}.

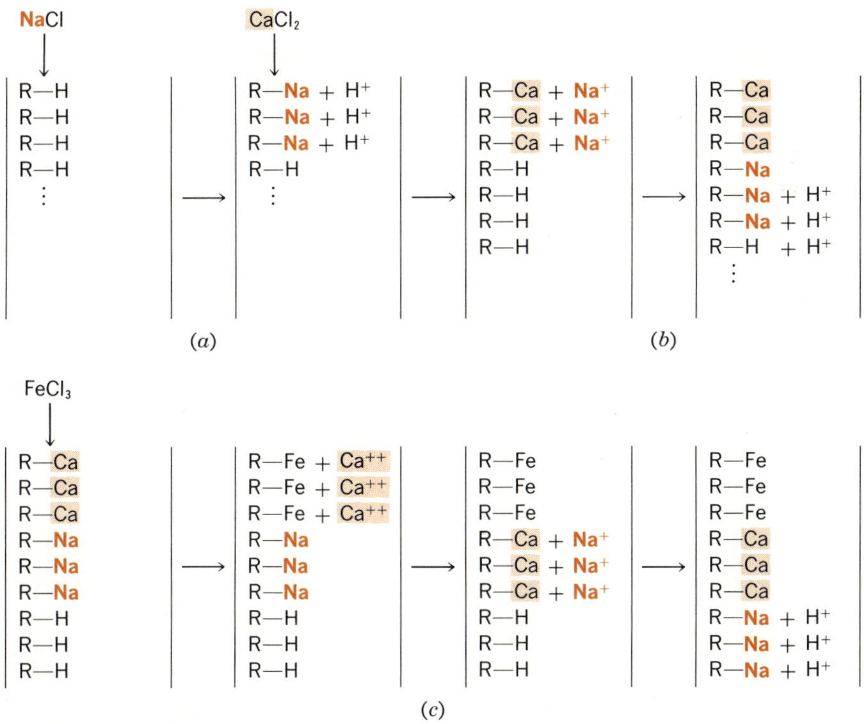

FIGURE 5.14 Diagram of ion-exchange chromatography.

The Ca^{++} now in solution moves down the column and displaces Na^+ from R^-—Na^+; the Na^+ in turn moves down the column and displaces H^+ from more R^-—H^+. The net result is that the three ions are bound to the resin in a definite order, from top to bottom of the column. The most strongly bound ion (Fe^{3+}) is followed by Ca^{++} and then the most weakly bound ion (Na^+).

Ion-exchange reactions are equilibrium reactions and therefore reversible under the appropriate conditions. The *law of mass action* (see Sec. 6.1) can be applied to reverse the above reactions of ion exchange. For example, bound ions such as Na^+, Ca^{++}, and Fe^{3+} can be exchanged for H^+ by passing a strong enough acid solution through the column containing these ions bound to the resin. This is called *eluting* the column. Since Na^+ is least strongly bound, it is the first ion eluted; following Na^+ would be Ca^{++} and then Fe^{3+}. By appropriate means of collecting the *eluate* (the eluting solution) from the column, volumes containing each ion can be separated as it comes off the column. Ion-exchange chromatography affords a convenient means of separating and collecting ions from complex mixtures.

Cation and anion exchange resins are used commercially to remove inorganic salts from water, i.e., to deionize the water. The reactions that take place if a solution of inorganic salts is passed first over a cation exchange resin in the hydrogen state and then over an anion exchange resin in the hydroxide state are given below, $CaCl_2$ representing the inorganic salts in solution. The passage of the salt water over a cationic exchange resin replaces the calcium ions with hydrogen ions.

Cation exchange reaction: $R^-—H^+ + Ca^{++} + 2\,Cl^- \longrightarrow R^-—Ca^{++} + H^+ + 2\,Cl^-$

The solution tends to become more acidic. The subsequent passage of the solution over an anion exchange resin removes the negative chlorine ions and replaces them with hydroxyl ion.

Anion exchange reaction: $R^+—OH^- + 2\,Cl^- \longrightarrow R^+—2Cl^- + OH^-$

Thus, the inorganic salts in the original solution have been replaced with H_2O, and the water has had all ions, except H^+ and OH^-, removed from it.

Net reaction: $\underset{\text{Soluble}}{CaCl_2} + \underset{\text{Insoluble}}{\text{resins}\begin{matrix} H^+ \\ OH^- \end{matrix}} \longrightarrow HOH + \underset{\text{Insoluble}}{\text{resins}\begin{matrix} Ca^{++} \\ 2\,Cl^- \end{matrix}}$

Commercial water softeners that make use of ion exchange are cation exchangers. Many of them are inorganic compounds, e.g., sodium *zeolite*. The sodium exchanges with the calcium, magnesium, and iron ions in hard water to form the corresponding insoluble zeolite salts.

$$\underset{\text{Insoluble}}{\text{Zeolite—Na}^+} + \underset{\text{Soluble}}{\text{Ca}^{++}} \rightleftharpoons \underset{\text{Insoluble}}{\text{zeolite—Ca}^{++}} + \underset{\text{Soluble}}{\text{Na}^+}$$

The Na^+ passes into solution, but since it does not form an insoluble scum with soap, the water has been softened. When the sodium zeolite has exchanged its sodium for calcium or other metallic ions in hard water, it cannot function again as an ion exchanger until the sodium salt has been regenerated. This is accomplished by passing a concentrated solution of NaCl over the calcium zeolite to elute the calcium.

$$\underset{\text{Insoluble}}{\text{Zeolite—Ca}^{++}} + \underset{\text{Excess}}{\text{NaCl}} \longrightarrow \underset{\text{Insoluble}}{\text{zeolite—Na}^+} + \underset{\text{Soluble}}{\text{Ca}^{++}\text{Cl}_2^-}$$

When the excess NaCl and the Ca^{++} released from the zeolite have been washed away, the sodium zeolite is ready for use again.

QUESTIONS

1. Fill in the following table:

Solute concentration, g/liter	*Molarity*	*Gram molecular weight Solute*
103	1.0	______
17	0.5	______
44	0.33	______
0.098	0.001	______
______	0.1	40
______	0.25	98

2. Complete the following table:

Compound	*Molarity*	*Normality*
NaOH	0.1	______
H_2SO_4	______	0.1
$AlCl_3$	1.0	______
$(NH_4)_3PO_4$	______	0.3
$Mg(NO_3)_2$	0.5	______

3. Find the normality of the following:
(*a*) 25 g NaCl in 200 ml solution.
(*b*) 140 g H_2SO_4 in 750 ml solution.
(*c*) A 1% (weight per volume) solution of NaOH.
(*d*) A solution containing 4.5 mg HNO_3 per 100 ml of solution.

4. How many milliliters of water should be added to 50 ml of 0.25 M H_3PO_4 to make a solution that is 0.25 N?
5. If 30 ml of 3 N KOH solution is diluted to 1.3 N, what volume of solution results?
6. What is the percentage by weight of $MgSO_4$ in a 0.12 N solution?
7. If it requires 20 ml of 0.1 N HCl to neutralize 50 ml of a NaOH solution, calculate (*a*) the normality of the NaOH solution and (*b*) the percent concentration (weight per volume) of NaOH.
8. How many milliliters of 1.75 N H_2SO_4 is required to react with 35 ml of 0.6 N NaOH?
9. How many milliliters of a 0.02% (weight per volume) solution of $CuSO_4$ contains 1.5 mg Cu?
10. It requires 8 ml of 0.15 M KOH solution to neutralize 10 ml of a HNO_3 solution.
    (*a*) What is the normality of the KOH solution?
    (*b*) How many grams of HNO_3 does 10 ml of the HNO_3 solution contain?
11. Define an unsaturated, a saturated, and a supersaturated solution.
12. Given a solution, how would you proceed to determine whether it is unsaturated, saturated, or supersaturated?
13. Use the solubility curves for compounds A, B, and C to answer the following questions:
    (*a*) Within what range of temperature is compound A less soluble than compounds B or C?
    (*b*) Where should one place an X on the graph to indicate the temperature and concentration at which a solution of compound B is saturated, A is unsaturated, and C is supersaturated?
    (*c*) Where should one place a Y to indicate the temperature and concentration at which a solution of compound C is unsaturated and compounds A and B are both supersaturated?
    (*d*) If 120 ml of a 30% (weight per volume) solution of A is at 60°, how many grams of A will precipitate out of solution if the temperature is lowered to 20°?

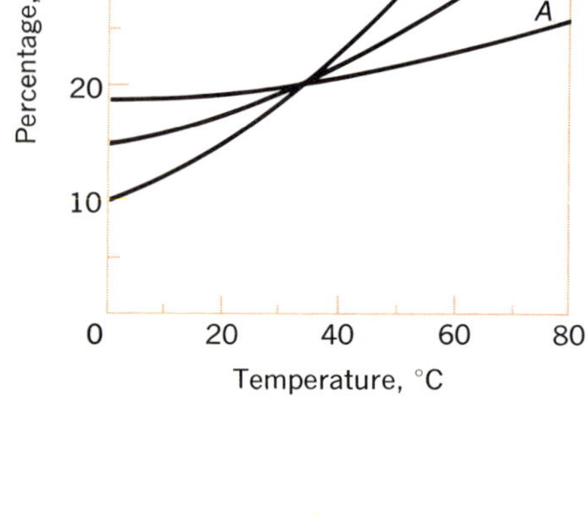

14. What equilibrium exists in a saturated solution? Is such an equilibrium also possible in an unsaturated solution? In a supersaturated solution?

15. Explain the difference in the procedure for preparing a *molal* and a *molar* solution.

16. If 5 g of NaOH is added to 100 g of H_2O, what is the molality of the resulting solution?

17. How would you prepare 150 ml of a 0.2 *m* solution of $CuSO_4$?

18. What is the mole fraction of acetic acid, CH_3CO_2H, in a solution that contains 30 g acetic acid in 100 g H_2O?

19. An investigator wishing to determine the purity of a product he has isolated finds the solubility of the preparation by adding increments of it to water and calculating the amount in solution. He obtains the following data. Is the product pure? Determine by plotting the data as in Fig. 5.6.

Preparation Added to 10 *ml* H_2O, *mg*	*Preparation Dissolved in* 10 *ml* H_2O, *mg*
0.5	0.5
1.0	1.0
1.5	1.5
3.0	3.0
4.0	4.0
4.5	4.4
5.0	4.8
6.0	5.7
6.8	6.3
7.0	6.5
7.5	6.5
8.5	6.5

20. From Fig. 5.7:
 (*a*) To what minimum temperature must a solution of 20% phenol and 80% water be heated to have a homogeneous solution?
 (*b*) To prepare a mixture of phenol and water such that the volumes of the two phases are equal at 45°C, what percentage of phenol and water should be mixed?

21. The distribution coefficients (water-phenol) of compounds X, Y, and Z are as follows: X = 1, Y = $\frac{1}{3}$, Z = 3. Roughly sketch the countercurrent distribution curves for these components after a nine-tube transfer.

22. In paper chromatography, what is meant by R_f?

23. With the aid of paper chromatography what solvent pairs would you use to effect the best separation of the components of the following mixtures of ions? It may be necessary to use two different solvents. Refer to Appendix E.
 (*a*) Ag, Cu, Fe^{3+} (*b*) Al, Cd (*c*) Cl, Br
 (*d*) Pb, Bi (*e*) Hg^{++}, Cd

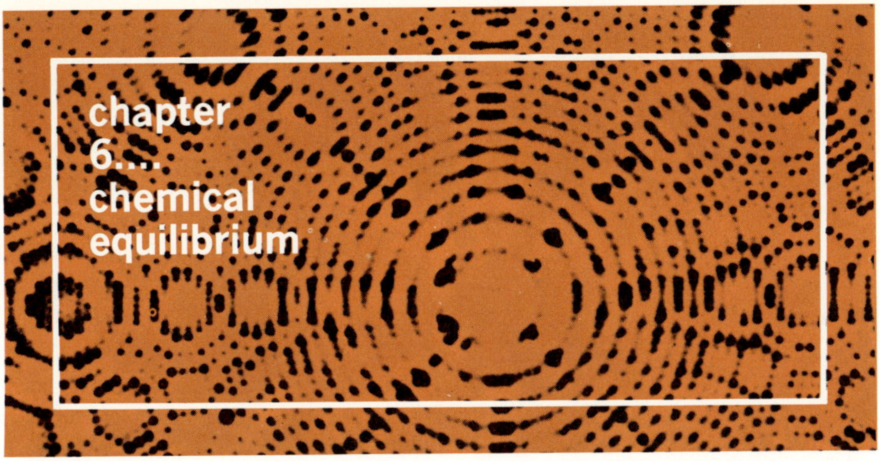

chapter 6.... chemical equilibrium

Chemical reactions may or may not go to completion. Many reactions do not go to completion because the reverse reaction also occurs. These are called *reversible reactions.* A state of equilibrium is reached when the forward reaction proceeds at the same rate as the reverse reaction. Most biochemical reactions do not go to completion in an isolated system outside the cell, but in the whole cell many do go to completion, especially those from which energy is derived. If these reactions remained in equilibrium, the cell would die. Since most inorganic and biological reactions take place in the presence of water, special attention must be paid to reactions in an aqueous medium and to factors affecting the rate of a reaction.

Before reading the rest of the chapter review the answers to the following questions to refresh your memory. What is equivalent weight? How does the electronegativity value of the elements vary within a group and within a period of the periodic table? What is molarity? What is the molarity of water? Is water a polar compound?

6.1 EQUILIBRIUM REACTIONS

One fundamental principle of chemical reactions is expressed in the *law of mass action: the rate of a chemical reaction in dilute solution is proportional to the molarity of the reacting substances.*

The law of mass action may be expressed by the following equation, in which brackets indicate molar concentration, v is the reaction rate, $\propto$ means is proportional to, and k is the proportionality constant for a particular reaction. A subscript to k or v defines which reaction is referred to.

$$\text{A} + \text{B} \xrightarrow{\text{Rate } v_1} \text{C} + \text{D}$$
$$v_1 \propto [\text{A}][\text{B}] \quad \text{or} \quad v_1 = k_1[\text{A}][\text{B}] \tag{6.1}$$

For the reverse reaction

$$\text{C} + \text{D} \xrightarrow{\text{Rate } v_2} \text{A} + \text{B}$$
$$v_2 \propto [\text{C}][\text{D}] \quad \text{or} \quad v_2 = k_2[\text{C}][\text{D}] \tag{6.2}$$

At equilibrium

$$\text{A} + \text{B} \underset{v_2}{\overset{v_1}{\rightleftharpoons}} \text{C} + \text{D}$$
$$k_1[\text{A}][\text{B}] = k_2[\text{C}][\text{D}] \tag{6.3}$$

Therefore

$$\frac{k_1}{k_2} = \frac{[\text{C}][\text{D}]}{[\text{A}][\text{B}]} = K_{eq} \quad \text{equilibrium constant} \tag{6.4}$$

In Eq. (6.1), A reacts with B to produce the products C and D. The velocity of this reaction is proportional to the product of the concentration [A] of A and [B] of B. Therefore we can write

$$v_1 = k_1[\text{A}][\text{B}]$$

As soon as they are formed, the products in a reversible reaction react with each other to produce the initial reactants, as shown in Eq. (6.2). Since these two reactions are going on at the same time, eventually a state is reached in which the velocity of the forward reaction is the same as that of the reverse reaction, Eq. (6.3). A state of equilibrium is said to exist when $v_1 = v_2$, and Eq. (6.4) is arrived at by collecting the two rate constants on the same side of the equation. Since k_1 and k_2 are both constants, they can be combined into a single constant, K_{eq}, the equilibrium constant.

If more than one molecule of each reactant is involved, such as

$$\text{A} + \text{B} \rightleftharpoons 2\,\text{C} + \text{D}$$

or

$$\text{A} + \text{B} \rightleftharpoons \text{C} + \text{C} + \text{D}$$

the equilibrium constant is

$$K_{eq} = \frac{[\text{C}][\text{C}][\text{D}]}{[\text{A}][\text{B}]}$$

Since [C] times [C] is equal to $[C]^2$, the final equation can be formulated as

$$K_{eq} = \frac{[C]^2[D]}{[A][B]} \qquad (6.5)$$

In the expression for the equilibrium constant for most reactions the molar concentration of each reactant and product is raised to the power of the number of molecules of that substance in the balanced equation.

Equilibrium is affected by temperature, pressure, concentration of reactants, and the chemical nature of the reacting substances. Changing one of these factors, such as temperature, tends to change the equilibrium of a reaction in such a manner that the added stress is offset. This is an expression of *Le Châtelier's theorem.* For example, the solubility of a solute eventually reaches equilibrium at saturation. If a substance absorbs heat while dissolving, the application of heat to the solution will shift the equilibrium so as to increase the solubility of the compound in the solvent.

EXAMPLE 1

Le Châtelier's theorem may be applied to the reaction between N_2 and H_2, in which a total of four volumes of gas combines to form two volumes of gas, NH_3.

$$\underset{3\text{ vol}}{3\,H_2} + \underset{1\text{ vol}}{N_2} \rightleftharpoons \underset{2\text{ vol}}{2\,NH_3}$$

An increase in pressure on this system would shift the reaction to the right, thus decreasing the pressure *by reducing the number of molecules.* Remember that the pressure of a gas is directly proportional to the number of molecules of the gas in the container.

EXAMPLE 2

If a product of an equilibrium reaction is removed, equilibrium can no longer be maintained and the reaction goes to completion. Consider the reaction for decomposition and formation of H_2CO_3.

$$H_2CO_3 \rightleftharpoons H_2O + CO_2\uparrow \qquad K_{eq} = \frac{[H_2O][CO_2]}{[H_2CO_3]}$$

If CO_2 is allowed to escape from the reaction container, the reaction will go to completion and all the H_2CO_3 will decompose. Removal of an end product can also be accomplished by a second reaction which uses the end product of the first reaction. For example, if A $\rightleftharpoons$ B is an equilibrium reaction and B + C $\longrightarrow$ D is not a reversible reaction, the net result of coupling these two reactions is the conversion of all A to D:

$$\begin{aligned} A &\rightleftharpoons B \\ B + C &\longrightarrow D \\ \text{Overall:}\quad A + C &\longrightarrow D \end{aligned}$$

6.2 IONIZATION OR DISSOCIATION

Ionization, or dissociation, is the process whereby compounds split into positively and negatively charged particles when dissolved. An ion is defined as an electrically charged atom or group of atoms (radical); hence the term ionization. Compounds that ionize are known as *electrolytes,* since a solution of ions conducts an electric current. Since most ionization reactions take place in an aqueous medium, we must consider the ionization of water as well as that of compounds dissolved in it. It is important to remember that *ionization is always reversible;* it is an equilibrium reaction.

TABLE 6.1 COMPLETE IONIZATION EQUATION FOR SEVERAL COMPOUNDS

Compound		Ions
$NaOH$	$\rightleftharpoons$	$Na^+ + OH^-$
KBr	$\rightleftharpoons$	$K^+ + Br^-$
$AlCl_3$	$\rightleftharpoons$	$Al^{3+} + 3\ Cl^-$
$(NH_4)_3PO_4$	$\rightleftharpoons$	$3\ NH_4^+ + PO_4^{3-}$
Na_3PO_4	$\rightleftharpoons$	$3\ Na^+ + PO_4^{3-}$
$MgSO_4$	$\rightleftharpoons$	$Mg^{++} + SO_4^{--}$
$Al(OH)_3$	$\rightleftharpoons$	$Al^{3+} + 3\ OH^-$
$KClO_3$	$\rightleftharpoons$	$K^+ + ClO_3^-$
$Ca(NO_3)_2$	$\rightleftharpoons$	$Ca^{++} + 2\ NO_3^-$
$Na(CH_3CO_2)$	$\rightleftharpoons$	$Na^+ + CH_3CO_2^-$
HCl	$\rightleftharpoons$	$H^+ + Cl^-$
H_2SO_4	$\rightleftharpoons$	$2\ H^+ + SO_4^{--}$
H_3PO_4	$\rightleftharpoons$	$3\ H^+ + PO_4^{3-}$
$Ca(OH)_2$	$\rightleftharpoons$	$Ca^{++} + 2\ OH^-$
NH_4OH	$\rightleftharpoons$	$NH_4^+ + OH^-$
$H(CH_3CO_2)$	$\rightleftharpoons$	$H^+ + CH_3CO_2^-$

Ionization is possible when atoms are united by ionic bonds (NaCl) or by a highly polar covalent bond (HCl). The greater the polarity of the bond, the greater the degree of ionization. The polarity of the solvent also influences ionization. Table 6.1 gives the ionization equation of a number of compounds.

6.3 WATER IN BIOLOGICAL REACTIONS

Water is very important to living organisms. The water content of different organisms may vary between wide limits. Seeds, for example, may contain only a few percent moisture; mammals are about 72% water, whereas a tadpole is approximately 89% water. Water is the medium in which biological reactions take place; it is a means of transporting nutrients in both animals and plants and also aids in regulating animal temperature. In addition, water enters into many chemical reactions of living organisms. For instance, water furnishes the hydrogen to reduce carbon dioxide during photosynthesis.

$$6\ H_2O + 6\ CO_2 \xrightarrow{\text{Light}} \underset{\substack{\text{Sugar} \\ \text{(carbohydrate)}}}{C_6H_{12}O_6} + 6\ O_2$$

The only chemical reaction associated with digestion is hydrolysis, i.e., the reaction of the food (carbohydrate, fat, protein) with water. Starch, for example, must be hydrolyzed by water to the sugar glucose before it can be utilized by the body.

$$\underset{\text{Complex carbohydrate}}{\text{Starch}} \xrightarrow[\text{(Hydrolysis)}]{\text{Digestion}} \underset{\text{Simple carbohydrate}}{\text{glucose}}$$

Water is also a reactant or a product of many reactions which occur inside cells. One important series of reactions is the reverse of photosynthesis, namely, the oxidation of sugar (glucose) to carbon dioxide, water, and energy

$$C_6H_{12}O_6 + 6\ O_2 \longrightarrow 6\ CO_2 + 6\ H_2O + \text{energy}$$

In this conversion the hydrogen in glucose reduces molecular oxygen to form water. The role of water in biological reactions will become apparent in subsequent chapters; of immediate interest are the properties of water associated with ionization.

6.4 HYDROGEN BONDING

Water is not a collection of isolated molecules but of aggregates (clusters) of molecules. The influence of these clusters on the physical properties of water, e.g., boiling point, is quite pronounced. Liquids whose molecules are not strongly attracted to each other have boiling points which increase as the molecular weight of the compound increases. For example, hydrogen sulfide, H_2S, has a molecular weight of 34 and boils at $-59.6°C$. Molecules of H_2S are not strongly attracted to each other. Water, with a molecular weight of about one-half that of H_2S, would be expected to boil at a temperature well below $-59.6°C$. However, the boiling point of water is almost $2\frac{1}{2}$ times that of H_2S because the water molecules are bonded together.

How are water molecules held together? Oxygen is more electronegative than hydrogen; water, H_2O, is a polar molecule because the oxygen pulls the bonding electrons of the hydrogen atoms toward it. This gives the oxygen a δ^- charge and the hydrogen atoms each a δ^+ charge (the δ^- and δ^+ mean a partial negative or positive charge on the atom). Because of their polarity, water molecules become oriented (Fig. 6.1) with a hydrogen atom between two oxygen atoms. The hydrogen atom can be considered to be associated with both water molecules, thus forming a hydrogen bond between the water molecules.

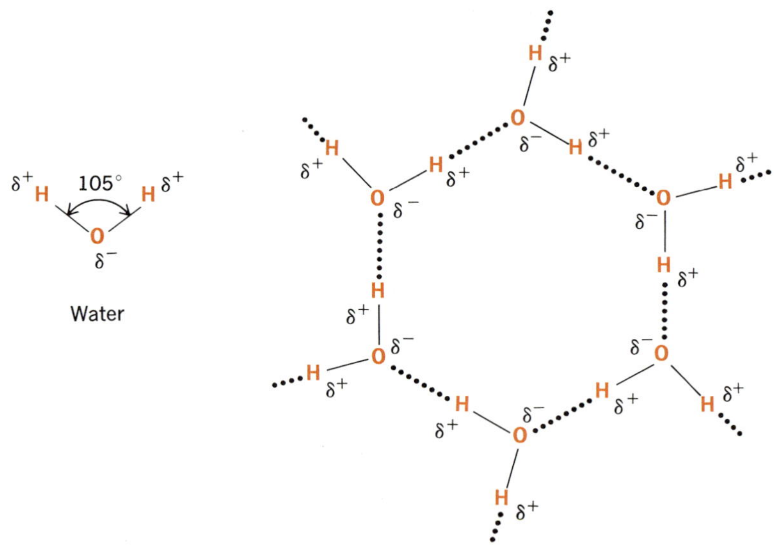

FIGURE 6.1 Hydrogen bonding between molecules. (*a*) A single water molecule; (*b*) a cluster of six water molecules in a cyclic arrangement.

The polarity of water molecules influences the ionization of ionic bonds or highly polar covalent bonds. A solvent that is able to neutralize the attraction between ions of an electrovalent bond or atoms of a polar covalent bond to the greatest extent causes the greatest amount of ionization, or dissociation. This so-called insulating property of the solvent, which keeps the ions separated, is known as the *dielectric constant* D of the solvent. The dielectric constant of water is quite high, $D = 80$, compared with that of vegetable oils, which is 2 to 3.

The highly polar nature of water and its high dielectric constant suggest that water itself should ionize, and it does.

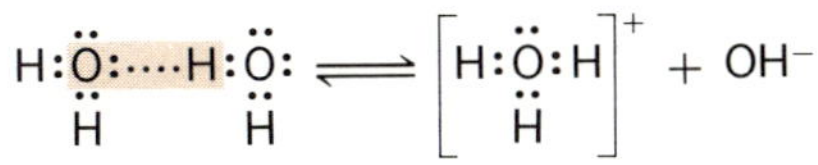

Hydrogen bonding Hydronium ion

or

$$2\ HOH \rightleftharpoons H_3O^+ + OH^-$$

Although the hydrated hydrogen ion, the *hydronium ion,* is actually the form in which the hydrogen ion occurs, we use H^+ for the sake of simplicity

$$HOH \rightleftharpoons H^+ + OH^-$$

The ionization constant for water is

$$K_{ion} = \frac{[H^+][OH^-]}{[HOH]} \tag{6.6}$$

At equilibrium the concentration $[H^+]$ of H^+ in moles per liter is 1×10^{-7}. This is an experimental fact. The concentration $[OH^-]$ of OH^- is also 1×10^{-7}, since ionization of 1 mole of water produces 1 mole of H^+ and 1 mole of OH^-. The concentration of water is 55.6 (the molarity of water, see Chap. 5) minus the amount that is ionized, or $55.6 - 1 \times 10^{-7}$ mole/liter. Therefore, the ionization constant for water can be written

$$K_{ion} = \frac{(1 \times 10^{-7})(1 \times 10^{-7})}{55.6 - 1 \times 10^{-7}} \tag{6.7}$$

However, the value $55.6 - 1 \times 10^{-7}$ is essentially constant and may be incorporated into the ionization constant represented by a new constant K_w

TABLE 6.2 CONCENTRATION OF IONS IN WATER

$[H^+]$	$[OH^-]$	pH	pOH	
1×10^{0}	1×10^{-14}	0	14	increasing acidity ↑
1×10^{-1}	1×10^{-13}	1	13	
1×10^{-2}	1×10^{-12}	2	12	
1×10^{-4}	1×10^{-10}	4	10	
1×10^{-6}	1×10^{-8}	6	8	
1×10^{-7}	1×10^{-7}	7	7	Neutral
1×10^{-8}	1×10^{-6}	8	6	increasing basicity ↓
1×10^{-10}	1×10^{-4}	10	4	
1×10^{-12}	1×10^{-2}	12	2	
1×10^{-13}	1×10^{-1}	13	1	
1×10^{-14}	1×10^{0}	14	0	

$$K_w = [H^+][OH^-] = (1 \times 10^{-7})(1 \times 10^{-7}) = 10^{-14} \quad \textbf{(6.8)}$$

Whenever water is present, the product of the molar concentration of hydrogen ions and hydroxyl ions, $[H^+][OH^-]$, in the solution must always equal 1×10^{-14} mole/liter. From this fact the values in Table 6.2 are obtained for the concentration of H^+ and OH^-.

6.5 pH AND pOH

The molar concentration $[H^+]$ of H^+ is used frequently. Because it is inconvenient to use the numerical values, such as 1.06×10^{-6}, in calculations or writing, the terms pH and pOH are used to express the $[H^+]$ and $[OH^-]$, respectively, where pH is defined as log $(1/[H^+]$ and pOH as log $(1/[OH^-])$. This relationship is

$$\text{pH} = \log \frac{1}{[H^+]} = -\log [H^+]$$

$$\text{pOH} = \log \frac{1}{[OH^-]} = -\log [OH^-]$$

Since

$$[H^+][OH^-] = 1 \times 10^{-14}$$

it follows that

$$-\log [H^+] + (-\log [OH^-]) = -\log 1 \times 10^{-14}$$

Therefore,

$$\text{pH} + \text{pOH} = -\log 1 \times 10^{-14} = 14 \quad \textbf{(6.9)}$$

The *sum of the* pH *and* pOH *must always equal* 14. At pH = 7, $H^+ = 1 \times 10^{-7}$; therefore, the concentration of the OH^- must also equal 1×10^{-7} [Eq. (6.8)]. A solution in which $[H^+] = [OH^-]$ is called a neutral solution (see Table 6.2). *A solution in which* $[H^+]$ *is greater than* $[OH^-]$ *is called an acidic solution, and a solution in which the* $[OH^-]$ *is greater than* $[H^+]$ *is called a basic solution.* Thus, at pH values less than 7 a solution is acidic, whereas at pH values greater than 7 the solution is basic. Since pH is a logarithmic function, a pH change of 1 unit represents a tenfold change in hydrogen-ion concentration. For example, a solution at pH 2 is 100 times more acidic (or less basic) than a solution at pH 4.

6.6 ACIDS AND BASES

Three definitions of acids and bases have been proposed at different times. The earliest defined acids as compounds that ionize to yield a greater number of H^+ than OH^- and bases as compounds that ionize to yield more OH^- than H^+:

Acid: $HCl \rightleftharpoons H^+ + Cl^-$
Base: $NaOH \rightleftharpoons Na^+ + OH^-$

Although this definition is satisfactory for many compounds, it is much too limited in its application because other ions react in a manner similar to H^+. The following two definitions of acids and bases are considerably more general than the first definition. The *Bronsted-Lowry* definition is that acids are proton donors and bases are proton acceptors. The *Lewis* definition (still more general) is that acids are electron-pair acceptors while bases are electron-pair donors.

	$H:\ddot{\underset{..}{Cl}}:$	$\rightleftharpoons$	H^+	+	$[:\ddot{\underset{..}{Cl}}:]^-$
	Acid		Acid		Base
Brønsted-Lowry:	Proton donor				Proton acceptor
Lewis:			Electron-pair acceptor		Electron-pair donor

Acids and bases can be divided into two main classes, strong and weak, according to the extent to which they dissociate. Strong acids and bases dissociate almost completely and therefore have a large K_{ion} value (see Table 6.3). For example, compare the K_{ion} for hydrochloric acid, HCl, and acetic acid, CH_3CO_2H. HCl is 5,780 times more highly ionized than CH_3CO_2H. HCl is a strong acid whereas CH_3CO_2H is a weak acid.

Acids with more than one ionizable hydrogen, such as H_2CO_3 and H_3PO_4, ionize in a stepwise manner; not all of the hydrogen atoms dissociate with the same ease. Usually the first hydrogen ionizes, K_1, to a larger extent than subsequent hydrogens, K_2, K_3, etc. Consider, for example, the ionization of H_2CO_3 and H_3PO_4:

$$H_2CO_3 \rightleftharpoons H^+ + HCO_3^- \qquad K_1 = 4.3 \times 10^{-7}$$
$$HCO_3^- \rightleftharpoons H^+ + CO_3^{--} \qquad K_2 = 5.6 \times 10^{-11}$$

$$H_3PO_4 \rightleftharpoons H^+ + H_2PO_4^- \qquad K_1 = 7.5 \times 10^{-3}$$
$$H_2PO_4^- \rightleftharpoons H^+ + HPO_4^{--} \qquad K_2 = 6.2 \times 10^{-8}$$
$$HPO_4^{--} \rightleftharpoons H^+ + PO_4^{3-} \qquad K_3 = 4.8 \times 10^{-13}$$

In general the ionization constant for the first hydrogen of acids is considerably larger than the ionization constant for the second hydrogen. The K_1 value for H_2CO_3 is low enough for this acid to be classified as a weak acid; however, the first hydrogen of H_3PO_4 ionizes to a large extent, almost completely, and H_3PO_4 is a strong acid. The second hydrogen of H_2CO_3 is very slightly ionized, much less than the second hydrogen of H_3PO_4. The K_2 value for H_3PO_4 is so small that $H_2PO_4^-$ is classified as a very weak acid. The third hydrogen of H_3PO_4 is almost completely un-ionized (see Table 6.3).

In sulfuric acid, H_2SO_4, both the hydrogens ionize to a large extent, $K_2 = 1.2 \times 10^{-2}$, and thus it is a very strong acid, yielding two hydrogen ions per mole of acid in dilute solutions. Thus a 0.01 M solution of H_2SO_4 contains 0.02 mole of H^+ per liter. By comparison, a 0.01 M solution of H_3PO_4 is 0.01 M with respect to H^+ since the second hydrogen of H_3PO_4 is only about 0.0006% ionized, a negligible amount, at the pH of 0.01 M H_3PO_4.

Organic acids are usually weak acids, with low ionization constants. They have as their acidic group the *carboxyl group,* —COOH. This group is usually part of a larger molecule which is indicated by an R—. The general formula for an organic acid is

$$R\!-\!\overset{\overset{\displaystyle O}{\|}}{C}\!-\!OH \qquad \text{or} \qquad R\!-\!COOH \qquad \text{or} \qquad RCO_2H$$

Carboxyl group

When R = H, the compound is formic acid

$$H\!-\!\overset{\overset{\displaystyle O}{\|}}{C}\!-\!OH \qquad \text{or} \qquad H\!-\!COOH \qquad \text{or} \qquad HCO_2H$$

TABLE 6.3 IONIZATION CONSTANTS FOR SOME ACIDS AND BASES

Ionization equation	Hydrogen ionized[a]	K_{ion}	$-\log K_{ion}$	pK[b]
	Strong acids			
$HCl \rightleftharpoons H^+ + Cl^-$		1×10^{-1}	1.0	1.0
$H_2SO_4 \rightleftharpoons 2\,H^+ + SO_4^{--}$	K_2	1.2×10^{-2}	1.9	1.9
$HNO_3 \rightleftharpoons H^+ + NO_3^-$		1×10^{-1}	1.0	1.0
$H_3PO_4 \rightleftharpoons H^+ + H_2PO_4^-$	K_1	7.5×10^{-3}	2.12	2.12
	Strong bases			
$NaOH \rightleftharpoons Na^+ + OH^-$		1×10^{-1}	1.0	13.0
$KOH \rightleftharpoons K^+ + OH^-$		1×10^{-1}	1.0	13.0
$Ca(OH)_2 \rightleftharpoons Ca^{++} + 2\,OH^-$	K_2	4.0×10^{-2}	1.4	12.6
	Weak acids[c]			
$H_2CO_3 \rightleftharpoons H^+ + HCO_3^-$	K_1	4.3×10^{-7}	6.37	6.37
$HCO_3^- \rightleftharpoons H^+ + CO_3^{--}$	K_2	5.6×10^{-11}	10.25	10.25
$H_2PO_4^- \rightleftharpoons H^+ + HPO_4^{--}$	K_2	6.2×10^{-8}	7.2	7.2
$HPO_4^{--} \rightleftharpoons H^+ + PO_4^{3-}$	K_3	4.8×10^{-13}	12.3	12.3
$CH_3CO_2H \rightleftharpoons H^+ + CH_3CO_2^-$		1.75×10^{-5}	4.76	4.76
	Weak bases[c]			
$NH_4OH \rightleftharpoons NH_4^+ + OH^-$		1.8×10^{-5}	4.75	9.25
$AgOH \rightleftharpoons Ag^+ + OH^-$		1.1×10^{-4}	3.96	10.04

[a] K_1 is the ionization constant for dissociation of the first hydrogen, K_2 for the second, and K_3 for the third.
[b] $pK = pK_{acid} = -\log K_{ion}$ of acid; $-\log K_{ion}$ of base $= pK_b$, but $pK + pK_b = 14$, therefore, pK (of base) $= 14 - \log K_{ion}$ (of base).
[c] Weak acids or bases, $K = 1 \times 10^{-3}$ to 1×10^{-8}; very weak acids or bases, $K < 1 \times 10^{-8}$.

When R = CH_3, the compound is acetic acid (vinegar)

$$CH_3{-}\overset{\overset{\displaystyle O}{\|}}{C}{-}OH \qquad \text{or} \qquad CH_3{-}COOH \qquad \text{or} \qquad CH_3CO_2H$$

The R chain in organic acids can be very long (see Chap. 11).

R—COOH is an acid because of the partial charge distribution in the carboxyl group. The electronegativity of the double-bonded oxygen makes the carbon partially positive, δ^+, which increases the pull of electrons from the oxygen of the hydroxyl group, thereby producing a strongly polar bond in the hydroxyl group and enhancing the ionization of the hydrogen.

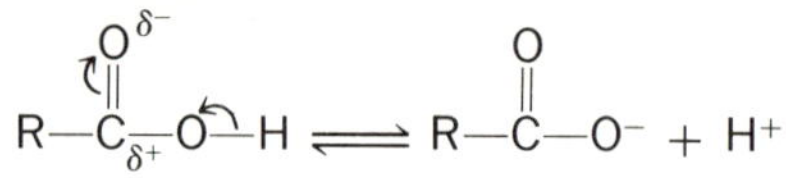

6.7 SOLUTIONS OF STRONG ACIDS AND BASES

Because a strong acid dissociates almost completely, the concentration of hydrogen ion in a dilute solution of a strong acid is equal to the normality of the acid (see Sec. 6.6).

EXAMPLE 1

In 1 liter of 1×10^{-4} N HCl there is 1×10^{-4} mole (or equivalent) of H^+ since the acid is considered to be completely ionized. The amount of H^+ from the dissociation of water (1×10^{-7} mole/liter) may be disregarded. The pH of this HCl solution is $-\log 1 \times 10^{-4}$, or pH = 4.0.

EXAMPLE 2

Calculate the pH of a 0.5×10^{-3} M solution of H_2SO_4. Since a 0.5×10^{-3} M solution of H_2SO_4 furnishes 1×10^{-3} mole of H^+ per liter, the solution would be 1×10^{-3} N. The pH of this solution is $-\log 1 \times 10^{-3}$, or pH = 3.0.

The pH of solutions of bases can be calculated by substituting $[OH^-]$ in the equation pH + pOH = 14. For strong bases the $[OH^-]$ in moles per liter is equivalent to the normality of the solution since strong bases are considered to be completely ionized. For example,

$$\begin{array}{ccccc} \text{NaOH} & \rightleftharpoons & \text{Na}^+ & + & \text{OH}^- \\ 1 \times 10^{-3}\ N & & & & \\ 1 \times 10^{-3}\ \text{mole/liter} & & 1 \times 10^{-3}\ \text{mole/liter} & & 1 \times 10^{-3}\ \text{mole/liter} \end{array}$$

$$\text{pOH} = -\log\,[OH^-] = -\log 1 \times 10^{-3} = 3.0$$

Since pH = 14 − pOH,

$$\text{pH} = 11.0$$

6.8 SOLUTIONS OF WEAK ACIDS

Since the dissociation constant of weak acids is small, only a small fraction of the total amount of acid present is ionized to form H^+; therefore, *the* pH *of solutions of weak acids is not equal to −log of the normality*. For example if an acid HA is 3% ionized, only 3 equiv of H^+ is produced from 100 equiv of the acid. At equilibrium there are 97 equiv HA, 3 equiv H^+, and 3 equiv A^-.

$$\underset{97\text{ equiv}}{\text{HA}} \xrightleftharpoons{3\%} \underset{3\text{ equiv}}{H^+} + \underset{3\text{ equiv}}{A^-}$$

The extent of dissociation of weak acids (and bases) is proportional to the concentration of the acid. Since the dissociation reaction is an equilibrium reaction, combining H^+ with A^- to form HA depends upon the concentration of the ions. As a solution of HA becomes more dilute, there is less chance that the ions will collide and form HA again because the ions are farther apart. Therefore, the extent of ionization of HA increases as the concentration of HA decreases (see Table 6.4).

The pH of solutions of weak acids can be calculated with the aid of the ionization, or dissociation, constant or from the percentage dissociation. Consider the ionization of acetic acid, CH_3COOH,

$$CH_3COOH \rightleftharpoons H^+ + CH_3COO^-$$

or

$$\text{AcOH} \rightleftharpoons H^+ + \text{AcO}^-$$

Where Ac is used to represent the *acetyl group,* $CH_3—\overset{\overset{\large O}{\|}}{C}—$,

$$K_{\text{ion}} = \frac{[H^+][\text{AcO}^-]}{[\text{AcOH}]}$$

For weak acids the concentration of H^+ is relatively small compared with the concentration of the undissociated acid present, [HOAc] in

TABLE 6.4 IONIZATION (DISSOCIATION) OF ACETIC ACID, CH_3COOH

Concentration		**Dissociation, %**	**pH**
CH_3COOH, *M*	**H^+, *M***		
1.0	4.2×10^{-3}	0.42	2.38
0.5	2.96×10^{-3}	0.59	2.53
0.1	1.3×10^{-3}	1.3	2.89
0.02	6×10^{-4}	3.0	3.22
0.01	4.2×10^{-4}	4.2	3.38
0.001	1.3×10^{-4}	13.0	3.89

this example. It is possible to make a close approximation of the amount of acid dissociated, or the concentration of H^+, by assuming that the concentration of the acid does not change during ionization, i.e.,

$$[\text{HOAc initially present} - \text{amount dissociated}] = [\text{HOAc initially present}]$$

This assumption is quite valid for slightly ionized acids but becomes increasingly less valid as the dissociation increases.

EXAMPLE 1

If a solution of acetic acid is 1.0 M and is 0.42% dissociated, there are 0.0042 mole of H^+ present in 1 liter at equilibrium and 1.0000 – 0.0042, or 0.9958, mole of undissociated acetic acid, which for our purposes is equal to the original concentration of acetic acid, namely, 1.0 M.

Taking into consideration the above approximation, if x is the moles of acid dissociated (and therefore the moles of H^+ or AcO^- formed),

$$K_{\text{ion}} = \frac{(x)(x)}{[\text{HOAc}]} = \frac{x^2}{[\text{HOAc}]}$$

Therefore,

$$x^2 = K_{\text{ion}}[\text{HOAc}] \tag{6.10}$$

The values for the hydrogen-ion concentration and percentage dissociation of acetic acid in Table 6.4 were calculated by appropriate substitutions in Eq. (6.10). For example, the values for 1.0 M acetic acid solution were calculated as follows:

$$[\text{HOAc}] = 1.0\ M$$
$$K_{\text{ion}} = 1.8 \times 10^{-5}$$

Substituting these values in Eq. (6.10),

$$x^2 = (1.8 \times 10^{-5})(1.0) = 18 \times 10^{-6}$$
$$x = 4.2 \times 10^{-3} = [H^+]$$

Since

$$\text{pH} = -\log[H^+] = -\log 4.2 \times 10^{-3} = 2.377$$

we get

$$\%\ \text{dissoc.} = \frac{4.2 \times 10^{-3}}{1.0} \times 100 = 0.42$$

The *Henderson-Hasselbalch equation* expresses the pH of a solution of a weak acid in terms of the K_{ion} of the acid and the ratio of the concentration of the negative ion (which is the same as the concentration of the dissociated species, or the *salt*) to the concentration of the undissociated acid. Let HA represent an acid; then

$$K_{\text{ion}} = \frac{[H^+][A^-]}{[HA]}$$

Solve for

$$[H^+] = K_{\text{ion}} \frac{[HA]}{[A^-]}$$

Take the $-\log$

$$-\log [H^+] = -\log K_{\text{ion}} - \log \frac{[HA]}{[A^-]}$$

If $-\log K_{\text{ion}}$ is defined as pK,

$$\text{pH} = \text{p}K - \log \frac{[HA]}{[A^-]}$$

Or, more generally,

$$\text{pH} = \text{p}K + \log \frac{[A^-]}{[HA]} = \text{p}K + \log \frac{[\text{salt}]}{[\text{acid}]} \quad \textbf{(6.11)}$$

Note that the expression log ([salt]/[acid]) is merely a ratio of the concentrations of the species in solution. This concentration is usually expressed in molarity; however, if the acid yields 1 mole of hydrogen ion per mole of acid, the number of equivalents of acid and salt may be used to express concentration. Molarity is the same as normality in this instance.

The Henderson-Hasselbalch equation will be used many times in future discussions. Apply this equation to the determination of the pH and percent dissociation of acetic acid solutions in the following examples.

EXAMPLE 2

Calculate the pH of a solution of acetic acid, HOAc, using Eq. (6.11).

$$K_{\text{ion}} \text{ for HOAc} = 1.75 \times 10^{-5}$$

The HOAc is 3% dissociated.

If a solution of acetic acid is 3% dissociated, from 100 moles of acid 3 moles of acetate ion are formed. This leaves 97moles of undissociated acid at equilibrium. Thus, the ratio of salt to acid is 3:97. This ratio is substituted in the Henderson-Hasselbalch equation.

$$\mathrm{p}K = -\log 1.75 \times 10^{-5} = 4.76$$

$$\mathrm{pH} = 4.76 + \log \tfrac{3}{97} = 4.76 + 0.477 - 1.987 = 3.26$$

EXAMPLE 3

What is the percent dissociation of an acetic acid solution with a pH of 3.55? The pK value for acetic acid (from Example 2) is 4.76. Substituting the values for pH and pK in the Henderson-Hasselbalch equation gives

$$3.55 = 4.76 + \log \frac{[\mathrm{salt}]}{[\mathrm{acid}]}$$

Solve for

$$\log \frac{[\mathrm{salt}]}{[\mathrm{acid}]} = -1.21 \qquad \text{or} \qquad \log \frac{[\mathrm{acid}]}{[\mathrm{salt}]} = 1.21$$

Take the antilog

$$\frac{[\mathrm{acid}]}{[\mathrm{salt}]} = \frac{1.62 \times 10}{1} = 16.2$$

There are 16.2 moles of acid for each mole of salt or dissociated acid. Therefore, 16.2 + 1, or 17.2, moles of acid dissociates to yield 1 mole of salt. The percentage dissociation is 1/17.2 × 100, or 5.81%.

6.9 SOLUTIONS OF WEAK BASES

Ammonia, NH_3, ammonium hydroxide, NH_4OH, and certain compounds that contain an amino, $—NH_2$, group are weak bases.

Ammonia is a base because of the unpaired electrons on the nitrogen atom. Thus ammonia can donate a pair of electrons to a proton or can accept a proton to form the ammonium ion, NH_4^+.

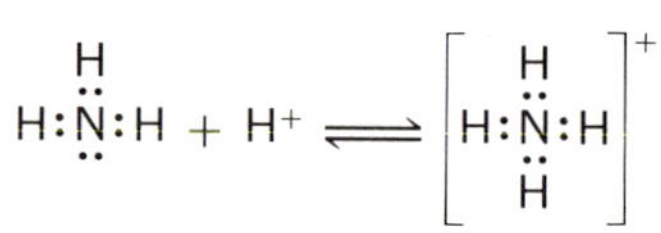

Base Acid

This reaction is reversible; hence, the ammonium ion is an acid since it is a proton donor. When ammonia is added to water, the following equilibrium reactions occur to form a solution of ammonium hydroxide, NH_4OH:

$$\begin{array}{ccccc} HOH & \overset{1)}{\rightleftharpoons} & H^+ & + & OH^- \\ & & + & & + \quad \overset{3)}{\rightleftharpoons} NH_4OH \\ & & NH_3 \overset{2)}{\rightleftharpoons} & & NH_4^+ \end{array}$$

Overall: $$NH_3 + HOH \rightleftharpoons NH_4OH$$

Ammonium hydroxide is a weak base, and the ammonium ion is a very weak acid. When a solution of ammonium hydroxide is heated, the ammonia is driven off as a gas, upsetting the equilibrium reactions (2) and (3). Therefore, the decomposition of ammonium hydroxide to give ammonia and water goes to completion. The ionization constant for ammonium hydroxide is practically the same as for acetic acid, 1.8×10^{-5}; therefore, ammonium hydroxide is as strong a base as acetic acid is an acid.

Amines may be considered to contain an ammonia or ammonium-ion group in which one or more hydrogens have been replaced with a carbon chain, —C—C—C (Chap. 7) represented by R:

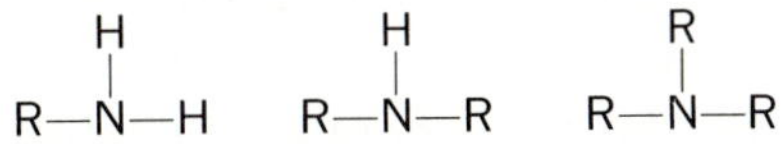

This class of compounds will be discussed in detail in Chap. 12. For the present our concern is only with the basic nature of amines. *Amines are basic; they accept a proton in a manner similar to* NH_3.

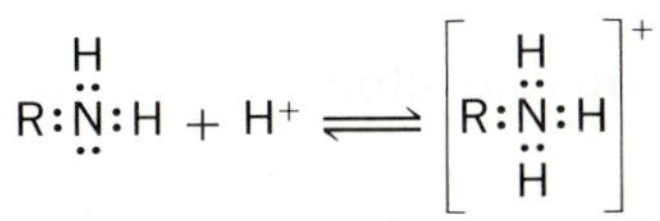

Amine Substituted ammonium ion

If the proton source is HCl, the product is a substituted ammonium chloride salt, or an amine hydrochloride.

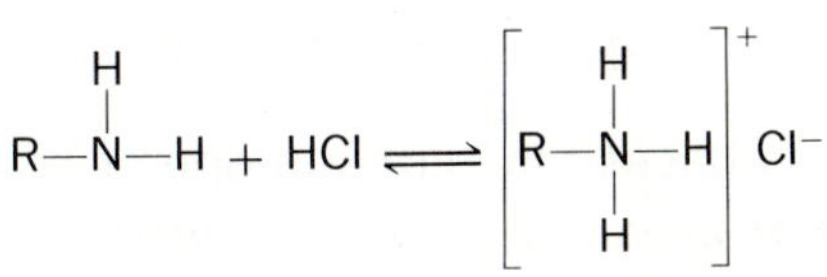

Substituted ammonium chloride

which can be written as

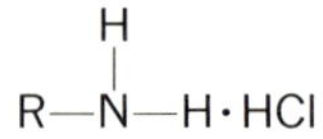

Amine
hydrochloride

The salts of amines are usually soluble in water, but most amines are insoluble in water.

6.10 DEGREE OF ACIDITY

Hydrochloric acid has been used as an example of a covalent-bond compound that ionizes to furnish H^+ and Cl^-. The ionization of HCl was explained on the basis of the greater electronegativity of chlorine than hydrogen, thus producing a very polar covalent bond or molecule easily split into ions by water. In general the greater the electronegativity of the element bonded to the hydrogen, the greater the dissociation constant of the acid. This generalization is true only within limits; e.g., it does not take into account the relative size or volume of the negative ionic species. As the volume of an ion increases, its stability also increases because the charge can be spread over a larger area (it is easier to catch a horse in a corral than in a 500-acre pasture). Since reactions go in the direction to form the most stable product, it follows that the greater the stability of the negative ion, the greater the ionization constant of the acid.

$$\underset{\text{Less stable}}{\textit{Reactants}} \longrightarrow \underset{\text{More stable}}{\textit{Products}}$$

In the halogen group the atomic radii, and consequently the atomic volume, increase from fluorine to iodine; therefore, the I^- should be more stable than the F^-. Consequently, one would expect HI to ionize to a greater degree than HF because of the greater stability of the I^- as compared with F^-. The HX acids (X indicates the halogens) can be arranged in order of their decreasing ionization:

$$HI > HBr > HCl > HF$$

Acids as well as bases may contain one or more OH groups. The *nature of the atom or group of atoms attached to the* OH *group determines whether the compound is an acid and yields* H^+ *or is a base and yields* OH^- *upon ionization.* An element of quite low electronegativity bonded to an OH group will yield its electron(s) to the more electronegative oxygen of the OH group; thus the compound would be expected to yield OH^- upon ionization. This is illustrated with potassium hydroxide, KOH,

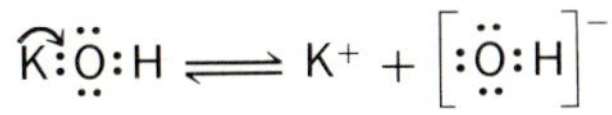

Electronegativity: K = 0.8 O = 3.5 H = 2.1

The electron of the potassium would be pulled toward the oxygen of the OH group—more so than the electron from the hydrogen. Consequently one would expect KOH to ionize into K^+ and OH^-.

On the other hand, if an element or group that attracts electrons from the oxygen of the OH group is attached to the OH group, it will cause a shift of electrons from the hydrogen to oxygen of the OH group:

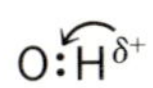

One would expect H^+ to be formed upon ionization and the compound to be classified as an acid. This situation is illustrated with hypochlorous acid, HOCl,

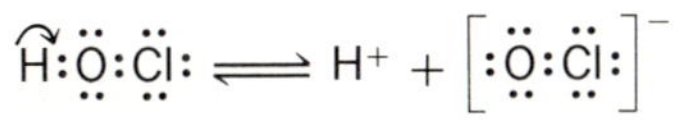

Electronegativity: H = 2.1 O = 3.5 Cl = 3.0

In a series of acids containing a common central element, the acidity of the acid increases as the oxidation number of the central element increases:

Increasing acidity →

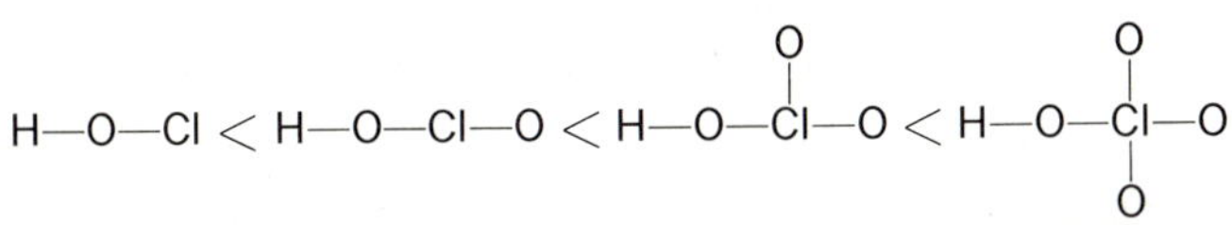

Acid:	Hypochlorous	Chlorous	Chloric	Perchloric
Ox. no. Cl:	+1	+3	+5	+7

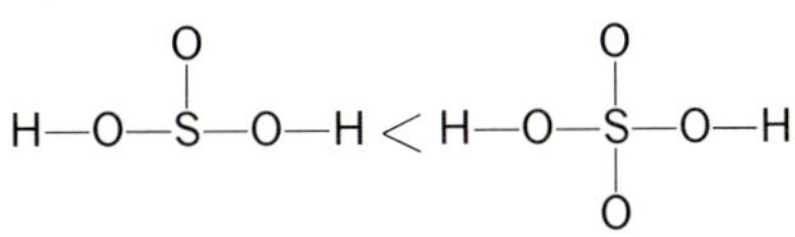

Acid:	Sulfurous	Sulfuric
Ox. no. S:	+4	+6

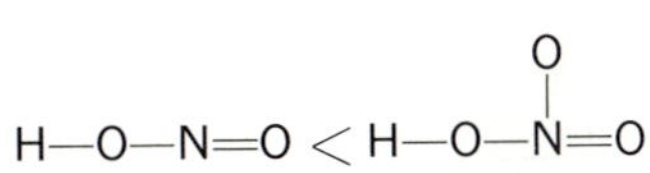

Acid:	Nitrous	Nitric
Ox. no. N:	+3	+5

When the central atom is not the same, the acidity of acids with similar structure increases as the electronegativity of the central atom increases:

Increasing acidity

	H—O—I <	H—O—Br <	H—O—Cl
	Hypoiodous	Hypobromous	Hypochlorous
Electronegativity:	I = 2.4	Br = 2.8	Cl = 3.0

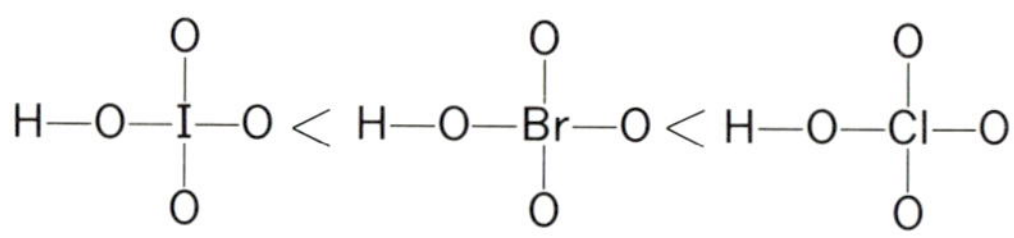

	Periodic	Perbromic	Perchloric
Electronegativity:	I = 2.4	Br = 2.8	Cl = 3.0

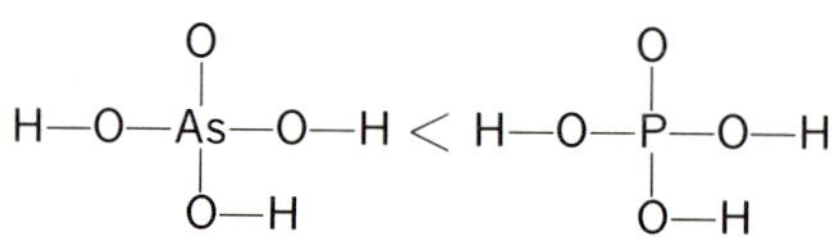

	Arsenic	Phosphoric
Electronegativity:	As = 2.0	P = 2.1

Elements in a given period of the periodic table *increase in electronegativity from left to right.* As hydroxides, the elements also increase in oxidation number from left to right. Consider the elements in the third period, sodium to chlorine: the electronegativity increases from 0.9 to 3.0. The oxidation number of the elements in their hydroxide form increases from +1 for sodium in NaOH to +7 for chlorine in $HOClO_3$. These two properties tend to cause an increase in the acidity of the hydroxides as one proceeds from sodium to chlorine. For example, NaOH is a very strong base, and $HOClO_3$ is a very strong acid. The hydroxides of the intermediate elements have intermediate strength as acids or bases.

Increasing acidity →

	NaOH	$Mg(OH)_2$	$Al(OH)_3$	$(HO)_2SiO$	$(HO)_3PO$	$(HO)_2SO_2$	$HOClO_3$
Electronegativity:	0.9	1.2	1.5	1.8	2.1	2.5	3.0
Ox. no.:	+1	+2	+3	+4	+5	+6	+7

← Increasing basicity

In organic acids, RCOOH, any strongly electronegative group on the carbon next to the carboxyl group, COOH, tends to pull electrons toward itself. This in turn would tend to pull the electron away from the carboxyl hydrogen and increase the tendency of the hydrogen to ionize, thus making the acid stronger. This can be illustrated by substituting chlorine for hydrogen in acetic acid:

Weak acid ——— Increasing acidity ———→ very strong acid

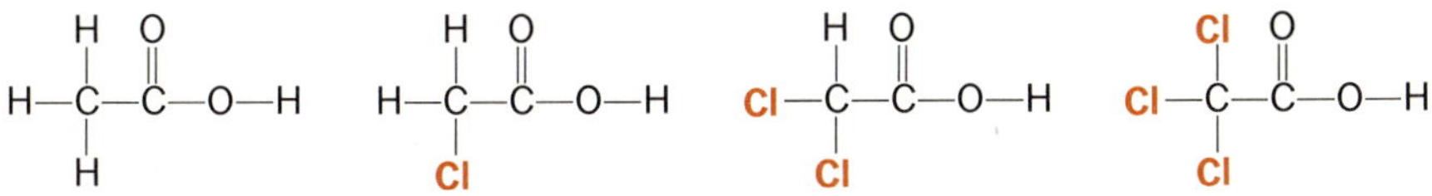

Acid: Acetic — Monochloroacetic — Dichloroacetic — Trichloroacetic

6.11 TITRATION

Titration is the process of determining the concentration of a substance in solution by adding a standard solution (one of known concentration) of a reactant. Usually the standard solution is added, in known volumes, to a measured amount of the solution of unknown concentration until enough reagent has been added to just react with the unknown present. For example, if a solution of HCl of known concentration is added to a definite volume of a solution of NaOH, the HCl will react with the NaOH as follows:

$$\underset{\text{1 equiv}}{\text{HCl}} + \underset{\text{1 equiv}}{\text{NaOH}} \longrightarrow \underset{\text{1 equiv}}{\text{NaCl}} + \underset{\text{1 equiv}}{\text{HOH}}$$

When the number of equivalents of HCl equals the equivalents of NaOH present in the unknown solution, the end point of the titration has been reached. In other words, sufficient HCl has been added to react with all the NaOH present: the NaOH has been neutralized. *The end point of any titration is when equal equivalents of reactants have been added, or* $V_1N_1 = V_2N_2$.

At the end point of titrating NaOH with HCl there is an aqueous solution of NaCl, the ionic species present being

$$\begin{array}{ccccc} \text{NaCl} & \rightleftharpoons & \text{Na}^+ & + & \text{Cl}^- \\ & & + & & + \\ \text{HOH} & \rightleftharpoons & \text{OH}^- & + & \text{H}^+ \\ & & \upharpoonleft\downharpoonright & & \upharpoonleft\downharpoonright \\ & & \text{NaOH} & & \text{HCl} \end{array}$$

Since NaOH is a strong base and HCl is a strong acid, they are both completely ionized and the concentrations of H^+ and OH^- are the same as they are in water, 1×10^{-7} M. The pH of the solution is $-\log 1 \times 10^{-7}$, or 7 (neutral). An aqueous solution of *any salt* of a strong acid and a strong base has a pH of 7.

EXAMPLE 1

To 20 ml of 0.1 N HCl are added increments of 0.1 N NaOH. The pH of the solution is calculated after each addition of base. Initially there are 2 meq of H^+ in the 20 ml of 0.1 N HCl. For each milliliter of 0.1 N NaOH added 0.1 meq NaOH is added that neutralizes the same number of milliequivalents of HCl. From the total volume after each addition of NaOH the concentration of H^+ remaining unneutralized and the pH of the solution can be calculated (Table 6.5). When the number of equivalents of NaOH added equals the number of equivalents of HCl initially present, the end point, or neutral point, of the titration has been reached. The H^+ and OH^- are equal to each other, namely, 1×10^{-7} M. The pH of the solution is 7.0. The pH of the solution upon adding increments of NaOH after the HCl has been neutralized is calculated as discussed in Sec. 6.7. From the data in Table 6.5 a titration curve is obtained by plotting the milliequivalents of excess H^+ or OH^- against pH (Fig. 6.2). Note the sharp change in the slope of the curve as the end point is reached.

Titration of a weak acid with a strong base

The end point of titrating a strong acid with a strong base is a pH of 7. This is not true when a weak acid is titrated with a strong base. Since the reaction of the salt formed during the titration with water (hydrolysis) does not produce a neutral solution, the pH at the end point must be calculated by taking into consideration the concentration of the salt formed during titration.

Hydrolysis of a salt of a weak acid and a strong base

All salts ionize completely; therefore, titrations of solutions containing salts must account for the presence of these ions. Consider the reaction of water with sodium acetate (a salt of a weak acid, acetic acid, and a

TABLE 6.5 TITRATION OF A STRONG ACID WITH A STRONG BASE

Ml 10^{-1} N NaOH added to 20 ml 10^{-1} N HCl	Final vol, ml	Meq H^+ in excess of OH^-	N (N = meq/V_{ml})	pH	pOH
0	20	2.0	10^{-1} [H^+]	1.00	13.00
2	22	1.8	8.18×10^{-2}	1.05	12.95
10	30	1.0	3.33×10^{-2}	1.74	12.26
18	38	0.2	5.26×10^{-3}	2.28	11.72
19	39	0.1	2.56×10^{-3}	2.59	11.44
20	40	0.0	1×10^{-7}	7.00	7.00
End point		All H^+ from HCl neutralized; ∴ same [H^+] as in H_2O			
		Meq OH^- in excess of H^+	N **[OH^-]**		
21	41	0.1	2.44×10^{-3}	11.39	2.61
22	42	0.2	4.76×10^{-3}	11.68	2.32
26	46	0.6	1.3×10^{-2}	12.11	1.89
34	54	1.4	2.59×10^{-2}	12.41	1.59

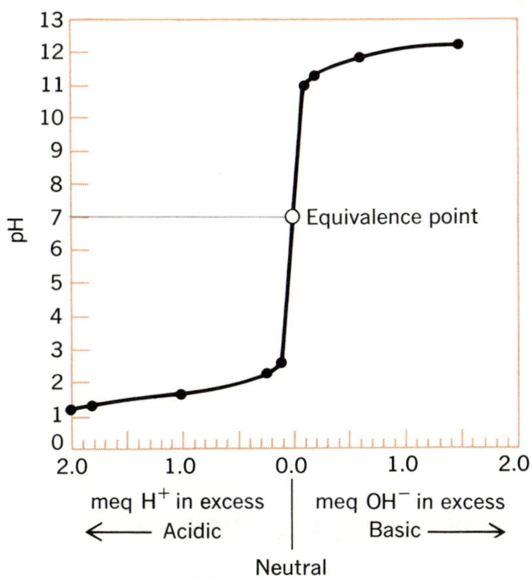

FIGURE 6.2 Titration curve for a strong acid and a strong base. Values for the curve were obtained from Table 6.5.

strong base, sodium hydroxide). When sodium acetate is dissolved in water, the ionized salt reacts in equilibrium reactions with the ions from water in a process known as hydrolysis.

$$
\begin{array}{ccccc}
\text{NaOAc} \rightleftharpoons & \text{Na}^+ & + & \text{OAc}^- \\
& + & & + \\
\text{HOH} \rightleftharpoons & \text{OH}^- & + & \text{H}^+ \\
& \updownarrow & & \updownarrow \\
& \text{NaOH} & & \text{HOAc} \\
& \text{100\% dissociated} & & \text{Weakly dissociated}
\end{array}
$$

The acetate ion, AcO^-, reacts with the hydrogen ion from the dissociation of water to produce the slightly dissociated acetic acid molecule, HOAc. This reaction removes hydrogen ions from solution. The sodium ions and the hydroxide ions are in equilibrium with sodium hydroxide, NaOH, but since this is a strong base, it is completely dissociated and does not remove hydroxide ions from solution. The net result of the hydrolysis of sodium acetate is to form a solution in which the $[OH^-]$ is greater than the $[H^+]$. The solution is basic and has a pH greater than 7. The pH varies (within limits) with the concentration of the salt.

Note that for each HOAc formed one OH^- remains in solution; therefore, $[HOAc] = [OH^-]$. Substituting [HOAc] for $[OH^-]$ in the dissociation equation for water, $[H^+][OH^-] = 1 \times 10^{-14}$, we arrive at

$$[\text{HOAc}] = \frac{1 \times 10^{-14}}{[\text{H}^+]} \qquad \textbf{(6.12)}$$

Substituting this value in Eq. (6.11), the Henderson-Hasselbalch equation,

$$\text{pH} = \text{p}K + \log \frac{[\text{OAc}^-]}{(1 \times 10^{-14})/[\text{H}^+]} = \text{p}K + \log \frac{[\text{OAc}^-][\text{H}^+]}{1 \times 10^{-14}}$$

Solve

$$\text{pH} = \text{p}K + \log \frac{1}{1 \times 10^{-14}} + \log [\text{OAc}^-] + \log [\text{H}^+]$$

$$\text{pH} - \log [\text{H}^+] = \text{p}K + \log 1 \times 10^{14} + \log [\text{OAc}^-]$$

Since $-\log [H^+] = \text{pH}$ and $\log 1 \times 10^{14} = 14$

$$2\text{pH} = \text{p}K + 14 + \log [\text{OAc}^-]$$

$$\text{pH} = \frac{\text{p}K + 14 + \log [\text{OAc}^-]}{2} \qquad \textbf{(6.13)}$$

EXAMPLE 2

What is the pH of a 0.01 M solution of sodium acetate? From Table 6.3 the pK for acetic acid is 4.76. Since salts are completely ionized in solution, the $[OAc^-] = 0.01\ M$, or $1 \times 10^{-2}\ M$. The values for pK and for $[OAc^-]$ are substituted in Eq. (6.13).

$$\text{pH} = \frac{4.76 + 14 + \log(1 \times 10^{-2})}{2}$$

$$= \frac{4.76 + 14 - 2}{2} = 8.38$$

A solution of 0.01 M sodium acetate is basic.

The titration curve for a weak acid and a strong base is shown in Fig. 6.3. Note how the hydrolysis of the salt formed during the titration causes the pH of the solution to change during titration. There is a gradual change in pH as base is added, the change as the equivalent point is reached being much less abrupt than when a strong acid and a strong base are titrated (Fig. 6.2). The pH at the end point of

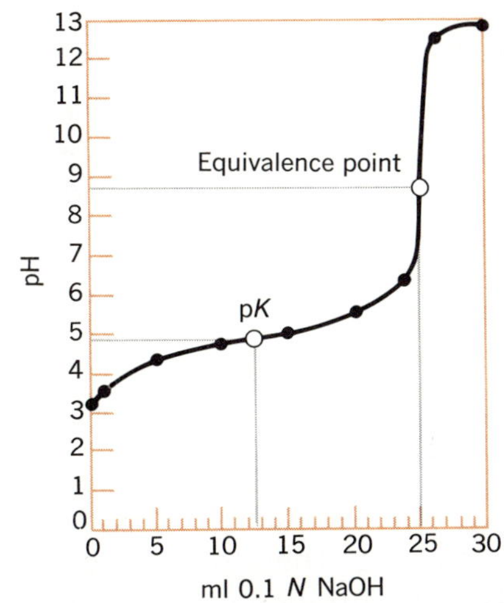

FIGURE 6.3 Titration curve for a weak acid and a strong base: 25 ml of 0.1 N acetic acid, CH_3COOH, is titrated with 0.1 N sodium hydroxide, NaOH. The equivalence point and the pK are indicated on the curve.

titration as shown in Eq. (6.13) and Fig. 6.3 depends upon two factors, namely, the concentration of the salt formed and the pK, or K_{ion}, of the acid. For example, the pH at the end point of titrating 25 ml of 0.2 N acetic acid, HOAc, with 0.2 N NaOH is calculated as follows. The pK for acetic acid is 4.76. The two solutions, HOAc and NaOH, are of the same normality; hence, the end point will be reached when equal volumes of NaOH and HOAc have been mixed. This would require 25 ml of 0.2 N NaOH to neutralize the HOAc in 25 ml of 0.2 N solution. Therefore, the final volume of the solution after reaching the end point is 50 ml (a twofold dilution), and the concentration of the salt is 0.2/2, or 0.1 M. Substituting the values for pK and the final salt concentration in Eq. (6.13) gives

$$\text{pH} = \frac{4.76 + 14 + \log(1 \times 10^{-1})}{2}$$

$$= \frac{4.76 + 14 - 1}{2} = 8.88$$

Carbonic acid is a weak acid and has two ionizable hydrogens; the stepwise ionization is

$$H_2CO_3 \underset{}{\overset{pK_1 = 6.37}{\rightleftharpoons}} H^+ + HCO_3^-$$

$$HCO_3^- \overset{pK_2 = 10.25}{\rightleftharpoons} H^+ + CO_3^{--}$$

The pK values of 6.37 and 10.25 for the two ionizable hydrogens indicate that the first hydrogen to ionize is classified as a weak acid and the second one is classified as a very weak acid. The neutralization of carbonic acid with a strong base is

$$H_2CO_3 \overset{pK_1 = 6.37}{\rightleftharpoons} HCO_3^- \overset{pK_2 = 10.25}{\rightleftharpoons} CO_3^{--} \tag{6.14}$$

0.5 equiv	0.5 equiv	0.5 equiv	0.5 equiv
pH = 6.37	pH = 8.31	pH = 10.25	

First H completely neutralized

When $\frac{1}{2}$ equiv of base has been added to 1 mole of H_2CO_3, the first hydrogen is half neutralized. Therefore, the concentration of H_2CO_3, [acid], is equal to the concentration of HCO_3^-, [salt], which when substituted in the Henderson-Hasselbalch equation with the appropriate pK value gives

$$\text{pH} = 6.37 + \log \frac{[HCO_3^-]}{[H_2CO_3]}$$

Since $[HCO_3^-] = [H_2CO_3]$,

$$pH = 6.37$$

The addition of another $\frac{1}{2}$ equiv of base (or a total of 1.0 equiv of base to H_2CO_3) completely neutralizes the first hydrogen of H_2CO_3 to form HCO_3^- as the negative ion. If HCO_3^- is half neutralized by the addition of another $\frac{1}{2}$ equiv of base, the pH of the solution is the same as the pK_2 of the acid, or 10.25. Show that this is so by substituting the correct values in the Henderson-Hasselbalch equation, as was done previously when the first hydrogen was half neutralized.

When the pK values for the acid hydrogens are far apart, as in H_2CO_3, the first hydrogen can be neutralized completely without appreciably neutralizing the second hydrogen. For example, the first hydrogen of carbonic acid would be neutralized at a pH value halfway between pK_1 and pK_2, or $(6.37 + 10.25)/2 = 8.31$. The reader should be able to show by appropriate substitutions in the Henderson-Hasselbalch equation that at this pH the second hydrogen of H_2CO_3 is only about 1% ionized.

Titration of a weak base with a strong acid

The pH of the solution at the end point of titrating a weak base with a strong acid reflects the hydrolysis of the salt formed during the titration. The pH at the end point will be less than 7.0, and the solution will be acidic because the salts of weak bases and strong acids hydrolyze to form acidic solutions.

Hydrolysis of salts of a weak base and a strong acid

Ammonium chloride, NH_4Cl, is the salt of a weak base, ammonium hydroxide, and a strong acid, hydrochloric acid, and is completely dissociated into ammonium ions, NH_4^+, and chloride ions, Cl^-. Therefore, there are the same number of ammonium or chloride ions present as molecules of salt initially present.

$$\begin{array}{ccccc} NH_4Cl & \rightleftharpoons & NH_4^+ & + & Cl^- \\ & & + & & + \\ HOH & \rightleftharpoons & OH^- & + & H^+ \\ & & \updownarrow & & \updownarrow \\ & & NH_4OH & & HCl \end{array}$$

Slightly dissociated Completely dissociated

Water ionizes to form hydroxide and hydrogen ions, which react with the ammonium and chloride ions, respectively, to form ammonium hydroxide and hydrochloric acid. Hydrochloric acid is a very strong acid and may be considered to be completely ionized; hence, there is no removal of hydrogen ions from solution. There is, however, a removal of hydroxide ions from the solution because ammonium hydroxide is only slightly dissociated. The net result is an excess of hydrogen ions over hydroxide ions thus making the solution *acidic* when equilibrium has been attained.

The equilibria to be concerned with in deriving an equation to calculate the pH at the end point of titration of a weak base with a strong acid are as follows, where $K_{\text{base}} = K_b = K_{\text{ion}}$ of NH_4OH:

$$[OH^-] = \frac{K_w}{[H^+]}$$

$$K_b = \frac{[NH_4^+][OH^-]}{[NH_4OH]}$$

Rearrange to

$$[OH^-] = K_b \frac{[NH_4OH]}{[NH_4^+]}$$

Since the two equilibria exist in the same solution, the $[OH^-]$ is the same for both equations. Hence, the right side of each equation can be equated to give

$$\frac{K_w}{[H^+]} = K_b \frac{[NH_4OH]}{[NH_4^+]}$$

Note that for each NH_4OH formed there is a H^+ produced *in excess* of OH^-; therefore, $[H^+]$ can be substituted for $[NH_4OH]$ to give

$$\frac{K_w}{[H^+]} = K_b \frac{[H^+]}{[NH_4^+]}$$

Solving for $[H^+]$,

$$[H^+]^2 = \frac{K_w}{K_b}[NH_4^+] = \frac{K_w}{K_b}[\text{salt}]$$

Take the $-\log$

$$2 - \log [H^+] = -\left[\frac{\log K_w}{\log K_b} + \log c\right] = -[\log K_w - \log K_b + \log c]$$

Since $-\log [H^+] = pH$, $-\log K_w = 14$, and $-\log K_b = pK_b$,

$$2pH = 14 - pK_b - \log [\text{salt}]$$

$$pH = \frac{14 - pK_b - \log [\text{salt}]}{2} \quad (6.15)$$

Note that $pK_w - pK_b$ expresses the strength of the base in terms of its acidity because $pK_{H^+} + pK_{OH^-} = pK_w$.

EXAMPLE 3

Calculate the pH at the end point of titration of 50 ml of 0.1 N NH_4OH with 0.1 N HCl. The K_{ion} of NH_4OH is 1.8×10^{-5}; therefore, $pK_b = 4.75$. At the end point the volume of solution is 100 ml (50 ml NH_4OH solution + 50 ml HCl solution). This solution contains 50×0.1, or 5 meq (or 5 mmoles), of salt; therefore, the molarity of this solution with respect to salt is $\frac{5}{100}$, or 5×10^{-2}. Substituting the value for pK_b and for the salt concentration in Eq. (6.15) gives

$$pH = \frac{14 - 4.75 - \log (5 \times 10^{-2})}{2}$$

$$= \frac{14 - 4.75 + 2 - 0.699}{2} = \frac{10.54}{2} = 5.27$$

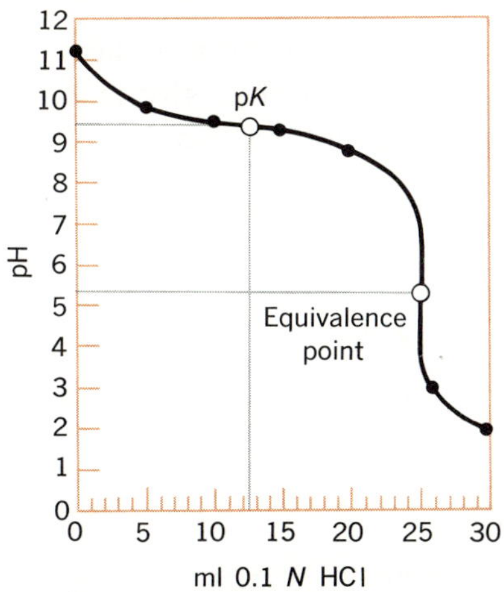

FIGURE 6.4 Titration curve for a strong acid and a weak base: 25 ml of 0.1 N ammonium hydroxide, NH_4OH, is titrated with 0.1 N hydrochloric acid, HCl. The equivalence point and the pK are indicated on the curve.

The titration curve for a weak base and a strong acid is shown in Fig. 6.4. The pH at the equivalence point is less than 7 because of the hydrolysis of the salt formed during titration. The value of pK, 9.6, is an expression of the acidic strength of the base, $pK + pK_b = 14$.

6.12 INDICATORS

Acid-base indicators are very weak acids or bases whose undissociated form has a color different from the dissociated form:

$$\underset{\text{Red}}{HA} \rightleftharpoons \underset{\text{Yellow}}{H^+ + A^-}$$

$$\underset{\text{Purple}}{BOH} \rightleftharpoons \underset{\text{Colorless}}{B^+ + OH^-}$$

Indicators change color over a limited pH range; e.g., bromocresol green is yellow at pH 3.8 and blue at pH 5.4. Between these pH values the color is shades of green. Phenolphthalein, another indicator, is colorless below pH 8.0 and red above pH 9.8. Between these pH values the color is shades of pink. A list of several indicators with pH range and colors is given in Table 6.6.

PROBLEM 1

Which indicator should be used to indicate the end point of titration of a weak acid with a strong base? As shown in Sec. 6.10, the end point pH would be approximately 9;

TABLE 6.6 COLOR CHANGE AND pH RANGE OF SOME COMMON INDICATORS

Name	Color		pH range
	Acidic	Basic	
Thymol blue	Red	Yellow	1.2–2.8
Methyl yellow	Red	Yellow	2.9–4.0
Methyl orange	Red	Orange-yellow	3.1–4.4
Bromophenol blue	Yellow	Blue-violet	3.0–4.6
Bromocresol green	Yellow	Blue	3.8–5.4
Methyl red	Red	Yellow	4.4–6.2
Bromophenol red	Yellow	Red	5.4–7.0
Bromothymol blue	Yellow	Blue	6.0–7.6
Phenol red	Yellow	Red	6.4–8.0
Thymol blue	Yellow	Blue	8.0–9.6
Phenolphthalein	Colorless	Red	8.0–9.8

therefore, the indicator to use would have to change color in the range of pH 8 to 9. Thymol blue or phenolphthalein would be satisfactory. On the other hand, if a weak base is titrated with a strong acid, the end point is below pH 7 and an indicator like methyl red or bromophenol red (pH range 5.4 to 7.0) should be used.

PROBLEM 2

What is the approximate pH of a solution which turns both phenol red and methyl red yellow? Phenol red is yellow at pH 6.4 or lower, and methyl red is yellow at pH 6.2 or higher. Therefore, the pH of the solution is between 6.2 and 6.4.

6.13 BUFFERS

Buffers are substances that tend to resist a change in pH. Titration curves in Figs. 6.3 and 6.4 show that there is a relatively small change in pH per increment of base, or acid, added in the region where the solution is *half neutralized.* At this region, pH $=$ pK, the solution is acting as a buffer.

Buffer solutions contain the *salt of a weak acid and a strong base plus some free acid* or the *salt of a weak base and strong acid plus some free base.* This definition suggests that the Henderson-Hasselbalch equation can be used to calculate either the pH or the composition of the buffer system.

EXAMPLE 1

What is the pH of a solution containing 1 meq each of sodium acetate and acetic acid in a total of 49 ml of solution? The pK of acetic acid is 4.76, and the pH of the solution is

$$\text{pH} = 4.76 + \log \frac{[\text{salt}]}{[\text{acid}]}$$

Since [salt] $=$ [acid],

$$\text{pH} = 4.76$$

If 1 ml of 0.1 N NaOH is added to this solution, what will be the change in pH? The milliequivalent of NaOH added is $1 \times 0.1 = 0.1$ meq. This would neutralize 0.1 meq of acetic acid and produce 1 meq more of salt. The new concentration of acetic acid would be $1.0 - 0.1$, or 0.9 meq, and the new concentration of salt would be $1.0 + 0.1$, or 1.1 meq. Substituting these values in the Henderson-Hasselbalch equation, the pH of the solution can be calculated as follows:

$$\text{pH} = 4.76 + \log \frac{1.1}{0.9} = 4.76 + \log 1.1 - \log 0.9 = 4.85$$

The change in pH is $4.85 - 4.76$, or 0.09 unit.

EXAMPLE 2

What would have been the change in pH if the NaOH had been added to 49 ml of water? The initial pH of the water is 7.0, and the final pH is calculated as follows:

$$0.1 \text{ meq NaOH in 50 ml} \simeq 2 \text{ meq/liter} = 0.002\ N \text{ or } M$$

with respect to NaOH, and also with respect to OH^-.

$$\text{pOH} = -\log [OH^-] = -\log 2 \times 10^{-3} = -(-3 + 0.301) = 2.7$$

$$\text{pH} + \text{pOH} = 14$$

$$\text{pH} = 14 - \text{pOH} = 14 - 2.7 = 11.3$$

The change in pH is $11.3 - 7.0 = 4.3$ units.

Compare this change of pH with that obtained with the buffered solution. Remember a change of 1 pH unit indicates a tenfold change in hydrogen-ion concentration; the change in hydrogen-ion concentration in the second solution is about 1×10^4 times greater than in the buffered solution.

6.14 AMPHOTERIC COMPOUNDS

Compounds that react both as an acid and as a base are called *amphoteric* compounds or *ampholytes*. Certain inorganic hydroxides and amino acids are examples of amphoteric compounds.

Inorganic hydroxides

The hydroxides of aluminum, zinc, chromium, silicon, and lead are amphoteric compounds. Their reactions with an acid and a base are

$$\underset{\text{Base}}{Al(OH)_3} + \underset{\text{Acid}}{3\ HCl} \longrightarrow \underset{\text{Salt}}{AlCl_3} + \underset{\text{Water}}{3\ H_2O}$$

$$\underset{\text{Acid}}{HOAl(OH_2)} + \underset{\text{Base}}{NaOH} \longrightarrow \underset{\text{Salt}}{NaOAl(OH)_2} + \underset{\text{Water}}{H_2O} \longrightarrow \underset{\text{Salt}}{NaOAlO} + H_2O$$

Aluminum hydroxide reacts as a base with a strong acid, HCl, to form aluminum chloride. When aluminum hydroxide reacts as an acid toward a strong base like sodium hydroxide, only one of the hydrogens of the hydroxide is replaced by sodium, the other two being much too weakly acidic to react. In fact the product of the reaction of NaOH with $Al(OH)_3$ is $NaOAl(OH)_2$, which is the hydrated form of $NaAlO_2$.

Amino acids

Compounds containing an amino group, NH_2, and a carboxyl group, COOH, in the same molecule are known as *amino acids.* Since the COOH group is a proton donor and the NH_2 group is a proton acceptor, these compounds have both acidic and basic properties. Proteins, which are made up of amino acids bonded together, also contain free carboxyl and amino groups and are amphoteric compounds.

When the amino acid $H_2N—CH_2—COOH$ is dissolved in water, the equilibria are

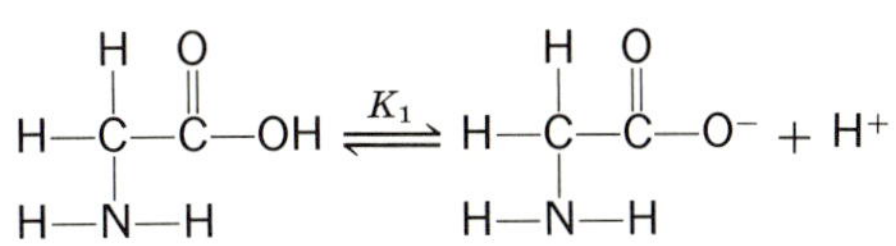

$pK_1 = 2.34$

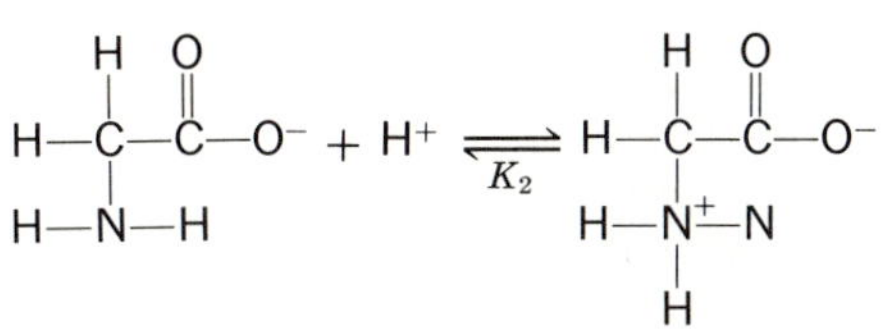

$pK_2 = 9.6$

Zwitterion

The intermolecular salt of an amino acid is called a *zwitterion;* it is a polar ion that has a net charge of zero. The internal negative charge is neutralized by the internal positive charge.

The presence of a proton acceptor (NH_2 group) on a carbon adjacent to the carboxyl group, COOH, increases the dissociation of the carboxyl group because some of the H^+ is removed from solution to form the NH_3^+ group; this shifts the equilibrium for the dissociation of the carboxyl group to the right. For example, compare the pK for the dissociation of acetic acid, $K_{\text{ion}} = 1.75 \times 10^{-5}$ or p$K = 4.76$, with the dissociation of the carboxyl group of the amino acid shown above, p$K_1 = 2.34$. The carboxyl group of the amino acid is about 260 times stronger an acid than the carboxyl group of acetic acid.

The NH_3^+ group in the zwitterion has weakly acidic properties and a pK_2 of 9.6. Thus a solution of this amino acid would have to be made to pH 9.6 before the NH_3^+ group of the amino acid would be one-half neutralized. When a compound has more than one acidic group, the pK's are indicated as pK_1, pK_2, pK_3, etc., in which the subscript numbers refer to the acid groups in order of their decreasing strength as an acid.

The titration of a completely protonated amino acid follows the same pattern as an inorganic acid with two replaceable hydrogens, like H_2CO_3 (see page 138). For our discussion consider the titration of an amino acid that contains one amino group and one carboxyl group,

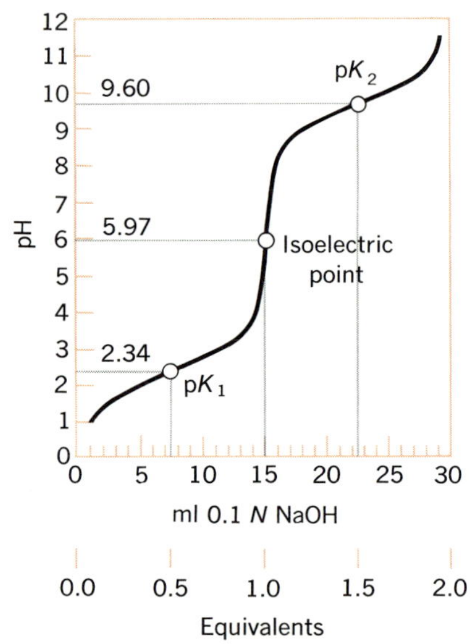

FIGURE 6.5 Titration curve for a monoaminomonocarboxylic amino acid: 15 ml of completely protonated 0.1 M glycine, $NH_3^+CH_2CO_2H$, is titrated with 0.1 N sodium hydroxide, NaOH. The isoelectric point, pK_1, and pK_2 are indicated on the curve.

a monoaminomonocarboxylic amino acid. Refer to Eq. (6.16) for the ionic species formed at different stages of the titration and to Fig. 6.5 for the titration curve.

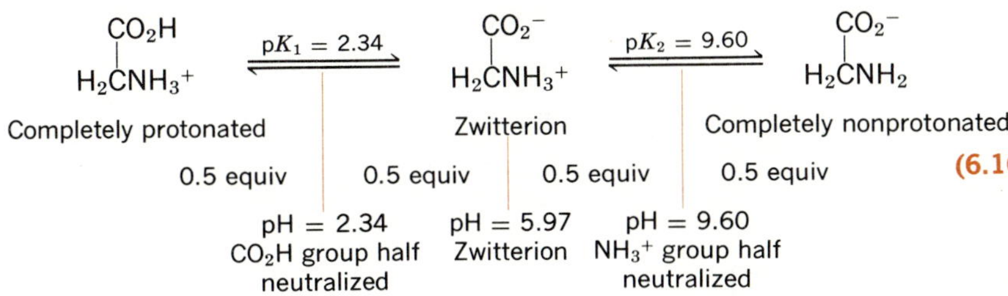

(6.16)

If $\frac{1}{2}$ equiv of base is added to the completely protonated amino acid, the carboxyl group will be half neutralized and the pH of the solution will equal the pK_1 of the amino acid carboxylic group. This is shown by appropriate substitutions in the Henderson-Hasselbalch Eq. (6.11):

$$pK_1 = 2.34 \qquad [\text{salt}] = \left[\begin{array}{c} CO_2^- \\ | \\ H_2CNH_3^+ \end{array}\right] \qquad [\text{acid}] = \left[\begin{array}{c} CO_2H \\ | \\ H_2CNH_3^+ \end{array}\right]$$

$$pH = 2.3 + \log \frac{\left[\begin{array}{c} CO_2^- \\ | \\ H_2CNH_3^+ \end{array}\right]}{\left[\begin{array}{c} CO_2H \\ | \\ H_2CNH_3^+ \end{array}\right]}$$

Since [salt] = [acid],

$$pH = 2.3 + \log 1 = 2.3$$

Upon the addition of another $\frac{1}{2}$ equiv of base, the carboxyl group is completely neutralized and the zwitterion, Eq. (6.16), exists in solution. This ion does not migrate to either the positive or negative electrode because its net charge is zero. Under this condition the amino acid is said to be at its *isoelectric point,* the pH at which the ionic species does not migrate under the influence of a direct current passed through the solution. The addition of another increment of $\frac{1}{2}$ equiv of base results in the neutralization of one-half the NH_3^+ groups present. The pH of the solution now is at the pK_2 of the amino acid:

$$pK_2 = 9.60 \qquad [\text{salt}] = \left[\begin{array}{c} CO_2^- \\ | \\ H_2CNH_2 \end{array}\right] \qquad [\text{acid}] = \left[\begin{array}{c} CO_2^- \\ | \\ H_2CNH_3^+ \end{array}\right]$$

$$pH = 9.60 + \log \frac{\left[\begin{array}{c} CO_2^- \\ | \\ H_2CNH_2 \end{array}\right]}{\left[\begin{array}{c} CO_2^- \\ | \\ H_2CNH_3^+ \end{array}\right]}$$

Since [salt] = [acid],

$$pH = 9.60 + \log 1 = 9.60$$

The addition of still another $\frac{1}{2}$ equiv of base completely neutralizes the amino acid, and the nonprotonated ionic species with a charge of -1 is formed. Because titration reactions are reversible, one could start with a completely nonprotonated amino acid and obtain the completely protonated ionic species by adding an acid. The titration curve for the amino acid in Eq. (6.16) is given in Fig. 6.5. Note that the curve is similar to the one obtained when H_2CO_3 is titrated with a strong base.

PROBLEM

To what pH would you make a solution of the above amino acid to have it exist as a zwitterion? Refer to Eq. (6.16). You could add 1 equiv of base to the completely protonated amino acid and take the pH of the solution; or you could add 1 equiv of acid to the completely nonprotonated species and take the pH of the resulting solution; or you could take the average of the two pK's of the amino acid because the zwitterion exists when the pH is halfway between the two pK values, or $(2.34 + 9.6)/2 = 5.97$.

6.15 SOME BIOLOGICAL CONSIDERATIONS

Acids, bases, and buffers play a vital role in biological systems because biological reactions must take place within a rather narrow pH range. A number of acids produced during biological transformations tend to decrease the pH of the cellular fluid markedly; however, such a change in pH is prevented by the counteracting effect of buffers. For example, the pH of blood is normally 7.4; a variation of about 0.2 pH unit on either side of this value may result in death. When hemoglobin, the red pigment in blood cells, unites with oxygen in the lungs, hydrogen ion is released to the cellular solution. This would easily reduce the pH of the red blood cells to a value lower than the limits of safety if buffers were not present. Enzymes are specific proteins that act as a catalyst, i.e., permit reactions to go at ordinary body temperature which would not go appreciably if the enzymes were not present. Most enzymes are very sensitive to pH changes, and their activity may be decreased markedly by a slight change in pH.

The main buffers in cellular fluids are HCO_3^- and H_2CO_3, HPO_4^{--} and $H_2PO_4^-$, and proteins. Since proteins contain amino and carboxyl groups, they have the same properties as amino acids with respect to acids and bases. Note in Fig. 6.5 that the pK values of an amino acid fit the prerequisites for a buffer.

QUESTIONS

1. (*a*) Write an equation for the equilibrium constant for

$$A + B \rightleftharpoons 2C + D$$

(*b*) If the reactants and products are all gases, which way would the equilibrium be shifted upon increasing the pressure on the system?

2. Write the stepwise ionization equation for H_2CO_3 and for H_3PO_4 and list the pK values for each step.

3. List the following acids in the order of their increasing acidity.

Acid	*pH*	*pOH*
A	7.5	
B	5.7	
C		13.0
D	9.5	
E		9.5

4. Why would you expect water molecules to hydrogen-bond with each other?

5. Explain why R—COOH is an acid whereas $Ca(OH)_2$ is a base.

6. What would be the pH of the following solutions?

(*a*) 0.003 N HCl (*b*) 0.01 N KOH (*c*) 0.025 M H_2SO_4
(*d*) 0.005 M NaOH (*e*) 0.01 M H_3PO_4 (*f*) 0.005 M HCl

7. What is the pH of a 0.001 N solution of an acid that is 4% ionized?

8. What is the percent dissociation of an acetic acid solution that has a pH of 3.76 (pK of acetic acid is 4.76)?

9. Why is periodic acid, HIO_4, a weaker acid than $HClO_4$?

10. State whether a solution of each of the following salts would be neutral, acidic, or basic and explain why:
(*a*) NaCl (*b*) K_2CO_3 (*c*) K_2SO_4
(*d*) Na_2HPO_4 (*e*) NH_4Cl

11. Explain why the pH of the end point of titrating a weak acid with a strong base is greater than the pH of the end point of titrating a weak base with a strong acid.

12. What is the pH at the end point of titrating 20 ml of a 0.01 N solution of acetic acid with 0.01 N KOH?

13. H_2CO_3 is transported by the blood from the cells to the lungs. If the pH of the blood is 7.4, what is the predominant ionic species of carbonic acid that exists in the blood ($pK_1 = 6.37$, $pK_2 = 10.25$)?

14. What is the function of a buffer? What constitutes a good buffer?

15. What is the pH of a solution which is 0.1 M with respect to sodium acetate and 0.05 M with respect to acetic acid ($pK = 4.76$)?

16. If a solution contains equal molar amounts of Na_2CO_3 and $NaHCO_3$ in 500 ml of solution, what is its pH? What will the pH of the solution be after it is diluted to 1,000 ml ($pK_1 = 6.37$, $pK_2 = 10.25$)?

17. To 49 ml of a solution 0.1 M with respect to both NaH_2PO_4 and Na_2HPO_4 is added 11 ml of a 0.01 M solution of HCl. What is the final pH of the solution?

SUGGESTED READINGS FOR PART ONE

Block, Richard J., Emmett L. Durrum, and Gunter Zweig: "A Manual of Paper Chromatography and Paper Electrophoresis," Academic Press Inc., New York, 1955.

Brescia, Frank, John Arents, Herbert Meislich, and Amos Turk: "Fundamentals of Chemistry: A Modern Introduction," Academic Press Inc., New York, 1966.

Brown, John F.: Inclusion Compounds, *Sci. Am.,* July, 1962, p. 82.

Buswell, A. M., and W. H. Rodebush: Water, *Sci. Am.,* April, 1956, p. 77.

Fowler, T. K., and Richard F. Post: Progress toward Fusion Power, *Sci. Am.,* December, 1966, p. 21.

Glasstone, Samuel: "Sourcebook on Atomic Energy," 2d ed., D. Van Nostrand Company, Inc., New York, 1958.

Hendricks, Sterling B.: How Light Interacts with Living Matter, *Sci. Am.,* September, 1968, p. 175.

Lederer, Edgar, and Michael Lederer: "Chromatography," Elsevier Publishing Company, New York, 1955.

MacInnes, Duncan A.: pH, *Sci. Am.,* January, 1951, p. 40.

Runnels, L. K.: Ice, *Sci. Am.,* December, 1966, p. 118.

Russell, John B.: "Study Guide to Accompany Sienko and Plane: Chemistry," McGraw-Hill Book Company, New York, 1966.

Sienko, Michell J., and Robert A. Plane: "Chemistry," 3d ed., McGraw-Hill Book Company, New York, 1966.

Van der Werf, C. A.: "Acids, Bases, and the Chemistry of the Covalent Bond," Reinhold Publishing Corporation, New York, 1961.

Whipple, Harold E. (ed.): Forms of Water in Biological Systems, *Ann. N.Y. Acad. Sci.,* **125**(2) (1965).

part two... organic chemistry

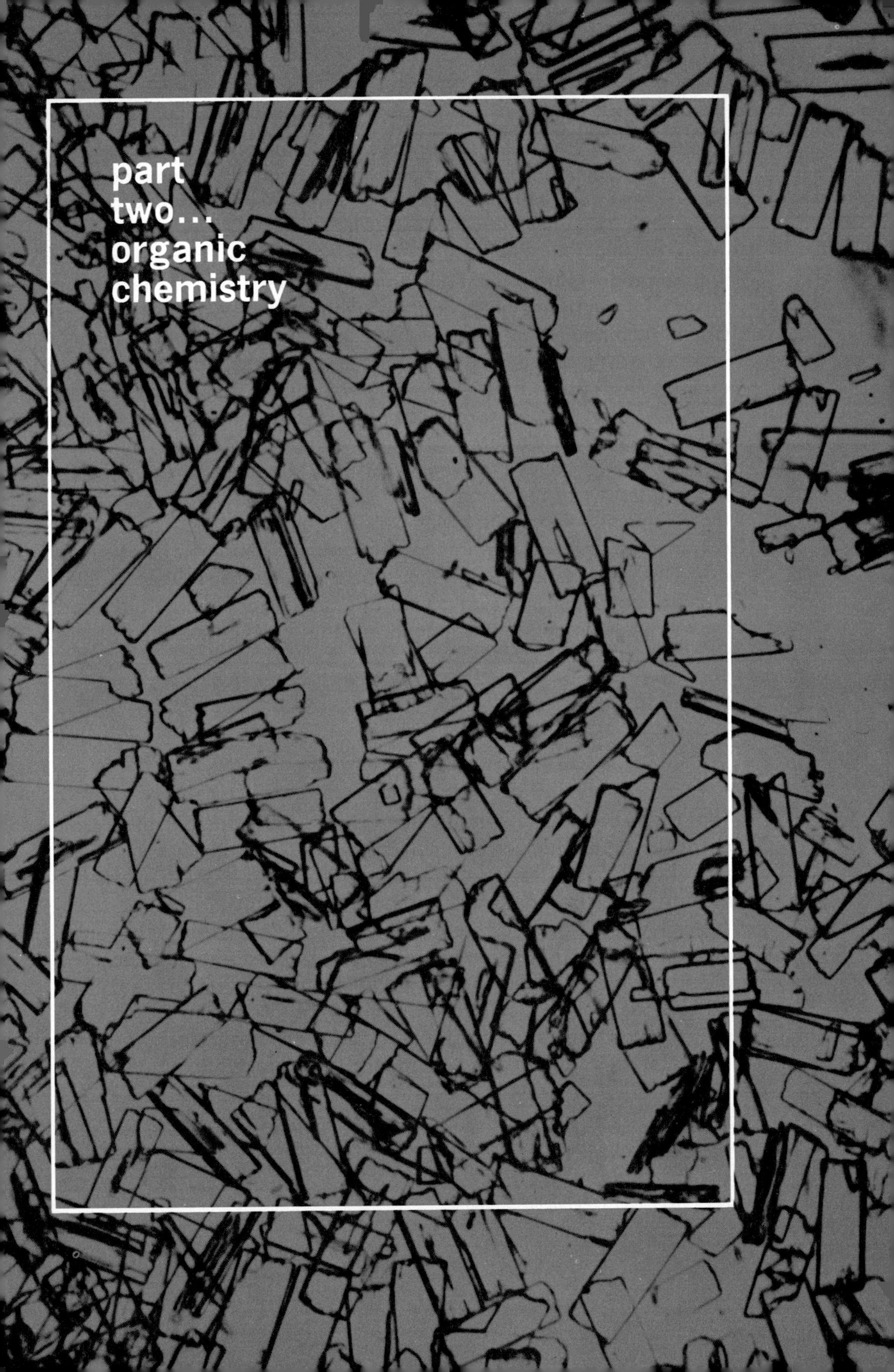

Organic chemistry was first defined as a study of the carbon compounds that occur in living matter. The distinction between inorganic and organic compounds dates back to the advent of the science of chemistry. This division was based upon observations of the differences between the chemical behavior of inorganic compounds and natural products from plant and animal sources. For example, organic compounds were unstable compared with most inorganic substances. Organic compounds were combustible, and many had no sharp melting point, charring and decomposing instead. Most reactions performed on inorganic compounds were carried out under nonbiological conditions, i.e., *high temperature or strongly acidic or basic solutions.* Such conditions *could not* be applied to the synthesis of organic compounds, for they were produced in living organisms at temperatures conducive to the integrity of the living cell. There was no plausible explanation for this phenomenon, and it was believed that a "vital force" inherent only in living cells was required for the synthesis of organic compounds.

One can easily imagine the impact on chemistry when in 1824 Wöhler announced the synthesis of the organic compound urea from the inorganic compounds ammonium sulfate and potassium cyanate.

$$\underset{\text{Ammonium sulfate}}{(NH_4)_2SO_4} + \underset{\text{Potassium cyanate}}{KOCN} \longrightarrow \underset{\text{Ammonium cyanate}}{NH_4OCN} \xrightarrow{\Delta} \underset{\text{Urea}}{H_2N\text{—}\overset{\overset{\displaystyle O}{\|}}{C}\text{—}NH_2}$$

This discovery was a turning point in the development of organic chemistry; a new horizon of unlimited synthesis possibilities appeared, and organic chemists were no longer limited to the isolation and elementary analysis of natural substances.

Slowly at first and then faster as the chemistry of carbon compounds became better understood, the synthesis of organic compounds was expanded. Today organic chemists synthesize many compounds which do not occur in nature but which enrich our lives, e.g., synthetic fibers, resins, rubber, drugs, and pesticides.

A study of the organic chemistry of natural compounds associ-

Crystals of L*-asparaginase from* Escherichia coli B. *Partially purified* L*-asparaginase has demonstrable antileukemic activity in animals and man. Crystals of this enzyme, prepared by a simple and efficient process, were effective in experimental tumor systems without adverse side effects (×500). (Peter P. K. Ho, The Lilly Research Laboratories, Eli Lilly and Company, Indianapolis, Indiana.)*

ated with biological reactions has made possible the advancement of biochemistry. For example, the chemical reactions associated with the digestion of carbohydrates, fats, and proteins and their subsequent conversion into body substances constitutes basic biochemistry. A general understanding of organic chemistry is a prerequisite for the study of biochemistry.

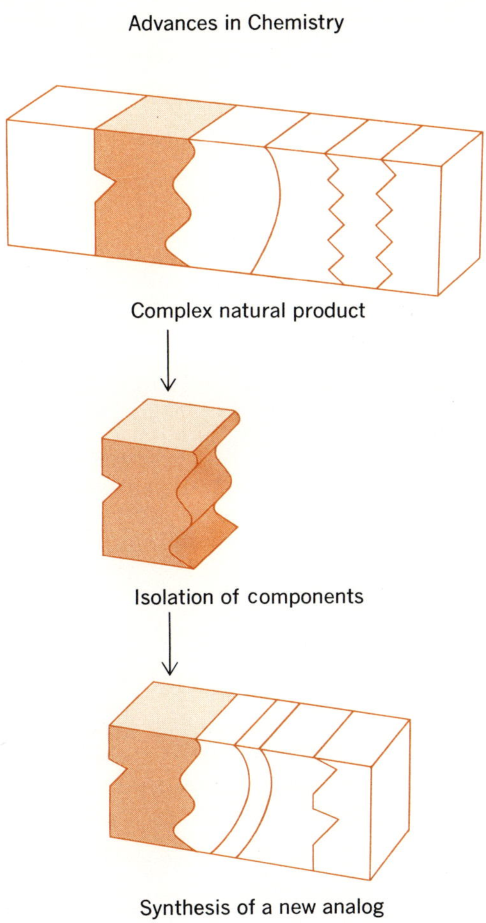

chapter 7.... carbon

Since carbon plays the central role in organic compounds, we begin our discussion of organic chemistry with a review of its electronic configuration. Be sure you can answer the following questions. What position does carbon occupy in the periodic table? Compare the electronegativity of carbon with that of hydrogen, oxygen, chlorine, nitrogen, and sulfur; would you expect carbon to form covalent bonds with these elements? What is the electron configuration of carbon? From the distribution of its electrons would you expect carbon to form four covalent bonds? This is an interesting question, and Chap. 2 began with a discussion of the answer.

7.1 ELECTRON DISTRIBUTION OF CARBON

As outlined in Chap. 2, the distribution of electrons about the carbon nucleus is

$$\underset{1s^2}{\underline{\uparrow\downarrow}} \quad \underbrace{\underset{2s^2}{\underline{\uparrow\downarrow}} \quad \underset{2p_x{}^1}{\underline{\uparrow}} \quad \underset{2p_y{}^1}{\underline{\uparrow}}}$$

First quantum energy shell (1s); Second quantum energy shell (2s, 2p)

The two electrons in each of the $1s$ and $2s$ orbitals are paired (↑↓); i.e., they are both in the same orbital and are spinning in opposite directions, whereas each electron in the two $2p$ orbitals is unpaired (↑). In the second quantum energy shell there are two kinds of electrons (s and p), but the difference in energy is not great. The electrons can all attain an excited state at temperatures well below room temperature and hence readily take part in reactions. The *most stable electronic*

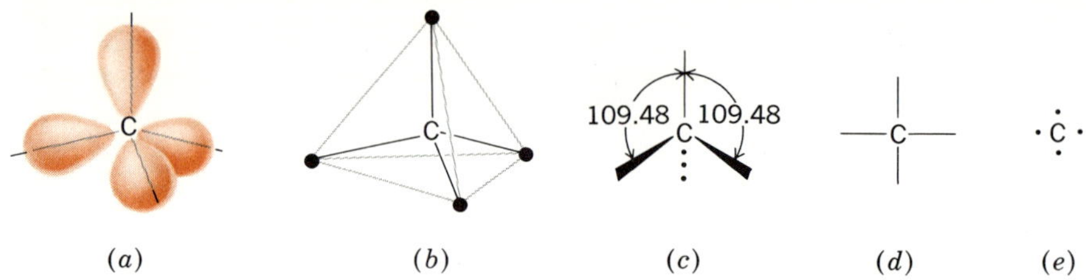

FIGURE 7.1 Five possible ways of depicting the electronic configuration of carbon. (*a*) The electrons are viewed as a dense cloud in a tetrahedral configuration around the nucleus; (*b*) the electrons are located at each of the points of a tetrahedron with the carbon nucleus at the center; (*c*) the bonds are drawn with two in front of the paper (thick bond), one in the plane of the paper (regular bond), and one behind the plane of the paper (dashed bond); (*d*) a one-dimensional diagram showing four bonds, but the angles are inaccurate; (*e*) vastly simplified diagram using dots for electrons in the outer shell with no attempt to portray the three-dimensional configuration.

configuration of carbon in compounds is for the four electrons in the second quantum energy shell to be equidistant from each other and to have the same energy, an *sp^3 hybrid* (Fig. 7.1). Since an electron is in continual motion, the volume in which it moves is often depicted as a cloud.

7.2 THE CARBON-TO-CARBON BOND

Carbon atoms are bonded to other carbon atoms by a single, double, or triple covalent bond. Since the maximum number of orbital electrons that can overlap between two carbon atoms is three, there can be no more than three bonds between adjacent carbon atoms.

Single bond

A single covalent bond between adjacent carbon atoms is a σ bond (see Chap. 3); the electron cloud encompasses both nuclei (Fig. 7.2). The covalent bond may be represented as C:C or C—C. The σ bond is quite stable.

Carbons joined by a single covalent bond are free to rotate about the bond unless hindered by the presence of large groups attached to the other valence electrons. Even such small atoms as hydrogen influence the relative position of the respective atoms. For example, hydrogens on adjacent carbon atoms may be in the same plane (eclipsed) or they may be staggered (Fig. 7.3). The term *conformation* refers to the various structures that can be written by rotation about a bond axis. In the eclipsed conformation the hydrogens are relatively close

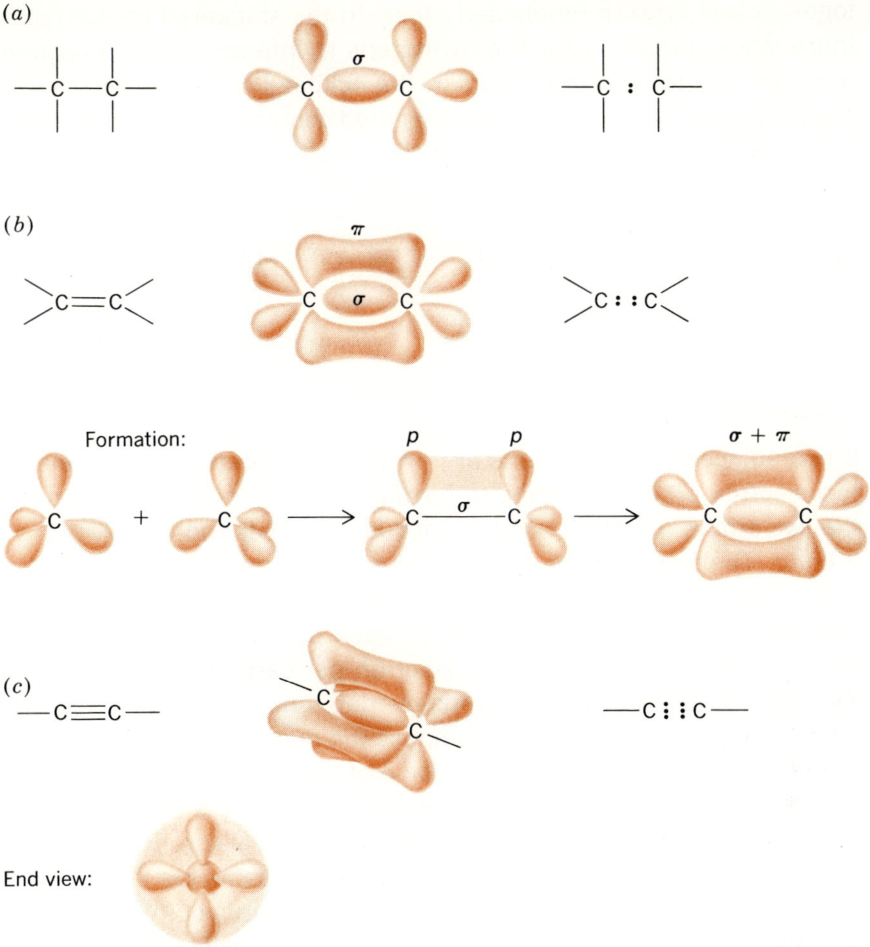

FIGURE 7.2 Carbon-carbon bonds: (*a*) σ bond, a single bond between adjacent carbon atoms; (*b*) π bond, a double bond between adjacent carbons; (*c*) triple bond.

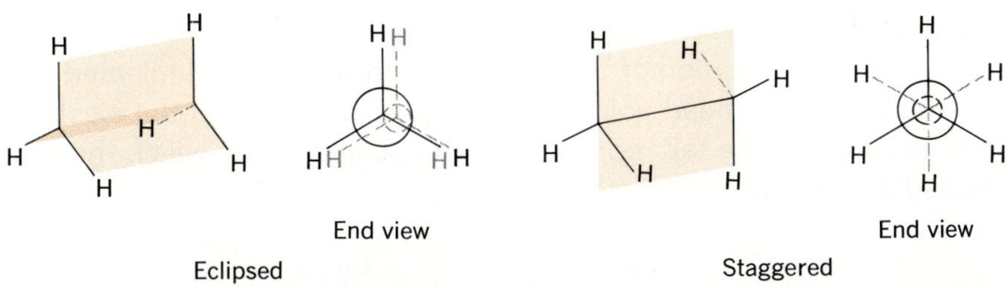

FIGURE 7.3 The eclipsed and staggered conformations of two substituted carbon atoms.

together and tend to repel each other. In the staggered conformation there is less tendency for the hydrogens to interact. Since it requires more energy to hold the atoms in the eclipsed conformation than in the staggered conformation, the staggered conformation is more stable.

The following compounds contain single C—C bonds:

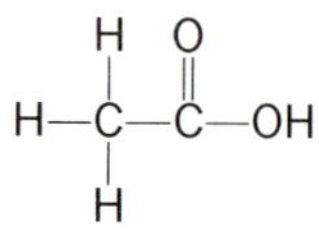

Acetic acid

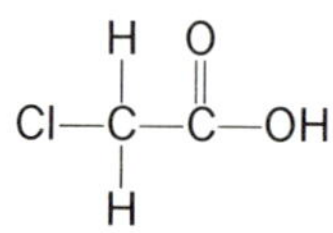

Monochloroacetic acid

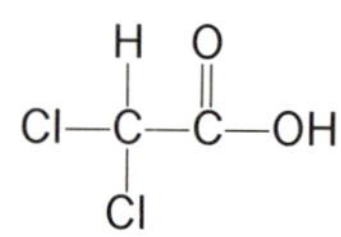

Dichloroacetic acid

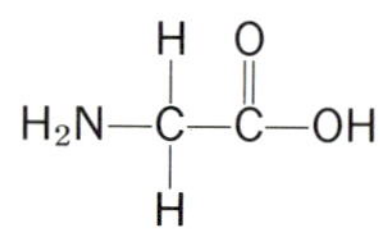

Amino acid

Double bond

If two adjacent carbon atoms are properly oriented, it is possible for two electron orbits to overlap to form a double covalent bond between the two atoms. The first bond formed is a σ bond, and the second is a π bond. The formation of a π bond can be envisioned as follows. Consider a two-carbon unit in which the carbons are attached by a σ bond and two hydrogens are attached to each carbon, also by a σ bond. There remains one unbonded p electron per carbon atom (Fig. 7.2b). The orbitals of these two electrons overlap, and a π bond results. In a π bond the electrons do not encompass the nuclei but are, as indicated in the figure, above or below the plane of the carbon atoms.

The π bond is not as strong as the σ bond. It is strongest when there is the greatest overlapping of the two p-electron clouds, i.e., when the two respective p electrons are in the same plane. Any tendency to twist the two carbons around the π bond would also tend to pull the π bond out of its plane and would be resisted. Hence, in a compound with a double bond between adjacent carbons there is no free rotation around the bond. This means that the hydrogens in the compound in Fig. 7.2b all lie in the same plane. The π-bond electrons are not as localized as the σ-bond electrons are, which accounts for many of the properties of C=C compounds.

The delocalization of the π-bond electrons can be indicated by a curved arrow, the head pointing to the position to which the electrons are moving and the tail representing the position from which the electrons have moved

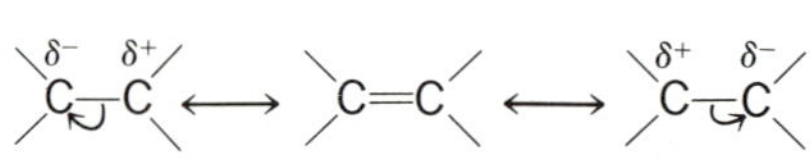

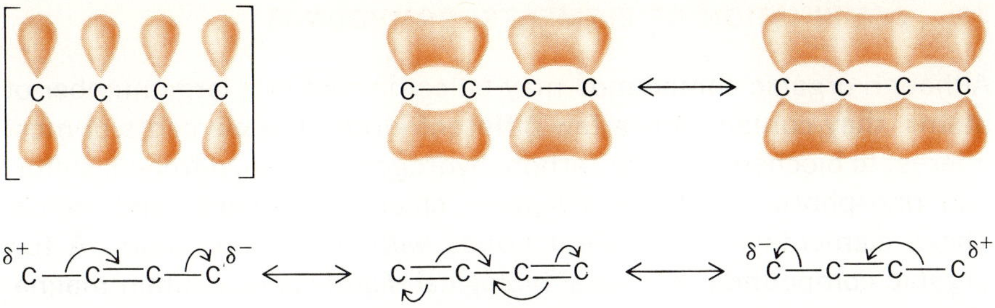

Resonance formulas

FIGURE 7.4 Two schematic ways of showing diene resonance and charge distribution.

These various electron configurations are known as *resonance forms* (resonance formulas involve only a change in electronic configuration). The double-headed arrow is used by convention to separate resonance forms; it *does not* represent a chemical reaction.

Dienes

Dienes are compounds that contain two double bonds, which are usually separated from each other by one or more single-bonded carbons. A *conjugated* diene has the two double bonds separated by a single-bonded carbon, C=C—C=C. A *nonconjugated* diene has the two double bonds isolated by more than one single-bonded carbon, C=C—C—C—C=C.

Conjugated dienes have a great delocalization of the p electrons associated with the double bonds. As seen in Fig. 7.4, the p electrons are in reality associated with all four carbon atoms making up the conjugated system.

Triple bond

The formation of a triple bond between two carbon atoms involves a σ bond and two π bonds. The two π bonds are at right angles to each other (Fig. 7.2*c*). Thus, in addition to the σ-bond cloud, the two carbon atoms are surrounded by the π-electron cloud. The delocalization of electrons in a triple bond is very great.

As the number of bonds between carbons increases, the distance between the carbon atoms decreases:

Bond type:

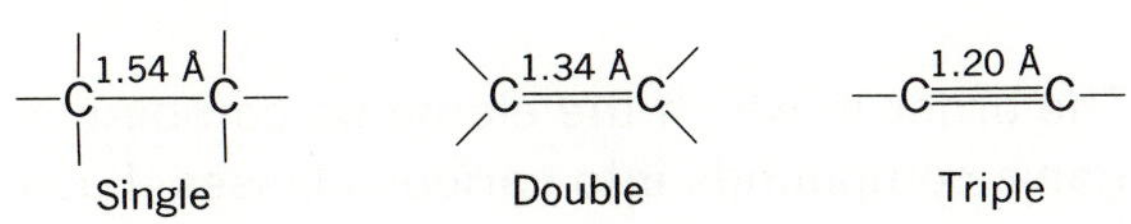

7.3 COMBINATION OF ELEMENTS: COMPOUNDS

Although organic compounds may be composed of a large number of atoms, they contain only a few different kinds. The elements of most interest to biochemistry are carbon, hydrogen, oxygen, nitrogen, sulfur, and phosphorus, and the halogens, chlorine, bromine, and iodine. These elements form covalent bonds with the carbon atom. A few organic compounds of special biological significance contain magnesium, iron, or cobalt. For example, iron is contained in hemoglobin, the red pigment of blood.

Many compounds may exist that contain varying numbers of the same kind of atoms, a simple example being methane, CH_4, and ethane, C_2H_6. There may also be a large number of different compounds that contain the same kind as well as the same number of atoms. For example, 150 different compounds have the formula $C_{10}H_{18}O$. Thus, to be certain of the compound in question the formula must be written in a way that shows the arrangement of atoms in the molecule. The several types of formulas are:

1. *Empirical:* Indicates only the *kind* of atoms and their *simplest ratio*, CH_2O.
2. *Molecular:* Indicates only the *kind* and *number* of atoms in a molecule. The molecular formula and the empirical formula for some compounds are the same, as for CH_2O, formaldehyde. But for $C_6H_{12}O_6$, glucose, the empirical formula is CH_2O.
3. *Structural:* Indicates the *arrangement* of the respective atoms in a molecule

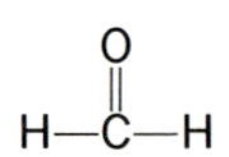

Formaldehyde

4. *Electronic:* Indicates the *distribution of electrons* in addition to that indicated by the structural formula

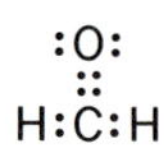

Formaldehyde

The order in which the elements combine is the basis for dividing organic compounds into various classes. Each class has a charac-

TABLE 7.1 CLASSIFICATION AND FUNCTIONAL GROUPS OF ORGANIC COMPOUNDS

Type	Formula[a]
I. C, H. Hydrocarbons	
A. Aliphatic	
1. Saturated, alkanes (single bond between C's)	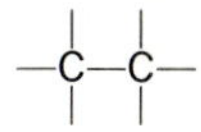
2. Unsaturated (multiple bonds between C's)	
(*a*) Alkenes, or ethylenes (double bond between C's)	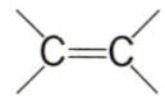
(*b*) Alkynes, or acetylenes (triple bond between C's)	—C≡C—
3. Cyclic, ring structures	
(*a*) Alicyclic (only C's in ring)	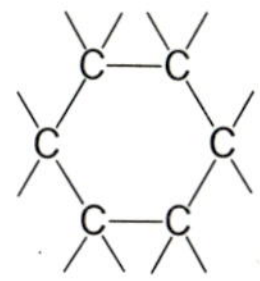
B. Aromatic (contain a benzene ring or the equivalent)	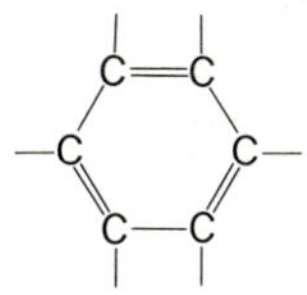
II. C, H, X (X = halide), alkyl halides	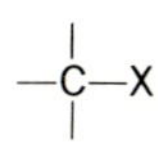
III. C, H, O	
A. Alcohols (organic hydroxides)	
1. Primary (not more than one C attached to the C holding the OH group)	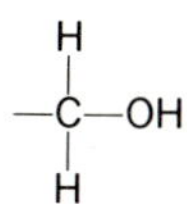 —CH_2—OH
2. Secondary (two C's attached to the C holding the OH group)	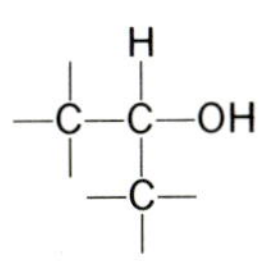 (≡C)$_2$CH—OH
3. Tertiary (three C's attached to the C holding the OH group)	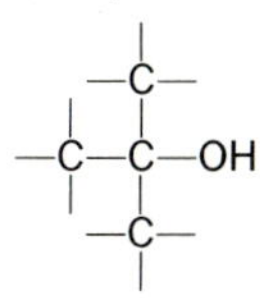 (≡C)$_3$C—OH
B. Aldehydes (carbonyl group with at least one H)	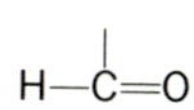 —CHO
C. Ketones (carbonyl group with no H)	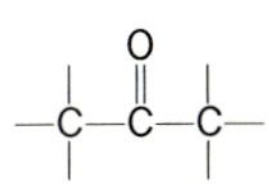
D. Acids (carboxyl group)	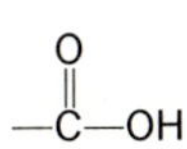 —COOH —CO_2H

TABLE 7.1 (*Continued*)

Type	Formula[a]
E. Acyl, e.g., acyl chloride, $H_3C-C(=O)-Cl$	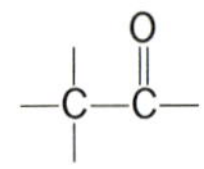
F. Esters	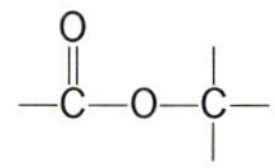
G. Anhydrides	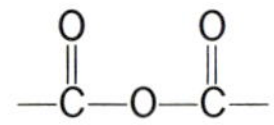
H. Ethers (organic oxides)	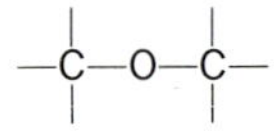
IV. C, H, N	
A. Amines (substituted ammonia)	
1. Primary (one C on the N)	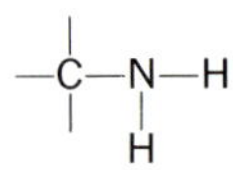 $\equiv C-NH_2$
2. Secondary (two C's on the N)	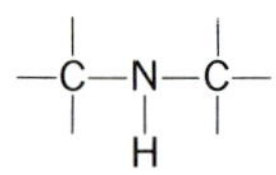 $\equiv C-NH-C\equiv$
3. Tertiary (three C's on the N)	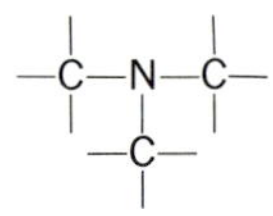 $(\equiv C)_3N$
4. Quaternary (four C's on the N)	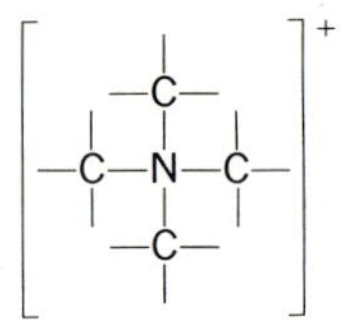 $(\equiv C)_4N^+$
B. Cyanides	$-C\equiv N$ $-CN$
V. C, H, O, N	
A. Amides	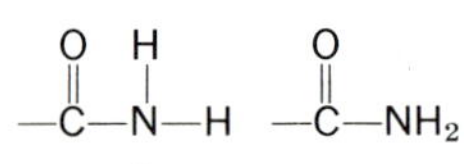
B. Nitro compounds	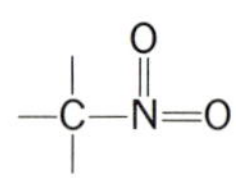 $\equiv C-NO_2$
VI. C, H, S	
A. Mercaptans (sulfides)	$\equiv C-S-C\equiv$
B. Disulfides	$\equiv C-S-S-C\equiv$
C. Thioalcohols, thiols (sulfhydryl group)	$\equiv C-S-H$

[a] The formulas show only the bonds involved in the functional groups.

teristic arrangement of atoms which is responsible for the chemical properties. The arrangement is called a *functional group.* Thus, by knowing the basic reactions of the various functional groups, one can understand and predict the chemical properties of many compounds. A partial list of functional groups with their formulas and names is given in Table 7.1. Because the valence electrons of the carbon atoms other than those involved in the functional group are left unattached, these formulas do not represent compounds.

There may be more than one functional group present in a compound. Such compounds have the properties of each functional group present.

Familiarity with the functional groups will increase as they are discussed and used in subsequent chapters. We are concerned here with identifying the functional groups, not with writing formulas.

7.4 ASYMMETRIC CARBON ATOM: OPTICAL ISOMERS

An *asymmetric* carbon atom is one with four different atoms or groups of atoms attached to it. Some compounds containing an asymmetric carbon atom are

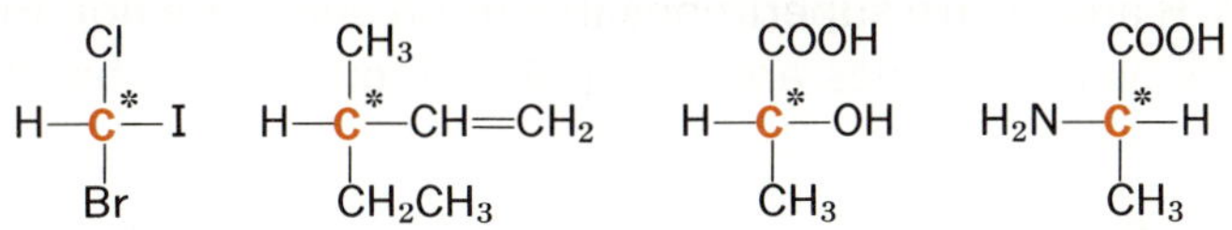

The asymmetric carbon is marked by an asterisk. An interesting feature of a compound that contains an asymmetric carbon atom is that it exists in two isomeric forms that are mirror images. Isomers are compounds that have the same molecular formula but different structural formulas. Because they rotate plane-polarized light, the name *optical isomers* has been applied to these isomers.

Consider the following example, in which a carbon atom is attached to four different groups, A, B, C, and D (Fig. 7.5). The four groups are shown in formula (*a*) attached to a carbon atom drawn as a tetrahedron. If this structure is placed in front of a mirror and its image drawn, formula (*b*) results. Particular note should be taken of the fact that in forming the mirror image the positions of only two groups in (*a*) were changed, namely, A and C. Formula (*a*) is the mirror image of (*b*).

The question to be considered now is this: Are formulas (*a*) and (*b*) identical? This is answered by applying the following reasoning. If the

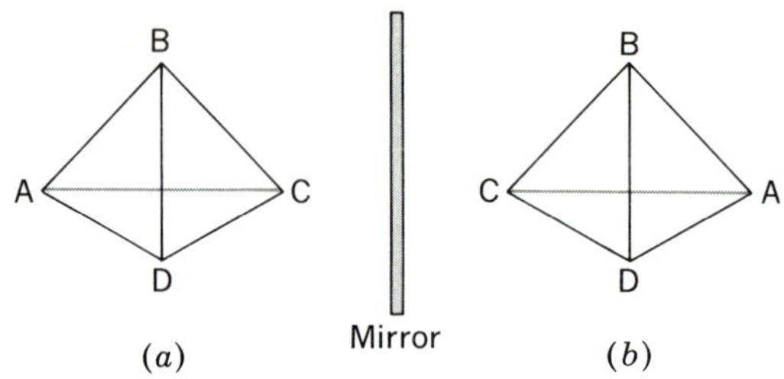

(*a*) (*b*)

FIGURE 7.5 Optical isomers (*a*) and (*b*) are mirror images of each other.

two molecules (*a*) and (*b*) are identical, i.e., just two ways of drawing the same compound, when the two images are superimposed upon each other, all four substituents will coincide. However, if the two formulas represent different molecules, only two substituents can be made to coincide. Figure 7.6 will clarify these statements.

If formulas 1 and 2 are slid together, each of the four substituents on the carbon atoms coincides; the molecules are superimposable and therefore identical. Note that 1 and 2 are not mirror images; they can be superimposed without taking them out of the plane of the paper. When we turn to (*a*) and (*b*) and slide these together, as in (*c*), two substituents, B and D, are superimposable but A and C are not. Regardless of how the two molecules are manipulated, only two substituents can

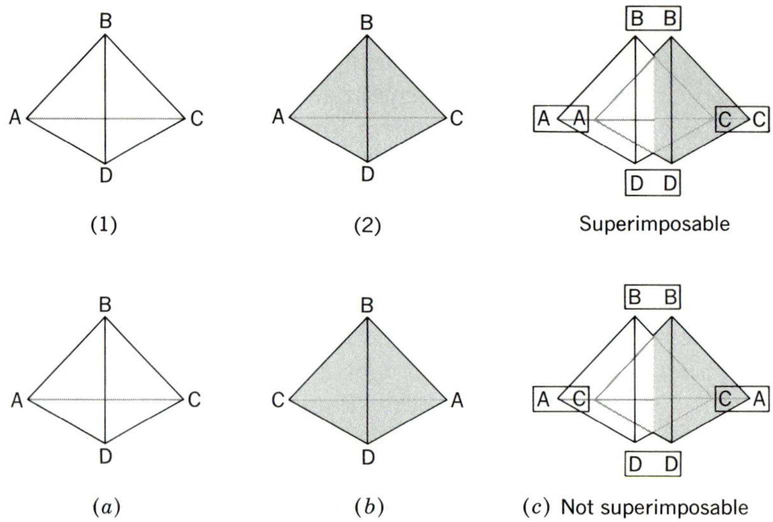

FIGURE 7.6 Optical isomers are not superimposable.

be superimposed. Molecules (*a*) and (*b*) are not identical. They are mirror images.

Some compounds contain more than one asymmetric carbon atom. The number of possible optical isomers is calculated by substituting in the formula 2^n, where n equals the number of asymmetric carbon atoms. Thus, if there are two asymmetric carbons in a molecule, the number of optical isomers is 2^2, or 4; for example,

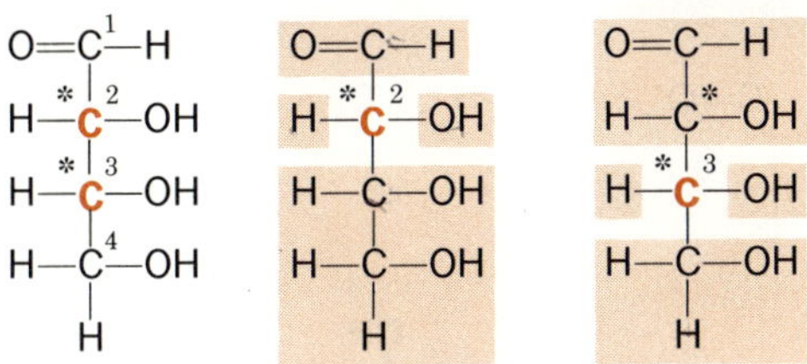

(I) (I′) (I″)

How many asymmetric carbons are there in compound I? The first carbon, the aldehyde carbon, is symmetric (not asymmetric) because two of its bonds are connected to the same atom, oxygen. The fourth carbon is also symmetric because two of its bonds are attached to hydrogen atoms. In formula I′ the four groups attached to the second carbon are shaded. No two groups are alike; therefore, the second carbon is asymmetric. Inspection of formula I″, in which the groups attached to the third carbon are shaded, shows that the third carbon is also asymmetric. Since there are two asymmetric carbons in compound I, there are 2^2, or 4, isomers:

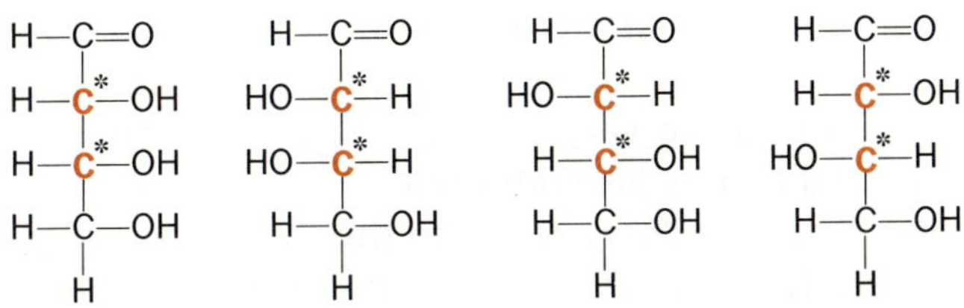

(I) (II) (III) (IV)

Mirror images (enantiomers) Mirror images (enantiomers)

Note that there are two sets of mirror images, I, II and III, IV. Compounds I and IV are optical isomers but not mirror images. The same is true of compounds II and III.

Enantiomer is another name for the mirror images of a compound. Compounds I and II are enantiomers, and so are compounds III and IV.

From this discussion it is apparent that the configuration about an asymmetric carbon atom is significant, whereas it makes no difference how the groups are arranged about a symmetrical carbon atom. For example, all the formulas in each group are identical:

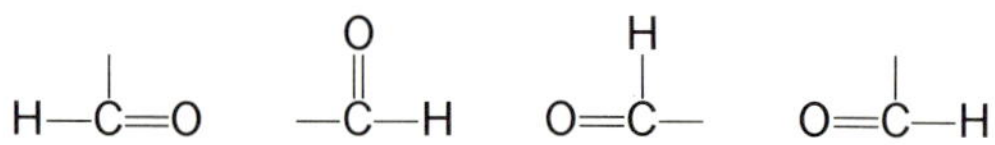

or

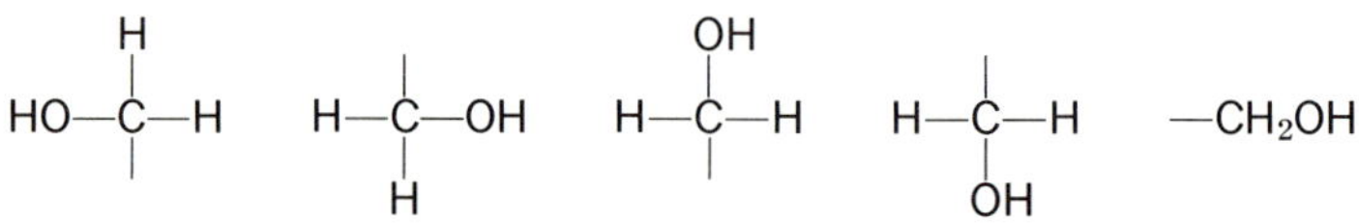

No carbon having a double bond or two identical substituents can be asymmetric.

7.5 PROPERTIES OF OPTICAL ISOMERS

Optical isomers that are mirror images of each other have the same chemical and physical properties, such as melting and boiling point and solubility, and differ only in the direction which they rotate plane-polarized light. Thus, compounds I and II have identical chemical properties and melting points. The same is true of compounds III and IV. Optical isomers that are not mirror images, e.g., compounds I and III, have different melting and boiling points and solubility characteristics. They often have very different biological activity, and their effect on polarized light is also different.

The term *optical activity* refers to the rotation of polarized light. As noted previously, this property is associated with the presence of an asymmetric carbon atom. Light emitted from any source is vibrating in all planes, from perpendicular to horizontal. By passing light through a specially cut prism called a Nicol prism (Fig. 7.7) all but one vibration can be removed, leaving light that is vibrating in only one plane, called plane-polarized light. When plane-polarized light is passed through a solution of an optically active compound, the plane of vibration of the light is rotated either to the right (clockwise) or to the left (counter-

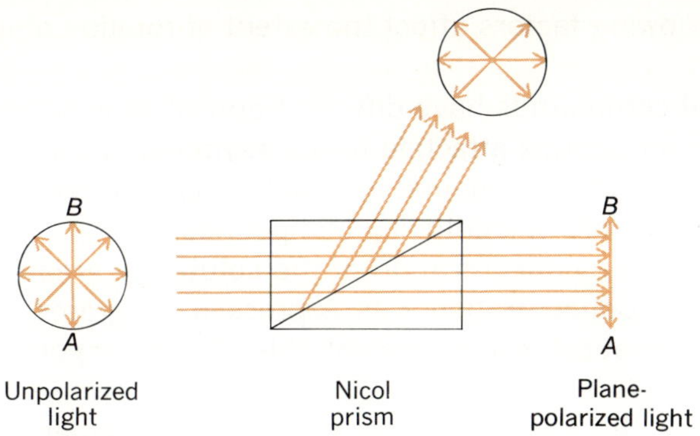

FIGURE 7.7 Schematic representation of the polarization of light.

clockwise). Compounds are called dextrorotatory or levorotatory, respectively, according to the direction of rotation they cause. The extent of this rotation α is expressed in angular degrees. If the light is rotated to the right, α is $(+)$; conversely, if the light is rotated to the left, α is $(-)$ (see Fig. 7.8).

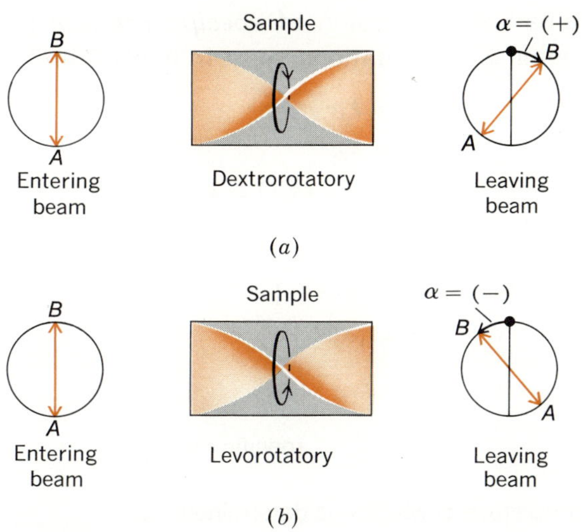

FIGURE 7.8 Optical activity. (a) Entering beam of plane-polarized light ($A \longleftrightarrow B$) is rotated clockwise as it comes *toward the viewer* through a solution of the optically active compound; α is the angle of rotation of the light. The sample is dextrorotatory, $\alpha = (+)$, if the light is rotated to the right. (b) Similar to (a) except that the light is rotated *counterclockwise* as it comes toward the viewer. The sample is levorotatory, $\alpha = (-)$, if the light is rotated to the left.

The following factors affect the extent of rotation of light:

1. Individual compounds have different optical activity because of the nature of the groups attached to the asymmetric carbon atom. This characteristic of a compound is called its *specific rotation* and plays an important role in identifying compounds.
2. The greater the concentration of the solute the greater the degree of rotation. Concentration c is expressed as grams per milliliter. The mathematical expression of this direct proportion is $\alpha \propto c$; $\alpha = k_1 c$.
3. The longer the light path the greater the angular rotation. The length l of the sample tube is expressed in decimeters (1 dm = 10 cm). This relationship is expressed as $\alpha \propto 1$; $\alpha = k_2 l$.
4. Because the wavelength of the light affects the degree of rotation, monochromatic light (light of only one wavelength) is used. This may be conveniently obtained from a sodium vapor lamp which emits a yellow light called the D line in the spectrum.
5. Temperature affects optical activity in different ways depending upon the compound.

A mathematical expression of *specific rotation* $[\alpha]$ can be obtained by combining the equations given in items 2 and 3.

$$\alpha = k_1 c$$
$$\alpha = k_2 l$$
$$\therefore \alpha = (k_1 c)(k_2 l) = k_1 k_2 (cl)$$

If

$$k_1 k_2 = K = [\alpha]$$
$$\alpha = [\alpha] cl$$

or

$$[\alpha]_{\mathrm{D}}^{T} = \frac{\alpha}{cl} = \text{specific rotation} \qquad \textbf{(7.1)}$$

where T = temperature at which α is determined, °C
D = yellow D line of sodium
c = concentration, g/ml
l = length of light path, dm
α = observed angular rotation
$[\alpha]$ = specific rotation

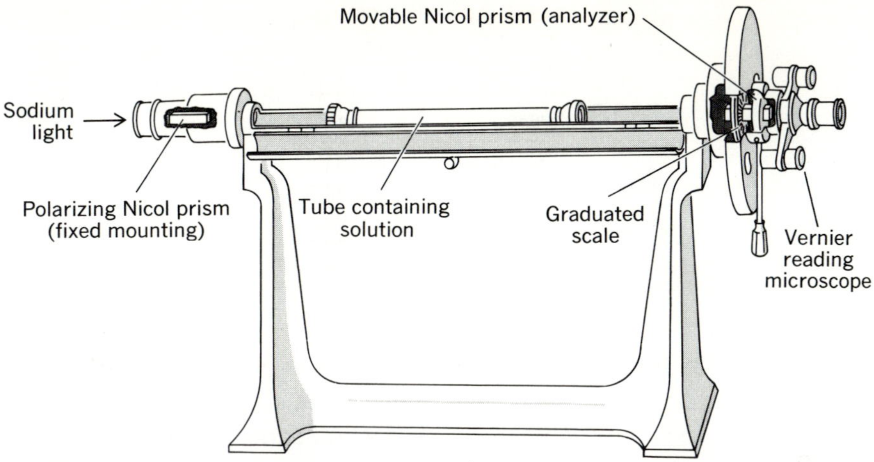

FIGURE 7.9 A simple polarimeter.

Specific rotation is defined as the angular rotation observed when plane-polarized light is passed through a solution one decimeter in length and at a concentration of one gram of solute per milliliter of solution. The D line of sodium is the wavelength usually used. The *polarimeter* (Fig. 7.9) is the instrument used to determine α.

The following examples illustrate the use of optical activity to identify a substance and to determine the concentration of an optically active solute.

EXAMPLE 1

You are given a 10% solution of an unknown sugar, a 2-dm sample tube, and a polarimeter. You are to determine which sugar is in your unknown. With the aid of the polarimeter, you make three observations of α in the 2-dm tube using the D line of sodium. Your observations are as follows: $+10.30$; $+10.20$; $+10.25$. The average is $\alpha = +10.25$. The concentration c of the solution is 0.1 g/ml. The specific rotation is then calculated using Eq. (7.1).

$$[\alpha] = \frac{\alpha}{c \times l} = \frac{+10.25}{0.1 \times 2} = +51.25$$

This specific rotation is now compared with that of known sugars (Table 7.2). The specific rotation of the unknown corresponds most closely with that of D-glucose. It can be

TABLE 7.2 SPECIFIC ROTATION OF SOME SUGARS

Sugar	[α]
L-Arabinose	+104.5
D-Galactose	+80.5
D-Fructose	−92
D-Glucose	+52.2
D-Mannose	+14.6

concluded that the unknown sugar was D-glucose and that the difference between your value and the correct one for $[\alpha]$ was due to experimental error. The significance of the D and L will be explained in a later section.

EXAMPLE 2

You are asked to determine the percent concentration of a solution of D-fructose. At your disposal is a polarimeter and a 2-dm tube. How would you approach this problem? First determine α; then substitute this value in Eq. (7.1) and solve the equation for c. The $[\alpha]$ for fructose is -92. Assume you obtained the following values for α:

$$\begin{aligned} \alpha &= -5.26 \\ & -5.21 \\ & -5.31 \\ \text{Av} &= -5.26 \end{aligned}$$

Rewrite Eq. (7.1) to isolate c, the unknown,

$$c = \frac{\alpha}{[\alpha] \times l} = \frac{-5.26}{-92 \times 2} = 0.0286 \text{ g/ml}$$

$$c \times 100 = \% = 2.86\%$$

Enantiomers (mirror images) differ with respect to their influence on polarized light, the $[\alpha]$ for each form being exactly the *same absolute numerical value but of opposite sign.* The terms dextrorotatory and levorotatory indicate compounds whose α values are (+) and (−), respectively. Usually just the sign (+) or (−) precedes the name of the compound to indicate the isomeric form. For example, lactic acid is a natural product occurring in two isomeric forms:

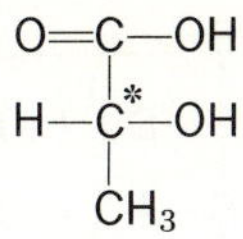

(−)-Lactic acid
$[\alpha] = -2.5$

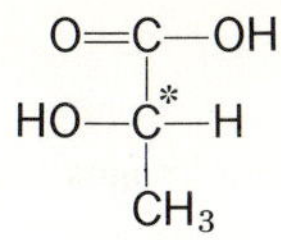

(+)-Lactic acid
$[\alpha] = +2.5$

Note that $[\alpha]$ for these compounds is numerically the same but of opposite sign.

Racemic mixture

A racemic mixture is one that contains equal amounts of enantiomers, the dextrorotatory and levorotatory forms. If a solution contains equal amounts of mirror images of a compound, α is zero because it is the summation of the optical activity of each asymmetric carbon in solution. In a racemic mixture, one enantiomer rotates polarized light to the right just as much as the other enantiomer rotates the light to the left. The net result is zero.

EXAMPLE 3

A solution contains a mixture of exactly 0.1 g/ml each of (+)-lactic acid and (−)-lactic acid. What is the α for this solution? Let $l = 1$. Calculate the α for each form of lactic acid in solution as follows:

(+)-lactic acid:	$[\alpha] = +2.5$	$\alpha = +2.5 \times 0.1 \times 1 = +0.25$
(−)-lactic acid:	$[\alpha] = -2.5$	$\alpha = -2.5 \times 0.1 \times 1 = -0.25$

The α for the mixture is equal to the sum of the respective α's for each component in the mixture. Thus, $\alpha = (+0.25) + (-0.25) = 0.0$. The solution will show no optical activity.

Meso compounds

If equal amounts of a pair of enantiomers are in solution, the optical activity of that solution is zero. It is possible to have a molecule containing asymmetric carbon atoms arranged in such a manner that the molecule is optically inactive. For example, if a molecule contains two asymmetric carbon atoms and one half of the molecule is the mirror image of the other half, the net optical activity is zero. The one half of the molecule would be dextrorotatory and the other half would be levorotatory to the same extent.

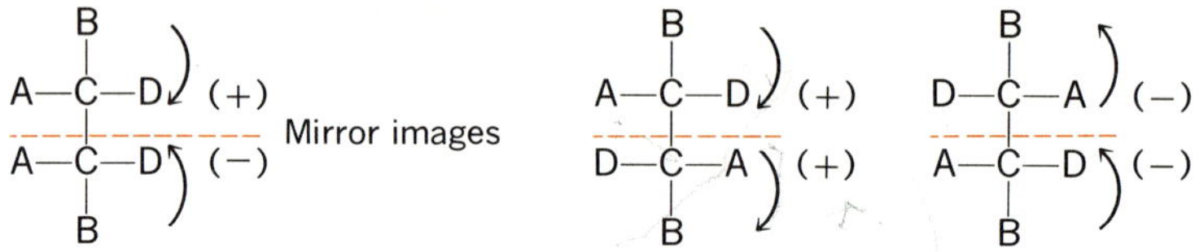

(I) Meso compound $[\alpha] = 0$ | (II) Dextrorotatory $[\alpha] = (+)$ | (III) Levorotatory $[\alpha] = (-)$

Mirror images

The optical activity of the respective asymmetric carbon atoms is indicated by an arrow. In formula I each half of the molecule is a mirror image of the other half. The molecule is internally compensated and the specific rotation is zero. In formula II the upper half is the same as in formula I, but the lower half is the mirror image of the corresponding part of formula I; therefore, the lower half of compound II is dextrorotatory and the upper and lower halves of molecule II rotate polarized light in the same direction. Formula III is the mirror image of formula II and therefore levorotatory. Compounds containing two asymmetric carbon atoms and capable of existing in the meso form have three optical isomers.

Tartaric acid is a classic example of a naturally occurring compound that can exist in the meso form. It occurs in grapes, mostly as the monopotassium salt, which precipitates out in the casks as grape juice is fermented. The pure salt is sold as cream of tartar and is used in making baking powder. The formulas for the various forms of tartaric acid are

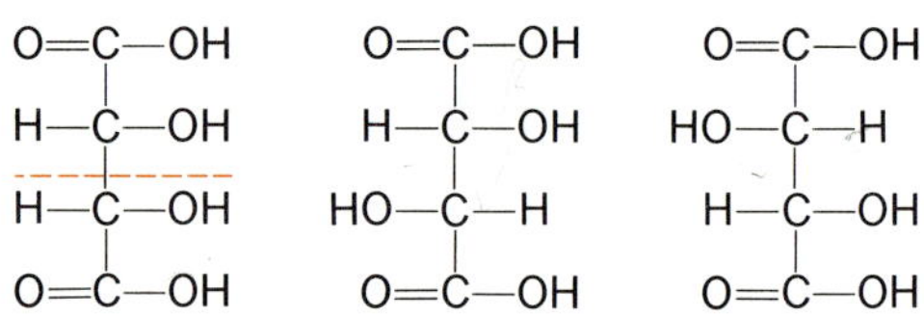

Meso $[\alpha] = 0$ | Dextrorotatory $[\alpha] = (+)$ | Levorotatory $[\alpha] = (-)$

Mirror images

Until 1848 only two forms of tartaric acid were recognized, natural (+)-tartaric acid and racemic acid, with no optical activity. Why the racemic acid was optically inactive was not understood. While examining some crystals of sodium ammonium tartrate under a microscope, Pasteur noted that there were two types, the facets of one type appearing to be the mirror image of the other. By hand Pasteur very carefully separated the two different crystals. A solution of one type of crystal rotated polarized light to the right, whereas a solution of the other rotated it to the left. The dextrorotatory crystals were identical with the natural product. Thus, the racemic acid was found to be a mixture of the two optically active forms of tartaric acid. However, it was not until 1874 that the theory of the asymmetric carbon atom was presented independently by van't Hoff and LeBel to explain optical activity.

7.6 BIOLOGICAL SIGNIFICANCE OF ASYMMETRY

From the viewpoint of cellular metabolism, the configuration on an asymmetric carbon atom is usually of great significance. It frequently is the factor deciding whether the cell can utilize a particular substance or not. In fact, the mirror images of some normal metabolic compounds are toxic. Although all organisms have many reactions in common, they frequently differ in their ability to make or to utilize a specific isomer. Compare wood (cellulose) with starch (amylose). The difference between them is the arrangement around one asymmetric carbon atom.

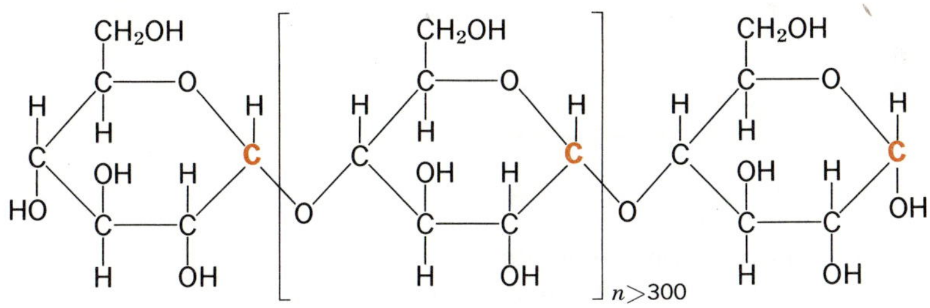

Amylose

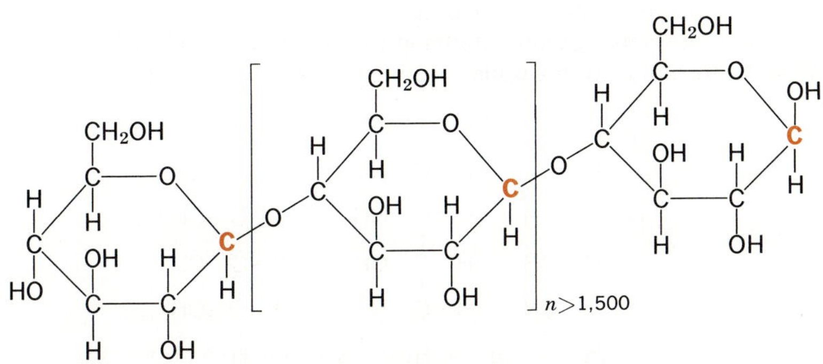

Cellulose

Both are made by plants; neither is made by animals. Furthermore, starch is a common main source of energy for man and of reserve energy in plants. Cellulose, on the other hand, is not utilized by man and to a very slight extent, if at all, by plants. Cellulose is the structural material of plants. It is used commercially in the building industry and in papermaking.

The full significance of the asymmetric carbon in biological reactions will become more apparent in subsequent chapters.

QUESTIONS

1. What is the structure of the carbon atom?
2. Are all four of the valence electrons of carbon equivalent?
3. What is meant by an eclipsed and by a staggered configuration?
4. How does a π bond differ from a σ bond?
5. What is the maximum number of bonds possible between two carbon atoms? Why?
6. Why is it necessary to write structural formulas for organic compounds rather than molecular formulas?
7. Circle and name each of the functional groups in the following compounds (refer to Table 7.1):

(*a*) 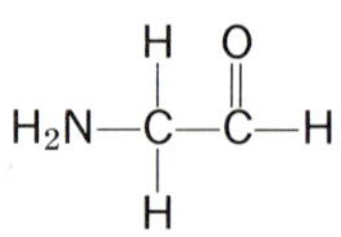

(*b*) 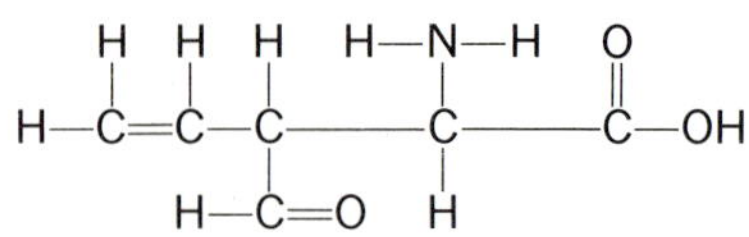

(*c*) 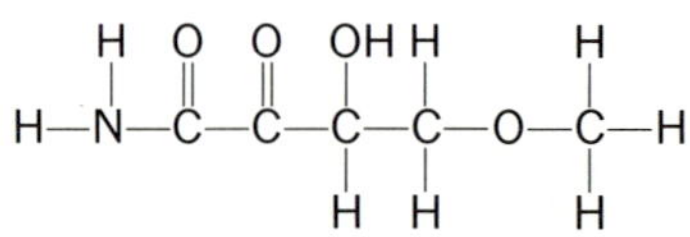

8. Define an asymmetric carbon atom.
9. What unique property does an asymmetric carbon have?
10. What are optical isomers?
11. Are optical isomers always mirror images of each other?
12. Star the asymmetric carbon atoms in the compounds in Prob. 7.
13. Write the formulas for the optical isomers of each of the following compounds.

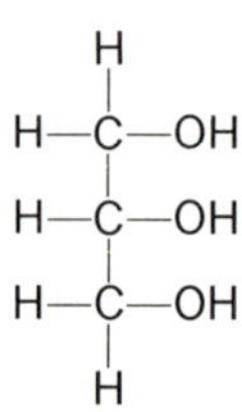

(I)

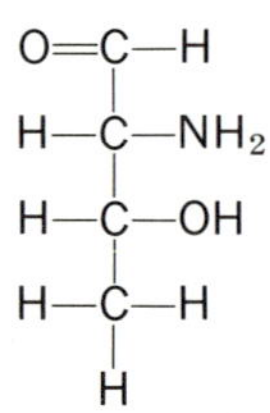

(II)

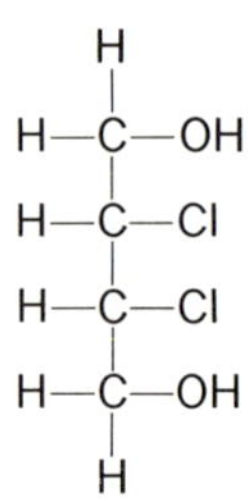

(III)

14. A 5% solution of a compound has an optical activity α of -12.8 when determined in a 2-dm tube. What is the specific rotation $[\alpha]$ of this compound?

15. (+)-A and (−)-A are mirror images. The specific rotation of (+)-A is $+62.3$. Given an 8% solution of (+)-A and a 4% solution of (−)-A; if equal volumes of these two solutions are mixed, what is the optical activity α of the resulting solution (assume no molecular interactions) measured in a 1-dm tube?

chapter 8.... hydrocarbons

Hydrocarbons are organic compounds containing only carbon and hydrogen. They are subdivided, depending upon their structure, into aliphatic or cyclic (or aromatic) and into saturated or unsaturated compounds (Table 7.1). Aliphatic compounds do not contain the benzene type of ring structure, whereas aromatic compounds do contain such a ring.

The various classes of organic compounds are made by substituting a functional group for one or more hydrogens in a hydrocarbon chain. Thus, hydrocarbons furnish the backbone structure for other classes of compounds. Furthermore, the name of the hydrocarbon serves as the basis for naming members of the various classes of organic compounds (see Sec. 8.3).

A discussion of the structure, nomenclature, and chemistry of the hydrocarbons involves the chemistry of the carbon-to-hydrogen bond and the carbon-to-carbon single, double, and triple bonds. These bonds, of course, occur in compounds other than hydrocarbons, particularly in biological compounds. Hydrocarbons as such are relatively unreactive biologically; however, the presence of certain functional groups may make the hydrocarbon part of the molecule quite active biologically. Some biological compounds are presented to illustrate reactions of the hydrocarbon portion of the molecule. This may seem somewhat of a departure from the classical presentation of hydrocarbon chemistry because the functional groups present would necessitate classifying the compounds into groups other than hydrocarbons. These biological examples will be discussed later from several different viewpoints and will therefore be familiar when they are encountered again.

It is important to remember that when there is more than one functional group in a compound, the chemical properties of the compound are determined by each functional group present.

8.1 ALKANES

Saturated hydrocarbons (alkanes) have only single bonds between carbons. Alkanes have the general formula C_nH_{2n+2}, where n is the number of carbon atoms. Because the carbon-to-carbon single bond is quite stable, members of this class are not particularly reactive chemically. For this reason they are frequently called *paraffin hydrocarbons,* a name derived from the Latin *parum affinis,* meaning "little affinity."

The structure of the carbon atom was discussed in Chap. 7. To review, carbon has its nucleus at the center of a tetrahedron with each of the four valence electrons at an apex. The valence electrons are all equivalent.

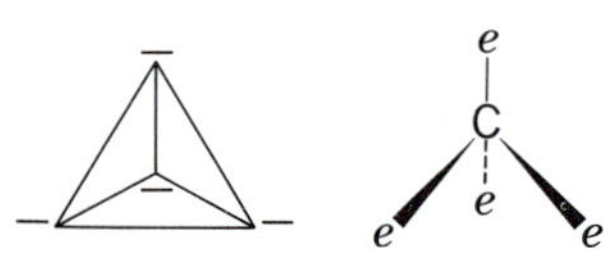

It must be remembered that the carbon atom has three dimensions and therefore cannot be adequately represented by writing a formula in a plane. In the formula on the right, the wedge-shaped bonds are used to indicate that the bond projects out of the plane of the paper toward the reader, and the dashed bond recedes behind the plane of the paper. When several tetrahedrons are drawn connected to each other by one corner, the carbons are not in a straight line as C—C—C—C but in a zigzag arrangement as

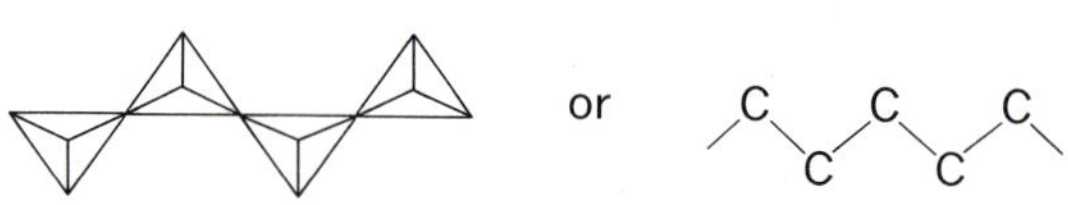

Several types of models are used to represent the carbon atom so that three-dimensional formulas can be visualized more easily. One type of model consists of different colored wooden balls representing different elements: the carbon atom is black and has four holes equidistant from each other and the hydrogen atom is shaded and has only one hole. Pegs representing bonds are used to join the different colored balls to form compounds (Fig. 8.1a). The simplest hydrocarbon

(*a*) (*b*)

FIGURE 8.1 Three-dimensional models of the carbon configuration: (*a*) ball-and-stick model; (*b*) more accurate atomic-radii model.

compound has four hydrogen atoms connected to a carbon, CH_4. A more accurate model is made to scale to indicate the respective atomic radii (Fig. 8.1*b*).

A compound containing four or more carbons can exist in more than one arrangement of carbon atoms. Consider for the present only the carbon atom with four single bonds attached to it.

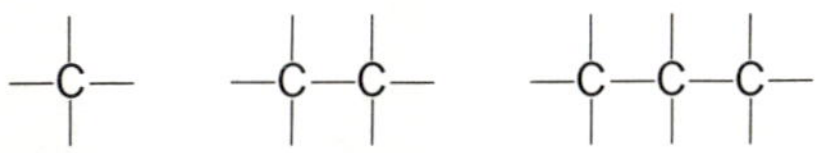

Only one arrangement possible

Look at the four-carbon chain. It is possible to have two arrangements of the carbons:

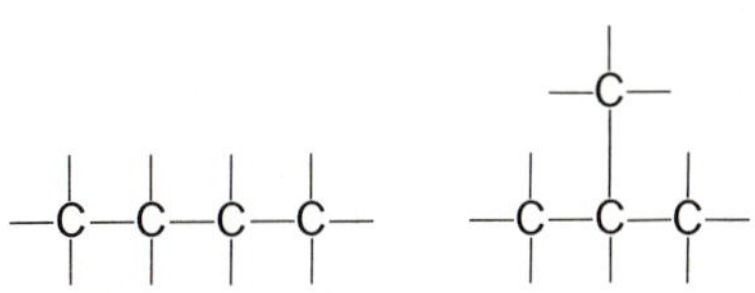

Straight chain Branched chain

If the valence bonds are satisfied by hydrogens, the two compounds have the same molecular formula, C_4H_{10}, but they have *different* physical and chemical properties.

Structural isomers are compounds with the same molecular formula but different spatial arrangements of the atoms. As the molecular weights of compounds increase, the number of possible isomers also increases. For example, there are 75 isomers of $C_{10}H_{22}$. As we have noted before, the possibility of isomers necessitates the use of structural formulas for organic compounds.

Before continuing with this subject it might be well to recall a very important property of the carbon-to-carbon single bond, namely,

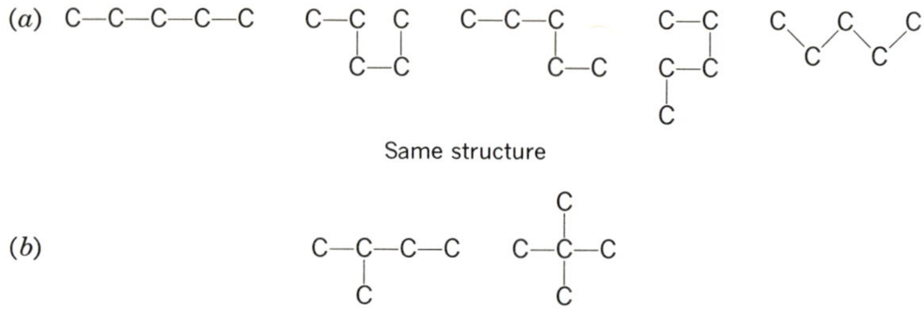

FIGURE 8.2 Straight and branched carbon chains. For simplicity only the carbons are shown; it is to be understood that each carbon has four single bonds.

the free rotation around a single bond in open-chain compounds. Because this fact is often forgotten, difficulties sometimes arise in naming or identifying isomers. For example, if free rotation were not possible, each of the following formulas would represent a different compound:

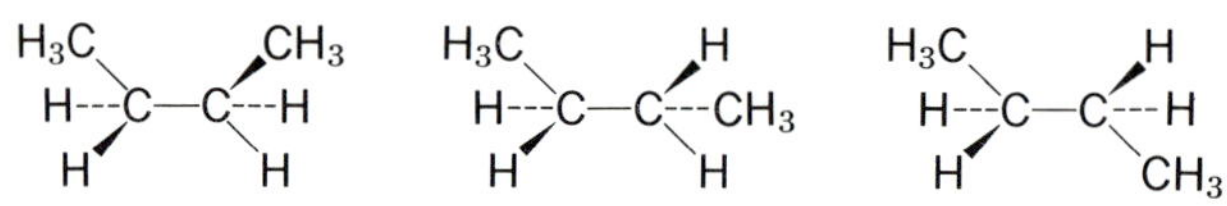

All the same compound

However, *only one* such compound exists; therefore, there must be free rotation about the carbon-to-carbon single bond.

The easiest way to determine branching is to start at one end of the carbon chain and draw a line through as many carbon atoms as possible without lifting the pencil or retracing the path. When this is done and the line includes all the carbons, there is no branching in the chain regardless of how the carbons are written. For example, all the formulas in Fig. 8.2*a* are identical, but formulas in Fig. 8.2*b* are different from each other and each is different from formula (*a*).

8.2 CLASSIFICATION OF CARBON AND HYDROGEN ATOMS

The fact that more than one carbon can be bonded to another carbon atom is the basis for the classification of the carbon and hydrogen atoms in organic compounds. This is illustrated in the formulas for pentane (a C_5 hydrocarbon) in Fig. 8.3. Examination of the formulas for pentane will reveal that:

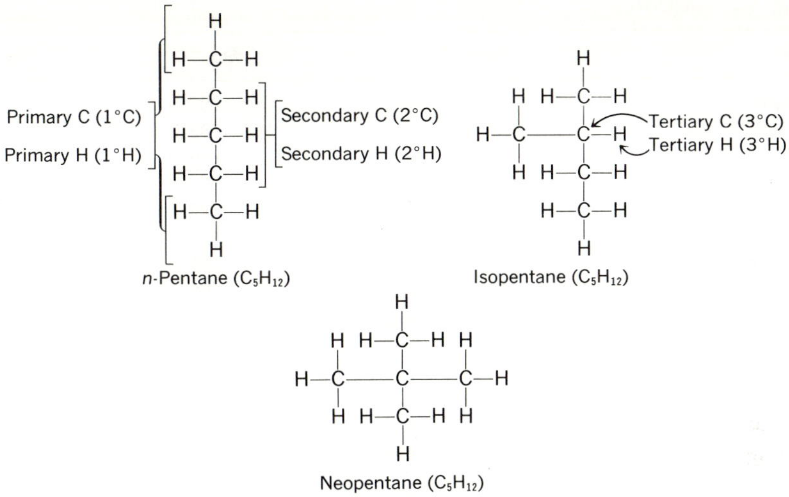

FIGURE 8.3 Classification of carbon and hydrogen atoms.

A *primary carbon* (1° C) is bonded to not more than one other carbon atom.
A *primary hydrogen* (1° H) is bonded to a primary carbon atom.
A *secondary carbon* (2° C) is bonded to two other carbon atoms.
A *secondary hydrogen* (2° H) is bonded to a secondary carbon atom.
A *tertiary carbon* (3° C) is bonded to three other carbon atoms.
A *tertiary hydrogen* (3° H) is bonded to a tertiary carbon atom.

The class of carbon and hydrogen has a direct bearing on the reactivity of the hydrocarbon.

Compounds whose formulas differ only by one or more CH_2 groups are called *homologs*. The formula and name for the homologous series of some alkanes are given in Table 8.1. The names of the compounds were developed by a committee on nomenclature of organic compounds (The International Union of Pure and Applied Chemists, IUPAC). The first four members are gases at room temperature, and those between C_5 and C_{17} are liquid. The higher-molecular-weight compounds are solids.

8.3 NOMENCLATURE

Referring to the name of hydrocarbons in Table 8.1, it will be noted that the first four alkanes are called by a common name that has no reference to the molecular structure. However, beginning with the C_5

TABLE 8.1 ALKANES

Name	Formula	Mp, °C	Bp, °C
Methane	CH_4	−182.5	−161.5
Ethane	$H_3C—CH_3$	−183.2	−88.6
Propane	$H_3C—CH_2—CH_3$	−189.9	−44.5
n-Butane	$H_3C—(CH_2)_2—CH_3$	−138.3	−0.5
n-Pentane	$H_3C—(CH_2)_3—CH_3$	−129.7	36
n-Hexane	$H_3C—(CH_2)_4—CH_3$	−95	68
n-Heptane	$H_3C—(CH_2)_5—CH_3$	−91	98.4
n-Octane	$H_3C—(CH_2)_6—CH_3$	−56.5	126
n-Nonane	$H_3C—(CH_2)_7—CH_3$	−51	150.8
n-Decane	$H_3C—(CH_2)_8—CH_3$	−29.7	174.1
n-Undecane	$H_3C—(CH_2)_9—CH_3$	−25	195
n-Dodecane	$H_3C—(CH_2)_{10}—CH_3$	−9.6	216.3
n-Tridecane	$H_3C—(CH_2)_{11}—CH_3$	−5.5	243
n-Tetradecane	$H_3C—(CH_2)_{12}—CH_3$	6	253.5
n-Eicosane	$H_3C—(CH_2)_{18}—CH_3$	36.8	343

alkane, pentane, the name is indicative of the number of carbons in the compound. The Greek prefix *pent-, hex-, hept-, oct-, non-,* etc., is added to the *-ane* of alkane to form the name of the hydrocarbon. For example, heptane is a saturated hydrocarbon containing seven carbons, and undecane is one containing eleven carbons. These two compounds have the molecular formulas C_7H_{16} and $C_{11}H_{24}$, respectively. The fact that there is no branching in the molecule is indicated by the term *normal,* abbreviated by *n* placed before the name of the alkane.

Simple branched-chain compounds are named according to their common name or, more informatively, according to the IUPAC nomenclature. The prefix *iso-* indicates a CH_3 group on the carbon next to the end.

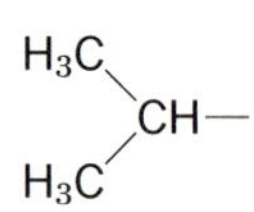

Iso group

When branched-chain compounds are named according to the IUPAC system, the longest straight chain of carbons is considered as the parent alkane, and the atoms or radicals substituted in it are named. The position of the substituents is indicated by giving each a number corresponding to the carbon to which it is attached. This raises the question of naming radicals and also from which end of the parent alkane one should start numbering the carbons.

Alkyl radicals

Alkyl radicals result when one hydrogen atom of an alkane is removed:

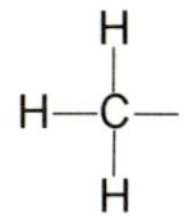

Radical, one carbon bond is free

The name of the radical is indicated by substituting *-yl* for *-ane* in the name of the alkane; hence, the general name *alkyl* radical. Some examples are given in Table 8.2.

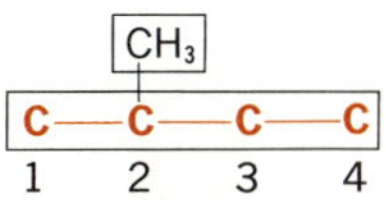

Isopentane
2-Methylbutane

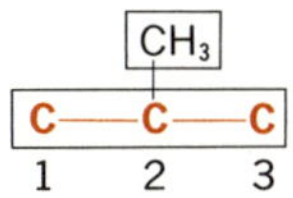

Isobutane
2-Methylpropane

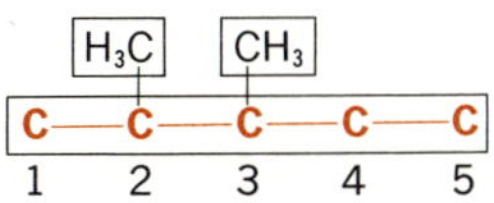

2,3-Dimethylpentane

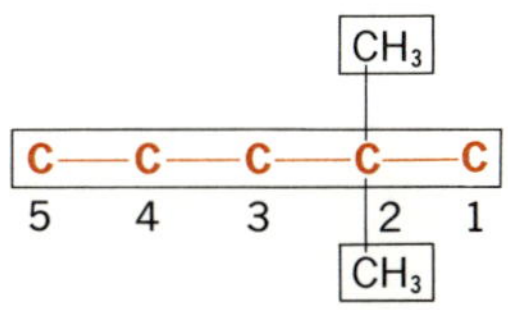

2,2-Dimethylpentane

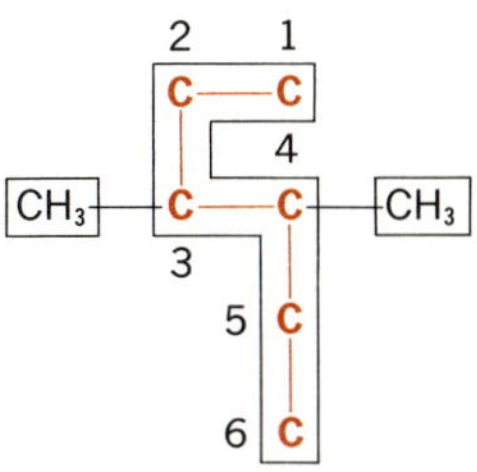

3,4-Dimethylhexane

What is the longest chain of carbons? It is six; therefore, the parent alkane is hexane.

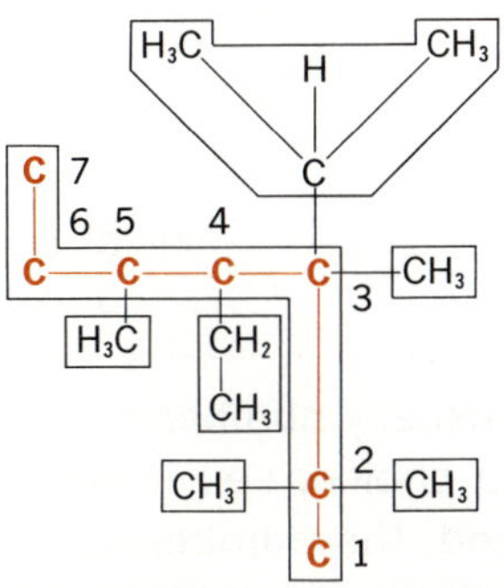

2,2,3,5-Tetramethyl-4-ethyl-3-isopropylheptane

FIGURE 8.4 Nomenclature of some hydrocarbons. The hydrogen atoms of the parent alkane are omitted for simplicity.

TABLE 8.2 SOME RADICALS

Name	Formula
Methyl	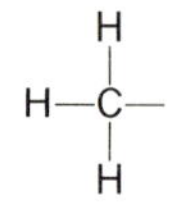
Ethyl	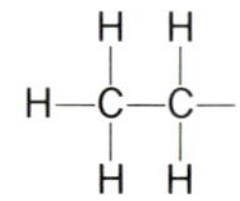
Normal propyl (*n*-propyl)	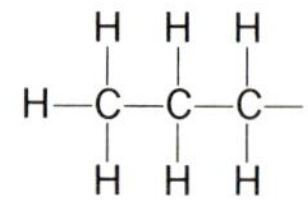
Isopropyl	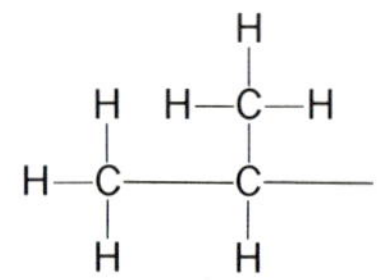
Isobutyl	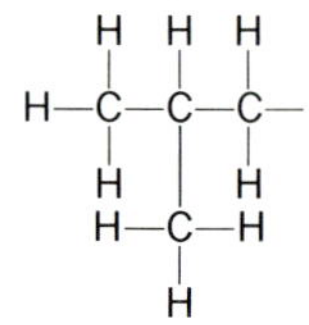
Tertiary butyl (*t*-Butyl)	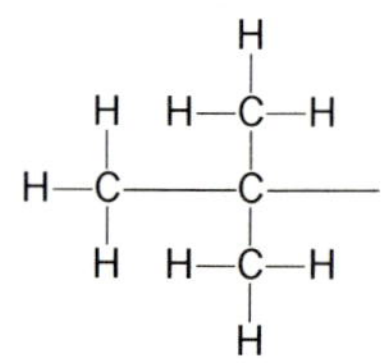
Radicals containing oxygen in addition to carbon and hydrogen:	
Hydroxymethyl	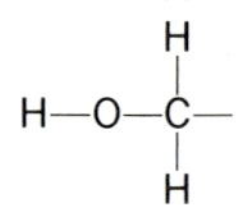
Methoxy	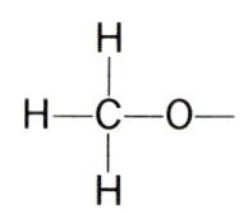
Ethoxy	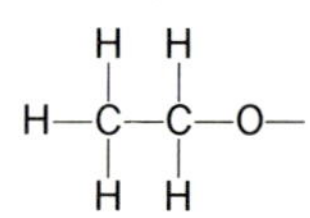

The carbons of the parent alkane are usually numbered from the end closest to the branching. If there is a choice, i.e., if branching occurs at the same distance from either end, the numbering is from the end that gives the lowest numbers for all the substituting groups.

Study the examples in Fig. 8.4 carefully. Note that each substituent group must be numbered according to the carbon to which it is attached. If more than one of the same group appears, the total is indicated by *di-*, *tri-*, *tetra-*, etc.

8.4 OCCURRENCE

The simplest hydrocarbon, methane, is the main constituent of natural gas, which is found above oil in oil wells. It is the sewer gas produced by bacteria in raw sewage and is also known as marsh gas and to miners as firedamp. Methane is used extensively as fuel for heating purposes, and the demand is so great that specially designed ocean tankers are being built to ship liquid methane (bp $= -162°C$).

Methane is produced in considerable volume by bacterial degradation of food in the rumen of animals. It cannot be utilized by the animal and is expelled into the atmosphere. If anything interferes with this process, the animal becomes bloated and may die if the pressure is not released. Hydrocarbons other than methane are not produced appreciably by living organisms.

The hydrocarbons propane and butane can be liquefied quite easily at moderate pressures. Portable tanks of these liquefied gases are used in rural areas for cooking and heating purposes.

Liquid hydrocarbons from pentane to decane are components of gasoline, a mixture. Mineral oil for lubricating purposes consists of hydrocarbons of higher molecular weight and boiling point. Paraffin wax, C_{20} to C_{34} hydrocarbons, is a mixture of solid alkanes.

8.5 REACTIONS OF HYDROCARBONS

The C—C and C—H σ bonds are quite stable, but the C—H bond can be broken more easily than the C—C bond. This accounts for substitution reactions, in which hydrogen can be replaced by other atoms, e.g., halogens. Although in general the alkanes are chemically unreactive compared with most functional groups, there are several reactions that involve the alkane part of the molecule.

Complete oxidation

The complete oxidation, or burning, of hydrocarbons is accompanied by the production of considerable heat. The equation for the oxidation of methane is

$$CH_4 + 2\,O_2 \longrightarrow CO_2 + 2\,H_2O + \text{energy (heat and light)}$$

Alkanes are not generally utilized by living organisms as a source of energy, but the oxidation of the alkane part of the molecule furnishes a substantial amount of cellular energy. For example, stearic acid, which occurs in fats, is oxidized as

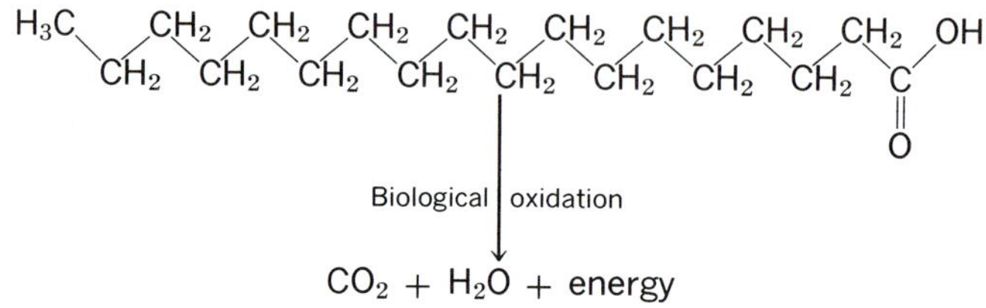

It will be noted that there are 18 carbons in stearic acid and that 17 of them represent the alkane part of the molecule.

If the carboxyl group, —COOH, were not substituted for a hydrogen in the C_{17} hydrocarbon chain, we would not be able to oxidize the hydrocarbon chain. There are differences between the mechanisms of biological and chemical oxidation. *Biological oxidation is not combustion.* Combustion would liberate too much energy at one time for the cells to dissipate. *Biological oxidation is a stepwise reaction which releases energy in manageable amounts.* In addition, *in most common biological oxidations of organic compounds the source of oxygen is water,* not molecular oxygen. Some of these reactions will be discussed in Sec. 9.5.

Substitution

Since alkanes contain only one σ bond between carbons and between the carbon and hydrogen, a chemical reaction must take place by removal of hydrogen and then substitution of another group or atom on the carbon in place of the hydrogen. Such reactions are called *substitution reactions.* The only reaction to be discussed here is with the halogens (X_2) to yield the corresponding R—X derivative and HX. This reaction is called *halogenation.* Note that the letter R is frequently used to denote a radical that does not take part in the reaction.

$$R{-}H + X_2 \longrightarrow R{-}X + HX$$

To make this reaction go at an appreciable rate the reactants must be exposed to light (ultraviolet rays) as a catalyst and heat. The mechanism of this reaction involves free radicals, i.e., a species without its full complement of eight electrons in the outer orbit. For example,

Methyl radical Chlorine radical

The reaction is general for hydrocarbons, the chlorination of methane being a specific example (see Fig. 8.5).

In this substitution reaction free radicals are formed as an intermediate product. These are exceptionally reactive radicals, and as

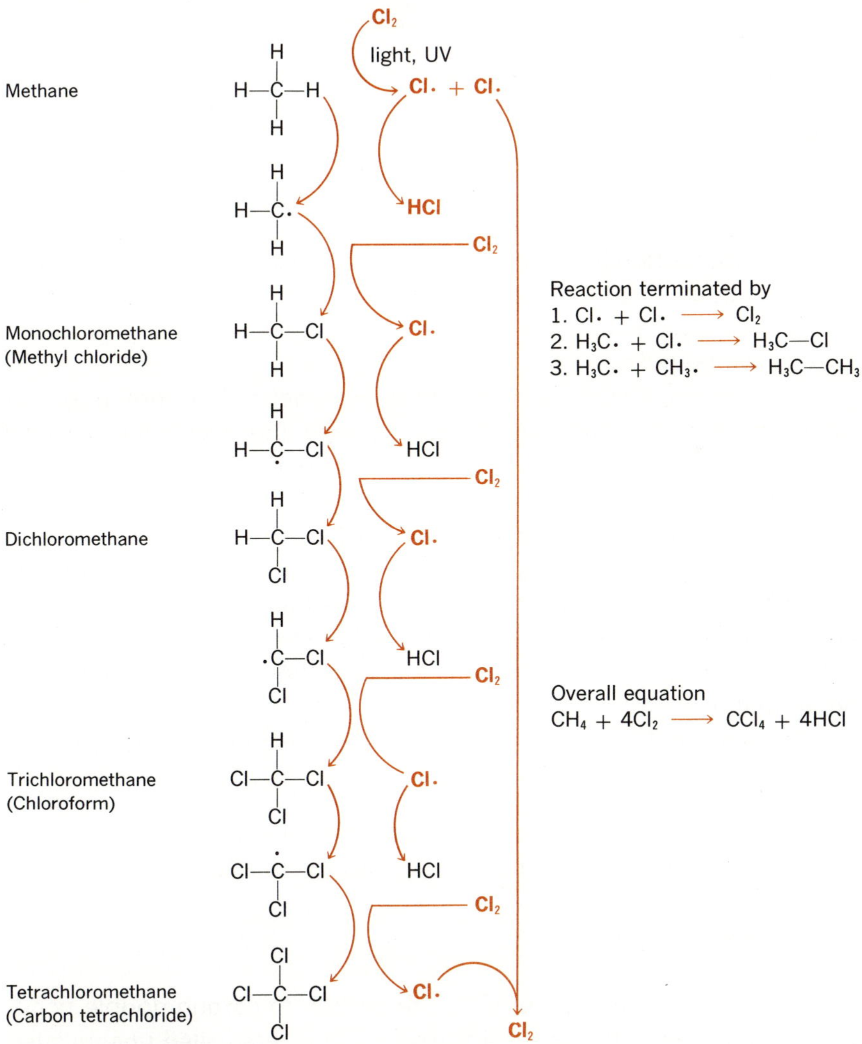

FIGURE 8.5 Free-radical chlorination of methane.

soon as they are formed they immediately react as indicated. Once a chlorine free radical (Cl·) is formed, each successive reaction produces another free radical and thus the reaction is propagated as a *chain reaction*. This chain reaction is stopped by combining any two free radicals, as shown in Fig. 8.5.

The ease with which the different classes of hydrogen atoms react to form free radicals varies:

$$3^\circ\ H > 2^\circ\ H > 1^\circ\ H$$

Most reactive   Least reactive

The easier it is to form a free radical, the easier it will be to carry out a substitution reaction. With the more complex hydrocarbons, many different products are possible.

Alkyl halides

Halogen substitution in alkanes results in a class of compounds known as alkyl halides. These compounds play an important role in organic chemistry because they are used in the synthesis of many different types of compounds.

Chloroform, $CHCl_3$, was the first anesthetic. It is not used for that purpose today because it may damage the liver when inhaled (even in small amounts) over an extended period of time.

Dibromomethane, CH_2Br_2, is used as a fumigant, e.g., in grain bins. The dichloro and the dibromo derivatives of ethane, 1,2-dichloroethane, or ethylene chloride, $CH_2Cl—CH_2Cl$, and 1,2-dibromoethane, or ethylene bromide, $CH_2Br—CH_2Br$, as well as other alkyl chlorides and bromides are used as soil fumigants. They must be used with caution under controlled conditions because they are very toxic to man and animal.

The alkyl halides are hydrolyzed in an aqueous alkaline solution to yield the corresponding alcohol.

$$\underset{\text{Alkyl chloride}}{R—CH_2Cl} + \underset{\text{Base}}{NaOH} \longrightarrow \underset{\text{Alcohol}}{R—CH_2OH} + \underset{\text{Salt}}{NaCl}$$

8.6 ALKENES

The functional group of alkenes is the carbon-to-carbon double bond, C=C. Alkenes, with the general formula C_nH_{2n}, are called unsaturated because they contain less hydrogen than the corresponding alkane, C_nH_{2n+2}. Alkenes are also called *olefins*.

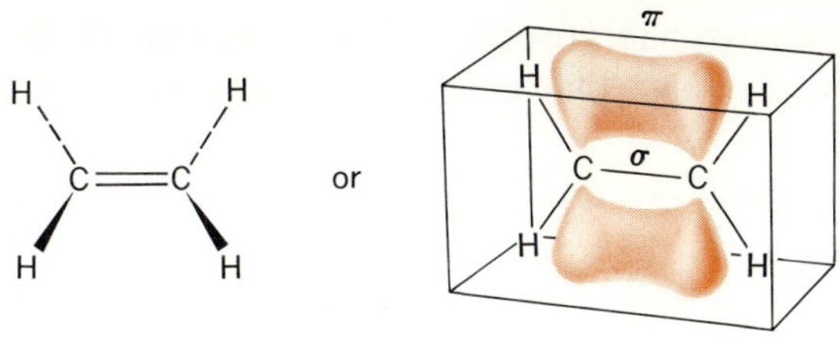

FIGURE 8.6 The configuration of the carbon-carbon double bond, as in ethylene.

The C═C group is responsible for the chemical properties of the alkenes. The π bond is much more reactive than the σ bond; it is very easily broken. Therefore, this class of compounds is very reactive chemically compared with the alkanes.

Many biological compounds contain one or more double-bonded carbons, but relatively few are hydrocarbons; most contain functional groups other than the C═C, such as hydroxyl or carboxyl. They are found in such diversified classifications as fats, hormones, antibiotics, and vitamins. Natural rubber is a good example of an unsaturated hydrocarbon produced by living cells.

The alkene group is characterized by one σ and one π bond between adjacent carbons, as in Fig. 8.6. Since the presence of the second bond between the carbons prevents free rotation about the bond, the groups on the two carbons are held in a fixed position and all lie in the same plane. In ethylene the four hydrogens and the two carbons are in one plane. This situation produces a new type of isomer.

Geometric isomers

The rigid structure associated with the double bond permits a spacial relationship among the groups attached to the two carbons. If two different groups are present, for example, H and CH_3, three possible structures, or *stereo isomers,* can be written

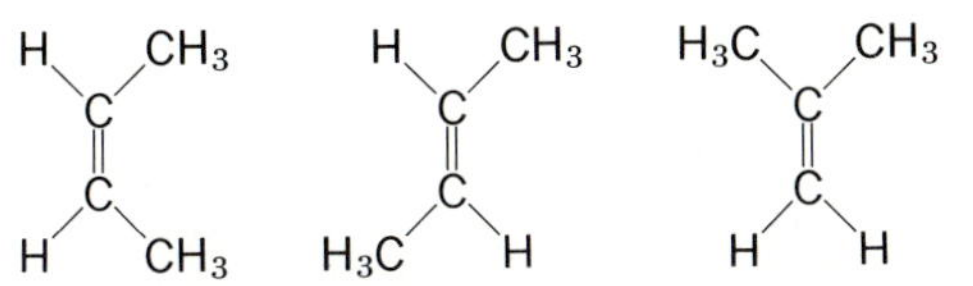

	(I) *cis*-2-Butene	(II) *trans*-2-Butene	(III) 2-Methylpropene
	Geometric isomers		No geometric isomers
Bp:	4°C	1°C	−7°C
Mp:	−139°C	−106°C	−140°C

Cis compounds have like groups on the same side of the double bond. Trans compounds have like groups on opposite sides of the double bond.

Geometric isomers are not limited to the alkenes but also occur in cyclic compounds. Examples are shown in the following formulas (only the carbons and the two groups associated with the cis and trans configuration are shown):

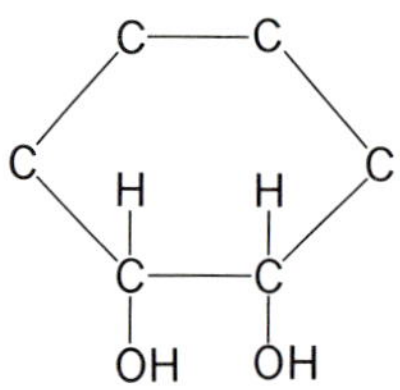

cis OH

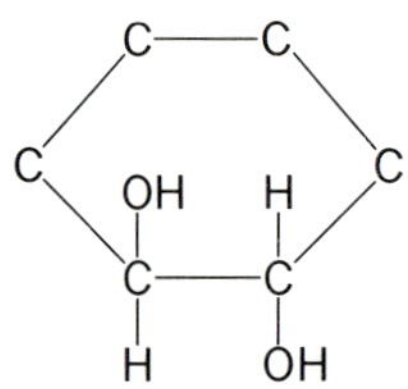

trans OH

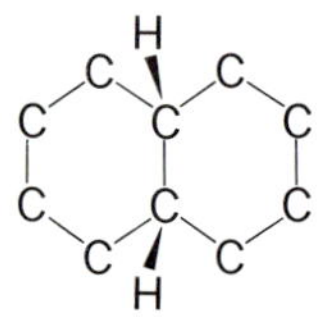

cis H

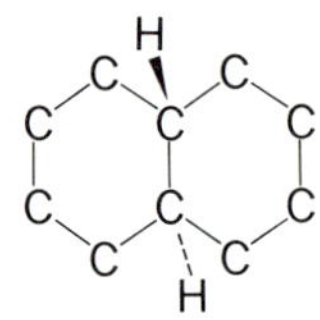

trans H

The geometrical structure may have a marked effect on the physical properties of the compound. For example, *cis*-2-butene has a higher boiling point and a lower melting point than *trans*-2-butene. The solubilities of the isomers also differ.

8.7 NOMENCLATURE

There are several common methods of naming alkenes. One is to consider them as derivatives of the simplest possible alkene, ethylene(I). Compound II is a *cis*-dimethyl derivative of ethylene.

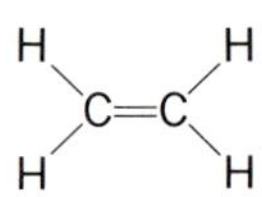

(I)
Ethylene
or
Ethene

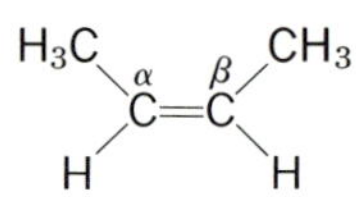

(II)
cis-Dimethylethylene
or
cis-α,β-Dimethylethylene
or
cis-2-Butene

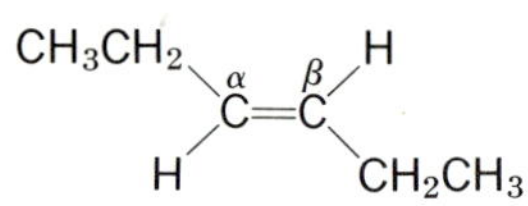

(III)
trans-Diethylethylene
or
trans-α,β-Diethylethylene
or
trans-3-Hexene

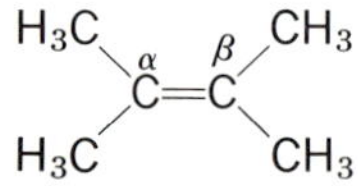

(IV)
Tetramethylethylene
or
2,3-Dimethyl-2-butene

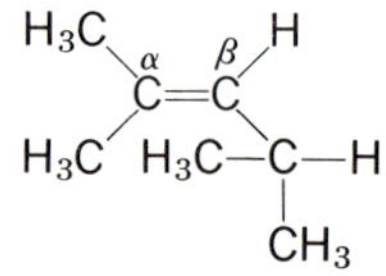

(V)
α,α-Dimethylisopropylethylene
or
2,4-Dimethyl-2-pentene

The cis indicates that the two methyl groups are on the same side of the double bond; therefore, one of these groups must be on each carbon of the ethylene unit. This can be specifically indicated by labeling the carbons associated with the double bond α and β. α,β-Dimethylethylene is another name for compound II. Formula III is a trans compound. The ethyl groups are on the opposite side of the double bond. In IV there can be only one structure for tetramethylethylene. There can be more than one geometrical isomer of dimethylisopropylethylene. One is shown in formula V. In this case it is necessary to indicate that the two CH_3 groups are on the same carbon, which is done by labeling the carbons associated with the double bond as α and β.

Because common names are sometimes difficult to remember, the more scientific method of nomenclature is preferred. The IUPAC method of nomenclature is based on the following rules.

1. Select as the parent alkane the one with the longest carbon chain containing the C=C.
2. Substitute the suffix *-ene* for *-ane* of the alkane.
3. Name the substituted groups and number them according to the carbon on which they occur. The carbon chain is numbered starting from the end closest to the C=C.

8.8 PREPARATION

Cracking

Olefins are formed chemically by several methods. One is by heating alkanes, usually a mixture such as petroleum, under conditions of high temperature and pressure in the absence of oxygen. This produces a number of different alkenes which are separated by distillation. This process is called *cracking*.

Elimination of HX

Alkyl halides can be treated with alcoholic KOH and converted into the corresponding olefin. The H and Cl trans to each other are eliminated.

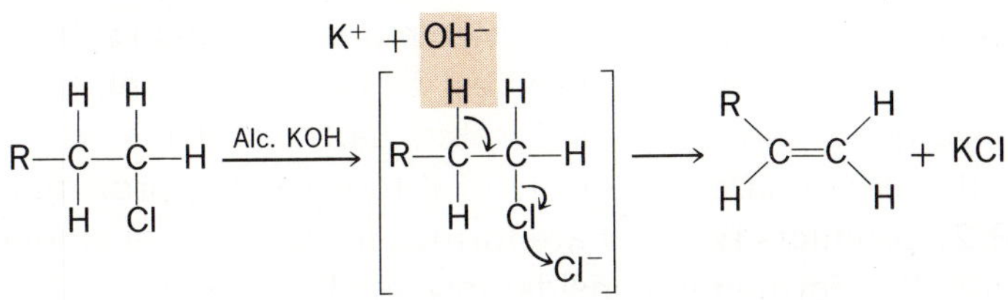

If the halide is on a secondary or tertiary carbon, more than one product is possible:

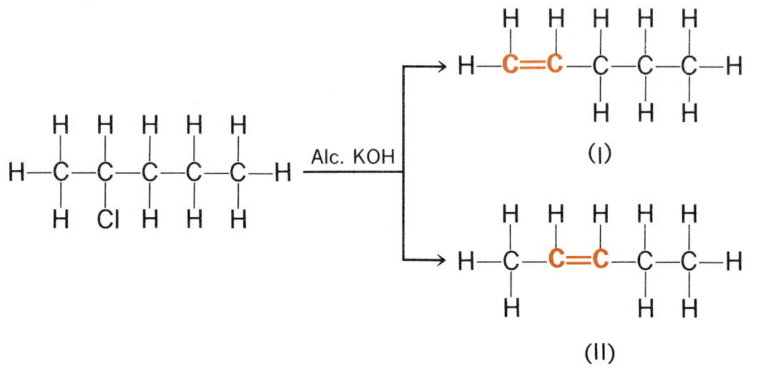

(8.1)

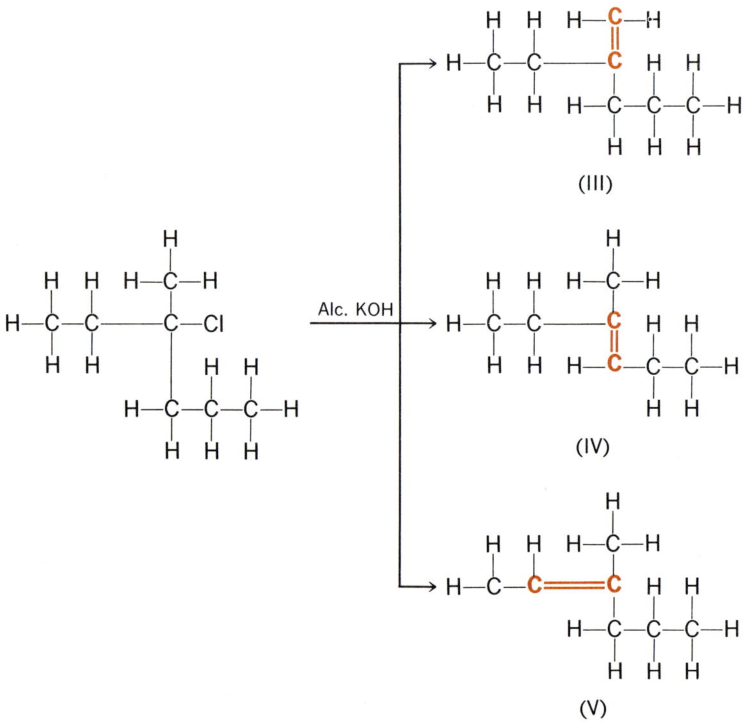

(8.2)

Which product is favored when more than one is possible? Remember that a secondary hydrogen is eliminated more easily than a primary one. Thus the olefin formed in the greatest amount is the one that has the fewest hydrogens on the unsaturated carbons. Thus in Eq. (8.1), product II predominates, the ratio of II to I being approximately 4:1. In Eq. (8.2), products IV and V are formed in about equal amounts, and compound III is formed in considerably smaller amounts.

Dehydration of alcohols

The elimination of 1 mole of water from 1 mole of alcohol to form an alkene will be discussed under reactions of alcohols. The overall reaction is

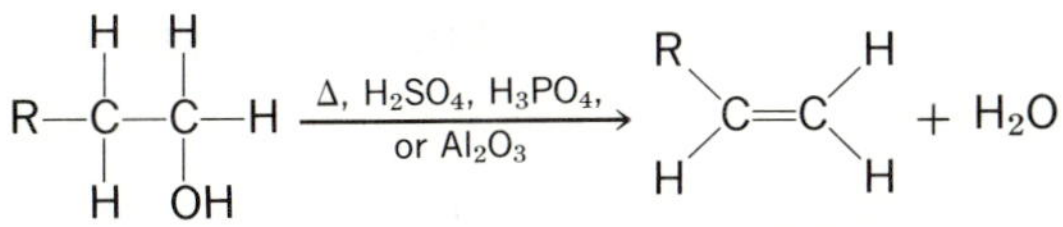

The structure of alcohols affects the rate of dehydration; they react in the decreasing order $R_3C—OH > R_2CH—OH > RCH_2—OH$.

8.9 REACTIONS OF ALKENES

Reactions of alkenes involve breaking the π bond and adding atoms or radicals to each carbon:

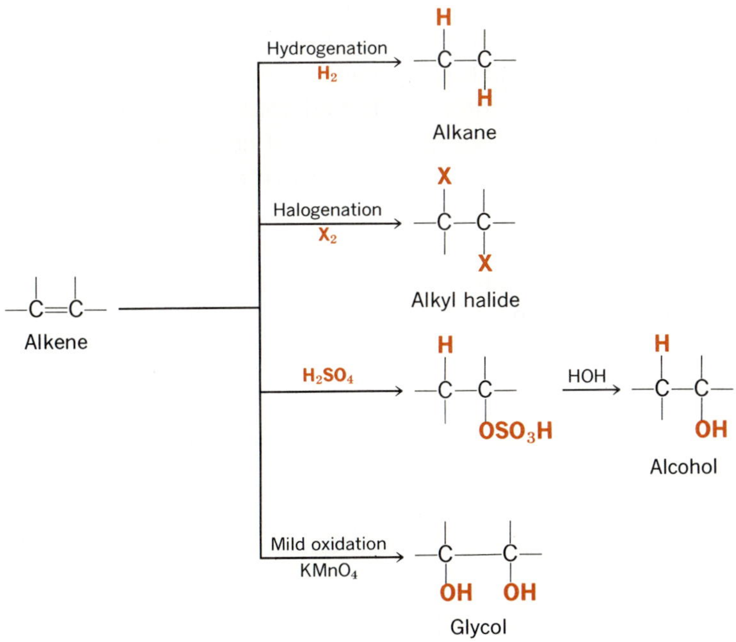

When we discussed formation of alkenes by the elimination of HX from alkyl halides, it was indicated that trans H and X are eliminated. Similarly here it is trans addition to the double bond. The actual mechanism of addition to a carbon double bond involves opening the π bond and adding the reactant in a stepwise manner with a carbonium ion as an intermediate

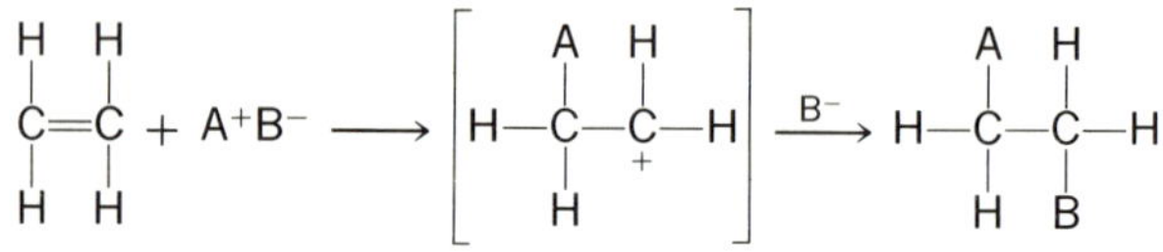

There is less interference with the addition of the second group if it adds in the trans position:

Cis product   Trans product

Hydrogenation

Chemical hydrogenation of an olefin group requires a catalyst, e.g., platinum or nickel, and elevated temperature and pressure. With the exception of ethylene, saturation or hydrogenation of a double bond increases the melting and boiling points of the hydrocarbon. This is also true of other classes of compounds that contain alkene groups.

Hydrogenation of alkene groups is of commercial importance in the preparation of lard and butter substitutes. Vegetable oils contain C=C groups in the hydrocarbon chain of the acid moiety of the oil. Three such acids, oleic, linoleic, and linolenic, contain one, two, and three double bonds, respectively, and all are C_{18} acids. These acids can be hydrogenated to the corresponding saturated C_{18} acid

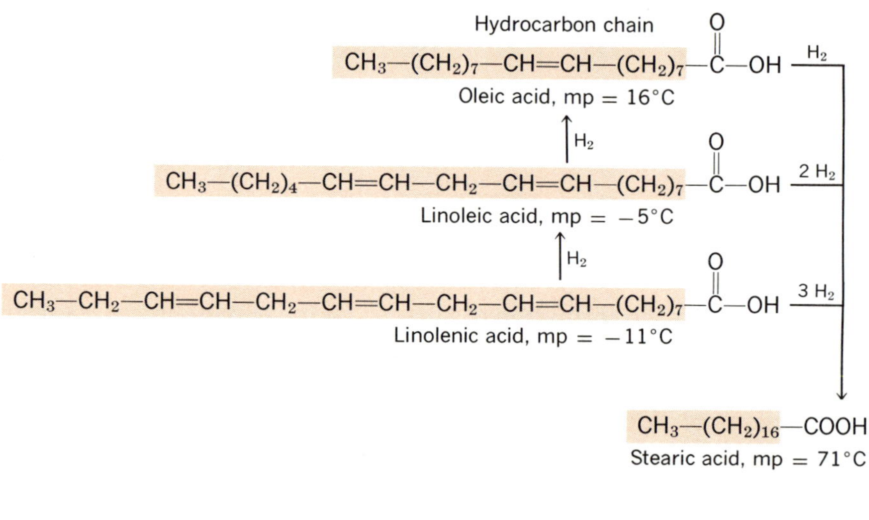

Note the increase in melting point as the compounds become more saturated. The addition of 1 mole of hydrogen to oleic acid (mp = 16°C) forms stearic acid (mp = 71°C). Stearic acid is also produced by the addition of 2 moles of hydrogen to linoleic acid (mp = −5°C) and 3 moles of hydrogen to linolenic acid (mp = −11°C). Stepwise hydrogenation of linolenic acid converts it successively into linoleic, oleic, and stearic acid. By controlling the extent of hydrogenation of the double bonds in the unsaturated-acid part of vegetable oils, a product with the desired melting point can be obtained.

Hydrogenation of alkene groups occurs in a number of biological reactions, but there are several differences to be noted between the biological and the chemical processes. Chemical hydrogenation requires an inorganic catalyst and high temperature and pressure. The biological process requires an organic catalyst (enzyme) and occurs at body temperature. Furthermore, molecular hydrogen is used in the chemical process but not in the biochemical one. As illustrated in the following reaction, the hydrogen atoms associated with the enzyme system are not in the molecular state:

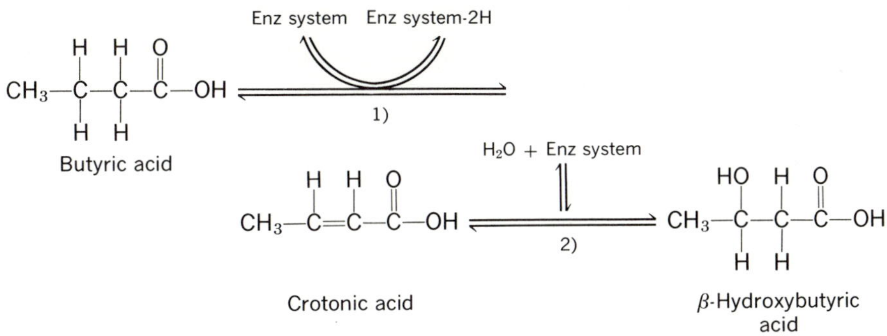

Two hydrogen atoms are removed from the alkane part of butyric acid to form an alkene group between the second and third carbons. This reaction is reversible and therefore also illustrates the biological hydrogenation of an alkene group. At this time we shall not pursue the fate of the two hydrogen atoms, merely noting that they become associated with an enzyme.

Hydration

Included in the previous series of equations is the hydration of the alkene bond to form the corresponding hydroxy acid, reaction (2). The reverse reaction, dehydration of the hydroxy acid to form an alkene group, is also of biological importance. At this point particular notice should be made of the net effect of combining reactions (1) and (2), namely, the conversion of the alkane part of butyric acid to a hydroxy derivative. This is an oxidation reaction. An oxygen atom has been inserted between the hydrogen and the third carbon of butyric acid. Where did this oxygen come from? It came from water, which is a very important compound biologically, not only because of its solvent characteristics but also because it plays a key role in many oxidation reactions. Note that most biological compounds contain *more than one* functional group; therefore, reactions of these compounds include reactions of *each functional group* present. This is easily forgotten under the stress of learning specific reactions of a specific group.

Chemical hydration of olefins occurs in two steps. First, the addition of H_2SO_4 to form the alkyl hydrogen sulfate, and second, the hydrolysis of this alkyl sulfate to the corresponding alcohol. The reaction illustrated with ethylene as the alkene is a convenient way of making many of the simple alcohols.

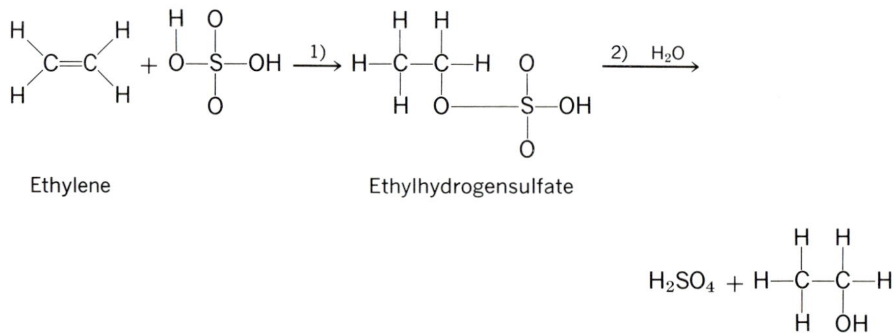

Ethylene — Ethylhydrogensulfate — Ethyl alcohol

Overall reaction:

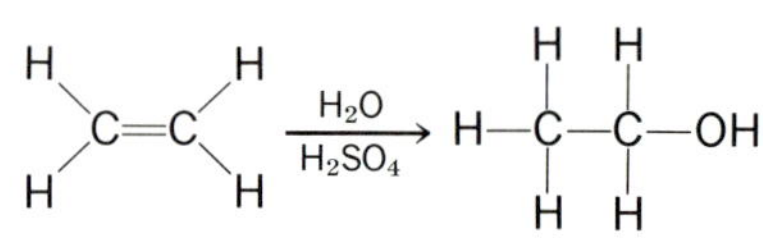

Ethylene — Ethyl alcohol

Hydrogenation of the alkene group involves the addition of like atoms to each carbon of the double bond. Regardless of whether or not the alkene is symmetrical, the same product results. However, in the case of hydration, two different units are added, a H^+ to one carbon and an OH^- to the other carbon. There is a relationship between the structure of the alkene and the order of addition of an unsymmetrical molecule, such as HOH, H_2SO_4, or HX.

Markownikoff's rule states that in the absence of peroxides, H_2O_2, the negative ion bonds to the carbon with the fewest hydrogens. Consequently, the H^+ of the adding molecule bonds to the carbon holding the *greatest* number of hydrogens.

Peroxides make the addition go in just the opposite manner. The negative ion adds to the carbon bonded to the most hydrogens, and the H^+ goes to the carbon bonded to the *smallest* number of hydrogens. Thus, by controlling the experimental conditions, the desired product can be obtained, as the following examples of the addition of H_2O, HCl, and HI will illustrate.

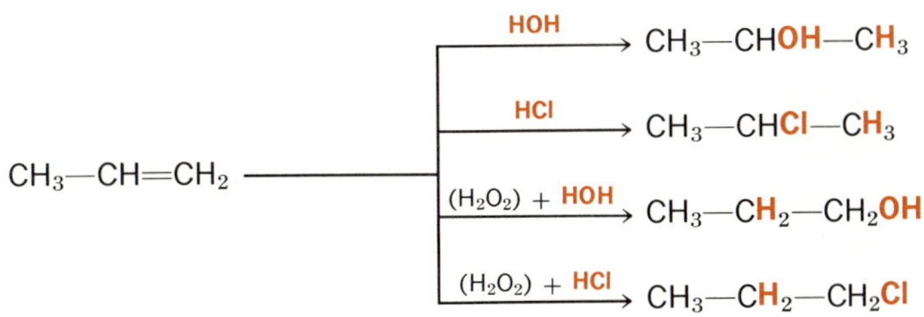

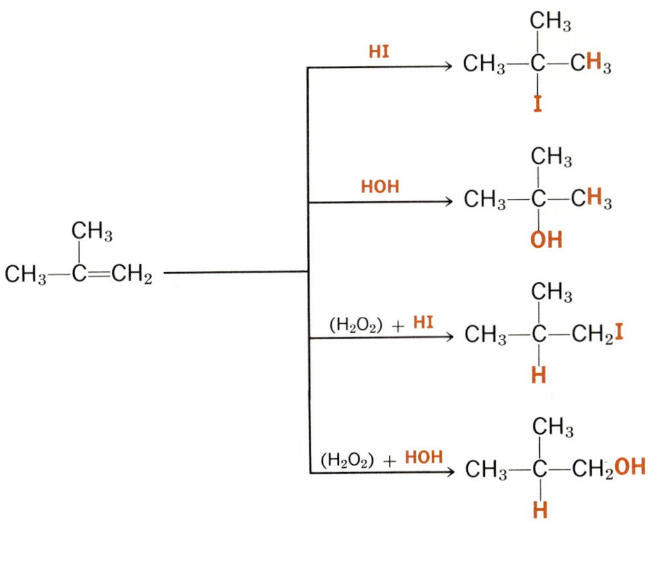

Halogenation

The halogens, X_2, add to the carbon double bond very easily and produce the dihalide derivative of the corresponding hydrocarbon. A positive ion intermediate is produced in this reaction, and, as mentioned before, trans addition occurs. The general reaction, which proceeds without a catalyst, can be represented as

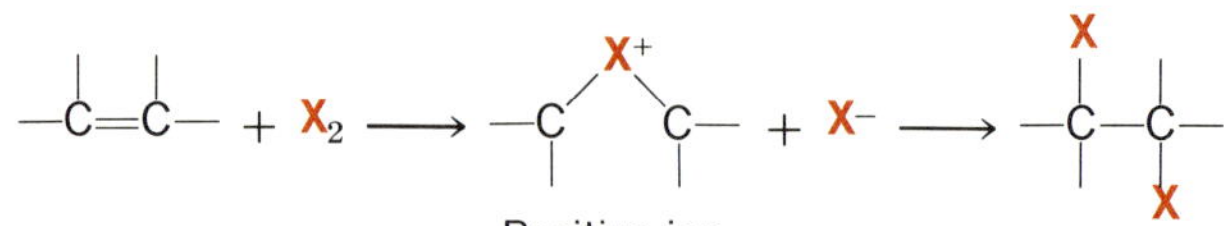

The addition of Br_2 or I_2 is used to find the number of double bonds in olefinic compounds. Iodine is used frequently because it is easy to determine quantitatively. A known amount of the sample (olefin) is treated with *excess* standard iodine solution (a standard contains a known concentration of I_2), and after the reaction has taken place, the excess iodine is determined by titration with a thiosulfate solution. The difference between the amount of I_2 added and the amount after the reaction has taken place is the amount of I_2 that has been added to the olefinic bonds in the sample. If 1 mole of I_2 is found to react with 1 mole of alkene, there is one double bond in the alkene. Similarly, if 2 moles of I_2 react with 1 mole of alkene, there are two C═C units in the substance.

The addition of halogens to a double bond takes place in the *absence of light* (without a catalyst). Because light catalyzes the substitution of X_2 in alkanes, if light were present during the halogenation of alkenes, the alkane portion of the molecule would also be halogenated. This is an example of how the chemist can choose conditions to favor the reaction at a selected functional group.

The *iodine number* (I no.) is the grams of iodine absorbed by 100 g of substance. The iodine number of fats is used to aid in their classification and characterization, since the I_2 will not react with other types of unsaturation, such as 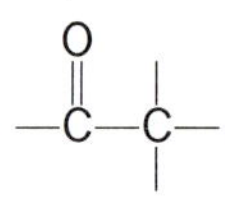. For example,

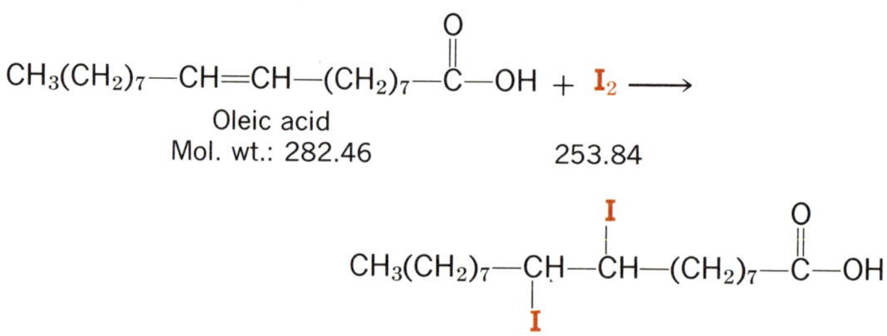

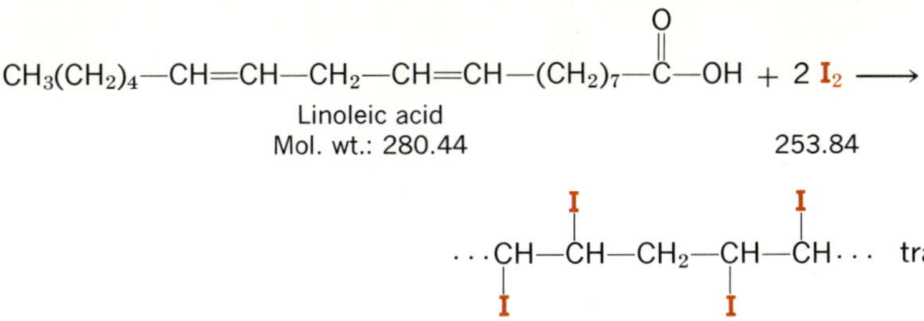

Linoleic acid
Mol. wt.: 280.44 253.84

trans addition

Oleic acid adds 1 mole of iodine per mole of compound, and linoleic acid with two double bonds adds 2 moles of iodine per mole of compound. To calculate the iodine number one finds the number of grams of iodine that reacts with 100 g of the olefin:

$$\frac{\text{Mol. wt.}}{\text{Mol. wt. I}_2 \times \text{no. moles}} = \frac{100}{\text{I no.}}$$

$$\text{I no.} = \frac{253.84 \times 100}{282.46} = 90.1 \text{ for oleic acid}$$

$$\text{I no.} = \frac{253.84 \times 2 \times 100}{280.44} = 181 \text{ for linoleic acid}$$

Therefore the difference of one double bond in a C_{18} acid makes a difference of about 90 in the iodine number.

Oxidation

Mild oxidation of the alkenes with neutral aqueous $KMnO_4$ forms the corresponding glycols, compounds with two OH groups on adjacent carbons. The general reaction is as follows (note that the cis glycol is formed):

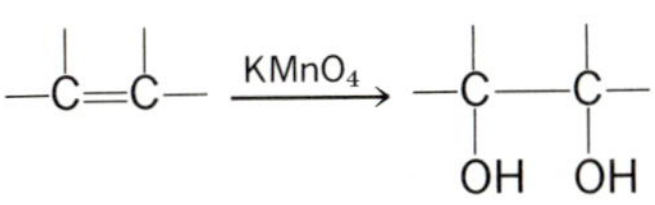

Alkene Glycol

This reaction proceeds easily at room temperature. Specific reactions are shown with ethylene and propylene to form the corresponding glycols.

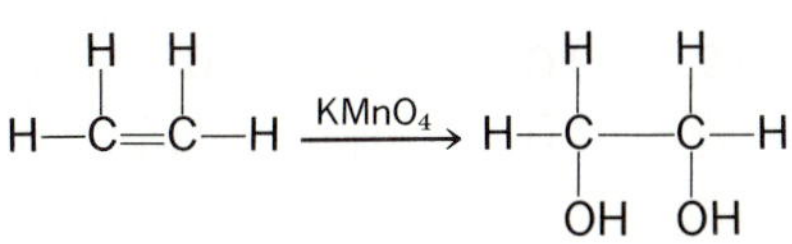

Ethylene Ethylene glycol

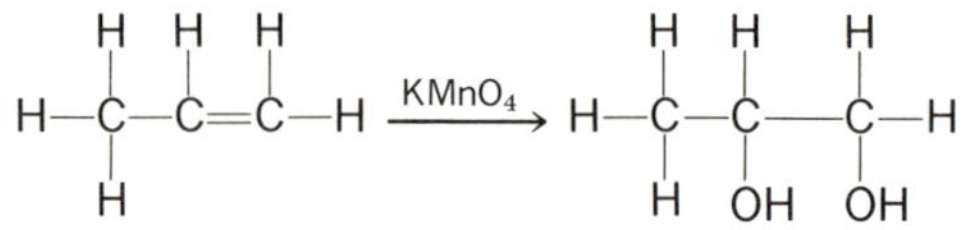

Propylene　　　　Propylene glycol

Potassium permanganate, $KMnO_4$, can be used to detect the presence of a double or triple bond. The manganese in $KMnO_4$ is reduced to MnO_2, which is an insoluble brown substance and precipitates out of solution as the reaction proceeds. A gaseous alkene may simply be bubbled into an aqueous solution of $KMnO_4$ and the precipitate of MnO_2 noted or filtered off. Very small amounts (microgram quantities) of a liquid or solid unsaturated compound can be spotted on filter paper and the $KMnO_4$ solution sprayed onto the paper with an atomizer. When the paper is rinsed with water to wash off the unreacted $KMnO_4$, a brown spot of MnO_2 remains where the alkene was applied. *This is not a specific test for unsaturation* since many easily oxidizable substances also give this test, e.g., aldehydes.

8.10 POLYENES

Polyenes are compounds that contain more than one double bond. A diene, triene, and tetraene contain two, three, and four double bonds, respectively. The fatty acid linoleic acid has already been mentioned; it is an example of a compound that contains two double bonds. Vitamin A and β-carotene are examples of naturally occurring compounds that are very highly unsaturated. Natural rubber is also a polyene.

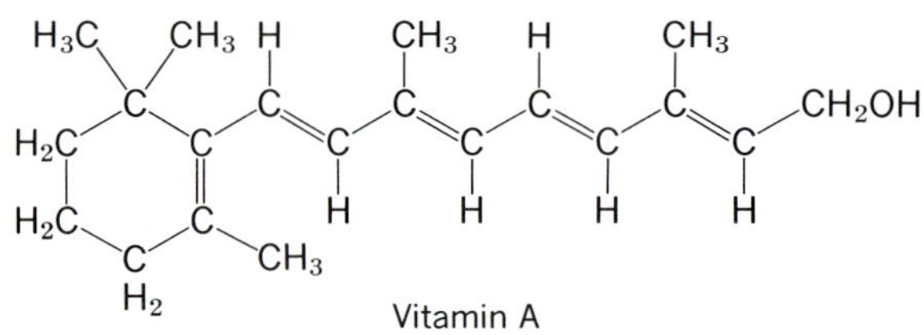

Vitamin A

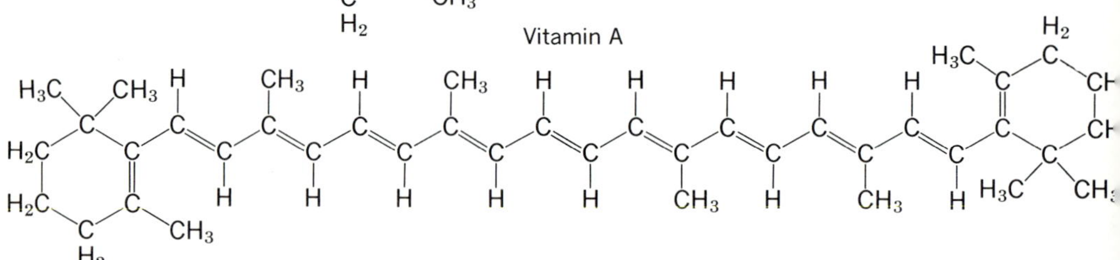

β-Carotene

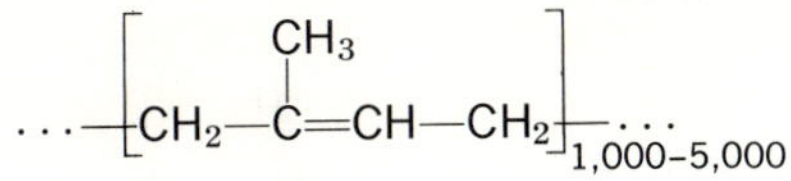

Natural rubber

Some of the compounds just mentioned are a class of compounds known as *polymers* (see Sec. 8.17). Polymeric substances may contain units of different types of molecules such as alkanes, alkenes, or the alkynes, which are discussed next.

The reaction of alkenes can be summarized as follows:

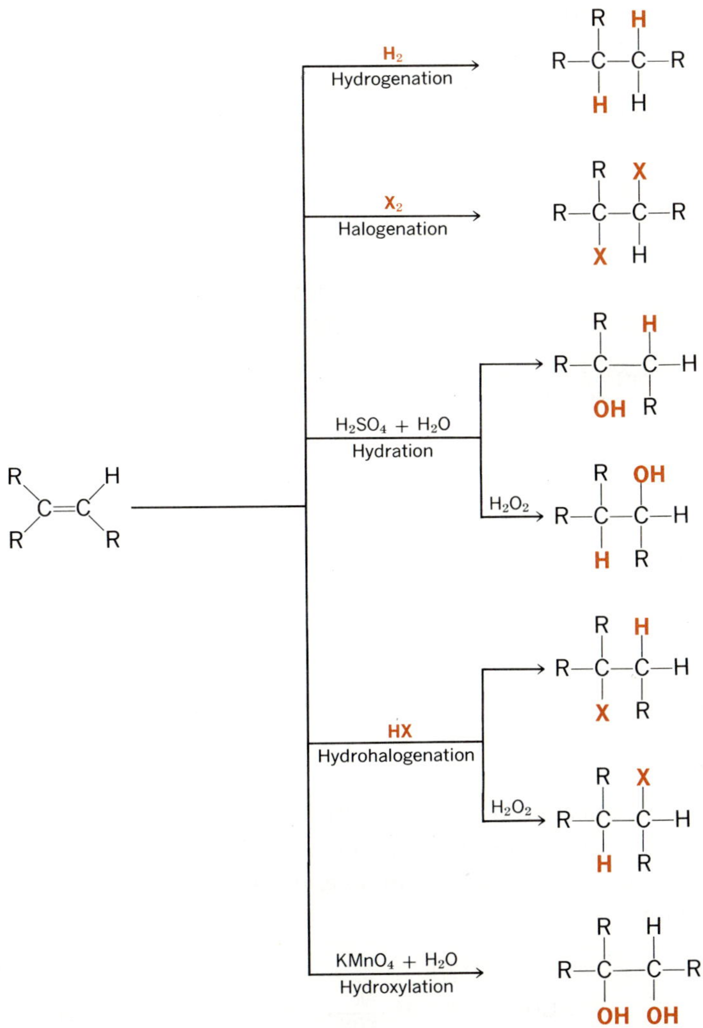

$$H—C\equiv C—H$$

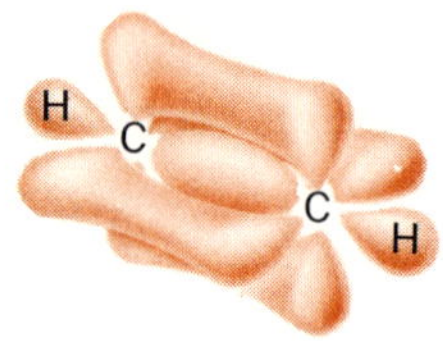

Acetylene

FIGURE 8.7 The carbon-carbon triple bond as in acetylene.

8.11 ALKYNES

Alkynes contain the C≡C group and have the general formula C_nH_{2n-2}. They are quite reactive chemically and many are so unstable they must be handled with special precautions or they may explode. The simplest alkyne, acetylene, is used in welding and in the synthesis of other chemicals (see Fig. 8.7). Alkynes do not enter into biological reactions in general, but certain microorganisms are able to produce molecules with several alkyne bonds. Many of these substances are antibiotic in nature. Just how the C≡C bond is formed biologically is not known. To show the unusual highly unsaturated nature of some natural compounds a few structures follow:

$$HC\equiv C—C\equiv C—CH=C=CH—\underset{\displaystyle OH}{\underset{|}{CH}}—CH_2—CH_2—COOH$$

Nimotinic acid

$$HOCH_2—C\equiv C—C\equiv C—C\equiv C—CO—NH_2$$

Agrocybin

$$HOCH_2—C\equiv C—C\equiv C—C\equiv C—CH=CH—COOH$$

Diatretyne-3

8.12 PREPARATION

Dihalide

Alkynes can be prepared by the elimination of 2 HX from alkyl dihalides when the halogens are on *adjacent carbons* (the dihalide can be formed by the halogenation of an alkene). Alcoholic KOH is needed for this reaction.

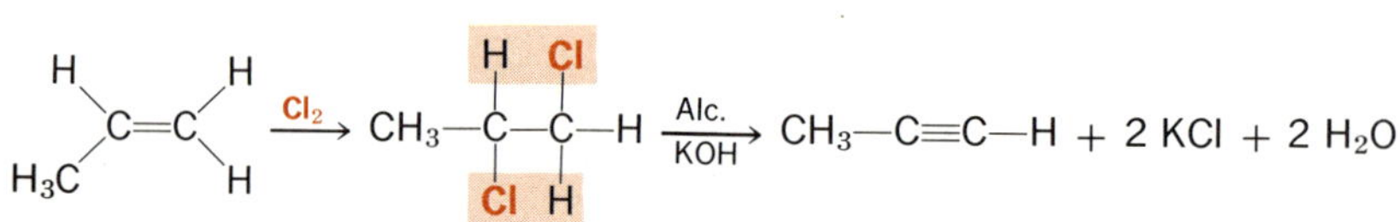

Tetrahalide

The removal of halide from a tetrahalide with zinc can also be used to prepare alkynes.

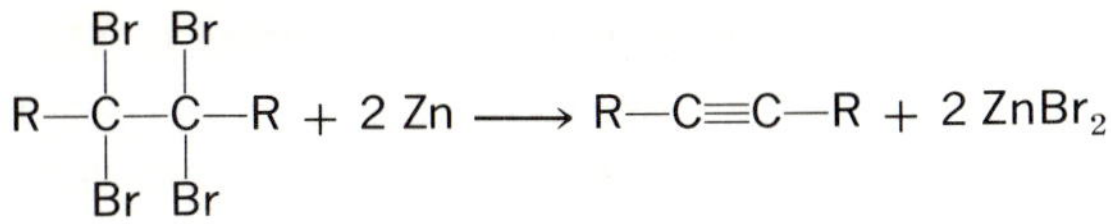

Calcium carbide

Acetylene is conveniently made by reacting calcium carbide, CaC_2, with water. Calcium carbide is produced by heating CaO and coke (carbon) in an electric furnace.

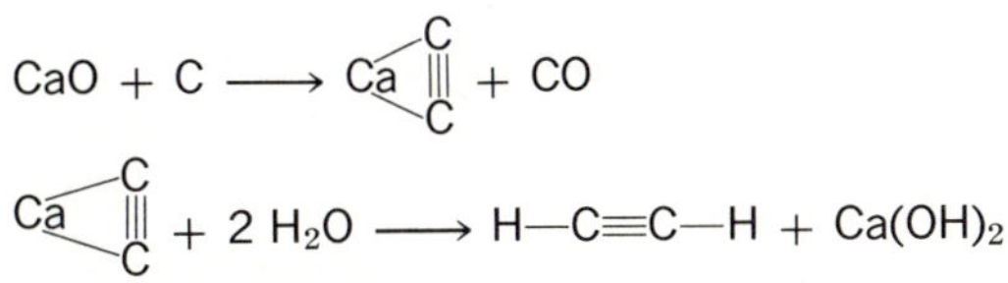

8.13 REACTIONS OF ALKYNES

The great chemical activity of the acetylenes can be attributed to the two π bonds which form a cylindrical electron cloud about the two carbons (Figs. 7.2 and 8.7). The reactions of the alkynes are similar to those of the alkenes inasmuch as they enter into addition reactions. They are, of course, specific reactions characteristic of the carbon-carbon triple bond. *Markownikoff's rule is followed when an unsymmetrical molecule adds to an unsymmetrical alkyne.*

The reactions of alkynes can be summarized as follows:

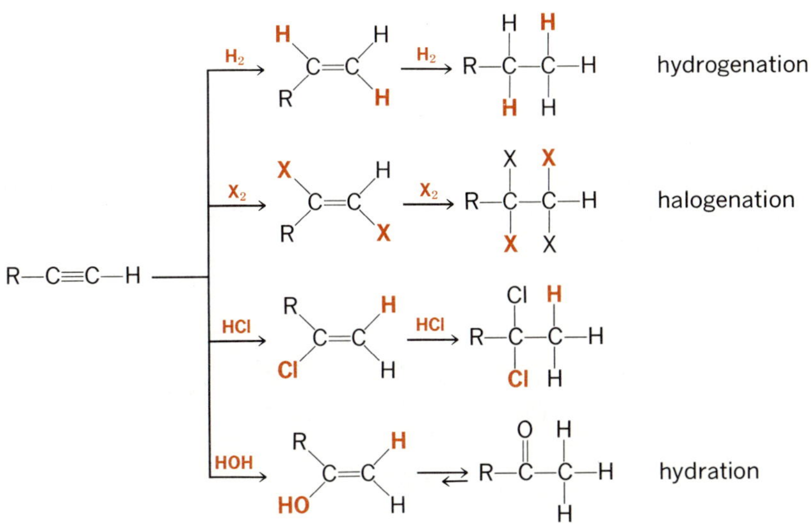

Hydration

The last reaction (hydration) deserves some comment. The addition of HOH to the triple bond forms a compound with an OH on an unsaturated carbon, an enol. These compounds are very unstable and immediately rearrange to form the corresponding carbonyl (C═O) compound. For example, the hydration of acetylene yields acetaldehyde:

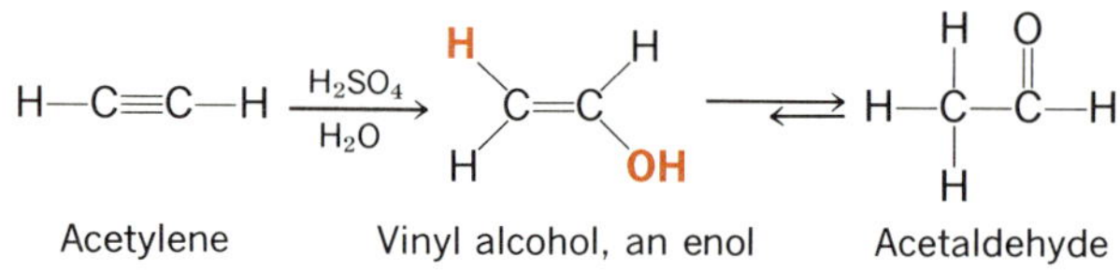

Vinyl alcohol is the *enol* form of acetaldehyde; the *en-* from alkene indicates a C═C, and the *-ol* from alcohol indicates an OH group. A general rule in organic chemistry is an OH on an unsaturated carbon atom is not stable. Only a few compounds with this structure are stable enough to exist as such. One is vitamin C, which is an enediol; it has two OH groups, one on each unsaturated carbon. Vitamin C is relatively unstable and is very easily oxidized.

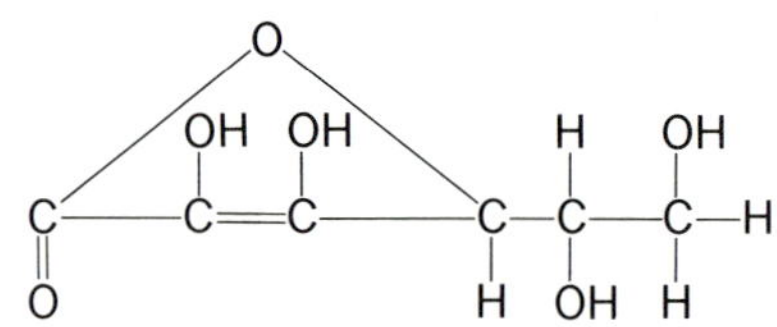

Vitamin C

8.14 AROMATIC COMPOUNDS

Compounds containing the benzene ring or the equivalent are classified as *aromatic compounds*. The benzene ring is composed of six carbons arranged in a ring having a degree of unsaturation equal to three double bonds. As shown in Fig. 8.8, the electrons form a π cloud above and below the ring and are not localized. For convenience, however, the structure is written in one of the following forms:

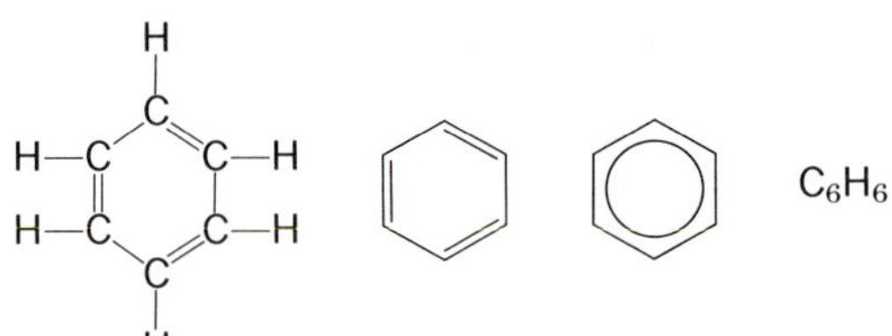

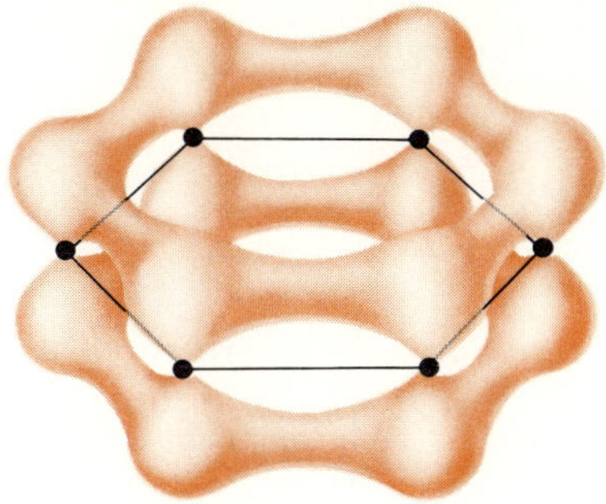

FIGURE 8.8 Delocalized cloud of π electrons in the benzene ring.

The cyclic structure prevents free rotation about the carbon bonds, and the carbon atoms are held in one plane. Benzene is therefore a planer molecule.

Compounds containing the benzene ring are studied separately from the aliphatic compounds because the benzene ring has some unique features which are not characteristic of unsaturated aliphatic compounds.

Aromatic compounds play a very important biological role. The benzene nucleus occurs in a wide variety of biological compounds, e.g., proteins, hormones, vitamins, and essential oils. Plants can make the benzene ring and all the aromatic compounds they need. Animals, on the other hand, cannot synthesize all aromatic nuclei and must therefore have in their diet certain aromatic compounds, which come ultimately from plants.

Aromaticity and the benzene nucleus

The benzene nucleus has six carbons in a cyclic structure containing alternate double bonds. What is the distribution of the π-bond electrons about the benzene ring? Refer back to the distribution of p electrons in a conjugated set of double bonds. There was an electron cloud above and below the four carbon atoms involved in the formation of the π bonds. This delocalization of electron distribution permits a shift of electron density back and forth, as indicated in Fig. 8.9*c* and *d*. This same situation tends to exist with respect to the benzene ring, but there the molecule has no end. Instead, there is a circular cloud of p electrons (making up the π bonds) above and below the ring of carbons with the carbon atoms sandwiched in between (see Fig. 8.10*b*).

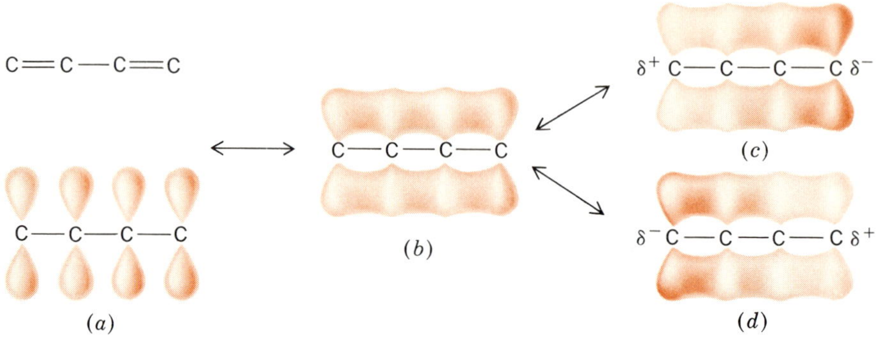

FIGURE 8.9 Electron configuration in a conjugated or diene bond, illustrating the idea of resonance.

The prerequisite for an aromatic ring is the presence of six p electrons to enter into the resonating structure. Each carbon atom of the benzene ring contributes one p electron (Fig. 8.10*a*). Thus the actual formula for the benzene nucleus cannot be written since the π bonds are associated with all the carbon atoms. The two formulas at the extreme ends of this resonating system would be

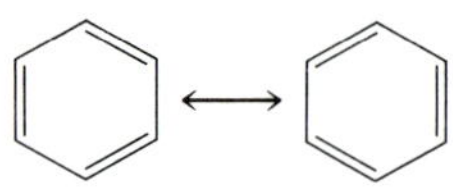

These two structures are identical and are known as resonance forms since they differ only in the distribution of electrons (remember the resonance forms of a conjugated diene).

The delocalization of electrons *stabilizes* the molecule to such an extent that the benzene ring does not react as a simple alkene but more like an alkane. Substitution reactions take place more easily than addition reactions.

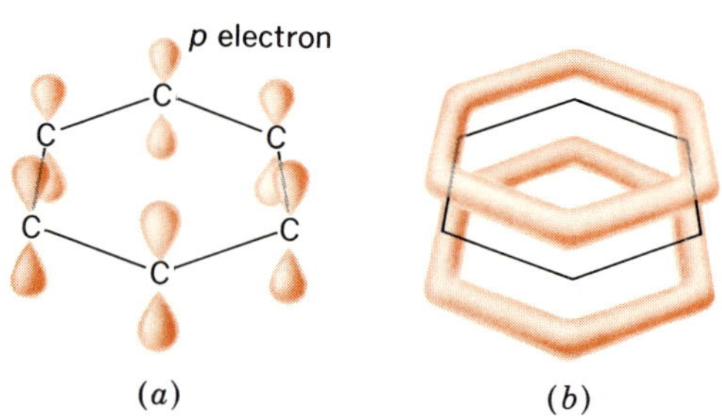

FIGURE 8.10 Distribution of p electrons about the benzene nucleus.

8.15 NOMENCLATURE

The relative position of groups substituted in the benzene ring is indicated by giving the number of the benzene carbon to which they are attached. Also the prefixes *ortho-* (*o*), *meta-* (*m*), and *para-* (*p*) are used to designate substituents that are on adjacent carbons, alternate carbons, or on opposite carbons (1 and 4), respectively.

An aromatic hydrocarbon, of course, contains only carbon and hydrogen, e.g., benzene and toluene.

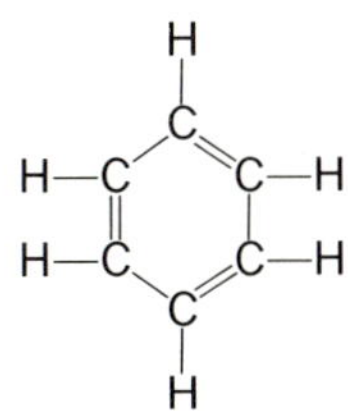

Benzene

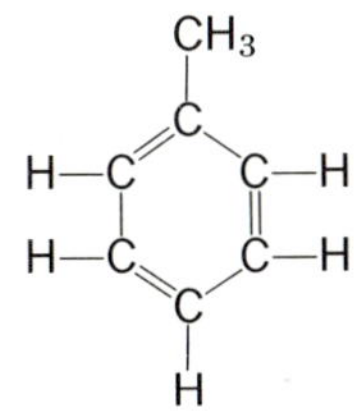

Toluene
or
Methylbenzene

The hydrogens of benzene can be substituted by groups, such as Cl, NO_2, SO_3H, and alkyl radicals (as in toluene). An aliphatic side chain may contain functional groups, e.g., alcohol, aldehyde, or acid; these compounds would have the chemical properties of the benzene ring as well as those of the functional group in the side chain. Hence, the benzene ring may occur in a large variety of classes of compounds other than hydrocarbons. For example,

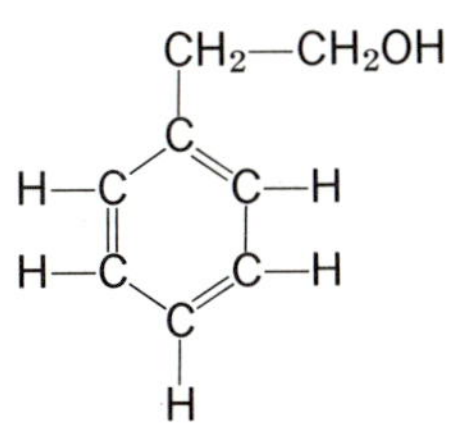

Aromatic alcohol
Phenyl ethanol

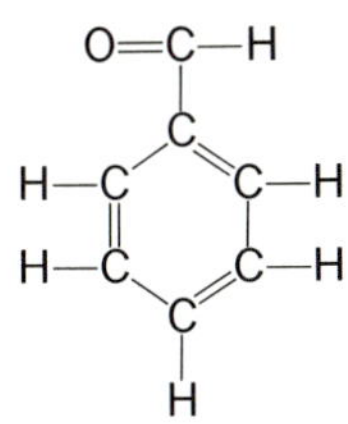

Aromatic aldehyde
Benzaldehyde

The phenyl radical is benzene less one hydrogen (C_6H_5—). The following examples will serve to illustrate the naming of benzene compounds.

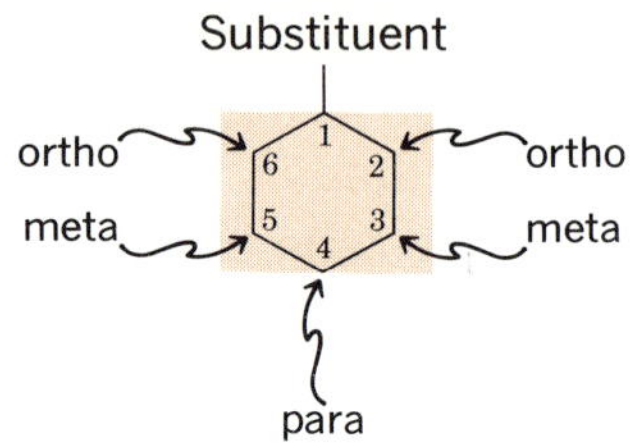

phenyl radical

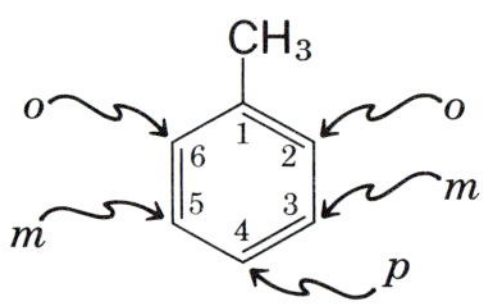

Toluene
Phenylmethane
Methylbenzene

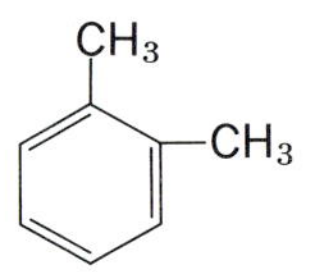

o-Xylene
1,2-Dimethylbenzene

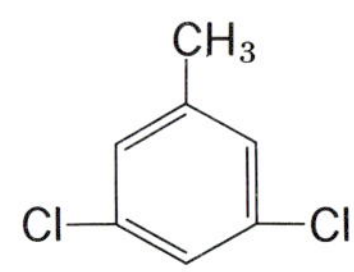

1-Methyl-3,5-dichlorobenzene
m-Dichlorotoluene

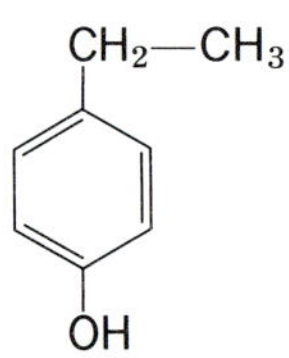

p-Ethylphenol
1-Ethyl-4-hydroxybenzene
p-Hydroxyphenylethane

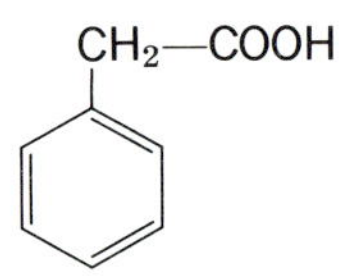

Phenylacetic acid

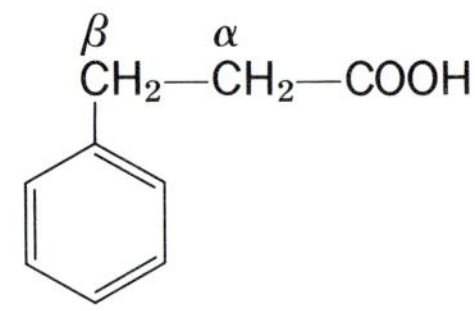

3-Phenylpropionic acid
β-Phenylpropionic acid

Some biochemical aromatic compounds are listed in Table 8.3.

TABLE 8.3 SOME BIOCHEMICAL AROMATIC COMPOUNDS

Compound	Comment
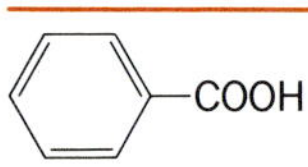 Benzoic acid	A weak acid, pK = 4.2, occurs in gum benzoin
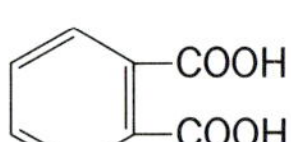 Phthalic acid	
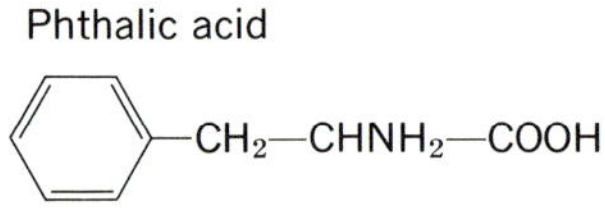 Phenylalanine	An amino acid needed in the diet
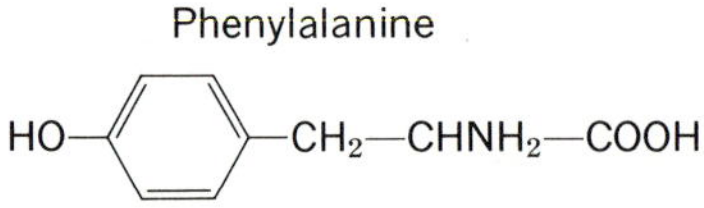 Tyrosine	An amino acid made biologically from phenylalanine

TABLE 8.3 *Continued*

Compound	Comment
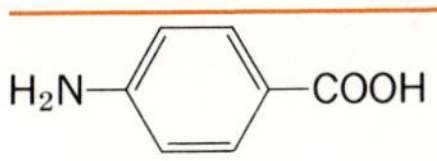 *p*-Aminobenzoic acid	Contained in the vitamin folic acid
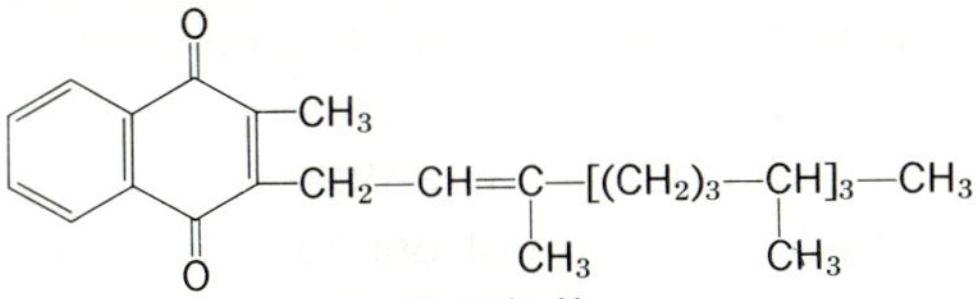 Vitamin K	A fat-soluble vitamin
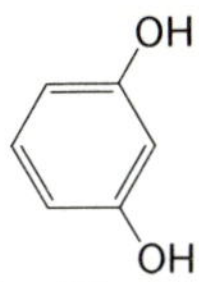 Resorcinol (*m*-Dihydroxybenzene)	
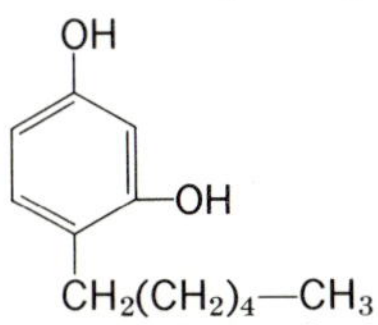 4-*n*-Hexylresorcinol (Hexylresorcinol)	Antiseptic
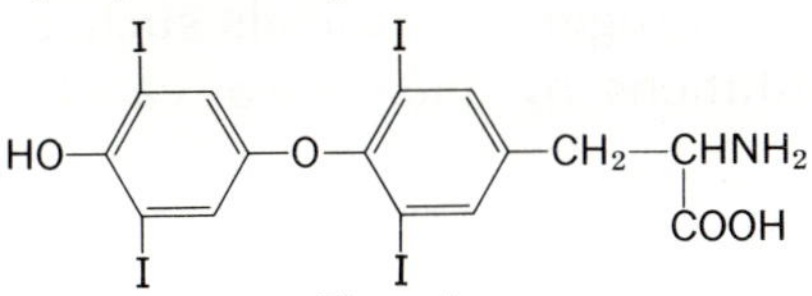 Thyroxine	A hormone made biologically from phenylalanine
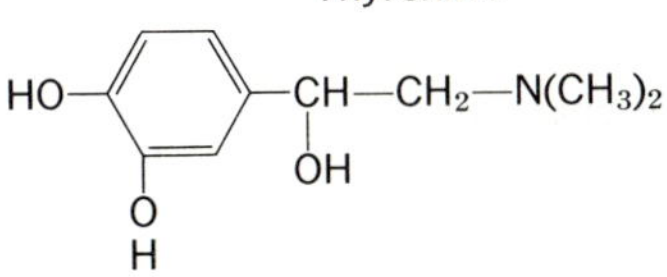 Adrenaline	A hormone made biologically from phenylalanine
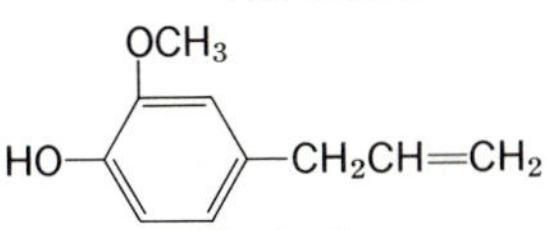 Euginol	Oil of cloves
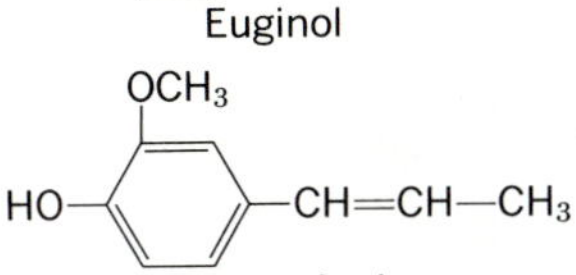 Isoeuginol	Oil of nutmeg
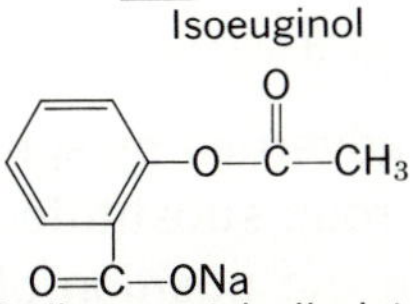 Sodium acetylsalicylate	Aspirin

8.16 REACTIONS OF AROMATIC COMPOUNDS

The stability of the benzene ring (because of resonance) decreases the reactivity of the double bonds. Only under rather strenuous conditions and with a very limited number of reagents do the double bonds in the benzene ring react like those in an alkene. For example, the cyclic nature of benzene was proved by hydrogenation of benzene to cyclohexane under much higher temperature and pressure than is needed to hydrogenate alkenes.

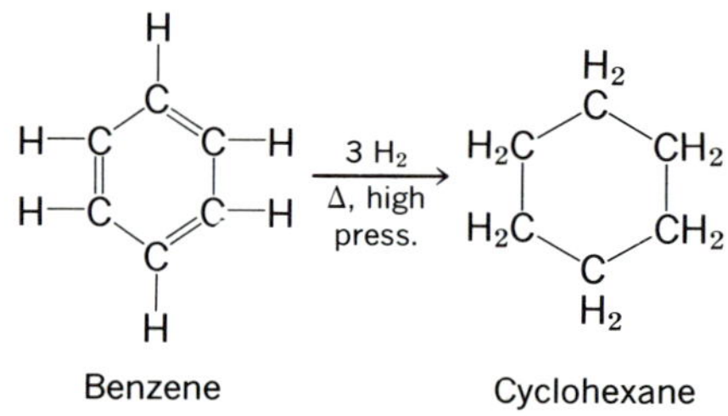

Substitution

Benzene reacts by substitution of the hydrogen for radicals such as halogens, nitrate, or sulfonate. Substitutions by one radical can be summarized as

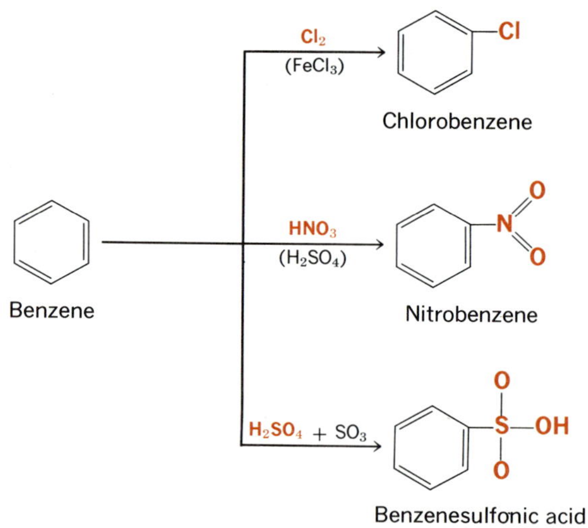

Multiple substitutions are also possible but are governed by specific rules. The electron-attracting nature of the first group substituted in the benzene nucleus has a directive influence on the position of entering additional groups. If the first group substituted in the benzene

nucleus causes a shift of electrons *from* the nucleus, substituents will enter at the positions meta to the first group; the group is *meta directing*. The *withdrawal* of electrons from the nucleus results in the ortho and para carbons becoming more positive. Consequently, they do not give up their hydrogen in substitution reactions as easily as the more negative meta carbons. For example, with nitrobenzene

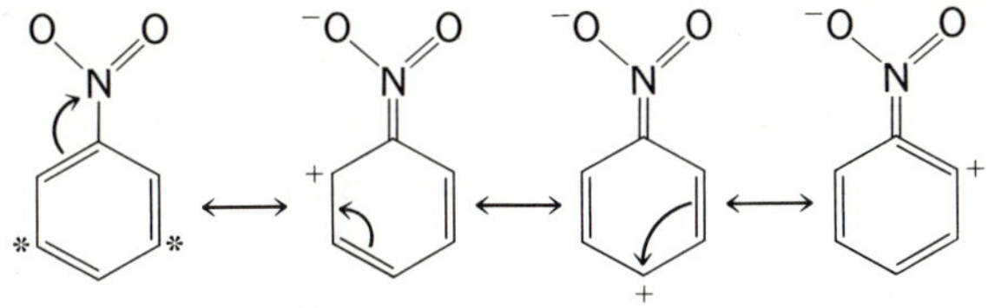

Nitrobenzene would substitute as follows:

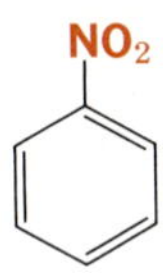

Nitrobenzene

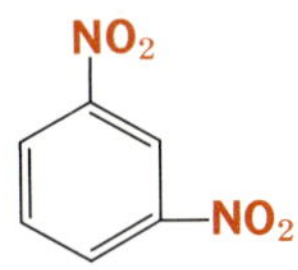

m-Dinitrobenzene
1,3-Dinitrobenzene

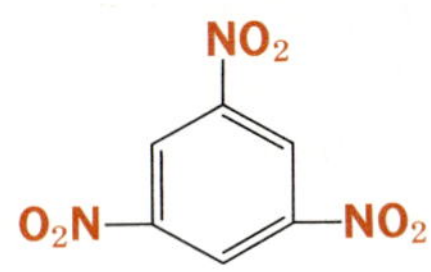

1,3,5-Trinitrobenzene

Meta-directing groups include CHO, COOH, and SO_3H.

If an *electron-donating* group is attached to the benzene ring, entering substituents will enter at the ortho and para positions; the group is *ortho-para directing*. In this instance it is the ortho and para carbons of the benzene ring that become more negative. Some electron-donating groups are NH_2, OH, CH_3, and Cl.

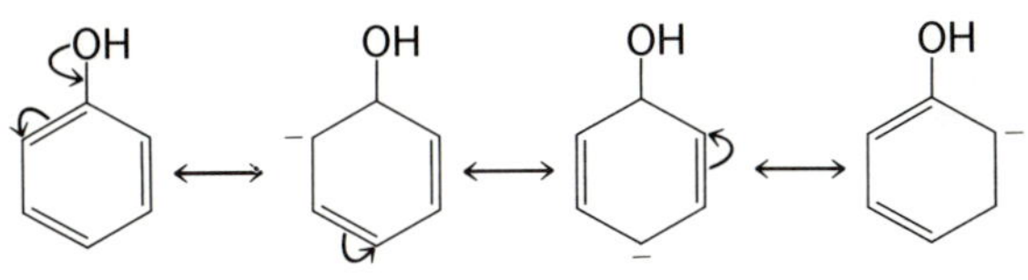

For example,

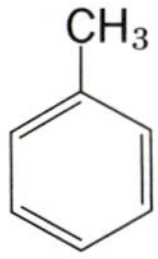

Toluene

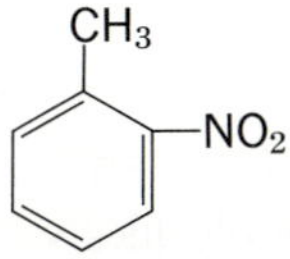

Nitrotoluene

2-Nitrotoluene

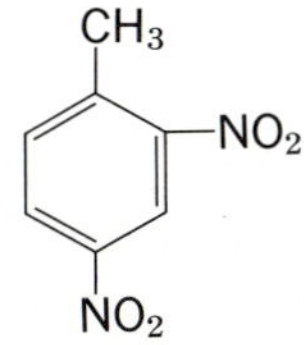

o,p-Dinitrotoluene
2,3-Dinitrotoluene

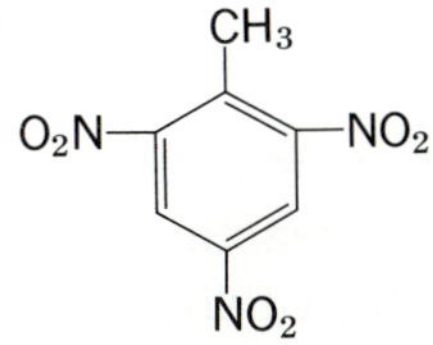

2,4,6-Trinitrotoluene
TNT

Not all aromatic compounds contain the benzene ring. A number of heterocyclic rings (rings that contain more than one kind of atom) have aromatic properties. The ones of most interest to us contain nitrogen in the ring structure and are discussed in Chap. 12 under nitrogen compounds.

8.17 POLYMERS

The study of polymers encompasses inorganic, organic, and biological chemistry. An example of an inorganic polymer is the silicate chain composed of SiO_3^{--} units. Butadiene and ribonucleic acid are examples of organic and biologically active polymers, respectively. A polymer that contains more than five of any one kind of unit (*monomer*) is called a *homopolymer,* represented by A—A—A—A—A— . . . . Special nomenclature is used for monomer chains having five or fewer monomers in the chain. A polymer that contains more than one kind of monomer, such as A—B—A—B—A—B— . . . , is called a *copolymer.*

Monomer:	A	Single unit
Dimer:	A—A	Two monomers bonded together
Trimer:	A—A—A	Three monomers bonded together
Tetramer:	A—A—A—A	Four monomers bonded together
Pentamer:	A—A—A—A—A	Five monomers bonded together
Polymer:	A—A—A—A—A— . . .	Many monomers bonded together
Copolymer:	A—B—A—C—A— . . .	Different monomers bonded together

The silicate homopolymer, which is composed of SiO_3^{--} monomers bonded together,

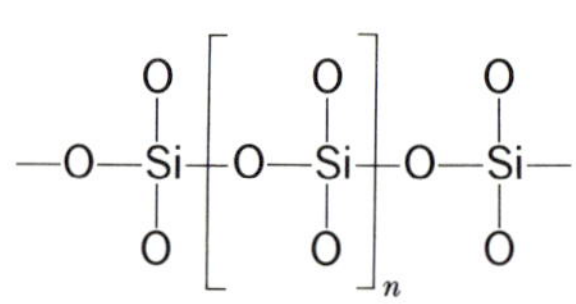

is an example of a *condensation polymer.* Condensation polymers are formed by reactions in which a small molecule like water is eliminated at the sites where monomers become joined together. Several organic derivatives of this polymer have been prepared. For example, the methyl derivative, methyl silicone,

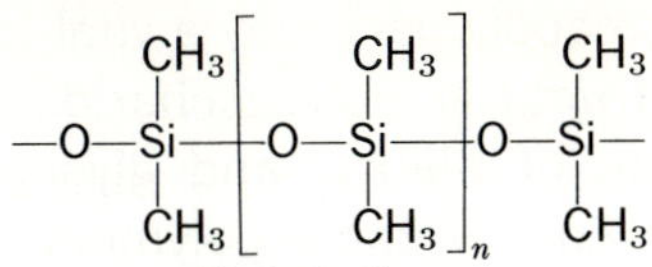

Methyl silicone

has lubricating properties. These silicones are quite stable at temperatures above which ordinary lubricating oils decompose. Phosphorus and boron form another polymer, called borophane, that contains both inorganic elements and organic groups.

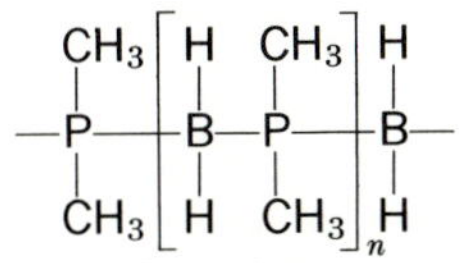

Borophane

Another inorganic polymer, polydichlorophosphonitrile, contains only phosphorus, nitrogen, and chlorine.

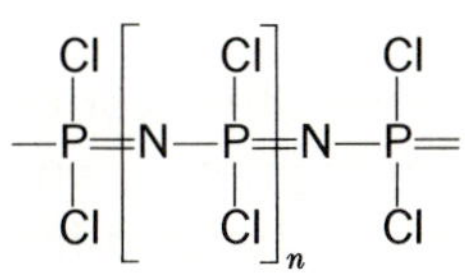

Polydichlorophosphonitrile

Natural rubber, according to its formula, may be considered as an addition polymer of the simple five-carbon isoprene unit:

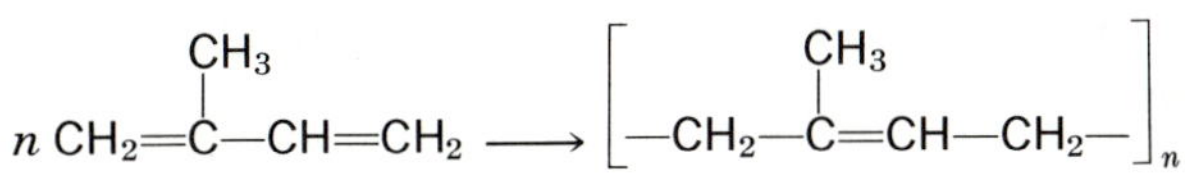

Isoprene Natural rubber

Isoprene is a conjugated diene (it contains two double bonds separated by a single bond), and polymerization occurs by way of a 1 to 4 addition. Compare the polymerization of isoprene units with the hydrogenation reaction

$$H_2 + CH_2{=}CH{-}CH{=}CH_2 \longrightarrow \overset{1}{C}H_3{-}\overset{2}{C}H{=}\overset{3}{C}H{-}\overset{4}{C}H_3$$

Butadiene
Conjugated diene

Butene-2
1–4 addition

The carotenoids, such as β-carotene, the precursor of vitamin A, are plant polymers of isoprene units.

Condensation homopolymers play a vital role in living organisms, e.g., cellulose, the structural polysaccharide of plants; starch, the reserve polysaccharide of plants; and glycogen, the reserve polysaccharide of animals. Each of these polymers contains glucose as the monomeric unit, and 1 mole of water is eliminated for each glucose unit incorporated into the polymer. In cellulose, glucose has the β configuration, whereas in starch and glycogen it has the α configuration. Complete hydrolysis of these polysaccharides yields only glucose. The structure and synthesis of these polymers are discussed in Part Three.

Two biologically important condensation copolymers are the proteins and the nucleic acids. Their importance is almost immeasurable. Proteins, polymers of amino acids, are not only components of the skin, tissue, and membranes of all living organisms but are also enzymes, the biological catalysts without which biological reactions could not take place. The nucleic acids, polymers of nucleotides, are the genes (DNA) which direct the synthesis of each particular protein in living cells and the ribonucleic acids (RNA), without which proteins could not be synthesized. These polymers are discussed in detail in subsequent chapters.

An application of polymers that contain either an acid or a base group is ion-exchange chromatography, where ions associated with the insoluble resin (polymer) are exchanged for ions in solution (Chap. 5).

A relatively new application of synthetic polymers is the *gel filtration* method for estimating the molecular weight of proteins. By appropriate techniques a polymer can be made which has a uniform pore size, but the sizes can differ in different polymers. This makes possible the molecular-sieve method of separating proteins of different molecular weight or size.

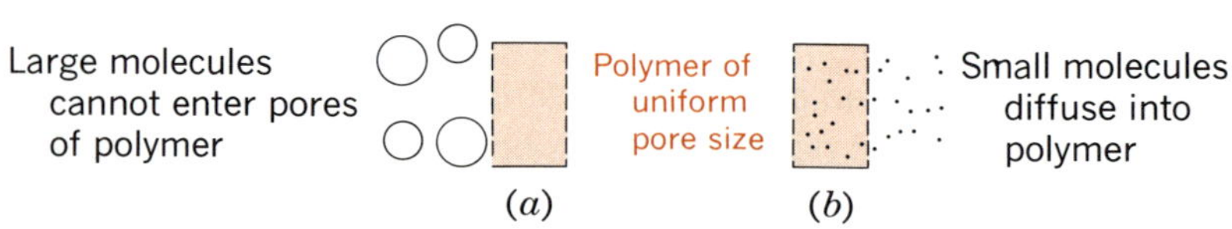

The polymer in (*a*) is exposed to a solution containing solutes of high molecular weight, which cannot penetrate the polymer because its pores are too small to allow the solute molecules to enter. In (*b*) the solute molecules are small enough to enter the polymer pores, and therefore they diffuse into the polymer. What would you observe with respect to the two conditions (*a*) and (*b*) if you passed some fresh solvent through a tube containing the polymer (*a*) and another tube containing the polymer (*b*)? In (*a*) the large molecules would be washed out of the tube with the solvent front and would not be held back by the polymer. In (*b*) the solute would not come out with the solvent front but would be held back because it has to diffuse out of the pores of the polymer before it can be washed out of the tube by the solvent. In this instance the solute would appear in the effluent some time after the solvent had started to pass through the column. Now consider a mixture of the large and small molecules. As above, the large molecules would be outside the polymer, whereas the small ones would be dispersed through it. By eluting the column with fresh solvent the large molecules are flushed out immediately; since the smaller molecules that penetrated the polymer appear in a later fraction of the effluent, a separation of large from small molecules has been effected.

Commercial polymers are made with a different uniform pore size so that they can be used as molecular sieves. For example, one product will not retard proteins of molecular weight 100,000 or greater, whereas other polymers will not retard proteins of molecular weight of, say, 50,000 or over or 40,000 or over. For example, if a protein is not retarded by a polymer retarding under 40,000 but is by a polymer retarding 50,000 or less, the protein has a molecular weight between 40,000 and 50,000.

QUESTIONS

1. What is the general formula for (*a*) alkanes; (*b*) alkenes; (*c*) alkynes?
2. Define a hydrocarbon.
3. What is a structural isomer?
4. Draw the structural formulas for all isomers of heptane that have five carbons as the longest chain. Name each isomer.
5. Label the hydrogens and carbons in the following compound as primary, secondary, or tertiary.

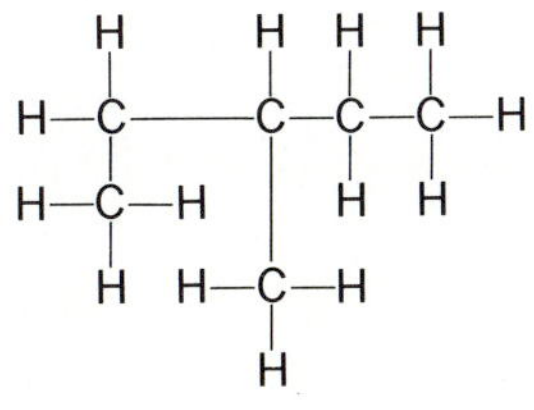

6. Which of the following group of compounds are isomers?

(*a*) 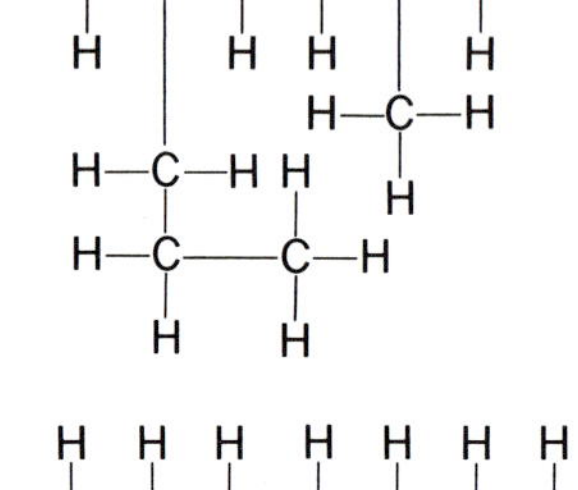

(*b*) 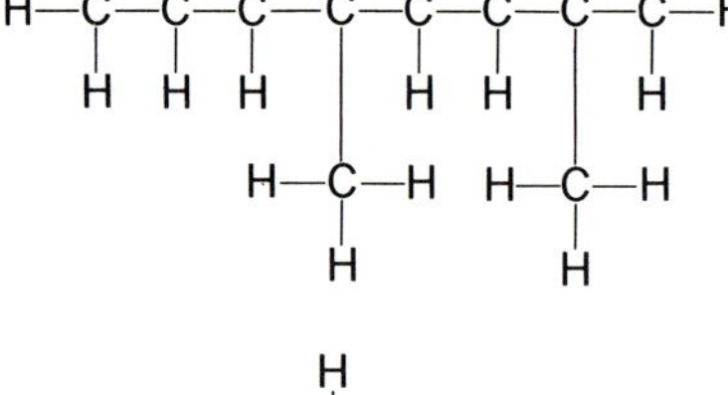

(*c*) 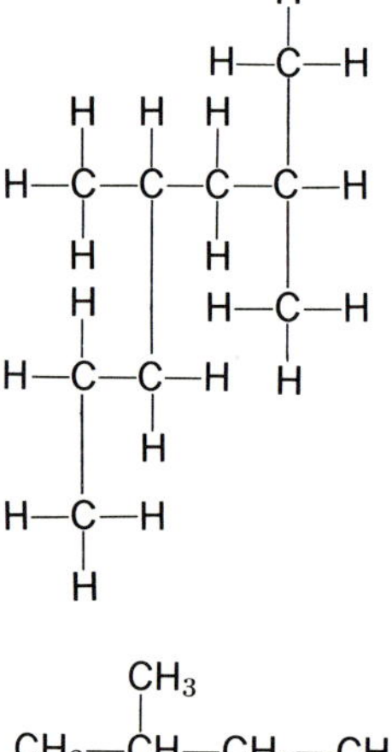

(*d*) 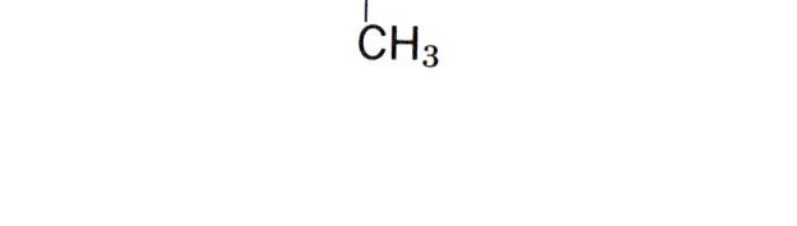

(*e*) 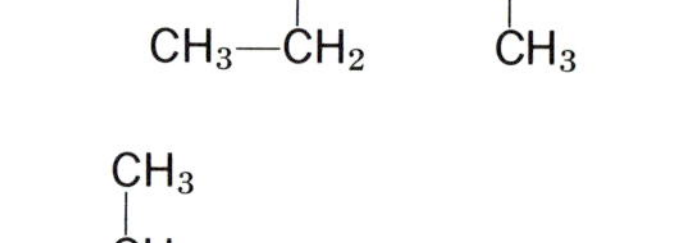

(*f*) 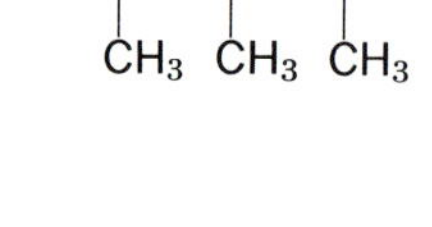

(*g*) 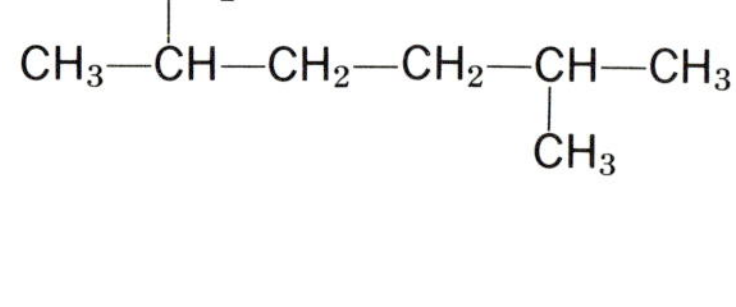

7. Write an equation to illustrate each of the following types of reactions:
   (*a*) Hydrogenation of an unsaturated hydrocarbon
   (*b*) Halogenation of an alkane
   (*c*) Oxidation of an alkene (partial oxidation)
8. Which group(s) in the list below when substituted for A in the following formula make the starred carbon asymmetric?

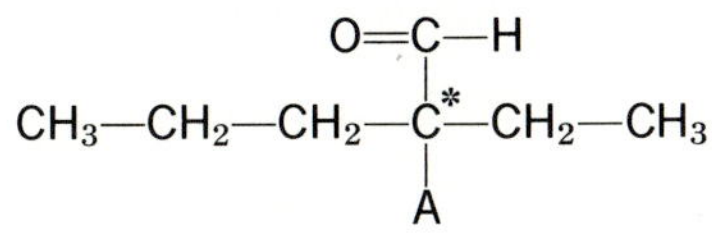

(*a*) Hydroxyl	(*b*) Isopropyl	(*c*) Cl
(*d*) Carboxyl	(*e*) Ethyl	(*f*) Methyl

9. Indicate how the electrons are distributed about a carbon-to-carbon double bond. Show the difference between a σ and a π bond.
10. Name the following compounds and indicate whether they are cis or trans:

(*a*) 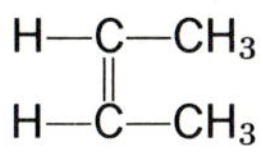 (*b*) 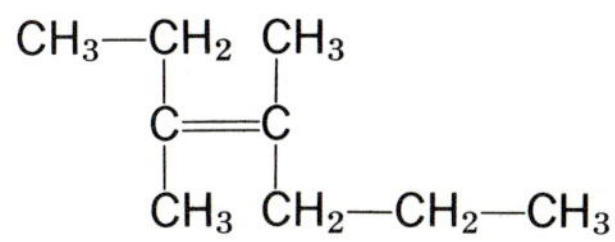

(*c*) 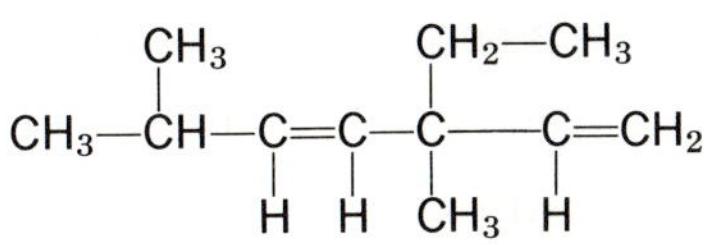

11. Write an equation for the formation of an alkene from an alkyl halide.
12. When a halogen, such as Cl_2, adds to an alkene, is it an oxidation-reduction reaction? Justify.
13. Write equations to show how $CH_3—CH_2OH$ and $CH_2OH—CH_2OH$ can be made from the same hydrocarbon.
14. What is meant by hydrogenation? Write an equation to illustrate this type of reaction.
15. How would you synthesize the following compound from an alkene?

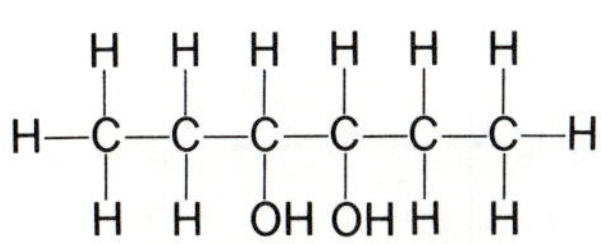

16. What is meant by iodine number? What is the iodine number of ethylene?
17. (*a*) A student named the following compound 3-*n*-butylhexane. Is this the correct name for the compound? If not, give the correct name.

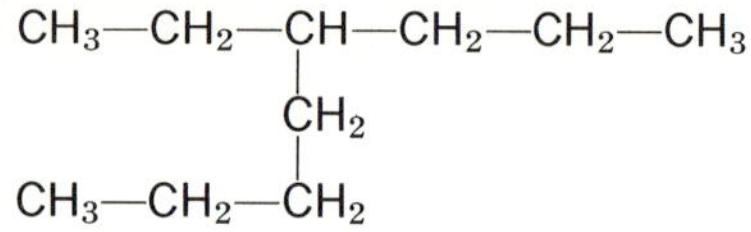

(*b*) A student named the following compound 2,3-*n*-propyl-4-methoxy-4-isopropylhexane. Is this the correct name? If not, what is the correct name? (Hydrogens are not given in the formula.)

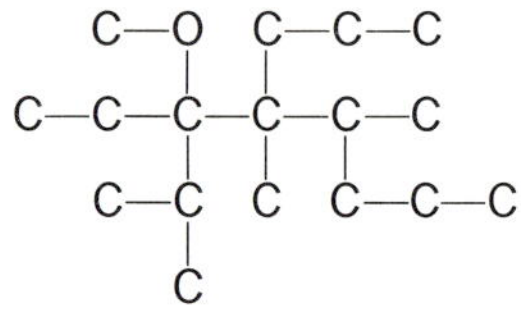

Ans: The correct name is 2,4,5-trimethyl-3-methoxy-3-ethyl-4-*n*-propyloctane.

18. Do alkynes exist in a cis and trans configuration? Explain.

19. A compound has the molecular formula C_6H_{12} and an asymmetric carbon atom. Write a possible structural formula for this compound.

20. Show how $CH_3—CH_2—CH(OH)—CH_3$ can be made from 1-butene.

21. Which compound of each of the following pairs would you expect to add H_2SO_4 more easily?

(*a*) Ethylene and propylene.

(*b*) Ethene and 2-butene.

(*c*) 2-Butene and 2,3-dimethyl-2-butene.

22. (*a*) From the following information write a possible formula for I, II, and III.

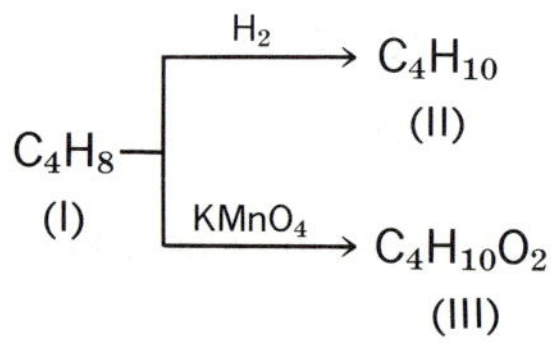

(*b*) How many isomers are there of I?

23. Compound I underwent the following reactions. From the products formed write a possible formula for each of the compounds I to IV.

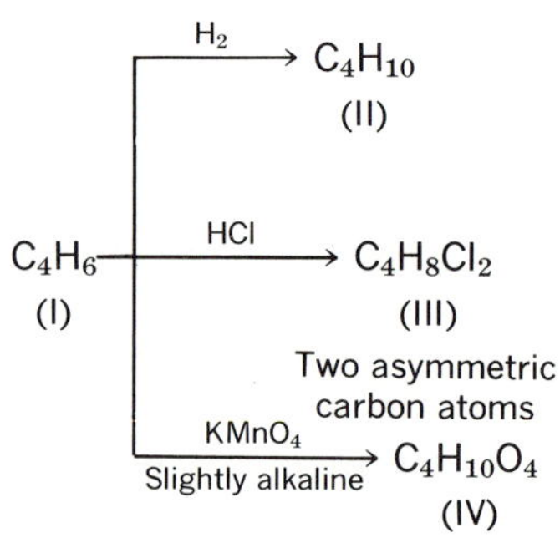

24. A hydrocarbon C_6H_{12} gave the following general reactions:

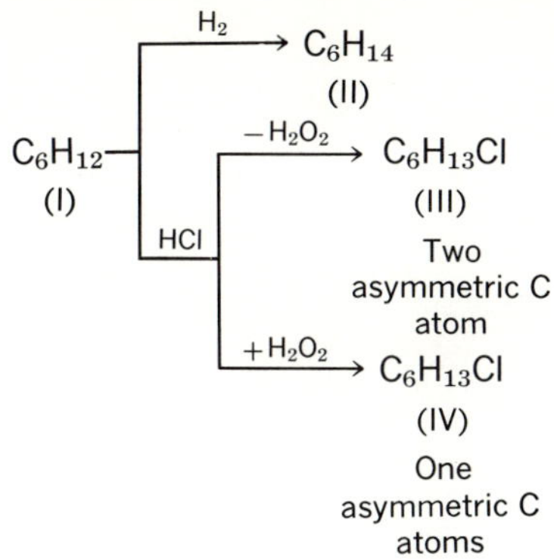

Write the structural formulas for I, II, III, and IV.

chapter 9.... alcohols

Alcohols are organic hydroxides of the general formula R—O—H, where R is an *alkyl or substituted alkyl radical.* The C—O—H group is responsible for the characteristic reactions of this class of compounds, but the structure of the R group as well as substituents in the R part of the molecule influence the chemical and biochemical activity of the OH group. They also influence the physical properties of the alcohols.

The following discussion stresses reactions of alcohols which are pertinent to biochemistry. Although there are numerous biological reactions involving alcohols, the types are relatively simple and few in number.

9.1 IMPORTANCE

Commercially, alcohols are important for their use as solvents and as starting materials for the synthesis of many other compounds, e.g., aldehydes, acids, esters, ketones, and alkenes.

Alcohols are also important biologically. The alcohol group occurs in a great variety of molecules associated with biological systems. Certain alcohols per se hold a prominent position in metabolic processes, but most biologically active alcohols also contain other functional groups. Many compounds contain one, two, or three hydroxyl groups, whereas others may contain more. Sugars, for example, all contain more than one OH; the simplest contains two. In contrast, a molecule of starch or cellulose contains thousands of hydroxyl groups (see formulas on page 175). Fats are derivatives of a simple trihydroxy alcohol, glycerol. Some vitamins are alcohols. For example, vision is partly dependent upon a reaction of the alcohol vitamin A (see formula on page 202).

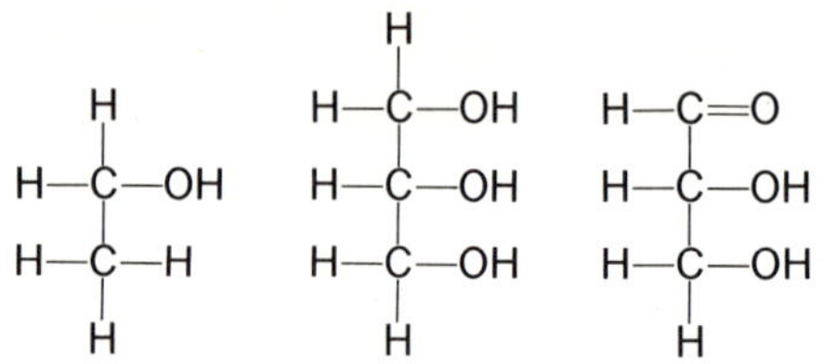

Ethyl alcohol Glycerol Glyceraldehyde

9.2 CLASSIFICATION

Alcohols are classified as primary, secondary, or tertiary depending upon the number of carbons bonded to the carbon holding the OH group. Primary alcohols have only one carbon (or three H) bonded to the carbon holding the OH group, secondary alcohols have two carbons bonded to the carbon holding the OH group, and tertiary alcohols have three carbons bonded to the carbon holding the OH group.

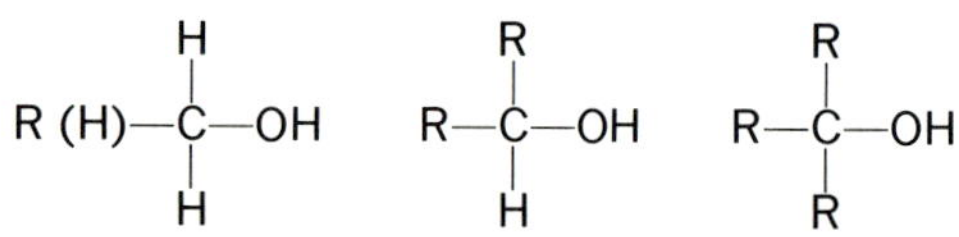

Primary Secondary Tertiary

9.3 NOMENCLATURE

As we have seen, there is more than one method for naming organic compounds. Today the common name for the alcohols is used only for a few of the simpler ones, e.g., ethyl alcohol (also known as grain alcohol because it is obtained from grain by fermentation) and methyl alcohol (also known as wood alcohol because it is obtained from wood by destructive distillation). This nomenclature system takes the name of the alkyl radical and adds the word alcohol to indicate the OH group. The methyl radical is CH_3, and the ethyl radical is CH_3CH_2; hence, their corresponding alcohols are called methyl alcohol and ethyl alcohol, respectively. The formulas for these alcohols and isopropyl alcohol are

CH_3—OH	CH_3CH_2—OH	$(CH_3)_2CH$—OH
Methyl alcohol	Ethyl alcohol	Isopropyl alcohol

Another common method of nomenclature is based on naming the C—OH group the *carbinol* group. Thus, the radicals which are attached to the carbinol group are named. For example, methylcarbinol has the formula CH_3—CH_2OH (ethyl alcohol), dimethylcarbinol is $(CH_3)_2$—CHOH (isopropyl alcohol), and trimethylcarbinol is $(CH_3)_3$—COH.

The *IUPAC system* of nomenclature is based upon the following rules:

1. The longest chain of carbon atoms is the parent substance (alkane) from which the name is derived.
2. The terminal *-e* of the parent alkane name is changed to *-ol* to indicate the OH group. If two OH groups are present, the suffix *-diol* is used, and if three OH groups are present, the suffix *-triol* is used.
3. The position of *each* OH is indicated by numbering the carbon on which it occurs, usually numbering the chain from the end *closest to the* OH *group.*
4. The other groups substituted in the parent chain are named and given a number, corresponding to the number of the carbon to which it is attached.

Some examples of the IUPAC nomenclature for alcohols are

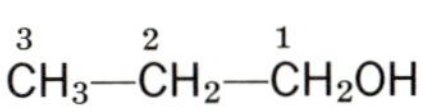

1-Propanol
Ethylcarbinol
n-Propyl alcohol

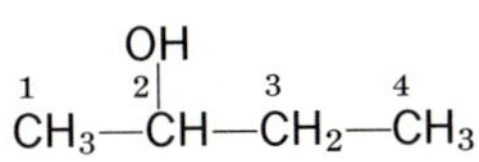

2-Butanol
Methylethylcarbinol

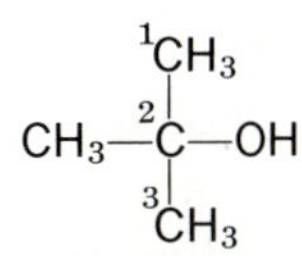

2-Methyl-2-propanol
Trimethylcarbinol

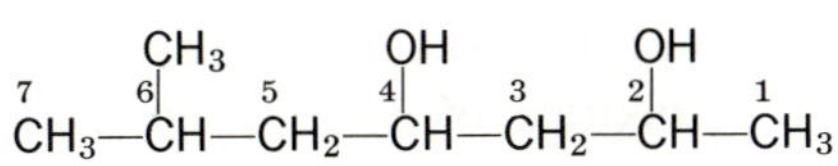

6-Methyl-2,4-heptanediol

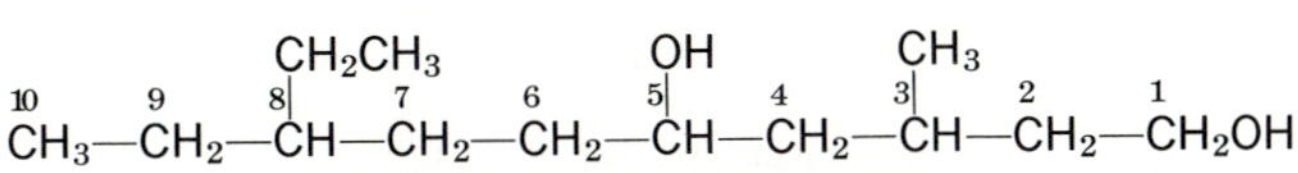

3-Methyl-8-ethyl-1,5-decanediol

Glycerol, CH_2OH—CHOH—CH_2OH, by the IUPAC system would be 1,2,3-propanetriol. Note that *each* of the OH groups must be numbered and the total number must also be indicated by *di-, tri-, tetra-,* etc.

9.4 FORMATION

Some reactions have already been discussed which yield alcohols as the end product, namely, (1) hydration of alkenes (Sec. 8.9); (2) hydrolysis of alkyl halides (Sec. 8.5); and (3) the oxidation of alkenes (Sec. 8.9).

Two major methods for the commercial preparation of low-molecular-weight simple alcohols are chemically by the hydration of alkenes and biologically by fermentation. The hydration of alkenes and the hydrolysis of alkyl halides have been discussed under the chemical reactions of these classes of compounds.

Fermentation

Although the production of alcoholic beverages by fermentation dates back to early man, it is only recently that the biochemical steps have been worked out. The stepwise reactions whereby sugar or starch is converted by yeast into ethyl alcohol are presented in Sec. 21.3. At this time only the overall equation for this process starting with glucose is considered.

$$\underset{\text{Glucose}}{C_6H_{12}O_6} \longrightarrow \underset{\text{Ethanol}}{2\ CH_3CH_2OH} + 2\ CO_2$$

Biologically, starch is not converted 100% to alcohol; a number of other products are formed, e.g., *n*-propyl alcohol, isobutyl alcohol, isopentyl alcohol, and amyl alcohol (the common name for pentyl alcohol). The mixture of these alcohols is known as *fusel oil*. The alcohols in this mixture can be separated by appropriate distillation procedures. By using a special medium in which to grow yeast, glycerol is produced as the main fermentation end product.

Organisms other than yeast are used to produce alcohols as the main fermentation product. For example, a specific bacterium is able to convert starch to a mixture of 10% ethyl alcohol, 30% acetone, $CH_3{-}C({=}O){-}CH_3$, and 60% *n*-butyl alcohol.

In summary,

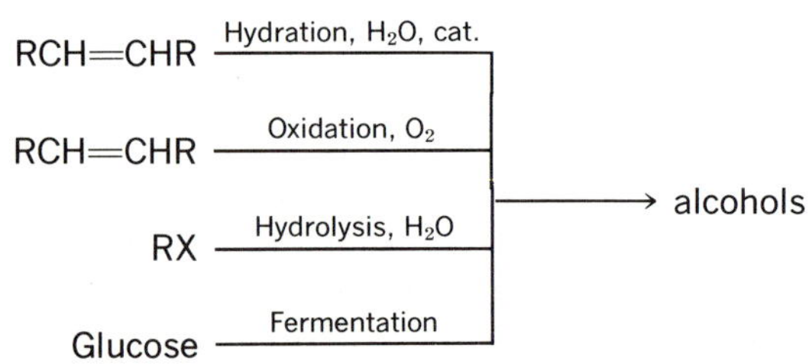

9.5 REACTIONS

Five types of reactions of alcohols will be discussed in this chapter: (1) the specific case of the oxidation of compounds that have two OH groups on adjacent carbons; (2) reactions in which the OH is removed and another group or element substituted for it; (3) dehydration reactions, in which the OH is eliminated along with a H from an adjacent carbon to yield an alkene; (4) reactions in which the H of the OH is replaced by another group or element; and (5) dehydrogenation reactions involving the removal of the H from the OH group and another H from the same carbon to which the OH is attached to form a C═O group.

Oxidation

Chemically the prerequisites for the oxidation of common compounds containing oxygen are that there must be a hydrogen on the carbon bonded to the oxygen and the oxygen must not be bonded to more than one carbon. These prerequisites are found in such compounds as

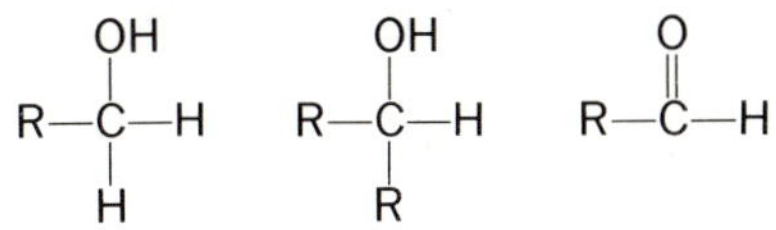

1° alcohol 2° alcohol Aldehyde

but not in such compounds as

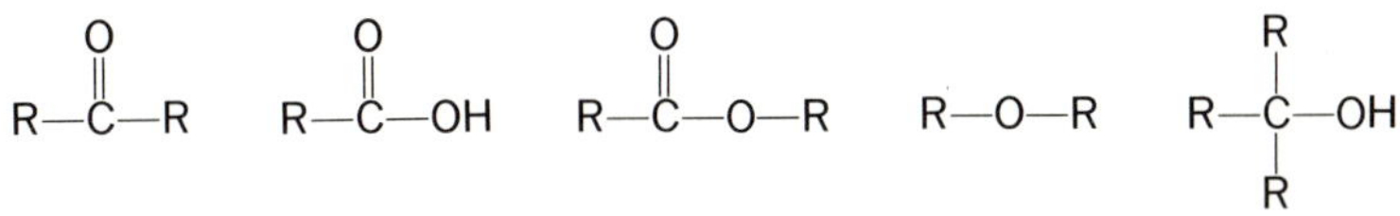

Ketone Acid Ester Ether 3° alcohol

The oxidation of alcohols is equivalent to the removal of a H from the OH group and a H from the C to which the OH group is attached.

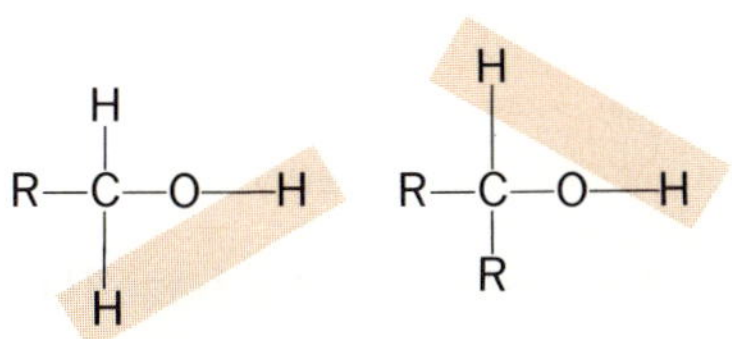

1° alcohol 2° alcohol

It must be remembered in this discussion that *oxidation of functional groups does not mean combustion of the compound.*

If a prerequisite for oxidation of the alcohols is the presence of a H on the carbinol C, would you expect tertiary alcohols to be easily oxidized? From the general mechanism of oxidation of alcohols, discussed next, it can be seen why a H on the carbinol C is essential.

The mechanism for the oxidation of alcohols is actually a dehydrogenation reaction, whereby two hydrogen atoms are removed, one hydrogen atom from the carbon holding the OH group and one hydrogen atom from the OH group.

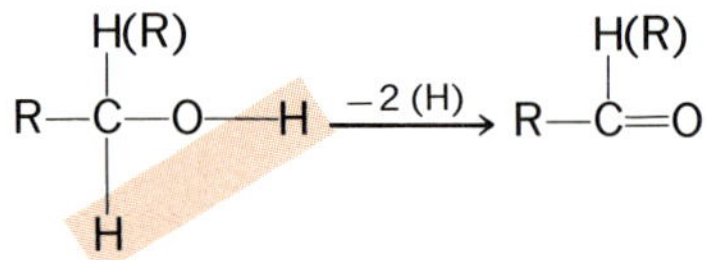

1° or 2° alcohol — Aldehyde or ketone

Tertiary alcohols, on the other hand, are not easily oxidized. There is no hydrogen atom on the carbon holding the OH group.

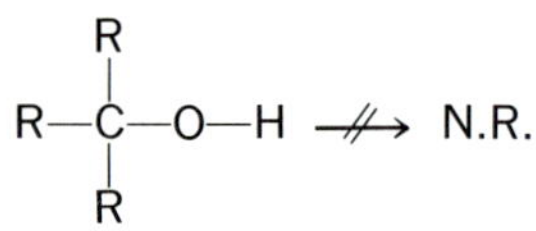

3° alcohol

The mechanism of oxidation of secondary alcohols has been thoroughly studied. With chromic acid, H_2CrO_4, in a slightly basic solution the oxidation proceeds as

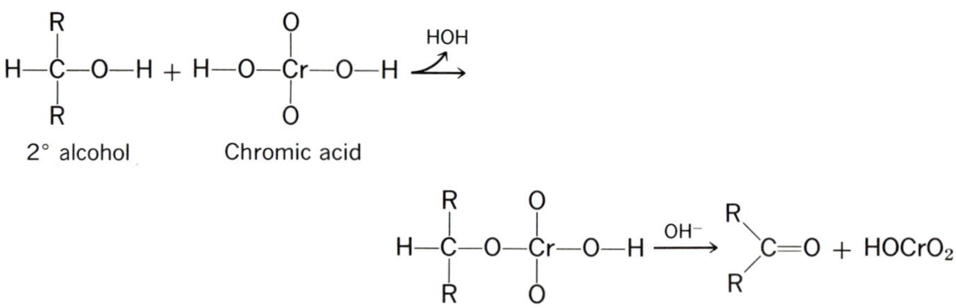

2° alcohol — Chromic acid — Alkyl chromate ester — Ketone — Chromous acid

The chromate ester in the presence of a base (OH^-) splits as shown in the equation to yield a ketone as the organic product. The overall reaction is the dehydrogenation of the alcohol.

The aldehyde resulting from the oxidation of a primary alcohol

contains a H on the C═O. This structure is necessary for an easily oxidizable substance. In fact, the aldehyde group is even more easily oxidized than the alcohol group. Therefore, if aldehydes are to be prepared by an oxidation reaction as above, it is necessary to use a very selective oxidizing reagent or to remove the aldehyde from the reaction mixture before it is oxidized further. The mechanism for the oxidation, which is somewhat different than for alcohols, will be given under aldehydes.

DICHROMATE OXIDATION. Primary alcohols can conveniently be oxidized to aldehydes with $K_2Cr_2O_7$. The resulting aldehyde can be removed by appropriate methods or allowed to be further oxidized to the corresponding acid. Secondary alcohols are oxidized to the corresponding ketone:

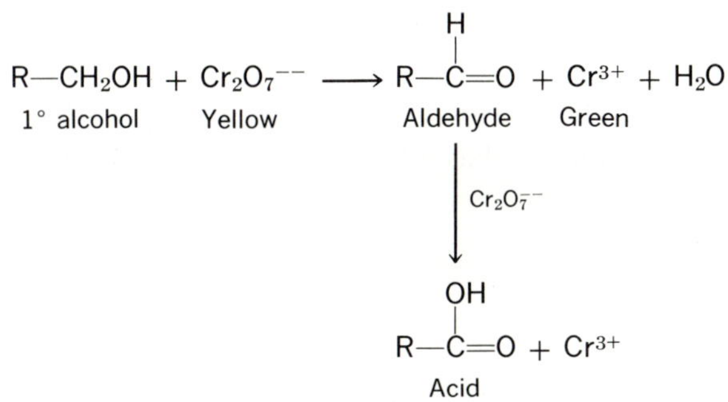

(9.1)

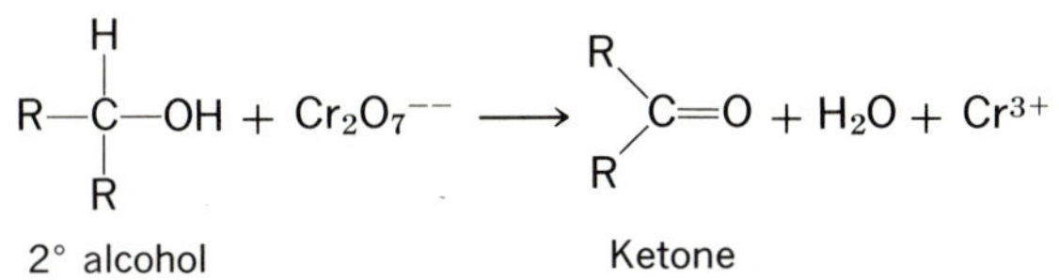

(9.2)

Examine the formula for the ketone produced in Eq. (9.2). It will be noted that there is no H atom on the C═O. Therefore, ketones are relatively difficult to oxidize, as breaking carbon-to-carbon bonds would be necessary.

CATALYTIC OXIDATION. Inorganic catalysts, e.g., copper, oxidize primary and secondary alcohols to the corresponding C═O compound by removing the H of the OH group and a H from the carbinol carbon. This can be accomplished by passing the alcohol vapors over metallic copper at 200 to 300°C:

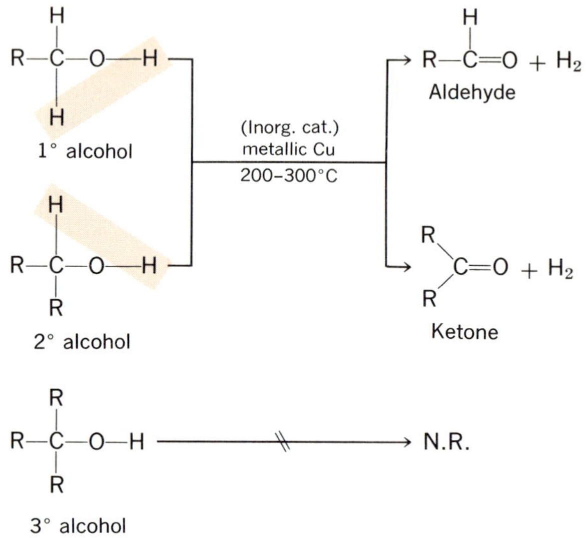

Biological catalysts (enzymes) also catalyze the oxidation of primary and secondary alcohols by the removal of two atoms of hydrogen:

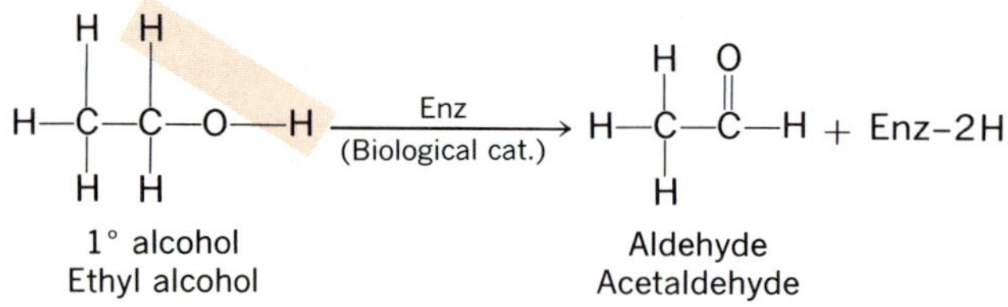

There are several differences to be noted between the chemical and the biological reactions. In the chemical reaction molecular hydrogen, H_2, is produced, whereas in the biological reaction the H atoms become associated with the enzyme (this is indicated as Enz–2H). The fate of this hydrogen will be discussed later. The chemical reaction is not specific; i.e., alcohols in general react as indicated, and only general formulas for alcohols were used to illustrate the reaction. The biological reaction, on the other hand, is quite specific. Only certain alcohols react in this manner, and for maximum oxidation a different enzyme is required for each alcohol. Inorganic catalysts function at elevated temperatures, but biological catalysts decompose with heat.

Replacement of OH by halogens

The OH group of alcohols can be replaced by halogens to form the corresponding alkyl halide. This is accomplished by reacting alcohols with the hydrogen halide, HX.

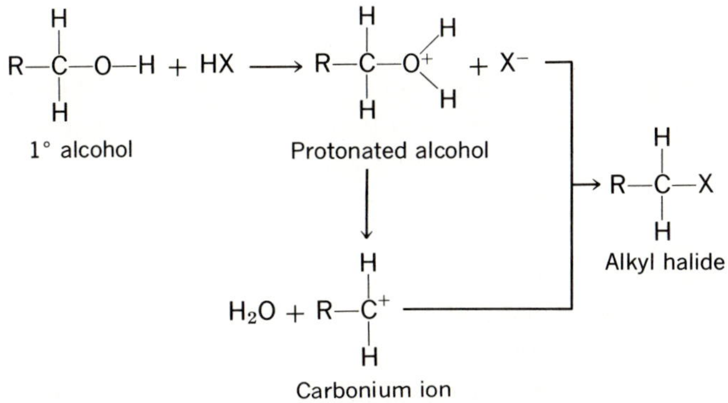

Net reaction: $RCH_2OH + HX \longrightarrow RCH_2X + H_2O$

The reaction is influenced both by the nature of the HX and by the structure of the alcohol. The order of reactivity is HI > HBr > HCl, and for the alcohols 3° > 2° > 1°. Hence, to carry out this reaction different conditions must be employed depending upon the acid and alcohol used. The conditions under which primary, secondary, and tertiary alcohols react with HCl are

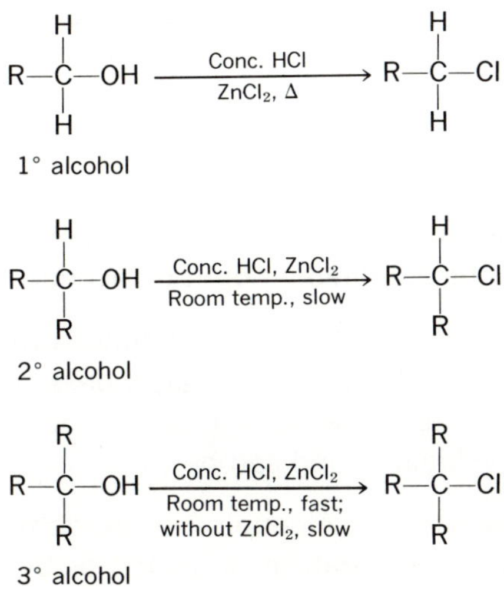

This reaction with HCl provides a simple means of differentiating between primary, secondary, and tertiary low-molecular-weight water-soluble alcohols (those containing five carbons or less). Since the resulting alkyl chlorides are insoluble in aqueous HCl, as the reaction proceeds two phases appear.

EXAMPLE

You are given a test tube containing a water-soluble alcohol and asked to determine whether it is a primary, secondary, or tertiary alcohol. How would you proceed? You could distinguish between the three classes of alcohols by adding at room temperature about 1 ml of the alcohol to about 10 ml of a solution of concentrated HCl containing some $ZnCl_2$ as a catalyst. The tertiary alcohols react immediately to give a cloudy solution, the secondary alcohols react after about 5 min, and the primary alcohols do not react at room temperature but must be heated to initiate the reaction.

Dehydration

The dehydration of alcohols proceeds (1) by the elimination of 1 mole of water between two molecules of alcohol to yield ethers or (2) by the elimination of 1 mole of water from 1 mole of alcohol to yield an alkene. The first reaction is an intermolecular dehydration, whereas the second reaction is an intramolecular reaction. The general equations are

$$\underset{\text{Alcohol}}{\text{R—O—H}} + \underset{\text{Alcohol}}{\text{H—O—R}} \xrightarrow{H_2SO_4} \underset{\text{Ether}}{\text{R—O—R}} + \text{HOH}$$

$$\underset{\text{Alcohol}}{\text{R—CH}_2\text{—CH}_2\text{—OH}} \xrightarrow[Al_2O_3]{H_2SO_4} \underset{\text{Alkene}}{\text{R—CH=CH}_2} + \text{HOH}$$

To form ethers there must be an excess of alcohol, whereas, to form alkenes there must be a limited amount of alcohol.

DEHYDRATION TO FORM ETHERS. A mixture of concentrated H_2SO_4 and the alcohol is heated to 140°C, and the alcohol is continuously added to keep it in excess as the ether forms. At the temperature of the reaction the ether volatilizes and can be condensed and collected.

DEHYDRATION TO FORM ALKENES. Alcohols can be dehydrated to form alkenes by H_2SO_4 at a temperature of 180°C or by passing vapors of the alcohol over Al_2O_3 at a temperature of 250°C. The nature of the alcohol governs the ease with which it is dehydrated. The order of reactivity of the alcohols toward dehydration is 3° > 2° > 1°.

MECHANISM OF DEHYDRATION WITH H_2SO_4. The dehydration of alcohols by H_2SO_4 involves several steps. First is the protonation of the alcohol to form a substituted hydronium ion [reaction (1)].

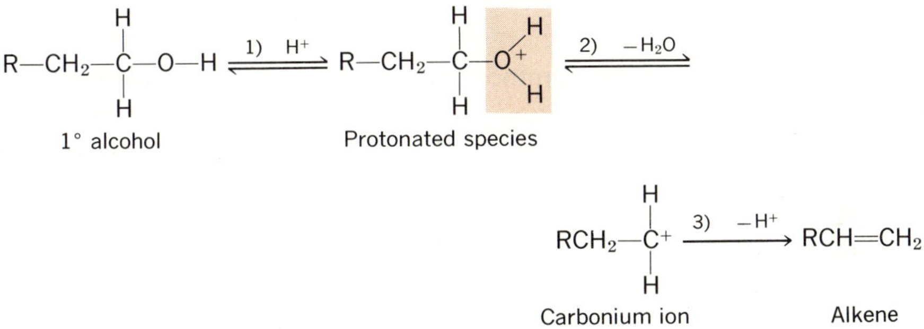

Step 2 involves the elimination of water to form a carbonium ion, an ion containing a carbon atom which has lost one electron, C^+. Step 3 involves the elimination of a proton from the carbon atom adjacent to the carbonium carbon to permit the formation of a C═C bond. In the formation of ethers the carbonium ion reacts with another mole of alcohol to form an ether

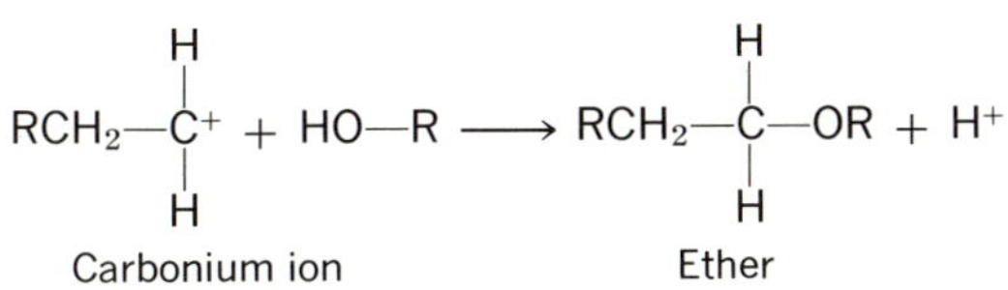

Summary equations:

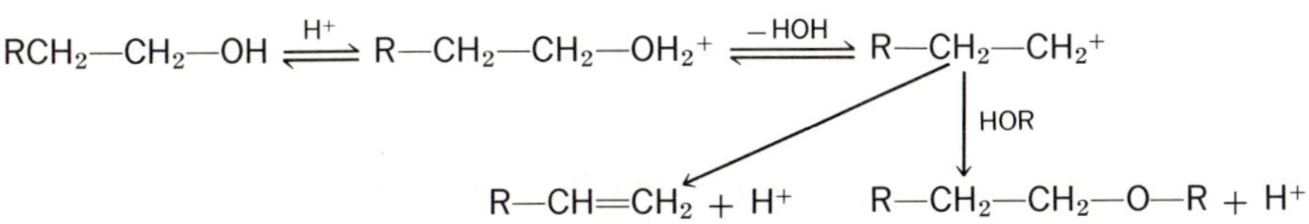

COMPARISON OF DEHYDRATION WITH SUBSTITUTION. It is to be noted that the order of reactivity of the alcohols (3° > 2° > 1°) is the same *for both dehydration reactions and substitution reactions by* HX; both reactions involve the elimination of an OH group. This suggests that the mechanism for these two reactions may involve a common intermediate, which it does. Compare the following equations with those for the dehydration of alcohols:

$$R{-}CH_2{-}OH + HX \overset{1)}{\rightleftharpoons} R{-}CH_2{-}\overset{+}{O}H_2 + X^- \overset{2)\ -H_2O}{\rightleftharpoons}$$

$$\underset{\text{Carbonium ion}}{R{-}CH_2^+} \overset{3)\ X^-}{\rightleftharpoons} R{-}CH_2X$$

The carbonium ion is the common intermediate. The relative reactivity of the alcohols in these two types of reactions is governed by the ease with which the carbonium ion is formed. The order of decreasing ease is

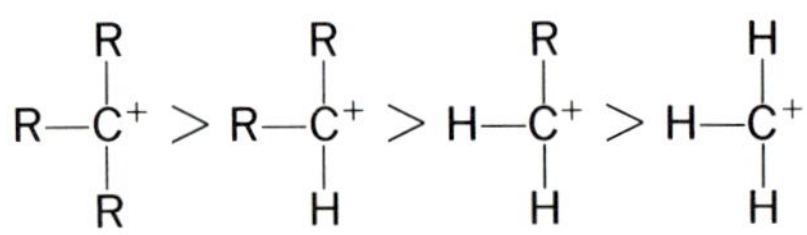

3° 2° 1°

Replacement of H in the OH group

The H in the OH group of alcohols can be replaced by active metallic elements, such as sodium or potassium, to form alkoxides. In this sense the alcohol is reacting as an acid similar to water.

$$H{-}O{-}H + Na \longrightarrow H{-}O{-}Na + \tfrac{1}{2}H_2$$

$$\underset{\text{Alcohol}}{R{-}O{-}H} + Na \longrightarrow \underset{\text{Sodium alkoxide}}{R{-}O{-}Na} + \tfrac{1}{2}H_2$$

In the presence of water, sodium alkoxides are converted to the original alcohol:

$$R{-}O{-}Na + HOH \longrightarrow R{-}O{-}H + NaOH$$

Alkoxides are used in a number of syntheses. One is the preparation of ethers by reacting an alkyl halide with an alkoxide:

$$R{-}O{-}Na + Cl{-}R' \longrightarrow R{-}O{-}R' + NaCl$$

The sodium and potassium alkoxides are very strong bases, stronger than the corresponding metallic hydroxides. It is frequently necessary to have a strong base in a solvent for water-insoluble compounds. This can be accomplished by reacting sodium with excess methanol or ethanol. The alkoxides formed are relatively soluble in the excess alcohol.

Some biological dehydrogenation, hydration, and dehydration reactions

Many reactions in living organisms involve the dehydrogenation of the alkane part of a molecule, the hydration of an alkene group, and the dehydrogenation of an alcohol group. These steps are key reactions in the synthesis of many biochemicals from food as well as in the metabolism of foods to water, carbon dioxide, and energy. Such reactions require enzyme catalysts because they take place at body temperature. A cell cannot survive the treatment with concentrated H_2SO_4 and high temperatures required for the chemical reaction of dehydration of alcohols.

Fatty acids, from fats, are utilized by living organisms as a source of energy by oxidizing the acids. The initial phase of fatty acid oxidation involves conversion of the acid to a keto acid. (The acid is written as the free acid for the sake of simplicity at this point, the R group representing a saturated hydrocarbon chain and Enz representing enzymes.)

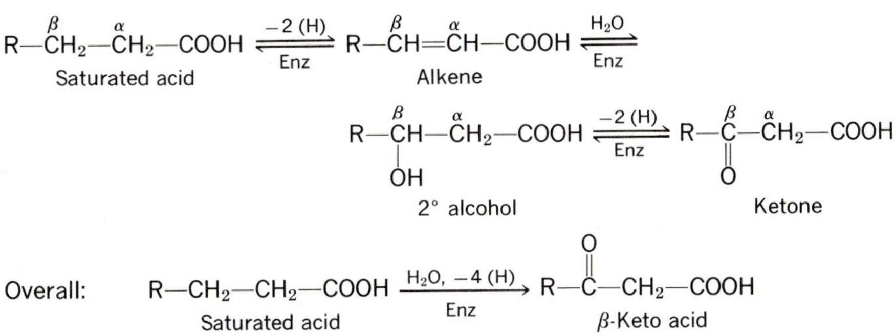

The first reaction is the dehydrogenation of the alkane part of the acid to yield an alkene group (discussed in Chap. 8). Water is then added to the alkene group to form a secondary alcohol, which is subsequently dehydrogenated to form a ketone group on the β carbon. The overall reaction is the oxidation of the β carbon of the saturated acid to a ketone group, the oxygen coming from water.

Two similar reaction series are the conversion of succinic acid to oxaloacetic acid and of citric acid to oxalosuccinic acid. Both these conversions play a major role in the complete oxidation of fatty acids, carbohydrates, and proteins. Without these pathways it would be impossible for us to oxidize food properly or to synthesize many biological compounds. To cite just one example, succinic acid is used in the synthesis of hemoglobin. These reactions will be presented in their proper perspective in future discussions; for the present it is essential to understand the basic chemistry of the several transformations.

The conversion of succinic acid to oxaloacetic acid is

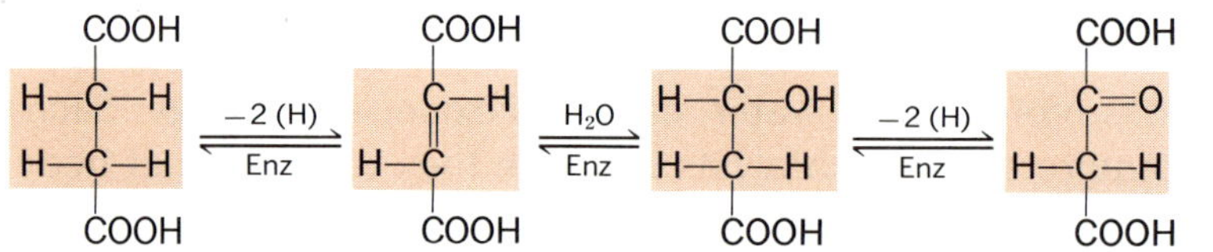

Alkane Succinic acid — Alkene Fumaric acid — 2° alcohol Malic acid — Ketone Oxaloacetic acid

Overall:

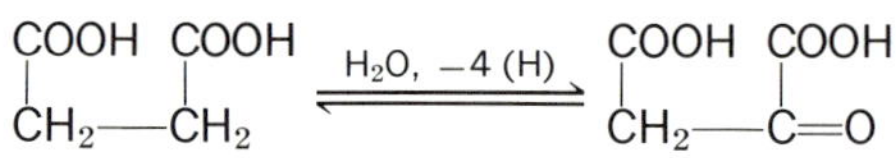

Succinic acid — Oxaloacetic acid

The conversion of citric acid to oxalosuccinic acid is

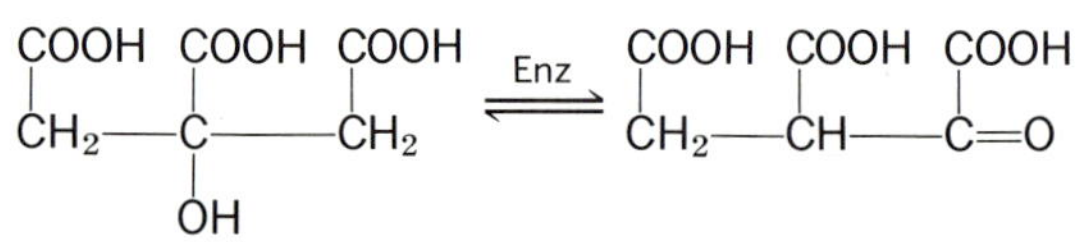

Citric acid — Oxalosuccinic acid

PROBLEM

To write an equation for this last conversion it is perhaps best to start by working backward. Based on previous reactions the following reasoning can be applied:

1. The C=O group can be derived by dehydrogenation of a secondary alcohol:

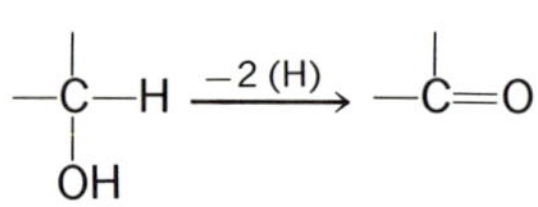

2. The secondary alcohol in turn can be produced by the hydration of an alkene group:

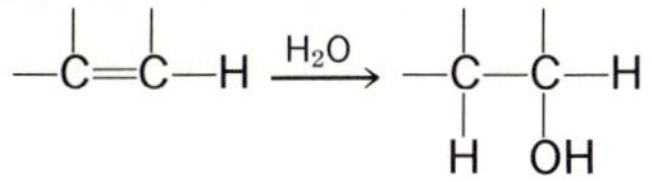

3. The alkene group can be produced by one of two reactions: (1) the dehydrogenation of an alkane group or (2) the dehydration of an alcohol group. The latter reaction seems more likely in this instance since citric acid itself contains an alcohol group.

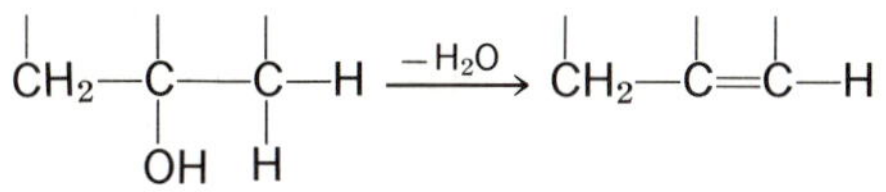

Starting with citric acid and combining these types of reactions gives

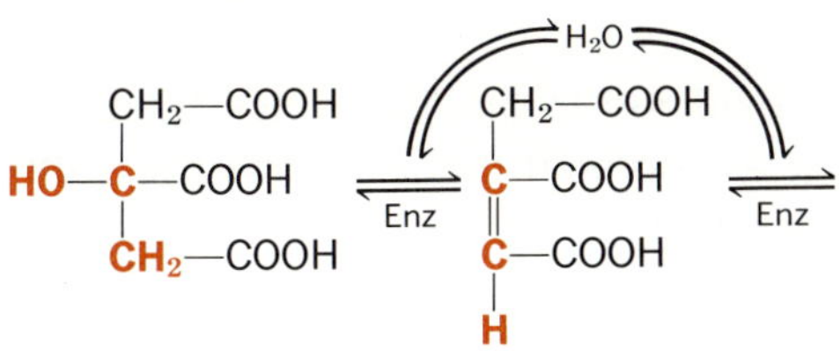

Citric acid · *cis*-Aconitic acid

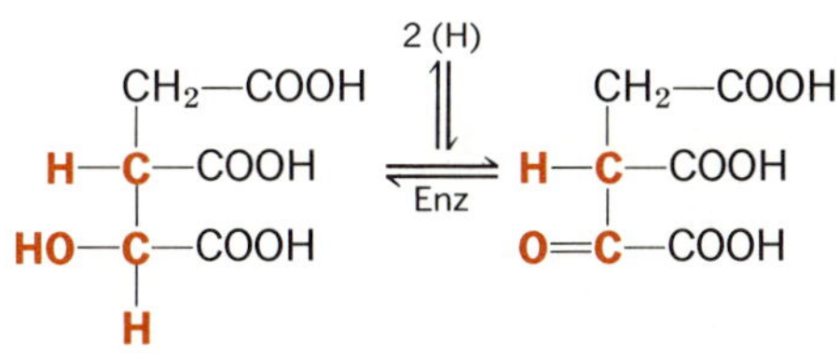

Isocitric acid · Oxalosuccinic acid

Citric acid is dehydrated to form *cis*-aconitic acid, which contains an alkene group. The addition of water to the alkene group in the manner indicated leads to the formation of an isomer of citric acid, isocitric acid. Subsequent dehydrogenation of the secondary alcohol group of isocitric acid forms the ketone group in oxalosuccinic acid.

These reactions illustrate the biological role of water. Water is a biologically active compound; it is not merely an inert solvent. The oxygen added to the compounds discussed above came from water; the oxidation was brought about by dehydrogenation. The hydrogens removed are associated with the enzyme; note that molecular hydrogen is not formed. This hydrogen is passed eventually to molecular oxygen, O_2, to form water and energy. A word of caution is needed about the use of the abbreviation Enz in these reactions: *each reaction requires a different and characteristic enzyme.*

9.6 THIOALCOHOLS

Oxygen and sulfur are in the same group of the periodic table. Both have six electrons in their outer orbit and form many similar types of compounds. For example, if the oxygen of water is replaced by sulfur, hydrogen sulfide, H_2S, results. The hydrogen attached to a sulfur atom is weakly acidic in nature. H_2S is a weak acid, $K_1 = 7.04$. If an alkyl group replaces one hydrogen in H_2S, a thioalcohol is formed. The prefix *thio-* indicates the sulfur. Thioethanol would have the formula $CH_3—CH_2—SH$.

Thioalcohols are easily oxidized to disulfides, R—S—S—R. Cysteine is a naturally occurring amino acid that contains the sulfhydryl, SH, group. A biological reaction involving the oxidation of the SH group of two cysteine molecules is

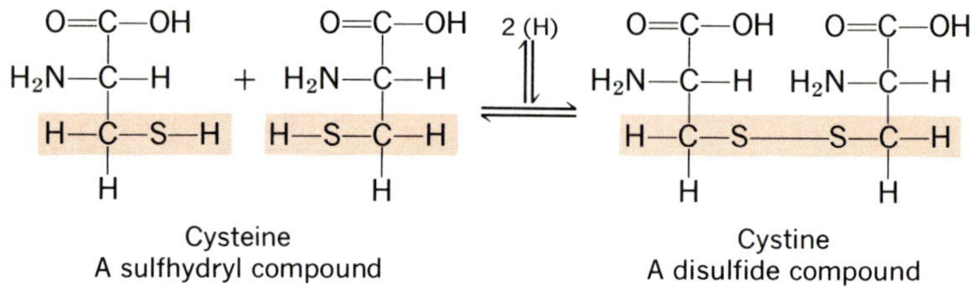

Thioalcohols are hydrolyzed by strong aqueous base and heat to the corresponding alcohol and the sulfide of the base:

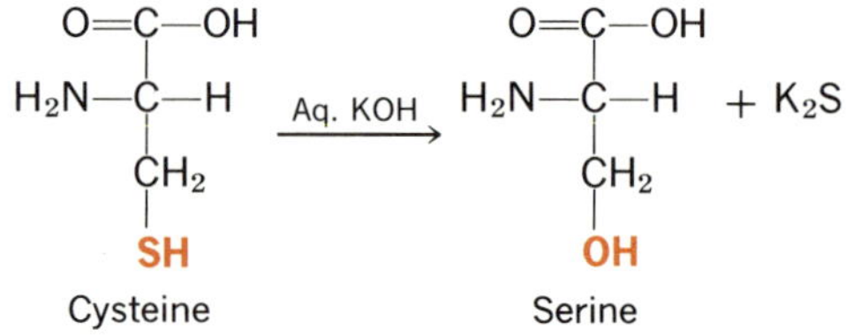

9.7 SUMMARY OF THE REACTIONS OF ALCOHOLS

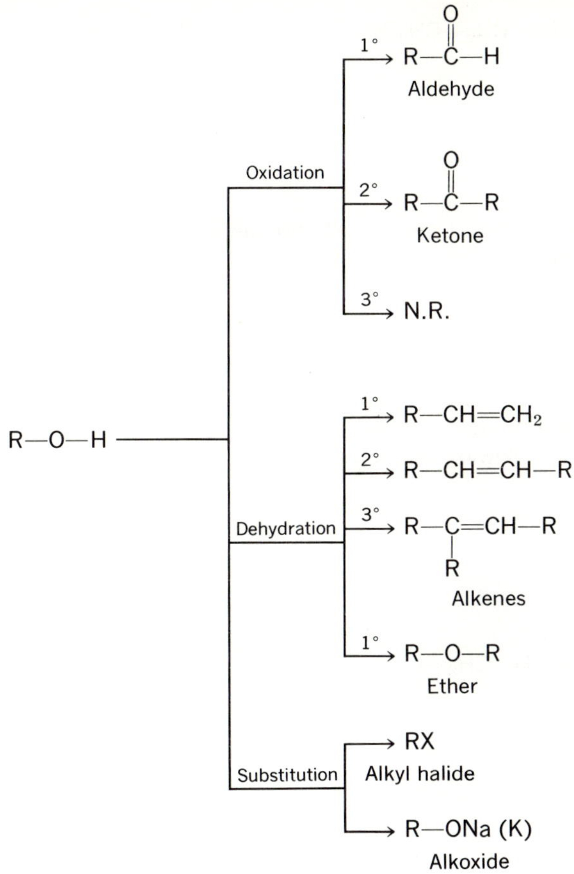

QUESTIONS

1. (*a*) Find and name the functional groups in the following compound:

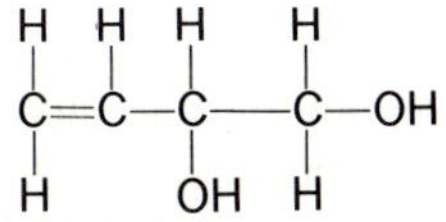

(*b*) Does it have an asymmetric carbon atom?

2. Write the formula for the following compounds:

(*a*) Isopropyl alcohol

(*b*) 2,2-Dimethyl-1-propanol

(*c*) 2-Ethyl-3-chloro-2-hexanol

3. Define and write the structural formula for an example of primary, secondary, and tertiary alcohol.

4. Name the following compounds:

(*a*) $CH_3—CH_2—CH_2—CH_2OH$

(*b*) 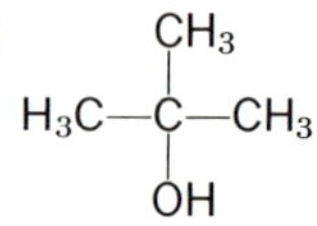

(*c*) $CH_3—CHOH—CHOH—CH_3$

(*d*) $CH_3—CHOH—CHOH—CH_2OH$

5. What is the structural prerequisite for the oxidation of an alcohol?

6. (*a*) Write the equation for the oxidation of

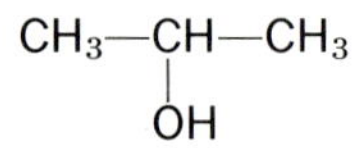

(*b*) Name the product formed.

7. Which of the following alcohols are easily oxidized?

(*a*) $CH_3—CH_2OH$

(*b*) 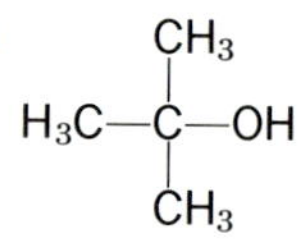

(*c*) $CH_3—CHOH—CH_2—CH_3$

(*d*) 2-Methyl-2-butanol

(*e*) 2-Methyl-2-propanol

8. What is the class of compounds formed when a primary alcohol is oxidized? A secondary alcohol?

9. (*a*) What is the order of reactivity of primary, secondary, and tertiary alcohols with concentrated HCl and $ZnCl_2$?

(*b*) Write an equation for one of the classes of alcohols to illustrate this reaction.

10. What two classes of compounds can be formed from the dehydration of primary alcohols? Write an equation to illustrate your answer.

11. What is a carbonium ion?

12. Will alcohols react with sodium? Write an equation for this reaction.

13. What is meant by dehydrogenation of an alcohol? Is this reaction an oxidation or a reduction reaction? Explain.

14. Is the dehydrogenation of alcohols a biologically important reaction? Give an example of such a reaction.

15. Write the equations for the conversion of succinic acid to oxaloacetic acid.

16. An unknown compound I is found to have the molecular formula $C_4H_{10}O$ and contains no asymmetric carbon atom. It undergoes the following general reactions, the molecular formulas of products being as given. What is the structural formula for each of the compounds?

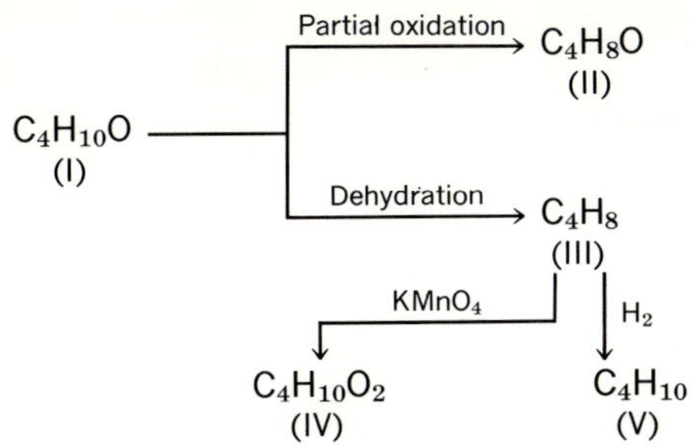

17. Write a possible structural formula for compounds I to VIII.

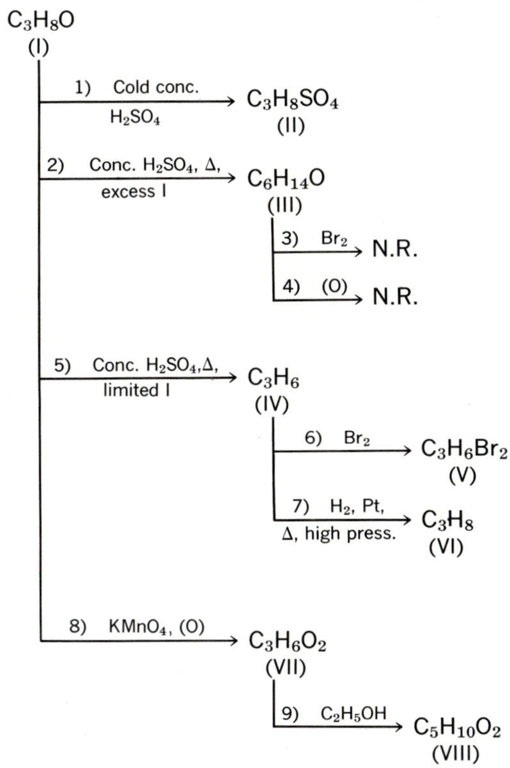

chapter 10.... aldehydes and ketones

Aldehydes and ketones constitute a wide variety of biologically important compounds. They are also intermediates in a number of biological reactions. Carbohydrates contain one or more hydroxyl groups and an aldehyde or ketone group. Vitamin A is an alcohol which plays a prominent role in the process of vision. The aldehyde group is also involved in this process because vitamin A alcohol is oxidized to vitamin A aldehyde. The series of biological oxidation reactions in Chap. 9 involved the formation of a ketone group. Reactions of the aldehyde and ketone group are prominent in many biological systems.

Our discussion of the chemistry of the aldehyde and ketone groups emphasizes biological compounds and reactions, but it should be remembered that aldehydes and ketones are important starting materials for a variety of chemical syntheses which neither time not space permits including in this chapter.

The reactions in the previous chapter illustrating the biological dehydrogenation of alcohols are reversible; therefore, they can also be used to illustrate the reduction of the carbonyl, C═O, group. What compounds would be formed upon the dehydrogenation of the alcohol group in malic acid, isocitric acid, and β-hydroxybutyric acid?

The classes of compounds known as aldehydes and ketones contain the carbonyl group, C═O:

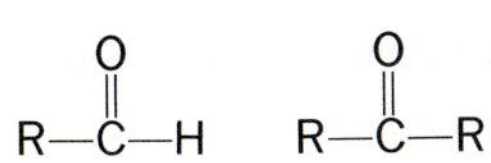

Aldehyde Ketone

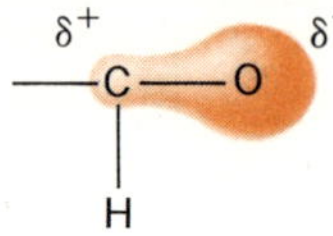

FIGURE 10.1
Electron polarization in the carbonyl group.

The electronegativity of the oxygen atom causes a greater electron density to be associated with it than with the carbon,

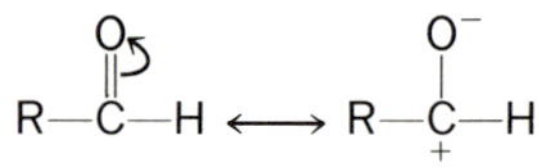

Perhaps a better representation is given in Fig. 10.1. There is not a complete shift of electrons to the oxygen atom, but the aldehyde group is sufficiently polar to allow for the possibility of the following addition mechanisms.

10.1 REACTIONS OF THE CARBONYL GROUP

Addition of water

Because of the polarity of the aldehyde group it is easily hydrated:

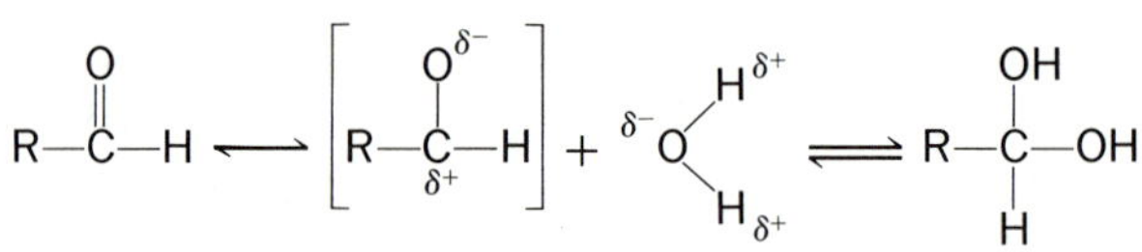

Aldehyde

Hydrated aldehyde (unstable)

Specific example:

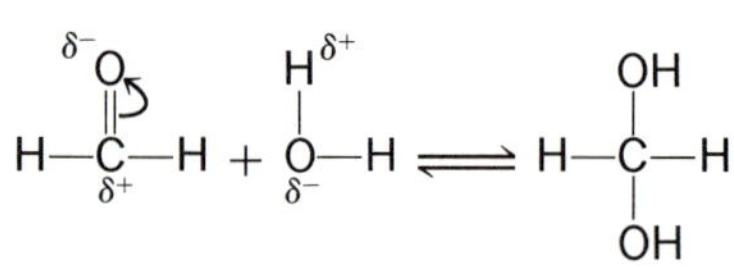

Formaldehyde

Hydrated formaldehyde

Formaldehyde in an aqueous solution is about 99% in the hydrated form.

Two points should be remembered with respect to the hydration of aldehydes: (1) the reaction is an equilibrium reaction, and (2) the hydrated aldehydes are very unstable. Only in a very few instances are the hydrates stable enough to be isolated.

A hydrated aldehyde has two hydroxyl groups on one carbon, which leads to instability. Water is eliminated to form the free aldehyde. A general rule to remember is that *two hydroxyl groups on the same carbon give an unstable arrangement.* A common compound already studied that illustrates this general rule is carbonic acid, H_2CO_3 (Chap. 6).

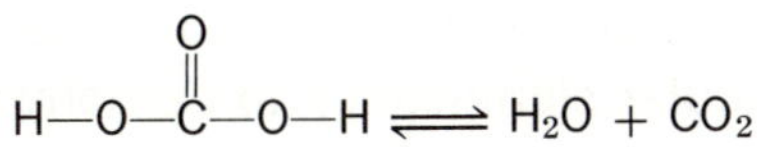

Carbonic acid is too unstable to exist alone; it occurs only in solution, where it is in equilibrium with carbon dioxide and water.

Reactions with alcohols

Hemiacetals and acetals are formed by reacting aldehydes with alcohols. In the presence of *dry* HCl, one molecule of alcohol adds to the aldehyde group to form a hemiacetal

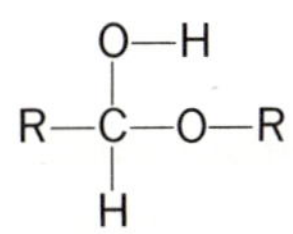

which may, in turn, form an acetal

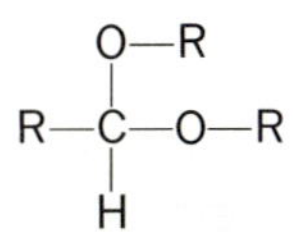

by reacting with another molecule of alcohol. The electronegativity of the oxygen atoms determines the direction of addition.

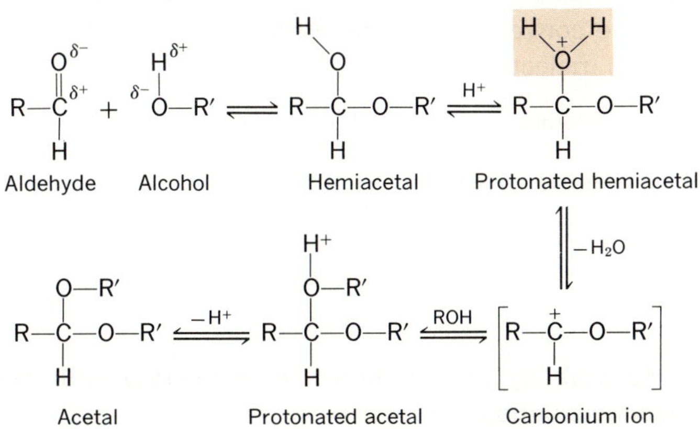

For example,

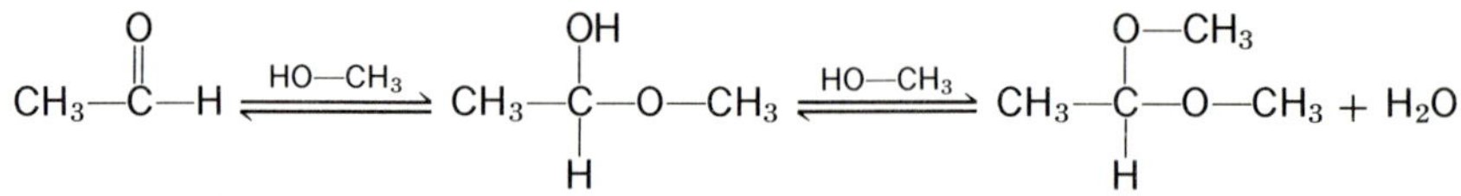

Acetaldehyde Methyl hemiacetal Dimethyl acetal

The stability of hemiacetals and acetals (also hemiketals and ketals, page 249) is of particular significance to carbohydrate chemistry and biochemistry. The acetal structure occurs in a number of important biological compounds, e.g., the compound that supplies energy for cellular reaction, the genetic material DNA, and a number of enzyme systems. Some sugars, such as glucose, exist in nature as a hemiacetal; other carbohydrates, like starch and cellulose, contain both acetal and hemiacetal groups. Digestion of carbohydrates involves the hydrolysis of the acetal linkage.

The synthesis of hemiacetals and acetals requires strict anhydrous conditions and H^+ as a catalyst. The hemiacetal group >C(OH)OR is *very labile (unstable) in an aqueous medium, and equilibrium* is reached between the respective aldehyde and alcohol and the hemiacetal. Acids and especially bases catalyze the attainment of equilibrium:

$$\text{Hemiacetal} \rightleftharpoons \text{aldehyde} + \text{alcohol}$$

The C—O—C group in hemiacetals does *not* have properties of an ether bond: the ether bond is stable in aqueous solutions. In hemiacetals the OH group on the "ether" carbon makes the C—O—C bond labile.

Acetals are also formed under anhydrous conditions in the presence of H^+. The conversion of a hemiacetal to an acetal involves the elimination of a molecule of water:

$$\text{Hemiacetal} + \text{alcohol} \rightleftharpoons \text{acetal} + \text{water}$$

In acetals there is no longer an OH group on the "ether" carbon:

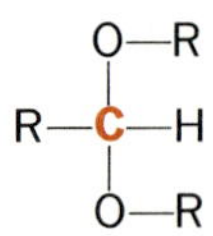

This leads to stability of the molecule, especially with respect to aqueous alkaline solutions.

The formation of acetals is reversible; the reverse reaction involves hydrolysis. (In the case of hemiacetals the reverse reaction was a decomposition reaction.) Hydrolysis of acetals proceeds under *aqueous acid conditions,* usually at elevated temperatures. Acetals are *not hydrolyzed by aqueous alkali.* For example, some are not hydrolyzed by boiling in 20% KOH for several minutes. Compare this alkali stability with the stability of hemiacetals in alkaline solution.

The formation and decomposition of acetals and hemiacetals may be summarized as follows:

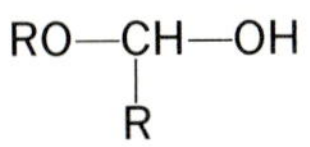
Hemiacetal

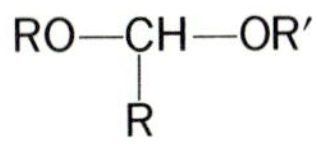
Acetal

Hemiacetal and hemiketal are formed in anhydrous acid. The reaction is reversed in aqueous acid or base.

Acetal and ketal are formed in anhydrous acid but the reaction is reversed in hot aqueous acid. These forms are stable in aqueous base.

The formation of hemiacetals and acetals was illustrated with an intermolecular reaction between two molecules, an aldehyde and an alcohol. It is important to remember that the reaction is between the two functional groups. Therefore, one may ask: If the aldehyde and the alcohol groups are in the same molecule, is it possible for an intramolecular reaction to take place? The answer: yes, provided there are a sufficient number of carbons between the aldehyde and the alcohol group to permit a stable ring formation. Five- and six-membered hemiacetal rings are stable. Consider the following equation:

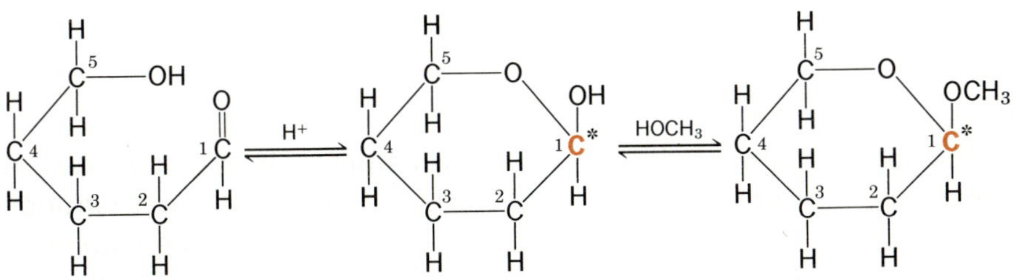

(I) Hydroxyaldehyde
(II) Cyclic hemiacetal
(III) Cyclic acetal

Are any of the carbon atoms in formulas I to III asymmetric? There is none in I, but note that C-1, which was the aldehyde carbon, becomes asymmetric upon ring closure. Therefore, both formulas II and III have an asymmetric carbon atom in their structure.

The cyclic hemiacetals and acetals are of importance in explaining the properties of sugars. The cyclization of glucose, an aldehyde sugar, is

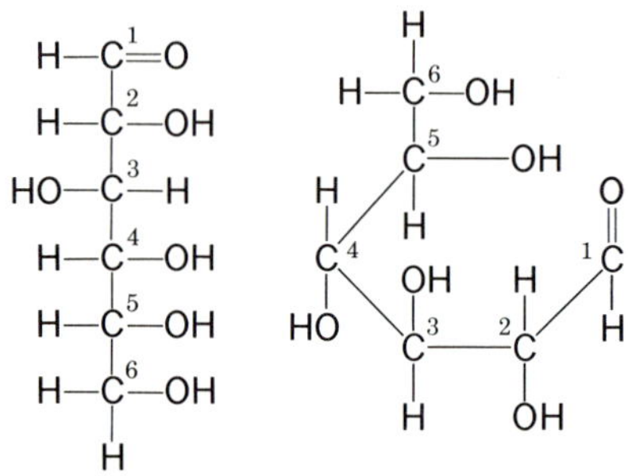

(I) (I′)

D-Glucose
Open-chain aldehyde

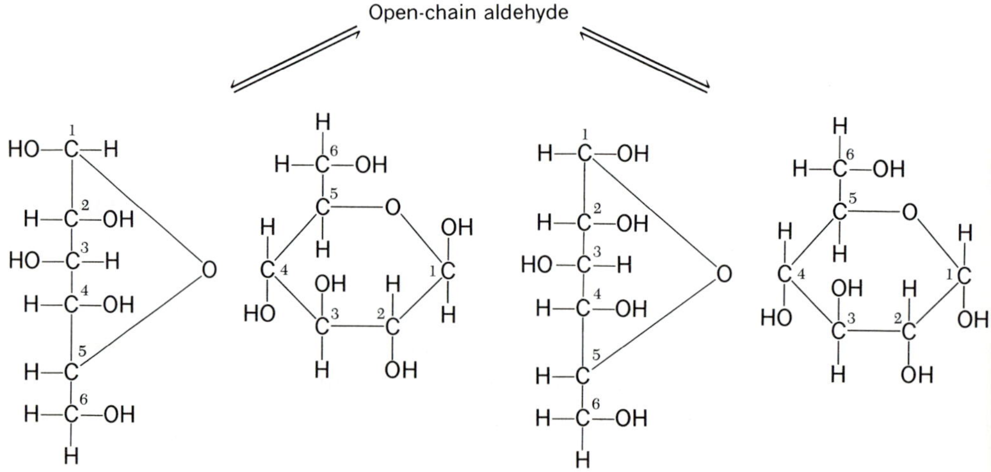

(II) (II′) (III) (III′)

β-D-Glucose α-D-Glucose

In the open straight-chain structure it is customary to write sugars with the aldehyde or ketone group at the top (see formula I). The OH groups are then placed on the left or right of the asymmetric carbon atoms in the chain to make the respective sugars. A more realistic method of writing the formula for glucose, shown in formula I′, brings the aldehyde group and the OH on the fifth carbon atom within close proximity of each other. To write this formula, C-1 is placed at the right, and the other carbons are written proceeding clockwise to the fifth carbon. If the OH appears on the right side of the carbon chain in formula I, it is placed below the carbon in the cyclic structure. The orientation about the C-5 may appear incorrect, but it is necessary to write it as shown to make the ring closure. The terminal CH_2OH group is written above the C-5 for this class of sugar. More will be said about writing these formulas and the significance of the letter D in a later chapter. For the present, we are interested in the formation of hemiacetals and acetals.

Note that formation of the cyclic hemiacetal produces a new asymmetric carbon, C-1. Consequently there are two possible optical isomers depending upon the configuration around the first carbon atom. These isomers are called the α and β form. When the OH group on C-1 of glucose is on the right side of the chain, it is called *α-glucose,* and when it is on the left side of the chain, it is *β-glucose.*

The cyclic hemiacetals of glucose are capable of forming acetals known as *glucosides.* One example is

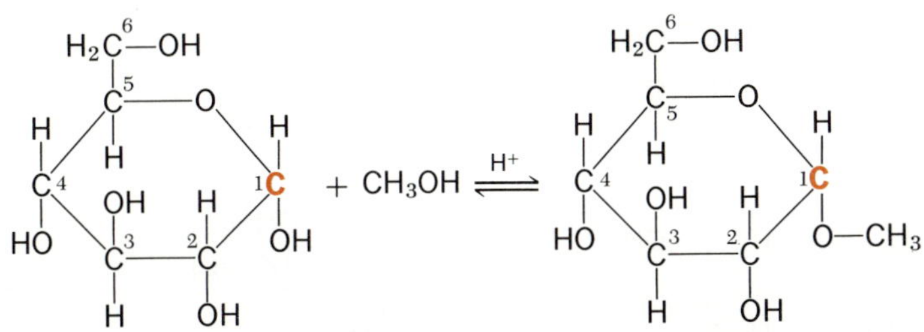

α-D-Glucose   Methyl alcohol   α-Methylglucoside

Hemiketals and ketals are produced by reacting ketones with alcohols. In general, ketones form hemiketals less readily than aldehydes form hemiacetals. However, the sugar fructose, a polyhydroxyketone, forms a cyclic hemiketal quite readily.

In the formation of hemiketals and ketals the mechanism of the reaction is the same as for acetal formation:

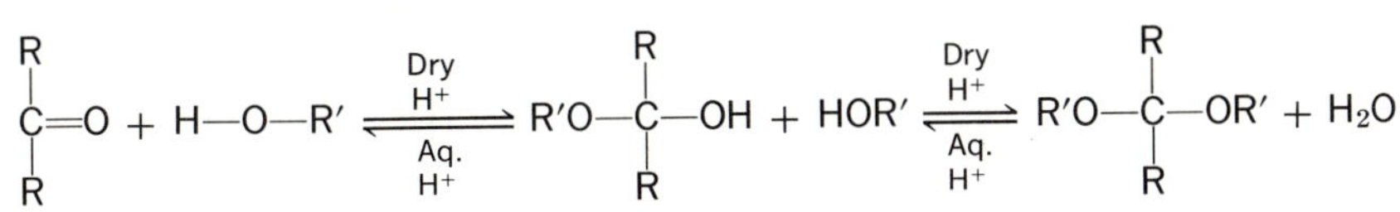

Ketone   Alcohol   Hemiketal   Ketal

A specific reaction with acetone and methyl alcohol is

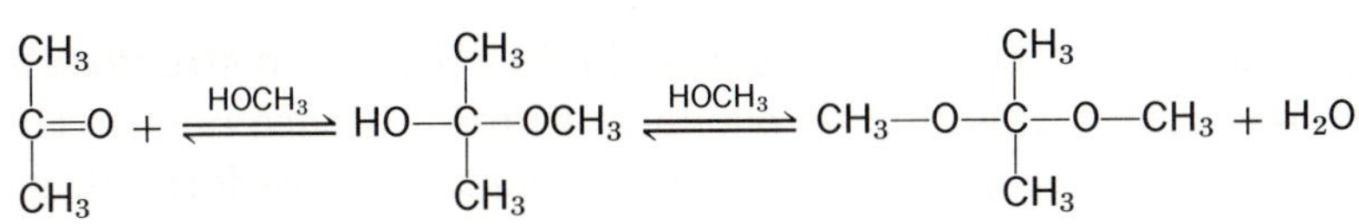

Acetone   Acetone methylhemiketal   Acetone dimethylketal

Ketones may form cyclic hemiketals or ketals if the molecule contains the hydroxy and the ketone groups on the appropriate carbons. The sugar fructose readily cyclizes to form a hemiketal containing a five-membered ring:

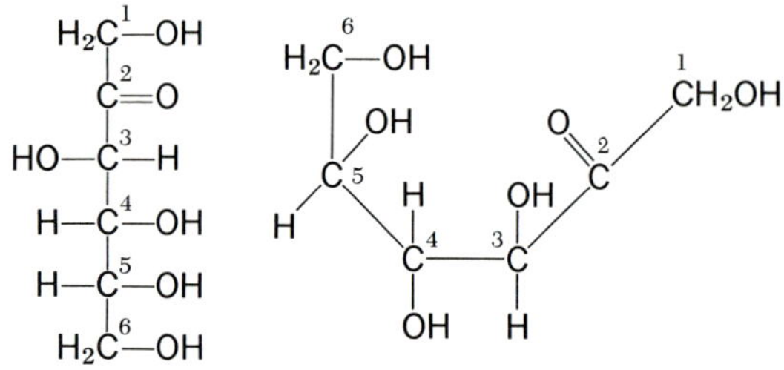

(I) (I′)

D-Fructose, open-chain ketone

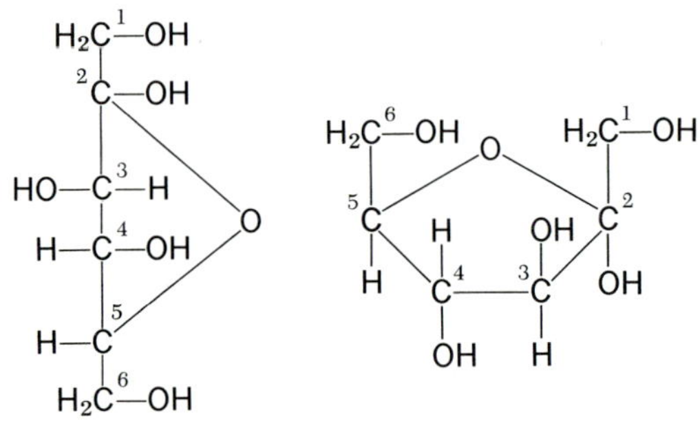

(II) (II′)

α-D-Fructose, cyclic ketone

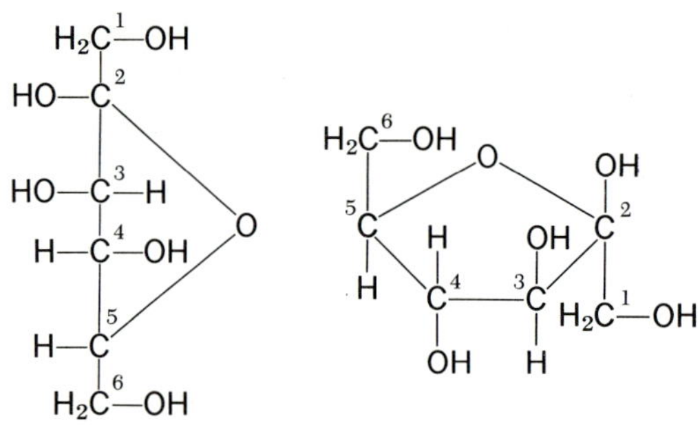

(III) (III′)

β-D-Fructose, cyclic ketone

The carbonyl carbon (C-2) of fructose in the open-chain structure is not asymmetric, but when the hemiketal is formed, it becomes so. Hence, as with glucose, two forms of the cyclic structure are formed, α- and β-fructose.

It is of particular significance to note that the open-chain aldehyde form of glucose and the ketone form of fructose are in equilibrium with the α and β forms of the respective cyclic structures. Since the α and β forms of D-glucose differ in configuration about the first carbon, the specific optical activity $[\alpha]$ of these two forms would be expected to have different values. The values are $+112°$ for the α form and $+19°$ for the β form. The α-glucose is obtained by crystallizing glucose from solution at temperatures below 87°C. If the temperature is maintained above this value during recrystallization, the β form crystallizes from solution. *In an aqueous solution the three forms of glucose are in equilibrium:*

$$\underset{[\alpha] = +112}{\alpha\text{-D-Glucose}} \rightleftharpoons \text{aldehyde form} \rightleftharpoons \underset{[\alpha] = +19}{\beta\text{-D-glucose}}$$

$$[\alpha] = +112 \xrightleftharpoons[{[\alpha] = +52.5}]{} [\alpha] = +19$$

Only a trace amount of the free aldehyde form is present.

Mutarotation refers to the change in optical activity observed if an aqueous solution of a sugar is allowed to stand. This change is a result of the attainment of equilibrium among the various forms. If α-D-glucose is dissolved in water and the optical activity of the solution determined immediately, $[\alpha] = +112°$. This value slowly decreases until $[\alpha] = +52.5°$. When β-D-glucose is treated in a similar manner, the $[\alpha]$ changes from $+19°$ to $+52.5°$ (see Fig. 10.2), which represents the optical activity of the equilibrium mixture of the α and β forms. At equilibrium there is about 35.5% α and 64.5% β-D-glucose present. The presence of a small amount of base, such as NH_4OH, permits attainment of equilibrium in a matter of minutes instead of hours.

Any sugar that forms a cyclic hemiacetal or hemiketal exhibits mutarotation. Would you expect a glycoside to mutarotate? The answer is no. A glycoside is an acetal (or ketal) and is stable in aqueous solutions. It requires H^+ and heat to hydrolyze an acetal group.

Addition of HCN

HCN adds to aldehydes and ketones to form *cyanohydrins.* This is a method whereby a carbon atom can be added to an aldehyde or ketone to produce a compound containing one more carbon atom. Cyanohydrins can be converted to the corresponding acid by hydrolysis:

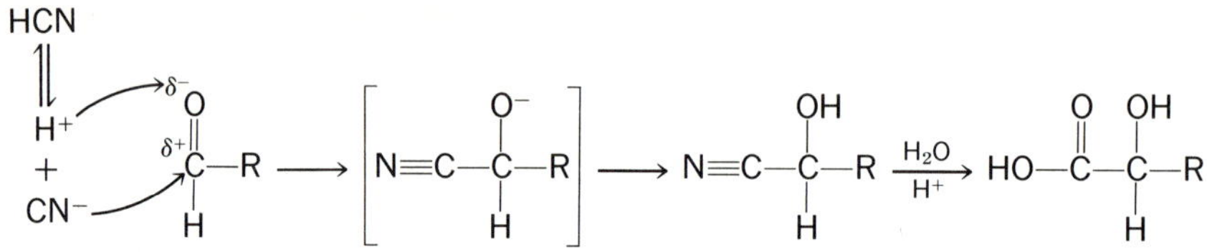

Aldehyde Cyanohydrin α-Hydroxy acid

Specific example:

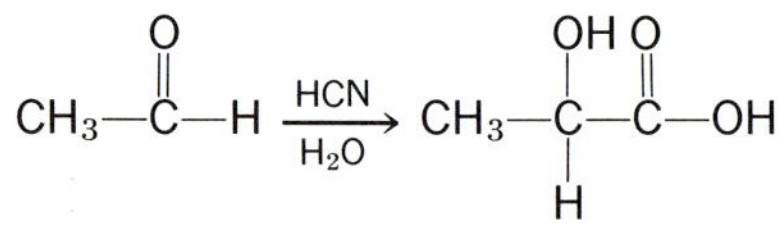

Acetaldehyde Lactic acid

Reaction with NH_4CN

If NH_4CN is reacted with aldehydes and the resulting product hydrolyzed, α-amino acids are formed.

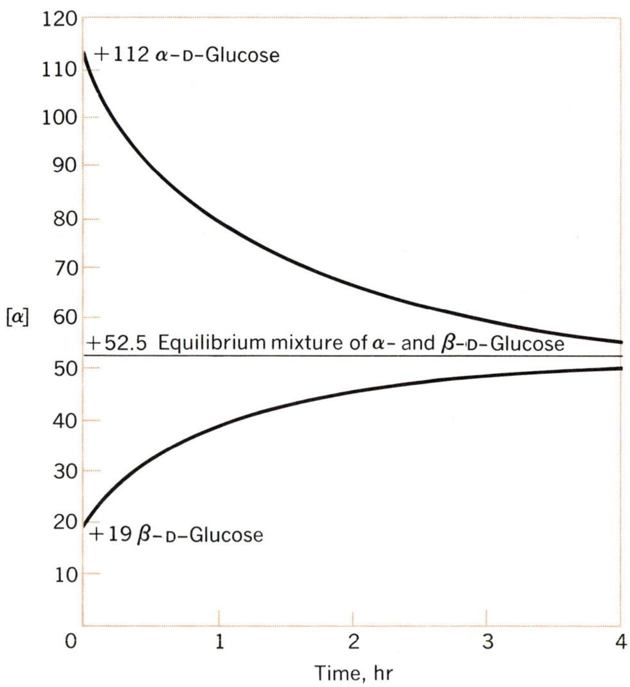

FIGURE 10.2 Mutarotation of D-glucose.

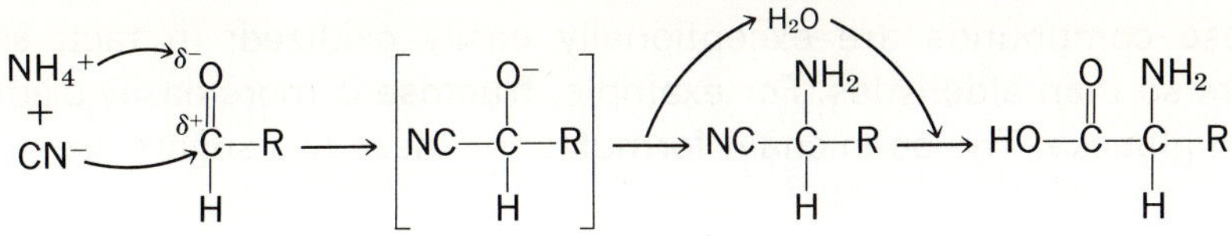

Aldehyde α-Amino acid

Specific example: 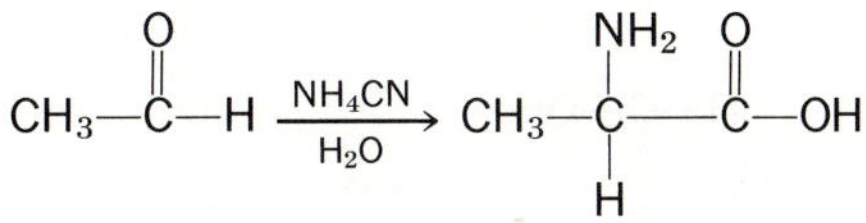

Acetaldehyde Alanine

Oxidation

Aldehydes are quite easily oxidized to the corresponding acid. Ketones are oxidized only under vigorous conditions which rupture the C—C bonds producing lower-molecular-weight compounds. The general equation for the oxidation of an aldehyde is

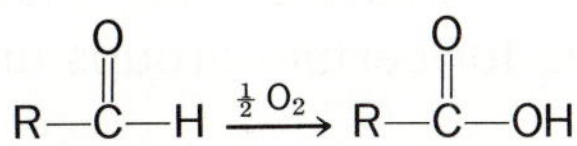

Aldehyde Acid

Oxidation of aldehydes may be visualized as dehydrogenation of the hydrated aldehyde:

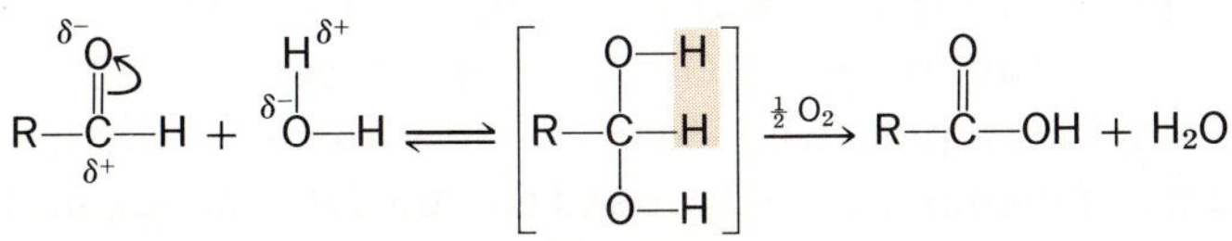

Aldehyde Hydrated aldehyde Acid

A notable exception to the general rule that ketones are not easily oxidized are compounds that contain a primary alcohol group adjacent to the carbonyl group,

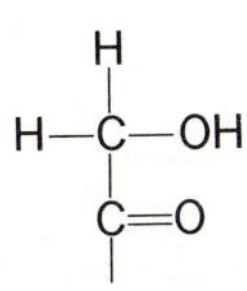

These compounds are exceptionally easily oxidized; in fact, some more so than aldehydes. For example, fructose is more easily oxidized than glucose. The open-chain formulas for these two sugars are

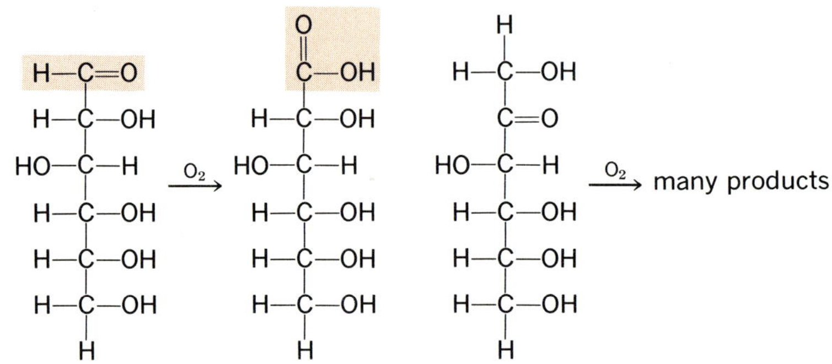

Glucose Aldehyde sugar — Gluconic acid — Fructose Ketone sugar

PERIODATE OXIDATION. The periodate ion, IO_4^-, is a rather selective oxidizing reagent. It is specific for certain groups on adjacent carbons; two of the groups are

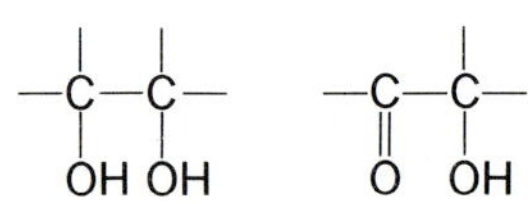

The C—C bond is broken, and the carbons are oxidized to a higher oxidation state. One mole of IO_4^- is required per C—C bond broken. The following examples will illustrate this reaction. In quantitative experiments, the amount of IO_4^- used can be determined, and thus the number of C—C bonds broken can be calculated. This reagent has played an important role in determining the structure of polyhydroxy compounds like the sugars.

Group:

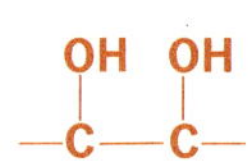

General reaction:

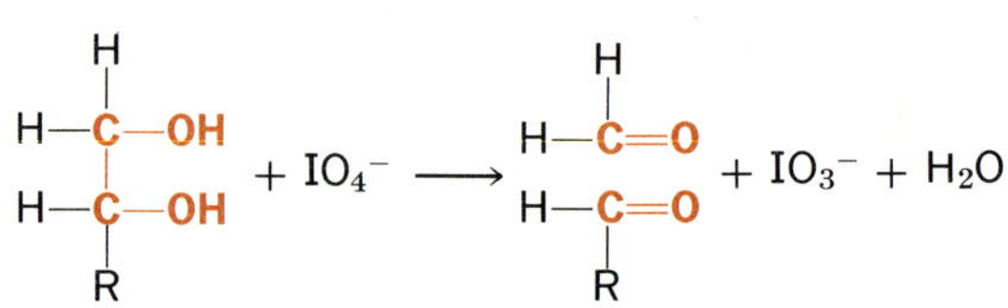

Specific reaction:

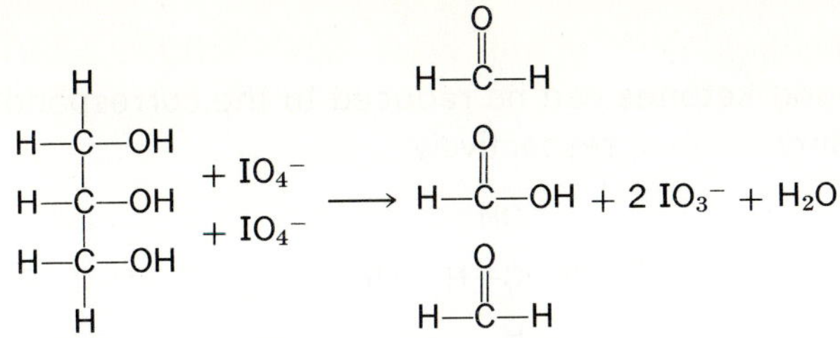

$$\text{Glycerol} + 2\ IO_4^- \longrightarrow 2\ HCHO + HCOOH + 2\ IO_3^- + H_2O$$

Glycerol is oxidized by 2 moles of IO_4^- to yield 2 moles of formaldehyde and 1 mole of formic acid.

Group:

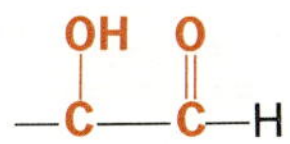

Oxidation of glyceraldehyde:

Stepwise:

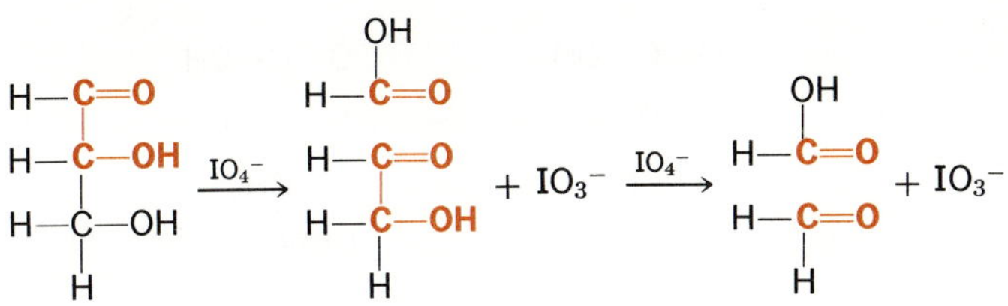

Overall:

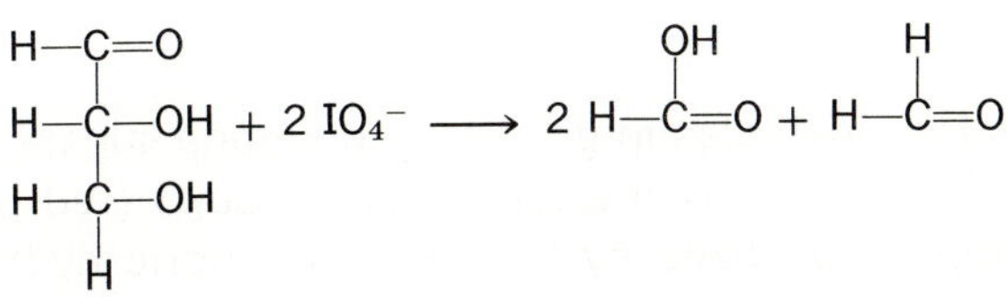

Glyceraldehyde   Formic acid   Formaldehyde

It requires 2 moles of IO_4^- to oxidize 1 mole of glyceraldehyde. The products of this reaction are 2 moles of formic acid and 1 mole of formaldehyde. The C-1 aldehyde group and the C-2 secondary alcohol groups of glyceraldehyde are oxidized to formic acid. The C-3 primary alcohol group yields formaldehyde as the end product.

Reduction

Aldehydes and ketones can be reduced to the corresponding primary and secondary alcohol, respectively:

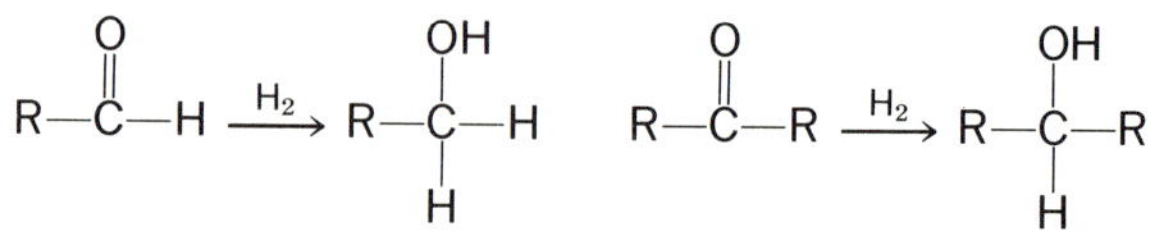

Aldehyde 1° alcohol Ketone 2° alcohol

The reduction of aldehydes and ketones is important both commercially and biologically for the synthesis of organic compounds. Acetaldehyde is reduced to ethyl alcohol in the fermentation of sugar by yeast:

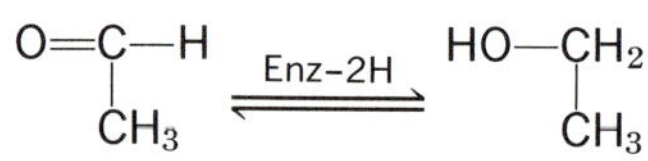

Acetaldehyde Ethyl alcohol

The reduction of pyruvic acid to lactic acid is a very important biological reaction. Pyruvic acid is an α-keto acid, and lactic acid is an α-hydroxy acid.

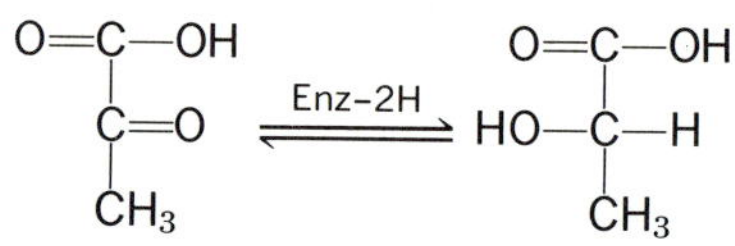

Pyruvic acid Lactic acid

10.2 TESTS FOR THE CARBONYL GROUP

Cupric Ion

Fehling's test is for aldehydes in general and ketone sugars. A positive Fehling's test is indicated by the formation of Cu_2O (red precipitate) when a compound is oxidized by the reagent, cupric hydroxide. The active oxidizing agent is cupric hydroxide, which is kept in solution as a complex with sodium potassium tartrate.

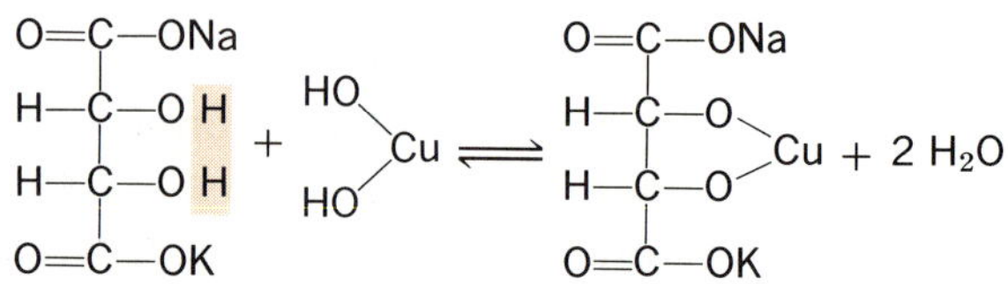

Sodium potassium tartrate

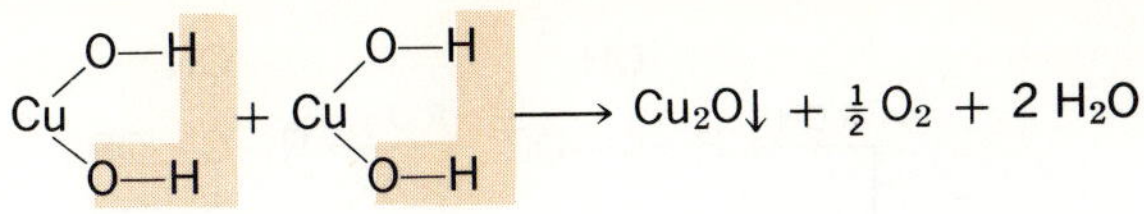

Cupric hydroxide   Cuprous oxide (red)

When a compound is oxidized by Fehling's reagent, the Cu^{++} is *reduced* to Cu^{+} and a red precipitate of Cu_2O is formed. Sugars containing an aldehyde or ketone group reduce Fehling's solution and are called *reducing sugars.* Sugars which do not reduce this reagent are called *nonreducing sugars;* they do not contain a free aldehyde or ketone group. Glucose and fructose are reducing sugars. A nonreducing sugar is sucrose (table sugar, cane sugar, or beet sugar).

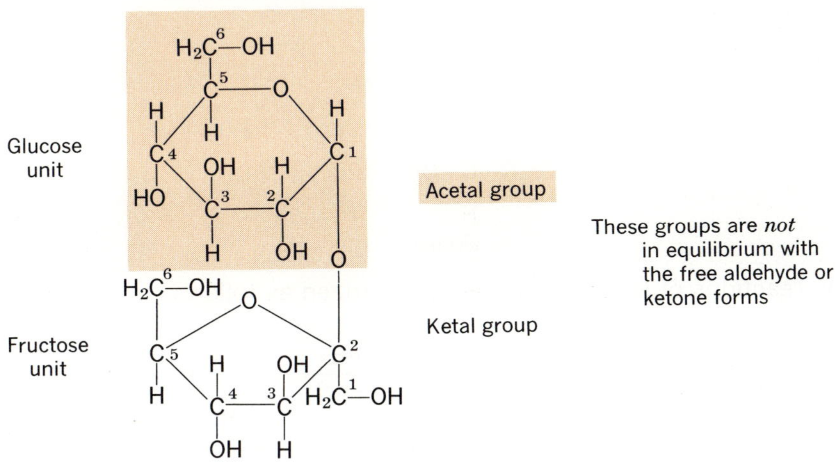

Sucrose

Benedict's reagent

Benedict's reagent also uses $Cu(OH)_2$ as the active oxidizing agent, but the solution is less basic than Fehling's reagent. It can be used in solutions containing a fair amount of NaCl, e.g., urine, whereas Fehling's solution cannot be used under these conditions.

10.3 SUMMARY OF REACTIONS

The reactions of aldehydes can be summarized as follows:

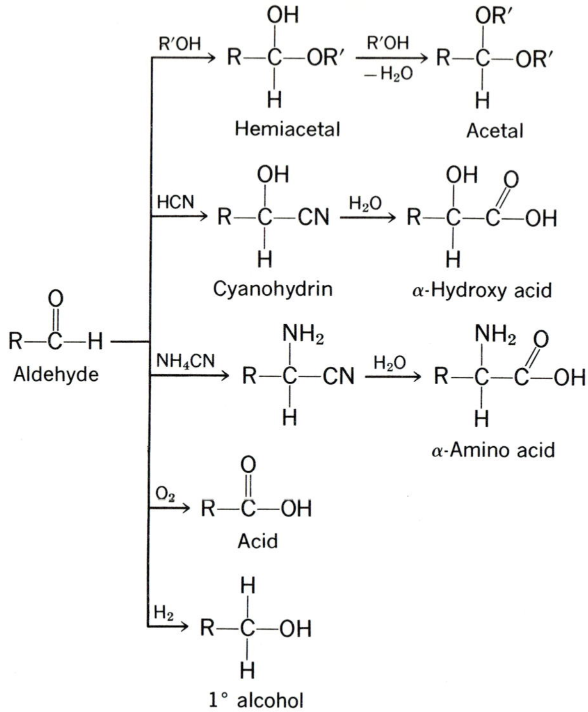

The reactions of ketones can be summarized as follows:

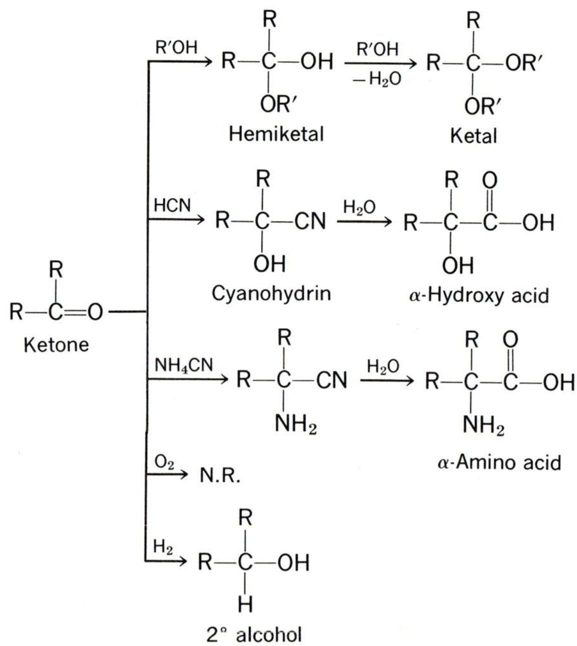

QUESTIONS

1. Write the general formula for an aldehyde and for a ketone.
2. When a primary or secondary alcohol is oxidized to the corresponding aldehyde or ketone, does the aldehyde or ketone contain more, the same, or less oxygen than the original alcohol? Can this reaction be classified as a dehydrogenation reaction?
3. What structure characteristic of aldehydes permits them to be easily oxidized whereas ketones are not?
4. What is the oxidizing reagent in Fehling's solution?
5. Write an equation for the formation of $CH_3—CH_2—CHOH—COOH$ from the appropriate aldehyde.
6. How are hemiacetals or hemiketals formed? Write an equation to illustrate your answer. Is water eliminated during this reaction?
7. Is water eliminated when an alcohol and a hemiacetal or hemiketal react to form an acetal or a ketal, respectively? Write an equation for this reaction.
8. Show the formation of a cyclic hemiacetal.
9. Are hemiacetals and acetals stable or labile toward aqueous alkaline treatment? Toward aqueous acid treatment?
10. How might one form an α-amino acid from an aldehyde? An α-hydroxy acid?
11. Given the two known compounds below, each in separate test tube marked *A* and *B*, respectively, how would you prove by a simple experiment that the markings on the tubes are correct?

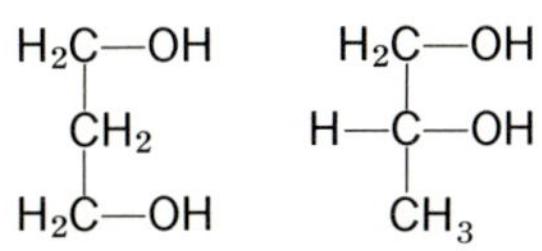

(*A*) (*B*)

12. (*a*) Illustrate the phenomenon of mutarotation as applied to glucose (use structural formulas).
    (*b*) In each structure, star the asymmetric carbon atoms.
13. Write a possible structural formula for compounds I to IV.

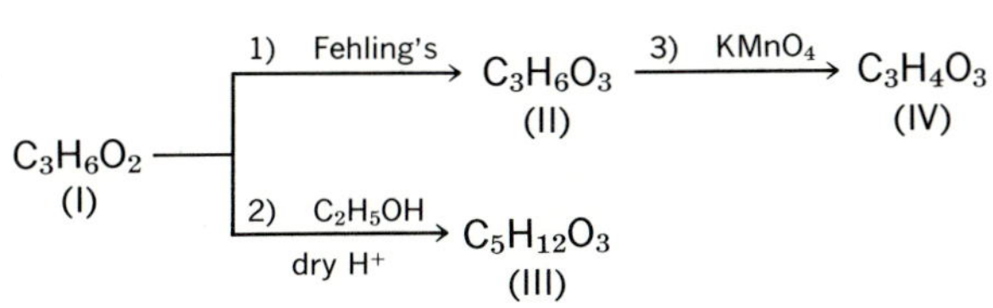

14. Write a possible structural formula for compounds I to VI.

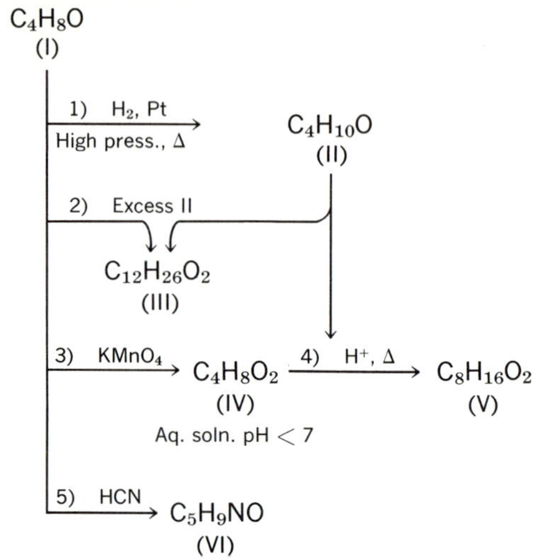

15. The reversible reactions presented in the previous chapter to illustrate the biological dehydrogenation of alcohols can also be used to illustrate the reduction of the carbonyl group. The first column is a list of carbonyl compounds. The second column is a list of the reduction products of the compounds in the first column. Match the members to each other.

Carbonyl compound	*Reduced product*
(*a*) Dihydroxyacetone	1. Isocitric acid
(*b*) Oxaloacetic acid	2. Malic acid
(*c*) Oxalosuccinic acid	3. Glycerol
(*d*) β-Keto acid	4. β-Hydroxy acid

chapter 11.... carboxylic acids

The functional group of carboxylic acids is the carboxyl group $-\overset{\overset{\displaystyle O}{\|}}{C}-OH$ (also written —COOH or —CO_2H). The general formula for acids is R—COOH, where R is C_nH_{2n+1}, for the saturated acids that contain only one carboxyl group. Although substituents in the carbon chain of acids may markedly affect the strength of the acid, the general reactions of members of this class of compounds are those of the —COOH group.

Acids and acid derivatives are important biologically. Some acids are incorporated into body constituents, e.g., fats, whereas others serve as intermediates in the synthesis or breakdown of other molecules. Two examples of the conversion of one acid into another were discussed in Chap. 9, namely, the conversion of succinic acid into oxaloacetic acid and citric acid into oxalosuccinic acid.

11.1 NOMENCLATURE

The common name for the acids will generally be used. Table 11.1 lists the common and IUPAC names and the formula of some acids.

The IUPAC system uses the longest straight chain of carbons as the parent compound. The terminal *-e* of the name of the alkane is replaced by the suffix *-oic* to indicate the carboxyl group. Ethanoic acid has the formula

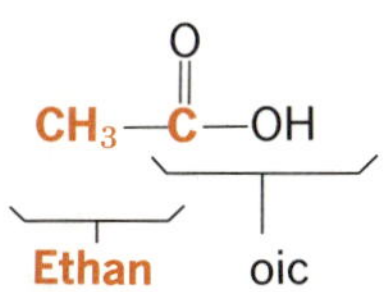

Acetic acid

TABLE 11.1 SOME IMPORTANT ACIDS

Common name	IUPAC name	Formula
I. Monocarboxylic (monobasic)		
A. Saturated		
Formic	Methanoic	H—COOH
Acetic	Ethanoic	CH_3—COOH
Propionic	Propanoic	CH_3CH_2—COOH
Butyric	Butanoic	$CH_3(CH_2)_2$—COOH
Valeric	Pentanoic	$CH_3(CH_2)_3$—COOH
Caproic	Hexanoic	$CH_3(CH_2)_4$—COOH
Caprylic	Octanoic	$CH_3(CH_2)_6$—COOH
Capric	Decanoic	$CH_3(CH_2)_8$—COOH
Lauric	Dodecanoic	$CH_3(CH_2)_{10}$—COOH
Myristic	Tetradecanoic	$CH_3(CH_2)_{12}$—COOH
Palmitic	Hexadecanoic	$CH_3(CH_2)_{14}$—COOH
Stearic	Octadecanoic	$CH_3(CH_2)_{16}$—COOH
B. Unsaturated		
1. One C═C, oleic acid series		
Myristoleic	*cis*-9-Tetradecenoic	$CH_3(CH_2)_3CH{=}CH(CH_2)_7$—COOH
Palmitoleic	*cis*-9-Hexadecenoic	$CH_3(CH_2)_5CH{=}CH(CH_2)_7$—COOH
Oleic	*cis*-9-Octadecenoic	$CH_3(CH_2)_7CH{=}CH(CH_2)_7$—COOH
2. Two C═C, linoleic acid series		
Linoleic	*cis,cis*-9, 12-Octadecadienoic	$CH_3(CH_2)_4CH{=}CHCH_2CH{=}CH(CH_2)_7$—COOH
3. Three C═C, linolenic acid series		
Linolenic	*cis,cis,cis*-9,12, 15-Octadecatrienoic	$CH_3(CH_2CH{=}CH)_3(CH_2)_7$—COOH
II. Dicarboxylic (dibasic)		
Oxalic	Ethanedioic	HOOC—COOH
Malonic	Propanedioic	HOOC—CH_2—COOH
Succinic	Butanedioic	HOOC—CH_2CH_2—COOH
Glutaric	Pentanedioic	HOOC—$CH_2CH_2CH_2$—COOH
III. Tricarboxylic (tribasic)		
Citric	3-Hydroxy-3-carboxy-pentanedioic	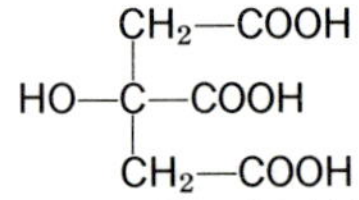
Isocitric	2-Hydroxy-3-carboxy-pentanedioic	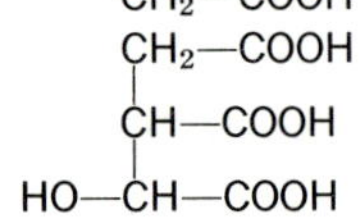
cis-Aconitic	*cis*-3-Carboxy-2-pentenedioic	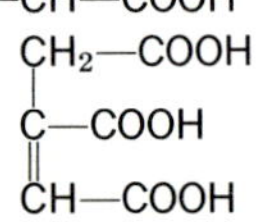
Oxalosuccinic	2-Keto-3-carboxy-pentanedioic	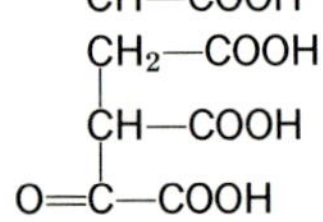

When there are two carboxyl groups in an acid (dicarboxylic acid) the terminal *-e* of the parent name is retained and the number of carboxyl groups indicated by the suffix *-dioic*. Since the two carboxyl groups must be on terminal carbons, it is not necessary to number the carbons on which they occur. Tricarboxylic acids may be considered as carboxy-substituted (COOH) alkanes. Consider the formulas and names of the following acids:

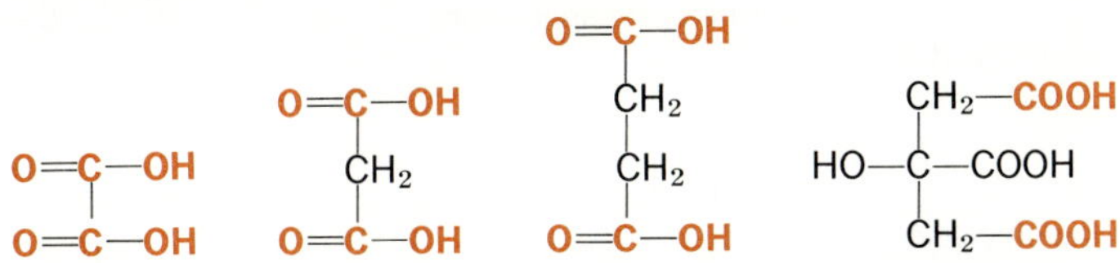

Ethanedioic Oxalic | Propanedioic Malonic | Butanedioic Succinic | 3-Hydroxy-3-carboxy-pentanedioic Citric

When there are substituted groups in the carbon chain, the name of the group and the number of the carbon to which it is attached are given. This is similar to the naming of compounds in previous chapters. The carboxyl carbon is always C-1, $—\overset{4}{C}H_2—\overset{3}{C}H_2—\overset{2}{C}H_2—\overset{1}{C}OOH$. Lactic acid is

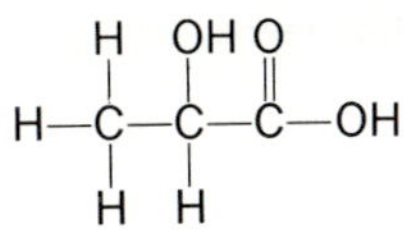

2-Hydroxypropanoic acid

The common way to indicate substituents in the carbon chain is to label the carbons α, β, γ, δ, etc. Since the word acid indicates the —COOH group, the α carbon of an acid is the first carbon next to the —COOH group. The β, γ, and δ carbons are the second, third, and fourth, respectively, from the carboxyl group.

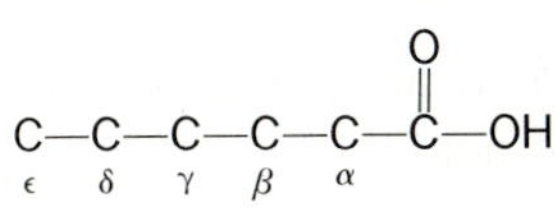

Some biologically important acid derivatives and their names are listed in Table 11.2. For example, tartaric acid is a dihydroxydicarboxylic acid found in grapes:

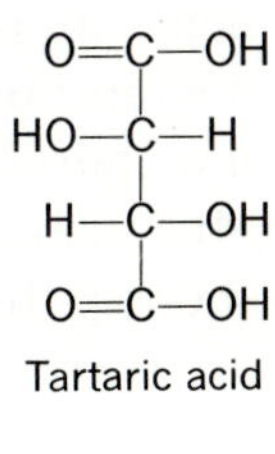

Tartaric acid

TABLE 11.2 FORMULA AND NOMENCLATURE OF SOME SUBSTITUTED ACIDS

Common name	IUPAC name	Formula
Isobutyric acid or α-methylpropionic acid	2-Methylpropanoic acid	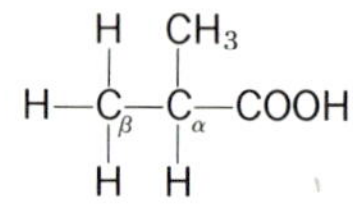
Alanine or α-aminopropionic acid	2-Aminopropanoic acid	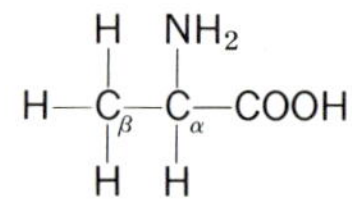
Lactic acid or α-hydroxypropionic acid	2-Hydroxypropanoic acid	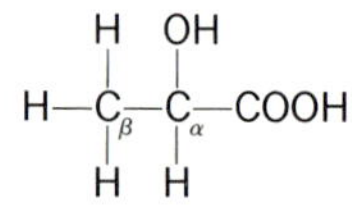
β-Hydroxybutyric acid	3-Hydroxybutanoic acid	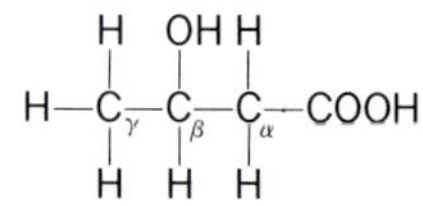
Pyruvic acid or α-ketopropionic acid	2-Ketopropanoic acid	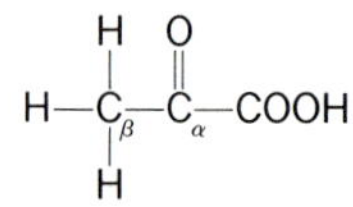
β-Ketobutyric acid or acetoacetic acid	3-Ketobutanoic acid	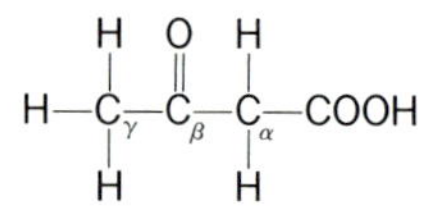
Serine or α-amino-β-hydroxypropionic acid	2-Amino-3-hydroxypropanoic acid	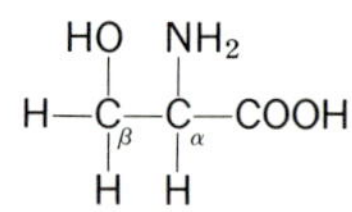
Methylmalonic acid	2-Methylpropanedioic acid	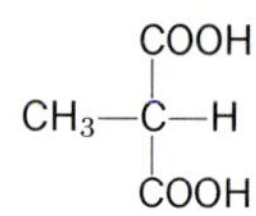

The name of this acid according to the IUPAC nomenclature is 2,3-dihydroxybutanedioic acid.

11.2 FATTY ACIDS

In the list of monocarboxylic acids in Table 11.1 it can be seen that those from caproic on have an even number of carbon atoms. These acids occur naturally in fats and therefore are known as *fatty acids.* The biological synthesis of fatty acids ultimately involves the condensation of two-carbon units; hence, the resulting acids have an even number of carbons. Only a relatively few naturally occurring acids have an odd number of carbon atoms. They are frequently intermediates in the metabolism or synthesis of body constituents and are not found appreciably in fat.

11.3 POLYCARBOXYLIC ACIDS

The formulas for several di- and tricarboxylic acids are given in Table 11.1. Oxalic acid is the simplest dicarboxylic acid. The radical HOOC—CO— is known as the *oxalyl* group. In oxalosuccinic acid the oxalyl group is substituted on the second carbon of succinic acid; hence the common name, oxalosuccinic acid. Another acid previously discussed that contains the oxalyl group is oxaloacetic acid.

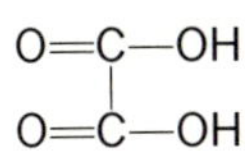
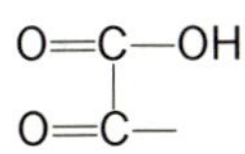

Common: Oxalic acid — Oxalyl radical acid

IUPAC: Ethanedioic acid

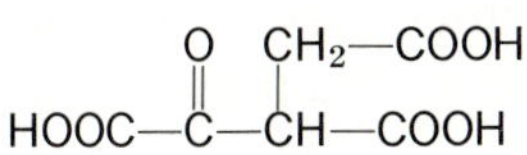
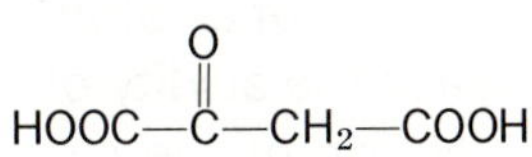

Common: Oxalosuccinic acid — Oxaloacetic acid

IUPAC: 2-Keto-3-carboxypentanedioic acid — 2-Ketobutanedioic acid

The name oxaloacetic acid indicates that the oxalyl group has been substituted in acetic acid. Tricarboxylic acids, according to the IUPAC system, are considered as derivatives of dicarboxylic acids in which a carboxyl group substitutes for a hydrogen (Table 11.1).

11.4 REACTIONS

Ionization

Before proceeding with the discussion of ionization, the student should review some of the material presented in Chap. 6. What is the significance of the term K_{ion}? What is the pK of an acid? Define zwitterion. How does a buffer function? Review the Henderson-Hasselbalch equation. How does electronegativity of elements vary within a period and within a group of the periodic table? Be sure you can answer these questions before proceeding with this discussion.

A characteristic of all acids is their ability to donate protons. The carboxyl group, COOH, is the functional group of the aliphatic acids, and it is *only* the hydrogen of the carboxyl group that ionizes. This is true regardless of the other groups present.

$$\mathrm{R{-}\overset{\overset{\displaystyle O}{\|}}{C}{-}O{-}H} \rightleftharpoons \mathrm{R{-}\overset{\overset{\displaystyle O}{\|}}{C}{-}O^-} + \mathrm{H^+} \qquad \mathrm{p}K = 4.7\text{–}4.9$$

The unsubstituted monocarboxylic acids, except formic acid, HCOOH, are all of about the same strength with respect to acidity. Their K_{ion} values lie between 1.8×10^{-5} and 1.3×10^{-5}, and their pK values are between 4.7 and 4.9. Formic acid is the only acid in this series which does not have a carbon attached to the carboxyl group. This difference in structure makes formic acid a *stronger* acid than the others. It has a K_{ion} of 1.77×10^{-4} and a pK of 3.75.

Substituted acids

Substituted acids may be stronger or weaker than the unsubstituted acid depending upon the type of substitution. In Chap. 6 it was noted that the substitution of an atom or element having high electronegativity tends to increase the acidity of the acid. The presence of such a group causes the electron of the carboxyl ion to be dissipated over a much larger area, increasing the stability of the ionic form (see Fig. 11.1). This is analogous to the increased stability of the halogen ions as the size of the halogen atoms increases (see Chap. 6).

Decreasing acidity

$$\mathrm{HI > HBr > HCl > HF}$$

The influence of substituted groups on the strength of acids is illustrated by comparing the K_{ion} or pK values of the acids in Table 11.3. It will be observed that the strength of acetic acid is increased about

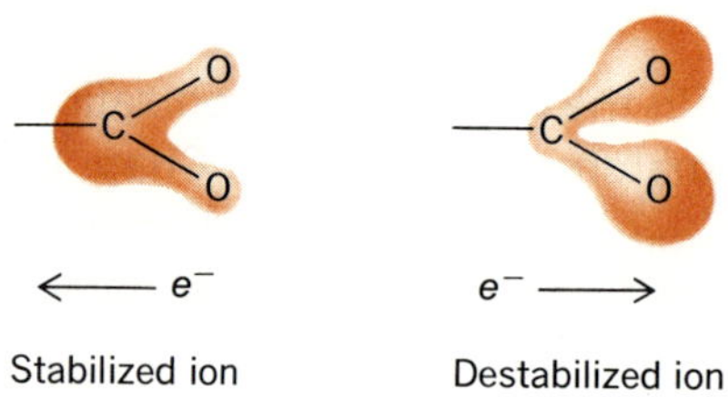

FIGURE 11.1 Stabilization of the carboxyl ion.

TABLE 11.3 DISSOCIATION CONSTANTS OF SOME ORGANIC ACIDS

Common name	Formula	K_{ion}	pK_1	pK_2
Formic	HCO_2H	1.77×10^{-4}	3.75	
Acetic	CH_3CO_2H	1.76×10^{-5}	4.76	
Chloroacetic	$ClCH_2CO_2H$	1.4×10^{-3}	2.85	
Dichloroacetic	Cl_2CHCO_2H	3.2×10^{-2}	1.5	
Trichloroacetic	Cl_3CCO_2H	2×10^{-1}	0.7	
Propionic	$CH_3CH_2CO_2H$	1.34×10^{-5}	4.87	
Pyruvic	CH_3COCO_2H	3.2×10^{-3}	2.5	
Lactic	$CH_3CHOHCO_2H$	1.39×10^{-4}	3.86	
β-Hydroxypropionic	$CH_2OHCH_2CO_2H$	3.1×10^{-5}	4.51	
Butyric	$CH_3CH_2CH_2CO_2H$	1.5×10^{-5}	4.82	
β-Hydroxybutyric	$CH_3CHOHCH_2CO_2H$	2×10^{-5}	4.70	
β-Ketobutyric	$CH_3COCH_2CO_2H$	2.62×10^{-4}	3.58	
Oxalic	$HO_2C—CO_2H$	5.9×10^{-2}	1.23	
		6.4×10^{-5}		4.19
Malonic	$HO_2CCH_2CO_2H$	1.49×10^{-3}	2.83	
		2.03×10^{-6}		5.69
Methylmalonic	$HO_2CH(CH_3)CO_2H$	1.17×10^{-3}	3.07	
Succinic	$HO_2CCH_2CH_2CO_2H$	6.89×10^{-5}	4.16	
		2.47×10^{-6}		5.61
Fumaric	$HO_2CCH{=}CHCO_2H$	9.3×10^{-4}	3.03	
		3.4×10^{-5}		4.47
Malic	$HO_2CCH_2CH(OH)CO_2H$	3.9×10^{-4}	3.4	
		7.8×10^{-6}		5.11
Glutaric	$HO_2CCH_2CH_2CH_2CO_2H$	4.5×10^{-5}	4.31	
		3.89×10^{-6}		5.41

eightyfold by substituting an electron-attracting chlorine atom for a hydrogen atom. As more chlorine atoms are substituted in acetic acid, the resulting acid becomes stronger. Trichloroacetic acid is a very strong acid.

Electron-attracting groups which appear in a number of organic acids produced by living organisms are the carboxyl, hydroxyl, and keto groups. Several such acids which have already been mentioned include oxalic, malic, succinic, citric, oxalosuccinic, and oxaloacetic. The closeness of the electron-attracting group to the carboxyl group has a direct bearing on the acidity of the acid. The dicarboxylic acids can be used to illustrate this.

The hydrogen of one of the carboxyl groups of oxalic acid ionizes to a large extent, pK_1 = 1.23, thus making it a strong acid. Compare its strength with that of formic and acetic acid. Once formed, the carboxylate ion $—\overset{\overset{\displaystyle O}{\|}}{C}—O^-$ has a negative influence on the ionization of a second carboxyl group if they are close together because the carboxylate ion is negatively charged and is an electron-donating group. This increases localization of electrons on the second carboxyl group, thus making it more difficult to release a proton. This depression of ionization is quite marked. The pK_2 of oxalic acid is 4.19, approximately 3 pK units less than the value for pK_1. By comparing the strength of formic and acetic acids it will be noted that the methyl group is a much more electron-donating group than the hydrogen atom is. Formic acid is about 10 times stronger than acetic acid.

Inserting a CH_2 group between the two carboxyl groups of oxalic acid (see malonic acid) and thus separating the two electronegative groups by one carbon decreases the ionization of the first carboxyl group fortyfold:

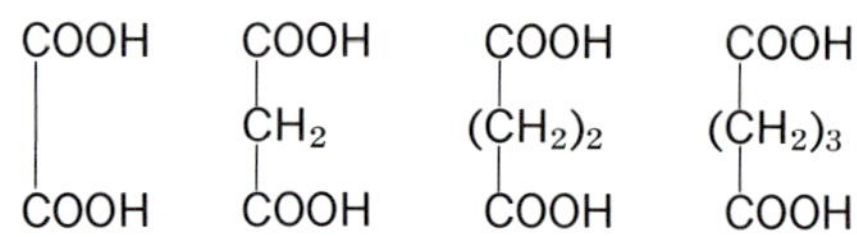

pK_1:	1.23	2.83	4.16	4.31
Acid:	Oxalic	Malonic	Succinic	Glutaric

The pK_1 for malonic acid is 2.83 as compared with 1.23 for oxalic acid ($\Delta pK = 1.6$). In succinic acid the two carboxyl groups are separated by two CH_2 groups. This causes a further substantial decrease in the strength of the acid. Succinic acid is about 20 times weaker than malonic acid and only approximately 4 times stronger than acetic acid ($pK = 4.76$). Thus, the second carboxyl group has only a moderate effect upon the ionization of the first carboxyl group of succinic acid. Further separation of the carboxyl groups has little effect, as noted when the pK_1 of succinic acid is compared with the pK_1 for glutaric acid (4.16 versus 4.31).

The conversion of the α carbon of an acid to a carbonyl group,

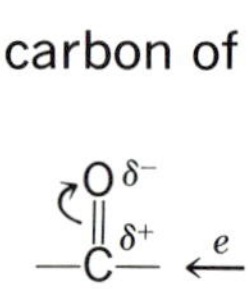

would be expected to greatly increase the acidity of the acid. This is substantiated by comparing the pK values of pyruvic and methylmalonic acid (2-methylpropanedioic acid). Pyruvic acid is about 4 times stronger than methylmalonic acid ($\Delta pK = 0.57$) and twice as strong as monochloroacetic acid ($\Delta pK = 0.35$).

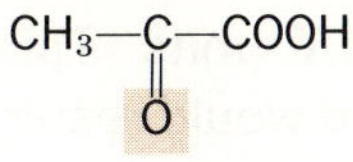

Pyruvic $\quad pK = 2.50$

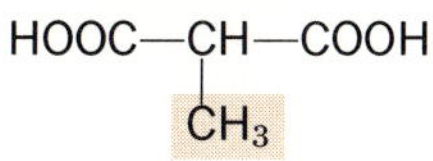

Methylmalonic $\quad pK = 3.07$

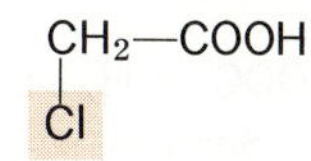

Monochloroacetic $\quad pK = 2.85$

Pyruvic acid is an intermediate of muscle glucose metabolism. One fate of pyruvic acid pertinent to the present discussion is its reduction to lactic acid (α-hydroxypropionic acid), a conversion that reduces the α-carbonyl group to an α-hydroxyl group.

$$\underset{\text{Glucose}}{C_6H_{12}O_6} \longrightarrow \underset{\text{Pyruvic acid, p}K = 2.5}{2\ CH_3\text{—}CO\text{—}COOH} \longrightarrow \underset{\text{Lactic acid, p}K = 3.86}{2\ CH_3\text{—}CHOH\text{—}COOH}$$

How would you expect this to change the acidity of the carboxyl group? The answer is based on the relative electron-attracting nature of the carbonyl and the hydroxy groups. Since the carbonyl group ($>C{=}O$) is more electron-attracting than the hydroxyl group, pyruvic would be a stronger acid than lactic acid. The pK values for pyruvic acid and lactic acid are 2.5 and 3.86, respectively. The $\Delta pK = 3.86 - 2.5 = 1.36$; hence, lactic acid is about 23 times weaker than pyruvic acid (antilog $1.36 = 23$). This conversion would cause the pH of the cellular fluids to increase *if no buffers were present.* Lactic acid is transported from the muscle to the liver via the blood. To what extent would lactic acid be neutralized at the pH of blood (pH = 7.4)? See the discussion in Chap. 6. Review the calculations. The answer is that 99.7% of the lactic acid has been neutralized.

An important series of biological transformations is the conversion of succinic acid to oxaloacetic acid.

	Succinic acid ⟶	fumaric acid ⟶	malic acid ⟶	oxaloacetic acid
pK_1:	4.16	3.03	2.83	1.7

As the reaction proceeds, the strength of the acids becomes progressively greater, the overall increase in acidity from succinic to oxaloacetic being 4.16 − 1.7, or 2.46 pK units. This is equivalent to a 288-fold increase in acid strength and would result in a marked change in pH of the cellular fluids if the acids were not neutralized.

Unsaturated acids

Fumaric acid is about 10 times stronger than succinic acid.

HOOC—CH=CH—COOH	HOOC—CH_2—CH_2—COOH
Fumaric, pK_1 = 3.03	Succinic, pK_1 = 4.16

How would you account for this when there is no appreciable electron-attracting effect of the C=C group? The answer is associated with the nature of the π bond of the C=C group. Note that the π-bond electrons of a C=C group are not localized but are associated with both carbon atoms. These electrons act as a conductor and permit the influence of one carbonyl to be transmitted to the other carbonyl. However, this influence is not as great as when the two carbonyl groups are attached to each other (oxalic acid, pK_1 = 1.23 versus fumaric acid, pK_1 = 3.03) (see Fig. 11.2).

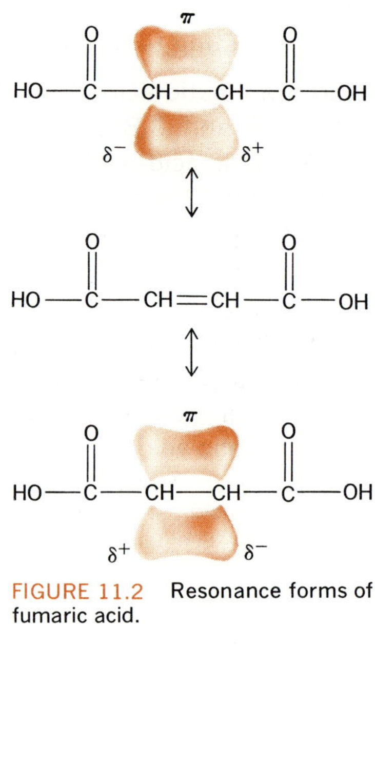

FIGURE 11.2 Resonance forms of fumaric acid.

Neutralization

The neutralization of an acid with a base yields a salt plus water. The neutralization of a monocarboxylic acid (one carboxyl group) requires 1 equiv of base per mole of acid, whereas a dicarboxylic and a tricarboxylic acid require 2 and 3 equiv of base per mole of acid. The general equation of the neutralization of a monocarboxylic acid and a specific example of a dicarboxylic acid are

Acid + base ⟶ salt + water

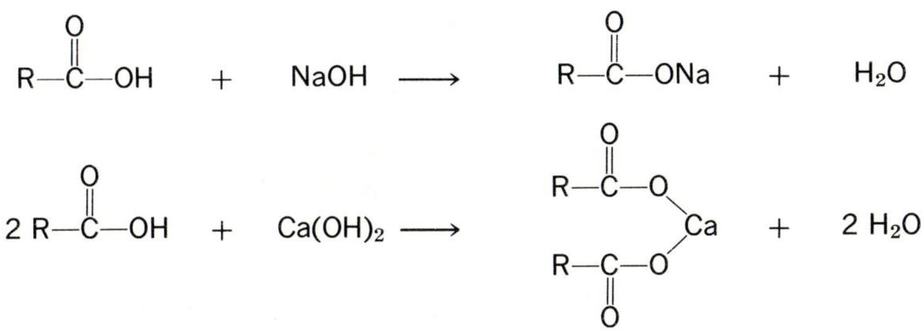

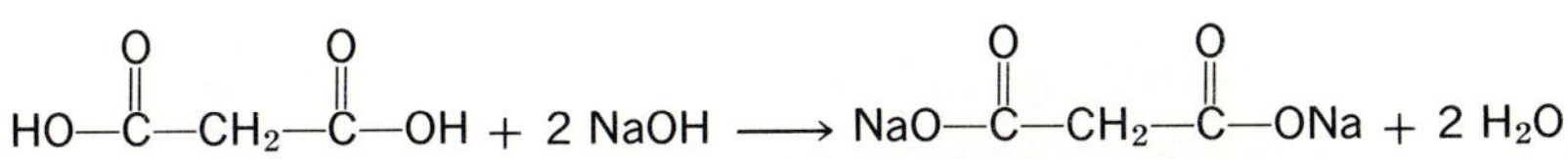

Malonic acid — Disodium malonate

When there is more than one replaceable hydrogen in an acid, a mixed salt can be obtained. For example, the formation of sodium potassium oxalate is accomplished as follows:

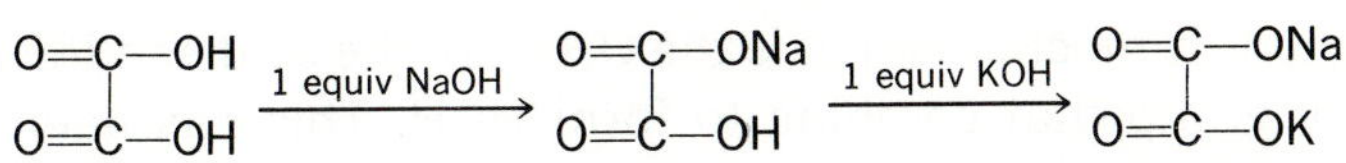

Oxalic acid — Monosodium oxalate — Sodium potassium oxalate

Esterification

The reaction of an acid with an alcohol, *esterification,* results in the formation of *esters*

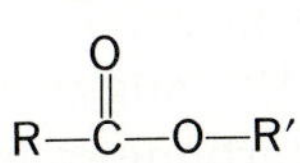

There are two classes of esters of biological importance, simple and mixed esters. Simple esters are formed by reacting an organic acid

with an alcohol, whereas mixed esters are formed when an inorganic acid, e.g., sulfuric or phosphoric, is reacted with an alcohol.

Esterification is an equilibrium reaction. The reverse reaction is called *hydrolysis*. The following is a general reaction mechanism for the formation and hydrolysis of esters:

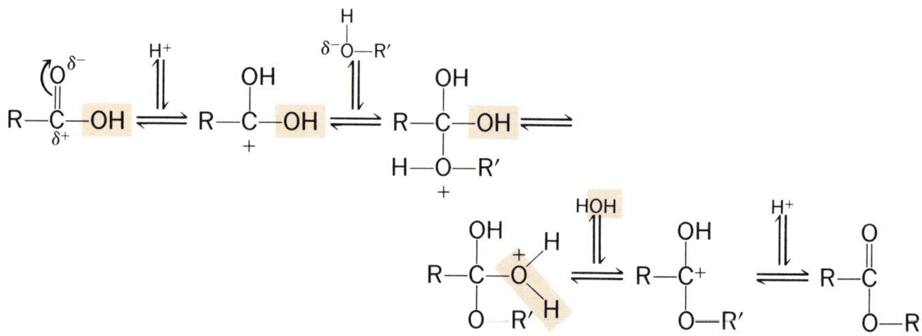

Overall reaction:

$$\underset{\text{Acid}}{R-\overset{\overset{\displaystyle O}{\|}}{C}-OH} + \underset{\text{Alcohol}}{HOR'} \overset{H^+}{\rightleftharpoons} \underset{\text{Ester}}{R-\overset{\overset{\displaystyle O}{\|}}{C}-O-R'} + \underset{\text{Water}}{H_2O}$$

Specific example:

$$\underset{\text{Acetic acid}}{CH_3-\overset{\overset{\displaystyle O}{\|}}{C}-OH} + \underset{\text{Ethyl alcohol}}{HO-CH_2-CH_3} \rightleftharpoons$$

$$\underset{\text{Ethyl acetate}}{CH_3-\overset{\overset{\displaystyle O}{\|}}{C}-O-CH_2-CH_3} + \underset{\text{Water}}{H_2O}$$

According to this mechanism, it is the OH from the acid and the H from the alcohol group that combine to form HOH. This has been established by the use of the isotope ^{18}O. If an acid containing ^{18}O in the carboxyl group, $O=\overset{|}{C}-{}^{18}O-H$, is esterified with an alcohol containing $-{}^{16}O-H$, the ^{18}O is found in the water eliminated in the reaction. Also, experiments with ^{18}O in the alcohol, $R-{}^{18}O-H$, show that the ester contains the oxygen isotope, $R-\overset{\overset{\displaystyle O}{\|}}{C}-{}^{18}O-R'$. Esterification is catalyzed by a mineral acid, H^+, suggesting the initial protonation of the organic acid to form the carbonium ion

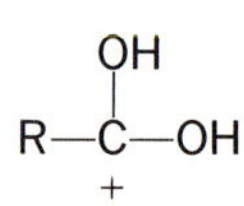

Note that these reactions are reversible and an overall equilibrium exists.

Reactivity of acids and alcohols

The simpler acids and alcohols react faster than the more complex ones. The substitution of groups on the carbinol carbon of alcohols and on the α carbon of the acid tends to hinder the formation of the intermediate addition products, thus preventing both the esterification and the hydrolysis reactions. The size of the substituted groups makes it more difficult for the carboxyl and alcohol groups to come into contact with each other. This is known as *steric hindrance.*

The order of reactivity of the alcohols is

$$CH_3OH > RCH_2OH > R_2CHOH \ggg R_3COH$$

For the acids the order is

$$HCOOH > CH_3COOH > RCH_2COOH > R_2CHCOOH \ggg R_3CCOOH$$

For example, the esterification of t-butyl alcohol with α,α-dimethylpropionic acid is quite difficult.

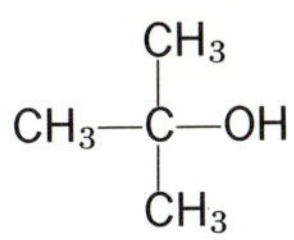

t-Butyl alcohol

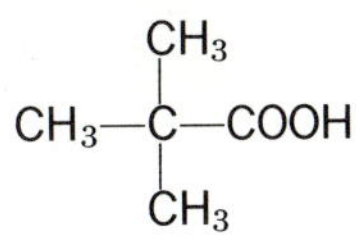

α,α-Dimethylpropionic acid

Mixed esters

Mixed esters are formed by reacting alcohols with inorganic acids containing oxygen. Esters of sulfuric acid, and especially those of orthophosphoric and pyrophosphoric acids, are of particular interest. Sulfate esters play an important role in certain detoxication reactions. Phosphate or pyrophosphate esters (or both) are associated with practically all biological reactions. For example, energy obtained from food is transferred to a phosphate compound and stored for future use. Phosphate esters are also associated with the basic structure of the gene (DNA) and with the transfer of nerve impulses. Most biological reactions are more or less directly dependent upon the presence of phosphate esters.

In common usage the term phosphoric acid refers to orthophosphoric acid. This also applies to esters of this acid; e.g., methyl phosphate is a methyl ester of orthophosphoric acid.

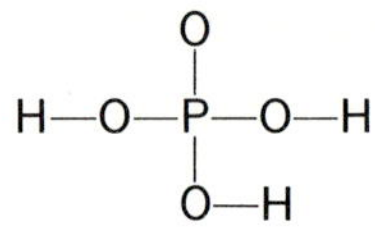

Orthophosphoric acid
Phosphoric acid

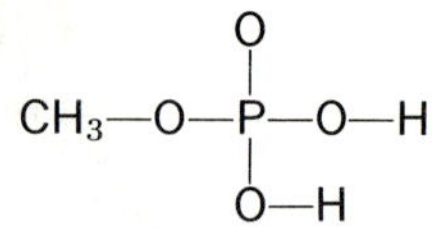

Monomethyl orthophosphate
Monomethyl phosphate

Phosphoric acid esters that have an ionizable hydrogen are strong acids.

	pK_1	pK_2	pK_3
H_3PO_4 Phosphoric acid	2.13	7.21	13.0
$CH_3OPO_3H_2$ Monomethyl phosphate	1.52	6.58	
$(CH_3O)_2PO_3H$ Dimethyl phosphate	0.76		

Phosphate esters produced during metabolism all have at least one highly ionizable hydrogen which is mostly dissociated at physiological pH. Several biologically active phosphate esters are

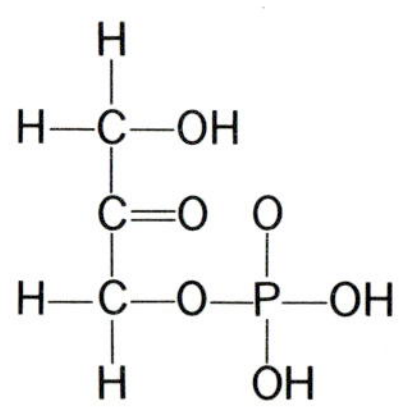

Dihydroxyacetone phosphate

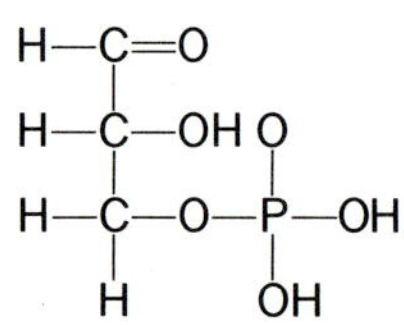

Glyceraldehyde-3-phosphate

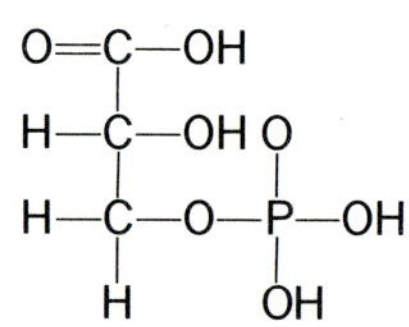

Glyceric acid–3-phosphate

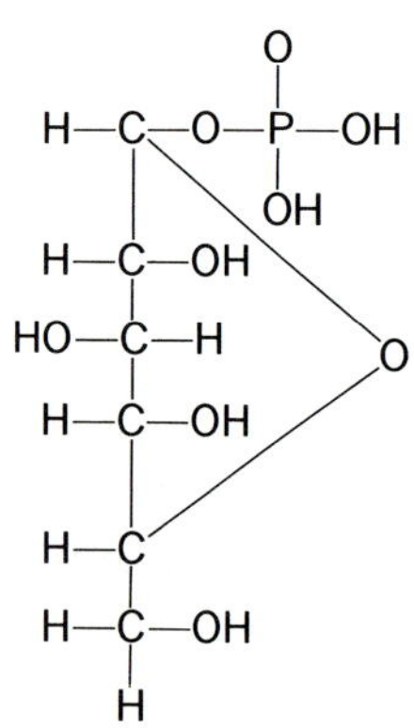

Glucose-1-phosphate

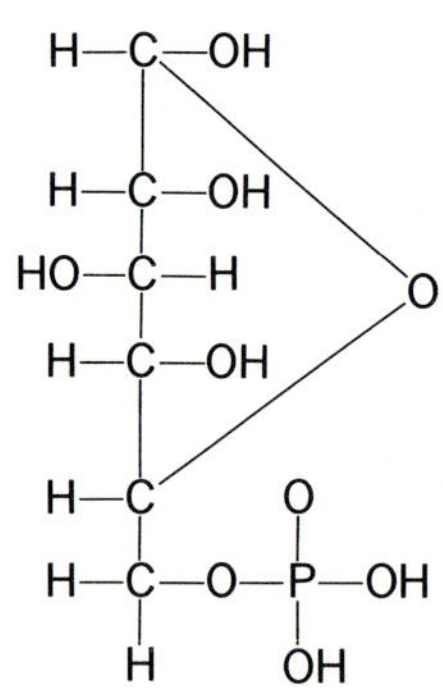

Glucose-6-phosphate

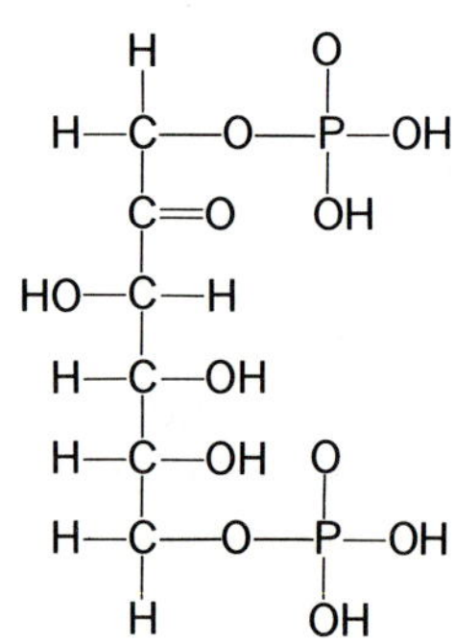

Fructose-1,6-diphosphate

PROBLEM

Assume that the pK values for glucose-6-phosphate are the same as for methyl phosphate (pK_1 = 1.52, pK_2 = 6.58). The first hydrogen would be completely dissociated at the physiological pH of 7.4. What would be the approximate percentage dissociation of the second hydrogen of the phosphate group at this pH? Substitute in the Henderson-Hasselbalch equation:

$$7.4 = 6.58 + \log\frac{[\text{salt}]}{[\text{acid}]}$$

$$\log\frac{[\text{salt}]}{[\text{acid}]} = 0.82$$

$$\frac{[\text{salt}]}{[\text{acid}]} = \frac{6.6}{1} = 87\% \text{ in the salt form}$$

Thus 87% of the second hydrogen of glucose-6-phosphate is dissociated at pH 7.4.

The monophosphate esters are more readily hydrolyzed in acid solution than in neutral or slightly alkaline solution. For example, between pH values of 0.5 and 7.5 methyl phosphate exhibits a maximum rate of hydrolysis at pH = 4, and glucose-1-phosphate is very rapidly hydrolyzed at pH values of 5 or lower.

The biological importance of a number of phosphate esters and their reactions is discussed in subsequent chapters dealing with metabolism. A few examples of mixed-ester formation are

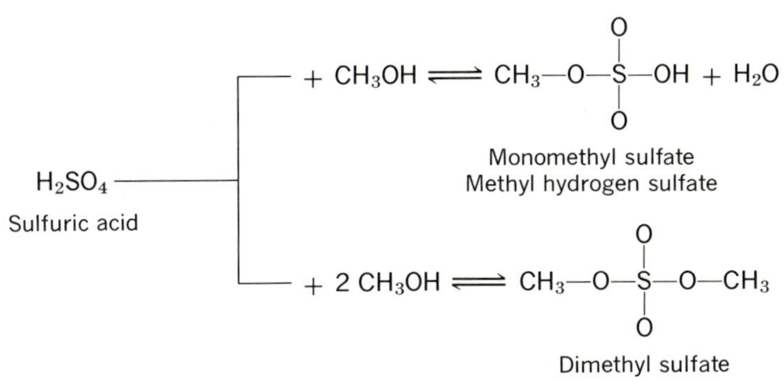

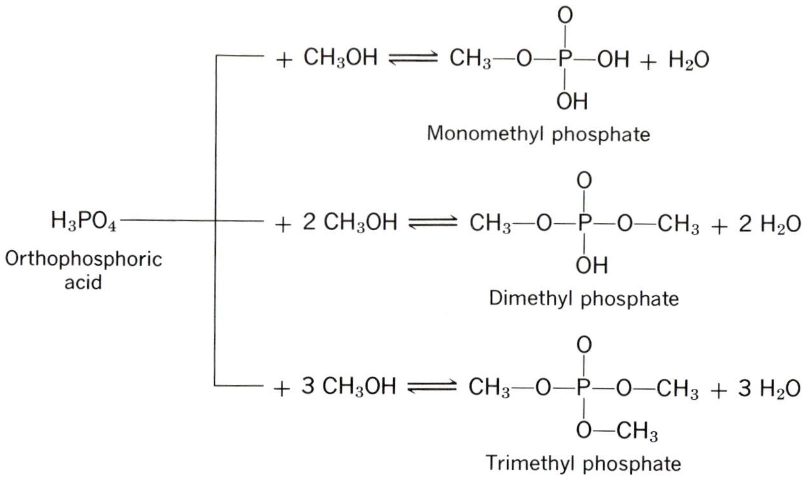

Saponification

It is often desired to hydrolyze an ester completely to its component parts. This cannot be accomplished in an acid medium, but in the presence of a base to neutralize the acid produced hydrolysis can be made to go to completion.

Saponification is the hydrolysis of an ester in the presence of a base. Since most esters are water-insoluble, the reaction is usually carried out in an alcoholic solution of the base, e.g., alcoholic KOH. The alcohol dissolves the ester and thus permits the reaction to be performed in a homogeneous medium. The simplified equations for the saponification of an ester are

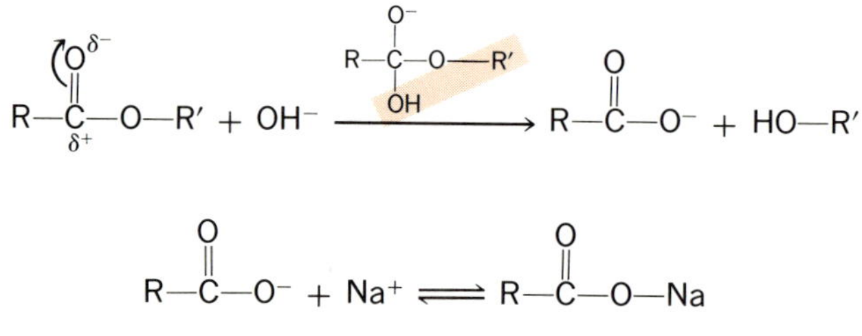

Overall:

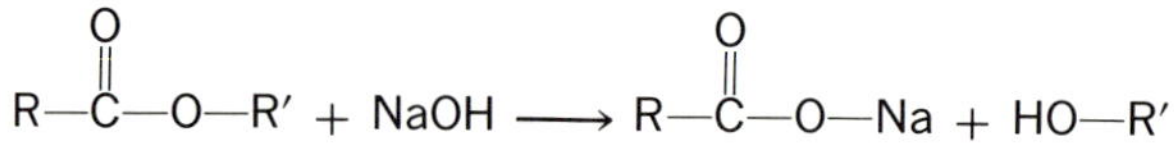

Specific example:

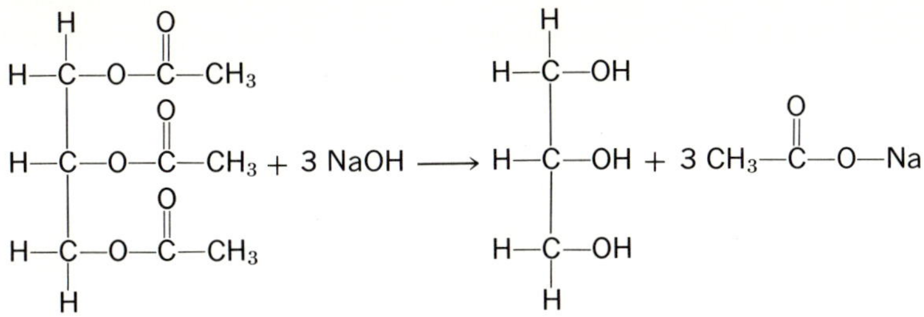

Glycerol acetate — Glycerol — Sodium acetate

11.5 ANHYDRIDES AND ACYL CHLORIDES

Anhydrides are formed by eliminating a molecule of water from two molecules of acid. Acyl chlorides are prepared by substituting a Cl for the OH of the carboxyl group of acids. For example, acetic acid forms acetic anhydride or acetyl chloride.

General equation:

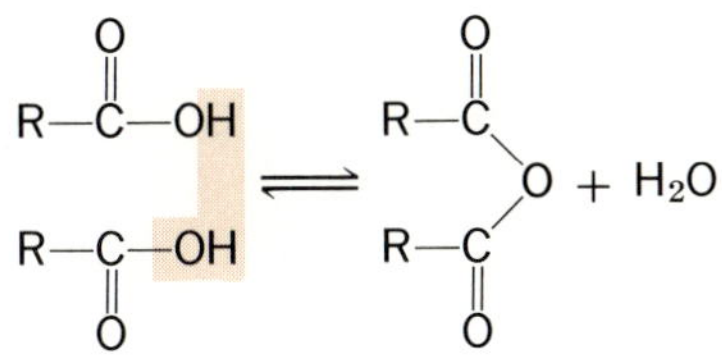

Acid — Anhydride

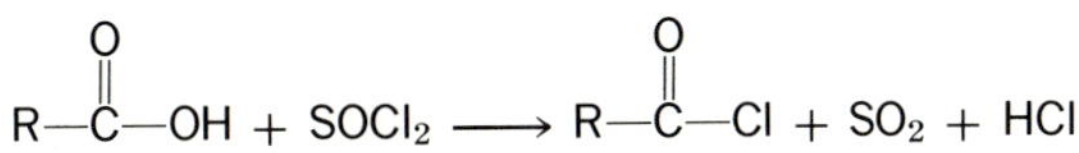

Acid — Thionyl chloride — Acyl chloride

Specific equation:

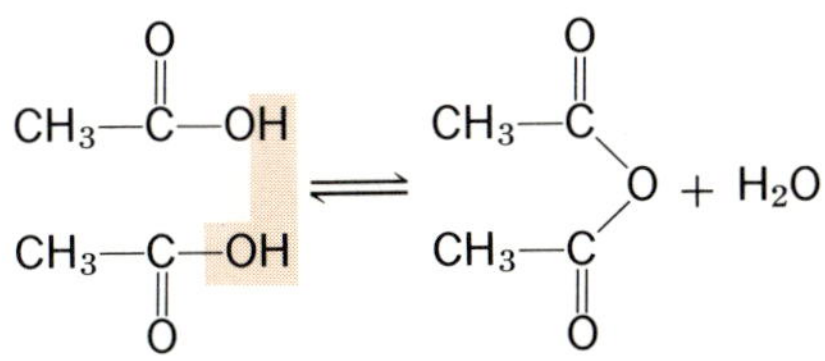

Acetic acid — Acetic anhydride

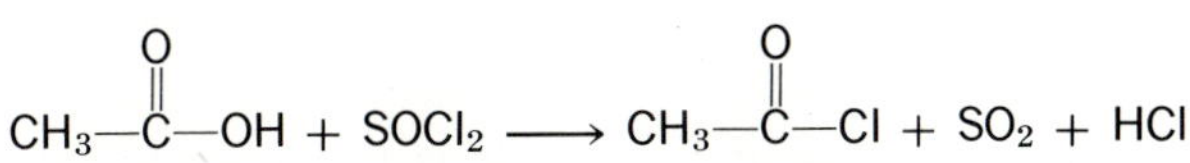

Acetic acid — Thionyl chloride — Acetyl chloride

Both anhydrides and acyl chlorides react with water to produce the respective acid, and they both produce esters when reacted with alcohols. Anhydrides and especially acyl chlorides are convenient reagents for the preparation of esters in general.

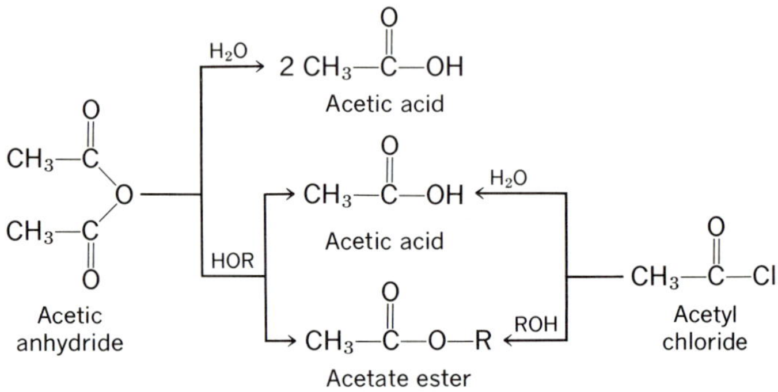

These reactions proceed by the following mechanism:

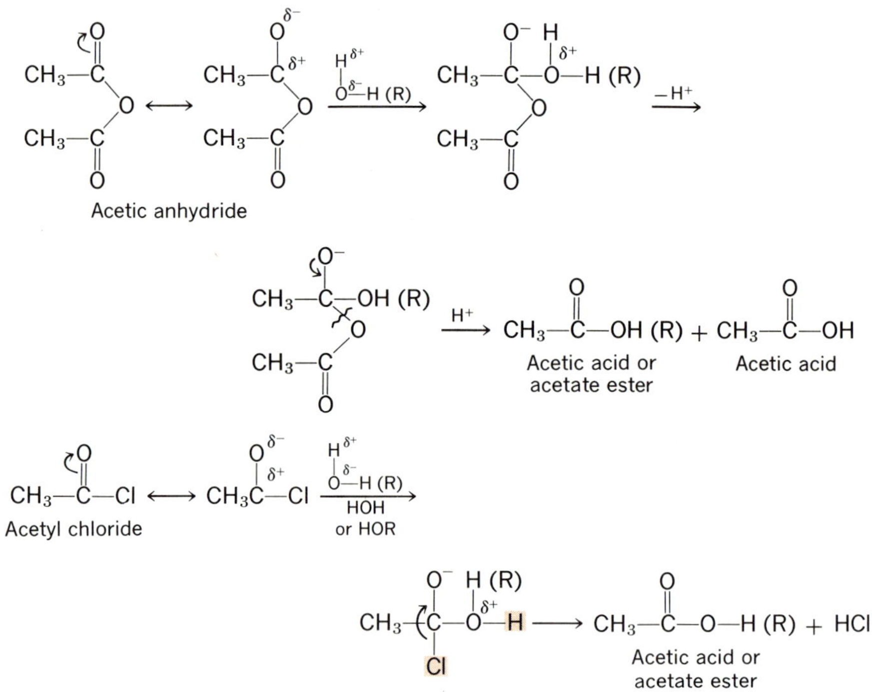

One mole of acid anhydride is required to esterify each OH group of an alcohol. Thus, if ethylene glycol (dihydroxyethane) is esterified with acetic anhydride, the reaction is

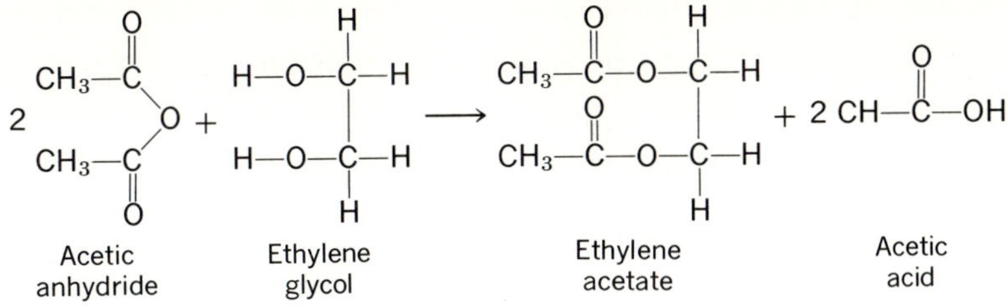

Inorganic anhydrides are formed when 2 moles of an inorganic acid react, with the elimination of 1 mole of water. The anhydrides of phosphoric acid are of interest from the biochemical standpoint. Two moles of orthophosphoric acid, H_3PO_4, react to form the anhydride called pyrophosphoric acid, which is an acid and therefore capable of forming mixed esters.

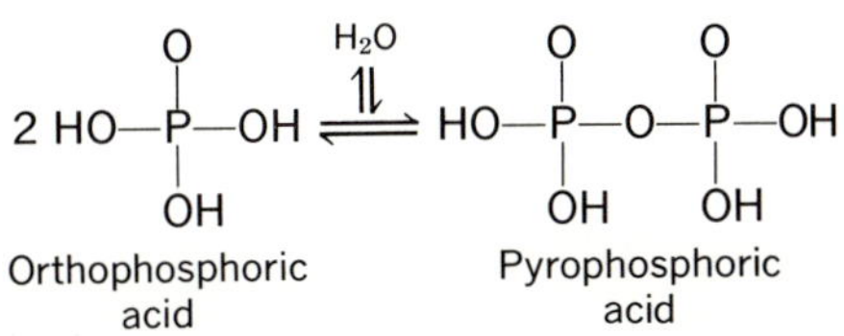

$$pK_1 = 0.85$$
$$pK_2 = 1.96$$
$$pK_3 = 6.54$$
$$pK_4 = 8.44$$

The biological compound adenosine triphosphate (ATP) contains both phosphate ester and anhydride bonds. This compound is the source of energy for many biological reactions.

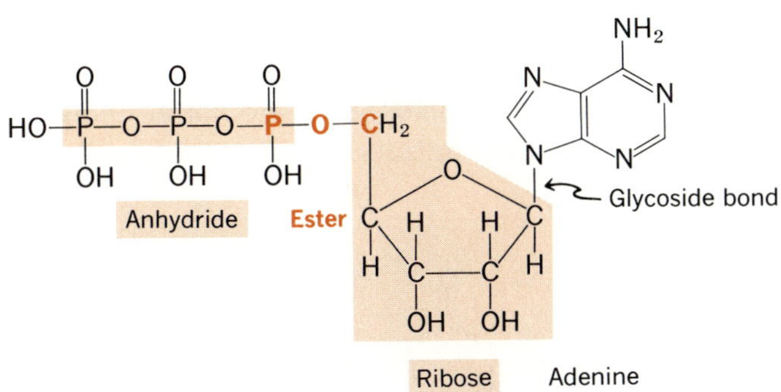

Adenosine triphosphate (ATP)

Mixed anhydrides are formed by eliminating 1 mole of water from an organic acid and an inorganic acid, e.g., phosphoric acid.

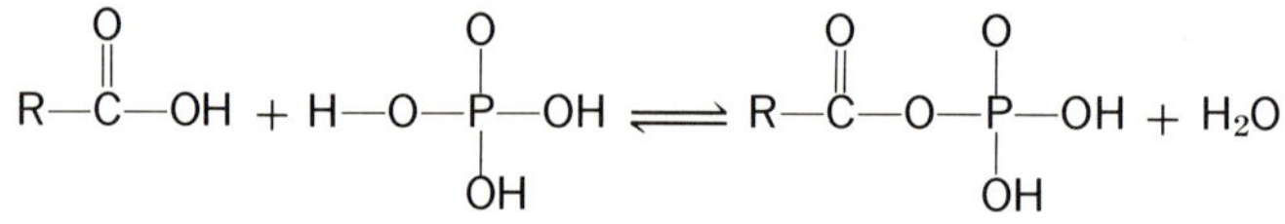

Organic acid Inorganic acid Mixed anhydride

A mixed anhydride, 1,3-diphosphoglyceric acid, is produced during carbohydrate metabolism. In this instance the anhydride is formed from glyceraldehyde-3-phosphate and H_3PO_4 according to the overall reaction

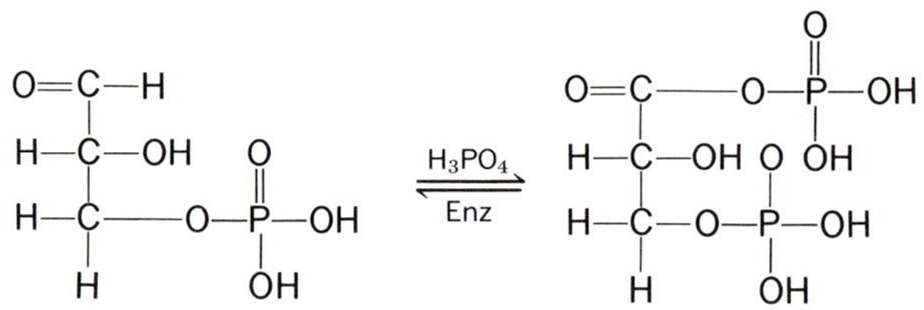

Glyceraldehyde-3-phosphate 1,3-Diphosphoglyceric acid

11.6 AROMATIC ACIDS AND SULFHYDRYLS

Aromatic acids may be divided into two classes, the phenols and the phenyl-substituted carboxylic acids. Sulfhydryl compounds, or thioalcohols, constitute another class of organic acids.

Phenols

Phenols are hydroxy benzene derivatives. The phenyl ring attracts the oxygen electrons of the OH group, thus permitting the hydrogen to ionize. Phenols, therefore, are acids; they are *not* alcohols. The following equation illustrates the ionization and neutralization of phenol (carbolic acid):

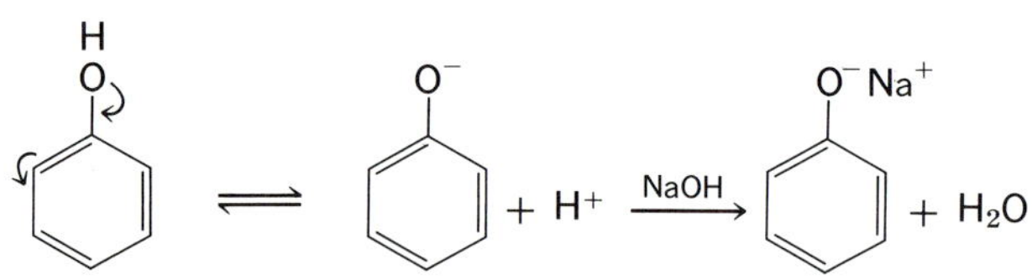

Phenol
Carbolic acid

Phenols are generally *weak* acids, but, by appropriate substitutions in the benzene ring the acidity can be strengthened. For example, substituting an electron-attracting group in the benzene ring tends to withdraw electrons from the ring, increasing the shift of the electrons from the oxygen in the phenol group. It is therefore easier for the proton to dissociate, and the phenol becomes more acidic. For example, by substituting nitro, NO_2, groups in phenol the pK value of the phenol group can be decreased from 9.95 to 1.02:

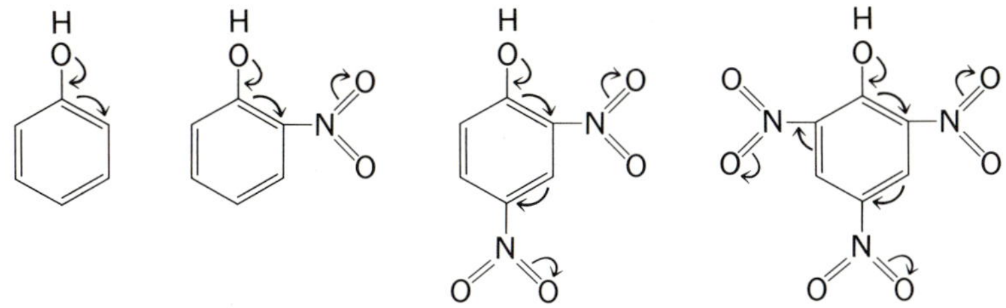

pK:	9.95	7.2	4.01	1.02
	Phenol	2-Nitrophenol	2,4-Dinitrophenol	2,4,6-Trinitrophenol
	Carbolic acid	*o*-Nitrophenol	*o,p*-Dinitrophenol	Picric acid

The insertion of one nitro group ortho to the OH group (2-nitrophenol) increases the acidity of the phenol group about 560-fold. The effect of two nitro groups in the 2,4 positions (2,4-dinitrophenol) increases the acidity about another 1,580 times. Trinitrophenol (picric acid) is a very strong acid with a pK of 1.02; it is 8.5×10^8 times stronger than phenol.

Aromatic carboxylic acids

The *phenyl* group as well as the phenol group can occur as a substituent in aliphatic carboxylic acids, such as benzoic, phenylacetic, and *p*-hydroxyphenylacetic acid.

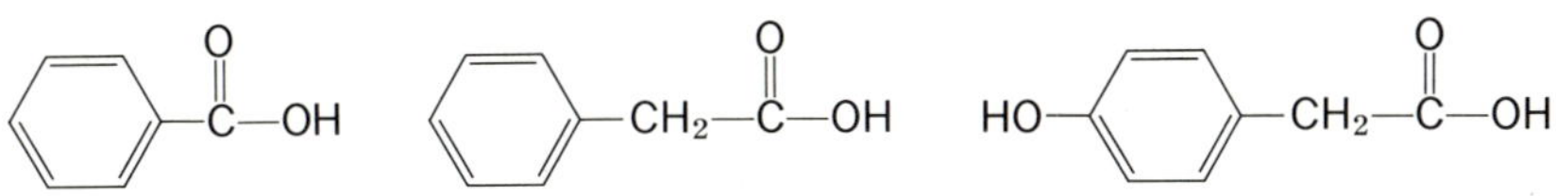

Benzoic acid   Phenylacetic acid   *p*-Hydroxyphenylacetic acid

p-Hydroxyphenylacetic acid has two pK values, one for the carboxyl group and one for the phenolic group.

Both phenyl and phenol groups occur in a number of biologically active compounds; some associated with the benzene ring have already been mentioned. Others will be discussed in subsequent chapters. The body is able to make some of the phenolic compounds required for proper metabolism. For example, the amino acid tyrosine (*p*-hydroxyphenylalanine) is synthesized from phenylalanine:

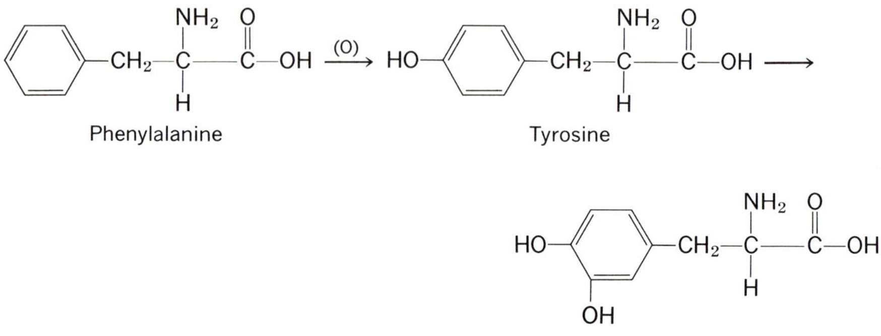

Phenylalanine → Tyrosine → Dopa

The browning of human skin upon exposure to ultraviolet light is caused by the oxidation of tyrosine to dopa.

Sulfhydryl compounds

Sulfhydryl compounds, R—S—H, or thioalcohols, may be considered derivatives of the acid hydrogen sulfide, H—S—H, in which one hydrogen has been replaced by a carbon chain. The sulfhydryl group has *weakly* acidic properties. The general ionization reaction is

$$\text{R—S—H} \rightleftharpoons \text{R—S}^- + \text{H}^+ \qquad \text{p}K \approx 8.0$$

Sulfhydryl compounds form salts of the SH group. For example, a general neutralization reaction is

$$\text{R—S—}\mathbf{H} + \mathbf{Na}\text{OH} \rightleftharpoons \text{R—S—}\mathbf{Na} + \mathbf{H}\text{OH}$$

The toxicity of heavy metals, such as lead, arsenic, and mercury, results from their ability to form stable salts with the SH group of enzymes. Some enzymes are biologically active only if they contain uncombined SH groups. The formation of a metallic salt destroys the free SH groups and thus inactivates the enzymes.

$$\underset{\text{Active enzyme}}{\text{Enz—SH}} + \mathbf{Hg}^{++} \longrightarrow \underset{\text{Inactive enzyme}}{\text{Enz—S—}\mathbf{Hg}\text{—S—Enz}} + 2\ \text{H}^+$$

Since enzymes are present in very small amounts, it requires only a trace of the heavy metals to inactivate many enzyme systems, which is lethal to the organism. This is the biological action of insecticides, such as Bordeaux mixture, which contain copper and lead. Mercurochrome is an organic mercury compound which has antiseptic properties because of its mercury content.

Oxidation

Organic sulfhydryl compounds are easily oxidized to a disulfide.

$$\underset{\text{Sulfhydryl}}{2\ \mathrm{R{-}S{-}H}} \xrightleftharpoons{2\,(\mathrm{H})} \underset{\text{Disulfide}}{\mathrm{R{-}S{-}S{-}R}}$$

This reaction of the SH group is of great biological significance. Consider, for example, the enzyme systems that require free SH groups for activity. If these SH groups are oxidized to a disulfide group, —S—S—, the enzyme is no longer active.

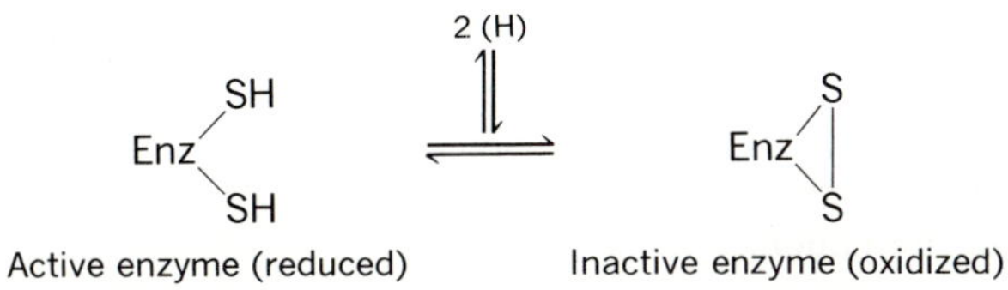

Sulfhydryl groups also play an important role in protein structure. Two protein strands A and B are shown in Fig. 11.3*a*. Each strand contains free SH groups. Upon oxidation of the SH groups to the disulfide group the two strands of protein A and B may be held together as in Fig. 11.3*b*.

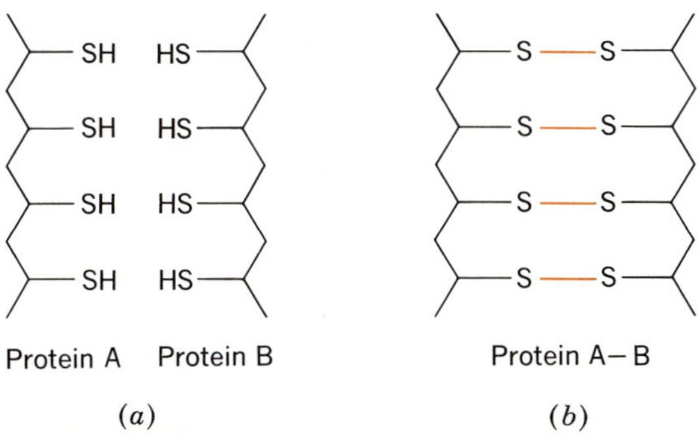

FIGURE 11.3 Formation of disulfide linkages between protein molecules: (*a*) two separate strands of protein, A and B; (*b*) protein A—B, composed of strand A and strand B held together by sulfide bonds.

QUESTIONS

1. What property is characteristic of all acids?
2. Why does a carboxyl group ionize appreciably whereas an alcohol group does not?
3. Write the formula for an example of each of the following:
   (*a*) Unsaturated acid (*b*) α-Hydroxy acid
   (*c*) β-Keto acid (*d*) Dicarboxylic acid
   (*e*) Tricarboxylic acid
4. What is the structure of a fatty acid?
5. Write the formula for each of the following:
   (*a*) Butyric acid (*b*) Pentanoic acid
   (*c*) Decanoic acid (*d*) α-Aminopropionic acid
6. Write the stepwise ionization of citric acid.
7. Arrange the following acids in order of their increasing strength:
   (*a*) $CH_3—COOH$ (*b*) $CH_3—CH_2—CHBr—COOH$
   (*c*) $CH_3—CHOH—COOH$ (*d*) $CH_2Cl—CH{=}CH—COOH$
8. If glucose is biologically converted only to lactic acid, how many moles of lactic acid are formed per mole of glucose?
9. Define an ester.
10. Write an equation for the formation of ethyl butyrate.
11. What are the end products of the saponification of an ester?
12. Arrange the following alcohols in the order of their decreasing reactivity toward ester formation:

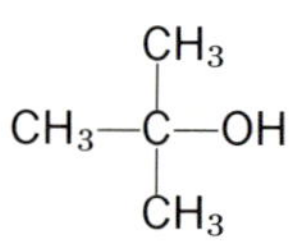

(I)

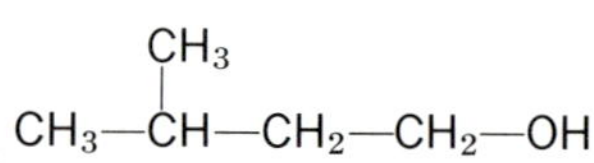

(II)

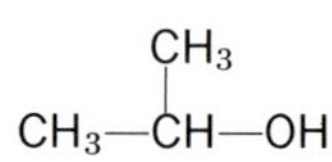

(III)

13. How does substitution of an alkyl group on the α carbon of an acid affect its reactivity toward esterification?
14. What is meant by the term anhydride? Write the formula for an organic anhydride and an inorganic anhydride.
15. What anhydride would you use to react with *n*-propyl alcohol to form *n*-propyl butyrate? Write the equation for this reaction.
16. What is a possible structural formula for compounds I, II, and III?

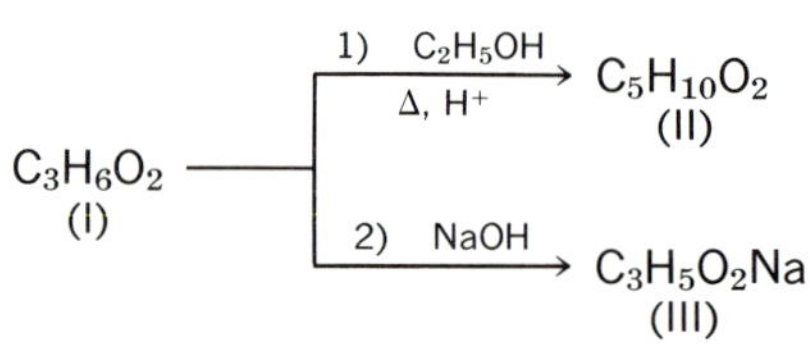

17. Write a possible structural formula for I, II, and III.

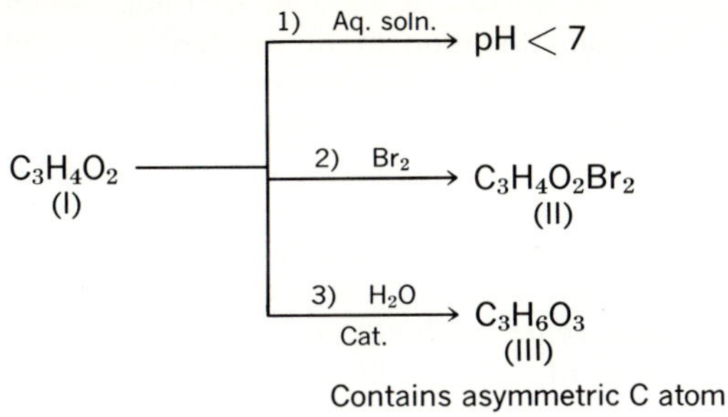

18. Write a possible structural formula for compounds I to V.

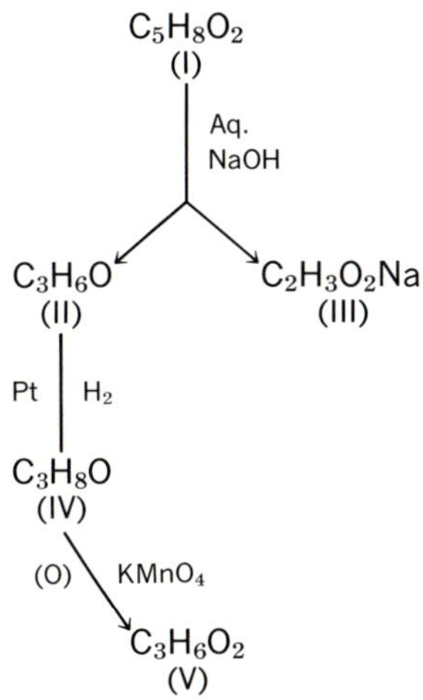

19. What is a structural formula for I to VI?

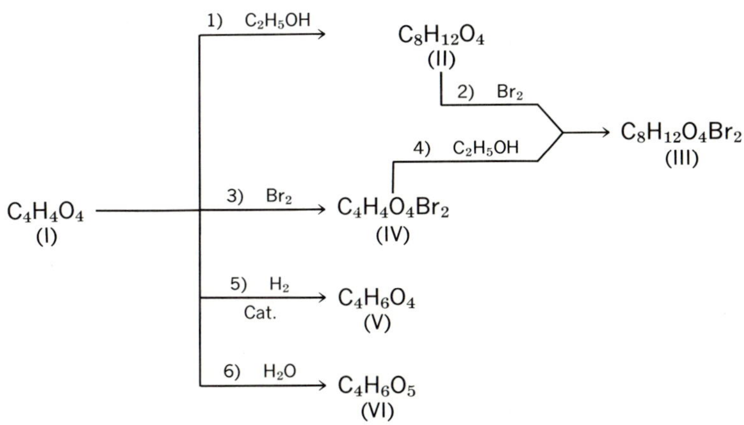

20. Compound I, $C_3H_6O_3$, gives a positive reaction with Fehling's reagent. I was reacted with acetic anhydride to yield II, $C_7H_{10}O_5$, which upon oxidation gave III, $C_7H_{10}O_6$. Saponification of III followed by acidification gave IV, $C_3H_6O_4$, and V, $C_2H_4O_2$. IV was also formed upon selective oxidation of I. What is the structural formula for each of the compounds?

21. An aqueous solution of I, $C_6H_8O_7$, has pH < 7. It requires 3 meq of NaOH to titrate 1 mmole of I to neutrality. I reacts with acetic anhydride to form II, $C_8H_{10}O_8$. It can be dehydrated to III, $C_6H_6O_6$, which adds Br_2. I contains an asymmetric carbon atom. What are the structural formulas of I, II, and III?

22. Compound I, $C_4H_4O_5$, reacts with ethanol to give II, $C_8H_{12}O_5$, and can be reduced with H_2 to give III, $C_4H_6O_5$. III can be dehydrated to give IV, $C_4H_4O_4$, which exists in either a cis or trans form. What are the structural formulas of I to IV?

chapter 12.... amines, amides, and heterocyclic nitrogen compounds

A large variety of biologically active compounds contain nitrogen, among which are all enzymes, some hormones and vitamins, and all genes. This chapter deals with organic nitrogen compounds classified as amines, amides, and heterocyclic nitrogen compounds.

Plants are able to make all their nitrogen-containing compounds from inorganic nitrogen, carbon dioxide, and water. Animals, however, are more restricted in their ability to synthesize organic compounds and require some preformed nitrogen compounds in their diet. These must come from plants.

Review Sec. 6.14, dealing with NH_3 and amino acid titration. Does the N of NH_3 have any unshared electrons? Can NH_4^+ be considered as being protonated NH_3? Are amino acids amphoteric compounds? Does the presence of an α-amino group, NH_2, of an amino acid affect the degree of dissociation of the carboxyl hydrogen? Can the amino group of an amino acid be protonated?

12.1 CLASSIFICATION OF AMINES

Amines are substituted ammonia, NH_3, in which one or more of the hydrogens have been replaced by a carbon chain. The amino group, NH_2, like the parent compound, ammonia, has an unshared pair of electrons on the nitrogen making the group basic. Amines are classified as primary, secondary, or tertiary, depending upon the number of carbon atoms bonded to the nitrogen. Quaternary nitrogen compounds are substituted NH_4^+ in which all hydrogens are replaced by carbon chains. They bear a positive charge and exist as salts or hydroxides. Table 12.1 illustrates these classifications.

TABLE 12.1 CLASSIFICATION OF AMINES

Class	General formula	Compound
Primary	H | R—N—H	H | CH_3—N—H Methylamine
Secondary	R | R—N—H	CH_2—CH_3 | CH_3—N—H Methylethylamine
Tertiary	R | R—N—R	CH_3 | CH_3—N—CH_3 Trimethylamine
Quaternary	[R | R—N—R | R]$^+$	[CH_3 | CH_3—N—CH_3 | CH_3]$^+$ Cl^- Tetramethyl- ammonium chloride

12.2 STRUCTURE OF AMINES

The electron configuration about nitrogen in ammonia and amines is such that the two unshared electrons of nitrogen are at one apex of a pyramid and one each of the three hydrogen atoms occupies one of the other three corners. The hydrogens are equidistant from each other, as shown in Fig. 12.1. The unshared pair of electrons on the nitrogen atom can be donated to positive ions, such as hydrogen, thus making the amines basic compounds.

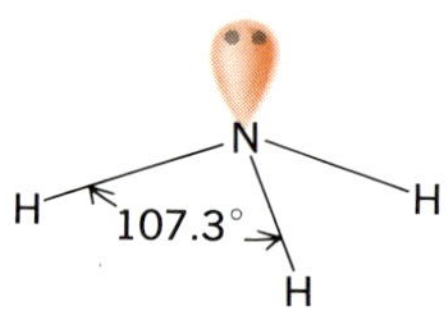

FIGURE 12.1 The electron configuration of the unshared pair on nitrogen in the ammonia molecule. The bond angle between two hydrogens is 107.3°, not the angle of 109.5° expected for a tetrahedral configuration.

12.3 NOMENCLATURE OF AMINES

Amines can be named according to the type and number of alkyl radicals attached to the nitrogen of the amino group (an *N* is used to indicate that substitution is on the nitrogen group) or as amino-substituted hydrocarbons.

$$CH_3—CH_2—CH_2—NH_2$$

n-Propylamine
1-Aminopropane

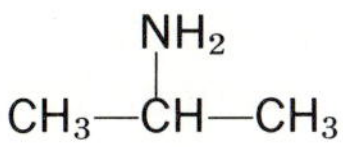

Isopropylamine
2-Aminopropane

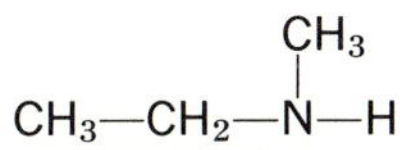

Methylethylamine
Methylaminoethane

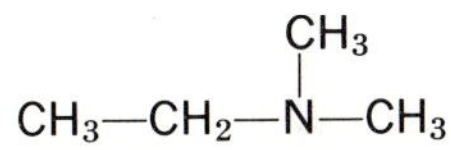

Dimethylethylamine
Dimethylaminoethane

The quaternary nitrogen compounds are named as derivatives of ammonium compounds:

$$(CH_3)_4N^+Cl^-$$

Tetramethylammonium chloride

$$(CH_3)_2(C_2H_5)_2N^+OH^-$$

Dimethyldiethylammonium hydroxide

12.4 PREPARATION OF AMINES

Only one method for the preparation of amines will be mentioned, namely, the substitution by alkyl halides on an amine nitrogen. The general reactions and examples are

$$RX + NH_3 \longrightarrow R—NH_2 \quad (1° \text{ amine}) \qquad R = CH_3 \longrightarrow CH_3—NH_2 \quad (\text{Methylamine})$$

$$RX + R—NH_2 \longrightarrow R_2—NH \quad (2° \text{ amine}) \qquad R = CH_3 \longrightarrow (CH_3)_2—NH \quad (\text{Dimethylamine})$$

$$RX + R_2—NH \longrightarrow R_3N \quad (3° \text{ amine}) \qquad R = CH_3 \longrightarrow (CH_3)_3—N \quad (\text{Trimethylamine})$$

$$RX + R_3N \longrightarrow R_4N^+X^- \quad (\text{Quarternary ammonium salt}) \qquad R = CH_3 \longrightarrow (CH_3)_4{}^+Cl^- \quad (\text{Tetramethylammonium chloride})$$

12.5 REACTIONS OF AMINES

Protonation

As shown in Chap. 6 a proton can bond with the two electrons of nitrogen in NH_3 to form the ammonium ion, NH_4^+ ($NH_3 + H^+ \rightleftharpoons NH_4^+$). This is also possible with the primary and secondary amines; hence, their basicity. The following reactions with HCl illustrate this property. Ammonia is included for comparison.

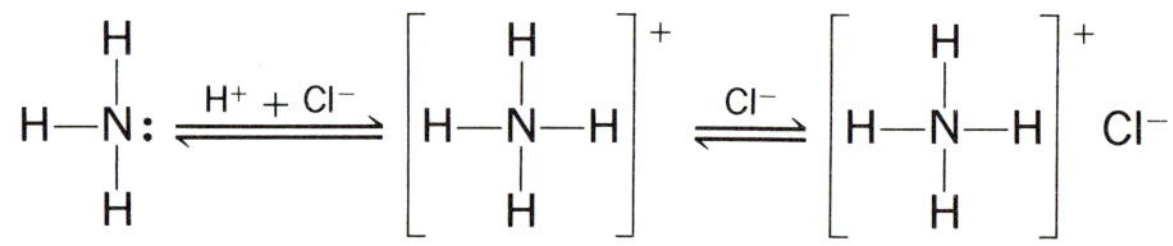

Ammonia Base | Ammonium ion Acid | Ammonium chloride Salt

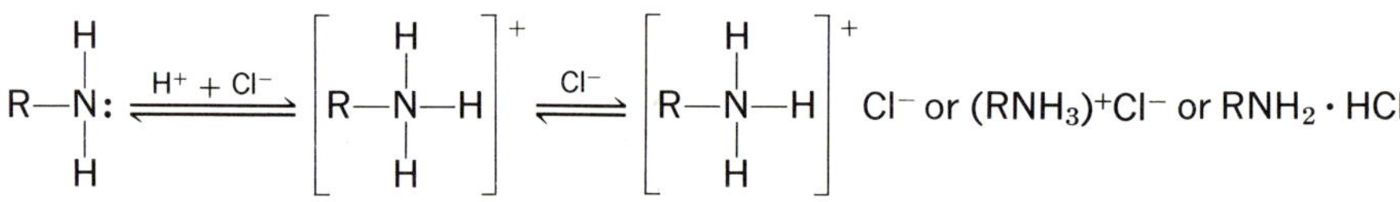

1° amine Base | Protonated amine Acid | Substituted ammonium chloride Salt | Amine hydrochloride

Specific example:

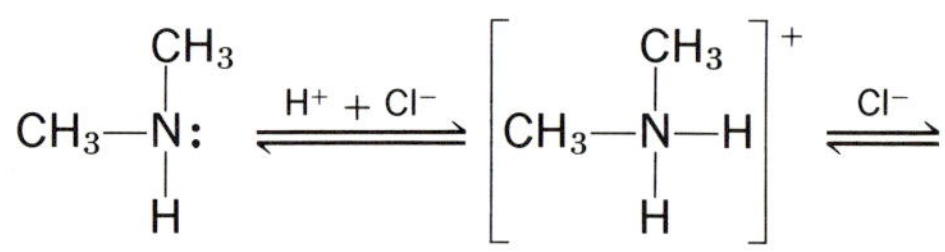

Dimethylamine Base | Protonated dimethylamine Acid

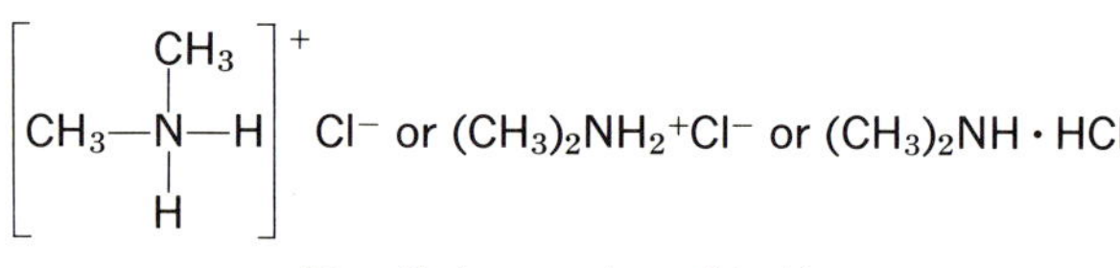

Dimethylammonium chloride
Dimethylamine hydrochloride
Salt

Amine salts can be named as substituted ammonium chlorides or, to keep the identity of the amine more intact, as amine hydrochlorides ($RNH_2 \cdot HCl$ or $RNH_3^+Cl^-$).

The primary, secondary, and tertiary amines are all of about the same basicity; however, as the two electrons of the nitrogen become more easily available, the amine becomes more basic. Quaternary hydroxides are very strong bases. Choline, a quaternary hydroxide, is necessary for the transport of nerve impulses.

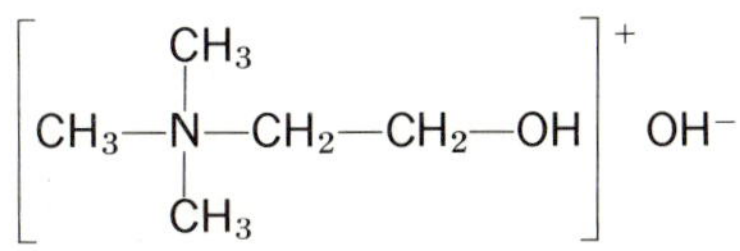

Choline
Trimethylethanolammonium hydroxide

Nitrous acid

Primary and secondary amines react with nitrous acid, whereas the tertiary amines do not. The primary amines yield alcohol as the final product, and the secondary amines yield *N*-nitrosamines. The reactions are as follows.

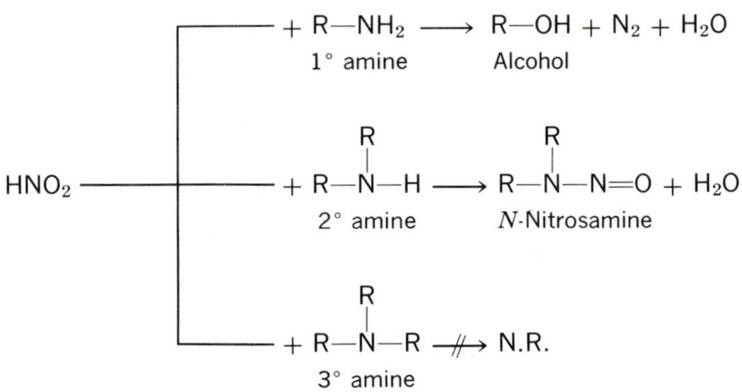

The reaction of primary amines with HNO_2 yields one mole of molecular nitrogen, N_2, for each mole of amine reacted. Hence, when the volume of N_2 eliminated in this reaction is determined, the quantity of amine initially present can be calculated. This is the basis of the Van Slyke determination of amino acids. The following problem will illustrate the calculations.

PROBLEM

Under the appropriate experimental conditions 2.25 mg of an amine was reacted with nitrous acid, and 1.12 ml of N_2 was collected at standard temperature and pressure (STP),

0°C and 1 atm pressure (see Appendix D). (*a*) What was the molecular weight of the amine? (*b*) Write a formula for the amine. The amine contains one NH_2 per formula weight.

(*a*) The amine was a primary amine, R—NH_2, because N_2 was eliminated. Since 1 mole of N_2 is eliminated from 1 mole of amine, if the number of moles of N_2 produced is known, the number of moles of amine represented by the 2.25-mg (0.00225-g) sample will also be known. Then the molecular weight of the amine can be calculated. The problem is worked as follows.

At STP 1 mole of any gas occupies 22.4 liters, or 22,400 ml. Thus, if 1 mole of N_2 occupies 22,400 ml, the number of moles contained in 1.12 ml of gas is found by the simple proportion

$$\frac{1 \text{ mole } N_2}{22{,}400 \text{ ml}} = \frac{x \text{ moles } N_2}{1.12 \text{ ml}}$$

$$x \text{ moles } N_2 = \frac{1 \times 1.12}{22{,}400}$$

Since moles N_2 is the same as moles of amine,

$$\text{Moles amine} = \frac{1 \times 1.12}{22{,}400}$$

but

$$\text{Moles amine} = \frac{\text{g amine}}{\text{mol. wt.}}$$

therefore

$$\frac{1.12}{22{,}400} = \frac{0.00225}{\text{mol. wt.}}$$

$$\text{Mol. wt.} = \frac{22{,}400 \times 0.00225}{1.12} = \frac{2.24 \times 22.5}{1.12} = 2 \times 22.5 = 45$$

The molecular weight of the amine is 45.

(*b*) The amine is a primary amine with a molecular weight of 45. The simplest primary amine is methylamine, CH_3—NH_2, with a molecular weight of 31. The difference between this value and 45 is 14, which is equivalent to one CH_2 group. Hence, inserting a CH_2 group into CH_3—NH_2 gives the amine CH_3—CH_2—NH_2, ethylamine, with a molecular weight of 45. The amine is ethylamine.

Fluorodinitrobenzene (FDNB)

Fluorodinitrobenzene reacts with primary amines.

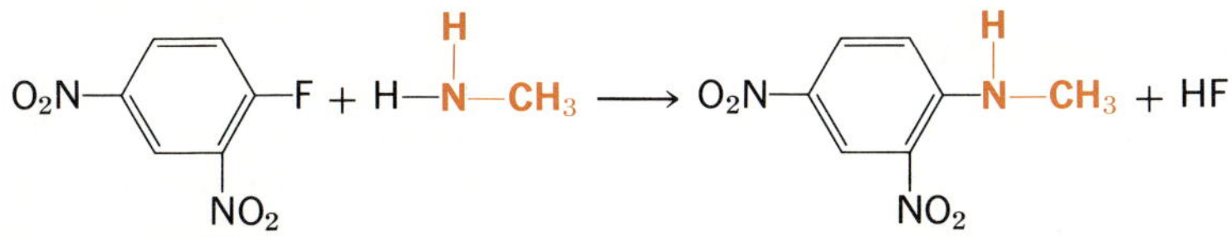

Fluorodinitrobenzene (FDNB)

N-Dinitrophenylmethylamine (DNP-methylamine)

It is useful in determining the free NH_2 groups in proteins. The reaction is illustrated with a dipeptide, a compound composed of only two amino acids (see page 297).

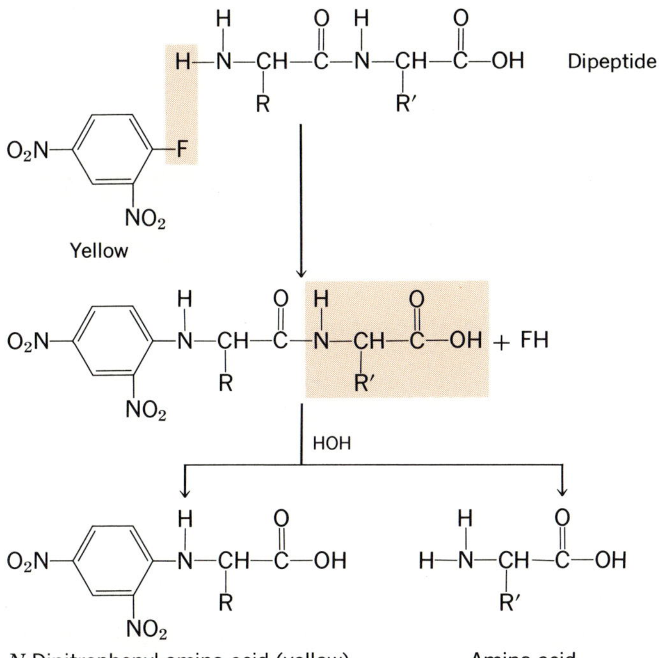

N-Dinitrophenyl amino acid (yellow)

Amino acid

After the reaction has taken place, the dipeptide is hydrolyzed. The dinitrophenyl group is attached to the amino acid that had the free NH_2 group in the dipeptide. The insertion of the large dinitrophenyl group into an amino acid changes its solubility characteristics so that its separation from the other amino acids can easily be effected by paper chromatography or by ion-exchange chromatography (see Chap. 5).

Aldehydes and ketones

With the exception of formaldehyde, aliphatic aldehydes and ketones react with ammonia to form an unstable imine (imines contain the C=N—H group and substituted imines the C=N—R group), which can be catalytically reduced to the corresponding primary amine. Primary amines react in a similar manner to yield secondary amines. The substituted imines are known as *Schiff's bases*. The mechanism of these reactions is as follows:

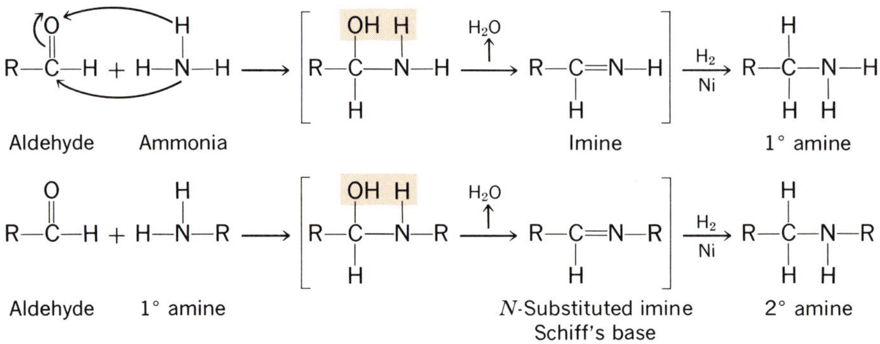

The simplified general equation for this reaction can be written

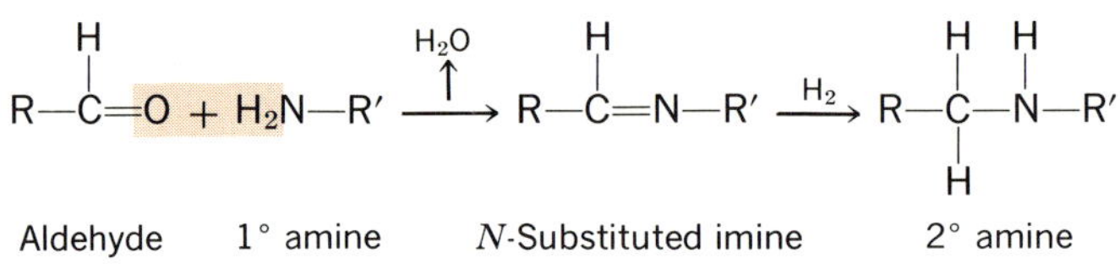

12.6 STRUCTURE OF AMIDES

Amides have the general formula $R-\overset{O}{\overset{\|}{C}}-NH_2$. Substituted amides have one or both hydrogens of the NH_2 group substituted by a carbon chain, R. The formulas for the monosubstituted and disubstituted amides are

Monosubstituted amide   Disubstituted amide

The R groups may or may not all be the same.

Amides can be considered as derivatives of acids in which the OH of the carboxyl group has been replaced by NH_2. The presence of the strongly electronegative carbonyl group, C=O, results in a delocalization of the nitrogen electrons and, as a result, makes them unavailable for bonding with a proton

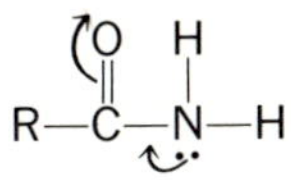

Thus, amides *are not* basic. For our discussion they will be considered neutral compounds.

12.7 NOMENCLATURE OF AMIDES

Amides are named by substituting the suffix *-amide* for *-ic* in the name of the acid. For example:

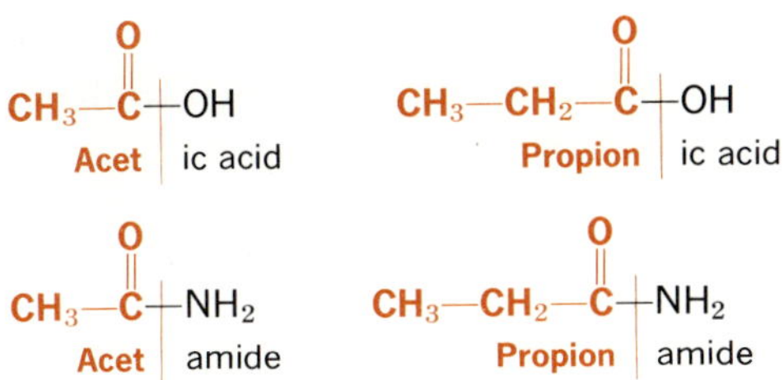

The name of a substituted amide must indicate the name of each substituent. The nitrogen atom of the amide group is not numbered; it is indicated by using the symbol *N*. In naming nitrogen compounds, *N* indicates that substitution takes place on the nitrogen group.

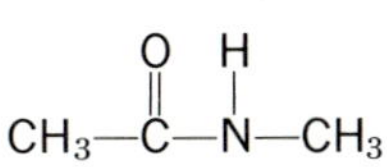

N-Methylacetamide

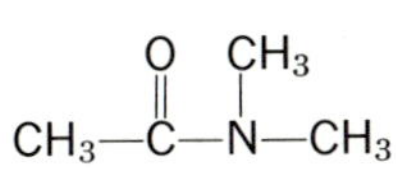

N,N-Dimethylacetamide

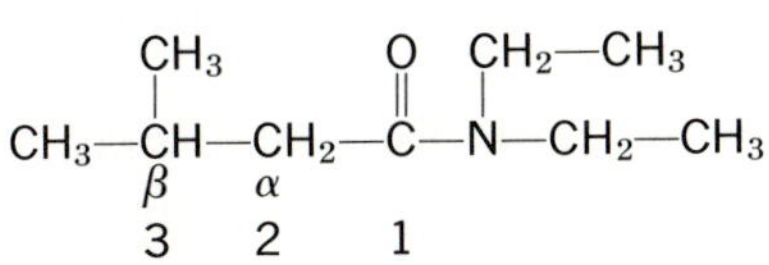

β-Methyl-*N,N*-diethylbutyramide
3-Methyl-*N,N*-diethylbutyramide

12.8 PREPARATION OF AMIDES

Acids

The amides of the aliphatic monocarboxylic acids are easily made by heat decomposition of the respective ammonium salts. For example, acetamide is made by heating ammonium acetate.

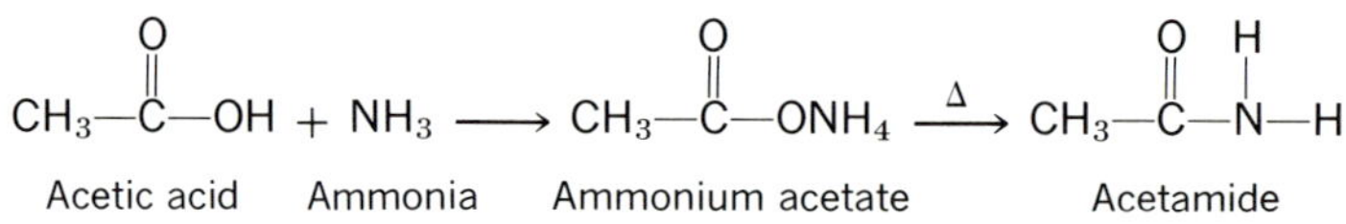

Some general reactions for the preparation of amides by this method are

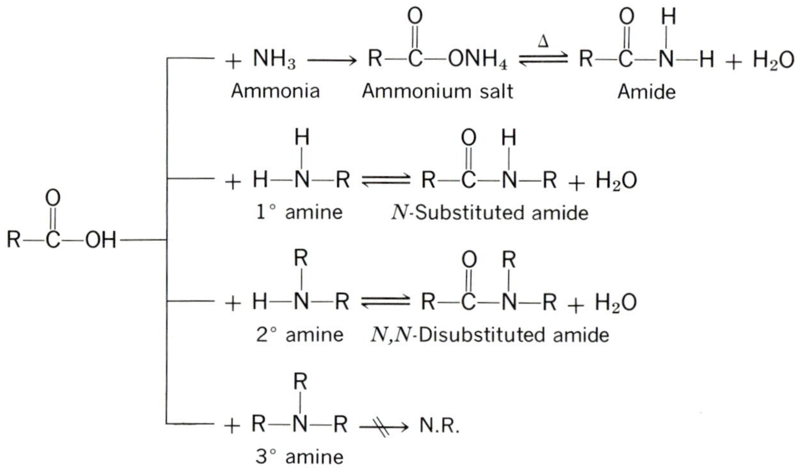

Since these reactions are reversible, the amides can be hydrolyzed to the corresponding acid and amine. The hydrolysis reaction proceeds faster in alkaline solution.

The mechanism of amide formation may be envisioned as follows. As in many previous reactions, the strong attraction of oxygen for electrons plays a key role.

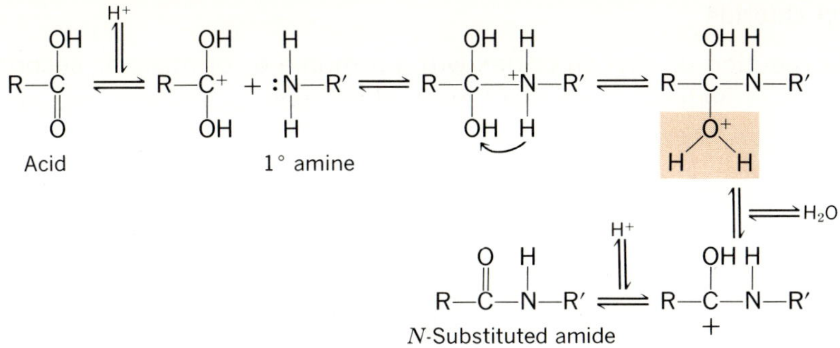

Net reaction: 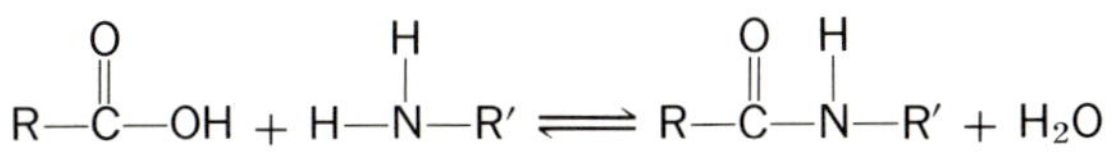

The hydrolysis of amides may be considered the reverse of the above reaction.

The formation of amides by the reaction of an acid and an amine catalyzed by an enzyme is of great biological importance. For example, two amino acids react to produce a substituted amide. The carboxyl group of one amino acid reacts with the amino group of another amino acid:

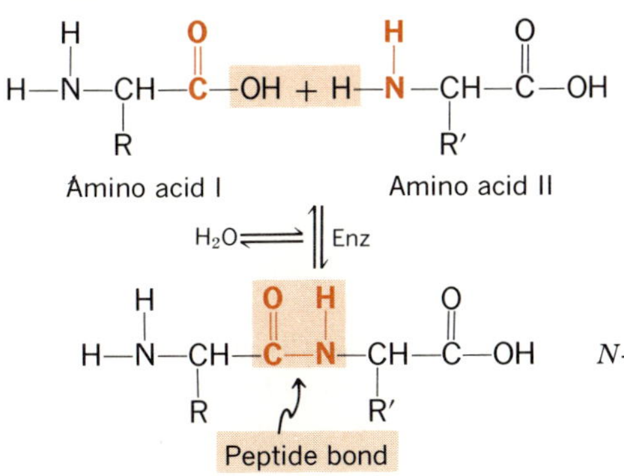

N-Substituted amide

The product is called a ***peptide,*** and the amide linkage between the amino acids is called a ***peptide bond.*** When two amino acids are linked together, as above, the compound is known as a ***dipeptide.*** Proteins are high-molecular-weight compounds which contain many amino acids linked together by peptide bonds. Amino acids and proteins constitute a topic in themselves and are discussed in more detail in a later chapter.

Acyl chloride

The reaction of acyl chlorides with ammonia or primary or secondary amines provides a convenient method for making amides. The general formula, $R-\overset{\displaystyle O}{\overset{\|}{C}}-Cl$, for acyl chlorides is used.

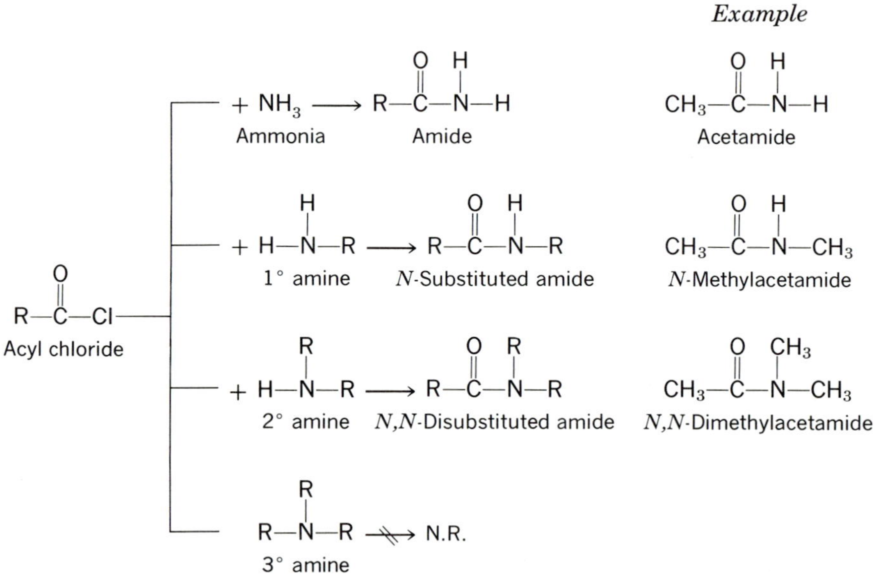

Specific examples:

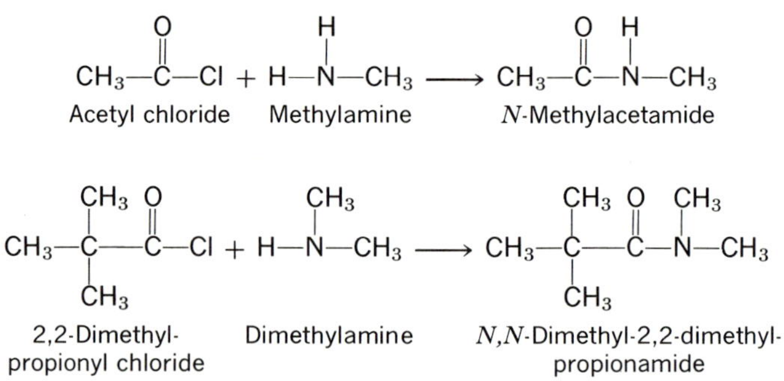

The reaction of benzenesulfonyl chloride with ammonia and amines serves to distinguish between primary, secondary, and tertiary amines.

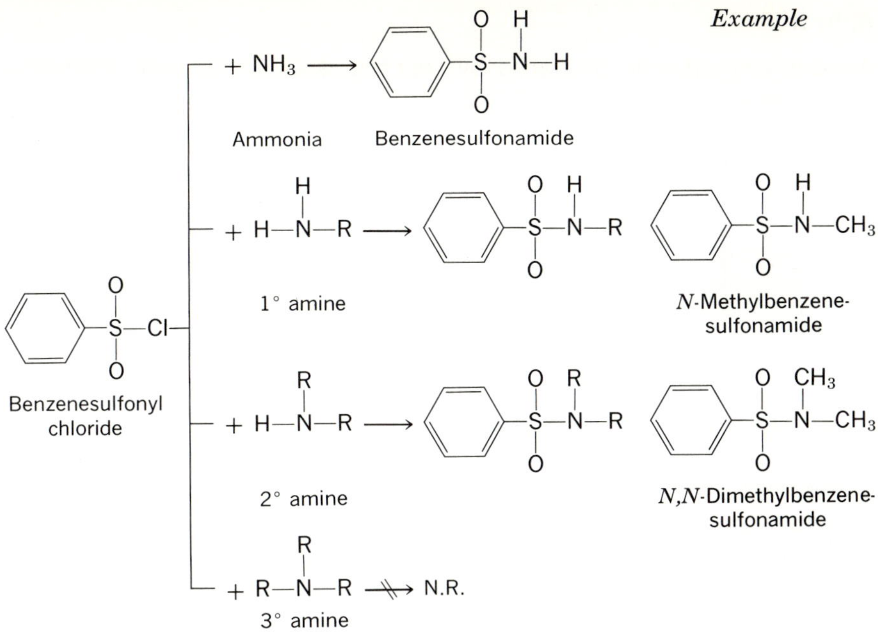

The sulfonyl derivative of primary amines is *soluble* in alkali, such as NaOH solution. This property is a result of the high electronegativity of the O—S—O group, which permits a shift in electrons equivalent to the enolization of aldehydes and ketones (see Sec. 8.13). This shift forms the S—OH group, which imparts acidic properties to the compound. The sodium or potassium salts of these acids are water-soluble. The equation for this reaction is

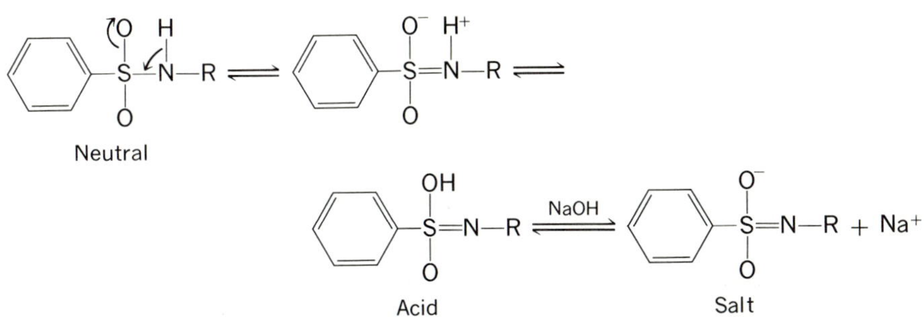

The secondary amines react with benzenesulfonyl chloride (see above), but the product is not soluble in alkali because there is no hydrogen on the amide nitrogen. The tertiary amines do not react with this reagent.

Anhydride

Amides can also be prepared by reacting anhydrides with ammonia and amines:

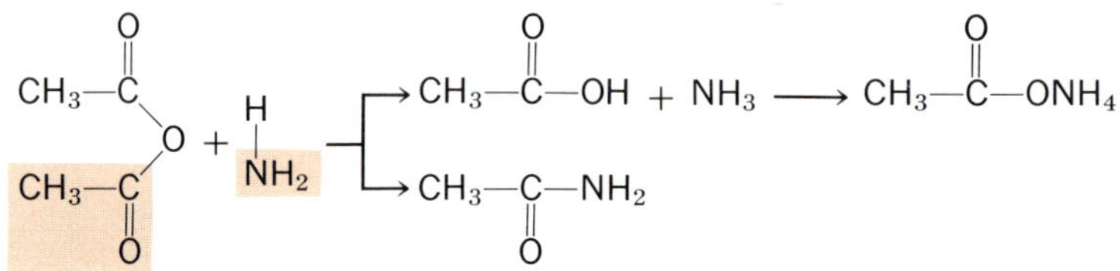

Acetic anhydride — Acetamide — Ammonium acetate

12.9 HETEROCYCLIC COMPOUNDS

Compounds that contain more than one kind of atom in a cyclic structure are called *heterocyclic compounds*. The elements most commonly found in heterocyclic structures are *carbon, oxygen, nitrogen,* and *sulfur*. The cyclic structure of glucose, the hemiacetal form, is a heterocyclic compound that has already been discussed. This cyclic structure contains carbon and oxygen in the ring.

Instead of attempting a thorough discussion of all heterocyclic compounds, this section is concerned only with the general structure of some that are of biological importance. The heterocyclic compounds, like cyclic compounds in general, may possess two distinctly different properties, depending upon the nature of the ring. Some react as aliphatic compounds, e.g., alkanes and alkenes, whereas others have properties of the aromatic (benzene) ring.

Electron distribution

The prerequisite for an aromatic ring is the presence of six p electrons capable of forming three resonating π bonds, as was observed with the benzene nucleus (see Fig. 8.8). For this to be so, there must be either a six-membered ring, which may or may not be heterocyclic (Fig. 12.2a), or a five-membered heterocyclic ring (Fig. 12.2b) in which one atom must be capable of furnishing two p electrons. These two electrons cannot be furnished by a carbon atom but can be furnished by a nitrogen, oxygen, or a sulfur atom because of their unbonded electrons.

In pyridine (Fig. 12.3) the nitrogen atom donates only one p electron (to make the π bond in the C=N group). This p electron plus one

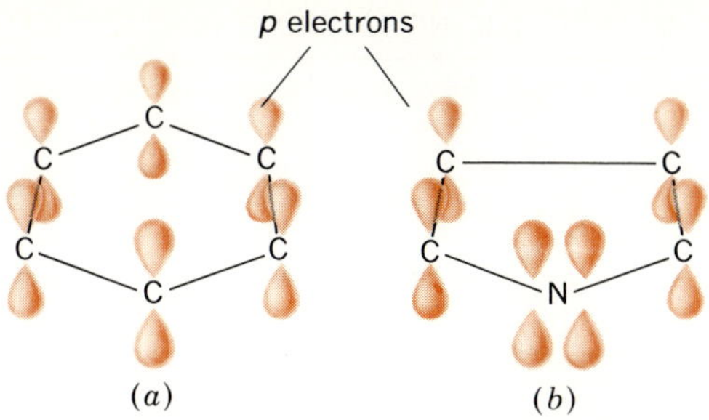

FIGURE 12.2 The aromatic ring configuration showing the p orbitals in (a) benzene and (b) pyrrole ring.

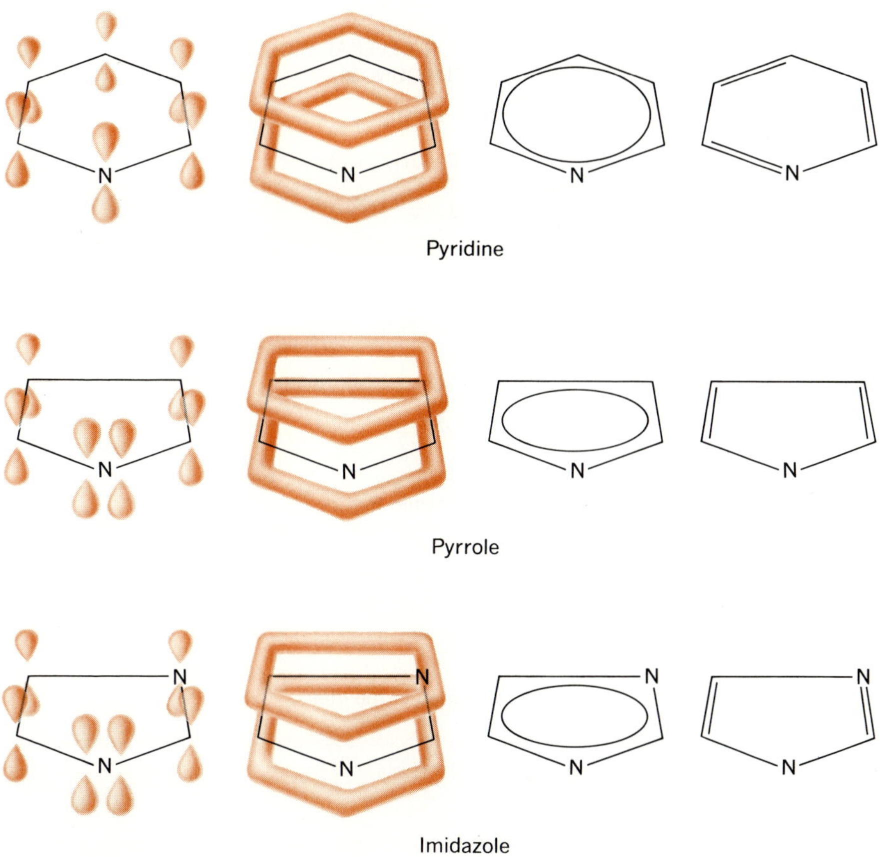

FIGURE 12.3 Aromatic ring structures showing the electron distribution.

TABLE 12.2 SOME HETEROCYCLIC NITROGENOUS RING STRUCTURES

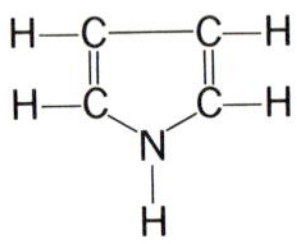

Pyrrole

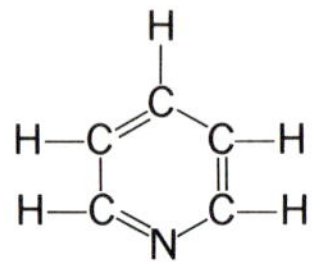

Pyridine

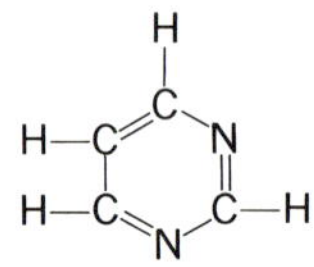

Pyrimidine

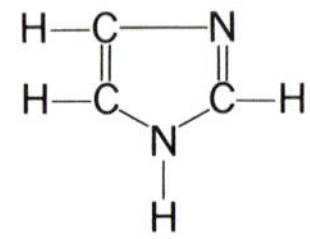

Imidazole

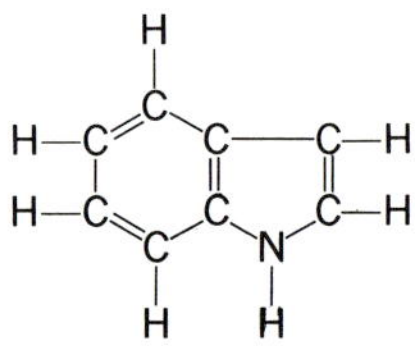

Indole

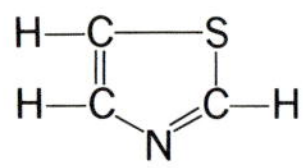

Thiazole

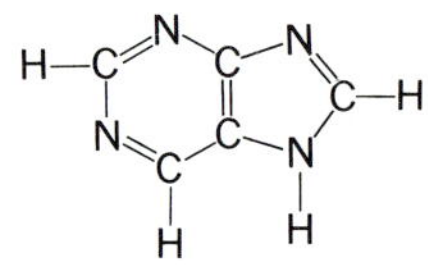

Purine

TABLE 12.3 FORMULAS FOR SEVERAL BIOLOGICALLY ACTIVE HETEROCYCLIC COMPOUNDS

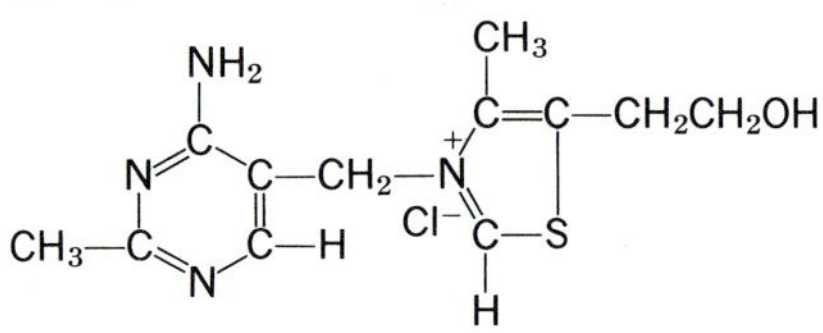

Thiamin, vitamin B_1

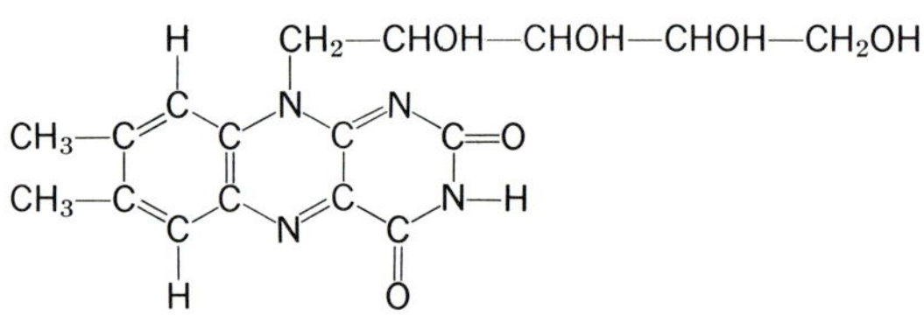

Riboflavin, vitamin B_2

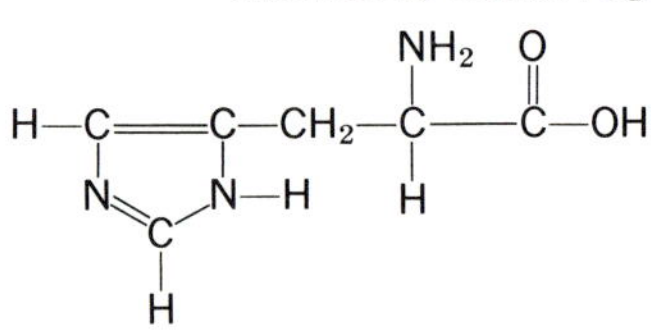

Histidine, an amino acid needed in the diet

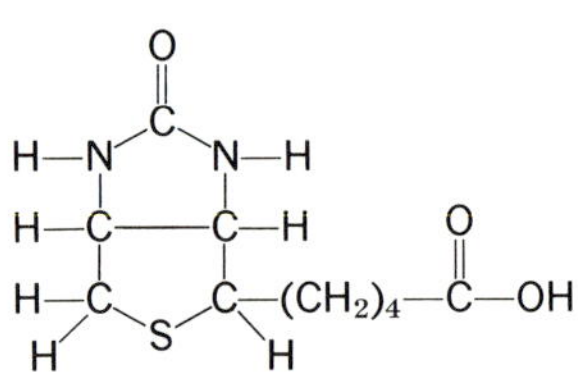

Biotin, a vitamin

TABLE 12.3 FORMULAS FOR SEVERAL BIOLOGICALLY ACTIVE HETEROCYCLIC COMPOUNDS (*Continued*)

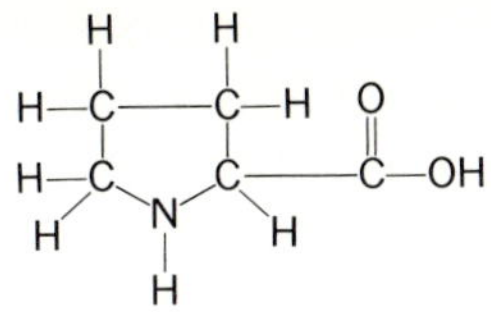

Proline, amino acid

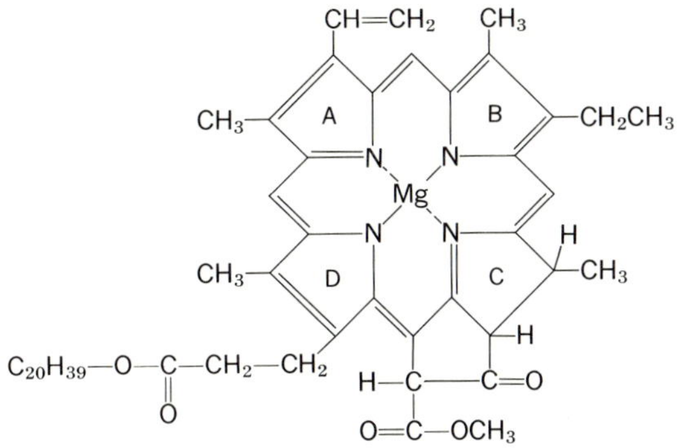

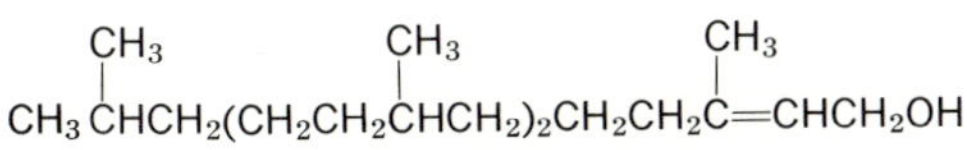

Phytol ($C_{20}H_{39}$—OH)

Chlorophyll *a*, a green pigment in plants associated with photosynthesis

from each of the five carbon atoms accounts for the six *p* electrons required to aromatize the ring. However, to aromatize the pyrrole ring, the nitrogen atom must furnish two *p* electrons.

A five-membered ring may contain two nitrogen atoms, as arranged in the imidazole nucleus. The tertiary nitrogen atom (C═N—C) furnishes one *p* electron, and the secondary nitrogen atom (C—N—C) furnished two *p* electrons to aromatize the ring. A similar structure is noted in the thiazole ring. It is the sulfur atom in this ring which furnishes the two *p* electrons to produce an aromatic ring.

12.10 SOME COMMON RINGS

Some heterocyclic nitrogenous ring structures are listed in Table 12.2. Note especially the structure of the indole and purine rings, which have two rings fused together. The indole nucleus is composed of the benzene ring fused with the pyrrole ring, each ring having two carbons in common with the other. The purine nucleus is composed of a pyrimidine ring fused with an imidazole ring.

Table 12.3 lists the formulas for several biologically active compounds that contain heterocyclic ring structures. Thiamin contains both the pyrimidine and the thiazole rings. Riboflavin contains three rings fused together, two of which are heterocyclic. Biotin, another vitamin, contains two fused five-membered rings neither of which is aromatic. Histidine, an amino acid, contains an imidazole ring attached to an α-amino acid chain. The ring structures mentioned cannot be synthesized by animals and therefore must be included in the diet. However, these compounds can be synthesized by plants. Chlorophyll is a green pigment of plants that is associated with photosynthesis. A glance at its formula reveals that it contains four pyrrole units (A, B, C, and D) connected to form a large heterocyclic ring.

Is proline classified as an aromatic or an aliphatic amino acid? This amino acid contains an aliphatic heterocyclic ring; therefore, the carbon and hydrogen react as in an aliphatic hydrocarbon. Proline is a secondary amino acid.

12.11 VITAMIN B_6 REACTIONS

A biological reaction associated with vitamin B_6 will illustrate the reaction of an aromatic amine with a carbonyl group and an aromatic aldehyde with an amine. Vitamin B_6 is a heterocyclic nitrogen compound containing the pyridine nucleus. It exists in the amine form, pyridoxamine, and in the aldehyde form, pyridoxal.

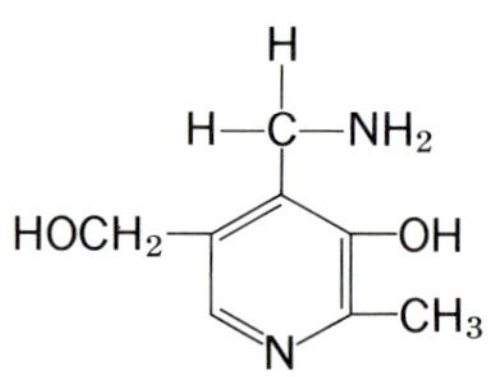

Pyridoxamine (B_6—NH_2)

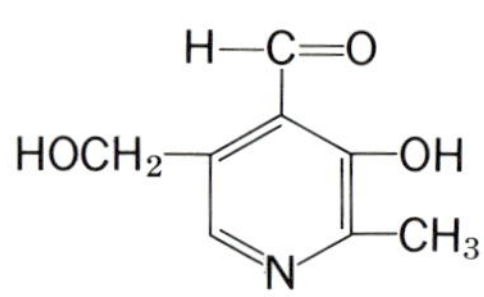

Pyridoxal (B_6—CHO)

The product of the reaction of an amine with a carbonyl group (Schiff's base) is relatively stable if either the NH_2 or the C=O group or both are attached to an aromatic nucleus. This structure exists in pyridoxamine and in pyridoxal. Reactions involving pyridoxamine with carbonyl compounds and pyridoxal with amine compounds are of great biological importance.

In the following reactions the ring structure of vitamin B_6 will be indicated as B_6. Thus pyridoxamine and pyridoxal will be represented as

respectively. The pyridoxamine reaction with pyruvic acid is brought about by the appropriate enzyme system.

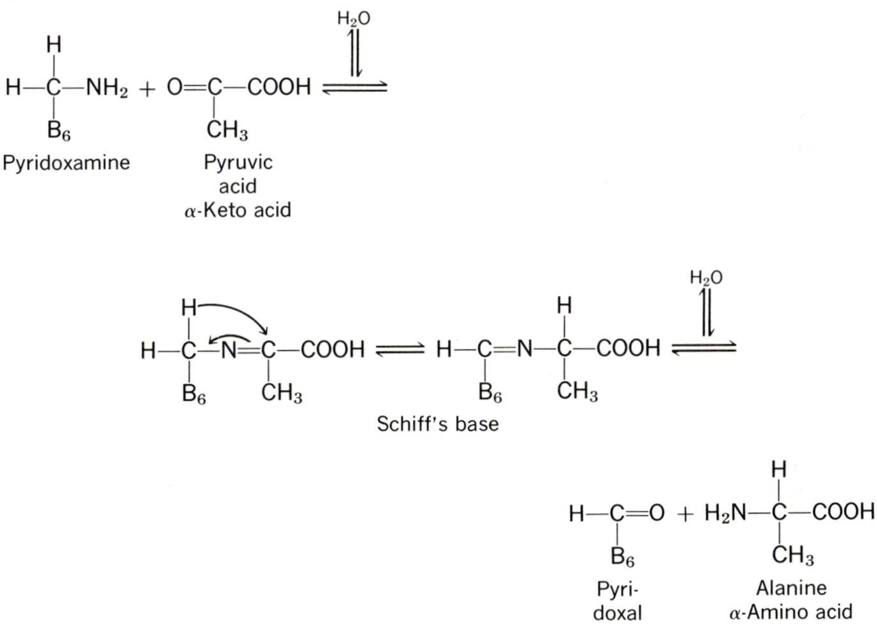

Overall:

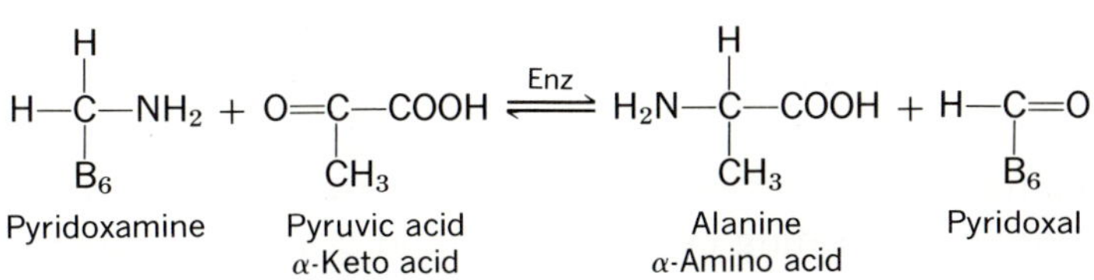

Amino acids in general and a number of α-keto acids enter into this reaction.

In this reaction the amino group of pyridoxamine and the keto group of pyruvic acid have been exchanged. It is of particular interest to note that the reactions are reversible. The reverse reaction with aspartic acid as the amino acid forms oxaloacetic acid:

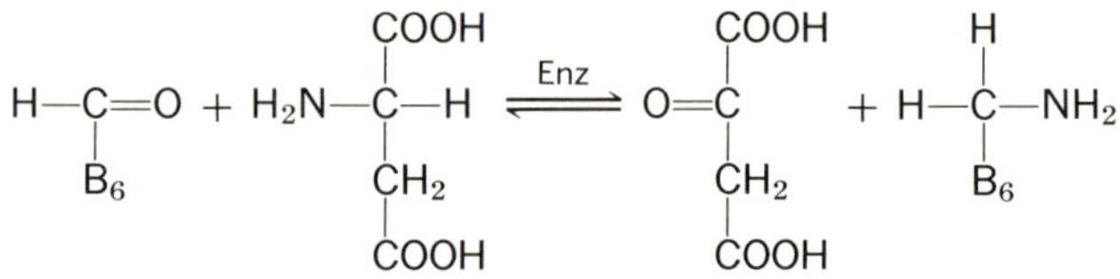

Pyridoxal | Aspartic acid | Oxaloacetic acid | Pyridoxamine

Thus, aspartic acid can be converted into oxaloacetic acid and vice versa.

The overall equation if the above two reactions are combined is

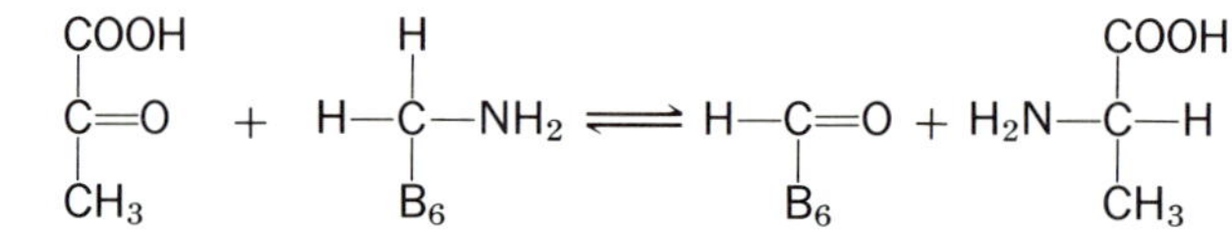

Pyruvic acid | Pyridoxamine | Pyridoxal | Alanine

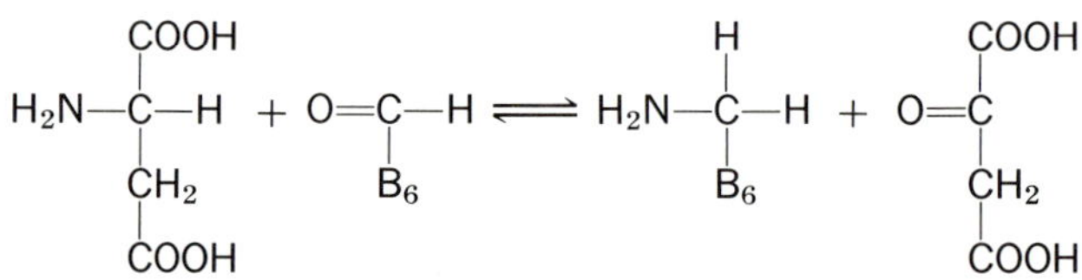

Aspartic acid | Pyridoxal | Pyridoxamine | Oxaloacetic acid

Net reaction:

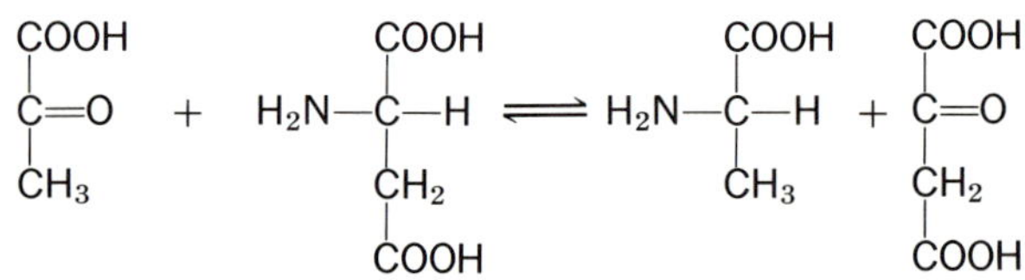

Pyruvic acid | Aspartic acid | Alanine | Oxaloacetic acid

The net reaction is equivalent to transferring an amino group from an amino acid (aspartic acid) to a keto acid (pyruvic acid). The amino acid is converted into the corresponding keto acid. Because of this net result, this reaction is called *transamination.* This is a key reaction in integrating carbohydrate and protein metabolism. Pyruvic acid is a product of glucose metabolism, and by transamination reactions it can be converted into alanine, which is a component of proteins. The reverse, of course, is also true. Alanine derived from proteins can be converted by transamination reactions into pyruvic acid. The significance of this reaction will become apparent in the subsequent discussions dealing with the interrelationship of carbohydrate and protein metabolism.

QUESTIONS

1. Write the general formula for a primary, secondary, and tertiary amine.
2. You are given three separate samples labeled 1° amine, 2° amine, and 3° amine. All three amines are water-insoluble. What tests would you perform to prove the amines were correctly labeled?
3. Write an equation for the preparation of each of the following compounds.

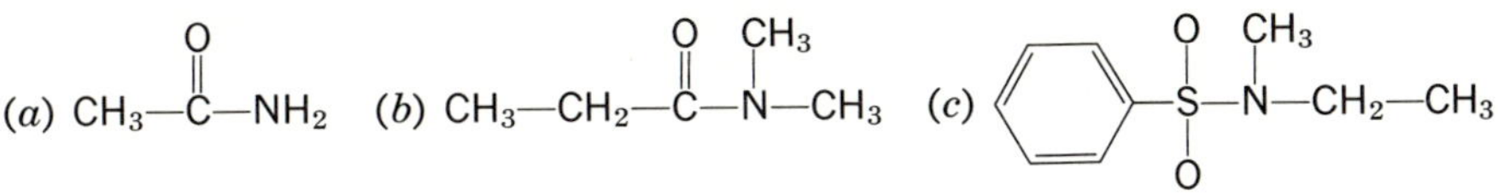

4. Write an equation for the hydrolysis of a dipeptide.
5. (*a*) What does transamination mean?
   (*b*) What vitamin is involved in this reaction?
6. The NH_2 group of amides reacts with HNO_2 as a primary amine group:

$$RCONH_2 + HNO_2 \longrightarrow RCOOH + N_2 + H_2O$$

   If 10 ml of a solution of acetamide, CH_3CONH_2, forms 50 ml of N_2 at STP when reacted with excess HNO_2, calculate the molarity of the acetamide solution.
7. Write the formula for the acid anhydride, acid chloride, and acid amide of isobutyric acid.
8. How would you distinguish between an acyl chloride and an alkyl chloride?
9. How many milliliters of 0.2 N NaOH are required to neutralize 20 ml of a 0.1 M aqueous solution of (*a*) acetic anhydride and (*b*) acetyl chloride?
10. Would a reaction occur if acetyl chloride is added to (*a*) ethyl alcohol, (*b*) benzene, (*c*) ethylamine, (*d*) trimethylamine?
11. What is the criterion for an aromatic nucleus? Explain.
12. What properties distinguish an aromatic nucleus from a triene?
13. Name a naturally occurring compound that contains:
    (*a*) A pyrimidine ring (*b*) A pyridine ring
    (*c*) An imidazole ring (*d*) A purine ring
    (*e*) A thiazole ring
14. Write a possible structural formula for compounds I to IV.

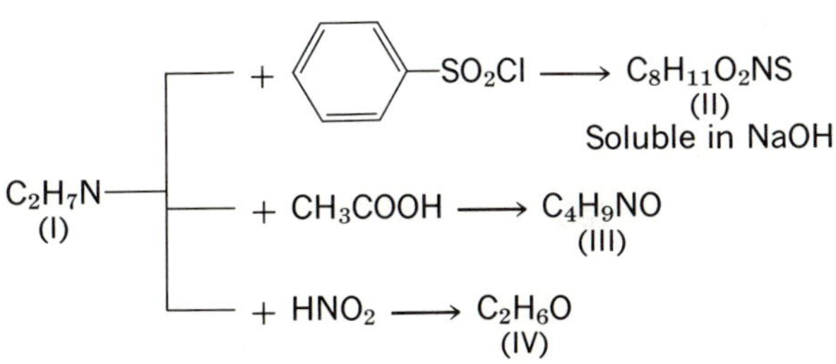

chapter 13.... intermolecular forces

Molecules, like people, do not exist in an isolated environment. On the contrary, everything is composed of aggregates of molecules or ions, each exerting a specific influence upon its neighbors. Remember that there are 6.02×10^{23} molecules in 1 g mole of a compound.

We have become accustomed to the fact that sugar and salt are soluble in water and that oil and water do not mix. Is there a logical explanation of these phenomena? Is there an explanation for the relative melting- and boiling-point characteristics of compounds?

Physical characteristics of aggregates of molecules are governed by the interaction of the molecules. The biological importance of intermolecular forces cannot be overemphasized. Reactions within each living cell take place in an aqueous environment. If the soluble cellular substances were to become insoluble, or vice versa, the cell would die. For plants to absorb such mineral elements as magnesium ion, the magnesium compound must be soluble not only in soil water but also in the cellular fluid so that it can be transported to various parts of the plant. At a particular place in the plant the water-soluble material may be incorporated or converted into a water-insoluble substance, such as chlorophyll. Bone, the structural substance of vertebrates, is insoluble in water and yet it is made by combining water-soluble compounds under appropriate conditions at definite locations of the body. Fats are a good example of a water-insoluble organic constituent of cells. Vitamins A, D, E, and K are insoluble in water and are localized in the fat of animals; they are fat-soluble. In contrast, some vitamins are water-soluble, e.g., vitamins C, B_1, B_2, and niacin.

Cellular components vary in their complexity from simple compounds like water to extremely complex substances like proteins, with molecular weights in the hundreds of thousands. Some proteins are soluble in water. What intermolecular forces exist between protein molecules and water molecules to make such high-molecular-weight substances water-soluble?

The chemistry of life involves not only extraordinarily complicated molecules but also common substances like water. Examples of biological reactions in which water is a reactant have already been discussed. Since water enters chemically into many biological reactions, the importance of its properties is stressed in this chapter.

13.1 TYPES OF INTERMOLECULAR FORCES

Intermolecular forces of most pure substances are ultimately based upon the formation of dipoles within the molecules and the subsequent attraction of oppositely charged areas of molecules for each other. The types of forces and their strength depend upon the extent to which dipoles are formed. Forces which play important roles in determining the physical state of matter in order of decreasing strength are:

1. Ionic
2. Metallic
3. Hydrogen bonding
4. Polar
5. Van der Waals

13.2 IONIC FORCES

Ionic bonds are easily recognized because of the great difference between the electronegativity of the atoms involved. What approximate difference in electronegativity is necessary to have an ionic bond (see Chap. 3)?

The crystalline structure of ionic compounds is characterized by the ordered arrangement of ions. For example, the crystalline structure of sodium chloride contains a lattice of positive ions surrounded by six negative ions (Fig. 13.1). Is there such a thing as a molecule of crystalline sodium chloride?

Compounds may contain ionic, covalent, or coordinate-covalent bonds or any combination of these bonds. The physical properties of

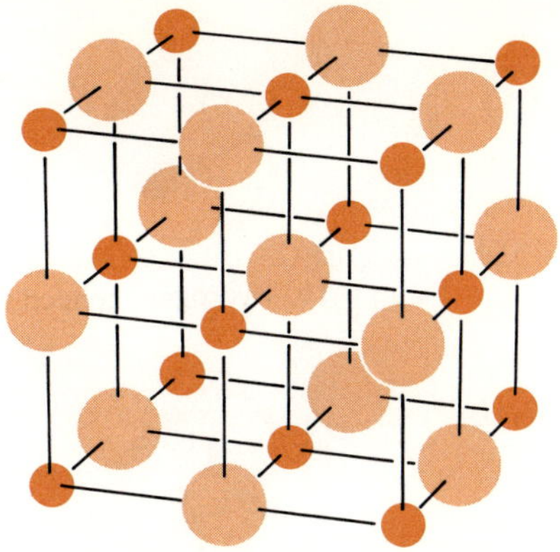

FIGURE 13.1 Arrangement of sodium and chloride ions in a sodium chloride crystal. The darker spheres are sodium ions, and the lighter spheres are chloride ions.

these compounds are governed by the strongest bond present—in this case ionic bonds. Thus, compounds such as NaCl, K_2HPO_4, and Na_2CO_3 have properties of ionic-bond compounds. Compounds containing ionic bonds have a high melting point and a crystalline structure.

13.3 METALLIC FORCES

Metals exist in definite crystalline structures. Current theories suggest that the crystalline pattern is formed by the regular arrangement of positive ions while the valence electrons are free to wander through the crystalline pattern of positive ions. This would account for the conductance property of metals. As electrons are added to one end of a wire, floating electrons are forced off the other end, producing a current of electricity.

13.4 HYDROGEN BONDING

Hydrogen bonding was introduced briefly in Chap. 6. Water readily forms this type of bond. Hydrogen bonds are formed between only three elements, nitrogen, oxygen, and fluorine. Hydrogen bonding involving nitrogen and oxygen is very important in biological systems. The electronegativity values for nitrogen, oxygen, and fluorine are 3.0, 3.5, and 4.0, respectively. It should be noted that these three elements have the three highest electronegativities of all the elements.

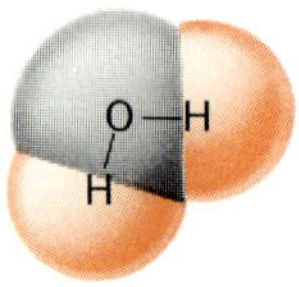

(*a*) Water molecule

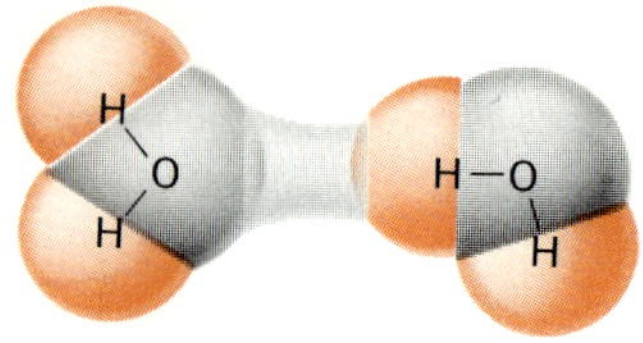

(*b*) Hydrogen-bonded water molecules

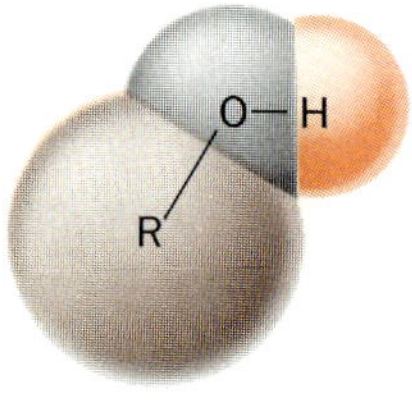

(*c*) R—O—H Alcohol

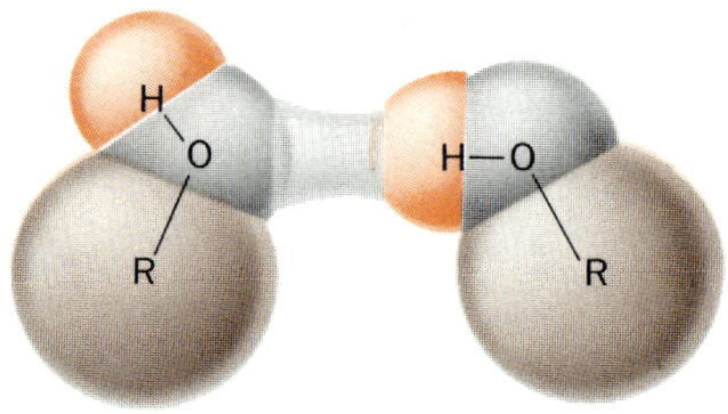

(*d*) Hydrogen-bonded alcohol molecules

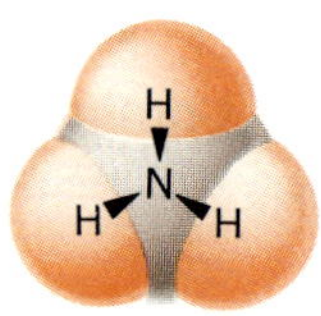

(*e*) NH_3 Ammonia

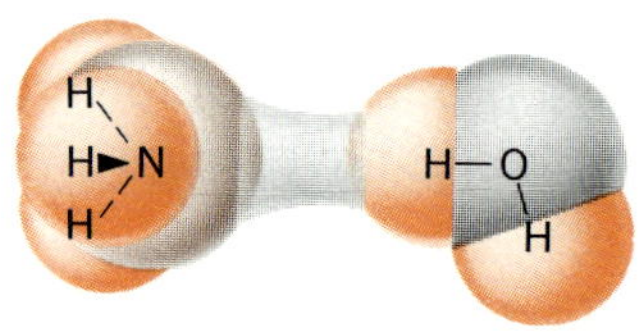

(*f*) NH_4OH Ammonium hydroxide

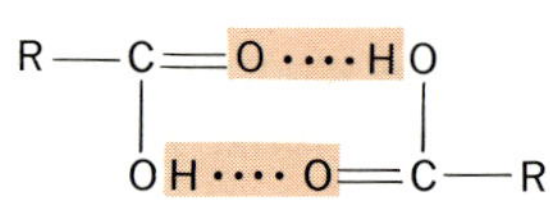

(*g*) Two molecules of acid hydrogen bonded

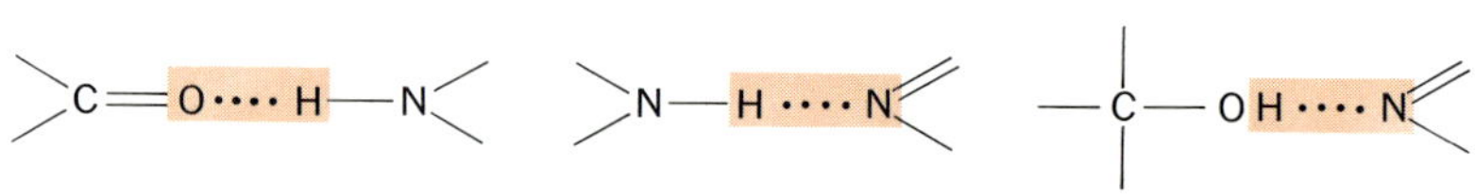

(*h*) Other examples of hydrogen bonding

FIGURE 13.2 Hydrogen bonding in a number of organic molecules.

The electronegativity of nitrogen and oxygen suggests a possible reason why hydrogen bonding occurs with these elements: because the electron of hydrogen would be pulled toward their nucleus. Therefore, if another highly negative element that has a pair of electrons available is close at hand, the proton could conceivably share these electrons.

$$\overset{\delta-}{:\ddot{\underset{\cdot\cdot}{O}}:}\overset{\delta+}{H}\overset{\delta-}{:\ddot{\underset{\cdot\cdot}{O}}:}$$

The molecules involved would be bonded together. The small size of the proton permits only two atoms of oxygen to be within bonding range.

The electron distribution about the oxygen atom is similar to that of the carbon atom; i.e., the electrons are located at the four corners of a tetrahedron having equal sides (Fig. 13.2*a*). The nitrogen atom may also be considered as having its outer electrons at the corners of a tetrahedron (Fig. 13.2*e*). The unshared pair of electrons of the nitrogen atom enter into hydrogen bonding. Other examples of hydrogen bonding are shown in Fig. 13.2*c*, *g*, and *h*. It will be noted that alcohols and acids associate by forming hydrogen bonds. Hydrogen bonds of the type shown in Fig. 13.2*h* play an important part in the structure of proteins.

13.5 POLAR FORCES

The extent to which a dipole is formed within molecules determines whether or not the molecules are attracted to each other by polar forces or by van der Waals forces. A polar molecule, such as HCl, may be considered as having the opposite charges more or less fixed in relation to the respective atoms rather than oscillating equally between them. This is because the bond electrons spend practically all their time under the influence of the atom with the greater electronegativity. Such molecules are said to have a dipole moment; they would be oriented under the influence of an electric field (Fig. 13.3). Molecules

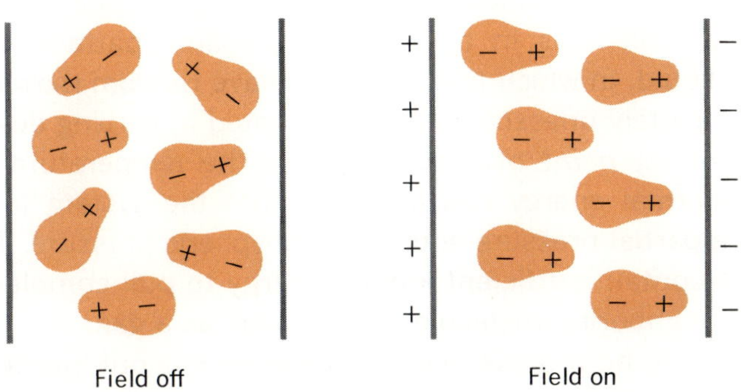

FIGURE 13.3 The orientation of a polar molecule in an electric field.

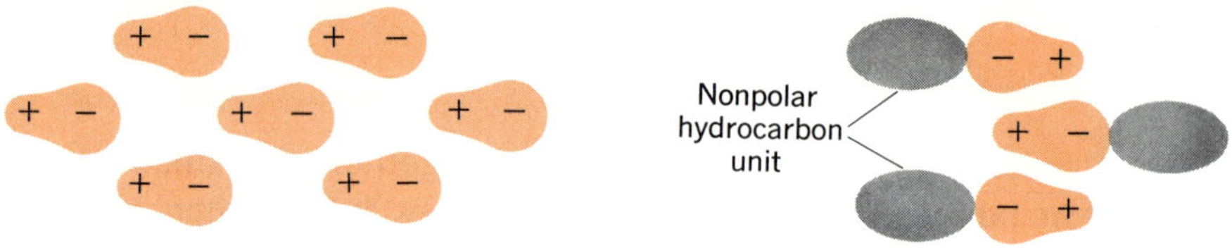

FIGURE 13.4 Aggregation of dipole molecules.

of this type are held together by a dipole-dipole interaction (Fig. 13.4). As expected, the size and shape of the molecule determine the extent to which the molecules are packed in a cluster.

13.6 VAN DER WAALS FORCES

The attraction of nonpolar molecules for each other is a result of van der Waals forces. Because the electrons of nonpolar compounds are equally distributed about the molecules, the molecules, unlike HCl, exhibit no net dipole. However, it must be remembered that bond electrons are not stationary; hence, at some instant a small dipole can exist due to a temporary unequal distribution of electrons (see Fig. 13.5). This initiates an unequal distribution of electrons in an adjacent molecule. A chain reaction like this results in a number of molecules being attracted to each other. Since these are not permanent dipoles, the charge is considered to be oscillating back and forth (Fig. 13.5). The partial charges in the figure are indicated arbitrarily on the surface of the molecules, with only two molecules associated.

13.7 MELTING AND BOILING POINT

The *melting point* of a substance is the temperature at which the thermal energy of the molecules is great enough to disrupt the forces holding the molecules in a definite crystalline pattern. The solid becomes a liquid, in which the molecules have random movement and may be completely unassociated or associated in varying degrees.

The *boiling point* of a substance is the temperature at which sufficient thermal energy has been put into the system to give the molecules a partial pressure equal to atmospheric pressure. The molecules must contain sufficient kinetic energy to pull completely away from their neighboring molecules and escape as a gas.

Because in both these phenomena energy is put into the system to overcome intermolecular force, these two physical properties are

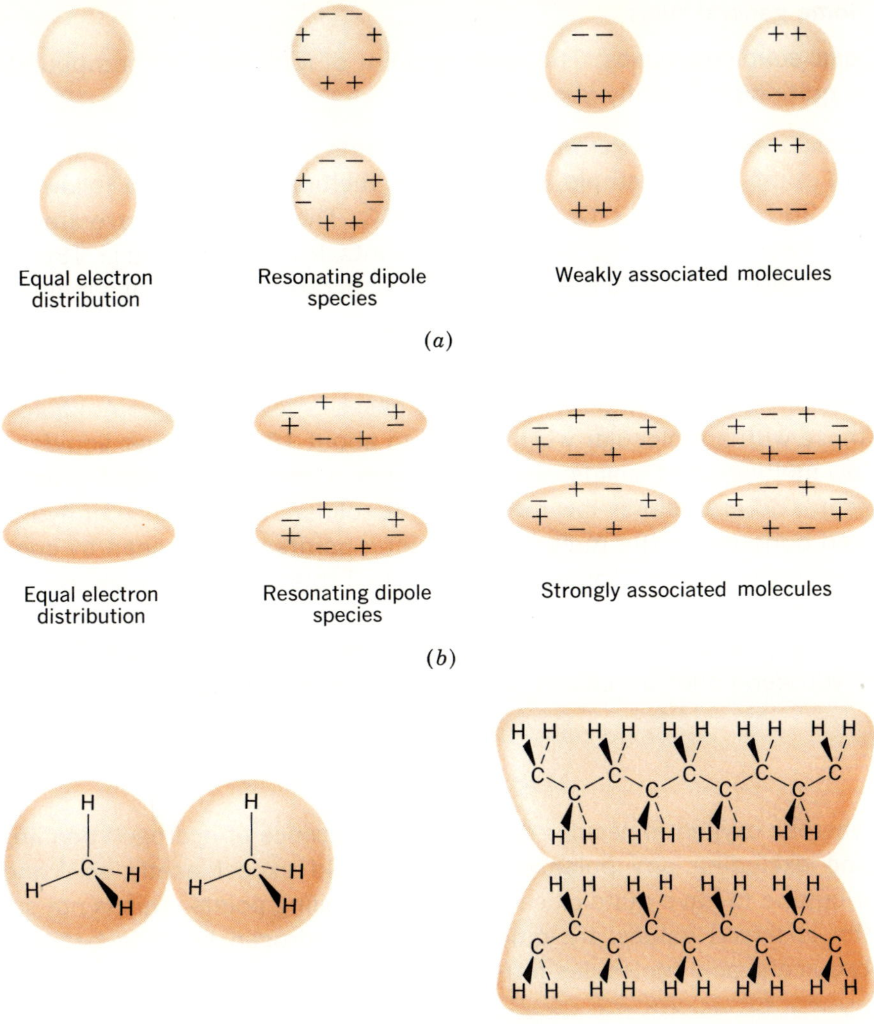

FIGURE 13.5 Van der Waals forces: (*a*) spherical molecules; (*b*) linear or flat molecules; (*c*) a more realistic representation.

discussed together. Solubility is also governed by the interaction of intermolecular forces of the solute and solvent and therefore is also included in this chapter.

The greater the intermolecular forces of a pure substance, the higher its melting and boiling point and vice versa. The greater the interaction between a solute and a solvent, the greater its solubility.

Some general rules

IONIC-BOND COMPOUNDS. The melting point of ionic-bonded compounds is quite high, usually above 600°C.

METALLIC-BONDED ELEMENTS. Metallic elements may have high or low melting points. For example, consider the melting point of the following: Cs, 28.5°C; K, 62.3°C; and Na, 97.5°C. These elements are among the lower-melting metallic bonded elements. Iron, 127°C, and Li, 186°C, are intermediate and tungsten, 3370°C, is representative of a metallic element with a very high melting point.

HYDROGEN-BONDED COMPOUNDS. Hydrogen-bonded compounds with a molecular weight up to 60 are generally gases or low-boiling-point liquids; those with a molecular weight of 60 to about 180 are either liquids or low-melting-point solids; and those with a molecular weight greater than 180 are usually low-melting-point solids. The influence of hydrogen bonding on the melting and boiling points may be seen by comparing the pairs of compounds listed in Table 13.1. Although oxygen and methyl alcohol have the same molecular weight, their melting and boiling points are markedly different. The CH_3OH molecules interact (hydrogen-bond formation) to form associated clusters

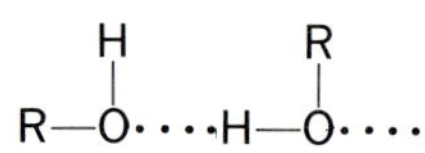

whereas oxygen molecules do not. In the absence of hydrogen bonding and excluding any polar effect, one would expect acetic acid and *n*-butane to have quite close melting and boiling points. This is not so; the observed values are quite different. Hydrogen bonding between two acetic acid molecules forms dimers

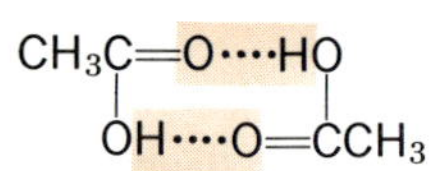

POLAR FORCES AND VAN DER WAALS FORCES. Generally speaking, compounds with either polar forces or van der Waals forces can be expected to be gases at molecular weights below 60 and liquids or low-melting-point solids at molecular weights from 60 to 180.

The physical characteristics of polar compounds depend not only upon the intensity of the polar force but also upon the group attached to the polar portion of the molecule. *The whole molecule must be considered.*

TABLE 13.1 INFLUENCE OF HYDROGEN BONDING ON MELTING POINT AND BOILING POINT OF SOME COMPOUNDS

Compound	Formula	Mol. wt.	Mp, °C	Bp, °C
Oxygen	O_2	32	−218.4	−183
Methyl alcohol	CH_3OH	32	−97.8	64.7
n-Butane	$CH_3(CH_2)_2CH_3$	58	−136	−0.5
Acetic acid	CH_3COOH	60	16.7	118

Although van der Waals forces are quite weak, they influence the physical properties of many compounds. The van der Waals forces are less between spherical molecules than between long or flat ones, as can be seen by comparing the melting and boiling points of methane, CH_4; *n*-hexane, $CH_3(CH_2)_4CH_3$; and isohexane, $(CH_3)_2CH(CH_2)_2CH_3$.

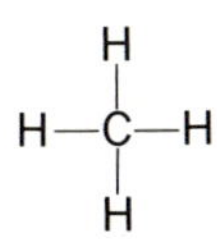

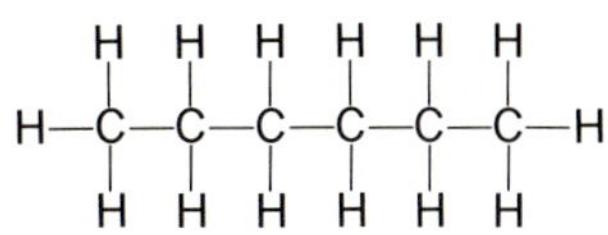

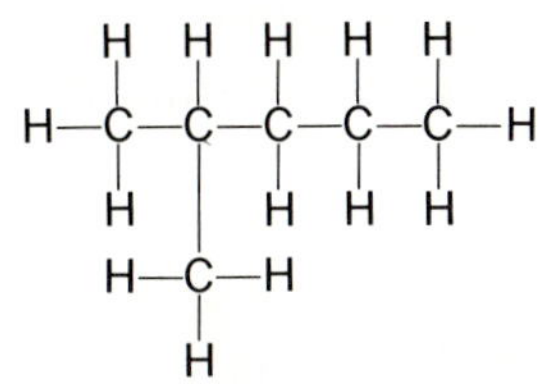

	Methane	*n*-Hexane	Isohexane
Mp, °C	−183	−95	−145
Bp, °C	−162	69	60

Methane is a small almost circular molecule, while *n*-hexane is a long molecule. Isohexane is somewhat intermediate between them as far as molecular shape is concerned; it cannot be packed as closely as *n*-hexane because of its branched-chain structure. Van der Waals forces are least in methane, intermediate in isohexane, and greatest in *n*-hexane. Does this correlate with the melting and boiling points of these compounds?

In benzene all the carbons of the ring and the hydrogen atoms lie in one plane; therefore, these units can easily stack on top of each other (Fig. 13.6). This fact is reflected in the relatively high melting and boiling points as compared with *n*-hexane.

Biological significance of polarity

The polar nature of low-molecular-weight compounds is of great biological significance. Many compounds produced during cellular metabolism have one or more strongly polar groups; many are capable of

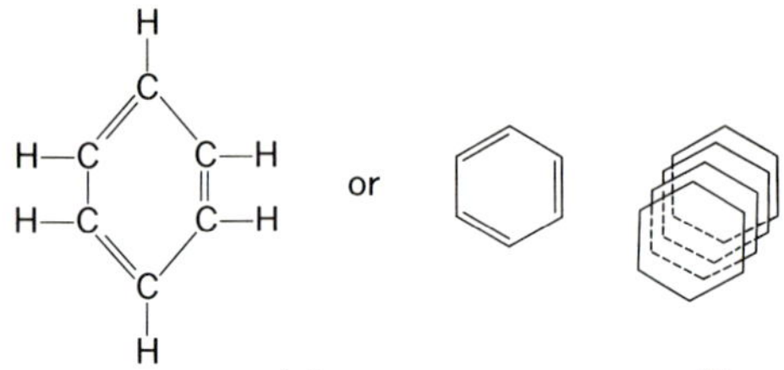

	(a)	(b)
mp, °C		5.4
bp, °C		80.1

FIGURE 13.6 Benzene molecules: (*a*) planar and (*b*) stacked.

hydrogen bonding. Why is this fact so important? Remember that as food material is metabolized to carbon dioxide and water, low-molecular-weight compounds must be formed. Also, complex body substances are made from compounds with relatively low molecular weights.

If these low-molecular-weight compounds had no polar groups, they would be gases or low-boiling liquids and would also be water-insoluble. This would be fatal: the cells could not metabolize them. The following few natural compounds are examples of low-molecular-weight polar compounds of biological interest:

HCHO Formaldehyde	CH_3—CO—COOH Pyruvic acid
CH_3—CHO Acetaldehyde	HOOC—CH═CH—COOH Fumaric acid
CH_3—CH_2OH Ethyl alcohol	HOOC—COOH Oxalic acid
CH_2OH—CHOH—CH_2OH Glycerol	H_2N—CH_2—COOH Glycine
CHO—CH_2OH Glycolic aldehyde	H_2N—CO—NH_2 Urea

The first compound, formaldehyde, is a gas, and acetaldehyde has a boiling point of 20°C. The other compounds are either liquids or solids. All are soluble in water. Apply the general rules given at the beginning of this chapter and determine what physical state you would expect each of these compounds to have if no polar groups were present.

13.8 STRUCTURE VERSUS BOILING POINT

The boiling point of organic compounds is more closely related to structure than the melting point is (see Fig. 13.7).

Nonpolar compounds

Among members of a particular homologous series of straight-chain compounds, the van der Waals forces increase with the length of the carbon chain but not necessarily by a constant value for each successive CH_2 group added to the molecule. Therefore, one would expect the boiling point to increase with molecular weight (see the curve for RH in Fig. 13.7).

The atomic weight of oxygen is higher than that of hydrogen and lower than that of chlorine. Therefore, if oxygen is inserted into a carbon-hydrogen chain in such a manner that it forms a nonpolar group, the boiling point would be expected to be higher than that of

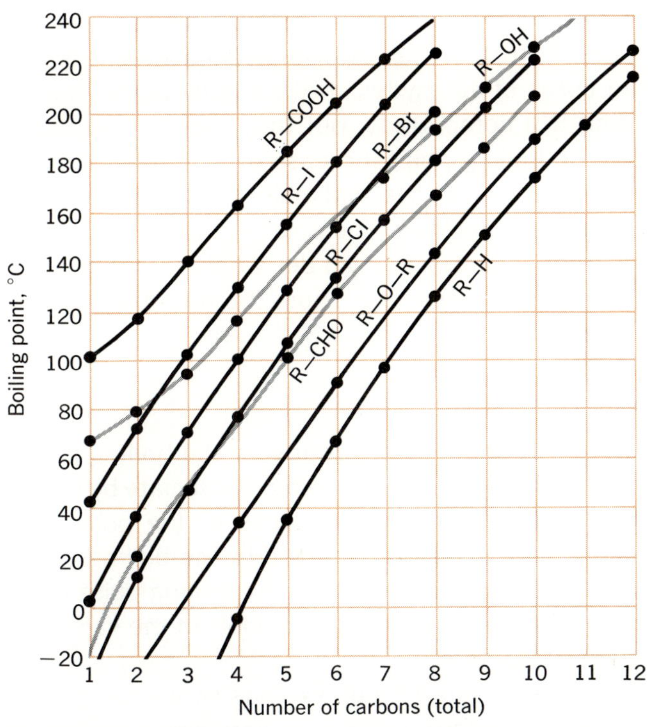

FIGURE 13.7 Comparison of the boiling points of several classes of organic compounds.

the corresponding RH compounds and lower than that of the corresponding RCl compounds. Remember that the atomic weight of oxygen is 16 times greater than that of hydrogen but about one-half that of chlorine. Oxygen inserted between two carbon atoms as in ethers, e.g., dimethyl ether, $CH_3—O—CH_3$, forms essentially a nonpolar bond. Follow the boiling point of ethers in Fig. 13.7. How do they compare with the boiling point of nonpolar compounds such as RH or with RCl?

Weakly polar compounds

The presence of a weakly polar bond in the hydrocarbon series, e.g., chlorine or bromine in place of a hydrogen, increases the molecular weight of the compound. As expected, the boiling point is increased accordingly. It will be noted that the curve for the hydrocarbons (RH) is parallel to those for RCl and RBr; this indicates that van der Waals forces rather than dipole-dipole interaction play the deciding role.

Oxygen may also occur in compounds in a relatively weak polar group such as CHO, the aldehyde group. These compounds would be expected to boil at a temperature higher than the corresponding ethers. This is especially true of the lower-molecular-weight aldehydes. The boiling-point curve of aldehydes starts slightly higher than that of the corresponding RCl compounds, but they rapidly approach the same value. Thus, the boiling-point increase of aldehydes containing more than four carbon atoms is governed mainly by the increase in molecular weight, i.e., the carbon chain. The polar nature of the aldehyde group remains constant, whereas the van der Waals forces are increasing as the molecular weight increases. Thus, the net effect of the aldehyde group is lessened.

Strongly polar compounds

The influence of strongly polar groups on the boiling point of liquids can be illustrated with the alcohols, which associate by hydrogen bonding (Fig. 13.2). The decreasing influence of the polar group as the molecular weight increases is reflected in the boiling-point curve for these compounds. If this were not so, the boiling-point curve for alcohols would parallel the curve for RI. Alcohols in the vapor phase are in the monomeric state.

Oxygen in the carboxyl group exhibits very strongly polar characteristics. Since the carboxyl group is capable of hydrogen bonding (Fig. 13.2*a*), straight-chain acids may be considered as dimers. The formation of dimers through hydrogen bonding is responsible for the high boiling point of this class of compounds. Interestingly enough, acids can also exist as dimers in the vapor phase. As with other polar compounds, the influence of the carboxyl group on the boiling point decreases with increasing number of carbon atoms in the compound.

More than one polar group

What effect would you expect the insertion of more than one polar group in a molecule to have on the boiling point of the compound? From the foregoing discussion one would deduce that it would be manifested by a decided increase in the boiling point of the compound. This is correct and is illustrated by the compounds in Table 13.2.

The insertion of one, two, and three OH groups into propane results in an increase in the boiling point from −45 to 290°C for the trihydroxy derivative, glycerol. There are two dihydroxypropanes. The one with the two OH groups on adjacent carbons has the lower boiling point. Does this indicate that intermolecular bonding is less for this compound than it is for the compound with the OH groups on the first and third carbons? How would you expect the boiling point of the compounds to change if the OH groups were changed to a less polar group?

Decreasing the polarity of glycerol by converting the OH groups into ether groups, R—O—R, would be expected to decrease the boiling point because the polar nature of the molecule is lessened. Examine

TABLE 13.2

	Compound	Bp, °C
	$CH_3—CH_2—CH_3$	−45
(I)	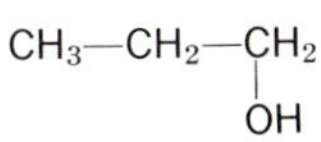	97
(II)	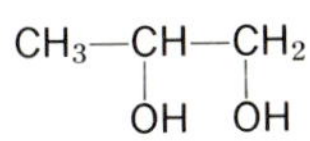	139
(III)	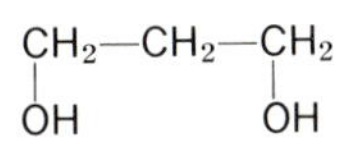	225
(IV)	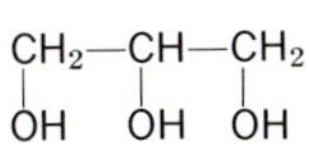	290
(V)	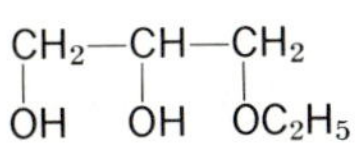	230
(VI)	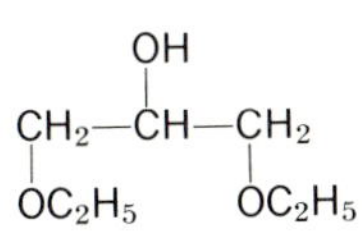	191
(VII)	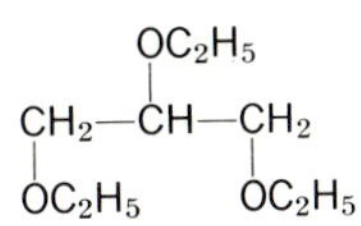	185

the boiling points of compounds V to VII. Why does compound III, with a molecular weight of 76, boil at a temperature higher than compound VII, which has a molecular weight of 176?

13.9 SOLUBILITY

A general rule for solubility is *like dissolves like.* A polar compound would be expected to be soluble in a polar solvent, and a nonpolar substance would be expected to be soluble in a nonpolar solvent. Water frequently has been called the *universal solvent.* It is the solvent for biological systems and is essential for all forms of life. We shall be interested mainly in aqueous solutions in the following discussion.

Ionic compounds

When an ionic compound dissolves in water, two steps may be envisioned: (1) the water must break the ionic bond, and (2) the water must become associated with the respective ions (see Fig. 13.8). The second phenomenon is known as *solvation;* if the solvent is water, the

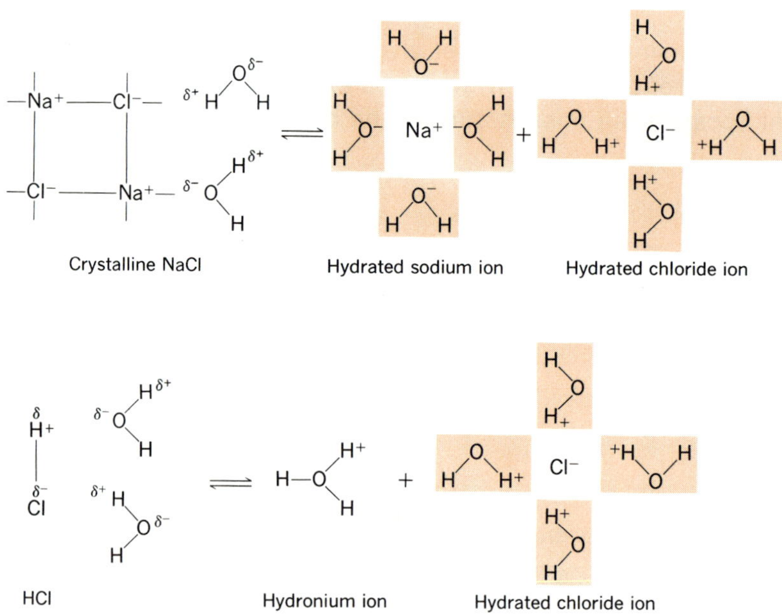

FIGURE 13.8 The effect of hydration on crystalline sodium chloride and hydrochloric acid.

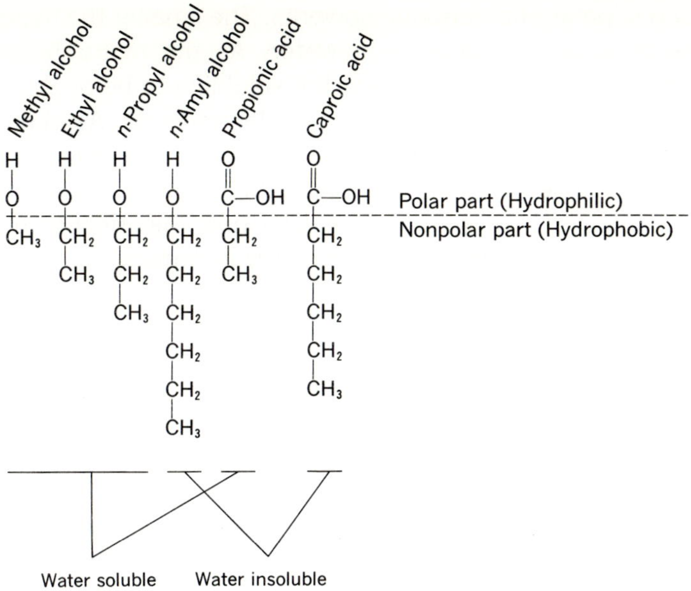

FIGURE 13.9 Solubility as a function of increasing molecular weight.

term *hydration* is used. The hydration of solutes is a biologically important phenomenon.

Highly polar solvents are also capable of rupturing weakly covalent bonds in highly polar compounds, such as in HCl and the carboxyl group, COOH, of organic acids, and subsequently hydrating the ions. Hydration of the ions tends to keep them separated and thus interferes with the interaction of the ions to form molecules again.

Organic compounds

For nonionic compounds, solubility is governed by the similarity of the solute and solvent. If the solvent-solvent molecular interaction is about the same as the solute-solute molecular interaction, the two will be miscible. As emphasized above, the entire molecule must be considered.

Most organic compounds have both a polar and a nonpolar portion (Fig. 13.9). The nonpolar part is repelled by the water, and the polar part is attracted to the water. The former is known as the hydrophobic (water-hating) group and the latter as the hydrophilic (water-loving) group. The relative size of the hydrophobic part compared with the hydrophilic part of the molecule influences the solubility of the compound in polar and nonpolar solvents. The smaller the hydrophobic part, the greater its solubility in water. As the hydrophobic group increases in size, one would expect the solubility to be proportionately more in nonpolar solvents and less in polar solvents. As illustrated in Fig. 13.9, methyl alcohol, ethyl alcohol, and *n*-propyl alcohol are completely miscible with water, whereas the six-carbon alcohol, *n*-amyl alcohol, is only 0.5% soluble in water. Also note the effect of increasing the carbon chain of acids on their solubility in water.

PROBLEM 1

How would you expect the molecules in a drop of a liquid water-insoluble organic acid, 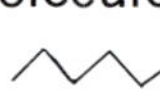COOH, to be oriented when added to a relatively large amount of water? Draw a schematic diagram to illustrate your answer. Consider the following questions based upon the previous discussion.

(*a*) What part of the acid is hydrophobic? Hydrophilic?
(*b*) Would you expect the hydrophilic group to interact strongly with the water molecules?
(*c*) Would you expect the hydrophobic part of the acid to be repelled by the water molecules?
(*d*) Would you expect the carbon chain of the acid to be pointed down into the water?
(*e*) Would you expect van der Waals forces of the hydrophobic part of the acid molecules to be greater than the interaction of water with the hydrophobic part?

If you had any doubts about the answer to any one of these questions, restudy this chapter. Now check your answers with the correct ones: (*a*) the carbon-hydrogen chain; the COOH group; (*b*) yes, the hydrophilic group, COOH, would be expected to interact very strongly with water molecules; (*c*) the hydrophobic part of the molecule would be repelled by the water molecules; (*d*) no; (*e*) yes.

A schematic diagram of the orientation of the acid molecules added to water would look like Fig. 13.10.

The molecules of acid would be one molecule thick with their COOH ends in the water and their carbon-hydrogen chains sticking above the surface of the water. Hydrogen bonding between water molecules and the COOH groups and the hydration of the COO^- ions

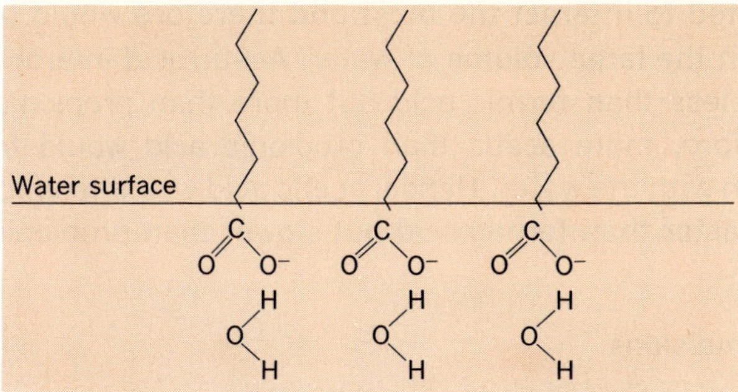

FIGURE 13.10 Orientation of acid molecules in water. Above the surface of the water, van der Waals forces exist between the hydrophobic groups; below the surface of the water, the hydrophilic groups associate with the water molecules.

account for the strong hydrophilic nature of this part of the acid molecule.

The cleansing action of soaps and detergents is a practical application of this phenomenon. Soap is composed of the water-soluble sodium salt of high-molecular-weight fatty acids (organic acids derived from fats). These salts stabilize oil-water emulsions.

Steam distillation involves the addition of a substance to a large amount of water, boiling the mixture in a suitable container, and collecting the liquid after the vapor is condensed. If the substance is steam-distillable, it distills over with the water. Low-molecular-weight acids are examples of such compounds.

PROBLEM 2

Consider the three acids formic, acetic, and propionic with the respective boiling points 10.18°, 118.1°, and 141.1°. If each of these acids is steam-distilled separately and 10-ml fractions of distillate collected, which 10-ml fraction will contain the greatest amount of acid, the first or subsequent fractions? In other words, which acid will steam-distill over to the greatest extent in the first 10 ml of distillate, which will be next, and which will be last?

The acid which would steam-distill the greatest amount in the first 10 ml of distillate is propionic acid. Acetic acid would be next and formic acid last.

To explain why the acids distill in this order, consider the interaction of the acids with water. Formic acid would be expected to interact the most and therefore would be held back in the large volume of water. Acetic acid interacts with water less than formic acid but more than propionic acid. Therefore, more acetic than propionic acid would tend to remain with the water. Hence, acetic acid would tend to distill over faster than formic acid but slower than propionic acid.

Oil-water emulsions

Let us now consider what happens if a few drops of oil are added to water. The oil floats on the surface (Fig. 13.11*a*). When the mixture is vigorously shaken, the oil is dispersed as small droplets throughout the water (Fig. 13.11*b*). If shaking is discontinued, the droplets coalesce and two layers soon form. However, if a little soap is added to the mixture and the mixture shaken, the oil forms very small droplets which do not readily coalesce to form two layers (Fig. 13.11*c*). Why?

The answer lies in the manner in which the soap molecules orientate themselves between the oil and water interface (Fig. 13.11*d*). The nonpolar part of the acid dissolves in the oil droplet, and the polar COO^- group is in the water phase. Thus, the surface of the respective droplets of oil literally has a coating of negatively hydrated ions surrounding it. Under these conditions the droplets cannot come together to form larger droplets and then separate out as another phase because the negative charge on the droplets tends to keep them apart. The oil has been emulsified in the water. Modern detergents act in a similar manner, but unlike the fatty acids they do not form insoluble salts with calcium and magnesium ions present in hard water. These insoluble salts form a scum when soap is used with hard water. Oil and grease in clothes are emulsified by the detergent during agitation in a washing machine and are then rinsed away. Another common emulsion is homogenized milk. The cream (fat) is finely dispersed in the liquid phase and does not separate out readily. Some salad dressings and mayonnaise are also emulsions.

Emulsions are the result of the suspension of one immiscible liquid in another liquid, the diameter of the emulsified particle being about 0.1 μ ($1\ \mu = 10^{-6}$ m.). *Emulsifiers* are the agents used to make an emulsion stable so that it will not readily separate into two phases.

The emulsification of fat is essential for normal digestion and absorption of this class of food. In the small intestine the fat is emulsified with aid of bile salts excreted by the gall bladder. These salts

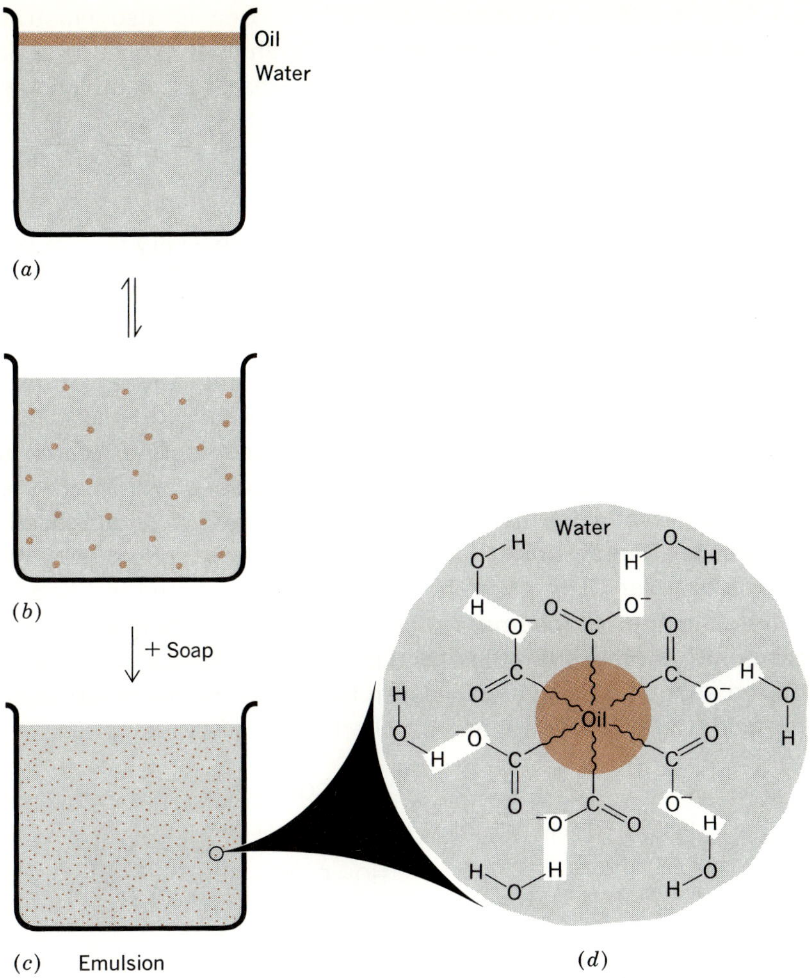

FIGURE 13.11 An oil-water emulsion, showing the effect of the addition of soap.

have a polar group, COOH, and a high-molecular-weight (cyclic) nonpolar group, the prerequisites for an emulsifier of oil in water. Fats that are liquid at body temperature are emulsified and thus present a large surface for the digestive enzymes to attack the fat. The greater the surface area, the more rapid and complete the digestive process.

The increase in surface area as a result of breaking a particle into very small pieces can be illustrated by considering a cube 1 cm on edge. Its area is 6 cm^2. If this cube is divided into smaller and smaller cubes until they are 0.01 μ on edge, the total surface area becomes about 600 m^2. The importance of surface area is also illustrated by red blood cells, which absorb oxygen through their exposed surface. It has been estimated that the total surface area of the red blood cells in an average man is about $\frac{1}{2}$ acre.

Polypolar compounds

Although one would expect compounds possessing a very high molecular weight, say 100,000, to be water-insoluble, many such biologically active compounds are water-soluble because they contain a large number of polar groups. The arrangement of the groups has a great deal to do with the water solubility of the compound, and the number of groups alone does not determine water solubility.

Consider the solubility of the two common compounds starch and cellulose. Could you build a waterproof house out of starch? Both these substances are of very high molecular weight and are composed of many glucose sugar units joined together. Both compounds have hundreds of polar OH groups; however, as will be noted in Chap. 21, they differ in the arrangement of atoms on one carbon atom of the glucose unit. Another example of the influence on solubility of the configuration about a carbon atom is seen by comparing the solubilities of mucic acid (water-insoluble) and saccharic acid (water-soluble).

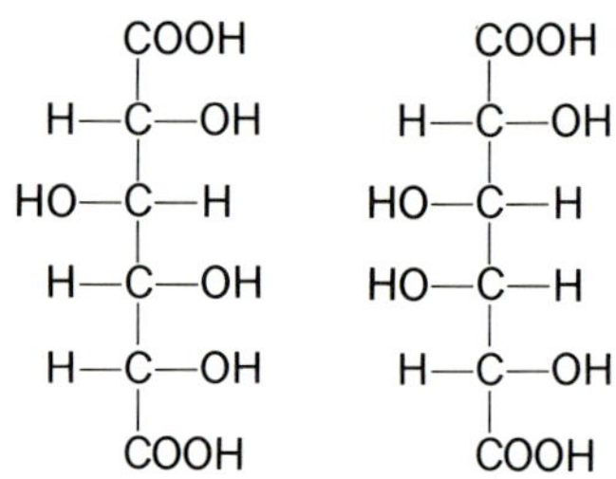

Saccharic acid Water-soluble

Mucic acid Water-insoluble

Apparently the arrangement of OH groups in saccharic acid is such that they can be readily hydrated, whereas in mucic acid this is not so.

Another notable example of a polypolar class of compounds is the proteins, flesh, hair, fingernails, casein of milk, and egg albumin (egg white). Proteins are composed of amino acids joined together in a long chain. Although all proteins contain the same kind of polar groups, they differ in the number of such groups and in the order in which the groups appear in the molecule. Each of these variables plays a role in

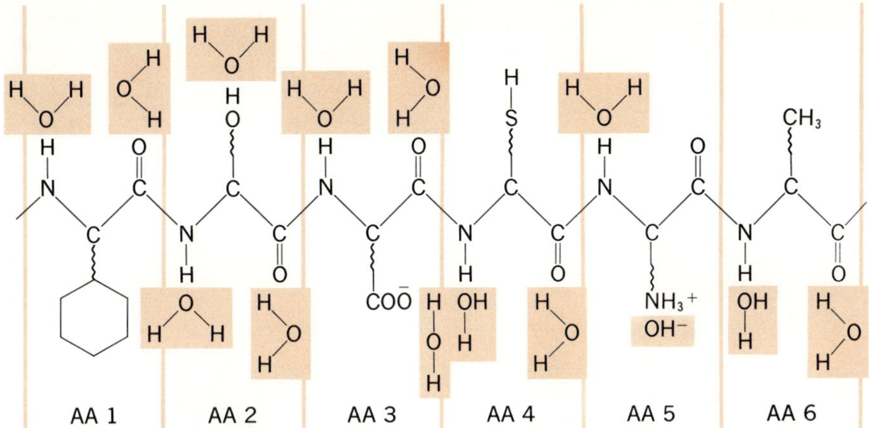

FIGURE 13.12 Hydration of protein; AA = amino acid.

determining solubility. Polar and nonpolar groups which occur in the amino acid side chain of proteins are illustrated schematically in Figure 13.12.

13.10 STRUCTURE OF WATER AND ICE

The structure of water itself and its influence on the solubility of solutes must also be considered. Liquid water has a certain structure due to hydrogen bonding between water molecules. The distribution of hydrogen atoms about the oxygen atom in water clusters is considered as at the four corners of a tetrahedron (Fig. 13.13*a*), similar to the hydrogen distribution about the carbon atom.

The smallest number of molecules of water that can be associated or clumped together in a regular pattern is a unit of five; six molecules can be clumped together to form a hexagonal ring structure (Fig. 13.13*b*). Water molecules arranged in the five-membered ring structure can cluster together to form a dodecahedron (12 sides), which would pack together as shown in Fig. 13.13*c*. Note that when the dodecahedrons pack together, there are holes in the structure which other molecules can enter. Water is known to form crystalline hydrates with some gases and low-molecular-weight water-insoluble compounds, e.g., xenon hydrate, chlorine hydrate, methane hydrate, ethane hydrate, and chloroform hydrate. These compounds fit into the holes in the packed dodecahedrons.

The hydrogen bonds between water molecules are not particularly

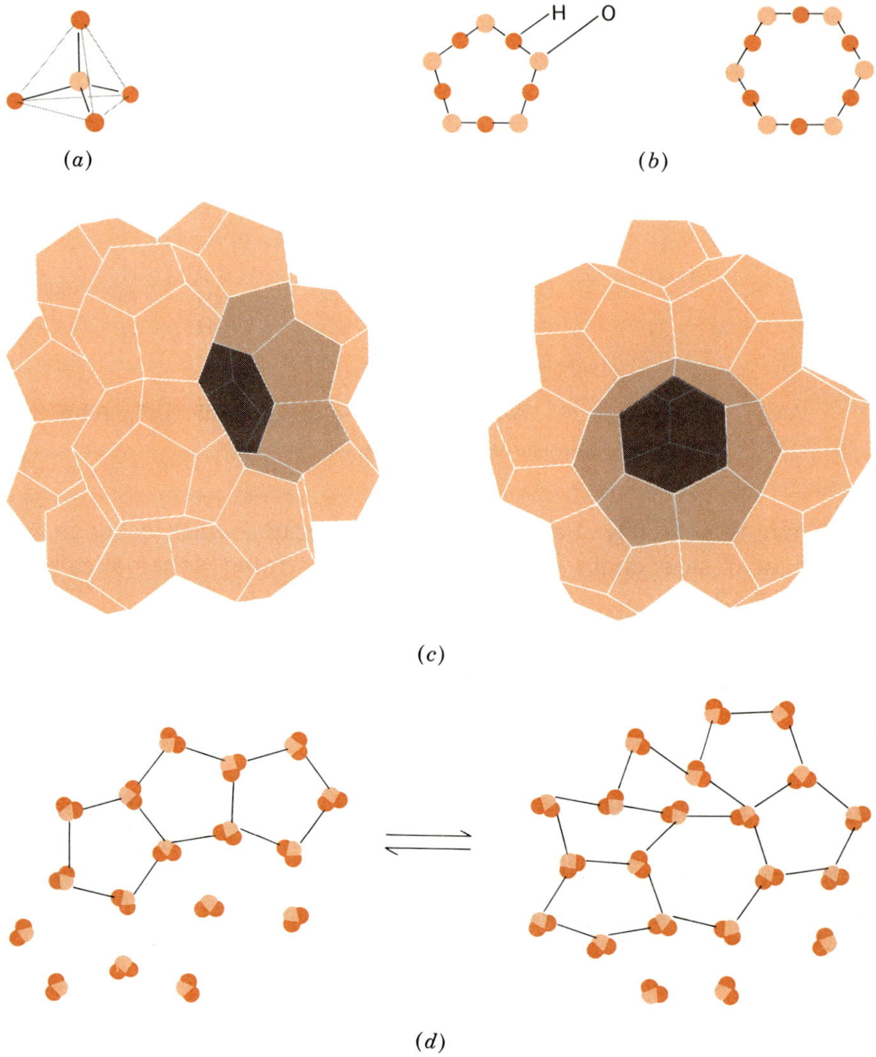

FIGURE 13.13 The structure of water: (*a*) oxygen atom bonded to four hydrogens; (*b*) clusters of water molecules; (*c*) dodecahedron water structure with the hydrogens omitted; (*d*) the dynamic state of the structure of water.

strong. Since the water molecules can rotate within the structure, the bonds are flickering from one oxygen atom to another on an adjacent water molecule. Thus, the structure of liquid water is often referred to as a *flickering structure*. A particular water molecule may be in structural form one second and a free molecule the next. An attempt to show this in a planar figure is Fig. 13.13*d*.

Ice floats on water because it is less dense, its mass per unit volume being less than that of water. The density of ice at 0°C is 0.917 g ml, whereas for liquid water at the same temperature it is 0.9998 g ml. This difference in density is related to the structure of water and ice. Water molecules, as noted above, are not all in a structural pattern; many occur as single molecules and, of course, occupy less space per gram than if they were held in a rigid structure like ice. To illustrate this, the density of water at 25°C is 0.997 g/ml. As water is cooled below this temperature, its volume becomes smaller because the water molecules have less kinetic energy. Consequently, the water becomes denser. At 4°C (actually 3.98°C) water reaches its maximum density of 1.000 g/ml, and then it becomes less dense until at 0°C it solidifies and has a density of 0.917 g/ml. At temperatures between 4 and 0°C more single water molecules are entering the structural form of water, in which they are held a certain distance apart. Fewer water molecules have sufficient energy to break away from the lattice structure. Therefore, 1 g of material occupies more space than when many single molecules of water are present and can crowd closer together. When all the water molecules have entered the rigid hexagonal structure of ice (Fig. 13.14), the density is minimum. When ice melts, the reverse of this process takes place.

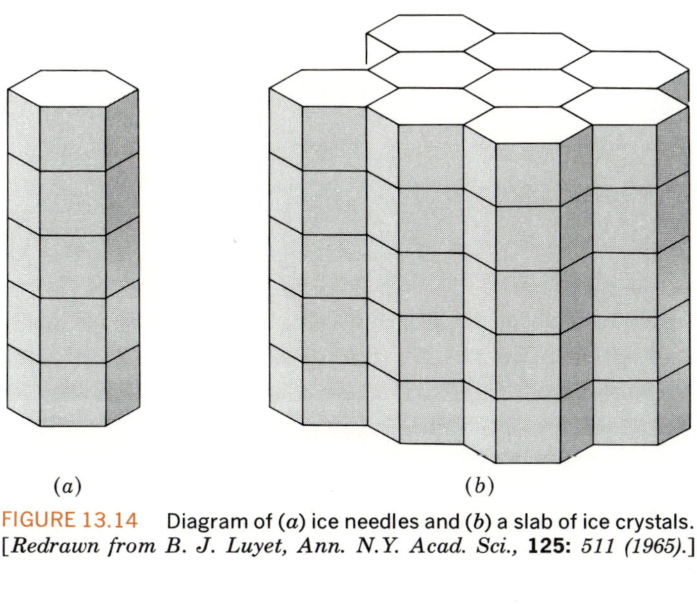

(*a*) (*b*)

FIGURE 13.14 Diagram of (*a*) ice needles and (*b*) a slab of ice crystals. [*Redrawn from B. J. Luyet, Ann. N.Y. Acad. Sci.,* **125:** *511 (1965).*]

13.11 HYDROPHOBIC BONDING

Hydrophobic bonding is the force holding nonpolar groups together in an aqueous environment. Since these nonpolar units do not associate with water, they must literally make a hole in the structure of water. This interference with the structure of water involves an energy change.

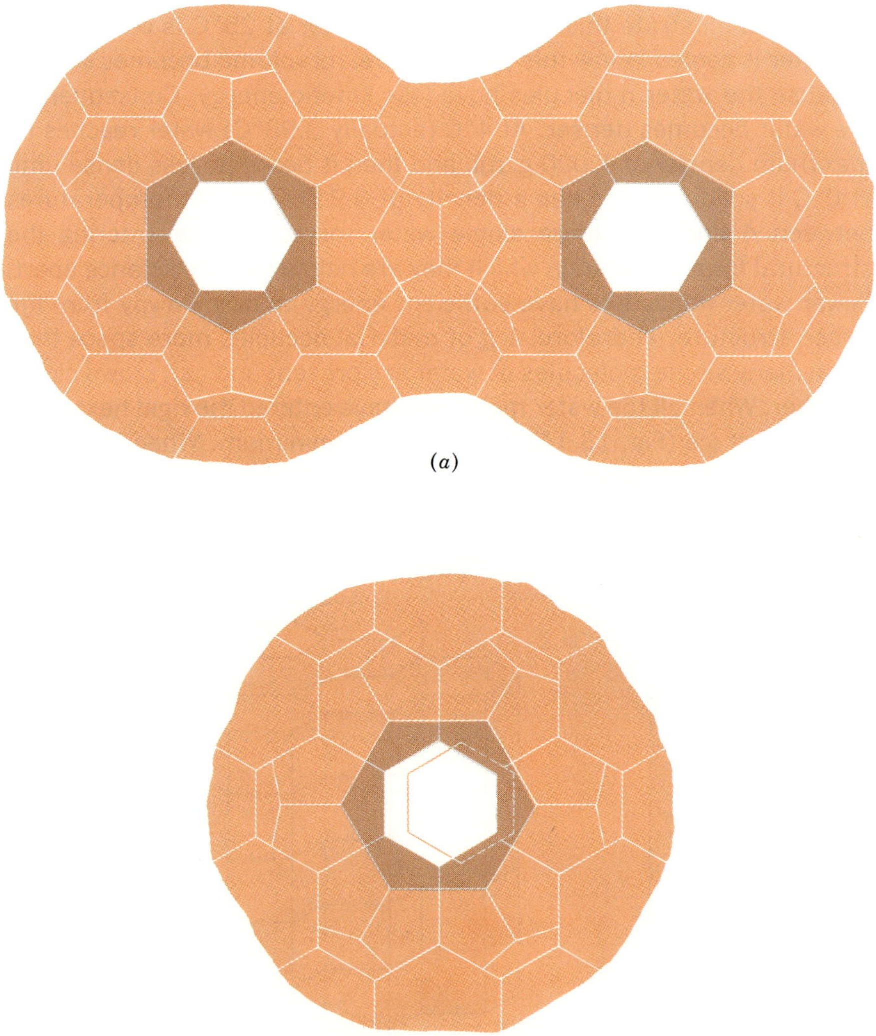

(*a*)

(*b*)

FIGURE 13.15 Hydrophobic bonding in (*a*) isolated hydrophobic units and (*b*) associated hydrophobic units.

Consider two such nonpolar groups isolated from each other, each creating a void volume in the structural pattern of water (Fig. 13.15). If these two nonpolar units are brought close together and put in the same hole, the void volume for the two is much less than when they were separated. There is less surface area surrounding the hydrophobic units if they are both in the same hole rather than separated; i.e., there is less interference with the structure of water. The difference in the energy relationship associated with the isolated cases and with the two-in-one hole may be considered as contributing to the energy of the hydrophobic bond. Another contributing factor is that the nonpolar groups are forced so close together that the water molecules between them are squeezed out and van der Waals forces can come into play.

When water and ethyl alcohol are mixed, the final volume is less than the sum of the volumes mixed. How can this fact be explained? Obviously the alcohol must cause some kind of collapse in the structure of water. The oxygen of the alcohol can hydrogen-bond to two hydrogen atoms in the water structure, causing a partial collapse of the ring structure of water immediately in the vicinity of the OH group of the alcohol. Consequently, there would be a partial decrease in the volume of the water. If the hydrocarbon part of the alcohol is considered as being within the center of the water structure, it would account for a decrease in the volume of water when alcohol is added.

QUESTIONS

1. How does an ionic bond differ from a covalent bond?

2. What is meant by a hydrogen bond? What elements enter into hydrogen-bond formation?

3. Arrange the following in the order of their decreasing bond strength:
(*a*) Polar forces (*b*) Metallic forces
(*c*) Ionic forces (*d*) Hydrogen bonding
(*e*) Van der Waals forces

4. From the data presented, predict the type of bond, if any, that exists in the following substances:
(*a*) A compound melting at 250°C
(*b*) A liquid of molecular weight 40
(*c*) A solid with a molecular weight of 60
(*d*) A liquid with a molecular weight of 72

5. Would you expect the following compounds to be solid, liquid, or gas?
(*a*) CH_4 (*b*) CH_3OH (*c*) C_7H_{16}

(*d*) $CH_3CH_2CH_2COOH$ (*e*) $\begin{matrix} COOH \\ | \\ COOH \end{matrix}$ (*f*) CH_2Cl_2

(*g*) CH_3CH_2Cl (*h*) $HOCH_2CH_2OH$ (*i*) $CH_3—O—CH_3$

(*j*) $CH_3CH_2—O—CH_2CH_3$

6. Arrange the following compounds in order of their increasing boiling point. Support your answer.
   (*a*) C_4H_9I (*b*) C_4H_9Cl
   (*c*) C_4H_{10} (*d*) $C_2H_5OC_2H_5$
   (*e*) C_3H_7COOH (*f*) C_4H_9Br
7. Why is water a good solvent for most biological substances?
8. Why is an emulsifying agent, such as soap, needed to make a stable oil-water emulsion?
9. What is the criterion for an oil-in-water emulsifier?
10. Would you expect a high-molecular-weight polypolar compound to be hydrated when placed in water? What is the name of a class of biologically important compounds with this property?
11. How could you effect a reasonable separation of formic acid and butyric acid from an aqueous solution?

SUGGESTED READINGS FOR PART TWO

Bennett, T. Peter: "Graphic Biochemistry," vol. I, The Macmillan Company, New York, 1968.

Cram, Donald J., and George S. Hammond: "Organic Chemistry," 2d ed., McGraw-Hill Book Company, New York, 1964.

Gould, E. S.: "Mechanism and Structure in Organic Chemistry," Holt, Rinehart and Winston, Inc., New York, 1961.

Noller, C. R.: "Chemistry of Organic Compounds," 3d ed., W. B. Saunders Company, Philadelphia, 1965.

Sykes, Peter: "A Guidebook to Mechanisms in Organic Chemistry," John Wiley & Sons, Inc., New York, 1961.

part three... biochemistry

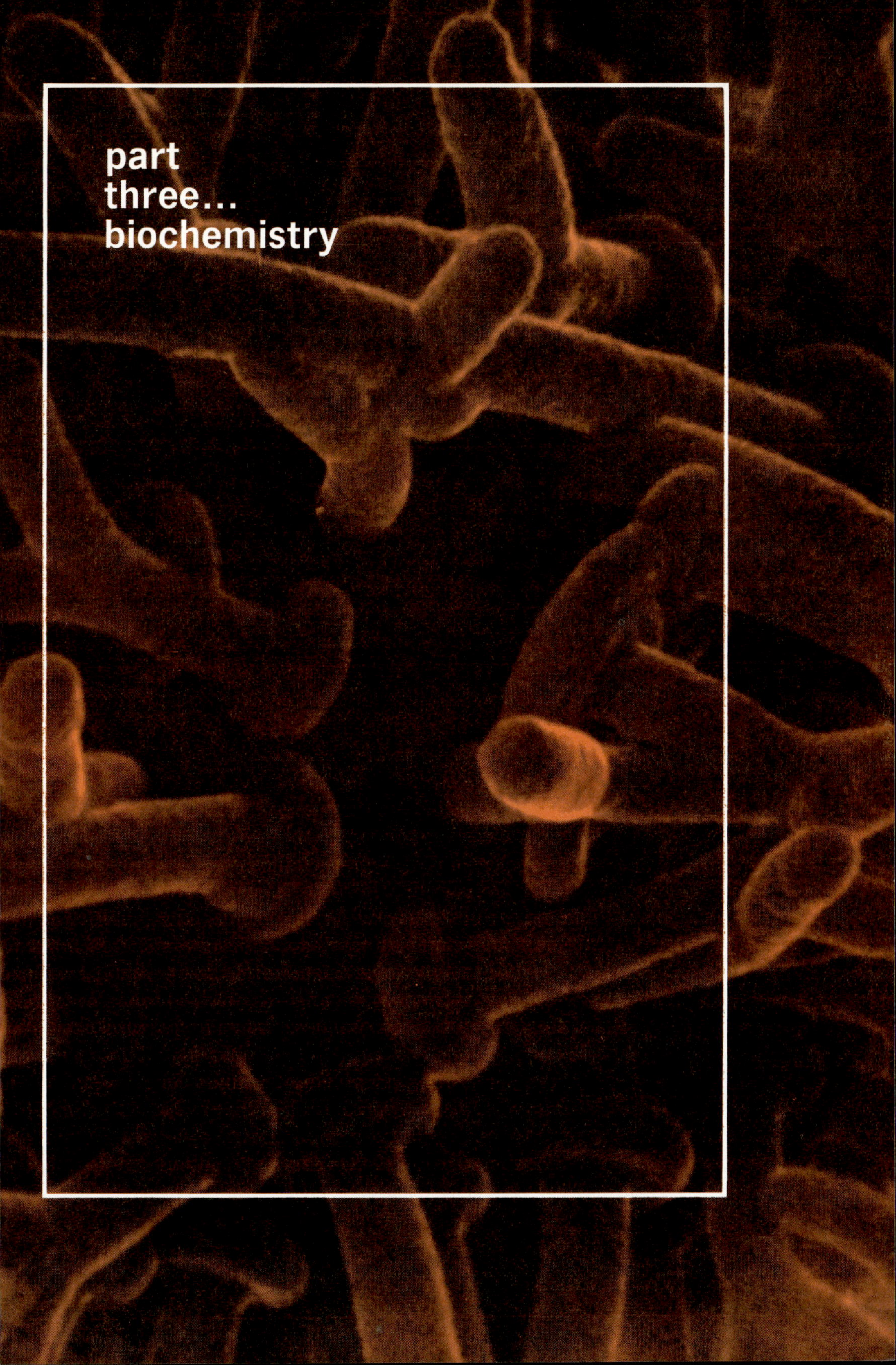

Biochemistry is the study of the chemistry of living matter and the substances related thereto. This definition includes the atmosphere, soil, plants, and animals, and the relationship of each of these to each other. Thus, the study of biochemistry includes inorganic and organic chemistry and the quantitative aspects of the dynamic state of living organisms as expressed in the concepts of physical chemistry.

This section discusses the chemistry of the main classes of compounds of biological importance and the major metabolic processes of living organisms.

The synaptic knobs or nerve endings of the sea hare (Aplysia californica) *magnified 20,000× in a micrograph taken with an electron scanning microscope. (E. R. Lewis, T. E. Everhart, and Y. Y. Zeevi at the University of California at Berkeley.)*

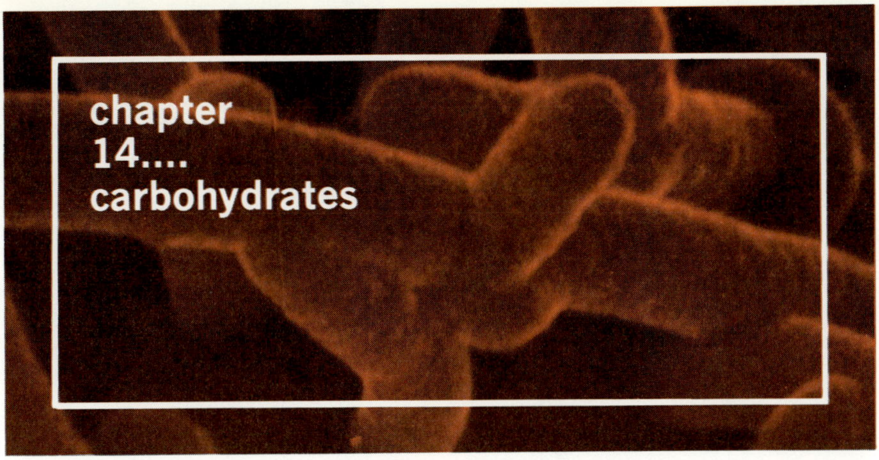

Carbohydrates are polyhydroxy aldehydes or ketones or substances that yield them upon hydrolysis. Since carbohydrates are too complex to define solely on the basis of chemical composition, their hydrolytic products must be included in the definition. Carbohydrates that are soluble in water and are sweet to the taste are called sugars.

Carbohydrates constitute the major material by weight in our diet and serve as a source of energy. They also play a major role in metabolic processes by furnishing the carbon chain for compounds synthesized by living organisms. In plants, certain carbohydrates constitute the structural material, wood or cellulose. Industries that depend upon the properties of carbohydrates for their success account for a large part of our economy, among them being those associated with wood (construction, paper), starch products, cotton, and fermentation.

14.1 CLASSIFICATION

The classification of carbohydrates is based upon one or more of the following: the chemical nature, the products of hydrolysis, and the number of carbons in the carbohydrate.

Monosaccharides, or simple sugars

Monosaccharides, or simple sugars, cannot be hydrolyzed to simpler sugars. Carbohydrates in this class are subdivided according to whether they contain the aldehyde or ketone group and according to the number of carbon atoms. For example, *ald-* indicates the aldehyde group, and the ending *-ose* indicates that the compound is a sugar. A three-carbon ketone sugar is classified as a *ketotriose*. The four-, five-, and six-carbon sugars are tetroses, pentoses, and hexoses, respectively.

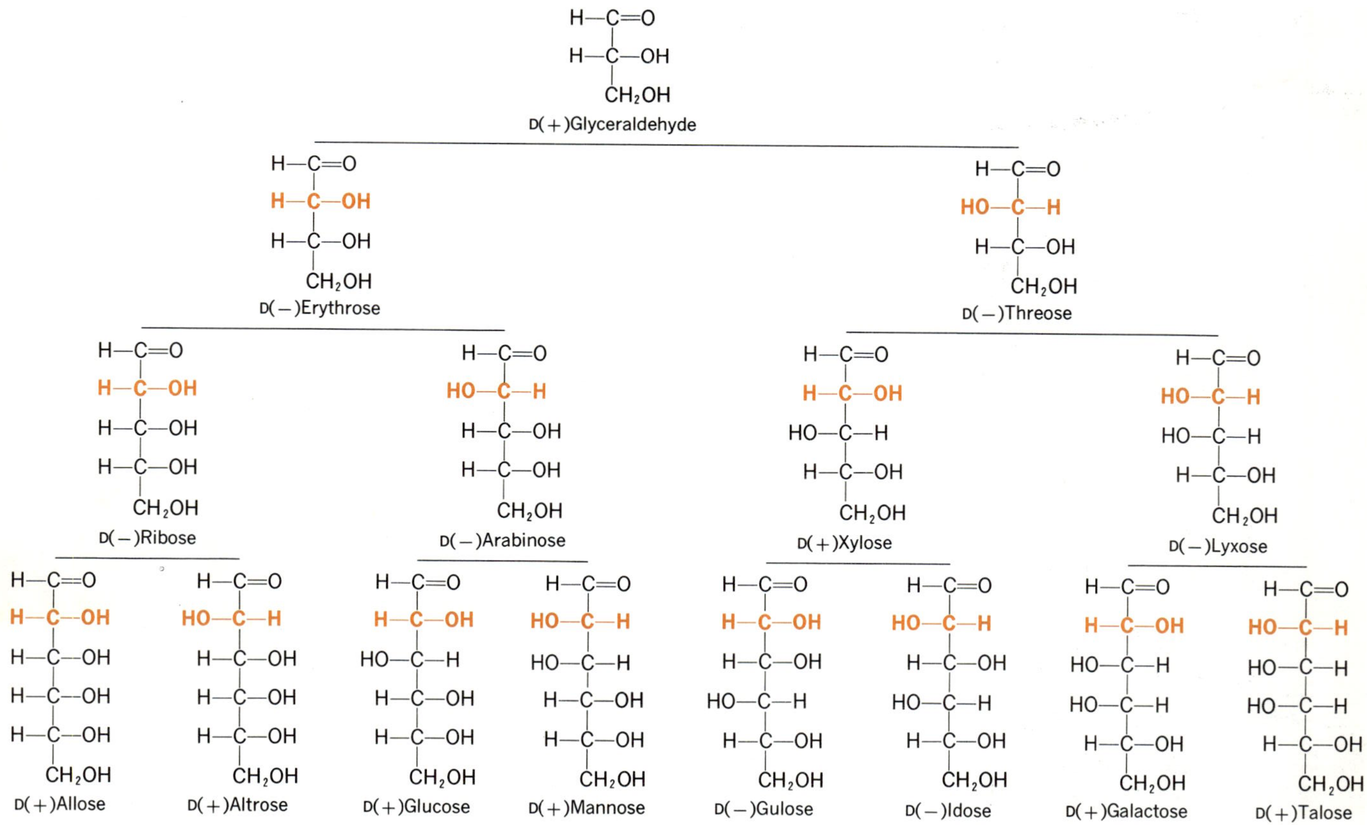

FIGURE 14.1 D family of aldose, open structure.

D and L family

There are two families of monosaccharides, the D family and the L family. In general, the D family of sugars occurs in nature. The D family of aldoses is derived from D-glyceraldehyde, whereas the L family is derived from L-glyceraldehyde, the mirror image of D-glyceraldehyde. Figure 14.1 illustrates a buildup of the D- family of aldehyde sugars from a triose through the hexoses. The D in carbohydrate chemistry means that the OH group on the *next to the last carbon* is on the right-hand side of the carbon chain if the aldehyde or ketone group is written at the top of the formula. *The* D *has nothing to do with indicating optical activity;* D sugars may be dextro- or levorotatory. For each member listed in Fig. 14.1 there is a mirror image; together they constitute the corresponding L family of sugars.

The formulas of the sugars in Fig. 14.1 were written by inserting a new CHOH between the first and second carbons of each previous member. For example, inserting a CHOH between C-1 and C-2 of glyceraldehyde gives two isomeric sugars (two tetroses) since the inserted carbon is asymmetric. Similarly, for each tetrose, two pentoses are formed by inserting a CHOH after the aldehyde group of the respective tetroses.

The number of optical isomers of a compound is equal to 2^n, where n is the number of asymmetric carbon atoms in the compound. Examine the formulas for the pentoses. Since there are three asymmetric carbon atoms, there are 2^3, or 8, optical isomers, 4 in the D family and 4 in the L family. There are 4 asymmetric carbons in the open-chain formula for aldohexoses; therefore, there are 2^4, or 16, total D and L family aldohexoses.

Are there an L family and a D family of dihydroxyacetone? To have a D or L family there must be an asymmetric carbon next to the primary alcohol group on the end. There is none in dihydroxyacetone; therefore, there is no D or L family of this ketose. The first member of a D or L family ketose must have four carbons. The formula of a D-tetrose and the formula of an L-tetrose are

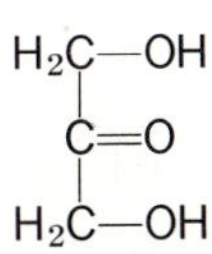
Ketotriose

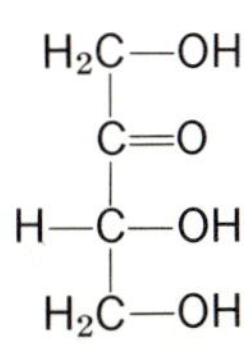
D-Ketotetrose

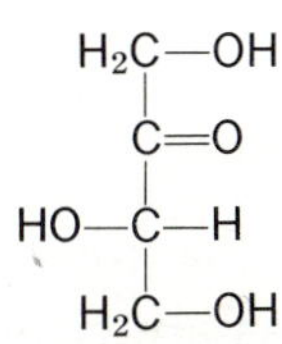
L-Ketotetrose

Oligosaccharides

Oligosaccharides are carbohydrates composed of two to ten monosaccharides linked together by an acetal or ketal linkage. These carbohydrates can be hydrolyzed to their respective monosaccharides. Carbohydrates which can be hydrolyzed to two molecules of simple sugar are called *disaccharides*. The simple sugars may be the same or different. *Trisaccharides* yield three molecules of monosaccharide when hydrolyzed. The monosaccharides may be the same or any combination of simple sugars.

Polysaccharides

As the name indicates, polysaccharides yield many molecules of simple sugar when hydrolyzed. They are of quite high molecular weight. The more complex ones contain other constituents, such as sugar acids or amino sugars, in their structure. A *homopolysaccharide* is composed of only one simple sugar, whereas a *heteropolysaccharide* is composed of more than one kind of monosaccharide.

14.2 MONOSACCHARIDES

Much of the chemistry of carbohydrates has already been presented in the chapters dealing with alcohols, aldehydes, and ketones. Since carbohydrates are hydroxy aldehydes or ketones, carbohydrates were used to illustrate a number of the reactions of these classes of compounds. Several reactions will be discussed again for emphasis, and a few additional reactions will be given. It is important that the previous material be reviewed. Consider the following questions. How can an alcohol be converted to an aldehyde? An acid? Compare the stability of acetals and hemiacetals toward acids and bases. How can the carbon chain be increased in length? Under what conditions are hemiacetals formed? Which group is more stable, a hemiacetal or an acetal? What is the reagent periodate used for? What equilibrium must be possible before mutarotation occurs?

14.3 STRUCTURE

The formulas of the free aldehyde form, aldehydo sugars, for the D family of aldoses are given in Fig. 14.1. The open-chain formula for the aldoses or ketoses is easily written by putting the aldehyde or ketone group at the top and the carbon chain below. The OH groups are placed on either the right or the left side of the carbon chain. However, when an aldehyde or a ketone group is in a molecule which also contains hydroxy groups, an intermolecular reaction may occur to form hemiacetal or hemiketal, respectively (see pages 247 and 250). This intermolecular hemiacetal or hemiketal is the basis for the cyclic structure of the sugars.

Two types of ring structures are possible, the five-membered, or *furanose,* ring and the six-membered, or *pyranose,* ring. These names are derived from the parent compounds furan and pyran:

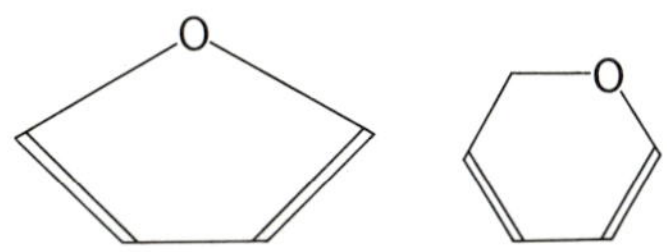

Furan α-Pyran

For example, the names glucopyranose and fructofuranose indicate that glucose and fructose contain a six- and a five-membered ring structure, respectively.

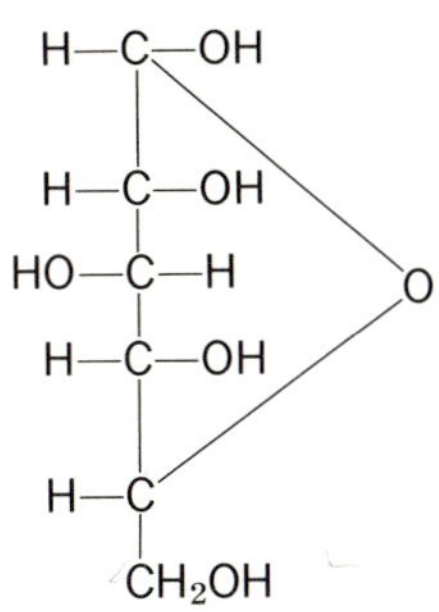

D-Glucopyranose
Most common

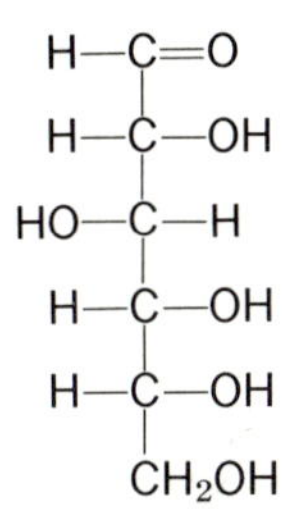

D-Aldehydoglucose

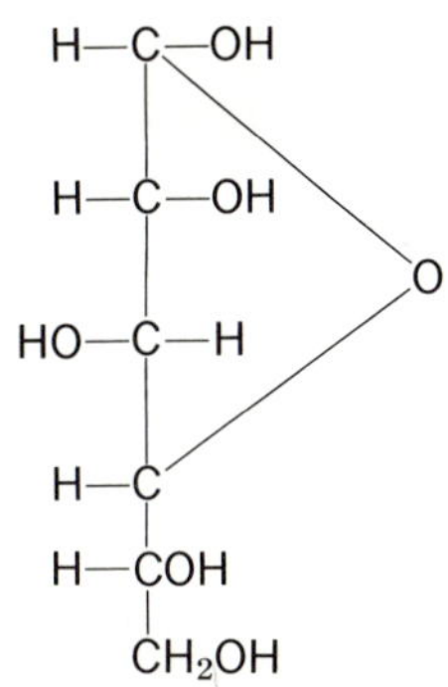

D-Glucofuranose
Less common

Carbons in ring structure

Never
Never
Less common
Most common
Least common

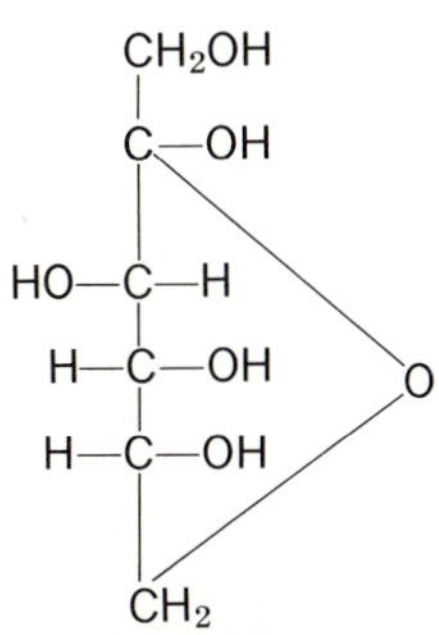

D-Fructopyranose
Most probable crystalline form

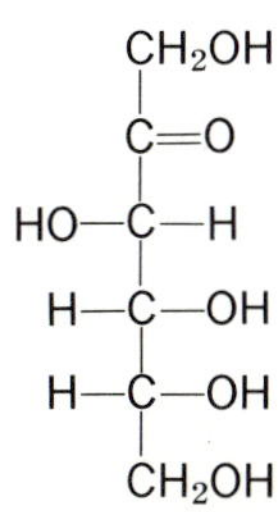

D-Ketofructose

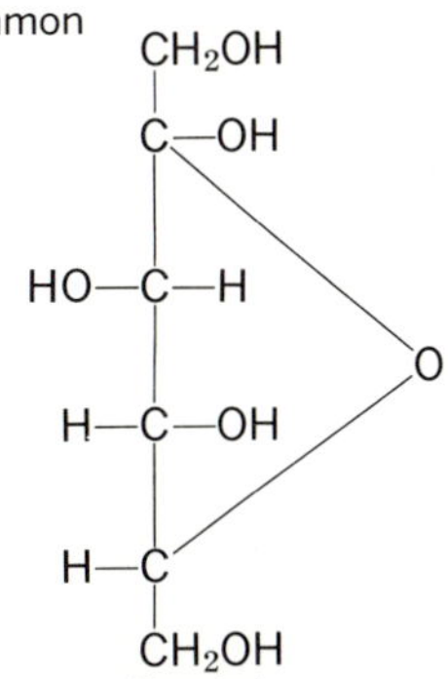

D-Fructofuranose
Common form in solution and in compounds

The second and third carbons of the aldehyde sugars never occur in the hemiacetal group. The most common ring structure for aldohexoses is the pyranose ring, which involves the first and fifth carbons of the sugars. The furanose ring structure is less stable than the pyranose structure and is not very common among the aldohexoses. Very seldom is a seven-membered ring formed. Fructose exists in solution and in compounds as a furanose; however, in the *crystalline* state only the pyranose ring is believed to exist. Ribose also occurs as the furanose structure in many important biological compounds.

When writing the cyclic, or Haworth, form of sugars (Fig. 14.2), it is sometimes difficult to judge whether an OH should be above or below the plane of the ring. A few simple rules can be followed for writing the Haworth formulas for carbohydrates. Glucose is used as an example.

Make a hexagon from a piece of board and stand it on one edge so that the ring projects out of the page as indicated in Fig. 14.2*b*. The shaded edge is closest to the eye. The five carbon atoms of the ring are not marked, but the oxygen atom is *always* shown. In a more abbreviated formula the hydrogens are also omitted and the OH groups are indicated merely by a line (Fig. 14.2*c*).

Rules for writing Haworth formulas

1. Write the oxygen at the upper right-hand corner of the ring structure and the carbons clockwise around the ring. At the fifth carbon it is necessary to rotate the bond 90° to make the ring closure.
2. For the D family of sugars the terminal CH_2OH is written above the plane of the ring. For the L family it is written below the plane of the ring.

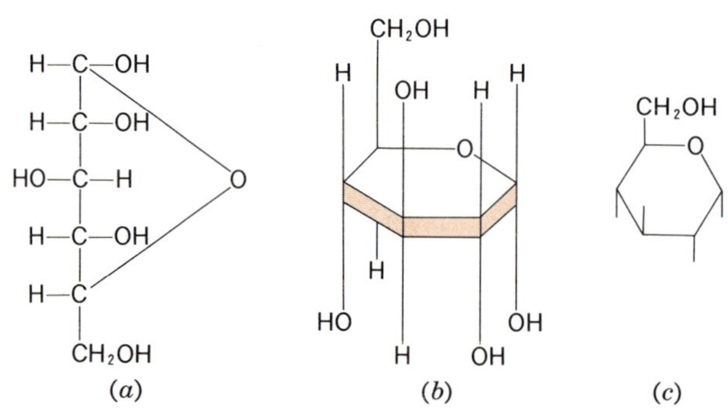

FIGURE 14.2 Cyclic structures of glucose: (*a*) straight-chain formula, (*b*) Haworth formula, (*c*) abbreviated Haworth formula.

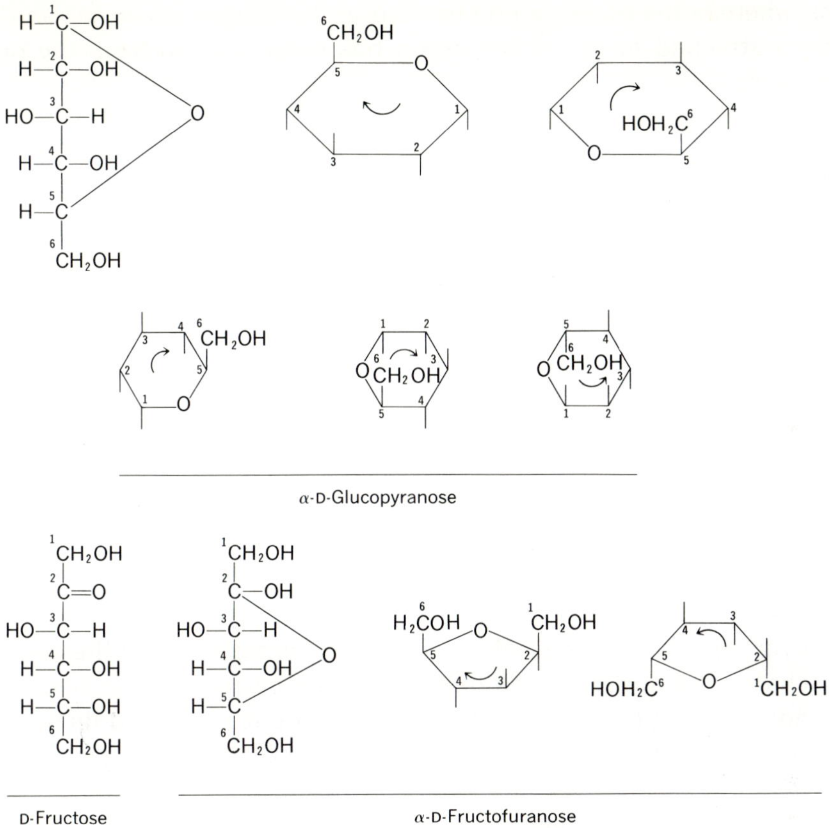

FIGURE 14.3 Cyclic formulas for glucose and fructose.

3. If the OH group occurs on the right-hand side of the carbon chain in the straight-chain formula, it is placed below the plane of the ring in the cyclic formula. Conversely, if the OH group is on the left side of the carbon chain in the straight-chain formula, it is placed above the plane of the ring in the Haworth formula.
4. Writing the formula with the carbons counterclockwise around the ring is equivalent to flipping the ring over. Consequently the arrangements about the carbons are the *reverse* of the above. The various ways of writing the formula for glucose and fructose are shown in Fig. 14.3.

It is not difficult to determine which is the hemiacetal or hemiketal carbon of the cyclic formulas when it is remembered that there is *always an OH group attached to the hemiacetal or hemiketal car-*

bon whereas the other carbon bonded to the oxygen *never* has an OH group attached to it; the OH group has been used to form the ring structure.

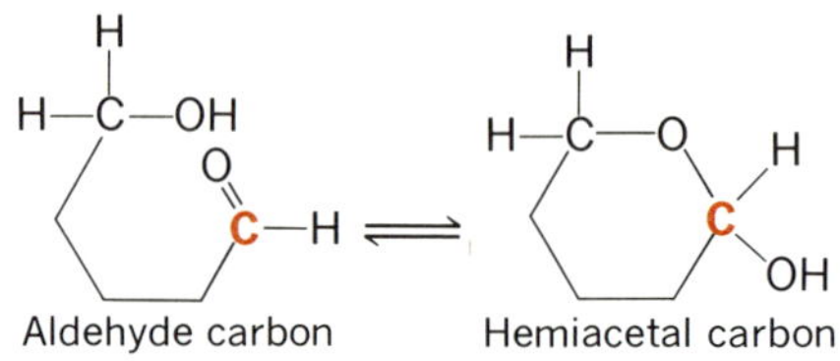

OH configuration

Each asymmetric carbon influences the overall optical activity [α] of a compound. However, with *sugar γ-lactones and amides* one asymmetric carbon atom is the factor deciding whether [α] is positive or negative. For our purposes, several general rules are used to determine the configuration of an OH about a carbon atom. The following rules are derived from Hudson's rules and apply *only* to sugar derivatives.

1. *The γ-lactone derivatives of the sugars are dextrorotatory* if the hydroxyl group in the lactone ring is on the right side of the carbon chain. The sugar can be oxidized to the corresponding acid with aqueous Br_2 (see page 350), the γ-lactone formed, and its optical properties determined. If α is (+), the γ-OH of the sugar was on the right side of the chain, and, conversely, if α is (−), the OH was on the left of the chain.

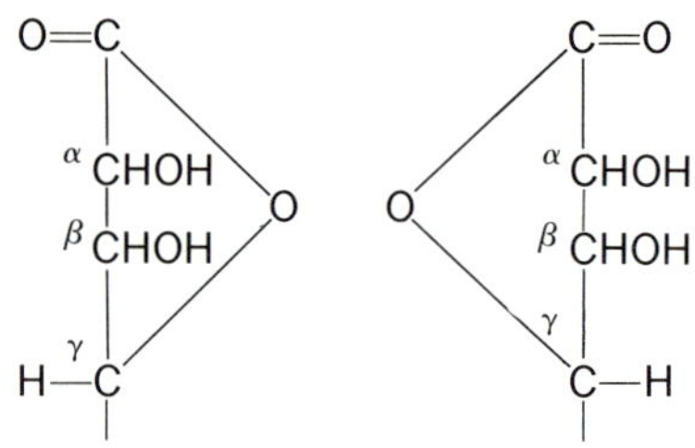

$\alpha = (+)$ $\alpha = (-)$

2. *The amide derivatives of the sugars are dextrorotatory* if the α-OH is on the right side of the carbon chain and levorotatory if the α-OH is on the left side of the chain. The amides can be made easily by treating the lactone with NH_3 (see page 351).

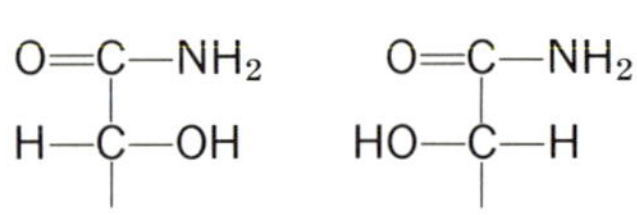

$\alpha = (+)$ $\alpha = (-)$

EXAMPLE

From the following information write the structural formula for the *pentose:*

(*a*) A pentose was oxidized with aqueous Br_2, and the γ-lactone of the resulting acid was formed. The lactone had a positive α.

(*b*) After treating the lactone with ammonia the resulting amide was found to have a negative optical activity.

(*c*) The original pentose was treated with HCN and the resulting cyanohydrin hydrolyzed to the acid. The γ-lactone was then formed, and it was found to have a positive optical activity.

Do you agree that the structure of the pentose is as follows?

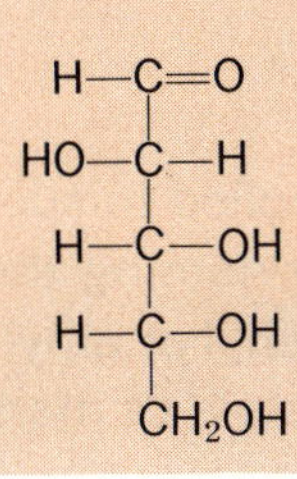

Mutarotation

Mutarotation is the change in optical activity of a freshly prepared solution of sugar that takes place as it is allowed to stand. For example, a fresh solution of one form of crystalline D-glucose has a specific rotation [α] of +112.4°, but upon standing the optical activity decreases until a value of +52.2° is reached. Another form of D-glucose has an initial [α] of +19°, which increases to the value of +52.2° if the solution is allowed to stand (see Fig. 14.3). This change in [α] is called mutarotation. The D-glucose form that mutarotates to a lower positive value is called the α form, and the D-glucose that mutarotates to a higher positive value is called the β form. This is true for all D-family sugars. The value for [α] of 52.2° represents the optical activity of an equilibrium mixture of α- and β-D-glucose, 36 and 64%, respectively.

Only sugars that form the hemiacetal or hemiketal structure mutarotate. The equation for the mutarotation of glucose was presented in Sec. 10.1 dealing with hemiacetal formation. In the following discussion only the hemiacetal group will be sketched. To obtain the

β form of glucose from the α form, or vice versa, it is necessary to pass through the free aldehyde structure:

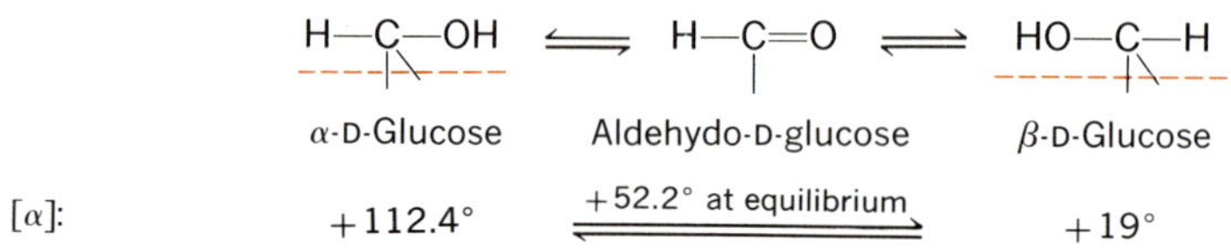

In doing so the asymmetry of the hemiacetal carbon is lost, and when the ring is formed again both isomers are produced. When α and β forms are in equilibrium with each other, $[\alpha] = +52.2°$.

14.4 REACTIONS

Glycoside formation

Glycoside is a general term applied to sugar acetals or ketals. The specific glycoside is indicated by naming the radical associated with the acetal or ketal group, e.g., methyl or ethyl glycoside. Acetals of glucose are called *glucosides,* and acetals of mannose are called *mannosides.*

Methyl glucoside is formed by reacting glucose and CH_3OH under anhydrous conditions with HCl as the catalyst:

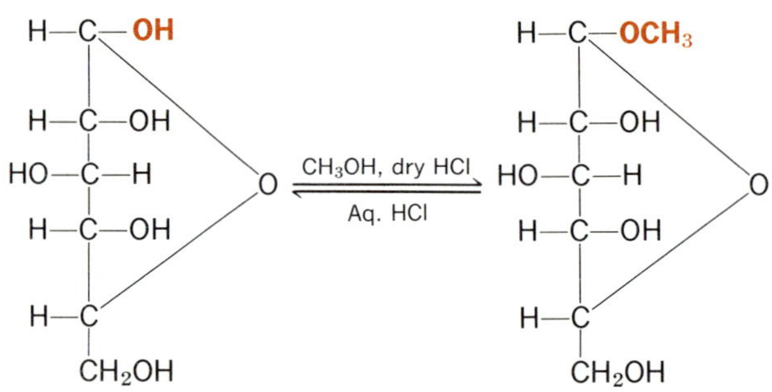

α-D-Glucopyranose
Glucose

Methyl-α-D-glucopyranoside
Glucoside

The formation of glycosides must be carried out under strict anhydrous conditions with H^+ as a catalyst because glycosides are hydrolyzed quite easily in aqueous acid solutions. The glycosides are, however, quite stable to alkali.

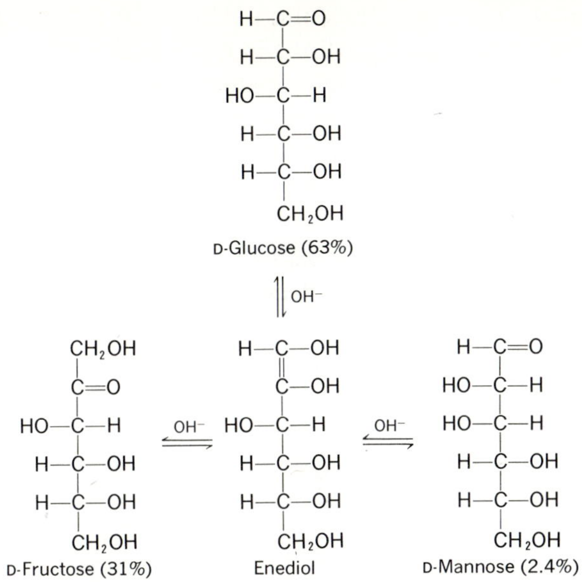

FIGURE 14.4 Interconversion of glucose, fructose, and mannose in the presence of alkali, which suppresses ring formation and favors enolization.

Alkali

Alkaline solutions favor enolization of aldehydes and ketones (CH_3—CHO $\rightleftharpoons$ CH_2=CHOH) but hinder hemiacetal or hemiketal ring formation. The sugars glucose, fructose, and mannose are all in equilibrium with each other in an alkaline solution because of enolization of the aldehyde and ketone groups and because these sugars have the same configuration below the second carbon atom (see Fig. 14.4). The second carbon atom of the intermediate compound (the enediol) is not asymmetric; therefore, when this carbon becomes asymmetric, the two isomers glucose and mannose are formed. The other isomer resulting from enolization is fructose.

Oxidation

Fehling's (page 256) and Barfoed's reagents contain copper as the oxidizing agent. Barfoed's reagent is a slightly acidic copper sulfate solution which, like Fehling's reagent, has oxidizing properties. Which of these solutions would be expected to be the stronger oxidizing reagent for sugars and why? The answer is Fehling's solution because alkali lessens the stability of the ring structure (favors the aldehyde form) and also increases enolization. The enediol form with the OH groups on the unsaturated carbons is quite susceptible to oxidation.

Sugars that reduce Fehling's reagent are called *reducing sugars.* A reducing sugar must have a free aldehyde or ketone group or exist as the hemiacetal or hemiketal. Therefore, all monosaccharides are reducing sugars. Disaccharides may or may not be reducing sugars, depending upon their structure (see page 361).

Bromine water is a weak oxidizing reagent that selectively oxidizes the aldehyde group of sugars to the corresponding acid. The active oxidizing substance in bromine water is HOBr [$Br_2 + HOH \longrightarrow HBr + HOBr$; $HOBr \longrightarrow HBr + (O)$]. The sugar acids are known as *-onic* acids; e.g., glucose yields *gluconic* acid, and galactose forms *galactonic* acid. The reaction for the formation of *-onic* acids from the aldehyde form of a sugar (aldehydo sugar) is illustrated with glucose:

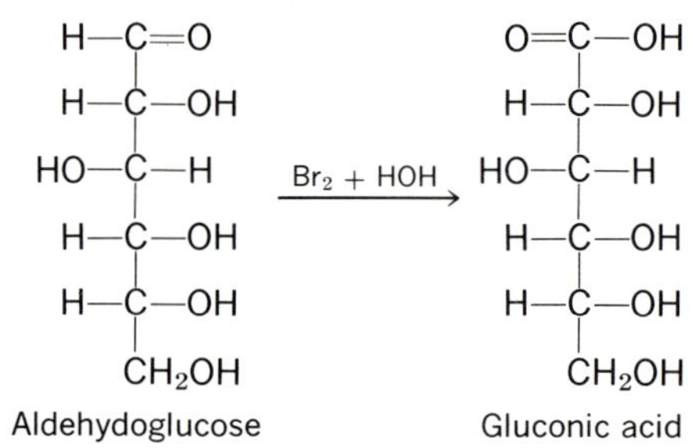

Gluconic acid is a polyhydroxy acid. Would you expect the acid group to react with an alcohol group within the same molecule? Yes; a lactone results when a carboxyl group reacts with a hydroxy group in the same molecule. The formation of the lactone of gluconic acid, gluconolactone, is

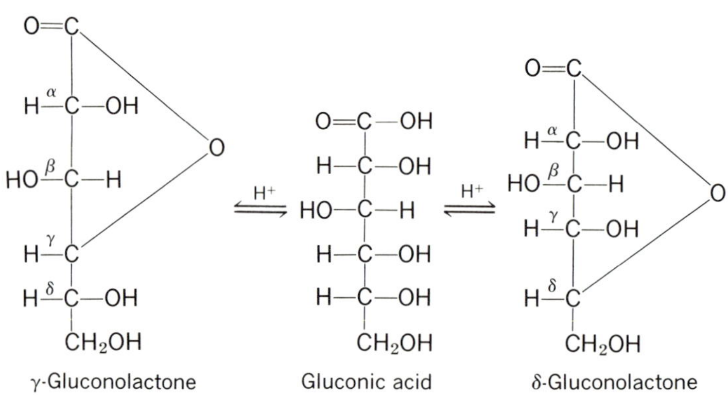

Acids catalyze the formation of lactones, which are intermolecular esters. Two lactones may be formed, either the γ- or the δ-lactone. The γ-lactone is more stable than the δ-lactone.

Two reactions characteristic of lactones should be mentioned. One is the reduction of lactones to the corresponding aldehyde by sodium-mercury amalgam (Na-Hg) plus water. Acids cannot be reduced, but esters (or lactones) can be reduced to the corresponding aldehyde. The second reaction is ammonolysis, reacting a substance with NH_3. If lactones or esters are reacted with ammonia, the corresponding amide is produced.

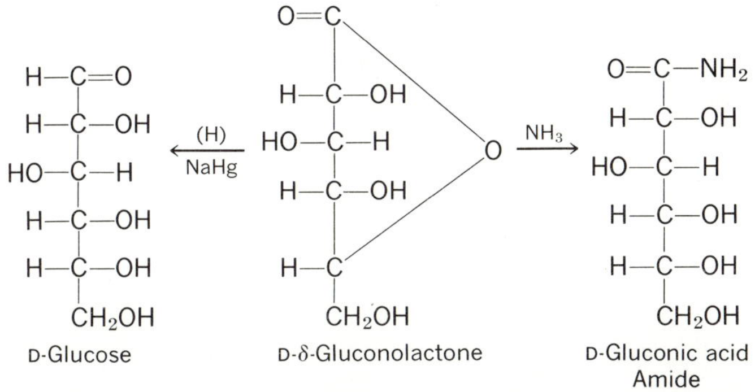

What methods have been discussed for the formation of amides?

The oxidation of certain sugars to the corresponding *-onic* acid is an essential step in metabolic reactions of carbohydrates. Two examples are the conversion of glucose-6-phosphate to the corresponding δ-lactone and the oxidation of glyceraldehyde-3-phosphate to the corresponding 3-phosphoglyceric acid:

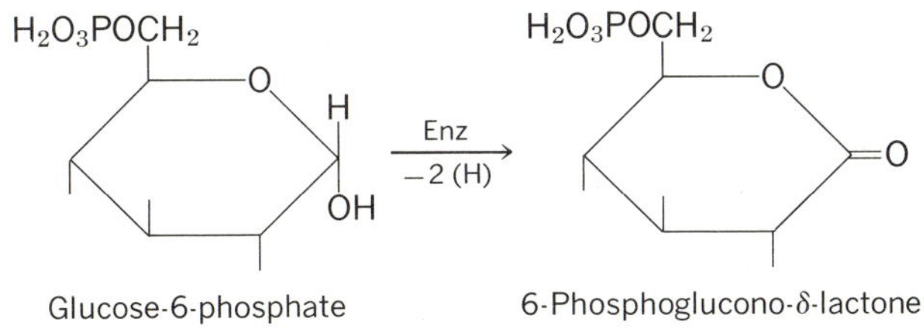

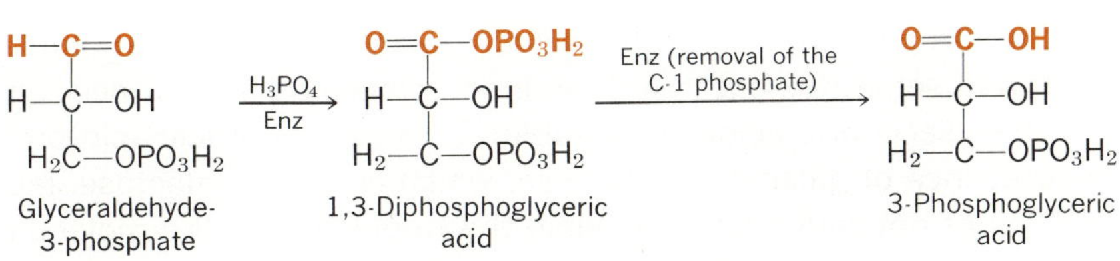

The formation of the δ-lactone upon oxidation of glucose-6-phosphate is easily visualized from the cyclic structure of glucose. The pyranose ring structure of glucose involves the δ carbon; thus upon the removal of two hydrogens from the groups attached to the first carbon, the δ-lactone results.

The oxidation of glyceraldehyde-3-phosphate to 3-phosphoglyceric acid involves the formation of the intermediary product 1,3-diphosphoglyceric acid. Subsequent removal of the 1-phosphate group yields 3-phosphoglyceric acid.

Nitric acid is a vigorous oxidizing reagent that oxidizes both the aldehyde group and the primary alcohol group of aldoses to the carboxyl group. Thus, dicarboxylic polyhydroxy acids are produced from the sugars. These are called *saccharic acids*. Glucose is oxidized to saccharic acid:

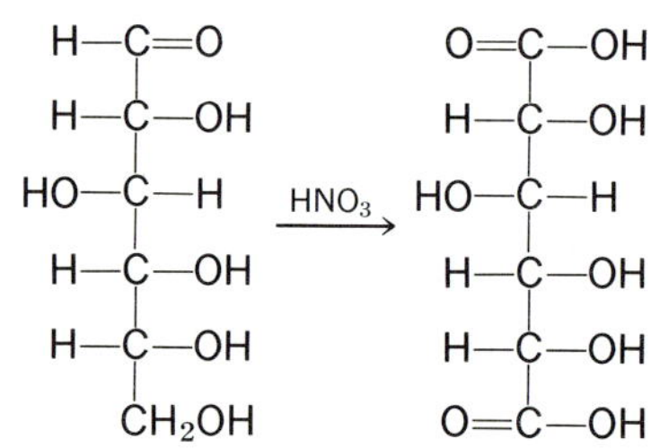

Glucose Saccharic acid

Galactose is oxidized by warm HNO_3 to galactosaccharic acid, commonly called mucic acid:

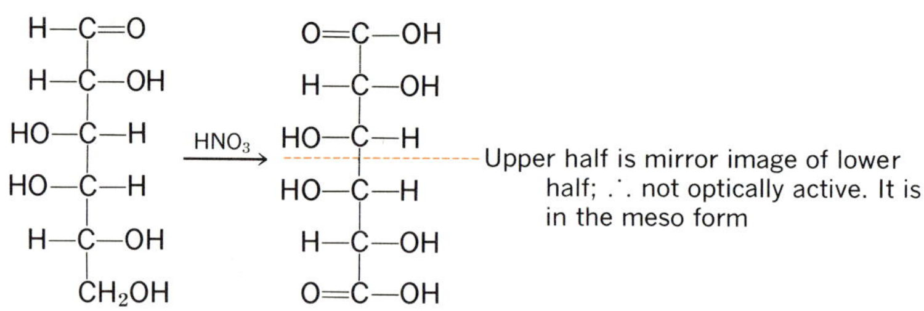

Galactose Galactosaccharic acid Mucic acid

Despite the polar nature of this acid, it is only very slightly soluble in water and separates out in well-defined crystals as it is formed. Since the other saccharic acids are soluble, formation of mucic acid proves the presence of galactose or lactose, which contains galactose. Mucic acid is not optically active because the upper half of the molecule is the mirror image of the bottom half.

Phenylhydrazine reacts with all reducing sugars to form the corresponding *osazone*. Three steps are involved in osazone formation: (1) the carbonyl group of the sugar reacts with 1 mole of phenylhydrazine to form the corresponding phenylhydrazone; (2) a second mole of phenylhydrazine oxidizes the phenylhydrazone to a keto derivative, and ammonia and analine are also formed; (3) the keto phenylhydrazone reacts with a third mole of phenylhydrazine to form the osazone.

Of particular significance is the fact that both glucose and fructose form the same osazone (Fig. 14.5). This is proof that both these sugars have the same configuration about the asymmetric carbon atoms, C-3, C-4, and C-5. Would you expect arabinose and ribose to

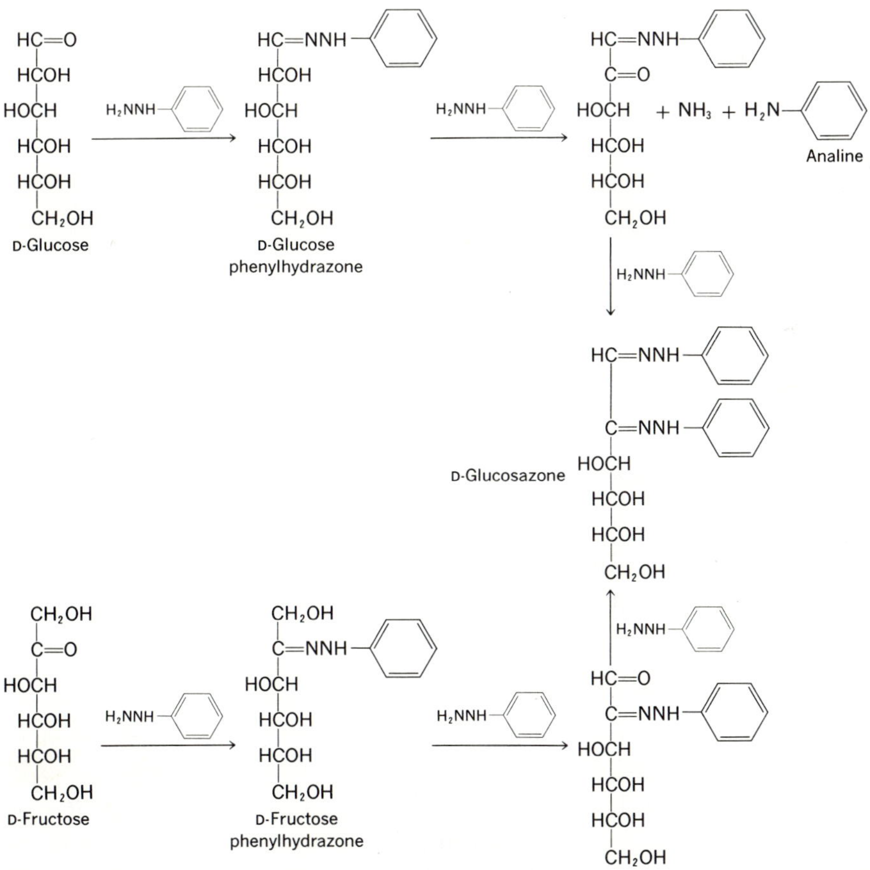

FIGURE 14.5 Formation of osazones.

form the same osazone? Would xylose and lyxose yield the same osazone?

The osazones are mostly insoluble and crystallize in well-defined characteristic crystalline structure. Also different sugars form osazones at different rates; e.g., fructosazone forms more readily than glucosazone. The characteristics of osazones are used in the study of carbohydrate structures.

Dehydration

When heated with strong nonoxidizing mineral acids pentoses are dehydrated to furfural, whereas the hexoses yield 5-hydroxymethylfurfural.

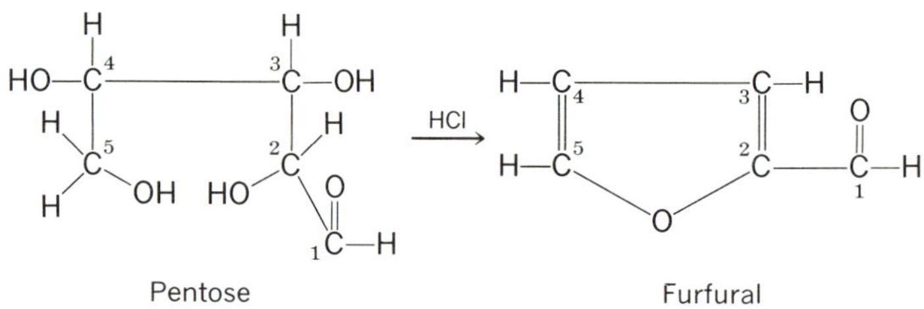

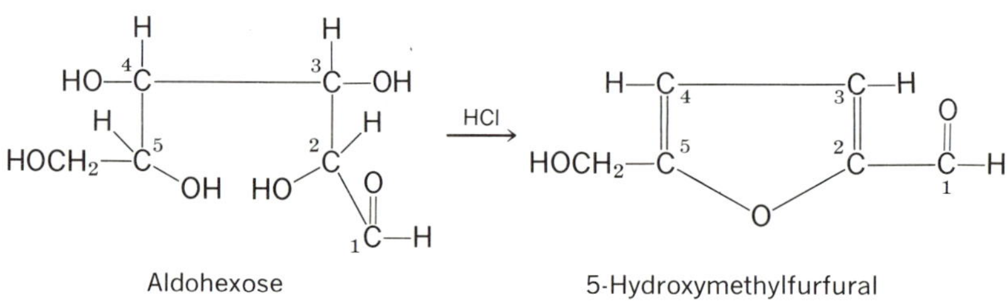

Furfural is volatile with steam, but hydroxymethylfurfural is not. Pentoses are dehydrated more easily than the hexoses.

14.5 INDIVIDUAL MONOSACCHARIDES

Trioses

Glyceraldehyde and *dihydroxyacetone* do not occur to any extent in the free form, but they are very important intermediary products (as the phosphate ester) in the metabolism of glucose. The phosphate ester of each of these trioses is formed from the cleavage of fructose-1,6-diphosphate

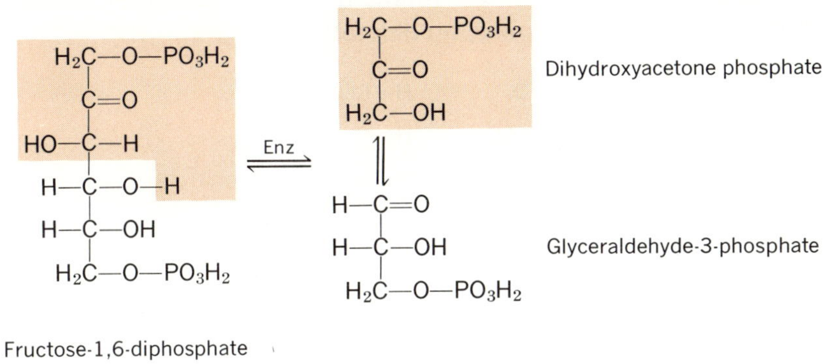

Enzymes in cells are able to cleave fructose-1,6-diphosphate into two C_3 sugars. Also in the cells is another enzyme which converts dihydroxyacetone phosphate to glyceraldehyde-3-phosphate. Since each of these reactions is reversible, it is possible for the body to synthesize fructose diphosphate from either of the two triose phosphates.

Dihydroxyacetone phosphate can be metabolically reduced to glycerol phosphate, and the phosphate can be hydrolyzed off to yield glycerol. The hydrogen for this process comes from the reduced coenzyme NADH + H^+. The reactions are

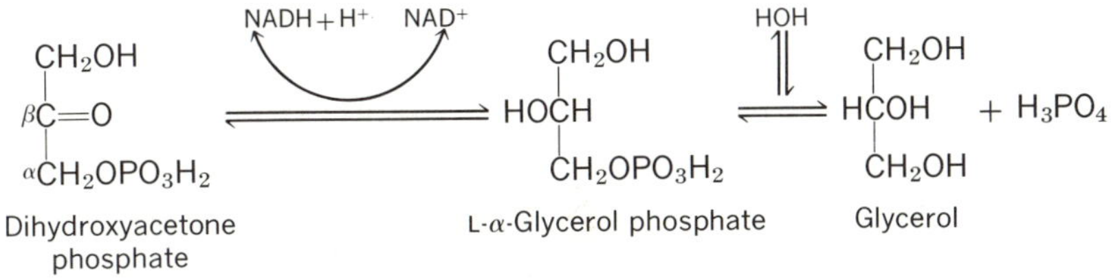

In the biological reduction of dihydroxyacetone phosphate, L-glycerol phosphate is formed. Note that the center carbon (β carbon) is asymmetric in glycerol phosphate but not in glycerol. If the acetone derivative were reduced chemically the reduction product would be a racemic mixture of the D- and L-glycerol phosphate. Enzymes are specific for the isomer produced, whereas chemically if a nonasymmetric carbon is converted into an asymmetric carbon, both isomers are formed in equal amounts to produce a racemic mixture.

Tetroses

Like the trioses, the tetroses do not occur free in nature, and some are intermediates in carbohydrate metabolism. The only tetrose to be mentioned here is D-*erythrose,* an aldotetrose.

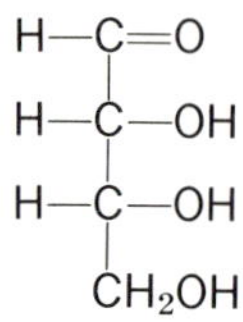

D-Erythrose

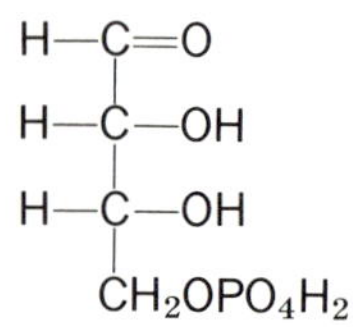

D-Erythrose-4-phosphate

Further like the trioses, erythrose occurs as the phosphate and is associated with carbohydrate metabolism.

Pentoses

None of the pentoses occur free in nature; however, several do occur as components of polysaccharides. The three pentoses, ribose, deoxyribose, and xylulose, play an important role in metabolism.

D-*Xylose* is found in plants in the polysaccharide xylan, as in seed hulls, straw, wood, and corncobs. It is not metabolized by animals but is fermented by certain classes of bacteria, a fact that is used in the classification of bacteria. Large amounts of xylose in the diet of certain experimental animals can produce a disorder of the eye known as cataracts.

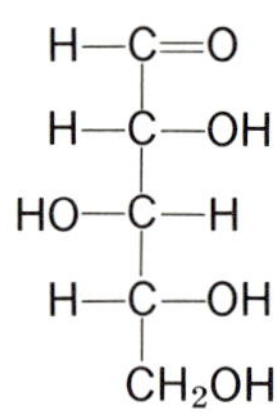

D-Xylose

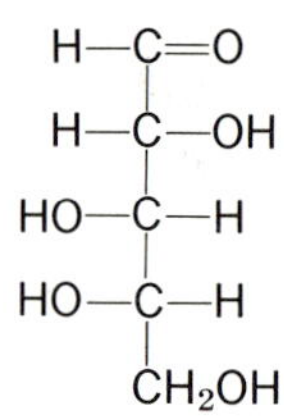

L-Arabinose

L-*Arabinose* is a component of complex plant polysaccharides called gums (cherry gum, gum arabic) and of wheat bran; it is also found in the free state in the heartwood of coniferous trees. Although arabinose is not utilized by animals, it is metabolized by certain microorganisms.

D-*Ribose* and *2-deoxy*-D-*ribose* are two pentoses whose biological importance cannot be overemphasized. They are found in every cell and are associated, for example, with many enzyme systems, with genes, and with protein synthesis. The *2-deoxy-* indicates that the oxygen on the second carbon is removed from the ribose skeleton.

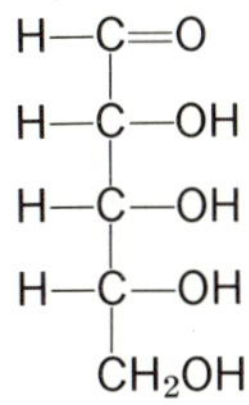
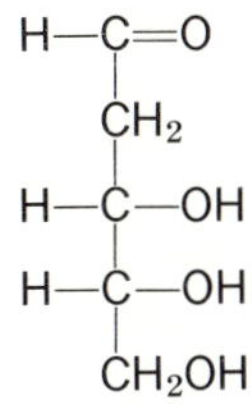

D-Ribose 2-Deoxy-D-ribose

D-*Xylulose* and L-*xylulose* are ketopentoses. D-Xylulose does not occur free in nature, but the 5-phosphate derivative is the only keto-pentose intermediate in glucose metabolism. L-Xylulose is an abnormal product of metabolism in man. In cases of pentosuria, or pentoses in the urine, L-xylulose is excreted in the urine. Pentosuria is an inherited trait. It apparently does not interfere with normal metabolism of other sugars and does not impair the health of the individual.

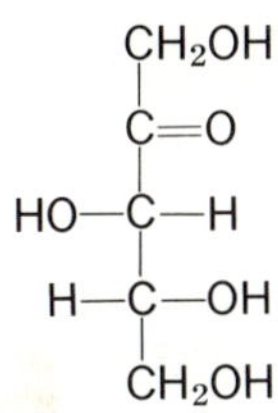
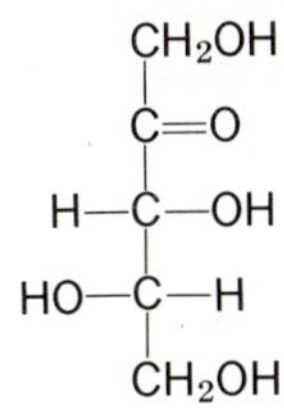

D-Xylulose L-Xylulose

Hexoses

Several hexoses occur in the free state in nature; some are components of di-, tri-, and polysaccharides.

Glucose, often called dextrose because it is dextrorotatory, is widespread in the free state. It is a constituent of most oligosaccharides and of the polysaccharides starch, glycogen, and cellulose. Normally glucose is found in the blood at a concentration of 80 to 90 mg per 100 ml of blood and in only trace amounts in the urine. But in diabetes a considerable amount of glucose is excreted in the urine. Glucose is often referred to as blood sugar because it is the only sugar transported in the blood to tissues for energy requirements and synthetic needs. All other sugars that are ingested and absorbed are first converted into glucose. In lactating mammary glands, the disaccharide lactose is made from glucose.

Glucose, fructose, and mannose all have the same configuration below the second carbon:

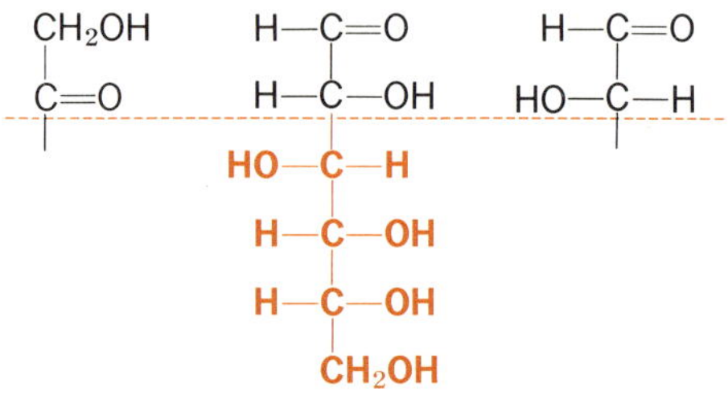

This similarity of structure accounts for the interconversion of these three sugars in the presence of alkali (see Fig. 14.4).

Fructose is a ketohexose widely found free in nature as well as in combination with other sugars. Also known as fruit sugar and levulose (because it rotates plane polarized light to the left), fructose occurs in fruits and is found with glucose in a ratio of 1 : 1 in honey. Fructose in combination with glucose is a component of common table sugar, sucrose. It also occurs in several trisaccharides and in the polysaccharide inulin.

As previously mentioned, in the crystalline state fructose occurs as a six-membered cyclic hemiketal (pyranose structure), whereas in the combined state fructose occurs only in the five-membered furanose ring structure.

Fructose as the diphosphate ester is an important intermediate in carbohydrate metabolism. As we saw in the discussion of trioses, fructose diphosphate is the precursor of dihydroxyacetone and glyceraldehyde phosphates.

Mannose, an aldohexose, is found in the vegetable kingdom combined in polysaccharides. It is a component of the polysaccharide mannan found in berries and in vegetable ivory (from the endosperm of palm nuts). When we compare the formula for mannose with that of glucose and fructose, we see that mannose, fructose, and glucose differ only in the arrangements of groups on the first two carbons; otherwise they have the same structure. The common form of mannose is α-D.

Galactose is an aldohexose that occurs in combination with other sugars but not in the free state. It is found in compounds called galactans (carbohydrates that yield galactose upon hydrolysis). Galactose is also called cerebose or brain sugar because it is found in cerebrosides and gangliosides in brain and nerve tissue. Galactose is an exception to the general rule that it is the D form of sugars that occurs in nature; the L form of galactose is found in a number of polysaccharides of plant origin. Cataracts have been produced experimentally by feeding high levels of galactose to animals. Galactose is a component of lactose, milk sugar.

14.6 OLIGOSACCHARIDES

Disaccharides

Disaccharides were classified as carbohydrates (sugars) that are hydrolyzed to two molecules of either the same simple sugars or two different ones. The disaccharides are classified by their component sugars and by the carbons through which the monosaccharides are attached. For example, the first carbon of one glucose molecule may be connected to the first, fourth, or sixth carbon atom of another glucose molecule to produce three different disaccharides, trehalose, maltose, and isomaltose, respectively. Several disaccharides and their component monosaccharides and linkage are shown in Table 14.1. The linkage between the respective carbons of the sugar units is indicated, for example, as α-sugar-(1—4) to show that the first sugar is in the α configuration and its first carbon is linked to the fourth carbon of the second sugar component. The formulas for six of these sugars are shown in Fig. 14.6.

TABLE 14.1 COMPOSITION OF SOME DISACCHARIDES

Name	Composition
Maltose	α-D-Glucose-(1—4)-α-D-glucose
Cellobiose	β-D-Glucose-(1—4)-β-D-glucose
Trehalose	α-D-Glucose-(1—1)-α-D-glucose
Isomaltose	α-D-Glucose-(1—6)-α-D-glucose
Gentiobiose	β-D-Glucose-(1—6)-β-D-glucose
Sucrose	α-D-Glucose-(1—2)-β-D-fructose
Lactose	β-D-Galactose-(1—4)-α-D-glucose

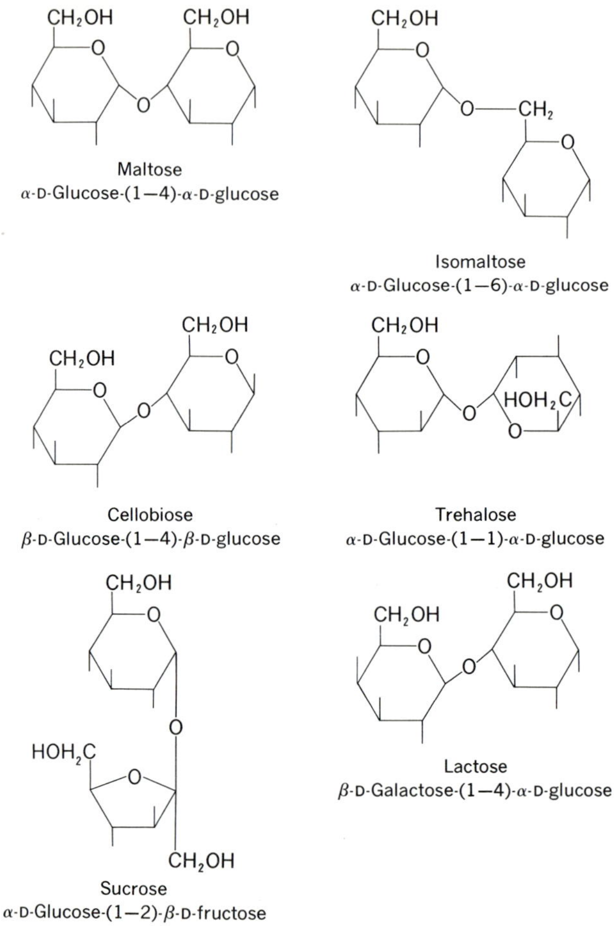

FIGURE 14.6 Structure of some disaccharides.

Maltose does not occur free in nature. It is an intermediate in the enzymatic hydrolysis of starch to glucose. During the sprouting of barley, starch is hydrolyzed to maltose; hence, the name maltose, or malt sugar. Maltose has two units of α-D-glucose connected by an acetal link from the hemiacetal carbon (C-1) of one glucose unit to the C-4 carbon of the second glucose unit α-(1—4). Thus the first glucose unit is in the acetal structure, and the second glucose unit is in the hemiacetal structure. The formula for maltose is shown in Fig. 14.6.

Cellobiose, like maltose, is composed of two glucose units connected in a 1—4 link. The only difference between these two sugars is the configuration about the acetal carbon. In cellobiose glucose is in the β form. Cellobiose can be considered a partial hydrolytic product of cellulose in the same manner that maltose is of starch.

Isomaltose is another disaccharide composed of two glucose units. Glucose in isomaltose is in the α form, and the two glucose molecules are connected by a 1—6 link. Isomaltose does not occur in the free state but is a partial hydrolytic product of glycogen.

Sucrose occurs free in the plant kingdom, especially in sugarcane and sugar beets, which contain up to 20% sucrose and constitute a commercial source of this sugar. As a result, sucrose is commonly known as cane sugar or beet sugar. Sucrose is composed of α-D-glucose and β-D-fructose joined by a glycosidic bond between the hemiacetal carbon of the glucose moiety and the hemiketal carbon of the fructose part. Thus the glucose is in the form of an acetal, and the fructose in the form of a ketal. Consequently there is no possibility of establishing an equilibrium between the hemiacetal or hemiketal and the aldehyde or ketone group, respectively. As a result sucrose does not reduce Fehling's solution; nor does it mutarotate.

Lactose, or milk sugar, occurs only in milk; it is not very sweet and is not fermented by yeast. It yields 1 mole each of galactose and glucose. The galactose is in the β-D configuration and is bonded to α-D-glucose by a 1—4 link, thus leaving the glucose moiety in the hemiacetal structure.

14.7 CHEMICAL PROPERTIES OF DISACCHARIDES

Oxidation

Disaccharides can be divided into *reducing disaccharides,* so called because they reduce Fehling's solution, and *nonreducing disaccharides,* which do not. Only disaccharides that contain a hemiacetal or hemiketal structure are easily oxidized and reduce Fehling's solution. Consider the formula for maltose (the two glucose units are labeled A and B)

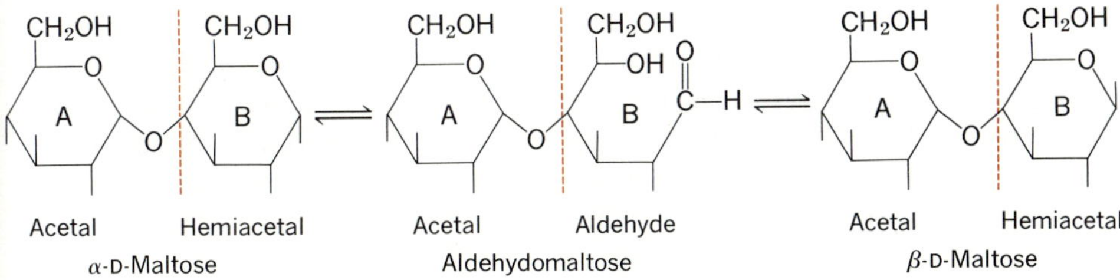

The acetal link between the two glucose units is relatively stable compared with the hemiacetal group in the B part of the molecule. The hemiacetal in the B part is in equilibrium with the free aldehyde form, aldehydomaltose. Opening the ring destroys the asymmetry of the hemiacetal carbon; therefore, when the ring recloses, α- and β-maltose are formed. Thus both α-D-maltose and β-D-maltose are in equilibrium with the free aldehyde form, and consequently both are easily oxidized by Fehling's reagent.

Is trehalose a reducing sugar (see formula, Fig. 14.6)? In trehalose the first carbon of each glucose unit is connected in an acetal link:

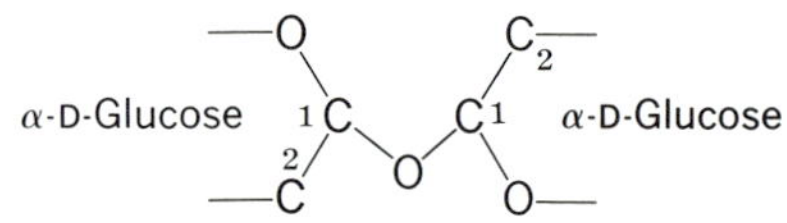

Acetal link

Remember that the acetal group is quite stable and will *not* be in equilibrium with the free aldehyde form of the sugars. Therefore, trehalose is not a reducing sugar.

Another nonreducing sugar listed in Table 14.1 is sucrose, in which the hemiacetal carbon of the α-D-glucose portion is linked to the hemiketal carbon of the β-D-fructose unit.

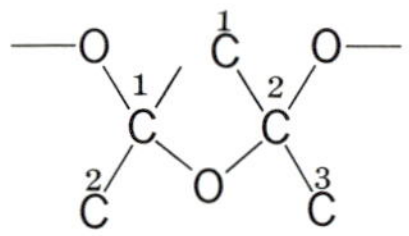

α-D-Glucose β-D-Fructose
Acetal Ketal

Thus glucose exists as an acetal and fructose as a ketal in sucrose. Since neither sugar component of sucrose can be in equilibrium with its corresponding noncyclic structure, sucrose is a nonreducing sugar.

Disaccharides are less easily oxidized than monosaccharides. As the number of sugar units in an oligosaccharide increases, the reducing power decreases. This fact serves as a basis for a chemical test to differentiate between the monosaccharides and the oligo- and polysaccharides. Barfoed's reagent oxidizes monosaccharides but not oligosaccharides.

Mutarotation

The prerequisite for mutarotation is the presence of a hemiacetal or hemiketal group. Which of the disaccharides listed in Table 14.1 will mutarotate? All except trehalose and sucrose.

The α and β forms of maltose are in equilibrium with each other. $[\alpha]$ is $+168°$ for α-maltose and $+118°$ for β-maltose; at equilibrium the value of $[\alpha]$ is $136°$.

$$\underset{[\alpha] = +168°}{\alpha\text{-D-Maltose}} \xrightleftharpoons{[\alpha] = +136° \text{ at equilibrium}} \underset{[\alpha] = +118°}{\beta\text{-D-Maltose}}$$

14.8 POLYSACCHARIDES

Polysaccharides are composed of repeating units of a monosaccharide and are of quite high molecular weight. Some contain only one kind of simple sugar, whereas others also contain such units as sugar acids, amino sugars, sugar sulfate, or a mixture of monosaccharides. Some are so complex that their structure has not been completely elucidated.

Polysaccharides are classified according to the component monosaccharide; the ending *-san* is reserved for polysaccharides. For example, a pentosan is a polysaccharide composed of pentoses; a specific pentosan is xylan, a polymer of xylose. (Polymers were discussed in Sec. 8.17.) Araban is a polymer of arabinose. Polymers of hexoses are called hexosans, the common ones being starch, glycogen, cellulose, and inulin. The first three are glucosans, but inulin is a fructosan. We shall be concerned mainly with the glucosans starch and glycogen. Although other polysaccharides such as those which contain hexosamine and uronic acid (sugar acid) are important from the biological standpoint, their discussion must be left to more advanced texts.

Starch

There are two forms of starch, amylose and amylopectin. *Amylose* consists of unbranched chains of α-D-glucose joined together in a 1—4 link, as shown in Fig. 14.7. The configuration of amylose is a helix with an open core down the center. The molecular weight of polysaccharides varies within a wide range. For example, the molecular weight of amylose ranges between 40,000 and 400,000 with a minimum of about 300 glucose units per amylose molecule.

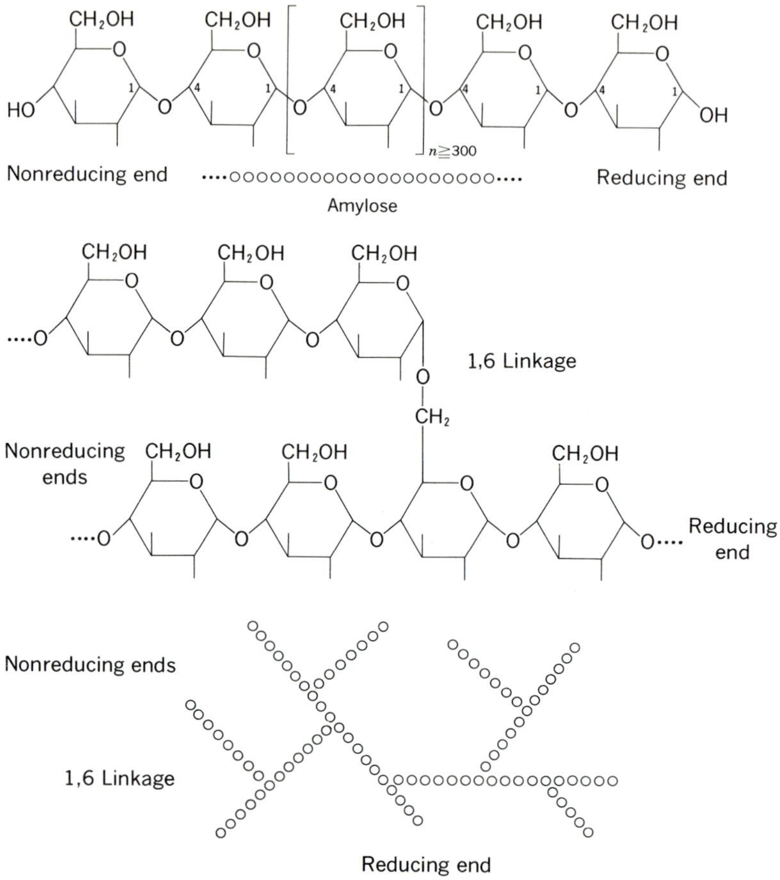

FIGURE 14.7 Schematic representation of amylose and amylopectin structures.

Amylose is insoluble in water but forms a solution of suspended particles. Soluble starch is amylose that has been partially hydrolyzed with HCl so that the starch units are of sufficiently low formula weight to be soluble.

Hydrolytic products of amylose are classified according to the color they give with a solution of I_2. Iodine complexes with amylose to give a deep blue color. The iodine fills up the hollow core of the amylose helix, which results in a colored complex; there is no chemical reaction between the starch and the iodine. As amylose is hydrolyzed, the chain becomes shorter and shorter and the color with iodine changes from blue to red to colorless, with purple as an intermediate color. The color depends upon the length of the amylose chain: the longer the chain the bluer the color. Purple results from a mixture of the blue iodine complex with the red complex. By following the change in color in the presence of I_2, the course of the hydrolysis of starch can be followed. The iodine color is more intense in cold solution. If the starch-iodine complex is heated, the color disappears and reappears when the solution is cooled. The partial hydrolytic products of starch are called dextrins (Table 14.2). When the iodine color disappears during hydrolysis, it is evidence that the starch has been hydrolyzed to achroodextrins or further.

Amylopectin contains chains of α-(1—4) glucose units, as in amylose, joined together by an α-(1—6) linkage from the aldehyde terminal end of one chain to the C-6 carbon of another chain (see Fig. 14.7). The molecular weight of amylopectin may be greater than amylose, up to about 1 million. Each chain in amylopectin contains on the average about 25 glucose units. Amylopectin gives only a faint reddish color with iodine.

Glycogen

Glycogen is a polysaccharide with a structure similar to that of amylopectin. Glycogen is the storage carbohydrate of animals as starch is the storage carbohydrate for plants. The structure for amylopectin can also be used for glycogen, the only difference being that glycogen has shorter chains and is therefore more branched. The chains contain 10 to 20 glucose units, averaging about 12 units in length. The molecular weight of glycogen is greater than 3 million. Glycogen gives a faint reddish color with iodine. Sometimes it is necessary to cool the solution in ice water to intensify the color.

TABLE 14.2 COLOR OF IODINE COMPLEX WITH CARBOHYDRATES

Color	Carbohydrate
Blue	Starch
Purple	Amylodextrin
Red	Erythrodextrin
Colorless	Achroodextrin
Faint red	Glycogen

Cellulose

This carbohydrate is another polysaccharide composed only of glucose units. It resembles amylose in having a linear structure but differs in that the glucose is in the β configuration (see Fig. 14.8). The molecular weight of cellulose is 200,000 to 2 million. Cellulose is sometimes referred to as a structural polysaccharide, compared with starch and glycogen, which are reserve polysaccharides, a reserve source of energy. Some difference in properties between the α- and β-linked glucose units are easy to visualize. Compare the properties of wood, paper, or cotton with those of starch. The β-acetal bond in cellulose is much more stable toward hydrolysis than the α-acetal bond in starch. Cellulose is not hydrolyzed by boiling for 30 min in 1.25% sulfuric acid, but starch is easily hydrolyzed under these conditions.

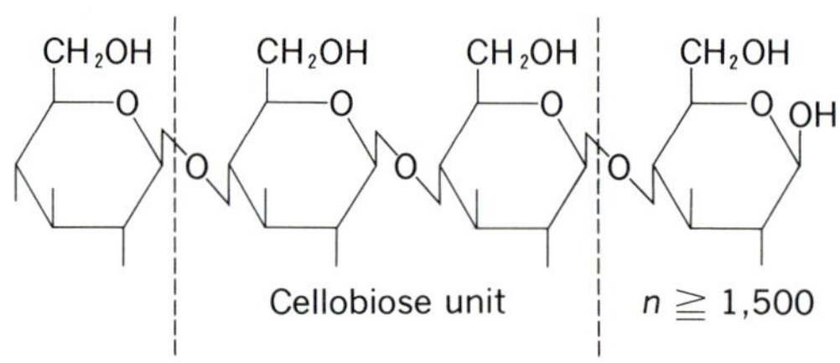

Cellulose

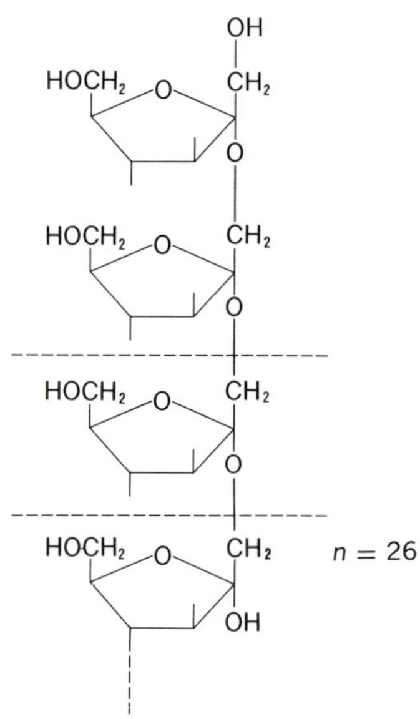

Inulin

FIGURE 14.8 The structure of cellulose and inulin.

Inulin

Inulin is the reserve polysaccharide of a number of plants such as the Jerusalem artichoke and dahlia tubers. Inulin is a fructosan in which the fructose units are joined together by a 2—6 link with fructose in the five-membered ring structure. Inulin gives no color with iodine. The structure of inulin is shown in Fig. 14.8.

QUESTIONS

1. Define carbohydrates.
2. What suffix is used to indicate that a carbohydrate is a sugar?
3. What is the significance of the terms "aldose" and "ketose"?
4. What is the significance of the letters D and L in carbohydrate chemistry?
5. Write the open-chain formula for an aldohexose. Label each functional group.
6. Write the formula for α-D-glucopyranose.
7. Explain why there are 16 aldehydoglucoses and 32 cyclic structures for glucose.
8. Which OH groups of glucose are never found in the hemiacetal structure? Which OH group is found most commonly in the acetal structure of glucose?
9. Explain mutarotation.
10. Write an equation for the formation of methyl-α-D-glucopyranoside. Will this glycoside mutarotate? Explain.
11. Are glycosides more stable under alkaline or acidic conditions?
12. In the parentheses preceding each name or statement listed in the left-hand column place the number of the appropriate item given in the right-hand column.

( ) Is hydrolyzed to two simple sugars	1. Simple sugar
( ) C-2 is involved in the cyclic structure	2. Disaccharide
( ) Fructose	3. Aldose
( ) Maltose	4. Ketose
( ) Amylose	5. None of the above
( ) Cellobiose	

13. Write the formula for 3-phosphoglyceraldehyde (glyceraldehyde-3-phosphate), fructose-6-phosphate, and fructose-1,6-diphosphate.
14. Indicate whether each of the following compounds would have a (+) or a (−) specific rotation [α].

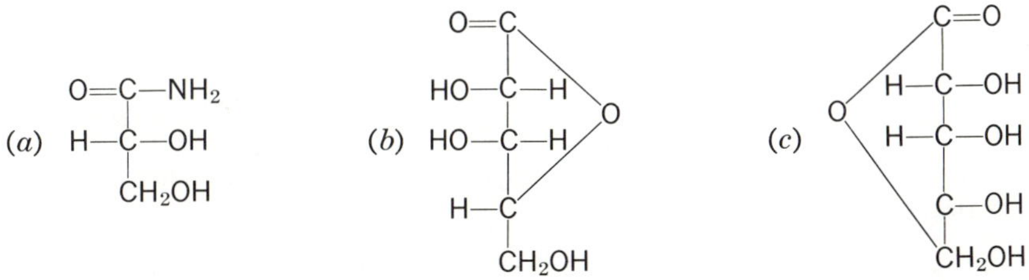

15. Write an equation for the following conversions:

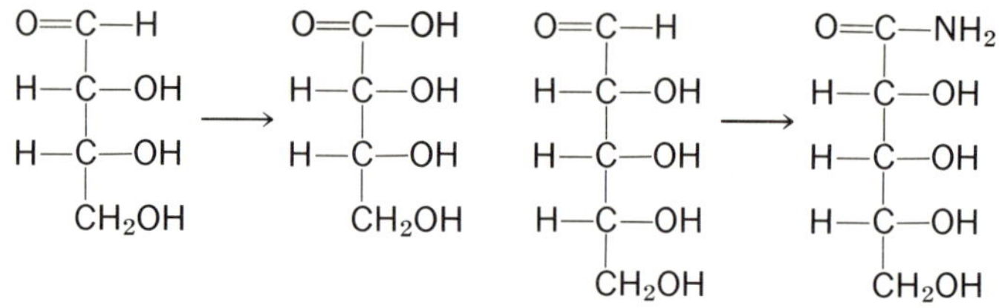

16. Why do glucose and fructose give the same osazone?

17. Why is maltose a reducing disaccharide?

18. The following tests are performed on an unknown sugar solution, the results being indicated as being positive or negative; e.g., a positive Fehling's test indicates that the sugar reduces the reagent. What is the unknown sugar?

Test	*Results*	*Conclusions*
Fehling's	Positive	______
Barfoed's	Negative	______
HNO_3 (conc.)	Crystals	______
I_2	Negative	______
Phenylhydrazine	Positive	______

19. How would you differentiate between a solution of maltose and a solution of sucrose?

20. Periodate oxidation can be used to determine the ring structure of simple sugar glycosides. Below are given all possible ring closures for glucose. Treat each compound with periodate and fill in the accompanying table.

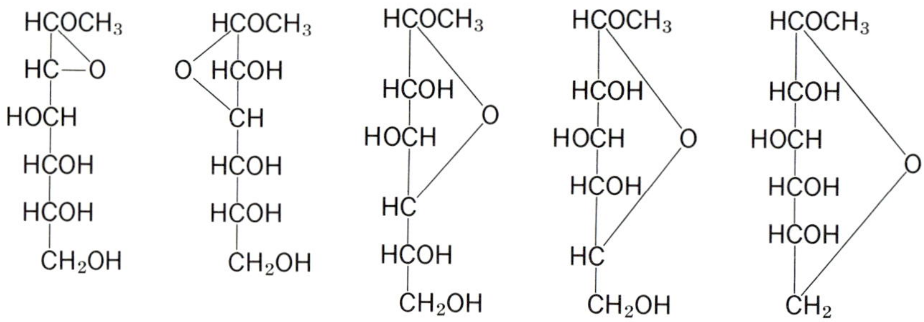

(I) (II) (III) (IV) (V)

Moles	*I*	*II*	*III*	*IV*	*V*
HIO_4 used	______	______	______	______	______
HCO_2H formed	______	______	______	______	______
HCHO formed	______	______	______	______	______

Could glucose rather than the methyl glucoside have been used to determine ring closure as above?

chapter 15.... lipids

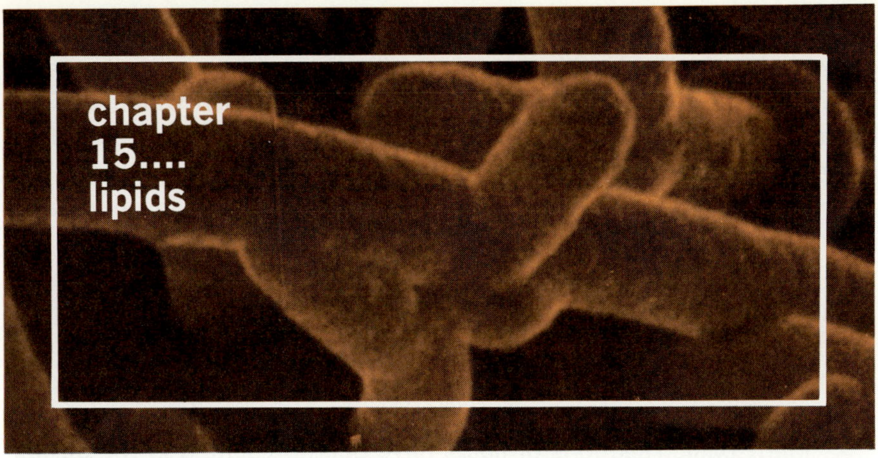

Because lipids are compounds of widely different composition and structure, as with carbohydrates, no concise definition can be made wholly on the basis of their chemical nature; some physical properties must be included in the definition. Lipids may be defined according to whether they are (1) esters or potential esters; (2) soluble in fat solvents, e.g., acetone, ether, and chloroform; and (3) biologically active.

A few words of explanation are needed to clarify this definition. Esters have been defined as products formed by reacting alcohols with acids. A potential ester is just a way of saying that compounds that can form esters and meet the other criteria are also classified as lipids, e.g., the fatty acids and some alcohols found in lipids. Biologically active means that lipids are made or utilized by living organisms; thus man-made compounds that do not occur naturally are excluded. Lipids are soluble in nonpolar solvents. This excludes such compounds as carbohydrates, amino acids, and minerals. Proteins are excluded by two requirements of the definition, esters or potential esters and solubility in nonpolar solvents.

Fats (lipids) constitute an appreciable amount of our diet by weight. Some of the fatty acids in fat are needed in the diet because the body cannot synthesize them. Fats carry vitamins A, D, E, and K; in fact, these vitamins are lipoid in nature.

The body stores fat as a reserve source of energy. Although glycogen is stored for this purpose too, there is a lower limit for glycogen storage than there is for fat. Excess carbohydrate is converted into fat for possible future needs. Actually it is more economical to store fat because its oxidation yields more energy than the oxidation of the same weight of carbohydrate, as can be seen when the formulas of fatty acids are compared with those of carbohydrates with the same number of carbon atoms. Consider a six-carbon acid which has the formula $C_6H_{12}O_2$ and a six-carbon sugar, $C_6H_{12}O_6$. Which is already in the more highly oxidized state? It is the carbohydrate.

Lipids are important biologically in more respects than just as a source of energy and a carrier of certain vitamins. They constitute a part of the cell wall and play a role in permeability of cellular membranes. They are also associated with proteins to form a class of compounds called lipoproteins and with carbohydrates to form glycolipids. Furthermore, some lipids are hormones which regulate enzymatic reactions. Overall, lipids play a very diversified role in metabolic processes.

Many basic reactions of lipids have been presented in the discussions of acids, alcohols, and alkenes; it remains to reorganize them as they apply to the various compounds classified as lipids. By way of preparation for this discussion it will be well to answer the following questions. What is the mechanism for the formation of an ester? What products are formed upon the saponification of an ester? Compare the physical properties of an acid with a high iodine number with an acid of low iodine number. Would you expect an acid with a high iodine number to be oxidized relatively easily?

15.1 CLASSIFICATION

Lipids are classified by their components into the following categories:

- Simple lipids
    - Triglycerides (neutral fats)
        - Fats
        - Vegetable oils
- Compound lipids
    - Phospholipids, phosphatides
        - Lecithins
        - Cephalins
        - Sphingosides
    - Glycolipids
- Miscellaneous
    - Sterols
    - Carotenoids
    - Fat-soluble vitamins

15.2 SIMPLE LIPIDS

Simple lipids are esters of fatty acids and alcohols. The alcohol is either glycerol or a high-molecular-weight monatomic (monohydroxy) alcohol. This class is further divided into two subclasses, the triglycerides (neutral fats) and the waxes.

Triglycerides

Triglycerides are esters of glycerol and fatty acids. All three alcohol groups are esterified, hence the prefix *tri-*. A general equation for the formation of a triglyceride by the esterification of glycerol is

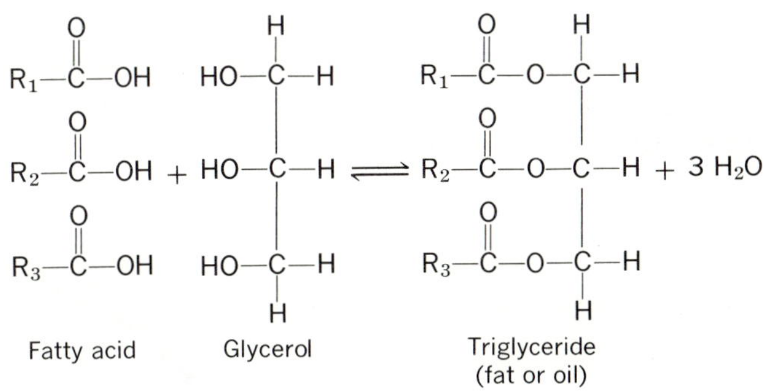

Natural triglycerides do not contain three molecules of the same fatty acid; there is always a mixture of acids. One of them is usually unsaturated. Thus, it is impossible to write a general molecular formula for the class triglycerides and it follows that it is not possible to give an exact molecular weight of natural triglycerides. Hence, we speak only of the average molecular weight or average composition of these compounds.

There is a wide variation of fatty acid composition in fats and oils from a given source. For example, the composition is related (within limits) to the diet of the animal and to the climatic conditions under which plants grow. This can be illustrated with the production of soft pork by feeding excessive amounts of vegetable oils to hogs. The fat in soft pork has a lower melting point than the fat deposited by animals on a normal diet because there are more unsaturated fatty acids in the vegetable oil fed and as a result the fat deposited has a somewhat higher degree of unsaturation than normal. Variations in the composition of fats and oils can be illustrated by comparing the range of their iodine number, which is indicative of the amount of unsaturated fatty acids in the triglyceride; e.g., butter fat, 26 to 38; lard, 50 to 70; coconut oil, 5 to 10; linseed oil, 170 to 200.

Fats and oils

Triglycerides, or neutral lipids, have been subdivided into two groups, fats and oils (vegetable). The difference between these groups is only one of melting points. Fats are solid and oils are liquid at room temperature. (Remember that oil refers to lipid and not to mineral oil. What is mineral oil?)

Two factors affect the melting point of fats and oils: (1) the molecular weight, or length of the carbon chain, of the fatty acids and (2) the degree of unsaturation of the acids. Beginning with butyric acid, as the molecular weight increases within any one series of acids, the melting point also increases. An odd situation is observed when the melting points of acids with even- and odd-numbered carbons are

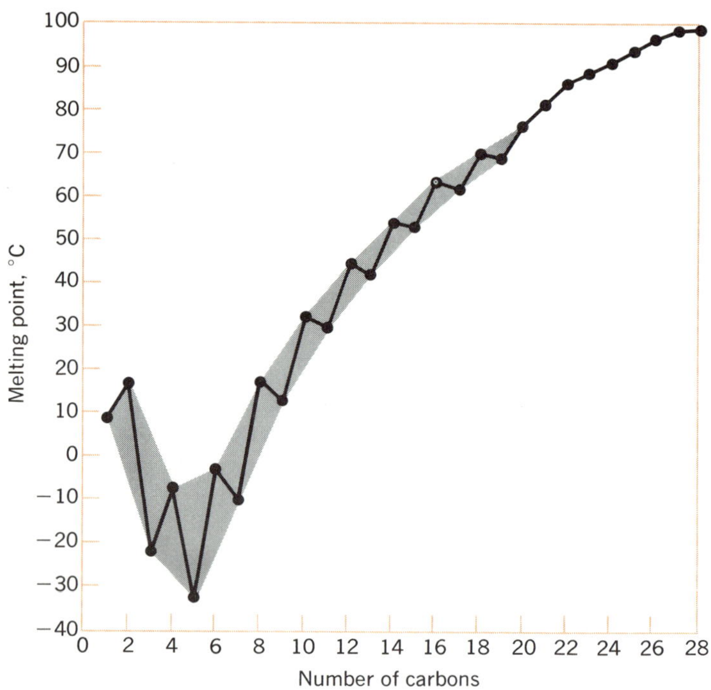

FIGURE 15.1 Melting point of alkanoic acids.

compared (see Fig. 15.1). The acids with an even number of carbons (up to C_{20}) melt at a higher temperature than acids with an odd number of carbons. There is also a striking difference between the melting points of saturated and unsaturated acids containing the same number of carbon atoms. For example, stearic acid melts at 70°C; the insertion of three double bonds to form linolenic acid decreases the melting point to about −12°C. Esters of the fatty acids follow the same general pattern with respect to melting points. The melting points of esters vary directly with the molecular weight of the acid and inversely with the degree of unsaturation of the acid. Thus, the difference between fats and oils is due merely to their composition with respect to the fatty acids.

The percentage composition of some triglycerides is given in Table 15.1. Butyric acid occurs only in butter. Caproic, caprylic, and capric acids seldom occur in fats and oils and then only in a relatively small amount. Lauric acid is not a common component of triglycerides; however, with the exception of butter, when it does occur, it is present in considerable amounts. For example, in babassu, coconut, and palm-kernel oil lauric acid constitutes from 44 to 47% of the acids. Coconut oil, with an iodine number of 10, is a good example of a triglyceride that is an oil because it contains low-molecular-weight acids rather than unsaturated acids. It contains only about 7.9% of acids with one double bond and 63% of lauric and myristic acids. For all practical purposes, butterfat and palm-kernel oil have the same iodine number. Why is butter a solid and palm-kernel oil a liquid at room temperature? An important difference between castor oil and the other triglycerides listed in Table 15.1 is that it is composed of 87% ricinoleic acid, the 12-hydroxy derivative of oleic acid, which is not readily absorbed from the intestinal tract.

Nomenclature

Glycerides are named to indicate the number and name of the acids and the carbon of glycerol to which they are attached. The glycerol carbons are labeled α, β, and α'. In naming the acid components, the *-ic* of the acid name is changed to *-o* except for the last acid in the name, whose *-ic* is changed to *-in*. For example, α-oleo-β-stearopalmitin would be a triglyceride with oleic acid, stearic acid, and palmitic acid on the α, β, and α' carbons of glycerol, respectively.

TABLE 15.1 TYPICAL COMPOSITION OF SOME TRIGLYCERIDES AND WAXES[a]

Triglyceride	Constituent fatty acid, grams per 100 g total fatty acids												
	Butyric	Caproic	Caprylic	Capric	Lauric	Myristic	Palmitic	Stearic	Palmitoleic	Oleic	Linoleic	Linolenic	I_2 no.
Butterfat	3.6	2.3	0.5	2.0	2.5	11.1	29.0	9.2	4.6	26.7	3.6		36
Depot fat						2.7	24.0	8.4	5	46.9	10.2		68
Tallow, beef						6.3	27.4	14.1		49.6	2.5		50
Mutton						4.6	24.6	30.5		36.0	4.3		40
Cod-liver oil						5.8	8.4	0.6	20.0	29.1	29.1		165
Menhaden oil						5.9	16.3	0.6	15.5			29.6	170
Babassu oil		0.2	4.8	6.6	44.1	15.4	8.5	2.7		16.1	1.4		16
Castor oil[b]					2.4	2.4	2.4	2.4		7.4	3.1		86
Coconut oil		0.8	5.4	8.4	45.4	18.0	10.5	2.3	0.4	7.5	trace		10
Corn oil						1.4	10.2	3.0	1.5	49.6	34.3		123
Cottonseed oil						1.4	23.4	1.1	2.0	22.9	47.8		106
Linseed oil							6.3	2.5		19.0	24.1	47.4	179
Palm-kernel oil			2.7	7.0	46.9	14.1	8.8	1.3		18.5	0.7		37

[a] From "Biology Data Book," compiled and edited by Philip L. Altman and Dorothy S. Dittmer, Federation of American Societies for Experimental Biology, Washington, D.C., 1964.

[b] 87% ricinoleic acid (12-hydroxy-*cis*-9-octadecenoic acid, 12-hydroxyoleic acid).

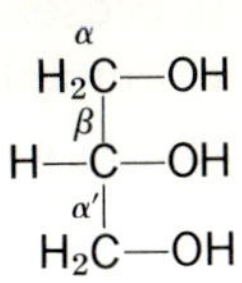

Glycerol

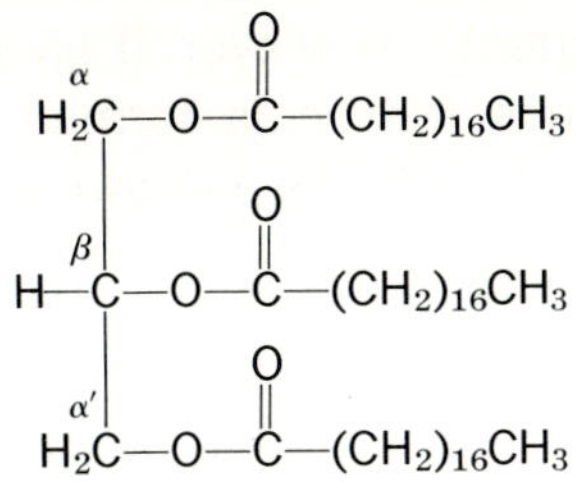

Tristearin

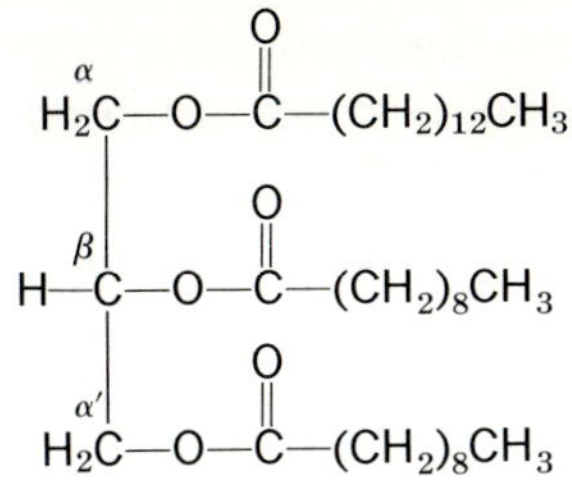

α-Palmitodicaprylin

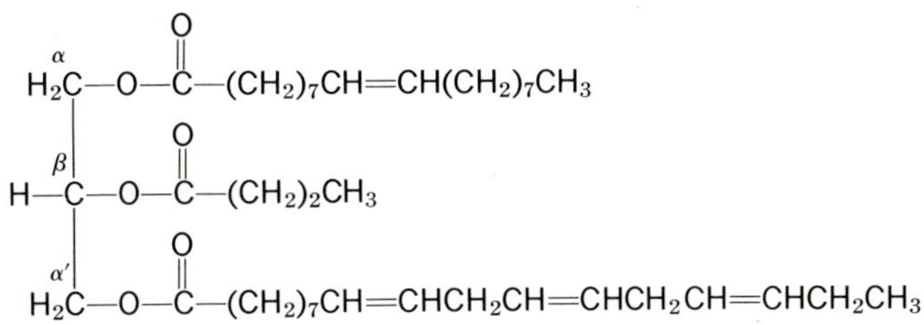

α-Oleo-β-butyrolinolenin

15.3 REACTIONS

Oxidation

The presence of one or more double bonds in the fatty acids of triglycerides makes them subject to oxidation. The greater the number of double bonds in an acid, the more susceptible it is to oxidation. Oxidation of the double bond is a rather complicated process which proceeds in several steps, one involving a free radical. The double bonds are eventually broken, and shorter-chain aldehydes, ketones, and acids result. If the acids are highly unsaturated, like linolenic acid, oxidation leads to the polymerization of the aldehydes and ketones to form a tough resinous film which is impervious to water. This is the basis for the protective nature of oil-base paints containing linseed oil.

Whereas the pure free fatty acids of C_{12} or higher are odorless and colorless, their oxidative products possess an off flavor and odor; the oxidized oil is said to be *rancid.*

Hydrolysis and saponification

Hydrolysis of triglycerides yields 3 moles of fatty acids and 1 of glycerol; this is the reverse of esterification (see Sec. 14.1). Esterification and hydrolysis are reversible reactions, and an equilibrium is established between the respective components. However, if alkali is present, the equilibrium is not attained since the acids formed are immediately neutralized. The salts of the fatty acids are known as soaps, and the reaction is saponification.

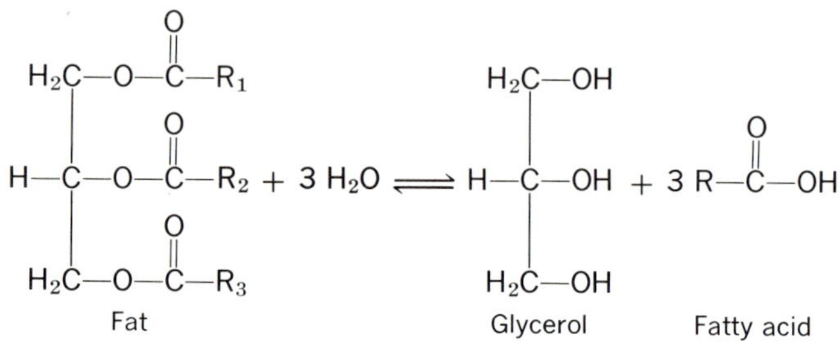

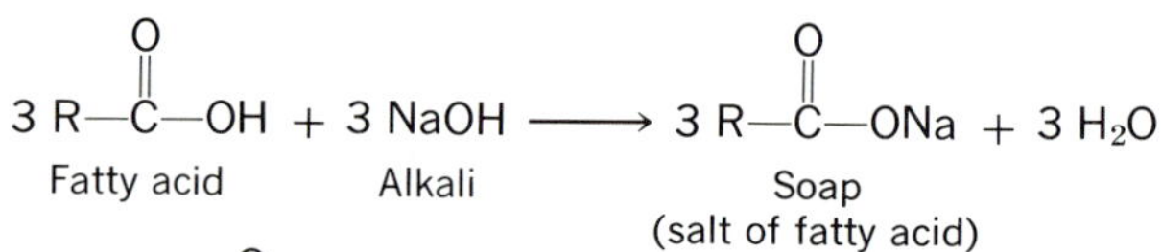

Net Reaction:

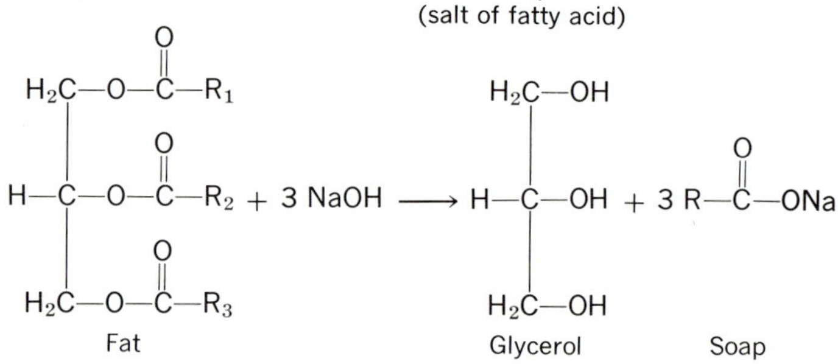

Soaps differ according to the base used, the sodium salt of the fatty acids being the most common. The potassium salt is soft soap; a mixture of the sodium and potassium soaps from linseed oil is green soap; and the heavy metal salts of calcium, iron, and cobalt are insoluble soaps. Calcium stearate is used in facial powders and greases, and iron, cobalt, and manganese soaps are used in certain paints to catalyze the oxidation reaction of the linseed oil.

The calcium, magnesium, and iron salts are formed when ordinary soap is used with hard water. Hard water contains the soluble salts of these metallic elements. Permanent hardness cannot be removed by heating and is caused by the chloride and sulfate salts of the heavy metals. Temporary hardness is caused by the bicarbonate salts of these metals and can be removed by boiling the water.

$$Ca(HCO_3)_2 \xrightarrow{\Delta} CaCO_3\downarrow + H_2O + CO_2$$

Most hard water is caused by soluble calcium salts. When a sodium soap is used in hard water, the calcium replaces the sodium and forms the insoluble calcium salt

$$2\ RCOONa + CaSO_4 \longrightarrow (RCOO)_2Ca + Na_2SO_4$$

Glycerol is dehydrated with reagents such as $KHSO_4$ and heat to acrolein, which has a characteristic acrid odor and is used as a test for the presence of fats. This is the odor that is noticed when fat is allowed to become overheated or burn. The reaction is

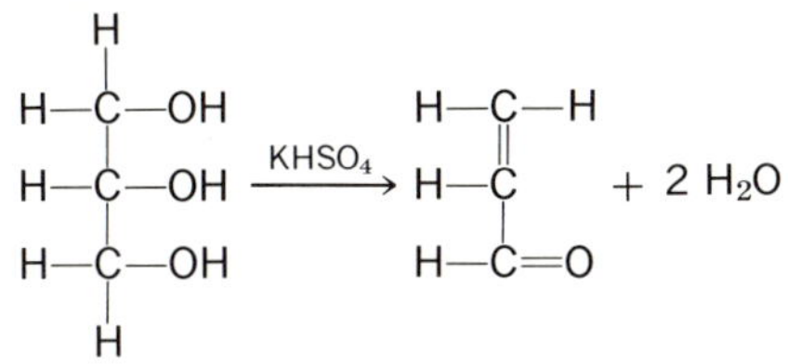

Glycerol Acrolein

Hydrolysis can also be brought about by enzymes (lipases) of the digestive juices or from bacteria. Hydrolytic rancidity is the off flavor in fats which have been partially hydrolyzed by enzymes of bacterial origin. The lower-molecular-weight acids, especially butyric, have a pronounced disagreeable odor and flavor.

The products of alkaline hydrolysis of lipids are saponification products. Any component of lipids that is soluble in the alkaline solution after saponification has occurred is *saponifiable* matter. Any material insoluble in the alkaline solution is *nonsaponifiable* matter. The saponification products are useful in the analysis of lipids.

Saponifiable matter from fats consists of glycerol and the salts of the fatty acids. Thus, all material derived from the saponification of triglycerides is saponifiable. Note that the solubility is determined *in the alkaline solution.* Of course, if after saponification the solution is acidified, the fatty acids will separate out as insoluble matter. Cholesterol (see page 383) is an example of nonsaponifiable matter which occurs to a large extent as an ester. After saponification the sterol is insoluble (nonsaponifiable matter), whereas the fatty acid to which it was esterified is saponifiable matter. Fat-soluble pigments and the fat-soluble vitamins are also nonsaponifiable matter. To obtain the nonsaponifiable fraction, the alkaline aqueous solution after saponification is extracted with a fat solvent such as ether. The nonsaponifiable material dissolves in the ether and can be recovered after evaporation of the solvent.

PROBLEM

Given a mixture of a triglyceride and cholesterol, how would you proceed to prove the presence of each component? Heat a sample of the unknown with $KHSO_4$, and smell it to determine whether acrolein is formed. If acrolein is liberated, glycerol must be present and can have come only from the triglyceride component of the mixture. This, of course, tells nothing about the other component present. Saponify the mixture and extract the alkaline solution with ether. After evaporation of the ether, a residue remains which is the cholesterol. Further evidence of the presence of the triglyceride is obtained by acidifying the ether-extracted alkaline solution to see whether a layer of fatty acids separates.

Hydrogenation

Hydrogenation of oils has become an important industry. The product of hydrogenation depends upon the conditions under which the reaction is carried out. When a lactone is reduced with NaHg amalgam the product is the corresponding aldehyde. Remember that lactones are intermolecular esters. Esters can be reduced in a similar manner. It is also possible to reduce esters to the corresponding alcohol of the fatty acid, and, of course, the alcohol of the ester is the other product. This requires rather vigorous conditions, e.g., temperatures of about 250°C, hydrogen under a pressure of 2,500 to 6,000 psi, and the presence of a catalyst. Less drastic conditions involve the use of sodium and ethyl alcohol as the reducing agent.

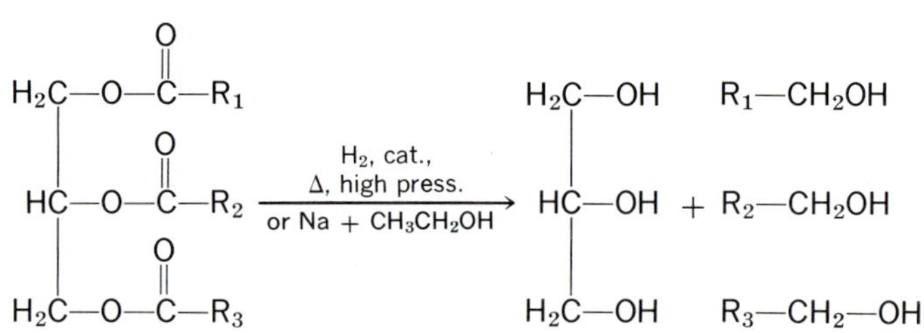

Triglycerids Glycerol Alcohols

For example, the reduction of coconut oil yields a high amount of lauryl alcohol, which can then be esterified with sulfuric acid to produce lauryl hydrogen sulfate.

$$\underset{\text{Lauryl alcohol}}{CH_3(CH_2)_{10}CH_2OH} \xrightarrow{H_2SO_4} \underset{\text{Lauryl hydrogen sulfate}}{CH_3(CH_2)_{10}CH_2OSO_3H} \xrightarrow{NaOH} \underset{\text{Sodium lauryl sulfate}}{CH_3(CH_2)_{10}CH_2OSO_3Na}$$

The sodium salt of this compound is a detergent. It has the characteristics of a good emulsifier, a hydrophilic and a hydrophobic part. The advantage of using a detergent is that since the calcium, magnesium, and iron salts are soluble in water, no scum forms in hard water. Also they are neutral in solution whereas soaps are alkaline. Other alkyl sulfates are used as detergents.

Hardening of oils is brought about by partial hydrogenation. Thus, only some of the double bonds of the unsaturated acids are hydrogenated. As hydrogenation proceeds, the melting point of the product increases. By controlling the extent of saturation, a product of the required consistency and melting point can be obtained. Crisco, Spry, Snodrift, and many other commercial shortenings are hydrogenated vegetable oils, as are margarine and butter substitutes.

15.4 WAXES

Waxes are naturally occurring esters of alcohols, other than glycerol, and fatty acids. The alcohols are of high molecular weight, some containing a double bond but generally not highly unsaturated. For example beeswax contains an ester of myricyl alcohol (a C_{30} saturated alcohol) and palmitic acid:

$$CH_3(CH_2)_{28}CH_2{-}O{-}\overset{\overset{\large O}{\|}}{C}{-}(CH_2)_{14}CH_3$$

Myricyl palmitate

Waxes are quite difficult to saponify. Their alcohols are insoluble in water and therefore are nonsaponifiable matter after saponification. Some natural waxes are:

Spermaceti, head oil of the sperm whale, cetyl palmitate, $CH_3(CH_2)_{14}COO(CH_2)_{15}CH_3$

Beeswax, myricyl palmitate, $CH_3(CH_2)_{14}COO(CH_2)_{29}CH_3$

Carnauba wax, a mixture of C_{28}, C_{30}, and C_{32} alcohols esterified with C_{32} and C_{34} acids

Cholesterol palmitate, the ester of cholesterol and palmitic acid, a wax found in blood plasma

Several ethers of glycerol have been found in waxes, one being batyl alcohol, which occurs in esters in wax from bone marrow.

$$\begin{array}{l} \quad H_2C{-}O{-}(CH_2)_{17}CH_3 \\ \qquad\; | \\ H{-}C{-}OH \\ \qquad\; | \\ \quad H_2C{-}OH \end{array}$$

Batyl alcohol

15.5 PHOSPHOLIPIDS

Phospholipids contain the phosphate group; they also contain another nonlipid component, choline, ethanolamine, amino acid, or inositol. Phospholipids are subdivided according to the alcohol which they contain.

Phosphatides

These compounds contain glycerol esterified with two molecules of fatty acids and one of phosphoric acid. The general formula of phosphatidic acid is

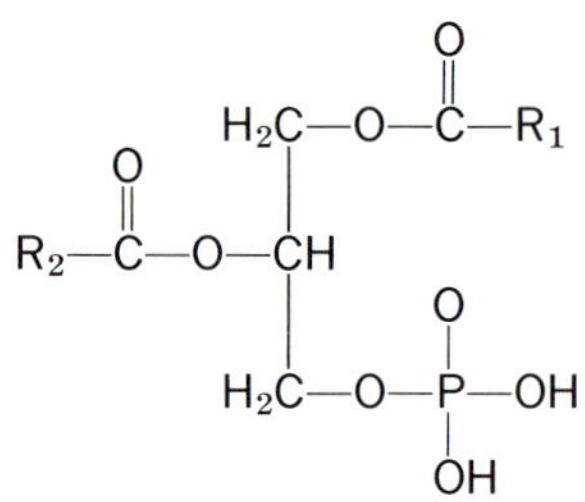

Phosphatidic acid

Phosphatidic acids contain an asymmetric carbon atom. The L configuration occurs in natural products. The fatty acid on the α carbon is usually unsaturated. Since the phosphate is esterified with the hydroxy group of an α carbon, the phosphatidic acid is an α-phosphatidic acid. The β-phosphatides have the phosphate group on the β carbon of the glycerol. They do not occur to any extent in natural phospholipids.

Lecithins

In lecithins choline is esterified with the phosphate of phosphatidic acid.

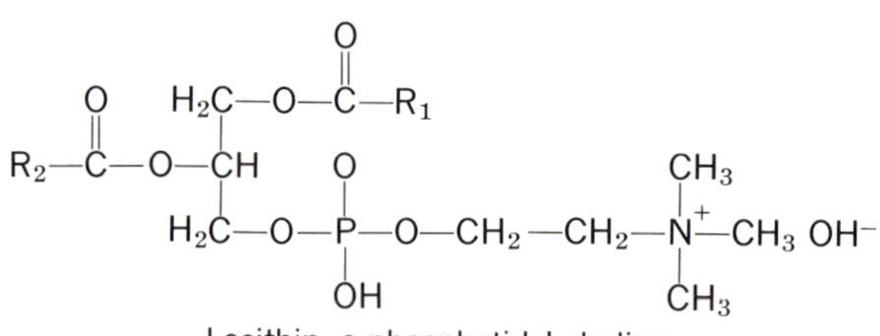

Lecithin, a phosphatidyl choline

There are many different lecithins due to the different combinations of fatty acids represented in the formula by R_1 and R_2. Choline is a quaternary nitrogen compound which exists as the base or salt. The formula of the base is

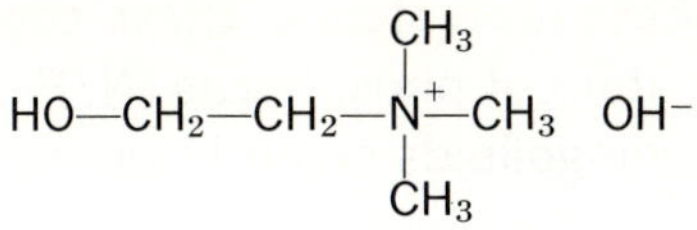

Choline hydroxide

Cephalins

Cephalins differ from lecithins only in the nitrogenous base. They contain ethanolamine or serine.

$$HO{-}CH_2{-}CH_2{-}NH_2$$

Ethanolamine

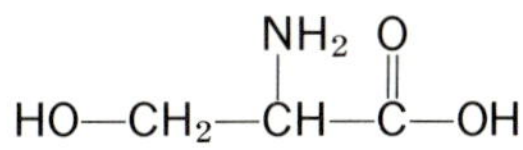

Serine

The formulas for two cephalins, phosphatidyl ethanolamine and phosphatidyl serine, are

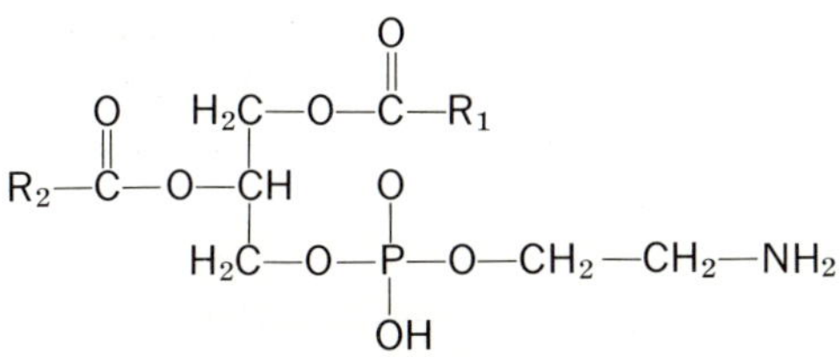

Phosphatidyl ethanolamine

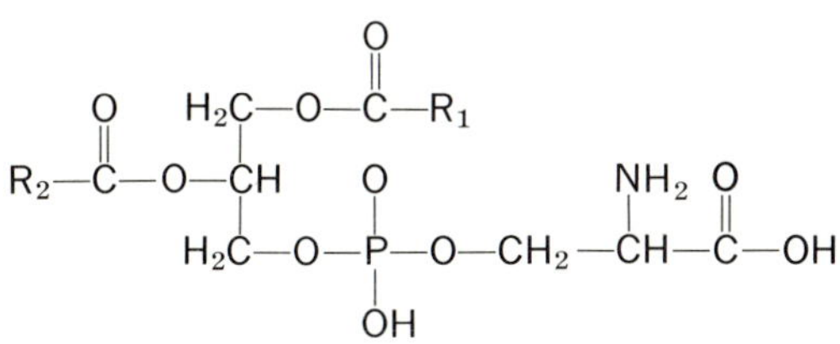

Phosphatidyl serine

Cephalins and lecithins are frequently symbolized as

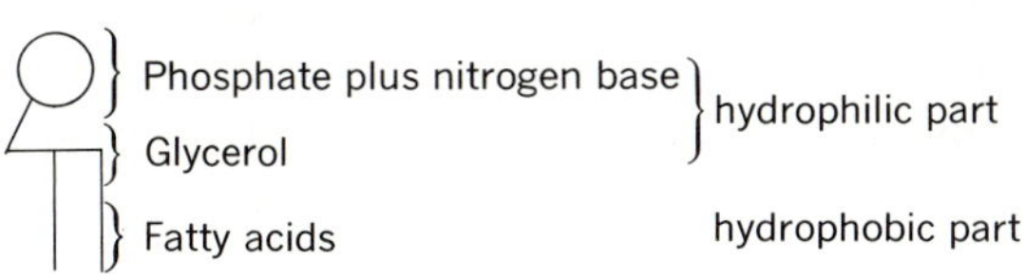

Sphingolipids

These compounds contain the high-molecular-weight nitrogenous alcohol sphingosine. They also contain a fatty acid and choline in addition to the phosphate unit. Hence, these compounds contain two atoms of nitrogen per atom of phosphorus (N/P = 2). In lecithins and cephalins N/P = 1. Sphingolipids occur in nerve tissue.

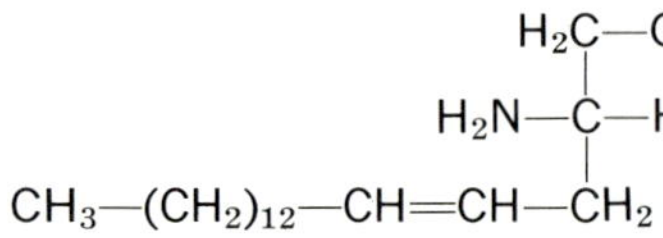

Sphingosine

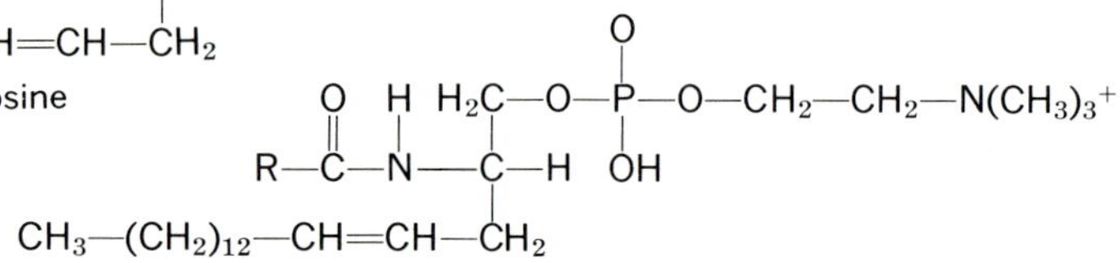

Sphingomyelin

15.6 GLYCOLIPIDS

Members of this class of compounds contain no phosphorus. Upon hydrolysis they yield a monosaccharide (usually galactose), fatty acid, and sphingosine or dihydroxysphingosine. Because glycolipids occur in nerve tissue, especially in the white matter of the brain, they are frequently referred to as *cerebrosides*.

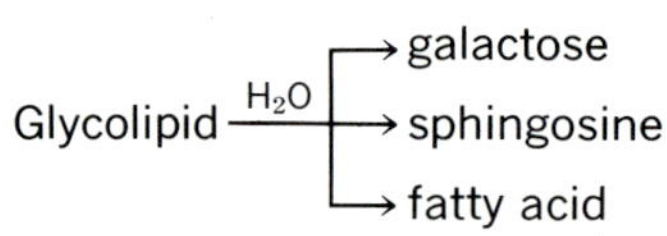

15.7 MISCELLANEOUS LIPIDS

A number of compounds that occur in plants and animals are classified as lipids but do not fall under any of the above categories.

Sterols

Sterols are solid cyclic compounds that contain three six-membered rings and one five-membered ring fused together. Steroids differ in the degree of saturation of the rings and in substitutions in the rings. Many have methyl groups at positions 18 and 19. Steroids are found among several classes of compounds, e.g., vitamins, hormones, and bile acids.

The formulas for several sterols are given in Table 15.2. Progesterone is a sex hormone, and vitamin D_3 is a steroid vitamin which is formed from 7-dehydrocholesterol by exposing it to ultraviolet light. This causes a rupture of ring B and the formation of a $=CH_2$ group at C-10. Unsaturation between C-7 and C-8 is a prerequisite for this

TABLE 15.2 FORMULAS OF SOME STEROIDS

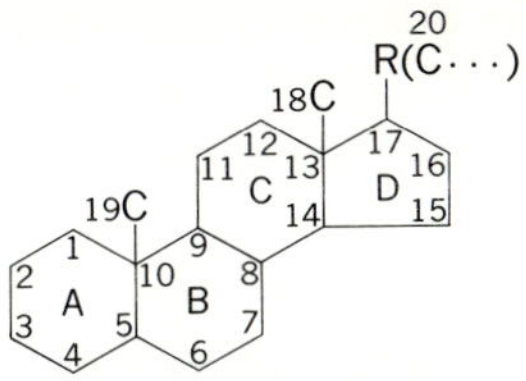

Steroid nucleus

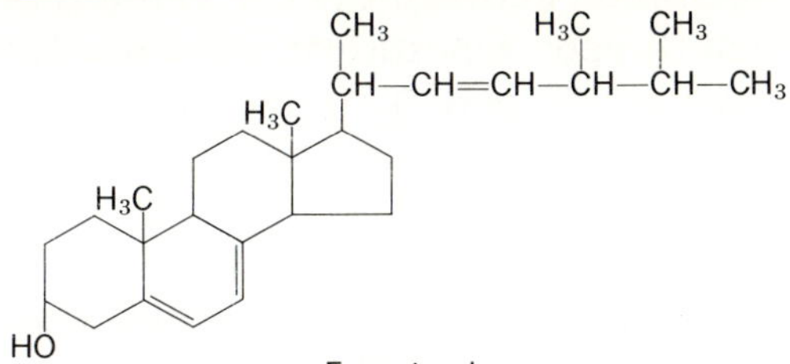

Ergosterol

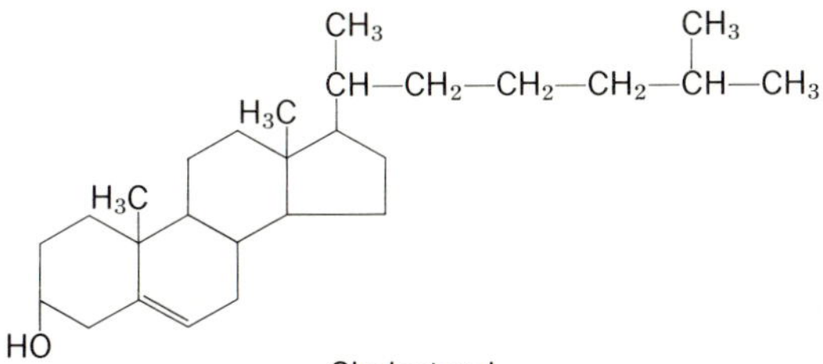

Cholesterol

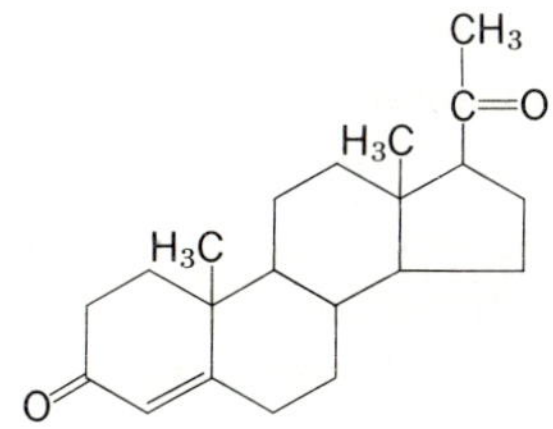

Progesterone

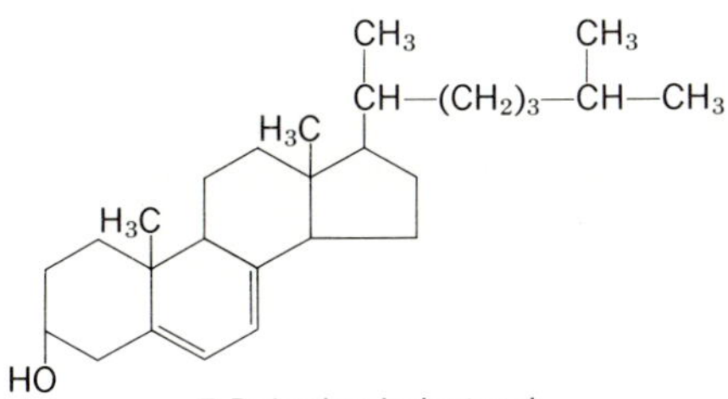

7-Dehydrocholesterol

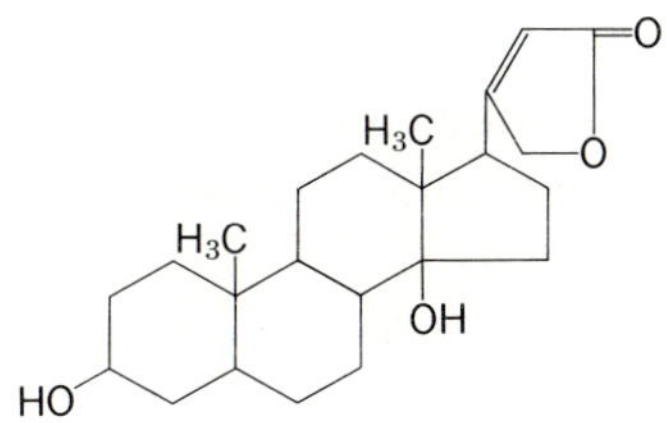

Digitoxigenin

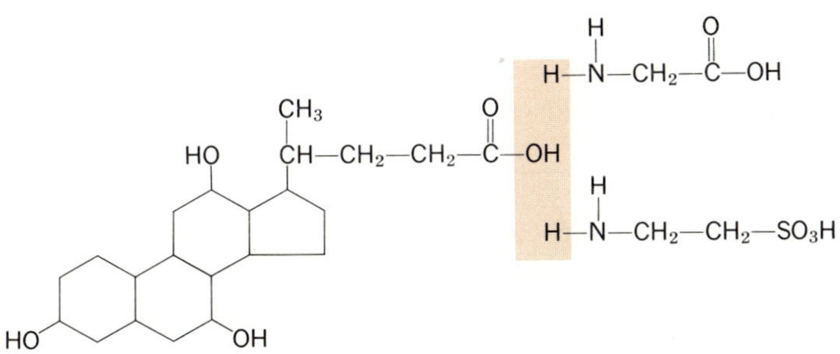

Glycine

Taurine

Cholic acid

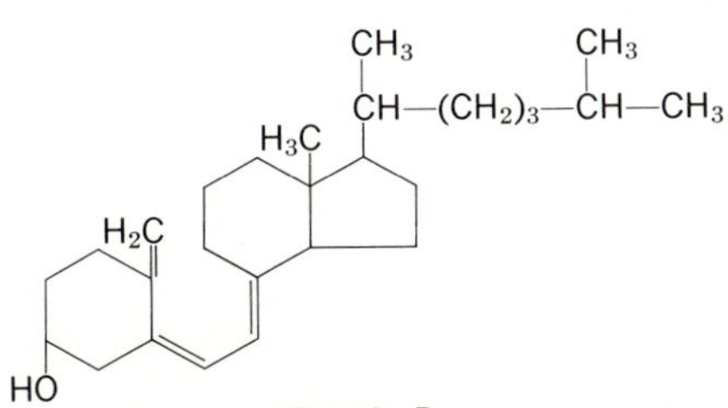

Vitamin D_3

transformation. Ergosterol, which, like 7-dehydrocholesterol, has unsaturation between the C-7 and C-8 of ring B, is converted into vitamin D_2 upon exposure to ultraviolet light. The only difference between vitamins D_2 and D_3 is the chain substituted on C-17 of the sterol nucleus. Ergosterol is of plant origin, whereas 7-dehydrocholesterol is of animal origin.

Cholesterol is a normal product of animal metabolism. Attention has recently been directed at this compound as a contributing cause of heart disease and hardening of the arteries. There is a considerable amount of cholesterol in eggs and meat fat. It is found in the blood at a level of up to about 200 mg per 100 ml of blood and is associated with lipoproteins. Cholesterol content is especially high in brain tissue, where it occurs to about 17% of the dry weight. Gallstones are chiefly cholesterol.

Examination of the formulas of sterols in Table 15.2 shows that there is an oxygen on the C-3, a characteristic of almost all naturally occurring sterols. The extent of ring unsaturation is a variable in this class on compounds. Some have no unsaturated rings, whereas others may have an aromatic ring A. For example, estrogens, female sex hormones, have an aromatic ring A, but androgens, male sex hormones, do not.

The bile acids, which occur in bile fluids, are not found free but as a peptide containing either glycine or taurine. The glycine derivative is called glycocholic acid, and the taurine derivative is taurocholic acid. The bile salts are good emulsifying agents. They are excreted into the intestines, where they aid in digestion and absorption by emulsifying the fats and oils.

Carotenoids

These compounds are related to β-carotene and vitamin A (see formulas page 202). Part of the green pigment in leaves is due to carotenes, as is the yellow color of carrots. The carotenoids are highly unsaturated compounds and therefore are easily oxidized. Several carotenoids are precursors of vitamin A. It is interesting to note that vitamin A itself does not occur in plants.

QUESTIONS

1. Define lipids.
2. (*a*) Write an equation for the complete esterification of glycerol with a fatty acid (R may be used).
   (*b*) To what class of lipids does this compound belong?

3. Why are fats insoluble in water?

4. (*a*) Name the products obtained upon saponification of a fat.
   (*b*) Are any of the products classified as nonsaponifiable? Explain.
   (*c*) What will be observed if the alkaline solution from (*a*) is acidified?

5. Write an equation for the formation of α-glycerol phosphate. Classify this compound.

6. What products are obtained upon the complete hydrolysis of each of the following classes of compounds?
   (*a*) Phosphatidic acid   (*b*) Lecithin   (*c*) Cephaline

7. Would you expect lecithin to be more soluble in water than fat? Explain.

8. Would you expect lecithin to be a good emulsifying agent? Explain.

9. Diagram the orientation of lecithin molecules in water (use the abbreviated structure for lecithin).

10. Diagram the basic ring structure of the steroid nucleus. Number the carbons.

11. (*a*) What is meant by 7-dehydrocholesterol?
   (*b*) What important transformation occurs upon irradiating 7-dehydrocholesterol with ultraviolet light? Does the same transformation occur upon irradiation of ergosterol?

12. What is the source (plant or animal) of cholesterol and ergosterol?

13. Label the functional groups in vitamin A.

14. (*a*) Does vitamin A occur in plants?
   (*b*) Name a precursor of vitamin A that occurs in plants.

15. Would you expect vitamin A to be associated with the fats or the proteins of animal tissue? Explain.

16. An unknown lipid was subjected to the following tests, the results of which tests are given. State the significance of each test and determine the classes of lipid to which the unknown could belong. What additional information would be needed to classify the unknown completely?

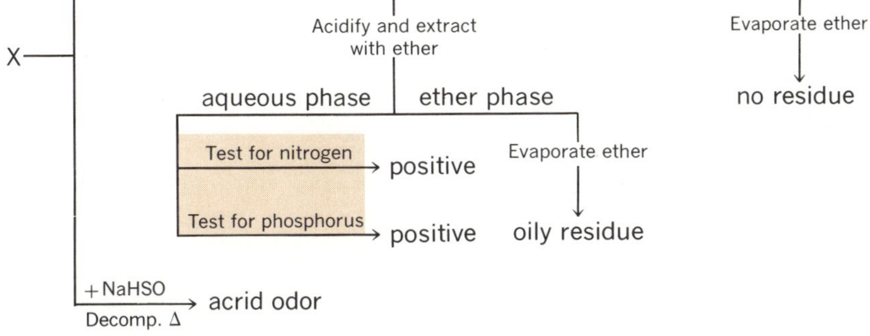

chapter 16.... amino acids and proteins

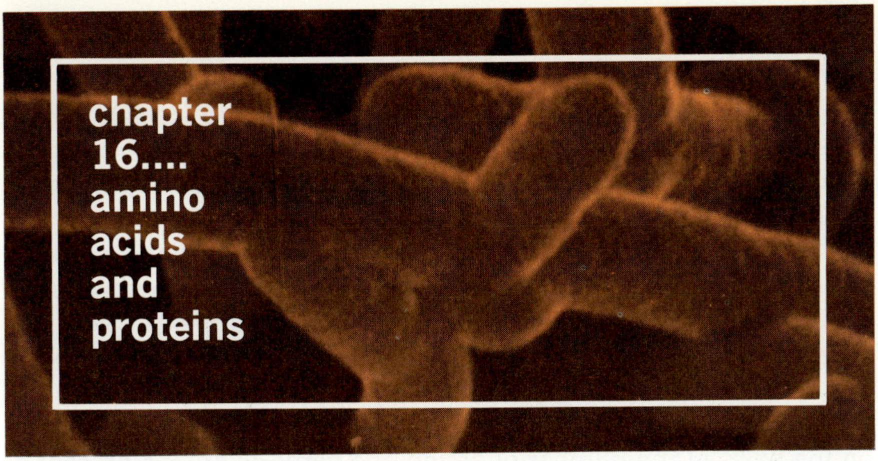

All living cells contain proteins as an essential component. Proteins are complex compounds that contain amino acids. The terms amino acids and protein have been used a number of times in previous discussions. For example, reference has been made to the fact that enzymes are proteins. This statement alone is in agreement with the derivation of the word protein, which comes from the Greek word meaning *preeminence* or *holding first place*.

Since the chemistry of proteins is the chemistry of the functional groups of the amino acids as they occur in the same molecule, the chemistry of the amino acids will be discussed first.

Some chemical properties of the amino acids have been discussed earlier under such related topics as reactions of acids, bases, and amines. In this chapter these properties will be correlated specifically with respect to amino acids and proteins. Hence, it is necessary to gather into a single chapter some properties of acids, bases, and amines for review. The answers to the following questions will provide this summation.

Define primary, secondary, and tertiary amine. Are all the valence electrons of nitrogen in primary, secondary, and tertiary amines in a bond formation? Are amines proton donors or proton acceptors? What is meant by a protonated amine? If an acid is half neutralized, how does one proceed to find the K_{ion} of the acid? Does a zwitterion react with acids as well as with bases? How does an amide differ from an amine? What is meant by a substituted amide? Which group of an amino acid does fluorodinitrobenzene (FDNB) react with? How can the Henderson-Hasselbalch equation be used to determine the pH of a buffer system of known concentration of salt and acid?

16.1 AMINO ACID CLASSIFICATION

The amino acid composition of proteins is of extreme biological interest. For example, the kind of amino acids in a protein determines its nutritive value, and the kind and sequence of amino acids determine the physiological role of a particular protein.

There are about 20 amino acids that commonly occur in proteins. A number of other amino acids are involved in metabolic pathways.

Table 16.1 lists the amino acids found in proteins. Since the amino acids are all α-amino acids, the side chain serves as a convenient handle for their classification. Amino acids can be divided into two main groups, aliphatic and cyclic. Further subdivisions are made on the basis of the number of amino and carboxyl groups present or the presence of an aromatic ring structure. A monoaminodicarboxylic acid would have an α-amino group and two carboxyl groups, as in glutamic acid. It would follow that a diaminomonocarboxylic acid would have a carboxyl group and an α-amino group plus another amino group in the side chain, e.g., lysine. Proline and hydroxyproline are heterocyclic aliphatic amino acids. Phenylalanine, tyrosine, tryptophan, and histidine are aromatic amino acids; the latter two are heterocyclic.

Amino acids that contain one amino and one carboxyl group are neutral. Acids containing more basic groups than acidic groups are called basic amino acids, and acids containing more acidic than basic groups are called acidic amino acids. Examples of the acidic class are cysteine, aspartic acid, glutamic acid, and tyrosine.

Cystine is actually a diaminodicarboxylic amino acid, but it is classified under the heading of monoaminomonocarboxylic acids because it is obtained from cysteine by oxidation. Asparagine and glutamine are listed after aspartic and glutamic acids because they are amides of these acids. The amide group is neutral in *most* common reactions.

Proline and hydroxyproline are not primary amines but secondary cyclic amines. Since the nitrogen is attached to the α carbon, they are logically classified as α-amino acids.

There are ten nutritionally essential amino acids. They are called *essential amino acids* because they cannot be made or made in sufficient amounts by metabolic reactions. They are shaded in Table 16.1.

Optical isomers

Only the simplest amino acid, glycine, contains no asymmetric carbon atom. The other amino acids in Table 16.1 are L-amino acids. Only the L-amino acids are incorporated into proteins.

TABLE 16.1 PROTEIN AMINO ACIDS

Acid	Abbreviation	Formula	pK	pI
		Aliphatic		
Monoaminomonocarboxylic:				
Glycine	Gly	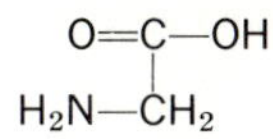	2.34 9.6	5.97
L-Alanine	Ala	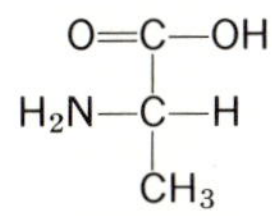	2.34 9.69	6.00
L-Serine	Ser	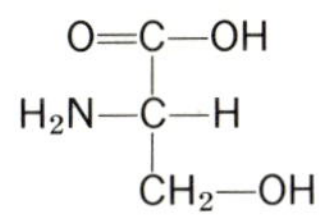	2.21 9.15	5.68
L-Cysteine	Cys	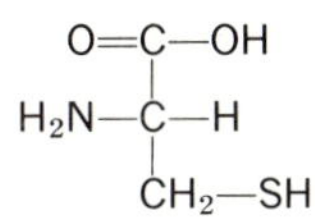	1.71 10.78 8.18	5.02
L-Cystine (oxidized cysteine)	Cys S S Cys	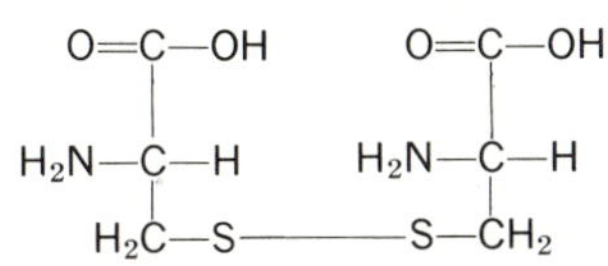	1.65, 2.26 7.85, 9.85	5.06
L-Methionine	Met	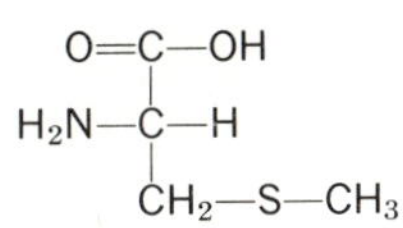	2.28 9.21	5.75
L-Threonine	Thr	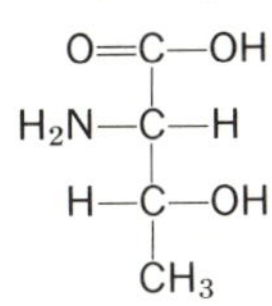	2.63 10.43	6.53
L-Valine	Val	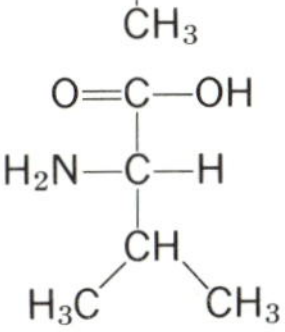	2.32 9.62	5.97
L-Leucine	Leu	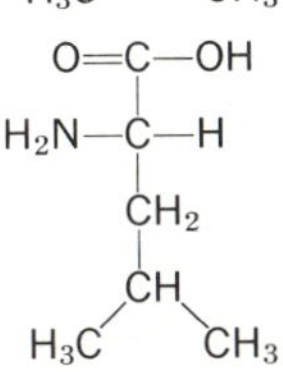	2.36 9.60	5.98
L-Isoleucine	Ile	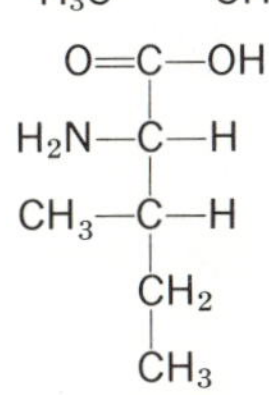	2.36 9.68	6.02

TABLE 16.1 PROTEIN AMINO ACIDS (*Continued*)

Acid	Abbreviation	Formula	pK	pI
Monoaminodicarboxylic (acidic amino acids):				
L-Aspartic acid	Asp	O=C—OH H$_2$N—C—H CH$_2$ O=C—OH	2.1 9.8 3.85	2.98
L-Asparagine (amide of aspartic acid)	Asn	O=C—OH H$_2$N—C—H CH$_2$ O=C—NH$_2$	2.08 8.8	5.41
L-Glutamic acid	Glu	O=C—OH H$_2$N—C—H CH$_2$ CH$_2$ O=C—OH	2.19 9.67 4.25	3.22
L-Glutamine (amide of glutamic acid, neutral)	Gln	O=C—OH H$_2$N—C—H CH$_2$ CH$_2$ O=C—NH$_2$	2.17 9.13	5.65
Diaminomonocarboxylic (basic amino acids):				
L-Lysine	Lys	O=C—OH H$_2$N—C—H (CH$_2$)$_3$ CH$_2$—NH$_2$	2.18 8.95 10.53	9.74
L-Hydroxylysine	Hyl	O=C—OH H$_2$N—C—H (CH$_2$)$_2$ H—C—OH CH$_2$—NH$_2$	2.13 8.62 9.67	9.15
L-Arginine (substituted diamino-acid, basic)	Arg	O=C—OH H$_2$N—C—H (CH$_2$)$_2$ CH$_2$ Guanidino group { NH C=NH NH$_2$ }	2.17 9.04 12.48	10.76

TABLE 16.1 PROTEIN AMINO ACIDS (*Continued*)

Acid	Abbreviation	Formula	pK	pI
		Cyclic		
Aliphatic:				
L-Proline (*N*-substituted) α-amino acid)	Pro	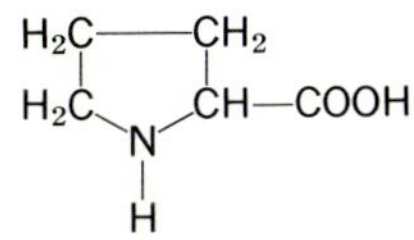	1.99 10.96	6.3
Hydroxyproline	Hyp	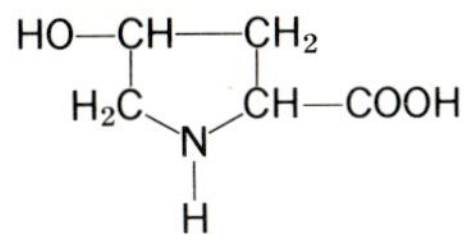	1.92 9.73	5.83
Aromatic:				
L-Phenylalanine	Phe	**Phenyl group** 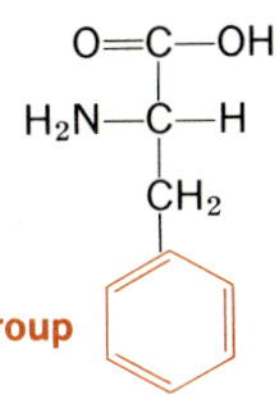	1.83 9.13	5.48
L-Tyrosine (*p*-hydroxyphenyl-alanine, a phenolic α-amino acid)	Tyr	**Phenol group** 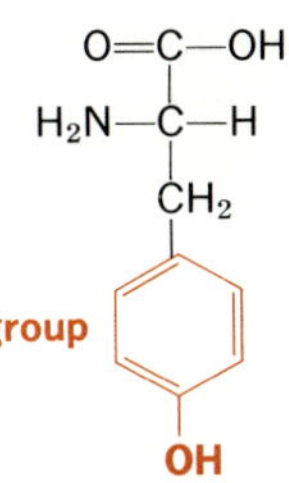	2.20 9.11 10.07	5.66
L-Tryptophan	Trp	**Indole group** 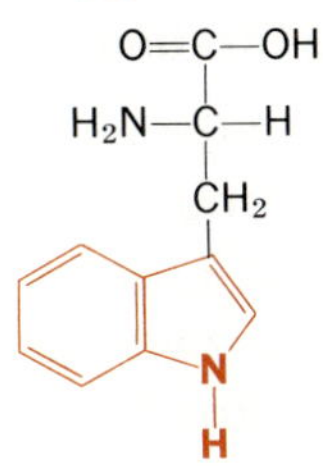	2.38 9.39	5.89
Histidine	His	**Imidazole group** 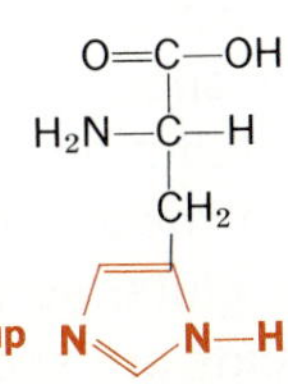	1.82 9.17 6.0	7.59

Shading indicates the essential amino acids.

In carbohydrate chemistry the letters D and L signify the position of the hydroxy group on the asymmetric carbon of the simplest aldehyde sugar, glyceraldehyde. Amino acids are also divided into D and L families. When the COOH group is written at the top and the carbon chain downward, if the NH_2 group on the α asymmetric carbon is on the right side of the carbon chain, the amino acid is in the D family. Conversely, if the NH_2 group is on the left of the carbon chain, the amino acid is in the L family. As with the sugars, the D *and* L *do not refer to the optical activity of the amino acid.* Optical activity is indicated by (+) or (−).

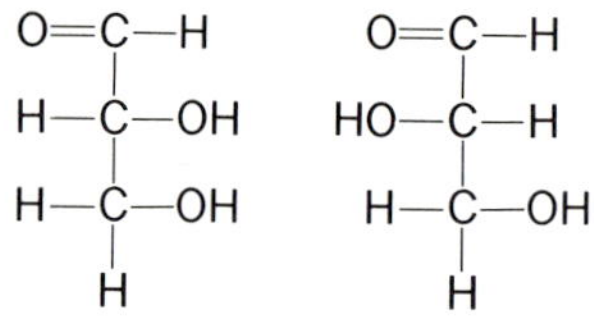

D L

Glyceraldehyde

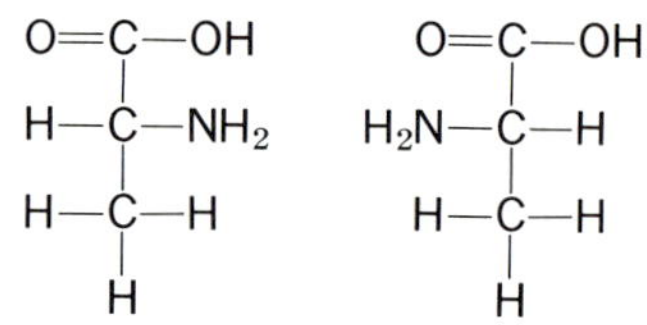

D L

Amino acid

Most amino acids contain only one asymmetric carbon atom, but cystine, threonine, isoleucine, hydroxylysine, and hydroxyproline contain two.

16.2 REACTIONS

Ionization and titration

The titration and the formation of a zwitterion was discussed earlier with respect to a monoaminomonocarboxylic amino acid (see page 145). The pI, isoelectric point, was defined as the pH at which the amino acid exists as a zwitterion. In this ionic state the net charge on the amino acid is zero; therefore, the acid would not migrate under the influence of a direct electric current.

What ionic species would be formed during the titration of a completely protonated monoaminodicarboxylic amino acid? How is the isoelectric point of these amino acids calculated? To answer these questions consider glutamic acid and lysine. The various ionic states of these amino acids are shown in the following equations, along with the number of equivalents of acid or base needed to effect the conversion from one ionic species to another.

Titration of glutamic acid:

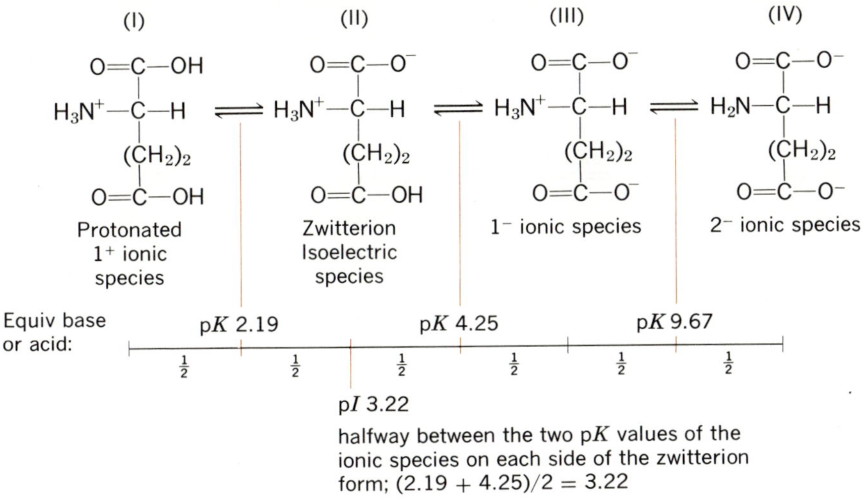

Titration of lysine:

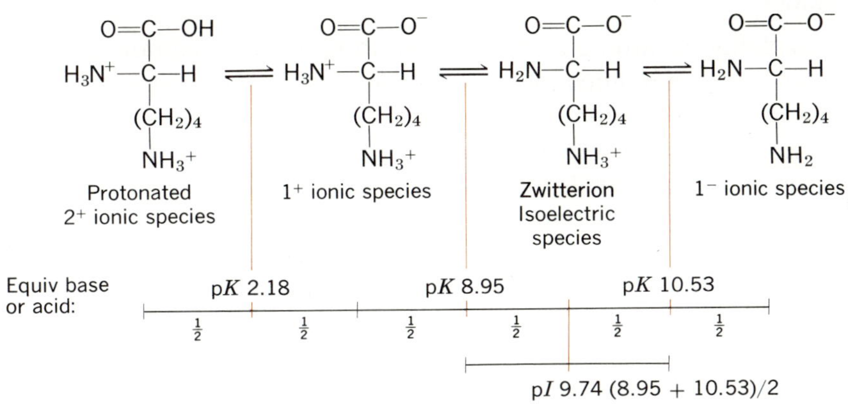

When amino acids combine to form peptides, it is the α-amino group of one acid and the carboxyl group of another that enter into the peptide-bond formation. There is, therefore, a free α-amino group at one end of the molecule and a free carboxyl group at the other. In addition to these two groups acidic or basic groups may also occur in the side chain of the respective amino acids. The ionic species of a dipeptide composed of glutamic acid and lysine illustrate the influence of side-chain groups in determining the pI.

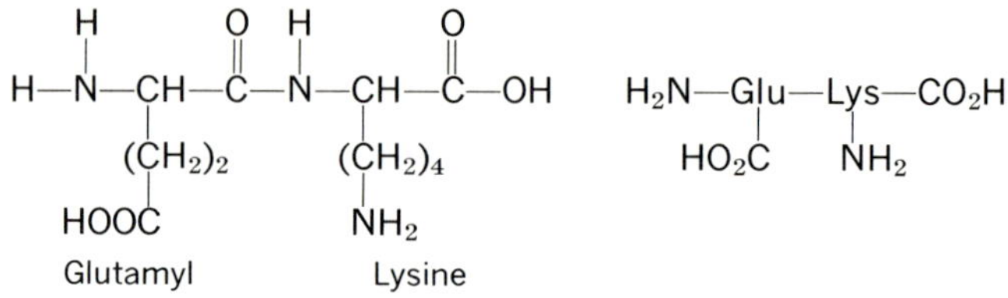

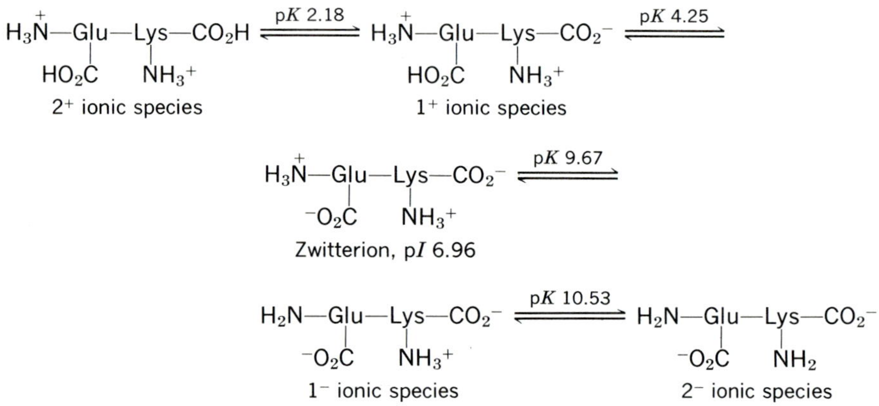

Oxidation

The oxidation of amino acids referred to here is not complete oxidation but that exemplified by, say, the conversion of cysteine to cystine or the oxidation by ninhydrin and nitrous acid. The conversion of cysteine to cystine is of importance with respect to protein structure and enzyme activity. Ninhydrin and nitrous acid are used in the quantitative determination of amino acids.

Two moles of cysteine are oxidized to form one mole of cystine according to the following equation. The disulfide bond in cystine is a covalent bond and is therefore quite stable.

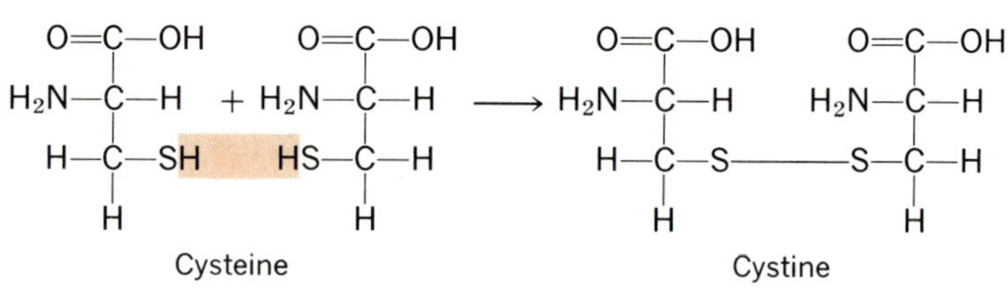

(16.1)

Sodium nitroprusside, $Na_2Fe(CN)_5$, in a dilute ammonium hydroxide solution is used to test for the sulfhydryl group. This reagent produces a red color in the presence of *free* SH *groups.*

Ninhydrin forms a purple-colored complex with NH_3.

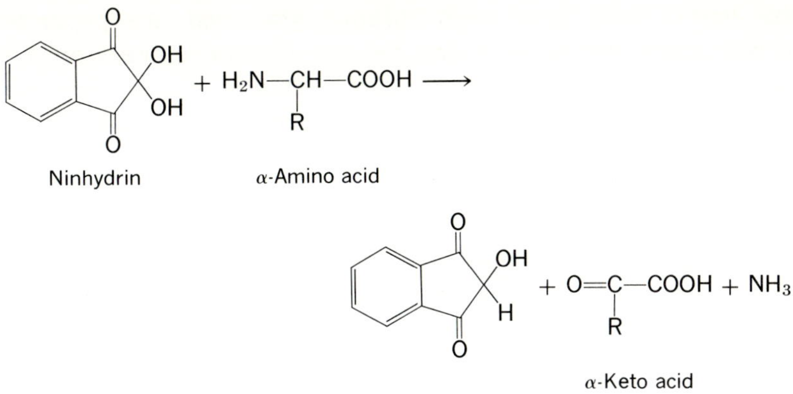

(16.2)

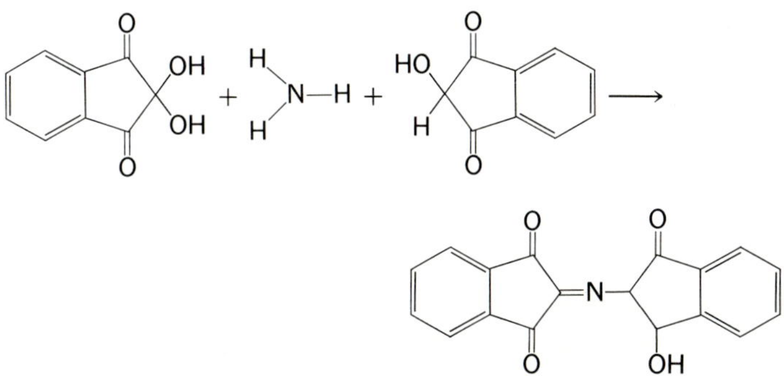

(16.3)

Purple

$$\underset{\alpha\text{-Keto acid}}{R—\overset{\overset{\displaystyle O}{\|}}{C}—COOH} \longrightarrow \underset{\text{Aldehyde}}{R—\overset{\overset{\displaystyle O}{\|}}{C}—H} + CO_2$$

Ninhydrin also produces a color with primary amines that have a hydrogen attached to the carbon holding the amine group; the —$CHNH_2$— is converted into =C=O and NH_3, which complexes with the ninhydrin. Thus, many amines as well as amino acids will give a positive ninhydrin test. α-Amino acids are converted into the corresponding α-keto acids and NH_3. The α-keto acids are unstable under the conditions of the test and are decarboxylated to the corresponding aldehyde and CO_2. Only α-amino acids yield CO_2; other primary amines do not.

Ninhydrin reagent has been used extensively for quantitative determination of the concentration of amino acids. The intensity of the color in the presence of this reagent is directly proportional to the concentration of the amino acid. Qualitatively ninhydrin is used to spot amino acids after they have been subjected to paper chromatography.

After the amino acids have been chromatogrammed, the paper is sprayed with ninhydrin reagent and heated. A color appears on the paper where the amino acids are located. This is a very sensitive method; as few as several tenths of a microgram of amino acid can be detected.

Nitrous acid reacts with primary amines to form the corresponding hydroxy acid and nitrogen. This reaction was discussed under amines in Chap. 12. The amount of nitrogen produced is a measurement of the number of primary amine groups present in the original substance.

$$\underset{\alpha\text{-Amino acid}}{R{-}\overset{NH_2}{\underset{H}{C}}{-}\overset{O}{\overset{\|}{C}}{-}OH} + HNO_2 \longrightarrow \underset{\alpha\text{-Hydroxy acid}}{R{-}\overset{OH}{\underset{H}{C}}{-}\overset{O}{\overset{\|}{C}}{-}OH} + N_2 + H_2O$$

16.3 PEPTIDES

Many compounds composed of amino acids linked together by peptide bonds are of too low molecular weight to be classified as proteins. These compounds are classified as peptides, and biologically they constitute a very important class. Some are antibiotics, whereas others are hormones (enzyme regulators). Unless otherwise stated, the L form of the amino acid is implied. Although the D form of amino acids is not normally found in proteins, it does occur in a limited number of peptides. Peptides have been thoroughly investigated with respect to biological activity, amino acid sequence, and synthesis. A number of low-molecular-weight peptides have been synthesized, a notable achievement being the synthesis of the hormone insulin, molecular weight 6,000. The formulas of several peptides are given in Table 16.2.

Oxytocin and vasopressin are both hormones, and both are octapeptides (contain eight amino acids). They have been synthesized in the laboratory. When the formulas of these two compounds are compared, they are seen to differ only in two amino acids.

Insulin is composed of two peptide chains, A and B, connected by two disulfide bonds. Another disulfide bond forms a loop in chain A. Reduction of one or more of these disulfide bonds produces a biologically inactive compound.

Gramiciden S and tyrocidin are antibiotic substances produced by *Bacillus brevis* and *B. licheniformis,* respectively. There are two

TABLE 16.2 AMINO ACID SEQUENCE OF SOME PEPTIDES

Bovine insulin

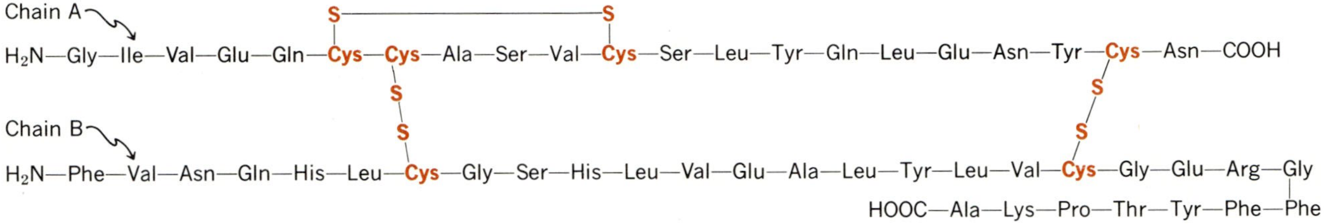

Beef oxytocin

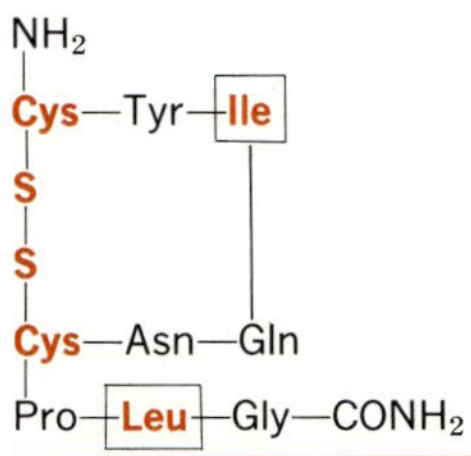

Beef vasopressin

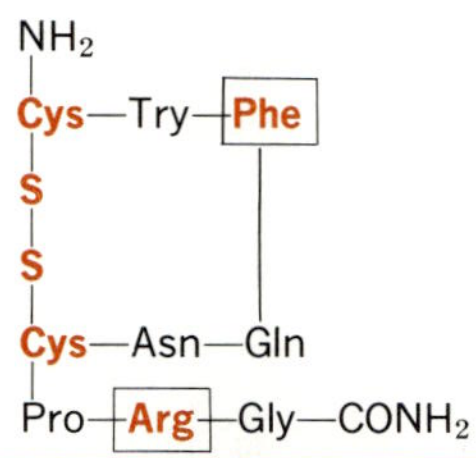

Gramiciden S

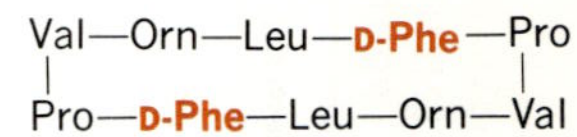

Tyrocidin S

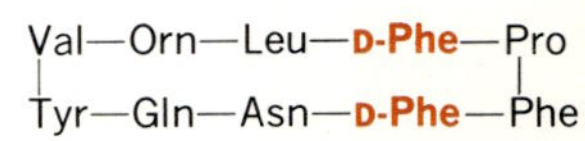

TABLE 16.3 AVERAGE ELEMENTAL COMPOSITION OF PROTEIN

Element	Percentage
Carbon	51–55
Hydrogen	6.7–7.3
Nitrogen	15–18
Oxygen	21–24
Sulfur	0.3–2.2
Phosphorus	0.0–1.5

noteworthy differences between the structures of these two antibiotics and peptides which are not antibiotic in nature: (1) the occurrence of ornithine, which normally is not a constituent of proteins but is important in metabolic reactions associated with the formation of urea, and (2) the occurrence of a D amino acid, namely D-phenylalanine. In tyrocidin both D- and L-phenylalanine are present.

16.4 PROTEINS

Proteins may be defined as high-molecular-weight compounds composed of amino acids linked together by a peptide bond. The elements carbon, nitrogen, oxygen, and sulfur occur in the amino acids of the protein (see Table 16.3), whereas the other mineral elements are found in non-amino acid groups attached to the protein unit. Such

TABLE 16.4 MOLECULAR WEIGHT OF SOME PROTEINS

Protein	Mol. wt.	p*I*
Insulin (hormone)	5,700	5.3
Ribonuclease (enzyme)	12,700	9.5
Lactoglobulin	15,500	
Trypsin (enzyme)	24,000	1.2
Ovalbumin	40,000	
Zein	40,000	
Hemoglobin	65,000	
Serum albumin, human	69,000	4.8
Edestin	310,000	
Catalase (enzyme)	250,000	5.5
Urease (enzyme)	480,000	5.0
Tobacco mosaic (virus)	40 million	
Vaccina, vertebrates (virus)	2 billion	

TABLE 16.5 PERCENTAGE ESSENTIAL AMINO ACIDS IN SOME FOODS[a]

Food	Total protein	Essential amino acids, edible portion, %									
		Phe	His	Ile	Leu	Lys	Met	Thr	Val	Arg	Tyr
Milk, cow, whole	3.5	0.17	0.092	0.22	0.34	0.27	0.086	0.16	0.24	0.13	0.049
Goat	3.3	1.72	0.94	0.087	0.28	0.31	0.065	0.22	2.44	1.30	0.39
Human	1.4	0.18	0.086	0.20	0.42	0.33	0.11	0.21	0.26	0.14	0.059
Casein	100	5.39	3.02	6.55	10.05	8.01	3.08	4.28	7.39	4.07	1.34
Gelatin	85.6	2.04	0.77	1.36	2.93	4.23	0.79	1.91	2.42	7.87	0.006
Barley	12.8	0.66	0.24	0.55	0.89	0.43	0.18	0.43	0.64	0.66	0.16
Corn, field	10.0	0.45	0.21	0.46	1.30	0.29	0.19	0.40	0.51	0.35	0.061
Zein	16.1	1.66	0.22	0.82	3.18		0.28	0.50	0.65	0.29	0.010
Rice, brown	7.5	0.38	0.13	0.35	0.65	0.30	0.14	0.29	0.52	0.43	0.081
Rye	12.1	0.57	0.28	0.52	0.81	0.49	0.19	0.45	0.63	0.59	0.14
Wheat, whole, durum	12.7	0.63	0.26	0.55	0.85	0.35	0.19	0.37	0.59	0.61	0.16
Bananas, common, ripe	1.2					0.055	0.011				0.018
Orange juice	0.8					0.021	0.002				0.003
Beans, lima, raw	7.5	0.39	0.25	0.46	0.61	0.47	0.080	0.34	0.49	0.45	0.10
Snap, raw	2.4	0.057	0.045	0.10	0.14	0.13	0.035	0.091	0.12	0.10	0.033
Peas, raw	6.7	0.26	0.11	0.31	0.42	0.32	0.054	0.25	0.27	0.66	0.056
Cabbage	1.4	0.030	0.025	0.040	0.057	0.066	0.013	0.039	0.043	0.11	0.011
Kale	3.9	0.16	0.062	0.13	0.25	0.12	0.035	0.14	0.18	0.20	0.042
Turnip greens	2.9	0.15	0.051	0.11	0.21	0.13	0.039	0.13	0.15	0.17	0.045
Potatoes, raw	2.0	0.088	0.029	0.088	0.10	0.11	0.025	0.079	0.11	0.099	0.021

[a] Data selected from M. L. Orr and B. K. Watt, Amino Acid Content of Foods, *USDA Home Econ. Res. Rept.* 4, December, 1957.

groups are called *prosthetic* groups, meaning additional. The mineral elements iron, phosphorus, magnesium, copper, and manganese are found in some proteins.

The molecular weight of proteins varies within wide limits (Table 16.4). The molecular formula of proteins is not constant because different proteins contain different numbers or kinds of amino acids (see Table 16.5). Although proteins may contain exactly the same number and kinds of amino acids, they may be entirely different proteins with different chemical and physiological properties because the amino acids are arranged in a different sequence. The number of formulas possible for a simple protein containing only 300 amino acids is enormous. For a tripeptide composed of amino acids designated as A, B, and C there are six different orders in which the amino acids could be linked together. The number of possible tripeptides that can be formed from 25 different amino acids is 15,625. Protein synthesis in each cell is carefully controlled; many proteins, each with a different sequence and kind of amino acid, can be synthesized continuously at the same time.

16.5 CLASSIFICATION OF PROTEINS

The complex nature of proteins excludes their classification simply according to structure or composition. Proteins have been classified in several different ways. One way is according to their physiological function, e.g., structural, enzyme, virus, or hormone, and another is according to solubility. The brief classification below is based upon composition and solubility.

Simple proteins

Simple proteins yield only amino acids upon hydrolysis. They are subdivided according to solubility and coagulability.

ALBUMINS. Albumins are soluble in both water and dilute salt solutions, and they are coagulated by heat. They are precipitated from solution by saturating the solution with ammonium sulfate. This ammonium sulfate precipitation is not unique to the albumins, for many proteins are precipitated under similar conditions. However, some are precipitated at lower concentrations of the salt.

Albumins occur in all forms of living organisms. Some examples of albumins are ovalbumin of egg white, serum albumin of blood, and lactalbumin of milk.

GLOBULINS. These proteins are insoluble in water and soluble in dilute neutral salt (NaCl) solution and are coagulated by heat. Whereas albumins are soluble in 50% saturated ammonium sulfate solution, globulins are insoluble at this concentration. Thus, these two proteins can be separated on the basis of their difference in solubility in ammonium sulfate solution. Another means of separating the albumins from the globulins is by the difference in their solubility in water.

Examples of some globulins are edestin of hempseed, legumin of legumes, α-, β-, and λ-globulins of blood serum, and ovoglobulin of egg white.

GLOBINS. Globins are a class of proteins high in histidine content. They are not basic and are soluble in ammonium hydroxide solution. The globins occur conjugated with heme, the iron-containing unit of hemoglobin of blood, and leghemoglobin of legumes.

HISTONES. Histones are quite basic proteins as a result of their high content of arginine. They are water-soluble but insoluble in ammonium hydroxide. They are associated with nucleoproteins.

PROLAMINES. Proteins in this class contain a high amount of the amino acid proline. They are insoluble in water and absolute ethanol but are soluble in 70% ethanol. Corn and wheat contain the two prolamines zein and gliadin, respectively.

ALBUMINOIDS. Albuminoids form the protective layer of skin on animals. These proteins are not found in the plant kingdom. Albuminoids are insoluble in water and solvents that do not bring about hydrolysis. Keratin is the albuminoid that occurs in hair, skin, fingernails, and horns. Connective tissue contains the albuminoid collagen, which is insoluble like the keratins. Upon partial hydrolysis of collagen a product known as gelatin is formed, which is soluble in water and is readily digested. Albuminoids are not easily digested.

Conjugated proteins

Conjugated proteins contain a nonprotein moiety, or *prosthetic group,* in their structure. The nature of the prosthetic group is the basis for subclassification of this group of proteins.

CHROMOPROTEINS. Proteins that contain a colored prosthetic group are called chromoproteins. Hemoglobin, for example, contains the prosthetic group *heme,* which gives this protein its red color. Copper complexes also occur in chromoproteins.

PHOSPHOPROTEINS. Phosphoproteins contain phosphate as the prosthetic group linked to the protein by an ester linkage. Casein of milk is a phosphoprotein, as is vitellin of egg yolk.

GLYCOPROTEINS. Proteins that contain a carbohydrate unit as the prosthetic group are called glycoproteins. Mucoprotein of the blood and mucin of the saliva are examples of this class of protein. Heparin, which inhibits the clotting of blood, is also a glycoprotein.

LIPOPROTEINS. These proteins contain a lipid, such as fatty acid. Lipovitellin of egg yolk is an example.

NUCLEOPROTEINS. Proteins that contain nucleic acids as their prosthetic group are called nucleoproteins. Nucleic acids contain a carbohydrate (ribose or deoxyribose), nitrogenous bases (purines and pyrimidines), and phosphoric acid. The proteins associated with nucleic acids are basic in nature and include the protamines and histones. Viruses are nucleoproteins. The nucleoproteins and nucleic acids are essential for the synthesis of proteins and will be discussed subsequently in more detail.

16.6 PRIMARY STRUCTURE OF PROTEINS

The structure of proteins is subdivided into primary, secondary, and tertiary according to the type of interaction of the amino acids.

The primary structure of proteins is limited to the peptide bond between amino acids and thus includes the sequence of amino acids in the protein (Fig. 16.1). The peptide bond is frequently referred to as the *backbone* of the protein molecule. The bonds involved in this link are *covalent* and quite stable. They can be disrupted by chemical or enzymatic hydrolysis but are not directly influenced by salt concentration, change in pH, or solvent.

Amino acid sequence

How can the sequence of amino acids of a peptide be determined? Basically the procedure involves the following steps: hydrolysis, identification of the products of hydrolysis, and then fitting the pieces together in the correct order. Hydrolysis of peptides can be effected by acid, alkali, or enzymes. Each method has its advantages and disadvantages, but together they make a formidable trio.

Acid hydrolysis is usually preferred to alkaline hydrolysis. Hydrochloric acid, 6 N, is a convenient acid to use in a sealed tube at 110°C

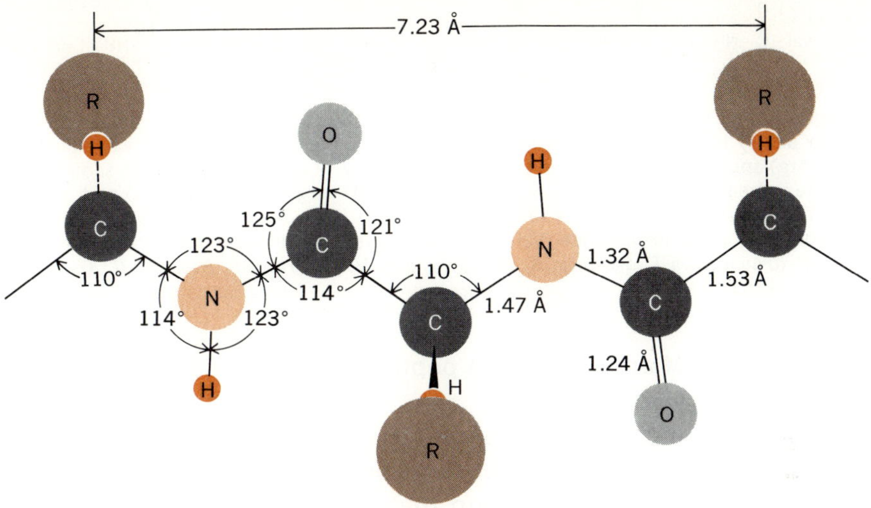

FIGURE 16.1 Primary structure, showing bond angles and distances.

for periods ranging from 10 to 100 hr, depending upon the nature of the peptide or protein to be hydrolyzed. Protein is completely hydrolyzed by this procedure, but tryptophan is destroyed. Some serine and threonine is also destroyed.

Alkaline hydrolysis is effected with about 4 N NaOH for periods ranging up to 10 hr at 100°C. Alkaline hydrolysis is used chiefly to obtain the tryptophan content. It has more disadvantages than acid hydrolysis, for example, arginine, cysteine, cystine, threonine, and serine are destroyed and some amino acids are partly deaminated. Also, alkali converts the L amino acids into a racemic mixture (a mixture containing equal amounts of the D and L forms).

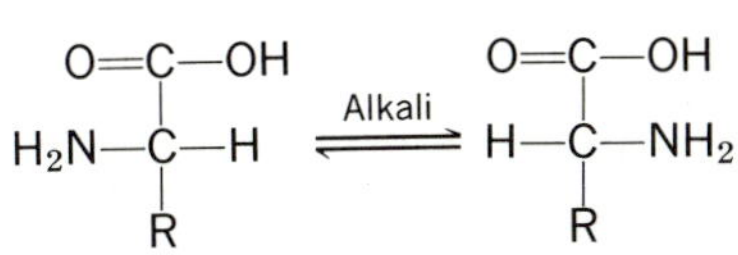

L amino acid D amino acid

Enzymatic hydrolysis of proteins by proteases (enzymes that hydrolyze proteins) is not a satisfactory means of *completely* hydrolyzing proteins since the reaction does not usually go to completion. Furthermore, proteases readily hydrolyze peptide bonds only between specific amino acids, a fact used to advantage in the determination of

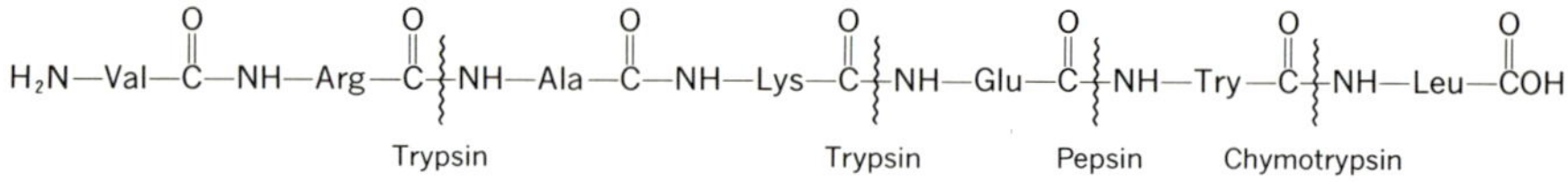

FIGURE 16.2 The protease action on peptides. If the heptapeptide (composed of seven amino acids) is treated with trypsin, the hydrolytic products would be two dipeptides (Val—Arg) and (Ala—Lys) and one tripeptide (Glu—Try—Leu). However, if the same heptapeptide is treated with a pepsin, a pentapeptide (Val—Arg—Ala—Lys—Glu) and a dipeptide (Try—Leu) are formed. The resulting fragments if the heptapeptide is treated with a chymotrypsin are a hexapeptide (Val—Arg—Ala—Lys—Glu—Try) and the free amino acid, Leu. [*Redrawn from L. Pauling, R. B. Corey, and H. R. Branson, Proc. Natl. Acad. Sci.*, **37**:206, (1951).]

the sequence of amino acids in peptides and proteins. The protease action is illustrated in Fig. 16.2.

Consider the three proteases *pepsin, trypsin,* and *chymotrypsin.* Although they hydrolyze peptides and proteins, each has a preference with respect to the amide group that is hydrolyzed. Other enzymes serve similar specific purposes.

Pepsin hydrolyzes most rapidly the peptide bond in which an aromatic amino acid (phenylalanine, tyrosine, tryptophan) furnishes the nitrogen.

Chymotrypsin preferentially hydrolyzes the peptide bond in which an aromatic amino acid furnishes the carboxyl group.

Trypsin hydrolyzes peptide bonds in which the carboxyl group is furnished by basic amino acids such as lysine and arginine; trypsin is more specific than chymotrypsin or pepsin.

Carboxypeptidase and *aminopeptidase* are two enzymes used in the determination of amino acid sequence of peptides.

Carboxypeptidase sequentially hydrolyses peptides *only* from the end that has the free carboxyl group.

Aminopeptidase sequentially hydrolyses peptides *only* from the end with the free amino group.

By following the increase in amino acids liberated during hydrolysis by these two enzymes it is possible to determine the sequence of amino acids in a peptide. Consider a time-course analysis (quantitative analyses of the free amino acids at regular time intervals) of the free amino acids in the hydrolysate upon treatment of the pentapeptide in Fig. 16.3. Hydrolysis of the pentapeptide with carboxypeptidase results

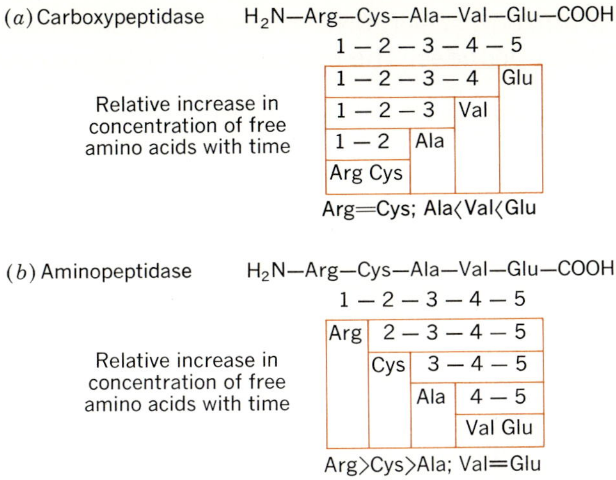

FIGURE 16.3 Enzymatic hydrolysis of the pentapeptide arginyl-cysteinyl-alanyl-valyl-glutamic acid.

in glutamic acid as the first amino acid to appear, and its concentration will be greater than the other amino acids in the hydrolysate. The second amino acid that will increase in concentration is valine. Alanine is the third. Arginine and cysteine both appear at the same time in equal amounts. Thus by the action of carboxypeptidase on the pentapeptide, the sequence of the first three amino acids from the carboxyl end of the molecule can be determined. The sequence of the last two amino acids cannot be determined with this enzyme but can be with the enzyme aminopeptidase.

A time-course analysis of the free amino acids in the hydrolysate of the pentapeptide in Fig. 16.3 with aminopeptidase gives the following results. Since this enzyme hydrolyzes the peptide starting at the end with the free amino group, the following order of increasing concentration of free amino acids is observed: arginine > cysteine > alanine; valine and glutamine appear simultaneously in the same amounts. Thus the sequence of the first three amino acids from the free amino end of the pentapeptide is arginine, cysteine, and alanine. Combining these results with those obtained by treating the pentapeptide with carboxypeptidase gives the primary structure of the amino acids in the pentapeptide.

Carboxypeptidase:	. . . . Ala—Val—Glu
Aminopeptidase:	Arg—Cys—Ala—. . . . . . .
Amino acid sequence:	Arg—Cys—Ala—Val—Glu

Fluorodinitrobenzene (FDNB) is used in the determination of free amino groups in peptides. The dinitrophenyl (DNP) derivative of the NH_2 is formed

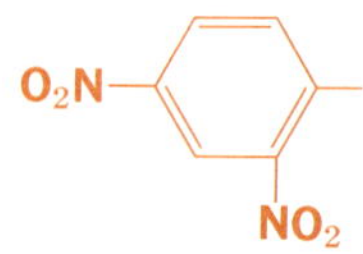

Dinitrophenyl radical (DNP)

The reaction of FDNB with amino acids was discussed in Chap. 12. The following examples serve as a review and illustrate further the methods used to determine amino acid sequence.

EXAMPLE 1

The hexapeptide Ser—Cys—Ala—Glu—Val—Lys is reacted with FDNB, the product is hydrolyzed, and the DNP-amino acid derivatives are separated.

(*a*) Which amino acids contain the DNP group?

(*b*) Which of the DNP-amino acids contain a free NH_2 group?

(*c*) Would DNP—Lys have been formed if the lysine were not a COOH-terminal amino acid in the peptide? Is this also true with respect to serine if it is not the NH_2-terminal amino acid?

(*d*) Write the formula for the DNP derivative of lysine if it had been the NH_2-terminal amino acid in the peptide.

Solution. Since FDNB reacts with free NH_2 groups, it may be best to rewrite the peptide in such a fashion as to show the free NH_2 groups. This is done below. There are two free NH_2 groups, the α-amino group of serine and the ϵ-amino group of lysine. The reaction with FDNB is

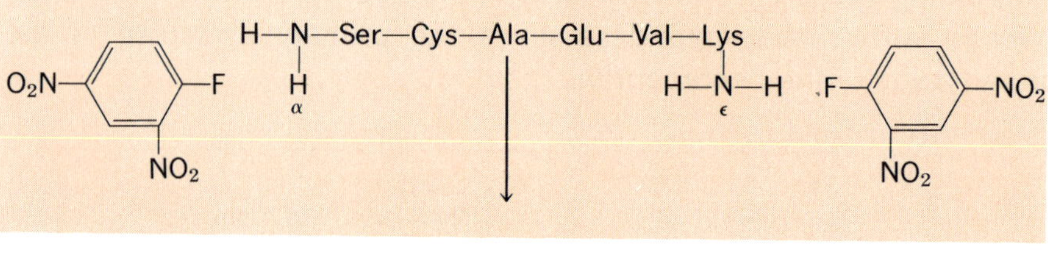

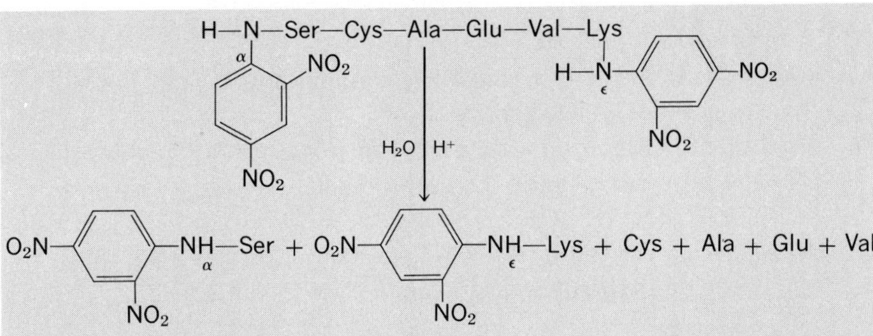

α-DNP—Ser ε-DNP—Lys Free amino acids

EXAMPLE 2

Determine the sequence of amino acids in a *heptapeptide* (I) from the following data. Amino acid analysis shows the heptapeptide to contain the following amino acids in the amounts indicated:

Amino Acid	*Moles*	*Amino Acid*	*Moles*
Arg	1	Ala	2
Phe	1	Leu	1
Glu	1	Lys	1

The following scheme shows the reactions performed and the results obtained.

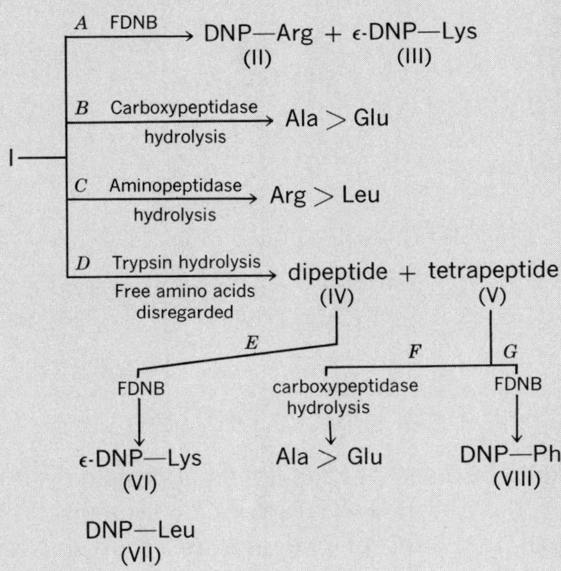

Interpretation of Data

Reaction A: Arginine is the NH_2-terminal amino acid; lysine has a free ϵ-NH_2 group. (I) Arg — — — — — — .

Reaction B: Alanine is the COOH-terminal amino acid, and glutamic acid is the next amino acid. (I) Arg — — — — Glu—Ala.

Reaction C: Arginine is the NH_2-terminal amino acid, and leucine is the next amino acid. (I) Arg—Leu — — — — Glu—Ala.

Reaction E: In dipeptide IV, leucine is the NH_2-terminal amino acid and lysine is the COOH-terminal amino acid. (IV) Leu—Lys. Since there is only 1 mole of leucine in I, the leucine of the dipeptide must be the leucine that is the second amino acid from the NH_2 end of I (see reaction *C*).

Reactions F and G: Tetrapeptide (V) contains alanine as the COOH-terminal amino acid and glutamic acid as the next amino acid. Phenylalanine is the NH_2-terminal amino acid. (V) Phe — Glu—Ala.

Correlation of Results

		1	2	3	4	5	6	7
	(I)	—	—	—	—	—	—	—
A, B, C		Arg—	Leu	—	—	—	Glu—	Ala
D, E		(IV)	Leu—	Lys				
F, G, only position for (5)					Phe	—	Glu—	Ala
One amino acid is missing, the second mole of alanine. It can only be in position 5.						Ala		
		1	2	3	4	5	6	7
	(I)	Arg—	Leu—	Lys—	Phe—	Ala—	Glu—	Ala

16.7 SECONDARY STRUCTURE OF PROTEINS

Hydrogen bonding between peptide bonds of proteins is the *only* bond responsible for the secondary structure of proteins. The hydrogen bond is between the C=O of one peptide group and the N—H of another.

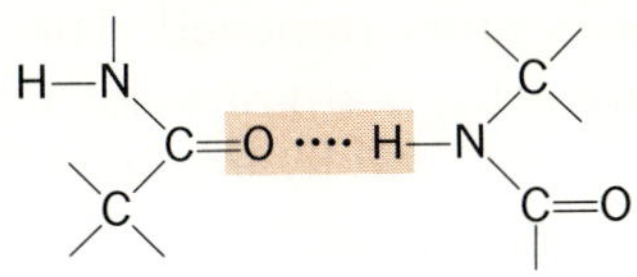

The secondary structure is divided into two classes, the helix (Fig. 16.4) and the pleated sheet (Fig. 16.5), depending upon whether the hydrogen bonding is intramolecular or intermolecular.

Helix structure

The helix structure of proteins is governed by intramolecular hydrogen bonding. X-ray studies not only have indicated the bond lengths and angles but have also furnished data for the spatial arrangement of amino acids in proteins. The α helix (Fig. 16.4) is the most stable helical arrangement of amino acids. In this structure hydrogen bonds are between the NH group of one peptide bond and the C═O group of

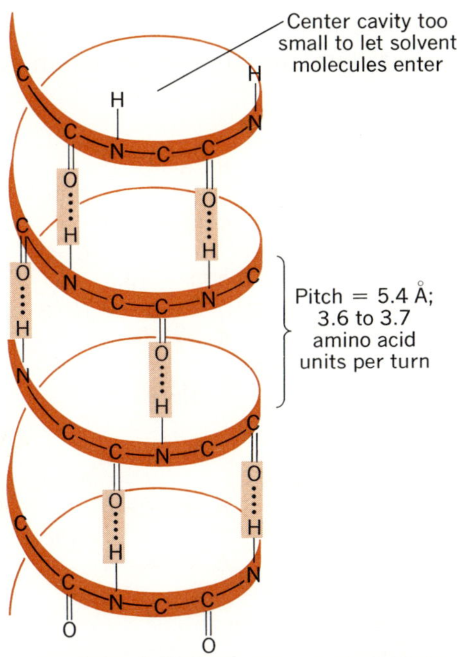

FIGURE 16.4 α helix. Only the amino acids in the front portion of the helix are shown. The carbons holding the H and R groups are shown as C. These H and R groups extend outward from the helix and for simplicity are not shown.

another three amino acid units removed. The L amino acids form a right-handed helix, whereas the D amino acids form a left-handed helix. The carbon, nitrogen, oxygen, and hydrogen of the peptide bond all lie in the same plane. The hydrogen and R groups on the α carbons stick out from around the helix and are therefore available for interactions with other groups of the same protein, solvent molecules, or groups of another strand of protein. The helix is so tightly wound that the space in the center is too small for solvent molecules to enter. Almost all proteins contain some α structure. There may be portions that are not helix (random structure). Hydrogen bonding cannot occur with proline in a peptide chain, and the helix structure is therefore broken where it occurs. A sharp bend in the protein structure may appear at this point (Fig. 16.6). Although the α-helix structure is more rigid than the random structure, it is not perfectly rigid and bends or folds around into various configurations.

Pleated-sheet structure

This protein structure is a result of intermolecular hydrogen bonding between protein chains to form a sheetlike arrangement. Proteins with this structure are crystalline and quite insoluble in aqueous solvents.

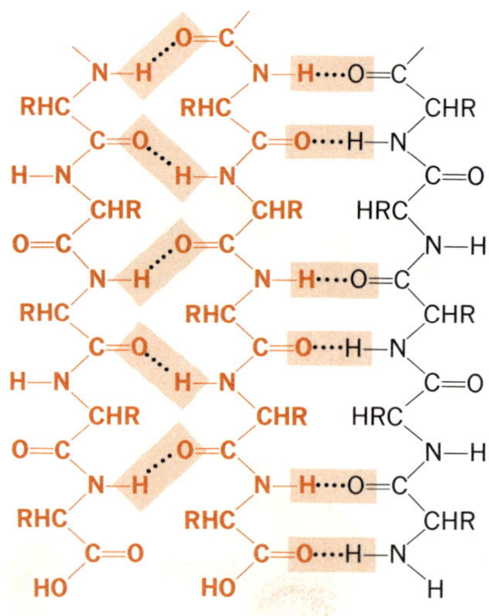

FIGURE 16.5 The pleated sheet structure of proteins. Two structures are possible, parallel, as represented by A—B, and antiparallel, as represented by B—C. Atoms of the peptide bond lie in one plane; the H and R groups are to the front and back of this plane.

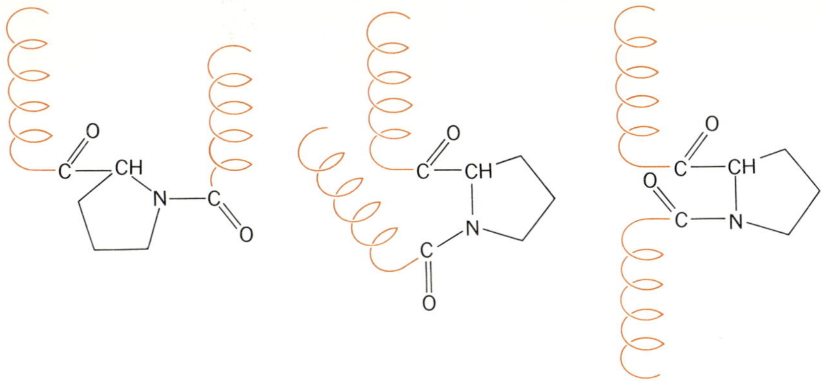

FIGURE 16.6 Proline disruption of the helix structure.

There are two ways in which protein chains can form the pleated sheet structure (Fig. 16.5). One is with the chains running in the same direction, the COOH and NH_2 ends of the proteins lying all at the top or all at the bottom of the sheet. This is called *parallel structure.* In the other type, known as *antiparallel* structure, the protein chains alternate in such a way that the COOH end of one is next to the NH_2 end of the other. In these structures the atoms of the peptide bond are in the same plane, but the hydrogen and R groups are in front or in back of the sheet.

16.8 TERTIARY STRUCTURE OF PROTEINS

The tertiary structure of proteins is associated with the interaction of groups in the side chain (R) of the amino acids. These interactions include the salt bond, hydrogen bond, disulfide bond, hydrophobic bond, and van der Waals forces. These bonds aid in holding the protein in a rather rigid structure or conformation (Fig. 16.7).

The disulfide bond is the only bond associated with the tertiary structure that is covalent; hence it is the strongest of the tertiary bonds. The other bonds are affected by solvent, pH, temperature, and salts, whereas the disulfide bond is not. The disulfide bond can be broken by reduction and can be reformed upon mild oxidation.

The tertiary structure is of special importance if the protein is an enzyme. To be active enzymes must bind a specific substrate. For this purpose the protein must have a certain shape and charges so arranged spatially that the substrate will bind in the proper position.

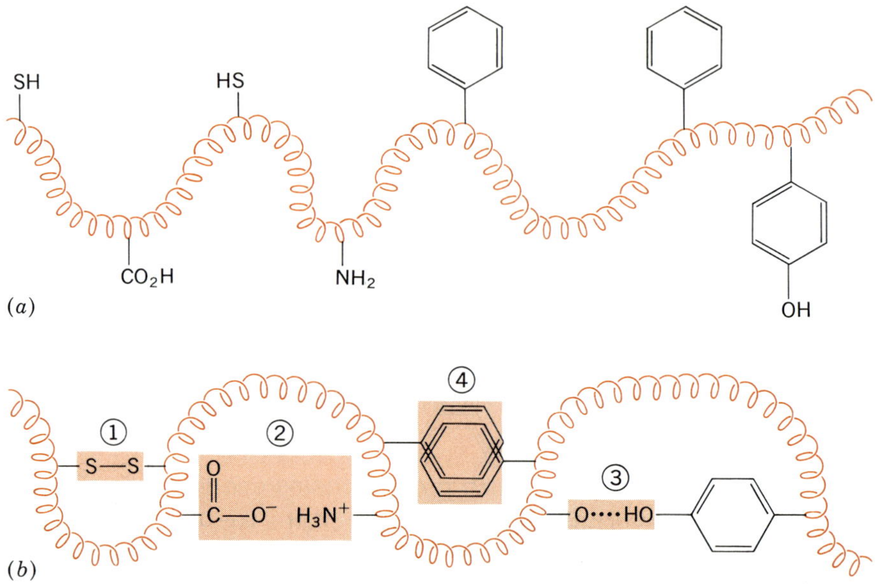

FIGURE 16.7 Interactions associated with the tertiary structure of protein: (*a*) random protein, no tertiary structure; (*b*) protein with a tertiary structure. Bonds: 1, disulfide; 2, salt; 3, hydrogen; and 4, hydrophobic.

16.9 QUATERNARY STRUCTURE OF PROTEINS

The quaternary structure of proteins refers to the number of protein units in the molecule, i.e., the state of polymerization. Some molecules of proteins contain more than one strand of protein, which may all be the same or all different. For example, hemoglobin is composed of two α strands and two β strands held together in a definite configuration. Collagen is composed of three strands of the same protein twisted together in a helical fashion like a piece of rope.

Quaternary structure may influence the activity of enzymes. Some enzymes are active only in their polymer state and become inactive if split into smaller units. Other enzymes are inactive in the polymeric state and are activated only upon being depolymerized. The enzyme phosphorylase *a* is a tetramer.

16.10 DENATURATION OF PROTEINS

Denaturation of proteins refers to the breaking of any bond except the primary bond (the peptide bond). Denaturation thus destroys the natural conformation of a protein. It can be caused by heat, oxidation, reduction, agitation, solvents, or x-rays.

There can be any number of degrees of denaturation. For example, varying the pH of a solution causes a disruption of salt linkages, hydrogen bonds, or hydrophobic bonds to varying extents depending upon how much the pH is changed. Urea causes a rupture of hydrogen bonds and at high enough concentrations causes an uncoiling of the α-helix structure to produce a random molecule with no secondary or tertiary structure. More vigorous reactions, such as reduction, disrupt disulfide bonds and produce a marked degree of denaturation. Oxidation may convert sulfhydryl groups to disulfide bonds and thus change the secondary structure of the protein.

The nature of the protein and the extent of denaturation determine whether or not the denaturation process is reversible. If the protein is composed of several chains, a polymer, there is less chance that the chains will return to their original conformation after denaturation. The precipitation of proteins by ammonium sulfate is a good example of a mild denaturation process which is reversible; in most cases the original protein is obtained upon redissolving in water. Many enzymes have been concentrated by precipitating them with ammonium sulfate. When the ammonium sulfate is removed, the enzyme redissolves and returns to its active conformation. Curling hair by permanent-wave lotion is a good example of denaturing a protein and reforming it into another conformation. In this process two reactions take place: (1) the disulfide bonds of the hair are reduced, and, after curling the hair around a suitable curler, (2) an oxidizing solution is applied which reforms the disulfide bonds in a different pattern, thus holding the hair in a wave. Egg-white proteins can be easily denatured by agitation, in the formation of meringue, or by heat to yield insoluble proteins. Some examples of denaturation of proteins are presented in Fig. 16.8.

16.11 COMPLETE AND INCOMPLETE PROTEINS

From the nutritional viewpoint proteins fall into two groups, complete and incomplete. A complete protein contains all the essential amino acids, whereas an incomplete protein lacks one or more of them. From the amino acid content of proteins in Table 16.5 it will be noted that most proteins are low or deficient in one or more of the essential amino acids. For example, zein is low in lysine and tryptophan, and gliadine is deficient in lysine. Fortunately our diet consists of a mixture of proteins which are not all deficient in the same way; thus the proteins supplement each other and together furnish a complement of the essential amino acids.

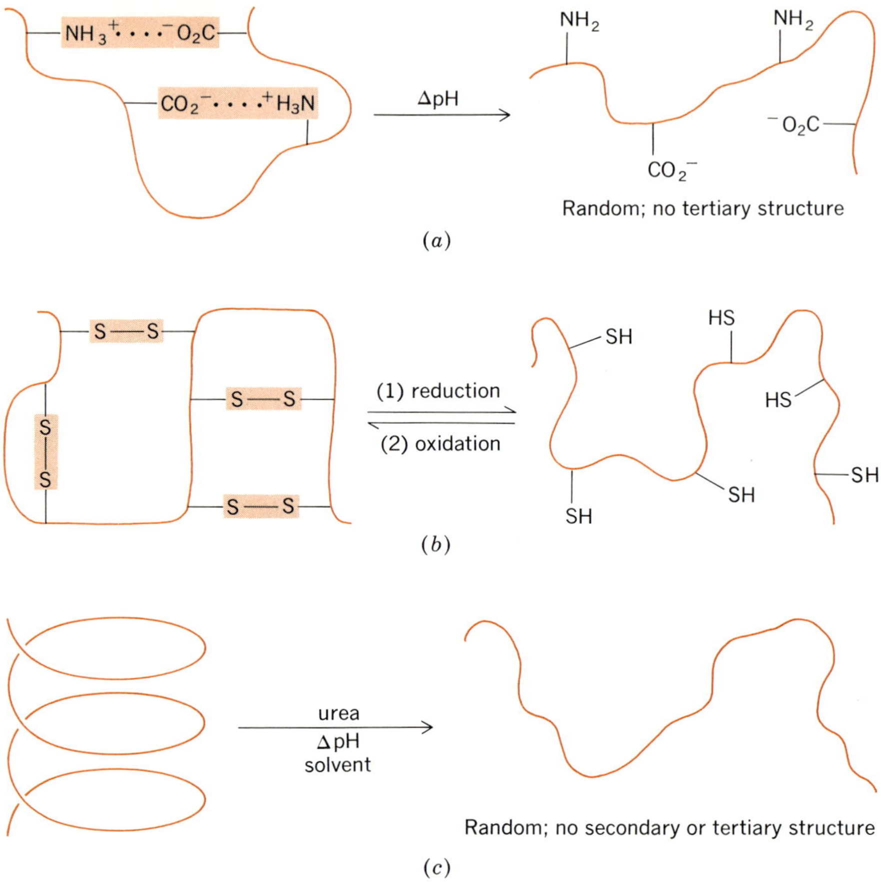

FIGURE 16.8 Protein denaturation (schematic): (*a*) salt-bond rupture; (*b*) disulfide-bond rupture and reformation in a different secondary structure; (*c*) hydrogen-bond rupture.

QUESTIONS

1. Write the formula of one amino acid for each of the following classes:
  (*a*) Aliphatic monoaminomonocarboxylic amino acid
  (*b*) Aromatic alicyclic amino acid
  (*c*) Aromatic heterocyclic amino acid
  (*d*) Basic amino acid
  (*e*) Acidic amino acid

2. What is the significance of the letters D and L in amino acid chemistry?

3. Would amino acids with a pI value of 3.2, 9.7, and 6.2 be considered as neutral, acidic, or basic amino acids? Explain.

4. Explain why proteins have isoelectric points.

5. Indicate whether the following compounds would be expected to give a positive or negative test with ninhydrin reagent. Which would liberate CO_2 during this test?

(*a*) 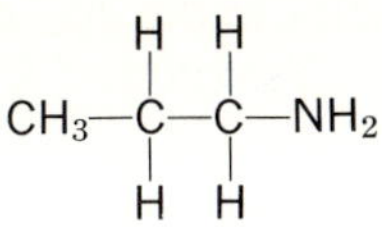 (*b*) 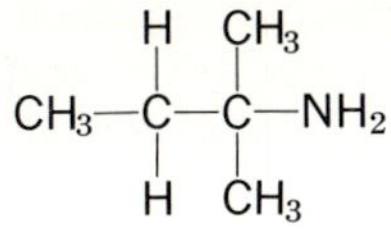 (*c*) 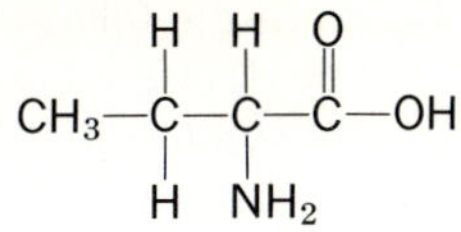

(*d*) 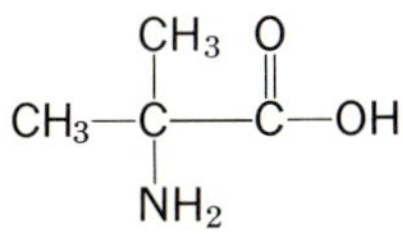 (*e*) 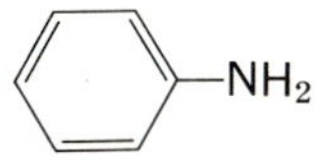 (*f*) 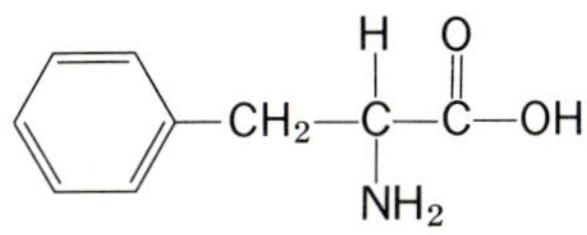

6. Given an unlabeled solution of cystine and one of cysteine, what simple test would you perform to distinguish between them?
7. Why are proteins not classified by their amino acid content alone?
8. Why is hemoglobin classified as a conjugated chromoprotein?
9. Label the following bonds and indicate their relationship to protein structure:

(*a*) 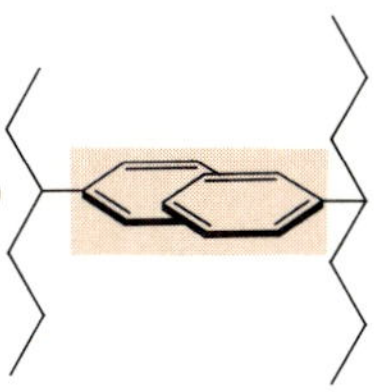 (*b*) 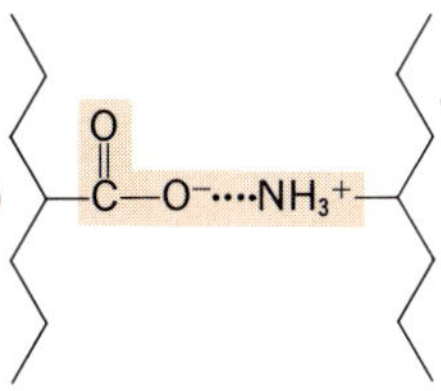 (*c*) 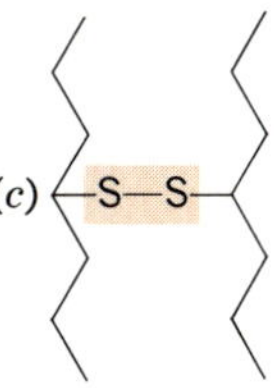

(*d*) 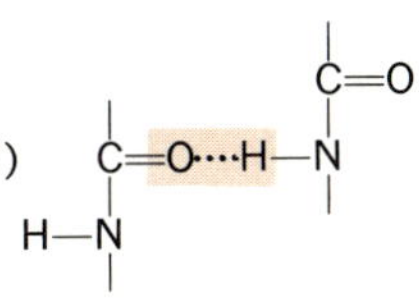 (*e*) 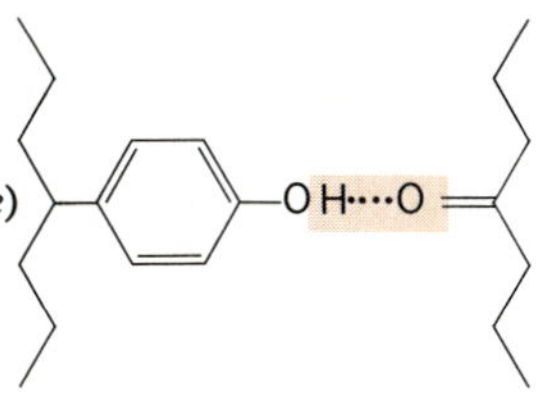

10. List the products formed upon hydrolysis of the following peptide with (*a*) trypsin, (*b*) chymotrypsin, and (*c*) pepsin:

Ala—Lys—Glu—Leu—Phe—Pro—Phe—Cys—Leu—Tyr

11. Time-course analysis of the hydrolysate of a protein with carboxypeptidase gave the following results. What conclusions can be drawn from the data?

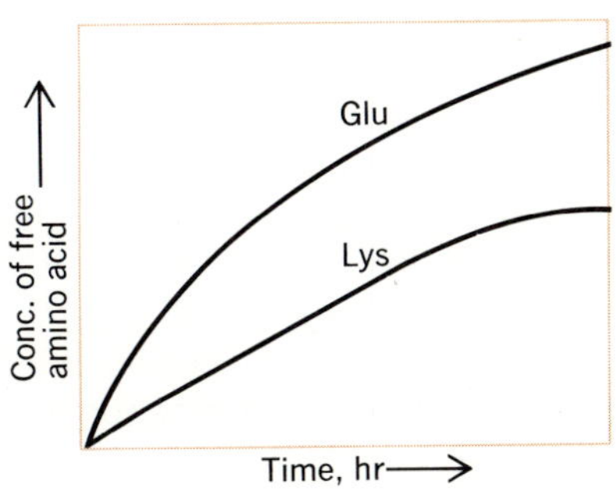

12. From the following data write the sequence of amino acids in the decapeptide (I) (10 amino acids). Any free amino acids formed are disregarded. *State the significance of each test.* Total amino acid content of I:

Amino acid	*Moles*	*Amino acid*	*Moles*
Ala	2	Phe	1
Ser	2	Tyr	1
Leu	1	Glu	1
Lys	1	Gly	1

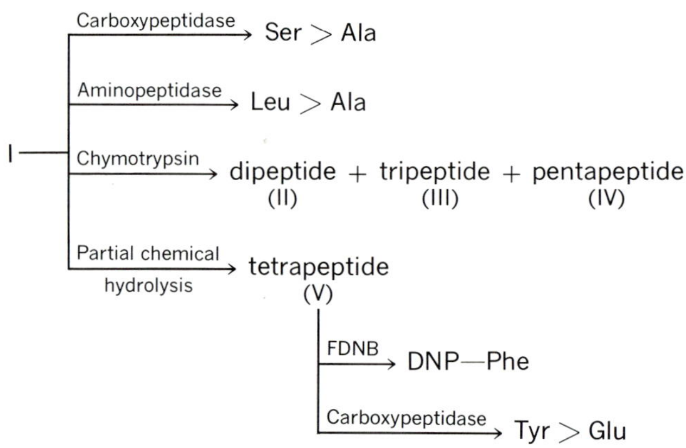

	Dipeptide II	*Tripeptide III*	*Pentapeptide IV*
Carboxypeptidase			Phe > Ser
Aminopeptidase			Leu > Ala
FDNB	DNP—Ala	DNP—Gly	DNP—Leu
			ε-DNP—Lys

	1	2	3	4	5	6	7	8	9	10
I	____	____	____	____	____	____	____	____	____	____

13. What is meant by the secondary structure of proteins?

14. Illustrate the bond responsible for the secondary structure of proteins.

15. What is meant by the helix and pleated-sheet structure as related to proteins?

16. Label the parts of the following schematic representation of a protein that illustrate the primary, secondary, and tertiary structure.

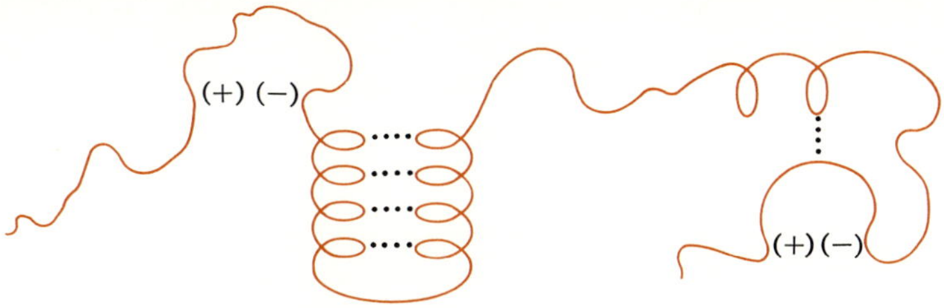

17. Why is the structure of an enzyme so important?

18. What is denaturation? Diagrammatically illustrate the denaturation of a protein.

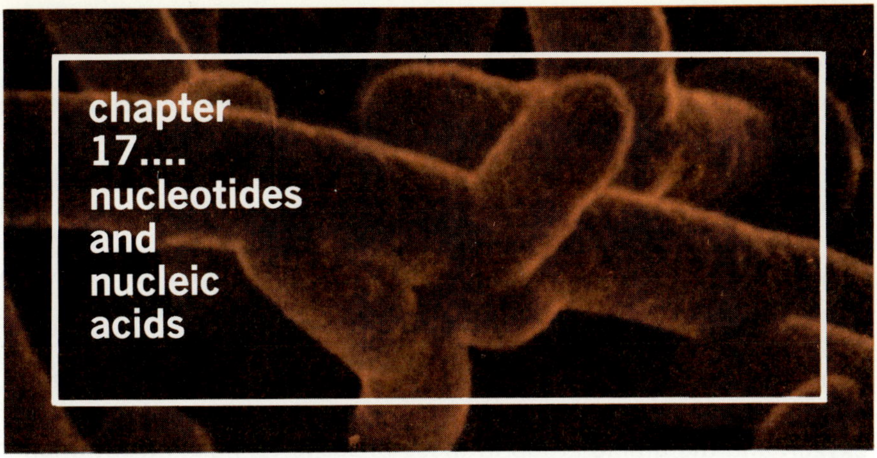

chapter 17.... nucleotides and nucleic acids

Nucleotides are substances which upon complete hydrolysis yield a nitrogen base, pentose, and phosphoric acid. The base components of the nucleotides are purine or pyrimidine derivatives, and the pentose is either ribose or deoxyribose. A mononucleotide consists of a base, pentose, and phosphoric acid; a dinucleotide consists of two nucleotides bonded together, and a polynucleotide consists of many nucleotides bonded together by phosphate ester bonds. The polynucleotides, called nucleic acids, are the prosthetic group of nucleoproteins.

Some nucleotides are coenzymes, and others are high-energy nucleotides associated with energy transfer. The nucleic acids are associated with genetic traits and protein synthesis. Nucleotides that are coenzymes are discussed later in the chapter dealing with enzymes and coenzymes. Here we shall discuss some of the chemistry of the high-energy nucleotides and nucleic acids. The following review questions are pertinent to this discussion.

What are the approximate pK values for phosphoric acid? Which ionic species of phosphoric acid would predominate at pH 7.4? Is pyrophosphoric acid an anhydride? Explain why HI dissociates to a greater extent than HCl. What is meant by resonance structures? What is the significance of *deoxy-* in deoxyribose? Why are purine and pyrimidine rings aromatic in nature?

17.1 NUCLEOSIDE AND NUCLEOTIDE COMPOSITION

Nucleosides contain only a nitrogen base and the sugar unit, whereas nucleotides also contain phosphate. A nucleotide, therefore, is a phosphorylated nucleoside.

Bases

The formulas for the common nitrogen bases found in nucleotides are given in Fig. 17.1. There are two main classes of bases, the pyrimidines and the purines. The nucleus of these compounds has aromatic properties. This is perhaps best shown for guanine, cytosine, uracil, and thymine by considering their enol formulas. Guanine is 2-amino-6-oxypurine, and adenine is 6-amino purine. The pyrimidines shown in Fig. 17.1 all contain oxygen in the C-2 position. In addition to this oxygen, cytosine contains an amino group on C-6, whereas uracil contains another oxygen on the C-6 position. Thymine is 5-methyl uracil, or 2,6-dioxy-5-methyl pyrimidine. As we shall see, the substitution of oxygen and the amino group in these bases is very significant when it comes to hydrogen bonding between bases.

Carbohydrates

There are only two pentoses contained in nucleotides, β-D-ribose and β-D-2-deoxyribose. In the latter sugar the 2-deoxy- indicates an oxygen

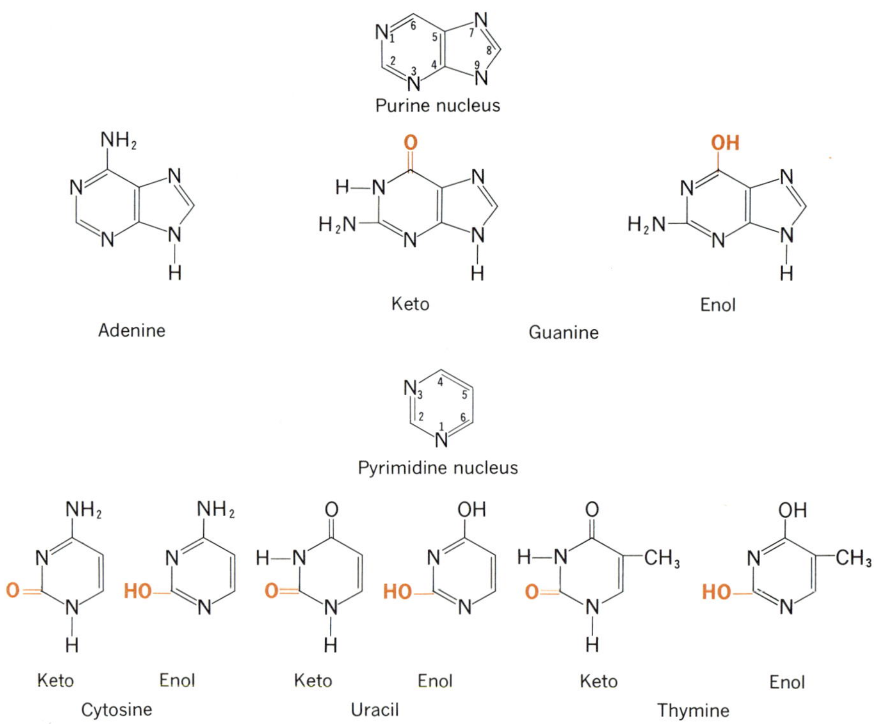

FIGURE 17.1 Common bases found in nucleotides.

atom has been removed from the second carbon; thus there are two hydrogen atoms on the C-2 of 2-deoxyribose. The formulas for D-ribose and 2-deoxyribose in the open-chain aldehyde form and in the cyclic furanose structure are

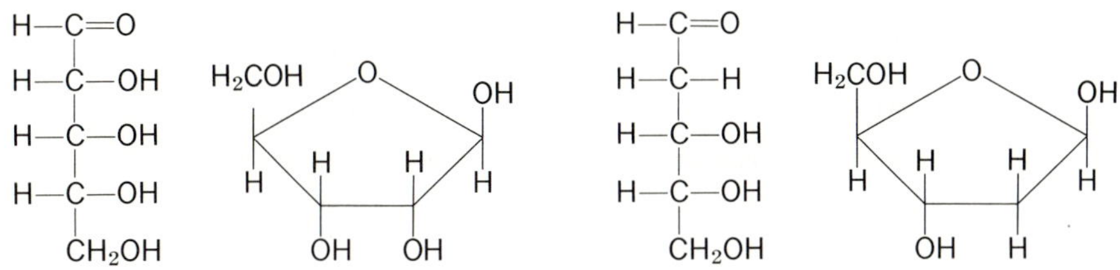

Aldehyde form | β cyclic form | Aldehyde form | β cyclic form

β-D-Ribose | β-D-2-Deoxyribose

The β form has the OH on C-1 trans to the OH on C-2. Hereafter these compounds will be referred to as ribose (R) and deoxyribose (D).

In both nucleosides and nucleotides the nitrogen bases are bonded to the pentose in a glycosidic link between the C-1′ (the carbon of the sugars are given prime numbers) of the pentose and the N-9 or N-1 of the purines and pyrimidines, respectively. Since this is an acetal linkage, the bond is alkali-stable and acid-labile with respect to hydrolysis.

Nucleosides

Compounds containing a nitrogen base bonded to the pentose are known as nucleosides.

Composition	*Compound*
Base–ribose	Riboside or ribonucleoside
Base–deoxyribose	Deoxyriboside or deoxyribonucleoside

As noted above, the pentose may be either ribose or deoxyribose. The formulas for several nucleosides are shown in Fig. 17.2. The nucleosides are named according to the base they contain. For example, if the base is adenine and the sugar is ribose, the nucleoside is called adenosine. If the sugar is deoxyribose, the name is deoxyadenosine. Table 17.1 lists the names of several nucleosides, and Fig. 17.2 shows several nucleoside structures.

Nucleotides

In addition to the base and ribose nucleotides have a phosphate group, which is bonded by an ester linkage usually to the C-5′ or C-3′ of the

Purines:

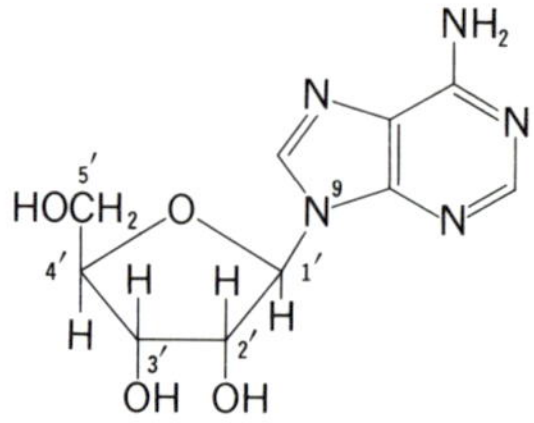

Ribose Adenine
Adenosine

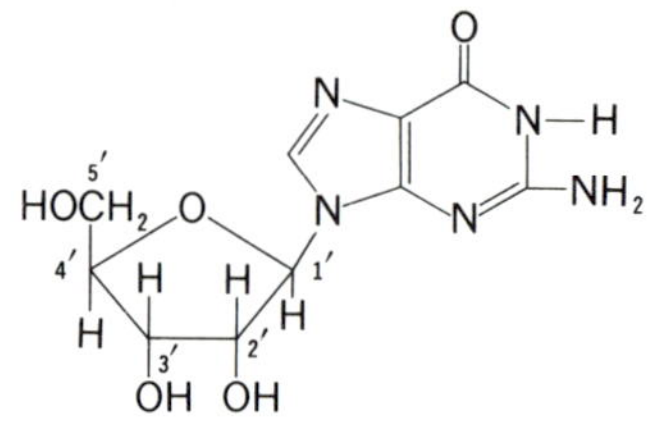

Ribose Guanine
Guanosine

Pyrimidines:

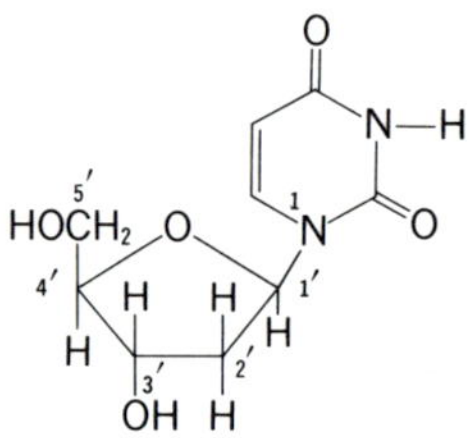

Deoxyribose Uracil
Deoxyuridine

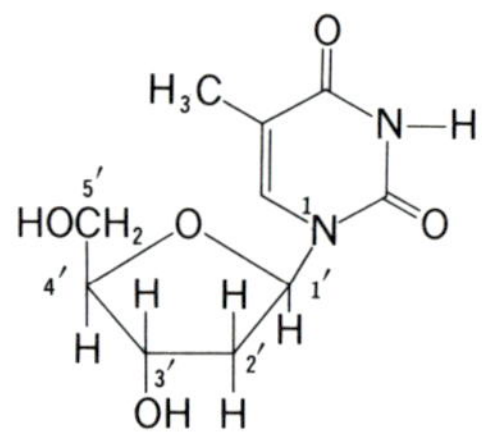

Deoxyribose Thymine
Deoxythymine

FIGURE 17.2 Formulas for some nucleosides.

ribose or deoxyribose moiety. They are therefore nucleoside phosphates. The phosphate ester group is hydrolyzed under basic conditions.

Composition	*Compound*
Base–ribose phosphate	Ribonucleotide, or riboside phosphate
Base–deoxyribose phosphate	Deoxyribonucleotide, or deoxyriboside phosphate

Base–ribose phosphate

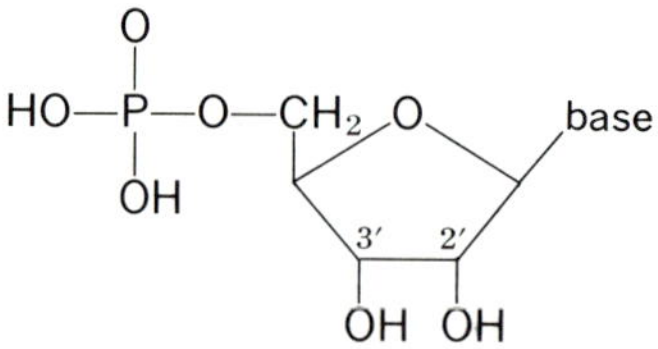

Ribonucleoside-5′-phosphate
Riboside-5′-phosphate

Base–deoxyribose phosphate

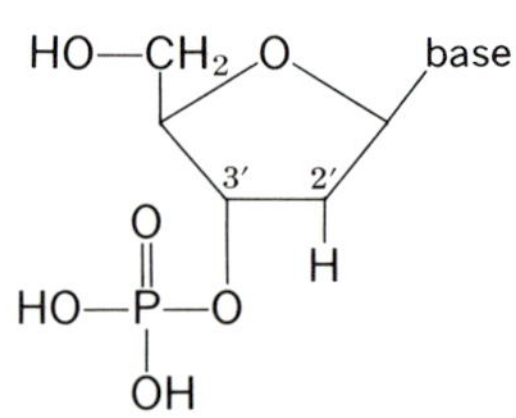

Deoxyribonucleoside-3′-phosphate
Deoxyriboside-3′-phosphate

These compounds may be named either as the nucleoside phosphate or as acids. For example, adenosine monophosphate is also

TABLE 17.1 NOMENCLATURE OF SOME NUCLEOSIDES AND NUCLEOTIDES

Base	Nucleoside		Nucleotide	
	Ribose	**Deoxyribose**	**Ribose**	**Deoxyribose**
Adenine	Adenosine	Deoxyadenosine	5′-Adenylic acid or adenosine-5′-monophosphate	5′-Deoxyadenylic acid or deoxyadenosine-5′-monophosphate
Guanine	Guanosine	Deoxyguanosine	5′-Guanylic acid or guanosine-5′-monophosphate	5′-Deoxyguanylic acid or deoxyguanosine-5′-monophosphate
Cytosine	Cytidine	Deoxycytidine	5′-Cytidylic acid or cytidine-5′-monophosphate	5′-Deoxycytidylic acid or deoxycytodine-5′-monophosphate
Uracil	Uridine	Deoxyuridine	5′-Uridylic acid or uridine-5′-monophosphate	5′-Deoxyuridylic acid or deoxyuridine-5′-monophosphate
Thymine	Thymidine	Deoxythymidine	5′-Thymidylic acid or thymidine-5′-monophosphate	5′-Deoxythymidylic acid or deoxythymidine-5′-monophosphate

called adenylic acid. The nomenclature of some nucleotides is given in Table 17.1, with only the 5′-monophosphates listed. They may also occur as a di- or triphosphate derivative. The formulas for the mono-, di-, and triphosphate derivatives of adenosine are

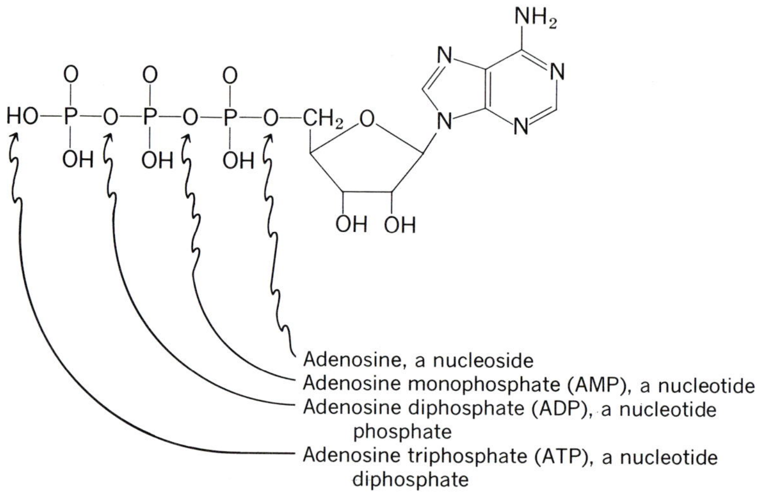

Some nucleotide phosphates trap energy from biological reactions and pass it on to energy-requiring reactions. These nucleotides are called ***high-energy compounds.*** The most common high-energy compound is ATP (adenosine triphosphate, a nucleotide diphosphate or nucleoside triphosphate). Several other triphospho nucleosides are high-energy compounds, but they function in more or less rather specific reactions whereas ATP is general. In biological systems the energy obtained from, or required by, reactions is frequently measured in terms of the number of moles of ATP formed or used. For example, the complete biological oxidation of glucose is equivalent to 38 ATP/mole of glucose oxidized. Other triphosphates will be discussed as they arise.

Looking more closely at the formulas for AMP, ADP, and ATP (see page 279), we can see that AMP contains a phosphate ester bond, ADP contains, in addition, a phosphate anhydride bond, and ATP contains both, together with a second anhydride bond. It is the anhydride bonding that imparts high energy to ATP. The energy is released when a phosphoric anhydride bond is broken. As there are two anhydride bonds present, there are two ways ATP can react, namely, by splitting off either 1 mole of phosphate or 1 mole of pyrophosphate:

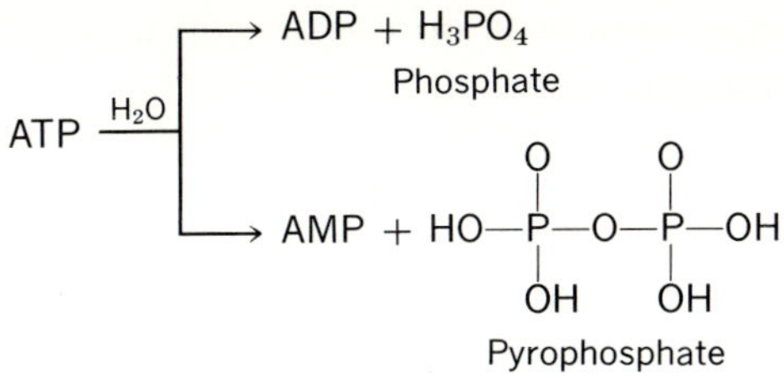

High-energy nucleotides

Why is ATP a high-energy compound whereas AMP is not? The energy referred to is the energy that is available for work, the *free energy G*. The amount of energy obtainable from a reaction is the difference between the free-energy content of the reactants and the final products, or the change in free energy ΔG of the reaction. Free energy released during a reaction is indicated as $-\Delta G$. If energy is required to drive a reaction forward, the ΔG is positive $+\Delta G$.

When a reaction proceeds with the liberation of energy ($-\Delta G$), it means that the products of the reaction are at a lower energy level than the reactants and therefore are more stable. A phenomenon that tends to stabilize an ion or molecule already discussed is the delocalization of electrons, which was illustrated by the dissociation of acids (such as the halogen acids) and the resonance formulas for organic acids. A notable example of the stabilizing effect of resonance is found in aromatic ring structures like benzene, purines, and pyrimidines. The greater the delocalization of electrons the more stable the compound. This principle can be applied to the discussion of phosphoric acid derivatives.

The pK values for H_3PO_4 are, in round numbers, 2, 7, and 12 for the three hydrogens. Thus, at body pH the first hydrogen is completely ionized, the second hydrogen is 60 to 70% ionized, and the third hydrogen is not ionized. When the first two hydrogens are completely ionized, HPO_4^{--} has three resonance formulas:

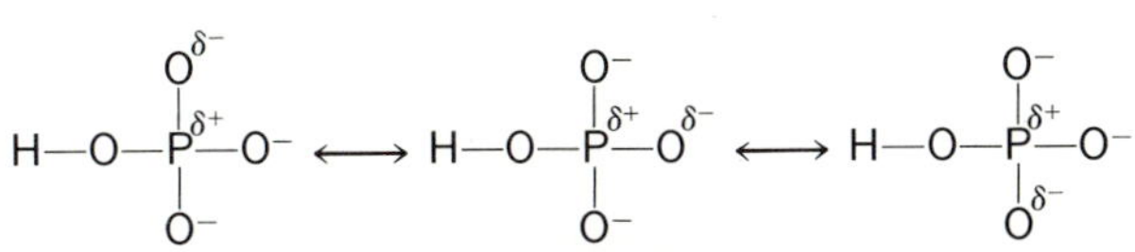

Resonance formulas for HPO_4^{--}

The resonance structures stabilize the ions because of delocalization of the electrons. Note the partial positive charge on the phosphorus and the partial negative charge on the oxygen due to the shift of electrons toward the oxygen atom,

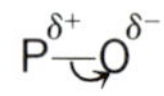

This shift is also seen in the resonance formulas of AMP, in which the adenosyl radical (R = adenine-ribose) replaces the H in HPO_4^{--}:

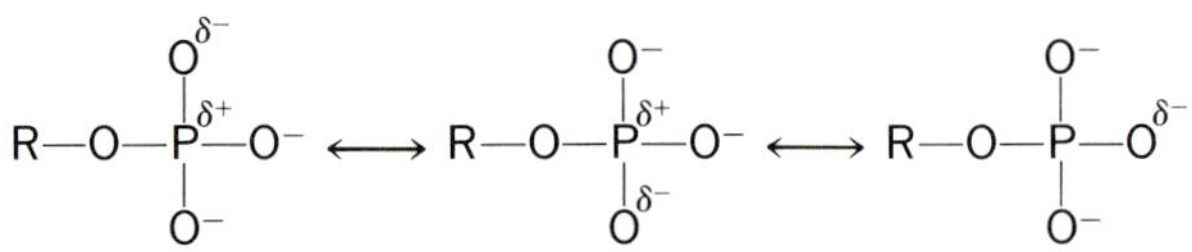

AMP

Now consider the resonance structures of ADP and ATP shown in Fig. 17.3 (see also Table 17.2). In addition to the three resonance formulas for the end phosphate of ADP there are two resonance formulas for the first phosphate group (the ester phosphate). This makes a total of $2 \times 3 = 6$ resonance formulas. In ATP there are four resonance structures for the first two phosphates and three for the terminal phosphate; this gives a total of $4 \times 3 = 12$ resonance formulas for ATP.

There are more resonance combinations for AMP + HPO_4^{--} ($3 \times 3 = 9$) than there are for ADP. This fact alone means that to unite HPO_4^{--} and AMP to form ADP would require energy to overcome resonance. Thus, in the reaction ADP $\longrightarrow$ AMP + HPO_4^{--} additional energy is released ($-\Delta G$). For ADP + HPO_4^{--} there are $6 \times 3 = 18$ combinations and for ATP there are only 12 resonance formulas; therefore, the reaction ATP $\longrightarrow$ ADP + HPO_4^{--} would release energy.

The energy released upon hydrolysis of the anhydride bond of

TABLE 17.2 NUMBER OF RESONANCE FORMULAS FOR HPO_4^{--} AND SOME NUCLEOTIDES

Compound	Number of resonance formulas
HPO_4^{--}	3
AMP	3
ADP	$2 \times 3 = 6$
ATP	$4 \times 3 = 12$
AMP + HPO_4^{--}	$3 \times 3 = 9$
ADP + HPO_4^{--}	$6 \times 3 = 18$

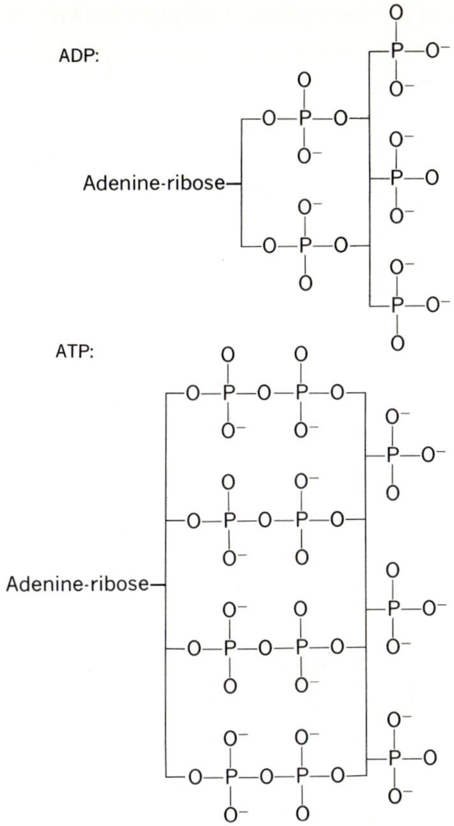

FIGURE 17.3 Resonance formulas for ADP and ATP.

ADP and ATP is greater than can be accounted for by the energy required merely to overcome resonance. As indicated in the resonance formulas, each phosphorus atom has a positive charge. These charges exert a repulsion on each other; therefore, it requires more energy to keep the phosphate anhydride bond intact than would be needed if the phosphorus atoms were not positively charged. Thus the high energy content of these nucleotides is a result of two factors: (1) the difference in the number of possible resonance formulas between the reactant and the products and (2) the repulsion of like charges. *The energy content of a compound is associated with the molecule rather than with just one single bond.* However, we commonly speak of the high-energy phosphate bond. These high-energy bonds are indicated by ∼ rather than a straight line:

Adenine—ribose—P∼P∼P

The difference in the energy content of AMP, ADP, and ATP can be compared by considering the energy released when the phosphate group is hydrolyzed off. The ester phosphate bond of AMP yields between 2000 and 3000 cal/mole of AMP hydrolyzed ($\Delta G = -2000$ to -3000 cal/mole). The hydrolysis of the anhydride bond of ADP to form AMP and HPO_4^{--} liberates 6000 to 7000 cal/mole ($\Delta G = -6000$ to -7000 cal/mole). ATP contains two phosphoric acid anhydride bonds and three $P^{\delta+}$ atoms; consequently its energy content is greater than that of ADP. When the terminal phosphate is hydrolyzed from ATP to yield ADP + HPO_4^{--}, about 8000 cal is liberated ($\Delta G = -8000$ cal/mole).

In summary, the useful energy obtainable by removal of phosphate from nucleotides is as follows:

$$\text{ATP} \longrightarrow \text{ADP} + HPO_4^{--} \qquad \Delta G = -8000 \text{ cal/mole}$$
$$\text{ADP} \longrightarrow \text{AMP} + HPO_4^{--} \qquad \Delta G = -6500 \text{ cal/mole}$$
$$\text{AMP} \longrightarrow \text{adenosine} + HPO_4^{--} \qquad \Delta G = -2200 \text{ cal/mole}$$

Other nucleotides that form di- and triphosphate derivatives have approximately the same energy content as indicated above.

Dinucleotides are composed of two nucleotides linked together by a phosphoric acid anhydride bond or by a phosphoric acid ester bond between the sugars of the respective mononucleotides.

Base—ribose—phosphate—**phosphate—ribose—base**
Nucleotide **Nucleotide**
Dinucleotide

Base—ribose—phosphate—**ribose—base**
|
phosphate
Nucleotide **Nucleotide**
Dinucleotide

Dinucleotides that contain the anhydride link occur as coenzymes, for example, NAD^+, $NADP^+$, and FAD; the formulas for these are discussed under coenzymes.

Dinucleotides linked together by a phosphate ester bond are partial hydrolytic products of the nucleic acids, just as maltose is a hydrolytic product of starch.

17.2 NUCLEIC ACIDS

Nucleic acids are the prosthetic group of conjugated proteins known as nucleoproteins. Cells contain two types of nucleic acids, which have received and are still receiving intensive investigation since their discovery. The importance of these acids is indicated by the fact that one nucleic acid, deoxyribonucleic acid (DNA), is the substance in the nucleus of cells which contains the genetic code. It is responsible for passing on hereditary traits from one generation to another. This compound is duplicated and passed on to each newly formed cell, where it is responsible for thousands of cellular reactions by virtue of its role in the synthesis of enzymes. DNA ultimately dictates the sequence of amino acids in proteins.

The nucleic acid in the cytoplasm of cells is called *ribonucleic acid* (RNA) because it contains the sugar ribose. Two types of RNA will be considered further with respect to protein synthesis, *transfer RNA* (*t*RNA), which forms a complex with amino acids and transports the amino acids to the *messenger RNA* (*m*RNA) for protein synthesis. The order of the nucleotides in both *t*RNA and *m*RNA is determined by the sequence of nucleotides in DNA. The role of nucleic acids in protein synthesis will be discussed in a subsequent chapter. Here some of the chemistry of the nucleotides and the structure of DNA and RNA are considered.

Base composition

DNA and RNA have three nitrogen bases in common, namely, the purines adenine and guanine and the pyrimidine cytosine. They differ, however, in their content of the bases uracil and thymine, both pyrimidines, uracil being found in RNA and thymine in DNA. The nitrogen base compositions of DNA and RNA are summarized in Table 17.3.

In DNA from any source there is a 1:1 molar ratio of adenine to thymine and of guanine to cytosine. Although the amount of these pairs may differ with the source of DNA, the ratio remains the same.

TABLE 17.3 COMMON BASES OF DNA AND RNA

	RNA			
		DNA		
Purines		Adenine (A) Guanine (G)		In DNA each pair of bases (A/T, C/G) occurs in a 1:1 molar ratio
Pyrimidines	Uracil (U)	Cytosine (C)	Thymine (T)	

This is a very important fact, for as we shall see shortly, it is the hydrogen bonding between these respective base pairs that gives DNA its helical structure.

Structure

Nucleotides are linked together to form a polymer by means of a phosphate ester bond between the C-5′ OH of the pentose of one nucleotide and the C-3′ OH of the pentose of another nucleotide unit. Thus, there is a phosphoric acid diester bond between two adjacent nucleotides, as shown for the two bases adenine and uracil and the pentose ribose. For nucleic acids containing deoxyribose, this same bonding occurs.

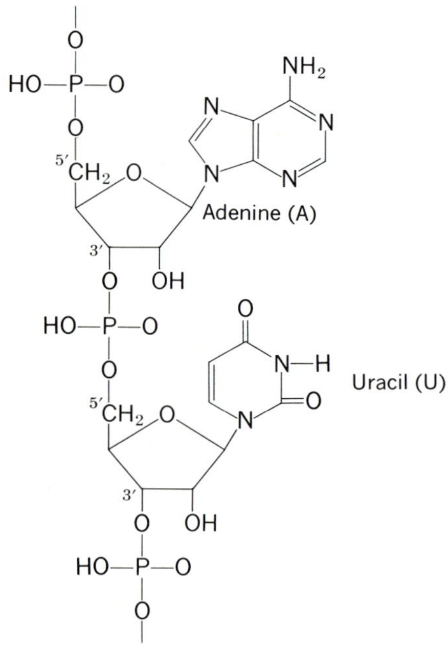

There is no doubt that writing a polynucleotide using structural formulas becomes cumbersome. A shorthand has been devised for writing the structure of nucleotides. The bases are indicated by their initial, the ribose by a straight line extending from the base, and the phosphate by P. The C-3′ and C-5′ of the ribose or deoxyribose are indicated by the fact that the C-5′ is at the end of the ribose line and the C-3′ is toward the middle of the line. Thus if adenosine and thymidine are joined by a phosphate from the C-3′ of adenosine to the C-5′ of thymidine, it can be represented as

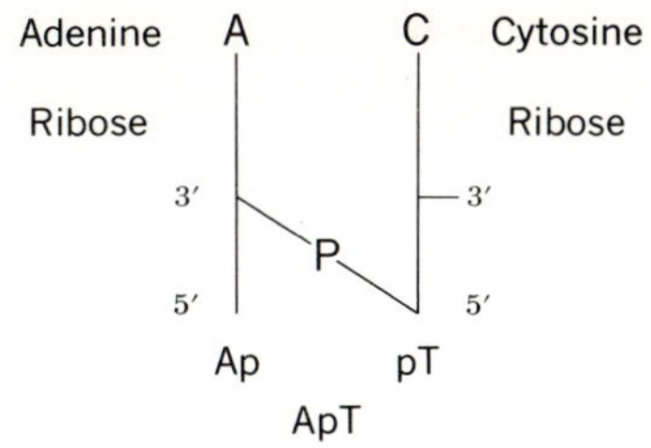

The shorthand for this compound is ApC (A = adenine, p = phosphate, C = cytosine). If the phosphate group P is to the right of the base (as in Ap), the phosphate is connected to the 3′ position of the ribose joined to that base. On the other hand, if the phosphate group is to the left of the base (as in pT), the phosphate is joined to the 5′ position of the ribose connected to the base.

The schematic representation of a portion of a nucleic acid is

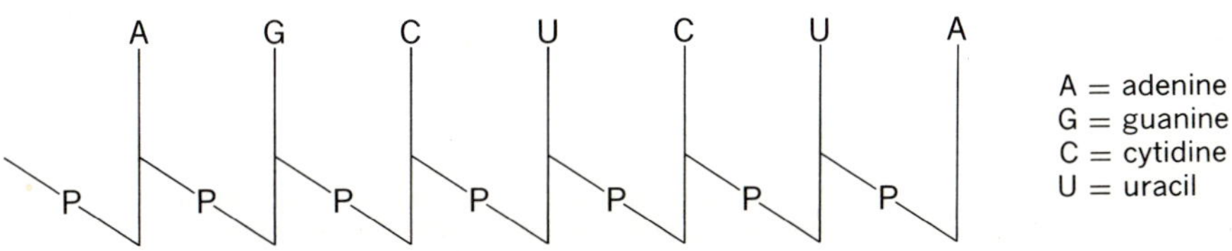

A polynucleotide (RNA), pApGpCpUpCpUpA (nucleic acid)

17.3 TRANSFER, OR SOLUBLE, RNA

The structure of RNAs may be illustrated by the polynucleotide above. Nucleic acids are divided into several categories according to their solubility, or function and occurrence, or both.

A group of RNAs has sufficiently low molecular weight (20,000 to 30,000) to be soluble in water; hence, the name soluble RNA (*s*RNA). As the function of these compounds became better understood, they were found to be involved in the transfer of amino acids from solution into the structure of protein and they became more appropriately known as transfer RNA (*t*RNA).

There are as many *t*RNAs as there are amino acids that occur in proteins because there is a specific *t*RNA for each amino acid. For example, the *t*RNA that complexes with alanine does not complex with other amino acids.

The complete sequence of bases of several *t*RNAs has been ascertained; Holley received a Nobel award in 1968 for determining the

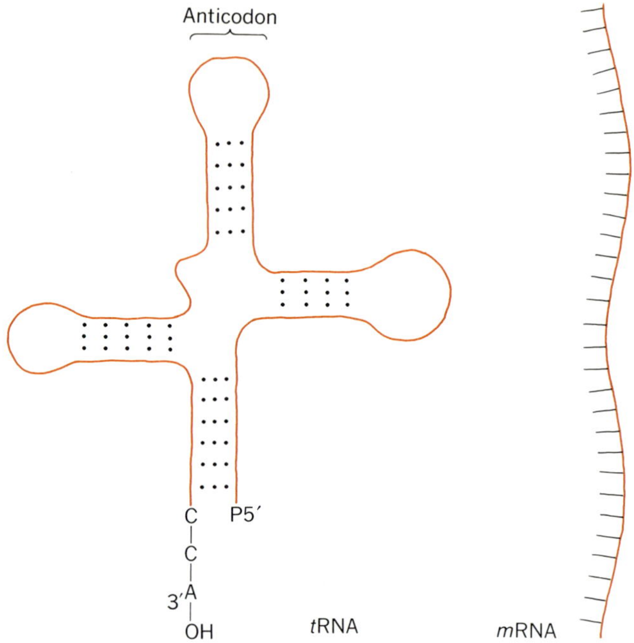

FIGURE 17.4 The structure of *t*RNA and *m*RNA. The solid line in *t*RNA represents the polynucleotide chain; dotted lines represent hydrogen bonding between complementary bases. The 3′ OH end of *t*RNA is the base sequence CpCpA, and the 5′ end is a 5′-phosphate. The anticodon is at the end of the loop indicated in the diagram. In *m*RNA the broad line represents the ribose-phosphate chain, and the small lines extending from it represent the bases.

sequence of bases in Ala—*t*RNA. The structure of Ala—*t*RNA is schematically represented in Fig. 17.4. Loops in the molecule are formed by hydrogen bonding between base pairs of the chain (see page 434).

Although each *t*RNA is specific for a particular amino acid, all *t*RNAs have cytidine-cytidine-adenine (C—C—A) as the terminal sequence of bases at one end of the chain. The *t*RNAs transport amino acids by forming an ester linkage with the C-3′ OH of ribose attached to the terminal adenine.

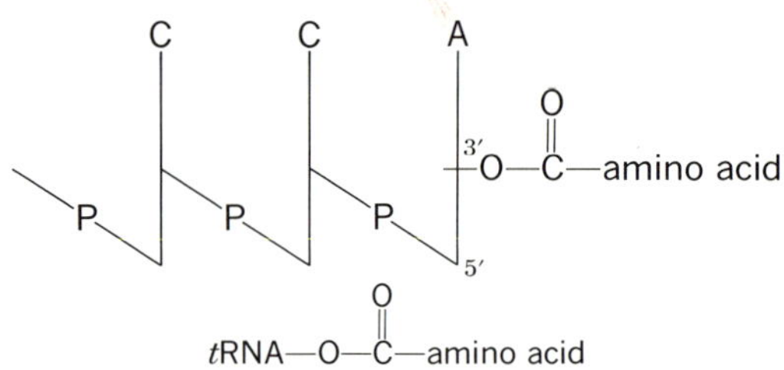

17.4 MESSENGER RNA

Other RNAs, of molecular weight between 200,000 and 50 million, are associated with both the *t*RNA and the ribosomes (particles of ribonucleoprotein) of the cells during the process of protein synthesis. Called messenger RNA (*m*RNA), this RNA is synthesized from nucleotides in the nucleus, and the order of bases in its structure is governed by the order of bases in DNA. Once formed, the *m*RNA migrates from the nucleus into the cytoplasm, where protein synthesis occurs. The order of bases in *m*RNA governs the sequence of amino acids in protein. Thus, the DNA has passed its information on to *m*RNA, which has acted as a messenger to take the information from DNA in the nucleus to the cytoplasm. This process will be discussed more in detail in the section dealing with amino acid and protein metabolism. For now the brief schematic representation of the function of these nucleic acids in protein synthesis in Fig. 17.5 will suffice.

The structure of *m*RNA is less well known than that of DNA. The RNA associated with protein synthesis is considered to be a ribbon of nucleic acids with the bases jutting out from the backbone of ribose-phosphate units, as indicated in Fig. 17.4. The *m*RNA is able to attach itself to the ribosomes during protein synthesis in a rather loose manner, for the ribosomes are thought to slide along the *m*RNA structure as proteins are being synthesized. Another structure found in some RNAs is a loop, where for a short distance there is a double-strand structure, as noted in Ala—*t*RNA. Some viruses are composed of a single strand of RNA, whereas others are a double-stranded helix similar to that of DNA (Fig. 17.7).

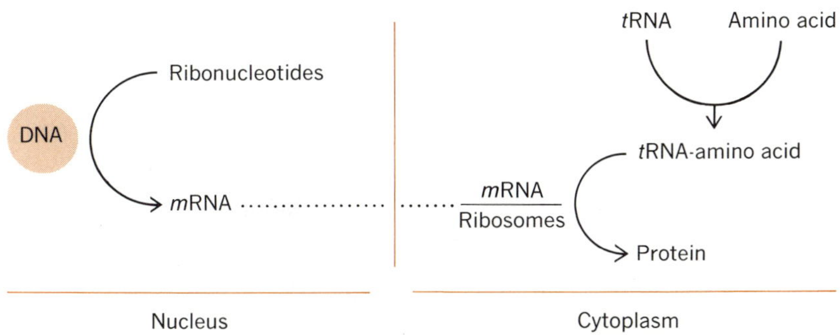

FIGURE 17.5 The role of DNA, *t*RNA, and *m*RNA in protein synthesis, greatly simplified.

17.5 DEOXYRIBONUCLEIC ACID

Deoxyribonucleic acid (DNA) is a high-molecular-weight polymer of deoxyribonucleotides and is composed of two polynucleotide chains each containing between 10^4 and 10^8 nucleotides. DNA occurs mainly in the nucleus of cells; however, it is also found to a limited extent in mitochondria and chloroplasts. DNA contains the bases adenine, thymine, guanine, and cytosine.

Examination of the bases of DNA from different sources has revealed a relationship between the amount of adenine-thymine and guanine-cytosine in the DNA. Adenine and thymine occur in a ratio of 1:1, and thymine and guanine are also found in this molecular ratio. Although the concentration of these bases varies in DNA from different sources, the ratios are consistent. The total amount of purines, adenine and guanine, has been found to equal the total amount of pyrimidines, thymine and cytosine. These relationships between the bases have a direct influence on the structure of DNA.

In 1953 Watson and Crick proposed the structure of DNA for which they later received the Nobel Prize. The deoxynucleotides are bonded together by a phosphate ester link from the C-5′ OH of one nucleotide to the C-3′ OH of another nucleotide. This is similar to the structure presented for RNA. DNA is double-stranded, two strands of deoxynucleotides being held together by hydrogen bonding between bases of the two chains. Hydrogen bonding between the chains is restricted to the pairs adenine and thymine and cytosine and guanine, thus accounting for the bases of each pair occurring in a molar ratio of 1:1. Also since each of these pairs consists of a pyrimidine and a purine, this structure also accounts for the fact that the total amount of pyrimidines equals the total amount of purines in DNA. Hydrogen bonding between adenine and thymine and between cytosine and guanine holds the two strands of nucleotides the same distance apart throughout the molecule.

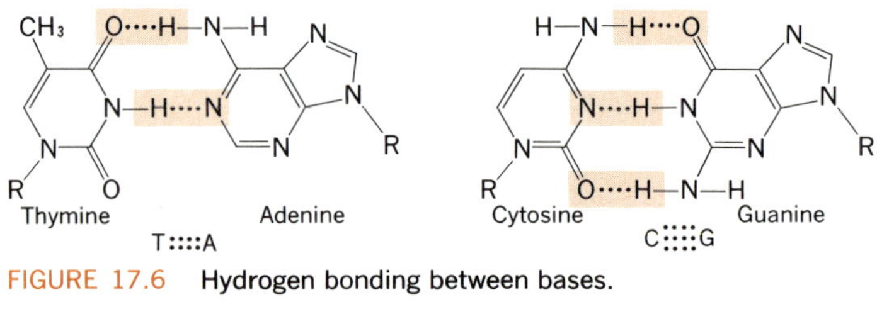

FIGURE 17.6 Hydrogen bonding between bases.

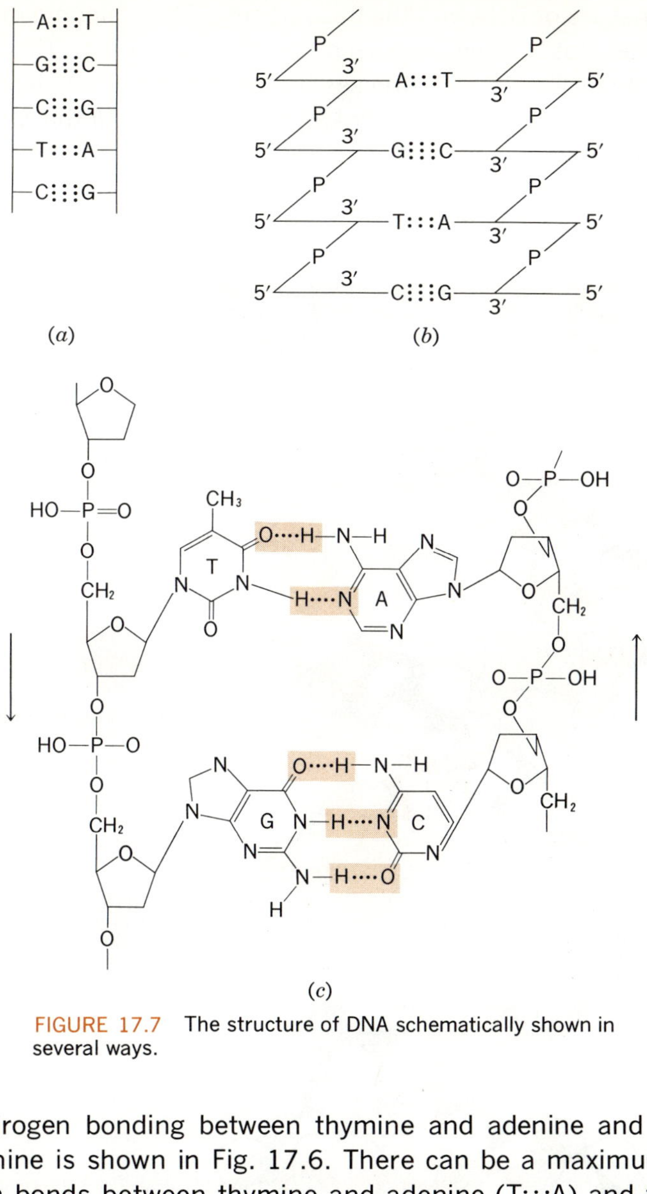

FIGURE 17.7 The structure of DNA schematically shown in several ways.

Hydrogen bonding between thymine and adenine and cytosine and guanine is shown in Fig. 17.6. There can be a maximum of two hydrogen bonds between thymine and adenine (T:::A) and three between cytosine and guanine (C⁝⁝⁝G). A portion of DNA in Fig. 17.7 shows the hydrogen bonding between two chains. As with RNA, shorthand or schematic methods are used to indicate the structure.

There are several ways of schematically indicating the double-stranded nature of DNA and the base pairing. They become more complicated the closer they approach the true situation. The simplest representation (Fig. 17.7*a*) stresses the base pairing but gives no indication of the sugar-phosphate linkages. In Fig. 17.7*b* the phosphate-deoxyribose linkage is indicated as well as the base pairing, while Fig. 17.7*c* shows the structural formula. It will be noted that the two strands

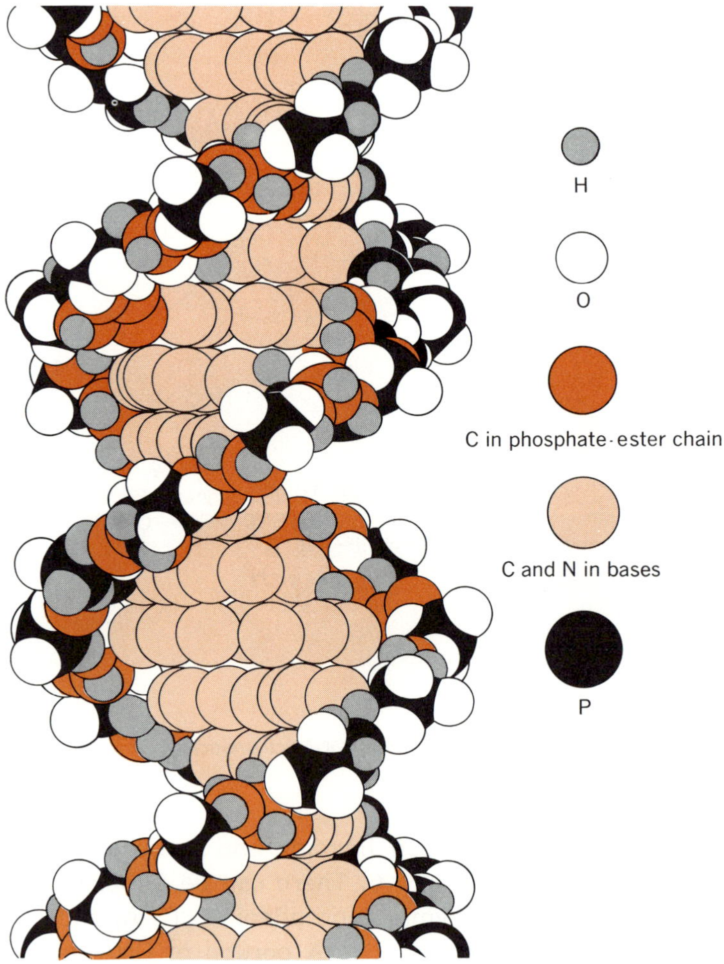

FIGURE 17.8 A molecular model of a portion of a DNA molecule. The size of the circles is proportional to their respective atomic radii. *(Redrawn from Gunther S. Stent, "Molecular Biology of Bacterial Viruses," W. H. Freeman and Co., San Francisco, 1963.)*

are running in opposite directions, i.e., antiparallel. It is as though a single chain was bent around in a U and base pairing occurred between the two side arms.

The two strands of DNA are twisted around each other to form a double-stranded helix. A model of this helix structure is shown in Fig. 17.8. DNA may occur as an open chain or, as in some bacteria, in a cyclic structure.

QUESTIONS

1. (*a*) What two classes of bases are found in nucleic acids?
   (*b*) Write the basic structure for the two classes of bases.
2. Classify each of the following compounds as pyrimidine or purine:

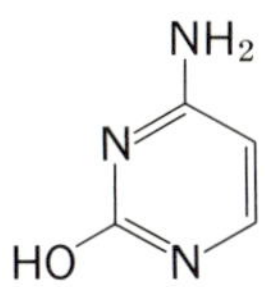

Cystosine

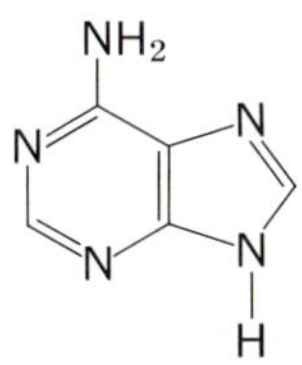

Adenine

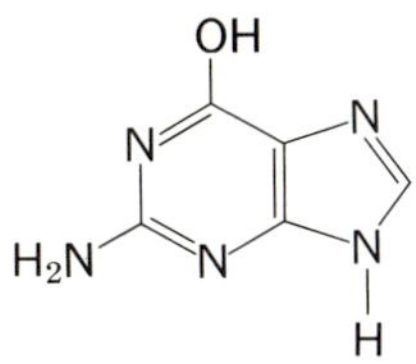

Guanine

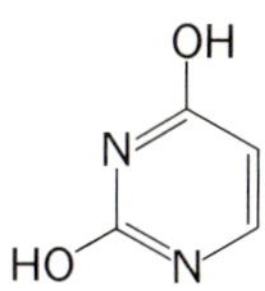

Uracil

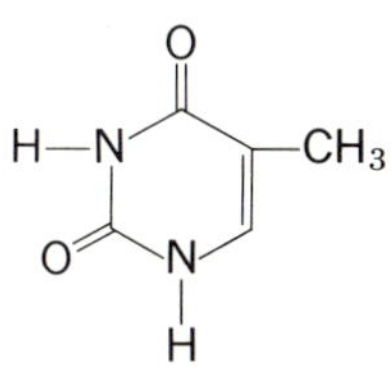

Thymine

3. Write the structural formulas for β-D-ribofuranose and β-D-deoxyribofuranose.
4. Name the products obtained upon complete hydrolysis of (*a*) adenosine and (*b*) deoxyadenosine.
5. Is adenosine-5′-pyrophosphate another name for adenosine diphosphate?
6. Explain why a compound like adenosine triphosphate (ATP) has high energy.
7. What is the difference between the composition of RNA and DNA?
8. What is the structural difference between *m*RNA and DNA?
9. What type of bond links the pentose with the bases in nucleic acids?
10. Following is a portion of a DNA molecule. The bases of one strand (A) are given.
    (*a*) Complete the diagram, showing the appropriate sequence of bases in strand B.
    (*b*) What would be the sequence of bases in the RNA strand (C) complementary to strand A?

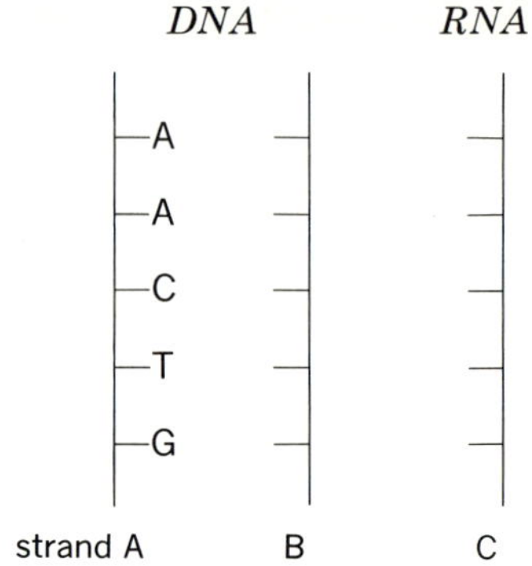

11. What is the biological significance of *t*RNA and *m*RNA?
12. Why is DNA important?

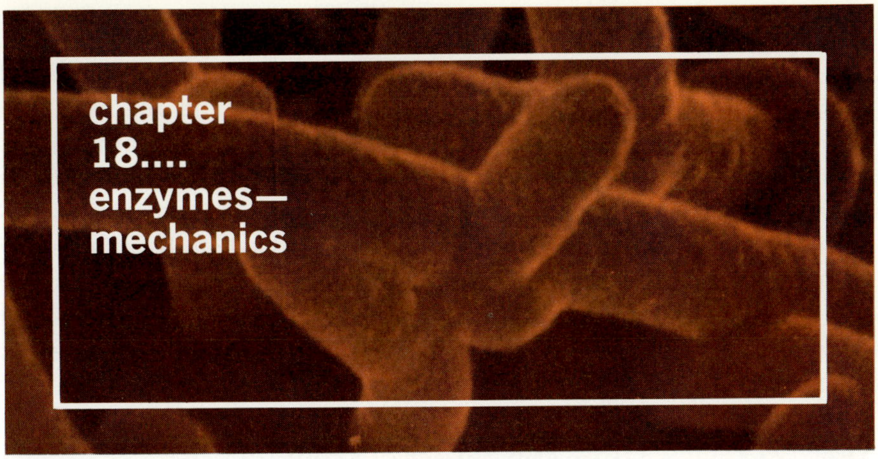

chapter 18.... enzymes—mechanics

Enzymes are a special class of proteins that are unique in that they are also catalysts. Produced by living cells, enzymes permit reactions to take place which normally would not do so at a reasonable rate.

The catalytic effect of enzymes is striking; for example, 1 mole of enzyme may catalyze the conversion of up to 1 million moles/min of substrate. Given the required components, cells can make compounds as complex as enzymes, which are so important in sustaining life. The cellular synthesis of enzymes becomes even more impressive when it is realized that many enzymes are so specific they react with only one compound. The many different enzymes in a cell are not located haphazardly. There is a definite location and arrangement of enzymes in the cell. Nor do all enzymes have the same activity. Enzyme activity is integrated with the need of the cell for the products of the reaction. For example, the need for large amounts of energy by organisms necessitates that the enzymes involved in the release of energy from metabolites be quite active to meet the demand for energy. Although many enzymes are the same in different organisms, the individual cellular control of them may differ quite markedly. Some enzymes are characteristic for a particular organism, e.g., enzymes associated with plant photosynthesis.

How does a catalyst increase the rate of a chemical reaction? Does it change the equilibrium constant of the reaction? Does a catalyzed reaction give off or absorb more energy than if it were noncatalyzed? Are there any similarities or dissimilarities between organic and inorganic catalyzed reactions? These are some questions that will be considered in the following discussion. First, however, answer the following review questions.

If equilibrium is attained when one-half the reactants have reacted (A = B), what is the value of K_{eq}? When would the reaction rate between A and B (the velocity of the forward reaction) be at its maximum—immediately upon mixing A and B or after A and B have been mixed for some time? How can equilibrium reactions be made to go to completion? Which structure of protein is most likely to be affected by a change in pH, primary, secondary, or tertiary? Does resonance stabilize a molecule or ion? What is the basic ring structure for purines and pyrimidines? Do aromatic aldehydes react with primary amines? Is pyrophosphate an anhydride?

18.1 CLASSIFICATION AND NOMENCLATURE

The enzymes that were discovered first received common names, which frequently had no relevance to their action; e.g., *ptyalin* is an enzyme found in the saliva which hydrolyzes starch. Later a more systematic classification and nomenclature was used, in which the enzyme is named after its substrate (the compound it acts upon). Accordingly ptyalin was renamed *amylase* because it acts upon amylose (starch). Other examples are *sucrase,* the enzyme that hydrolyzes sucrose, and *protease*, an enzyme that hydrolyzes proteins. The suffix *-ase* is reserved for enzyme names.

In 1961 a commission was confronted with the task of systematically naming the enzymes. They classified all enzymes into six main groups, which will be given with only one or two subclasses as examples.

1. *Oxidoreductases* are associated with oxidation-reduction reactions.
   (*a*) *Dehydrogenases* remove hydrogen from a substrate and pass it on to another organic compound. For example, succinic dehydrogenase and lactic dehydrogenase oxidize succinic acid to fumaric acid and lactic acid to pyruvic acid, respectively, by the removal of hydrogen from the substrate.
   (*b*) *Oxidases* pass hydrogen from the substrate to molecular oxygen.
   (*c*) *Oxygenases* incorporate oxygen into an organic molecule.
2. *Transferases* bring about the transfer of a group from one molecule to another molecule.
   (*a*) *Transaminases* transfer the amino group from an α-amino acid to the α carbon of an α-keto acid.
   (*b*) *Glucokinase* transfers a phosphate group from adenosine triphosphate (ATP) to glucose to form glucose-6-phosphate and adenosine diphosphate (ADP).

3. *Hydrolases* hydrolyze the substrate and include a large number of enzymes, e.g., those which hydrolyze proteins, carbohydrates, and fats.
   (*a*) *Carbohydrases* hydrolyze carbohydrates, starch to maltose, or sucrose to glucose and fructose.
   (*b*) *Phosphatases* hydrolyze off phosphate groups, glucose-6-phosphate to glucose and phosphoric acid.
   (*c*) *Lipases* hydrolyze fats to the corresponding fatty acids and glycerol.
4. *Lyases* split covalent links and thus bring about the removal of a group from the substrate other than by hydrolysis; the reactions *are reversible.*
   (*a*) *Decarboxylases* remove CO_2 from a carboxyl group, as in the conversion of oxaloacetic acid to pyruvic acid.
   (*b*) *Fumarase* removes water from malic acid to form fumaric acid.
5. *Isomerases* react on a substrate intramolecularly to form an isomer of the substrate.
   (*a*) *Phosphohexoisomerase,* e.g., *glucosephosphate isomerase,* which converts glucose-6-phosphate to fructose-6-phosphate.
   (*b*) *Intramolecular transferases,* or *mutases,* transfer a group from one carbon to another carbon in the same molecule to make an isomer. Such an enzyme is responsible for the conversion of 3-phosphoglyceric acid to 2-phosphoglyceric acid.
6. *Synthetases* or *ligases* join two molecules.

18.2 ENERGETICS

The energetics of a reaction has to do with the amount of energy released or absorbed during a chemical reaction. It is a very important concept in biochemistry because cellular processes must proceed so as to obtain energy from food and use it in subsequent reactions for maintenance and growth.

Free energy

The energy referred to in this discussion is energy that is available for work, called *free energy* or *useful energy* and indicated by the symbol G. The energetics of a reaction is considered from the viewpoint of the reaction; therefore, *if energy is given off* during a reaction, that amount of energy is lost from the system and is a *negative value.* On the other hand, *if energy is applied* to a chemical reaction, this amount of energy is taken into the system and is a *positive value.* The energy available from, or required by, a chemical reaction is the *difference* ΔG between the useful energy of the reactants and the products.

$$\text{A} + \text{B} - \text{energy} \longrightarrow \text{C} + \text{D} \qquad \Delta G = \text{negative}$$

The energy content of the reactants is greater than the energy content of products; useful work is obtainable from the reaction.

$$\text{C} + \text{D} + \text{energy} \longrightarrow \text{A} + \text{B} \qquad \Delta G = \text{positive}$$

The energy content of the reactants is less than the energy content of the products; useful work is not obtainable from the reaction.

Consider the following exergonic (energy-liberating) equilibrium reaction:

$$\text{A} - \text{energy} \rightleftharpoons \text{B} \qquad K_{eq} = \frac{[\text{B}]}{[\text{A}]}$$

The further the equilibrium lies toward the right, the greater will be the energy available from this reaction. Therefore, the extent to which equilibrium reactions go to completion, that is, [B]/[A], is directly related to the free energy for the reaction. However, since the extent to which reactions go to completion is reflected in the value of the equilibrium constant, K_{eq}, the following mathematical expression, not derived here, can be written.

$$\Delta G^\circ = -RT \ln K_{eq}$$

where ΔG° = standard-free-energy change when concentration of reactants is 1 mole/liter and 1 mole of reactants has reacted
R = const = 1.987
T = absolute temperature, °C + 273
$\ln K_{eq}$ = natural logarithm of equilibrium constant

A more convenient equation to use is obtained by converting to base 10 logarithms and incorporating R into this conversion. The equation then becomes

$$\Delta G^\circ = -4.576T \log K_{eq}$$

The ΔG° for a large number of reactions is given in various advanced chemical and biochemical texts.

Under ordinary laboratory conditions reactions are not carried out under conditions to obtain ΔG°; therefore, the ΔG of such a reaction would be different from ΔG°. An equation that expresses the change in free energy, ΔG, under any given conditions is

$$\Delta G = \Delta G^\circ + 4.576T \log \frac{[\text{products}]}{[\text{reactants}]}$$

Thus far in our discussion of ΔG the forward reaction A $\longrightarrow$ B + energy has been considered. The conditions for reversing this reaction, that is, B $\longrightarrow$ A would be

$$\text{B} + \text{energy} = \text{A}$$

It would require the same amount of energy to drive the reaction from B to A as was liberated in the forward reaction, A to B. The ΔG° value for this reverse reaction would have the *same absolute numerical value as* $-\Delta G^\circ$ for the forward reaction *but would be positive.*

$$\text{Large negative } \Delta G^\circ = \text{large amount of energy liberated}$$
$$\text{Large positive } \Delta G = \text{large amount of energy required}$$

At equilibrium the forward and reverse reactions are proceeding at the same rate; hence, the same amount of energy is being released in the forward reaction as is being used in the reverse reaction. *The net amount of energy available for work would be zero* ($\Delta G = 0$). *No useful energy can be obtained from a reaction that has reached equilibrium.* Thus, if energy is to be obtained from an energy-yielding reaction, equilibrium must not have been attained. At equilibrium then,

$$0 = \Delta G^\circ + 4.576T \log \frac{[\text{B}]}{[\text{A}]} \qquad \text{or} \qquad \Delta G^\circ = -4.576T \log K_{eq}$$

and ΔG° can be calculated by determining the value of K_{eq}.

All chemical reactions tend to proceed in the direction to form products of a lower energy content than the reactants. Another expression of this statement is that reactions having a negative ΔG° tend to go spontaneously and reactions having a positive ΔG° do not.

Energy of activation

Because a reaction has a tendency to go spontaneously is no reason to believe that the reaction will proceed at a rate fast enough to be measured unless it is started and carried out under the correct conditions. For example, hydrogen and oxygen gas can be kept mixed together at room temperature without a reaction occurring. But touch a lighted match to the mixture and a violent explosion results with the liberation of a large amount of energy. How did the match initiate the reaction, and why did the reaction not go initially?

A spontaneous reaction proceeds immediately upon mixing the reactants. This can be represented by the energy-change curve in Fig. 18.1*a*, in which A + B $\longrightarrow$ AB. The asterisk above the reactants indicates that they contain sufficient energy to react, i.e., are in an *activated* state.

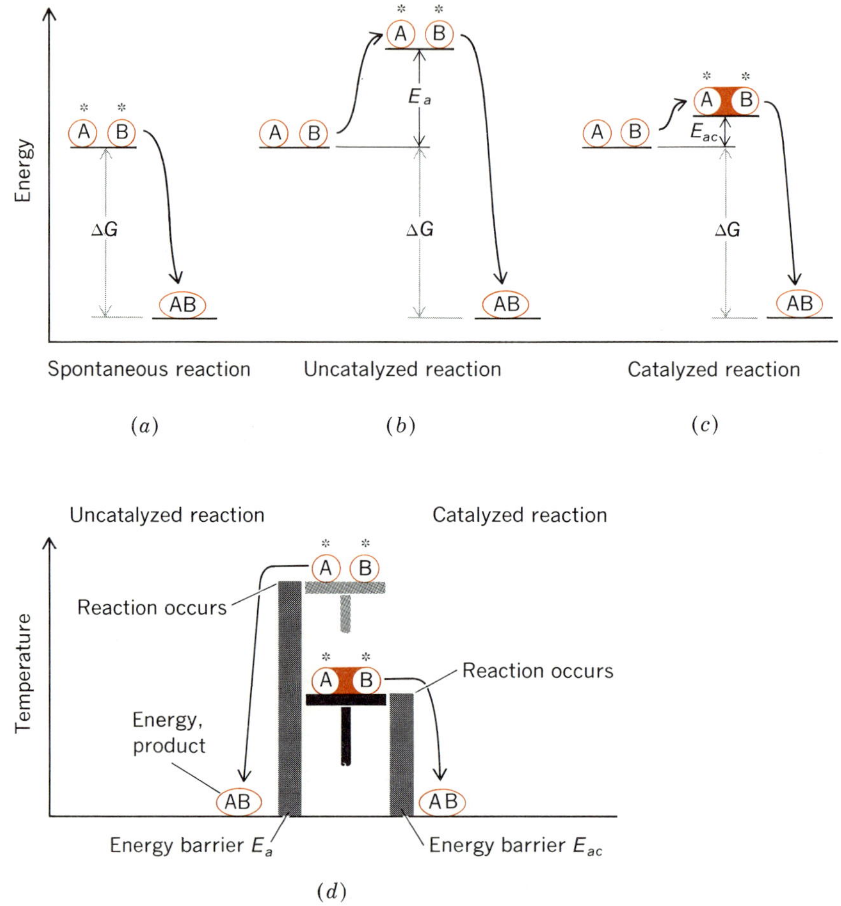

FIGURE 18.1 The energetics of a chemical reaction. E_a = energy of activation of an uncatalyzed reaction; E_{ac} = energy of activation of a catalyzed reaction; (A) (B) = reactants; (A■B) = catalyst complex with the reactants; (AB) = products; ΔG = change in free energy; starred molecules are in an activated state.

There are thousands of reactions that have a tendency to go spontaneously, but since the reactants do not have sufficient energy to be activated at room temperature, there is no measurable reaction rate. This situation is depicted in Fig. 18.1*b* and *c*. Reactants A and B do not react because they do not have sufficient energy. However, if by increasing the temperature of the system the energy content of A and B is increased until they are activated, the reaction will proceed. The amount of energy needed to raise the reactants to the activated state is known as the *energy of activation* E_a. Once an exergonic reaction has started, the energy given off may be sufficient to keep the reaction going. In the case of hydrogen and oxygen, the match supplied the energy to activate some of the molecules in the mixture, and the large amount of energy released when the first few molecules of hydrogen and oxygen reacted to form water was sufficient to activate many more molecules. The reaction then proceeded at such a fast rate that an explosion resulted.

An enzyme can assist in a chemical reaction by combining in some manner with the reactants to form an enzyme-reactant complex with the result that less energy is required to activate the molecules of the reactants in the enzyme-reactant complex form. Consequently, the reaction goes at a lower temperature (Fig. 18.1*c* and *d*). By comparing parts *b* and *c* of the figure it will be noted that the value for ΔG is the same; the only difference is that the energy barrier is much lower in the catalyzed reaction, E_a versus E_{ac}. Energy for activation is usually supplied in the form of heat. Thus, the molecules in a catalyzed reaction have sufficient energy at a lower temperature to surmount the energy barrier (Fig. 18.1*d*).

How would you explain the following observation? An enzymatic reaction was performed at 10, 25, and 37°C. There was no measurable activity at 10°C. The rate of the reaction was considerably faster at 37°C than at 25°C. Why? It must be remembered that when one speaks of a chemical reaction it is not with reference to only one or two molecules (remember Avogadro's number?). Not all molecules have the same energy at a given temperature; some have considerable energy while others have much less. Hence we must think in terms of average energy. At the temperature of 10°C no molecules have enough energy to be activated (Fig. 18.2*a*). At 25°C the average energy of the

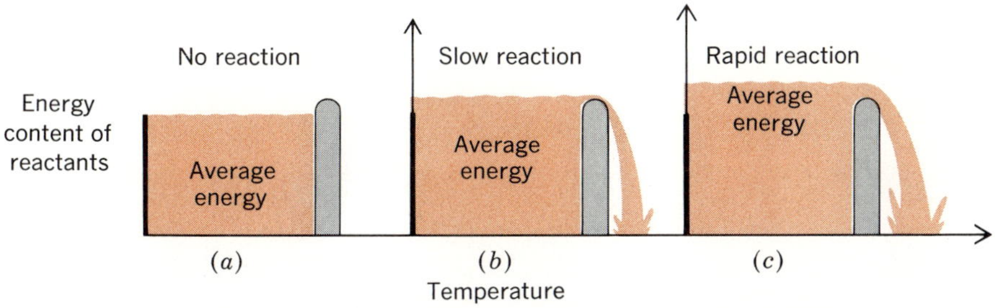

FIGURE 18.2 Energy content of reactants versus the speed of a chemical reaction: (*a*) water level below the top of the dam: energy content of the reactants is too low for a reaction to occur; (*b*) water level slightly above the top of the dam: energy content of only a fraction of the reactants is high enough for a reaction to occur; (*c*) water level well above the top of the dam: average energy content of the reactants is above that required for a reaction to occur, hence, a very fast reaction. The dam represents the energy of activation of the reaction mixture.

molecules is below that required for activation of all molecules, but some molecules have sufficient energy to be activated and will react. The reaction will proceed but not at a very fast rate (Fig. 18.2*b*). At 37°C most of the molecules have sufficient energy to be activated, and the reaction will proceed at a rapid rate compared to that at 25°C (Fig 18.2*c*).

It is very important to remember that *catalysts do not change the equilibrium constant but merely speed up the attainment of equilibrium at a given temperature.*

Active site

The part of the enzyme molecule to which the substrate binds is called the *active site*. For a number of enzymes, the active site has been quite

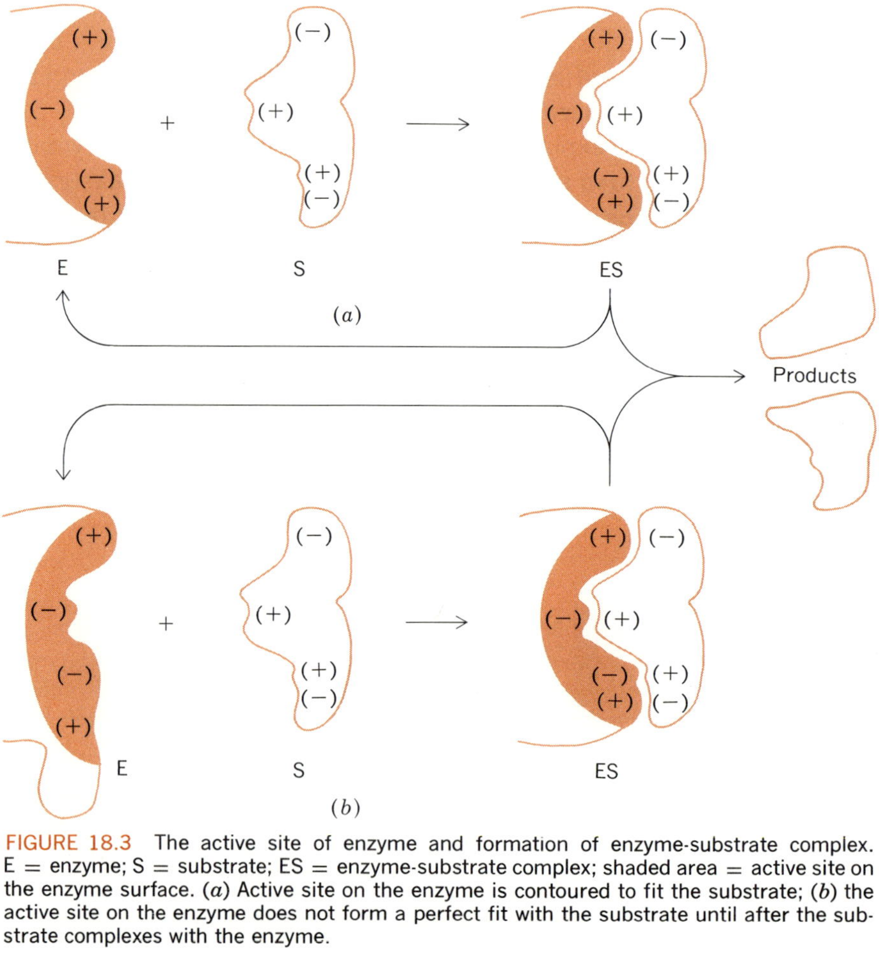

FIGURE 18.3 The active site of enzyme and formation of enzyme-substrate complex. E = enzyme; S = substrate; ES = enzyme-substrate complex; shaded area = active site on the enzyme surface. (*a*) Active site on the enzyme is contoured to fit the substrate; (*b*) the active site on the enzyme does not form a perfect fit with the substrate until after the substrate complexes with the enzyme.

well characterized with respect to the amino acids and charges. A diagrammatic illustration of the binding of an enzyme with its substrate is shown in Fig. 18.3. The active site on the surface of the enzyme is so contoured and charged that it attracts only the substrate and forms an enzyme-substrate complex ES which has a lower activating energy than the substrate alone (Fig. 18.3*a*). The surface of the enzyme in this example is already contoured to fit the substrate before complexing takes place and is said to have a perfect fit.

It is not always necessary for the enzyme to have a completely preformed conformation or perfect fit for the substrate. The first point of attachment of the substrate to the enzyme can sometimes change the conformation of the protein so that it bends into a perfect fit (see Fig. 18.3*b*).

18.3 FACTORS AFFECTING ACTIVITY

Temperature

The rate of a chemical reaction *varies directly* with the temperature. Usually the rate constant doubles for each 10°C increase in temperature. This is true whether or not the reaction is aided by a catalyst. If the catalyst is an enzyme, the increase in activity is not valid over a very large temperature range because of the adverse effect of heat on the enzyme.

Figure 18.4*a* illustrates the change in reaction rate with tempera-

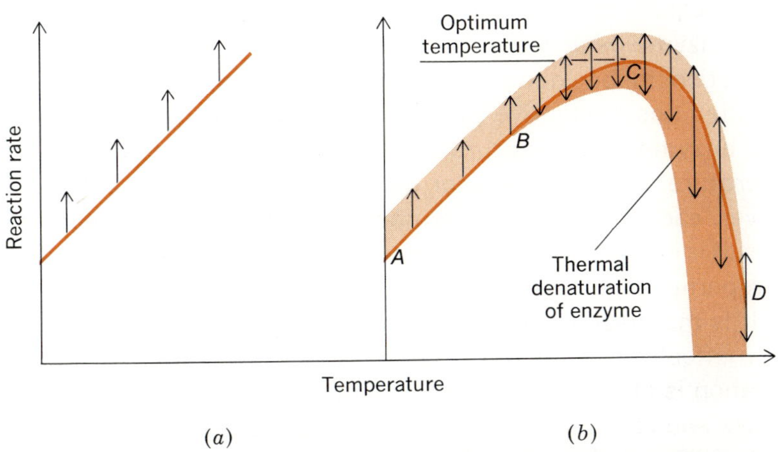

FIGURE 18.4 The effect of temperature on the rate of chemical reactions: (*a*) no catalyst or an inorganic catalyst; (*b*) enzyme catalyst. ↑ = thermal increase in reaction rate; ↓ = a decrease in the reaction rate due to enzyme destruction (denaturation by heat).

ture for a noncatalyzed reaction and for a reaction catalyzed by an inorganic catalyst. Figure 18.4*b* illustrates an enzyme-catalyzed reaction. There is a decided difference between these two curves. In the enzyme-catalyzed reaction, the *initial* increase in velocity follows the rule of doubling the reaction rate for each 10°C increase in temperature, AB, the increase in rate being indicated by an upward arrow. At B there is a departure from the straight-line function, and the curve starts to bend downward. As the temperature is increased, this decrease in reaction rate becomes more pronounced and eventually at C a maximum rate is obtained, after which the rate decreases rapidly. How can this curve be explained?

Enzymes are proteins and therefore subject to denaturation by heat. For an enzyme to act as a catalyst, it must have a definite specific conformation. In some cases a specific quaternary structure is also required. Depending upon the enzyme, the tertiary structure is disrupted to varying extents as the temperature is increased, thus inactivating the enzyme. In Fig. 18.3*b* there is no detectable destruction of the enzyme at temperatures from A to B; hence the curve is a straight line. Usually enzymes are not denatured at temperatures below about 40°C. Between B and C the reaction rate is still increasing, but at a slower rate, because two opposing forces are coming into play: (1) the increase in reaction rate because of the raise in temperature and (2) the decrease in the reaction rate because of the greater destruction of the enzyme as the temperature is raised. The first condition is the greatest until point C is reached. At C the two opposing rates are equal; hence a maximum in the curve. Between C and D the rate of enzyme destruction is so great that it offsets the tendency of the reaction to be increased because of temperature change. The curve then has a negative slope, usually quite a sharp one. Because not all enzymes are denatured to exactly the same extent at a given temperature, not all curves for different enzymes look exactly alike. However, each enzyme has a temperature at which maximum activity occurs. It is called the *optimum temperature* and is useful in describing the properties of an enzyme.

It is the inactivation of enzymes by heat that makes pasteurization an effective means of preserving foods. The temperature of pasteurization is about 60°C, which is sufficiently high to denature many enzymes and result in the death of the microorganisms.

At 0°C or below an enzyme has little if any activity, a fact that has changed our daily habits through the frozen foods industry. By quickly freezing and storing them at temperatures below zero, enzyme reactions are slowed down to such an extent that foods can be kept for a reasonable length of time without spoiling.

It is interesting to note the differences between organisms with respect to their ability to survive at low temperatures. Animals can exist only within a rather narrow limit of temperature; man cannot tolerate body temperatures appreciably lower than 37°C, but pine trees stay green and are able to survive temperatures far below zero, and many bacteria can be quick-frozen to the temperature of Dry Ice, −80°C, kept in the frozen condition for a considerable length of time (measured in months), and still be alive after they are thawed.

pH

In Chap. 15, in which intermolecular forces and protein structure were discussed, the highly polar nature of proteins was illustrated. The effect of pH was considered in relationship to changing the tertiary structure of proteins. Since each enzyme must possess a specific tertiary structure as well as primary structure, it follows that enzyme activity is sensitive to a change in pH. Due to the difference in conformation and the number and location of charges on enzymes, the effect of pH on different enzymes varies. Furthermore, *a change in pH may alter the charge on a substrate as well as the charge on the enzyme,* which would affect the binding of the substrate to the enzyme surface. For example, if the substrate is an amino acid, a change in pH can easily cause a change in the ionic species present in solution.

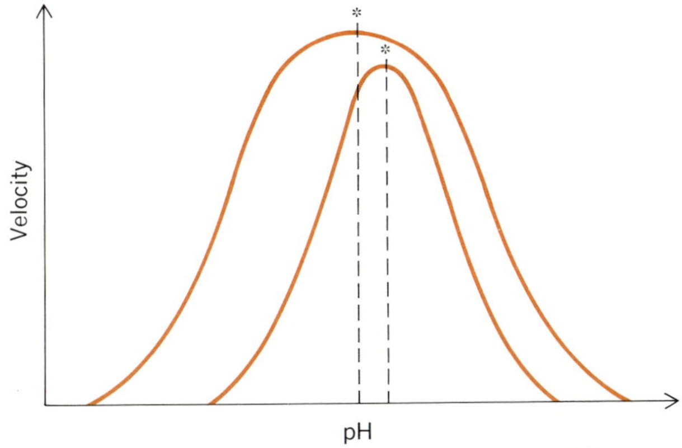

FIGURE 18.5 The influence of pH on enzymatic reactions. The optimum pH is indicated by an asterisk.

There are, however, certain general influences of pH on all enzymes.

Every enzyme has an optimum pH, that is, one at which the reaction rate is maximum (Fig. 18.5). At this pH the conformation of the enzyme is such that the maximum number of molecules are acting as catalysts. Some enzymes have a rather sharp maximum pH, whereas others have quite a broad maximum. As the pH is decreased or increased from the optimum pH, the charges on the protein are changed and the enzyme is no longer as active as a catalyst. This is a progressive situation: the greater the difference between the pH of the solution and the optimum pH of the enzyme, the greater the number of molecules of enzyme denatured and consequently the lower the reaction rate.

Substrate concentration

In the presence of adequate enzyme, if the velocity of an enzymatic reaction is measured against time, the reaction is linear at first and then begins to slow down as the substrate is used and end products increase in concentration. Eventually the velocity becomes zero. To study the effect of substrate concentration on the rate of an enzymatic reaction conditions must be such that the substrate is the *only* limiting factor; the enzyme and all other necessary components must be present in adequate amounts.

Figure 18.6 shows a series of curves for the velocity of a reaction

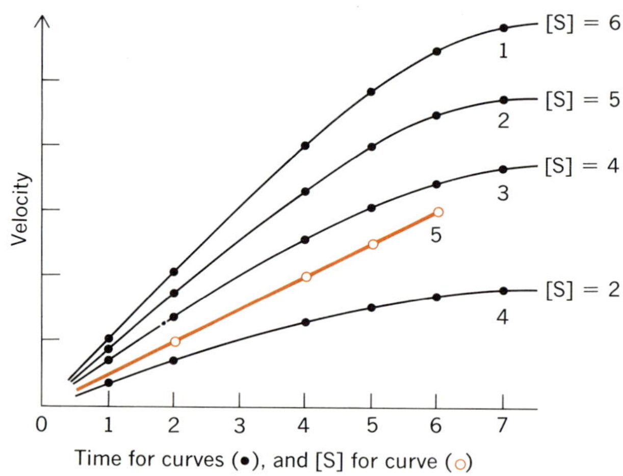

FIGURE 18.6 Velocity of an enzymatic reaction: black points = velocity versus time at different substrate concentrations; colored points = velocity versus change in substrate concentration at one specific time, $t = 3$.

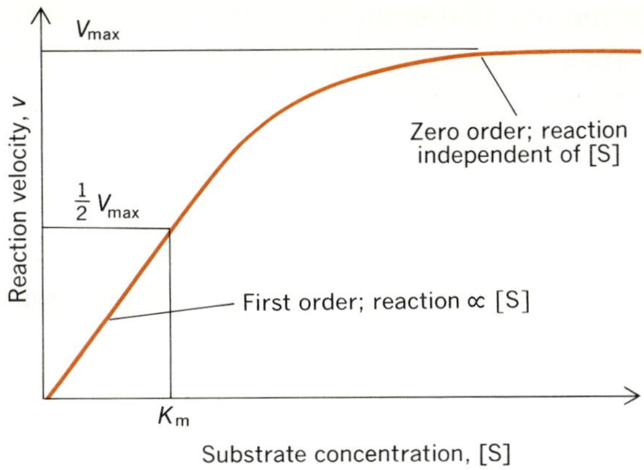

FIGURE 18.7 The effect of substrate concentration on reaction rate of an enzymatic reaction. V_{max} = maximum velocity of the reaction; K_m = Michaelis-Menten constant.

with several different substrate levels [S] as plotted against time. To prepare a curve for velocity versus [S] it is necessary to choose a time such that the velocity for the various [S] values lies on the linear portion of the respective time-course curve. Such a situation is met at time $t = 3$. Replotting the velocity at time $t = 3$ versus the various [S] produces curve 5, which is linear with respect to concentration of substrate. This linearity does not hold for high concentrations of substrate.

Figure 18.7 shows a Michaelis-Menten curve for velocity versus [S] to high levels. The first part of the curve is linear; it then decreases in slope until a zero slope is obtained, after which increasing the concentration of substrate has no effect. When this situation exists, the reaction is called a *zero-order reaction, a reaction in which the velocity is independent of* [S]. The part of the curve which is linear is referred to as a *first-order reaction, one in which the velocity is directly proportional to* [S].

How can one account for the fact that after reaching a certain substrate concentration, there is no increase in the velocity of the reaction when more substrate is added? Consider the mechanism of an enzyme-catalyzed reaction in which a complex is formed between the enzyme and the substrate:

$$S + E \longrightarrow ES \longrightarrow P + E$$

At constant enzyme level, there will be a certain level of substrate at which substrate is combining with enzyme at a maximum rate. The limiting part of the reaction is therefore the rate at which the enzyme-substrate complex is decomposing, thus freeing enzyme to combine with more substrate. Any increase in the concentration of substrate above this value will cause no increase in the rate of the reaction. The enzyme is saturated with respect to the substrate.

As an analogy, consider a shallow tank with an outlet at the bottom (Fig. 18.8). The maximum amount of water that can flow out is fixed by the size of the outlet. The size of the outlet is likened to enzyme concentration, the water to the enzyme-substrate complex, and the rate of the water flowing through the outlet to the rate of

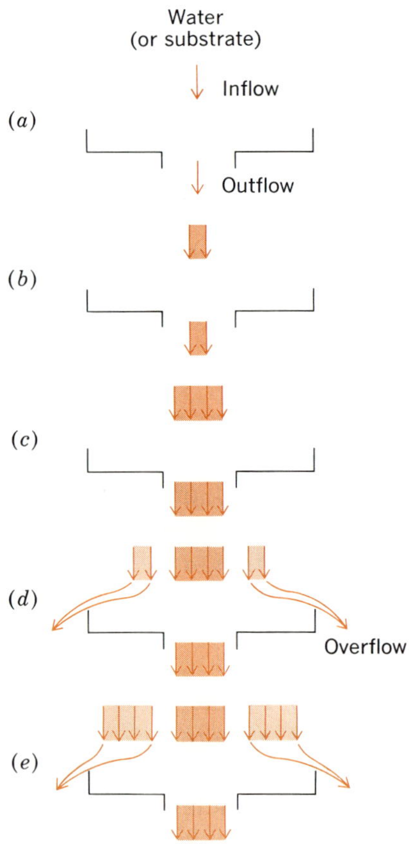

FIGURE 18.8 Analogy for the Michaelis-Menten plot of an enzyme reaction. Rate of inflow = substrate concentration; size of the outlet = enzyme concentration; rate of outflow = rate of enzyme reaction; overflow = excess substrate.

enzymatic reaction. When water is poured slowly into the tank, it flows out at the same rate as it is flowing in (Fig. 18.8*a* and *b*). The rate of water flowing out is directly proportional to the rate of water flowing in, until the maximum rate of outflow is reached (Fig. 18.8*c*). In our analogy this represents a first-order reaction: the rate of a reaction is directly proportional to the concentration of enzyme-substrate complex. If the water input rate equals the maximum outflow rate, the maximum output is reached. This is analogous to the maximum rate or velocity of an enzyme-catalyzed reaction. If the water input rate is increased, the outflow rate remains the same because water begins to spill over the side of the tank (Fig. 18.8*d* and *e*). The amount of water flowing through the outlet pipe is independent of the amount of water flowing into the tank. This situation represents a zero-order reaction in which the enzyme is reacting with the substrate as fast as possible and any extra substrate is left unreacted. Since the enzyme is saturated with respect to the substrate, the rate is independent of the concentration of substrate.

18.4 MICHAELIS-MENTEN CONSTANT K_m

Michaelis and Menten were the first to express a reversible enzymatic reaction in mathematical terms. From their mathematical treatment of the following equation is derived the *Michaelis-Menten constant* K_m:

$$\mathrm{E} + \mathrm{S} \underset{k_2}{\overset{k_1}{\rightleftharpoons}} \mathrm{ES} \underset{k_4}{\overset{k_3}{\rightleftharpoons}} \mathrm{P} + \mathrm{E}$$

where E = enzyme
S = substrate
ES = enzyme-substrate complex
P = product
k's = respective velocity constants for reactions indicated by arrows

If one considers the reaction rate when the products have not formed in sufficient quantity to cause the reverse reaction to occur to any appreciable extent ($k_4 = 0$), the forward rate is directly proportional to the substrate concentration at constant enzyme concentration; i.e., the initial velocity depends upon [S], and it is a first-order reaction. At maximum velocity all the enzyme is complexed with the substrate; hence the rate of the reaction is determined by the rate at which this complex decomposes into product and enzyme, a zero-order reaction.

The value K_m is of specific interest. It is the *concentration of substrate at which the reaction is half maximum* (Fig. 18.7). By appropriate calculations from the above rate constants k it can be shown that at this concentration of substrate the rate is dependent only on [S] and is independent of the concentration of the enzyme. There is a characteristic K_m value for each enzyme. K_m values are expressed in moles of substrate per liter.

18.5 INHIBITION

Inhibition of enzyme activity may occur at various metabolic levels. Some compounds inhibit enzyme activity by interfering with the synthesis of the enzyme, while others combine directly with the enzyme and thus decrease its activity. Such compounds as antibiotics and insecticides are enzyme inhibitors. Also some normal metabolic products are of significance in regulating cellular metabolism by acting as enzyme inhibitors when present in sufficiently high concentrations. Some inhibitors are quite specific and inhibit only one particular enzyme, while others may be rather general and inhibit all enzymes with a particular group at the active site, e.g., the SH group.

Product inhibition

One method by which enzyme reactions are inhibited is by product inhibition. The products of an enzyme reaction combine with the enzyme to form an inactive complex (Fig. 18.9). Thus when a certain

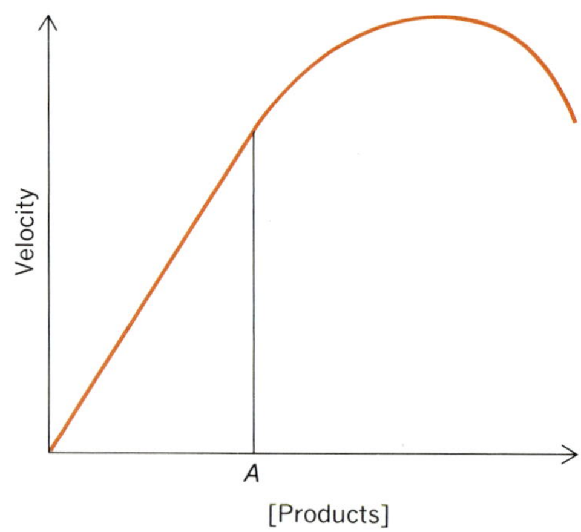

FIGURE 18.9 Product inhibition of an enzyme reaction. The reaction is not inhibited by products up to concentration A but is inhibited by products at concentrations greater than A.

concentration of product is reached, the activity of the enzyme to produce more product is decreased until the product is removed from the reaction medium and the enzyme system functions again.

Feedback inhibition

This type of inhibition occurs when the product of a sequence of reactions inhibits an enzyme toward the beginning of the sequence. For example, in the following equation the sequence of reactions is from A to E, but product E is inhibitory to the enzyme that converts A to B.

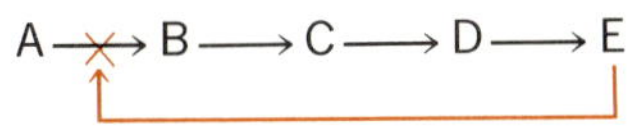

Thus as product E accumulates it slows down the entire sequence of reactions.

Competitive inhibition

There is a class of enzyme inhibitors whose activity depends upon the ratio of the inhibitor to substrate. At a given inhibitor concentration, if the substrate concentration is increased sufficiently, the inhibition will eventually be overcome (see Fig. 18.10). In this type of inhibition the inhibitor I is competing with the substrate to form an enzyme complex:

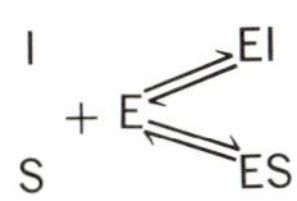

The enzyme-inhibitor complex dissociates; hence there is an equilibrium established between E, EI, and ES.

If the velocity of a reaction is plotted against substrate concentration [S] for several levels of a competitive inhibitor, the series of curves is as shown in Fig. 18.10. Note that the *same maximum velocity* is eventually attained if sufficient substrate is present. Another point of interest is that there is a *different* K_m *value* for each of the curves. This is one criterion of a competitive inhibitor.

Noncompetitive inhibition

Noncompetitive inhibition is not overcome by increasing the amount of substrate. The inhibitor combines *reversibly* with the enzyme to produce an inactive complex. This is equivalent to removing some of the enzyme from solution:

$$\text{E} + \text{I} \rightleftharpoons \text{EI} + \text{S} \rightleftharpoons \text{EIS (inactive)}$$

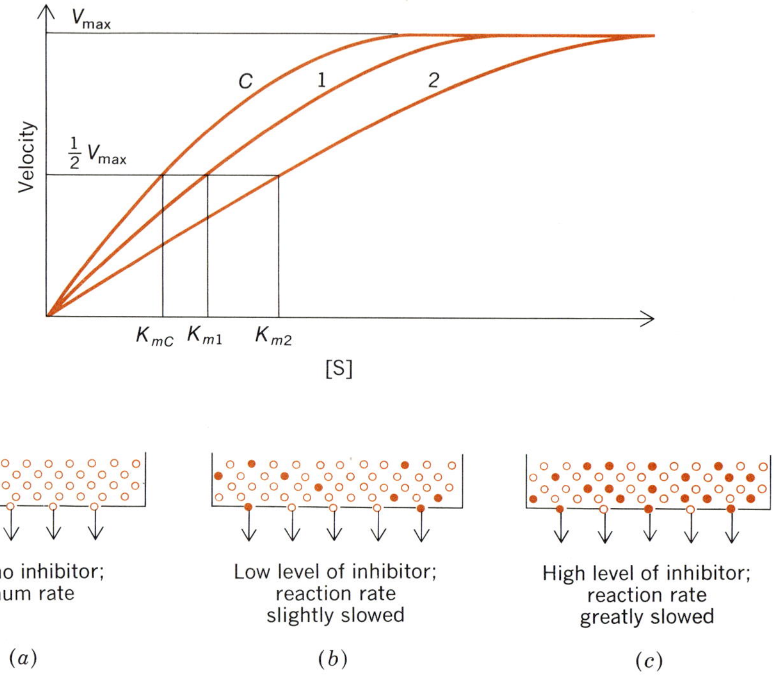

FIGURE 18.10 Competitive inhibition. Curve C is a control; i.e., no inhibitor is present; curves 1 and 2 have increasing amounts of inhibitor present. An analogy is shown below. Compare the rate at which light sand passes through a sieve in the absence and in the presence of varying amounts of dark sand. In (a) only light sand, or substrate, is present; in (b) and (c) the ratio of light sand to dark sand decreases. The more dark sand, or inhibitor, present, the less substrate goes through the sieve.

Consequently as the concentration of the inhibitor is increased, the amount of active enzyme decreases and there is a proportionate *decrease in the maximum velocity* of the reaction (Fig. 18.11). Note that *the K_m value remains constant,* whereas for a competitive inhibitor the K_m value varies.

The method of plotting enzyme data shown in Fig. 18.12 is called a *Lineweaver-Burk plot.* When $1/v$ is plotted against 1/[S], a straight line results which crosses the $1/v$ axis at $1/V_{\max}$ and intercepts the 1/[S] axis at $-1/K_m$. This plotting method is very convenient because it is often quite difficult to obtain experimentally such ideal curves as shown in Fig. 18.7 from which to determine $V_{\max}$ and K_m accurately. From a Lineweaver-Burk plot it is quite easy to determine whether an inhibitor is competitive or noncompetitive. For a competitive inhibitor $V_{\max}$ is the same as for the uninhibited reaction, but K_m values differ

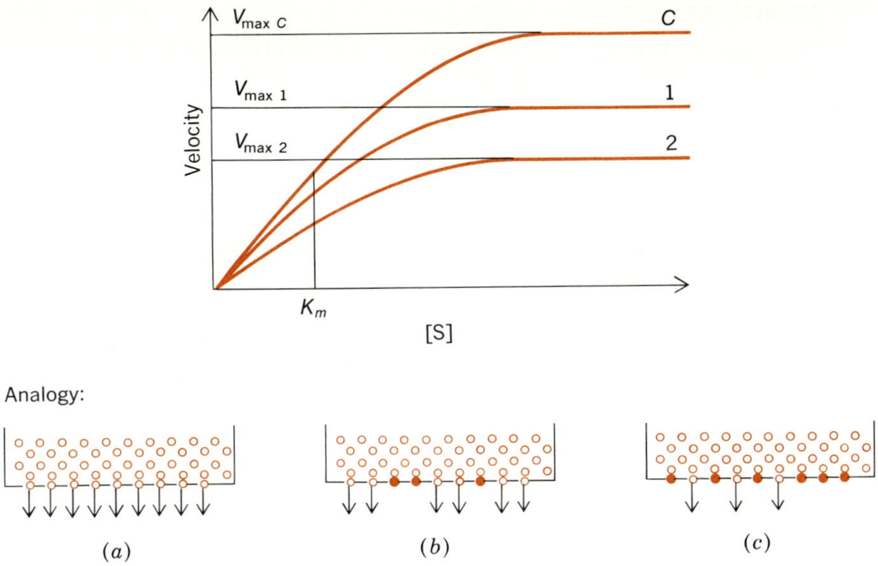

FIGURE 18.11 Noncompetitive inhibition. Curve C is the control; i.e., no inhibitor is present; curves 1 and 2 have increasing amounts of inhibitor present. Underneath the graph is an analogy, using light sand as the substrate and dark pebbles which block the sieve as the inhibitor. In (a) there is no pebbles, or inhibitor, present. In (b) and (c) increasing amounts of inhibitor plug the sieve; therefore, the rate of flow through sieve (a) is never equaled by that of sieve (b) or (c).

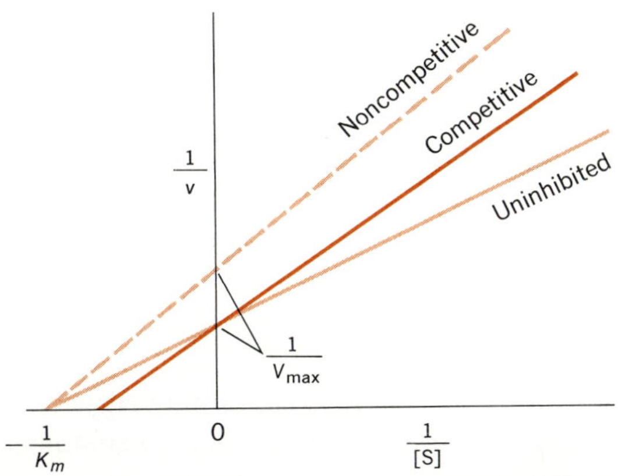

FIGURE 18.12 Lineweaver-Burk plot of uninhibited, competitive inhibited, and noncompetitive inhibited enzyme reactions. [S] = substrate concentration, V_{max} = maximum velocity, and K_m = Michaelis-Menten constant.

for the inhibited and uninhibited reactions. For a noncompetitive inhibitor the $V_{\max}$ for the uninhibited and the inhibited reactions are different, but the K_m values are the same.

18.6 INHIBITOR STUDIES

One of the tools used to determine the order of a sequence of reactions is enzyme inhibitors. The following example will illustrate the technique.

PROBLEM

Assume for this discussion that (1) all the appropriate enzymes are present and active, (2) it is known that oxaloacetic acid is required for the utilization of pyruvic acid, (3) pyruvic acid, after decarboxylation, combines with oxaloacetic acid to form citric acid as shown below, and (4) other acids (malic, fumaric, succinic, isocitric, and *cis*-aconitic) are also associated with the utilization of pyruvic acid. The following experiments were performed and the observations and conclusions made as indicated.

Experiment 1. Oxaloacetic acid alone was found to be required in only catalytic amounts to sustain the utilization of pyruvic acid. This indicates that oxaloacetic acid must be reformed during the reaction.

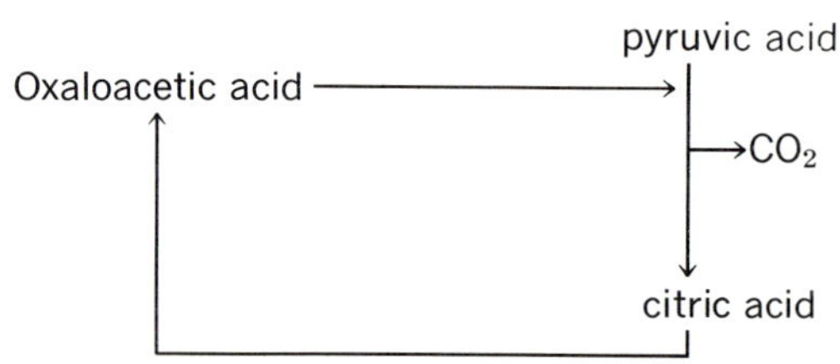

Experiment 2. The addition of only citric acid or any of the other acids in catalytic amounts also sustained the utilization of pyruvic acid. This indicates that these acids can be converted into oxaloacetic acid, but there is no indication of the order in which the reactions occur.

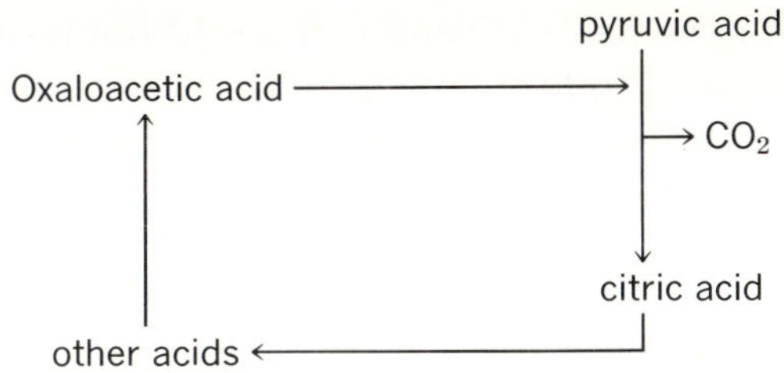

Experiment 3. After the addition of the enzyme inhibitor fluoroacetic acid, utilization of pyruvic acid was sustained (with a concomitant buildup of citric acid) only as long as oxaloacetic acid was present. The addition of citric acid caused no utilization of pyruvic acid. The addition of any one of the other acids, however, produced an accumulation of citric acid as long as the acid was present. It can be concluded that fluoroacetic acid inhibits an enzyme that utilizes citric acid as substrate, and the above observation that the other acids can be converted into oxaloacetic acid is confirmed.

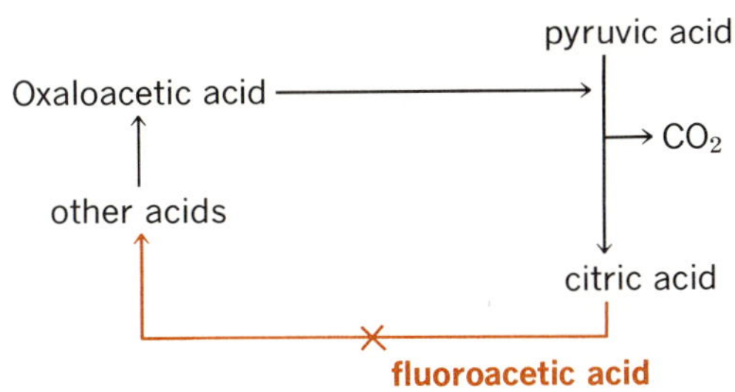

Experiment 4. When malonic acid, another enzyme inhibitor, was added together with fluoroacetic acid, an accumulation of citric acid accompanied the utilization of pyruvic acid only as long as oxaloacetic acid was present. The addition of either malic or fumaric acid also resulted in the utilization of pyruvic acid (and formation of citric acid) only as long as these respective acids were present. The addition of *cis*-aconitic, isocitric, or succinic acid did not result in the utilization of pyruvic acid. Succinic acid accumulated when either *cis*-aconitic or isocitric acid was added. The conclusions from these observations are that malic acid inhibits the enzyme using succinic acid as a substrate; otherwise succinic

acid would have been utilized and would not have accumulated. Malic acid and fumaric acid lie between succinic acid and oxaloacetic acid; and *cis*-aconitic acid and isocitric acid lie between citric acid and succinic acid

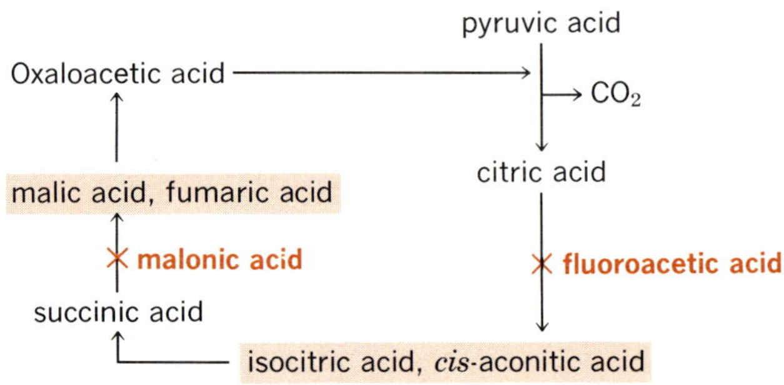

Thus using only two enzyme inhibitors considerable information about this series of reactions was obtained.

18.7 ACTIVATION

Some enzymes are active as the pure protein, whereas others are inactive until a nonprotein unit, e.g., mineral elements or coenzymes, is also present. Still other enzymes may exist in an inactive form (zymogen) until after the protein has been changed slightly, e.g., partially hydrolyzed. Coenzymes constitute a separate chapter; activation by mineral elements will be mentioned at appropriate places in the discussion.

QUESTIONS

1. How can one differentiate between a reaction catalyzed by an inorganic catalyst and an enzyme-catalyzed reaction?
2. (*a*) What does energy of activation mean?
   (*b*) Is the energy of activation greater or less for a catalyzed reaction than for a noncatalyzed reaction?
3. (*a*) In what form is energy usually applied to a chemical reaction?
   (*b*) Is this true for biological reactions? Explain.
4. What simple everyday observation can you give to illustrate the fact that not all molecules at a given temperature have the same kinetic energy?
5. In molecular reactions the reactants must collide with each other with sufficient force to cause an overlapping of electrons of the respective functional groups. Is this statement in line with your answer to part (*a*) of question 3? Explain.

6. Explain why enzymatic reactions exhibit an optimum temperature and an optimum pH.
7. What is meant by the active site of an enzyme?
8. Illustrate how pH might influence the charges at an active site of an enzyme.
9. Outline an experiment and show the results you would expect to obtain to indicate that enzymes form an enzyme-substrate complex.
10. What is meant by a first-order reaction and a zero-order reaction?
11. If an inhibitor changes the maximum velocity of an enzymatic reaction, is it a competitive or a noncompetitive inhibitor?
12. What is the difference between a zymogen and an enzyme?
13. What is the significance of the term ΔG? What is the relationship of ΔG to K_{eq}?
14. Arrange the following reactions in order of their increasing tendency to go spontaneously.

	ΔG
(*a*) A $\longrightarrow$ B	+1628
(*b*) C $\longrightarrow$ D	+562
(*c*) E $\longrightarrow$ F	0
(*d*) G $\longrightarrow$ H	−1425
(*e*) I $\longrightarrow$ J	−8000

15. Does a competitive inhibitor increase, decrease, or cause no change in the K_m value of an enzymatic reaction?
16. Does a noncompetitive inhibitor increase, decrease, or cause no change in the value of V_{max}?
17. What is meant by the term negative feedback with respect to enzyme reactions?

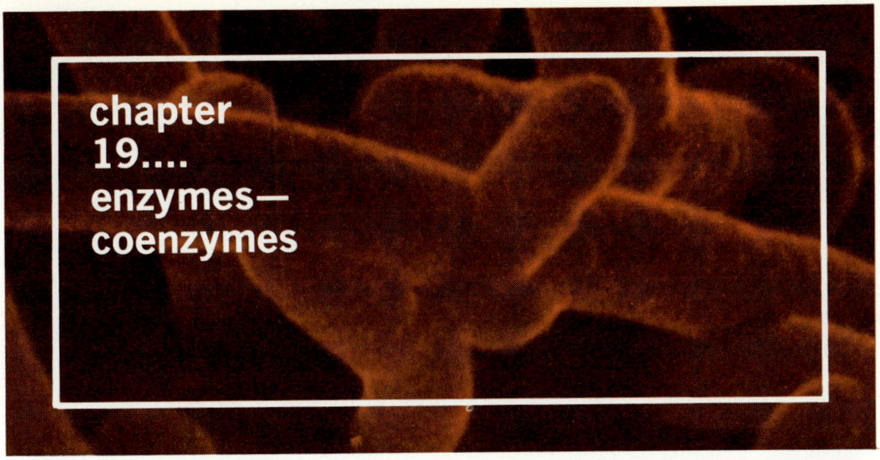

chapter 19.... enzymes—coenzymes

A coenzyme is the organic nonprotein part of the enzyme system. Coenzymes are soluble, heat-stable moieties and are often easily separated from the *apoenzyme* (the protein part of the enzyme) by precipitation of the protein portion. This discussion introduces a number of specific coenzymes and their functions and emphasizes the importance of vitamins as components of coenzymes.

19.1 CLASSIFICATION

Vitamins may be defined as organic compounds required in small amounts which cannot be synthesized by an organism. Thus, vitamins are differentiated from the essential amino acids, which are also organic compounds needed in the diet but in much greater amounts than the vitamins. The fact that vitamins are organic compounds differentiates them from mineral elements, some of which are also required in very small amounts. The importance of the vitamins that function as coenzymes stems from their involvement in enzyme systems. Each coenzyme contains a vitamin.

Another important characteristic of coenzymes is the role of phosphorus, as phosphate or pyrophosphate esters, and the five-carbon sugar ribose, which is in the structure of many coenzymes. It should also be noted that the mineral cobalt is in vitamin B_{12}.

VITAMIN B_6. The structure and one function of vitamin B_6 coenzymes *pyridoxal phosphate* and *pyridoxamine phosphate* have already been presented (see page 304). The reaction referred to is associated with transamination, the transfer of an amino group from an amino acid to an α-keto acid.

$$\underset{\substack{|\\ \mathrm{R}}}{\mathrm{H_2N{-}CH{-}COOH}} + \underset{\substack{|\\ \mathrm{R'}}}{\mathrm{O{=}C{-}COOH}} \xrightleftharpoons{\mathrm{B_6\ Enz}} \underset{\substack{|\\ \mathrm{R}}}{\mathrm{O{=}C{-}COOH}} + \underset{\substack{|\\ \mathrm{R'}}}{\mathrm{H_2N{-}CH{-}COOH}}$$

There are, however, a number of other reactions involving amino acids in which vitamin B_6 coenzymes function. One is the decarboxylation of amino acids, as in the conversion of histidine to histamine.

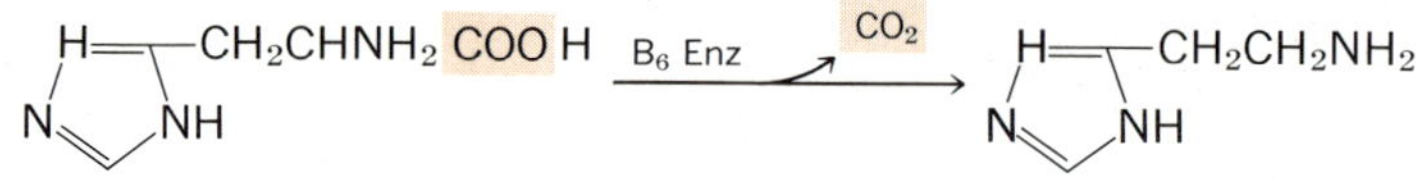

Histidine Histamine

Histamine is very active biologically; it depresses blood pressure quite strikingly.

NIACIN, NICOTINAMIDE. Niacin is the amide of nicotinic acid.

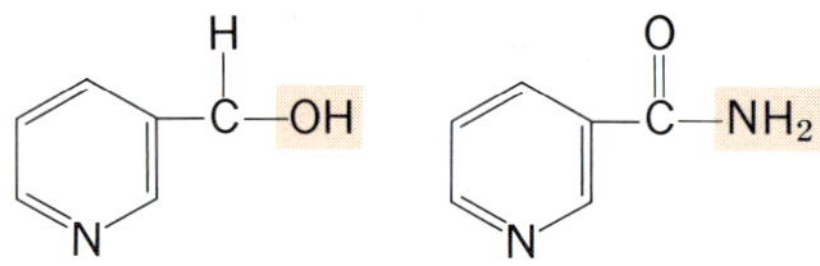

Nicotinic acid Nicotinamide Niacin

It contains the pyridine nucleus. Nicotinamide is a component of coenzymes associated with oxidation of substrate by the removal of hydrogen and, of course, the reverse reaction. Two examples of such reactions are the oxidation of lactic acid to pyruvic acid and the reduction of acetaldehyde to ethyl alcohol.

The two coenzymes that contain nicotinamide are abbreviated NAD^+ (nicotinamide adeninedinucleotide) or DPN^+ (diphosphopyridine nucleotide) and $NADP^+$ (nicotinamide adeninedinucleotide phosphate) or TPN^+ (triphosphopyridinenucleotide).

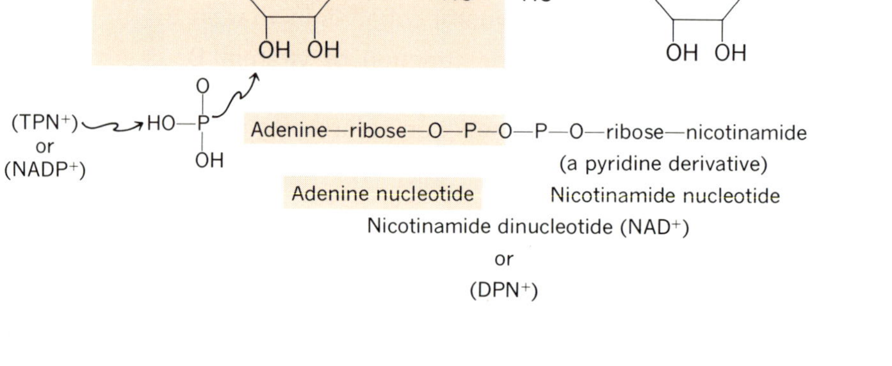

Each molecule of NAD^+ and $NADP^+$ coenzyme is capable of reacting with two atoms of hydrogen. In this reaction one hydrogen atom with its electron is added to the C-4 of the pyridine nucleus to form a covalent bond. The other hydrogen atom is stripped of its electron; the electron enters the pyridine nucleus to neutralize the quaternary nitrogen and the proton remains free in solution. The reaction showing only the essential part of the coenzyme is

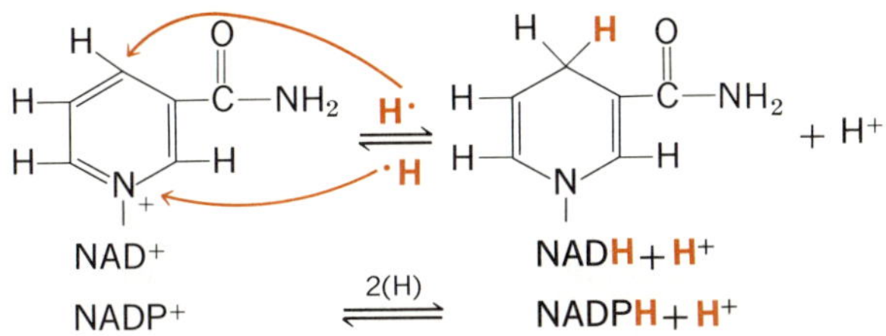

VITAMIN B_2, RIBOFLAVIN. Riboflavin phosphate complexes are coenzymes which with NAD^+ and $NADP^+$ play an important role in the transport of hydrogen from substrate eventually to oxygen.

There are two forms of B_2 coenzymes, the *flavin mononucleotide,* which contains flavin-ribityl-phosphate (FMN) and a *dinucleotide* composed of flavin-ribityl-phosphate-phosphate-ribose-adenine (a flavin-adenine dinucleotide) FAD:

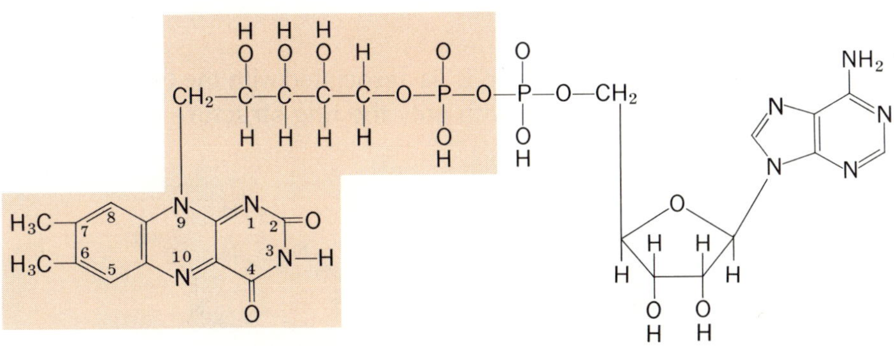

Flavin—ribityl—phosphate—phosphate—ribose—adenine

Flavin mononucleotide—Adenine mononucleotide

FMN

Flavin adenine dinucleotide

FAD

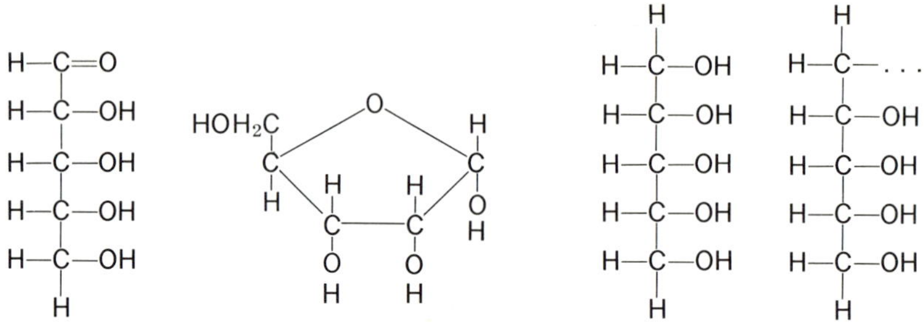

Ribose Ribitol Ribityl radical

Ribitol, the alcohol derived from the reduction of ribose, rather than the pentose ribose is contained in vitamin B_2 and consequently in FMN and FAD.

The functional part of B_2 coenzyme responsible for the transportation of hydrogen is the conjugated system of double bonds (=C—C=). Additions to a conjugated system occurs first at the 1 and 4 positions, and a double bond is formed between the 2 and 3 positions:

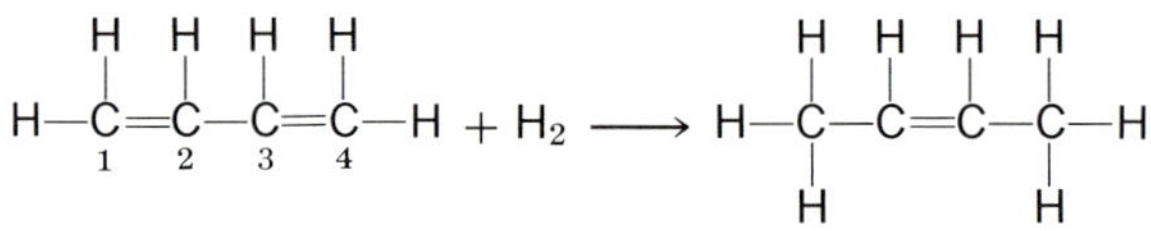

The conjugated system in vitamin B_2 is associated with the two nitrogen-containing rings shown below with only the ring structure given.

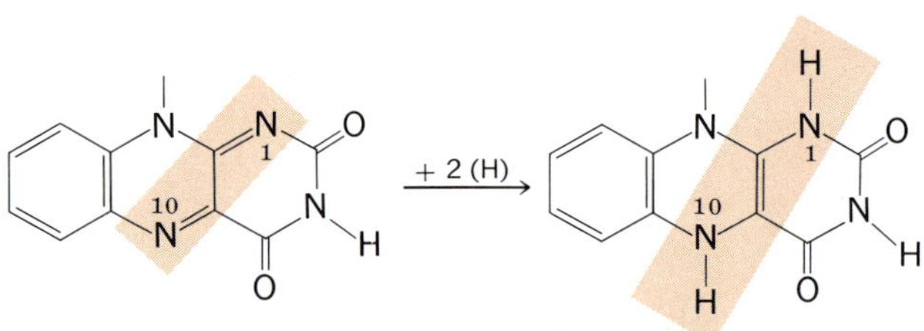

Oxidized FMN or FAD Reduced FMN or FAD

The coenzyme for succinic dehydrogenase is FAD. This enzyme catalyzes the conversion of succinic acid to fumaric acid:

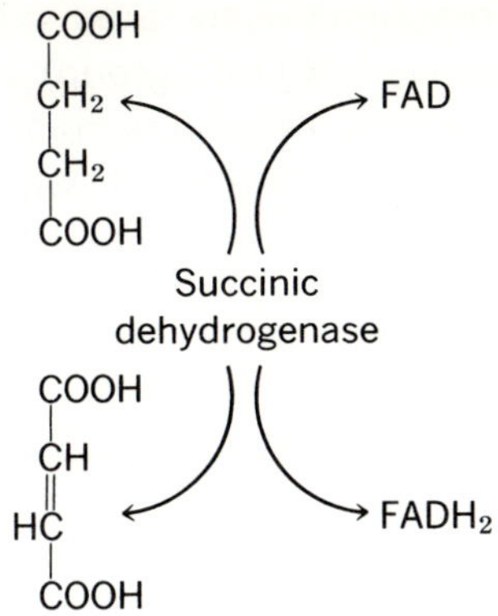

To aid in following a series of reactions frequently only the coenzymes are shown. However, it must be remembered that the protein part of the enzyme (apoenzyme) is also required. The following series of reactions with lactate as the substrate shows only the coenzymes involved to illustrate the important role of niacin and flavin coenzymes in passing hydrogen from a substrate eventually to oxygen.

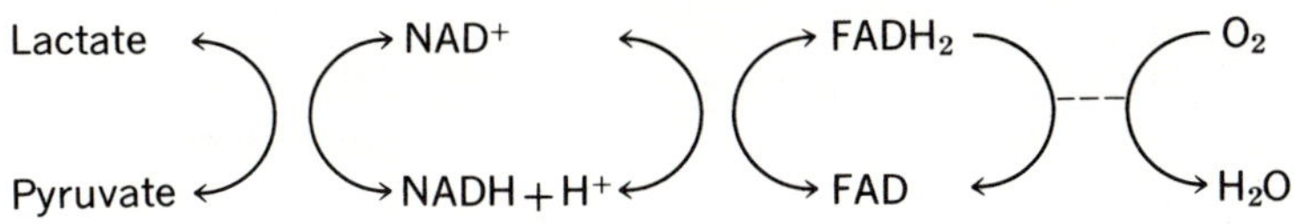

THIAMIN, VITAMIN B$_1$. Thiamin is another water-soluble nitrogenous vitamin that functions in coenzymes. Here too phosphate is important since it is thiamin pyrophosphate (TPP) that is the coenzyme. Also called *cocarboxylase,* TPP contains a pyrimidine and a thiazole nucleus. The latter is in the form of a quaternary ammonium salt NR_4^+ Cl^-:

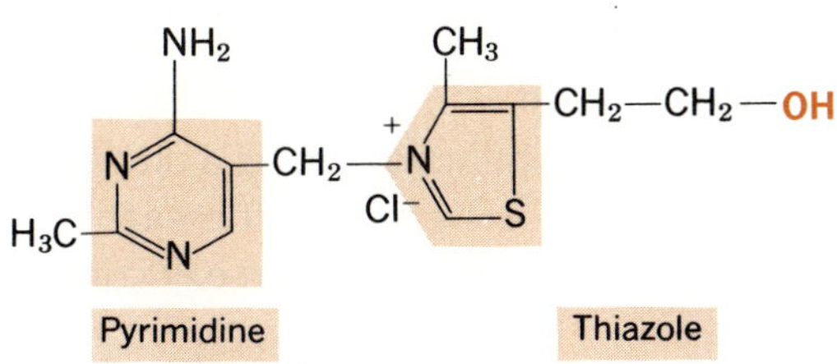

Thiamin

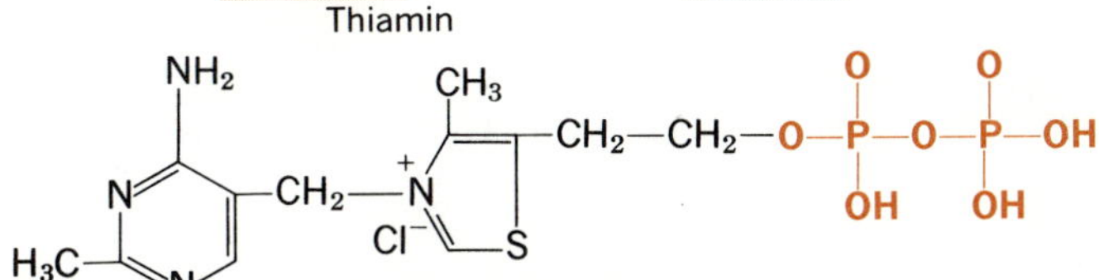

Thiamin pyrophosphate (TPP)

TPP functions in reactions associated with the decarboxylation (splitting out of CO_2 from a —COOH group) of α-keto acids and in reactions involving the transfer of groups such as $HOCH_2$—CO— and CH_3—CO—. The thiazole part of TPP bonds with the acyl unit, as illustrated with the acetyl group in the reductive decarboxylation of pyruvic acid to acetaldehyde:

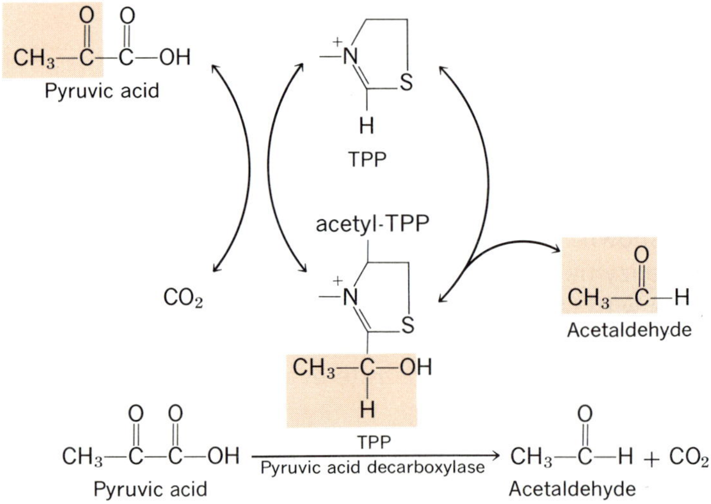

In the fermentation of glucose to ethanol, NADH + H^+ reduces the acetaldehyde formed from the decarboxylation of pyruvic acid to ethanol

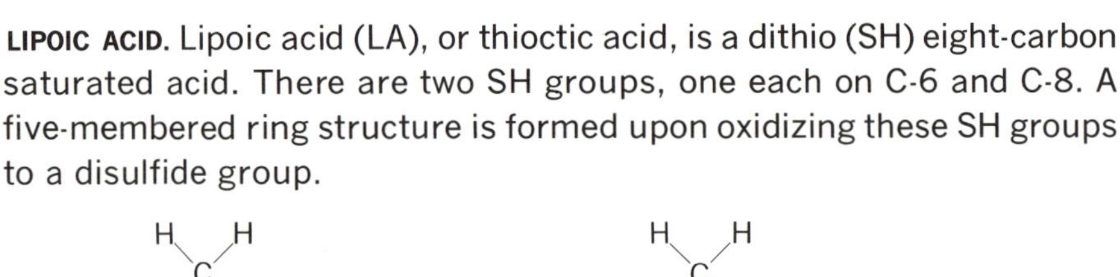

LIPOIC ACID. Lipoic acid (LA), or thioctic acid, is a dithio (SH) eight-carbon saturated acid. There are two SH groups, one each on C-6 and C-8. A five-membered ring structure is formed upon oxidizing these SH groups to a disulfide group.

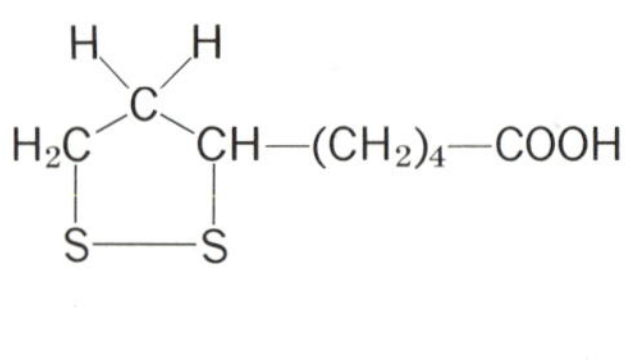

Lipoic acid (oxidized) (LA)

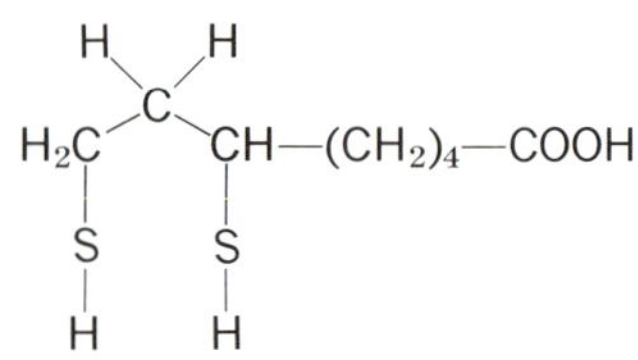

Lipoic acid (reduced) (LAH_2)

By virtue of its ability to be oxidized and reduced, lipoic acid plays an important role in hydrogen-transport reactions and in the transport of acyl groups from acyl-TPP obtained in the decarboxylation of acids such as pyruvic and α-ketoglutaric. In the following equation, which illustrates the chemistry of the transfer of the acetyl group from TPP to lipoic acid, only the reactive portion of TPP is shown.

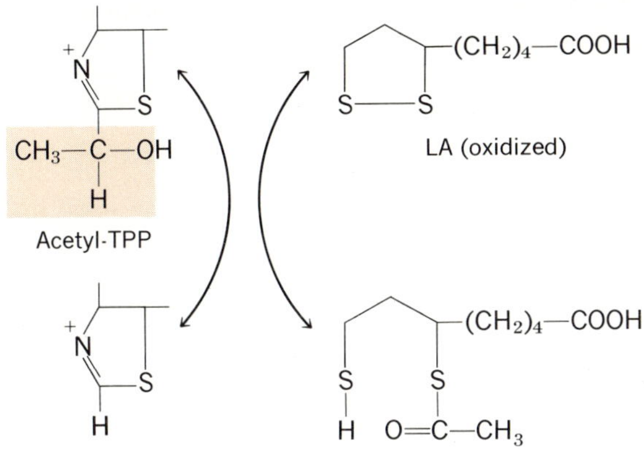

TPP Acetyl-LA

PANTOTHENIC ACID. Pantothenic acid is a simple compound, composed of β-alanine, $H_2N—CH_2—CH_2—COOH$, linked to 2,4-dihydroxy-3,3-dimethylbutyric acid, $HO—CH_2—C(CH_3)_2—CHOH—COOH$, by a peptide bond.

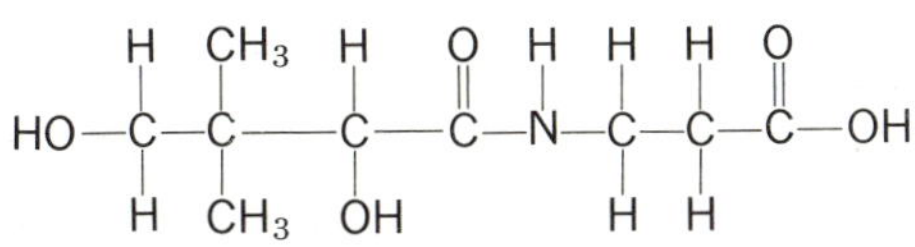

Pantothenic acid

Pantothenic acid is associated with a large number of reactions because it is a component of coenzyme A, CoASH, which also contains thioethanolamine, $HS—CH_2—CH_2—NH_2$, phosphate, ribose, and adenine.

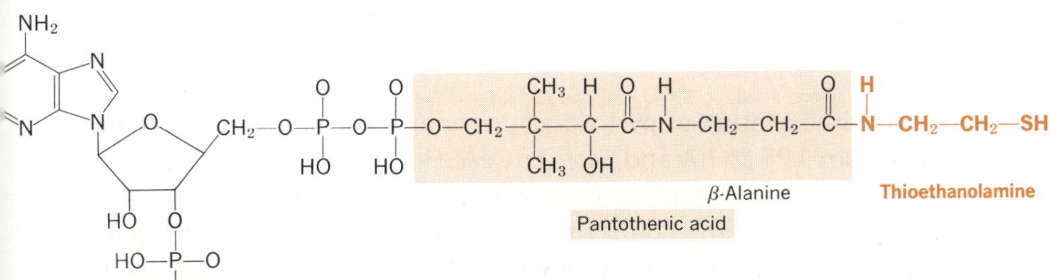

Coenzyme A

Coenzyme A plays a central role in the interrelationships between carbohydrate, fat, and amino acid metabolism as well as in reactions associated with transmission of nerve impulses and vision. In carbohydrate metabolism CoASH accepts the acetyl unit from acetyllipoic acid to form acetyl-SCoA and LAH_2. Coenzyme NAD^+ is also involved in this reaction; it oxidizes the reduced lipoic acid to oxidized lipoic acid, which is then free to pick up another acyl group.

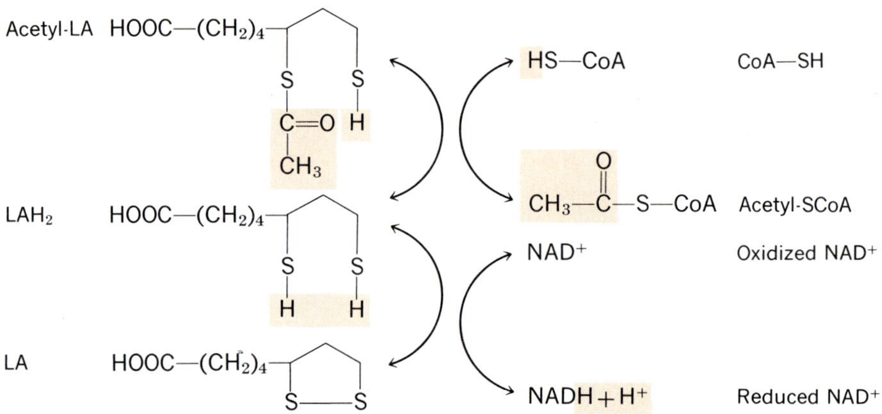

19.2 REACTIONS

Reactions of pyruvic acid have been used to illustrate the importance of coenzymes containing vitamins B_1, lipoic acid, and pantothenic acid. These reactions can be summarized in the following sequence:

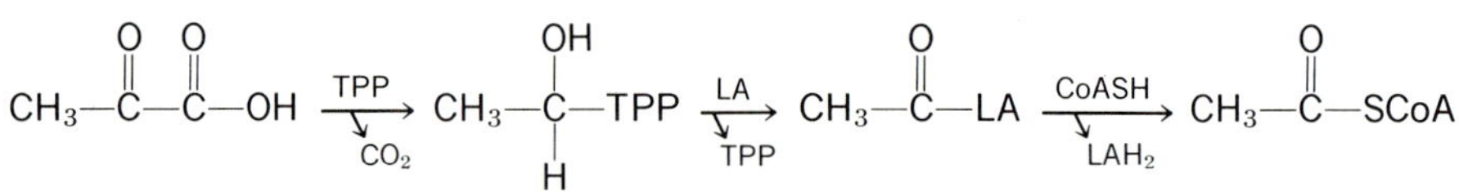

Pyruvic acid Acetyl-TPP Acetyl-LA Acetyl-SCoA

Upon decarboxylation of pyruvic acid, the acetyl group is subsequently passed from TPP to LA and then to CoASH.

α-Ketoglutaric acid is decarboxylated in a manner similar to pyruvic acid; the same coenzymes are involved. α-Ketoglutaric acid is converted into succinyl-SCoA by way of succinyl-TPP and succinyl-LA:

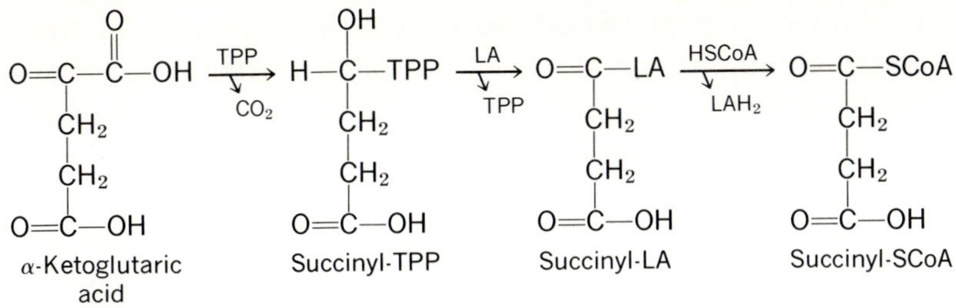

One reaction involving acetyl-SCoA is the transfer of the acetyl group to oxaloacetic acid to form citric acid:

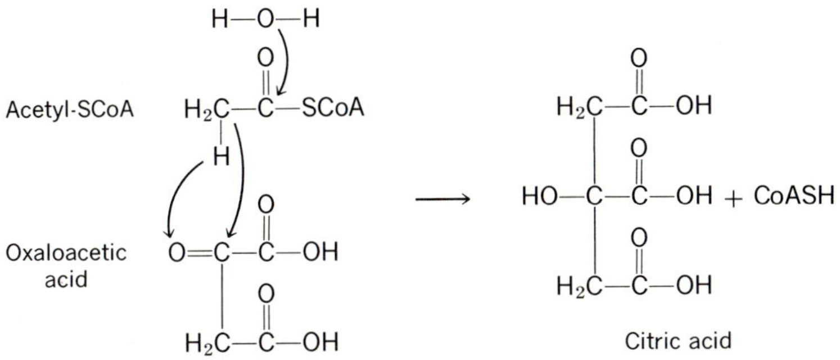

PROBLEM

Consider the following equation for the hydrolysis of acetyl-SCoA:

$$CH_3-\overset{O}{\overset{\|}{C}}-SCoA + H_2O \longrightarrow CH_3-\overset{O}{\overset{\|}{C}}-OH + CoASH$$

Would you expect the ΔG of this reaction to be (*a*) negative or positive and (*b*) to be small or relatively large?

To answer this question examine the product, acetic acid. Since an acid exists in resonance forms, to form acetyl-SCoA from CoASH and acetic acid sufficient energy must be supplied to the reaction to overcome the resonance of the acetate unit. This energy is released when acetyl-SCoA is hydrolyzed. Since reactions tend to proceed to form products of lower energy level than the reactants, this reaction would tend to go spontaneously in the direction indicated, with a concomitant release of energy. The ΔG for the hydrolysis of acetyl-SCoA would be negative. In view of the extra energy contained in acetyl-SCoA, to overcome resonance of the acetyl group one would expect the value of ΔG to be a relatively large negative value; it is -8000 cal/mole. *Acetyl-SCoA is classified as a high-energy compound.*

The condensation of the acetyl group with oxaloacetic acid to form citric acid requires energy to drive the reaction forward. The energy for this reaction comes from the hydrolysis of the acyl-SCoA bond. Note that water is involved in the reaction to hydrolyze off CoASH.

Coenzyme A is also required in the first step of fatty acid oxidation. The fatty acids are converted into the fatty acid acyl-SCoA derivatives, which are often referred to as *activated acids*.

$$\underset{\text{Fatty acid}}{R{-}\overset{\overset{\displaystyle O}{\|}}{C}{-}OH} + HSCoA \xrightarrow{ATP} \underset{\text{Acyl-SCoA}}{R{-}\overset{\overset{\displaystyle O}{\|}}{C}{-}SCoA} + HOH$$

The acyl-SCoA derivative of the fatty acid is a high-energy compound, similar to acetyl-SCoA; therefore, energy (in the form of ATP) must be supplied to drive the above reaction forward.

A number of reactions have been given so far to illustrate the function of several coenzymes, but when they are thus presented separately, it is not easy to see where they fit into the overall picture of biochemistry. It is like an artist painting a picture; the first few strokes are meaningless to a viewer who does not know what the artist has in mind. At this time it is therefore a good idea to take stock of several reactions thus far presented to obtain an indication of their correlation and importance. This is the aim of the following exercise.

The equation for the Krebs cycle, also known as the tricarboxylic acid cycle (TCA cycle) or the citric acid cycle, is given in Fig. 19.1. Take a large piece of wrapping paper and write the formulas for the compounds of this cycle in the sequence indicated. Note that *once around the cycle is equivalent to the oxidation of 1 mole of pyruvic acid.* The hydrogens from $NADH + H^+$ and $FADH_2$ produced in these reactions are eventually passed on to oxygen to form water, as was indicated in part under B_2 coenzyme. Pyruvic acid is formed during the metabolism of glucose, and the oxidation of this acid by way of the Krebs cycle is one of the main pathways in the metabolism of carbohydrates to carbon dioxide and water (Fig. 19.1).

TCA cycle (the Krebs cycle)

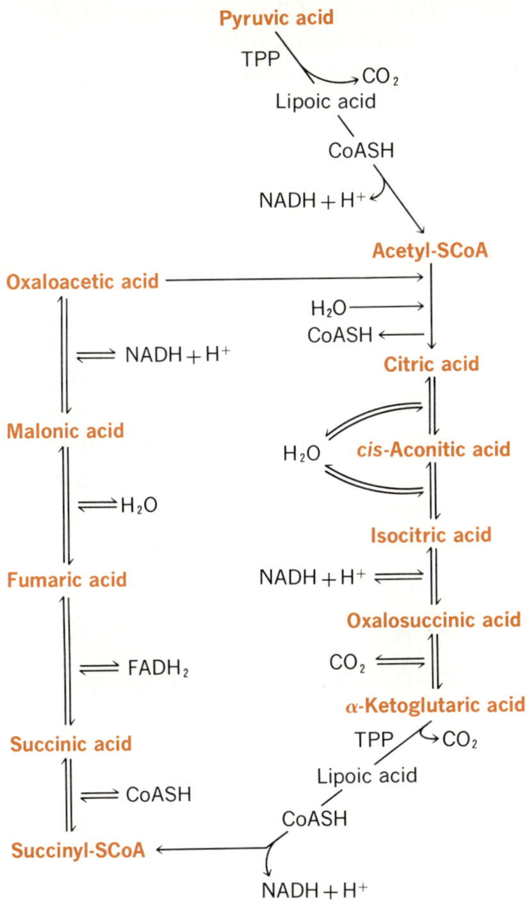

Overall reaction:

$$CH_3{-}CO{-}COOH + 3\ H_2O \longrightarrow 3\ CO_2 + 10\ (H) \qquad (4\ NADH + H^+ + FADH_2)$$

Subsequent oxidation:

$$10\ (H) + 2\tfrac{1}{2}\ O_2 \longrightarrow 5\ H_2O$$

$$\text{Net:}\quad CH_3{-}CO{-}COOH + 2\tfrac{1}{2}\ O_2 \longrightarrow 3\ CO_2 + 2\ H_2O$$

FIGURE 19.1 The tricarboxylic acid cycle or Krebs cycle.

BIOTIN. Biotin is a water-soluble vitamin that contains both sulfur and nitrogen and functions as a coenzyme. Its structure is shown below.

Whereas vitamin B_1 functions in enzyme reactions that eliminate carbon dioxide from the COOH group of α-keto acids (decarboxylation), biotin functions in the reverse manner, namely, in certain enzyme systems that form a COOH group by the addition of carbon dioxide to an organic compound. This is called *carbon dioxide fixation.* Biotin first forms an addition product with carbon dioxide and then "fixes" it in organic compounds nonphotosynthetically (light is not needed).

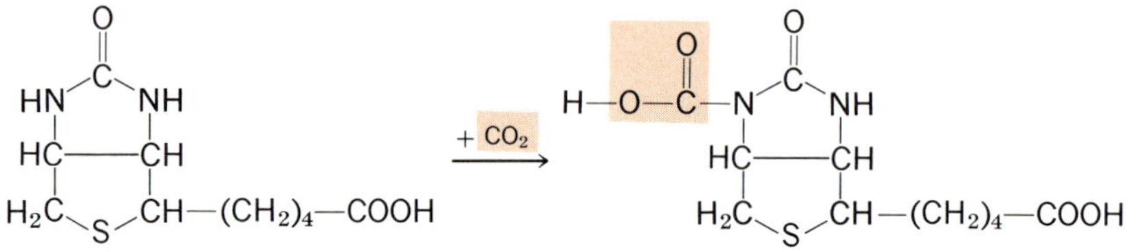

Biotin Biotin–carbon dioxide complex

A number of reactions in living organisms depend upon carbon dioxide fixation, e.g., in carbohydrate and fat metabolism and in the biosynthesis of purines. Several specific examples follow.

Reference was made earlier to the fact that oxaloacetic acid is required to metabolize acetyl-SCoA via the Krebs cycle (Fig. 19.1). One source of oxaloacetate is by the addition of CO_2 to pyruvic acid, a reaction which requires a biotin enzyme.

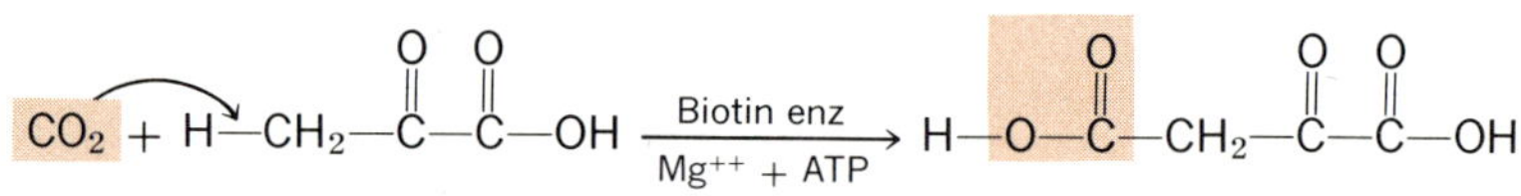

Pyruvic acid Oxaloacetic acid

It is all too apparent that eating carbohydrates can result in the deposition of fat. In the conversion of carbohydrates to fat, carbon dioxide is fixed into acetyl-SCoA to form an intermediate product, malonyl-SCoA, the CoA derivative of malonic acid, $HO_2C—CH_2—CO_2H$:

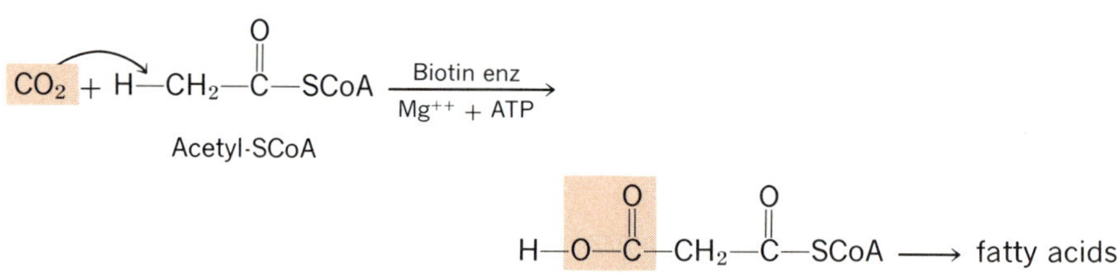

Acetyl-SCoA

Malonyl-SCoA
(intermediate in fatty
acid biosynthesis)

Energy for nonphotosynthetic carbon dioxide fixation is supplied by ATP, whereas for photosynthetic carbon dioxide fixation it is supplied by light.

In the above two reactions Mg^{++} ion is required. Mg^{++} is an activator for enzyme systems that utilize ATP. A mere trace is all that is needed. This is a good example of the importance of mineral elements in biological systems.

FOLIC ACID. Folic acid is a vitamin with a rather complex function. It is associated with enzymes that transfer a one-carbon unit, such as HCHO and CH_3. The chemistry is quite intricate and will not be presented in detail.

Folic acid is composed of pterin, *p*-aminobenzoic acid, and glutamic acid. The last two units are linked together by a peptide bond. There may be one or three glutamic acid residues in the natural coenzyme.

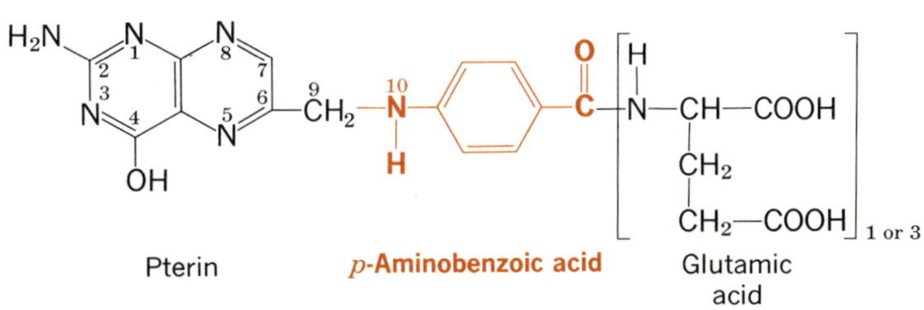

Folic acid

It is the reduced form of folic acid (tetrahydrofolic acid) that picks up the one-carbon unit, which becomes bonded to a nitrogen atom of the folic acid moiety of the molecule. The formulas for tetrahydrofolic acid and the complex with a one-carbon unit (formyl and methyl) attached to it are

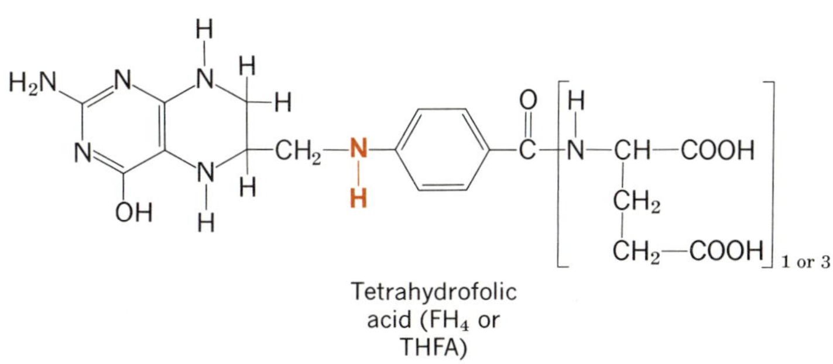

Tetrahydrofolic acid (FH_4 or THFA)

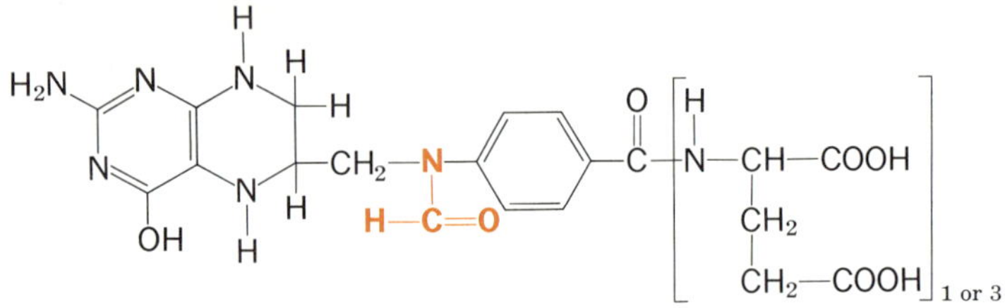

Formyl (N-10)–FH_4

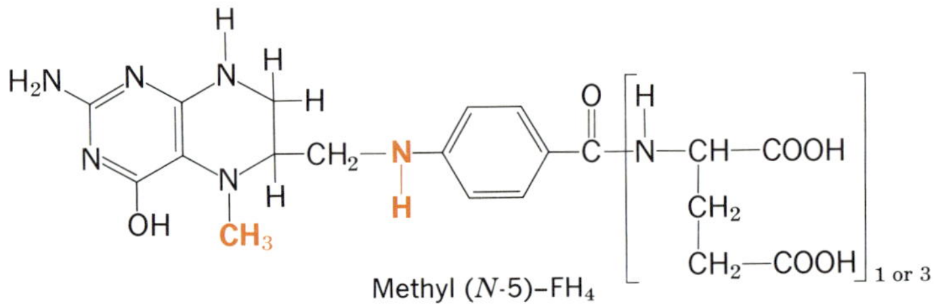

Methyl (N-5)–FH_4

The transfer of a one-carbon unit is associated with the biosynthesis of purines and pyrimidines and with the metabolism of the amino acids serine, glycine, and methionine. Thus folic acid enzymes ultimately play an important role in metabolism in general. For example, consider only the importance of the purine adenine as a component of NAD^+, $NADP^+$, FAD, and ATP, which is the immediate energy source for most biological reactions.

VITAMIN B_{12}. Vitamin B_{12} (cyanocobalamin) is the compound previously known as the *animal protein factor,* a lack of which causes pernicious anemia (low hemoglobin content of red blood cells). The structure of vitamin B_{12} is shown in Fig. 19.2. The isolated vitamin contains the CN group (R = CN), which is an artifact obtained during the isolation procedure. A number of compounds are known in which different groups are substituted for R, for example, hydroxycobalamin, in which R = OH. In the natural B_{12} coenzyme, however, R = 5′-deoxyadenosyl; it is called 5′-deoxyadenosylcobalamin. The formula for 5′-deoxyadenosyl is

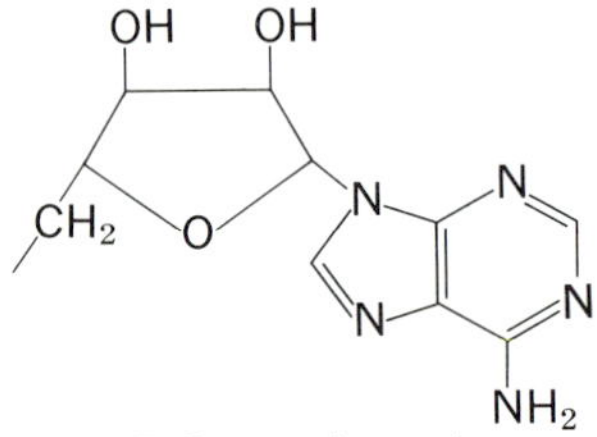

5′-Deoxyadenosyl

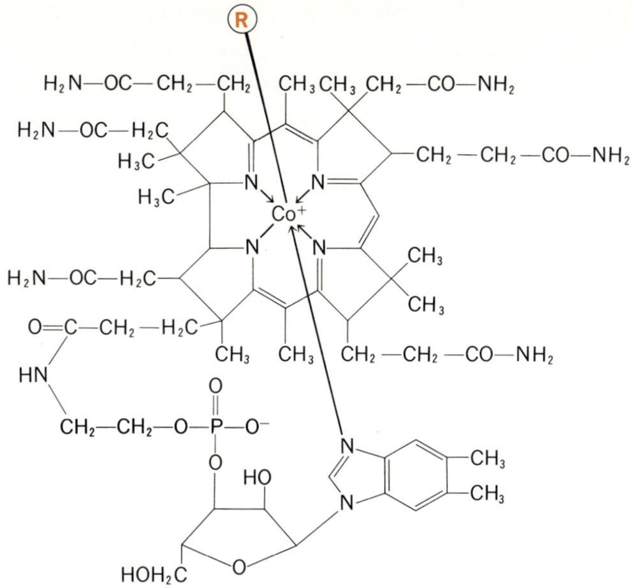

R—cobalamin

If	Then structure is
R = CN	Cyanocobalamin, vitamin B_{12}
R = OH	Hydroxocobalamin
R = 5′-deoxyadenosyl	B_{12} coenzyme

FIGURE 19.2 The general structure for cobalamin.

Vitamin B_{12} is an example not only of a vitamin associated with enzyme systems but also of the importance of the mineral element cobalt. No other element can replace cobalt in this coenzyme.

The function of vitamin B_{12} enzymes has not been completely elucidated, but they are known to be linked with the formation of the methyl groups of methionine and choline. Choline is a component of

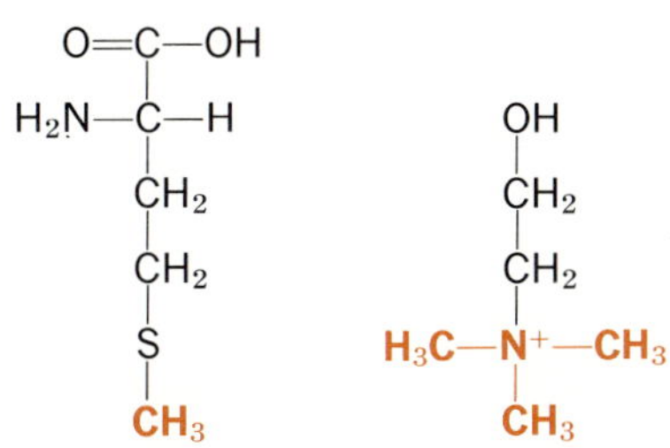

Methionine Choline

CH_3 groups from B_{12} enzyme

phospholipids and is also important in reactions associated with propagation of nerve impulses. Vitamin B_{12} coenzyme is also active in the enzymatic conversion of methylmalonyl-SCoA to succinyl-SCoA. Methylmalonyl-SCoA is formed by the fixation of CO_2 into propionyl-SCoA, a reaction that requires a biotin enzyme. The equation for the conversion of propionic acid to succinyl-SCoA is

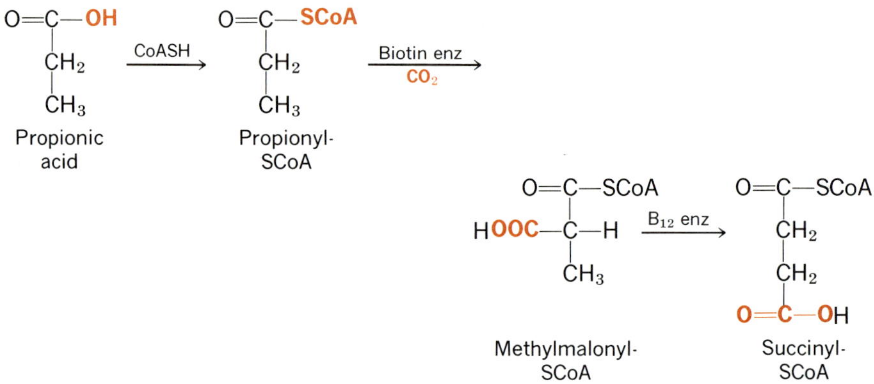

Propionic acid is produced in considerable quantity by bacterial fermentation in the rumen of ruminants. Propionyl-SCoA or methylmalonyl-SCoA or both are intermediary products of the metabolism of a number of different compounds, such as the amino acids valine, isoleucine, threonine, and methionine and the odd-numbered fatty acids. The series of reactions above affords a means whereby these products can enter into the Krebs cycle via succinyl-SCoA and be further metabolized.

COENZYME Q. Coenzyme Q (CoQ) is a fairly recent addition to the group of coenzymes associated with the passage of hydrogen from substrate to oxygen. Also called *ubiquinone* (UQ), it has the formula

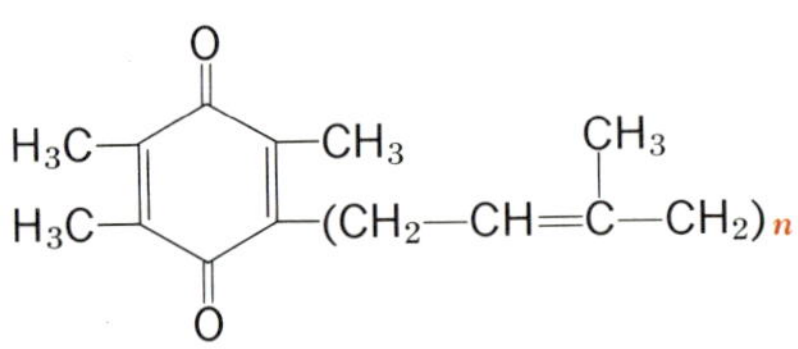

where $n = \begin{cases} 6 & CoQ_6 \text{ or } UQ_{30} \\ 10 & CoQ_{10} \text{ or } UQ_{50} \end{cases}$

Coenzyme Q (CoQ)
Ubiquinone (UQ)

CoQ accepts hydrogen from reduced B_{12} enzymes (flavoproteins, $FADH_2$) and passes the hydrogen to the cytochrome system

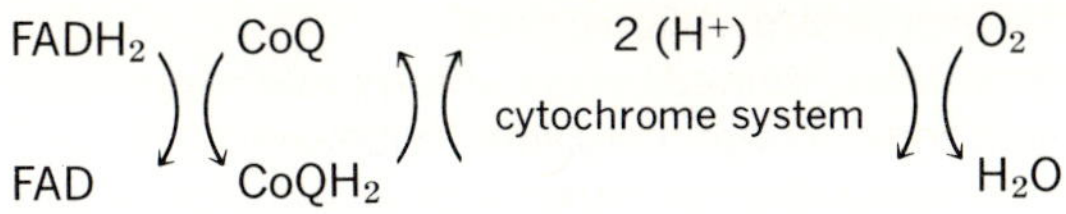

CoQ is easily oxidized and reduced by virtue of its quinone structure; the reaction is written for the simpler compound quinone.

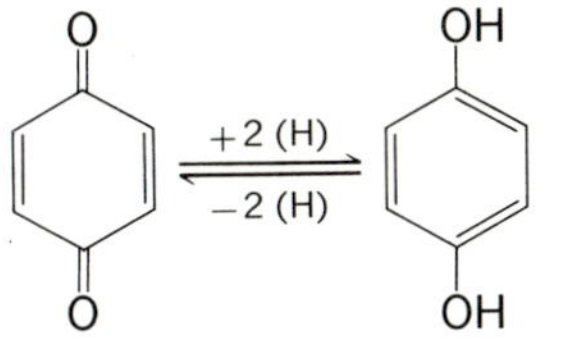

Quinone Hydroquinone

QUESTIONS

1. What is the significance of the term "coenzyme"?
2. Define vitamin.
3. Name three vitamins that contain nitrogen.
4. Name the products obtained upon complete hydrolysis of NAD^+ and $NADP^+$.
5. What atom in NAD^+ or $NADP^+$ is responsible for the positive charge?
6. Name three coenzymes that are associated with dehydrogenation reactions.
7. Place the letter of the compound in the first column in the parentheses of the applicable statements in the second column.

   (*a*) Thiamin
   (*b*) Riboflavin
   (*c*) Niacin
   (*d*) Pyridoxal or pyridoxamine
   (*e*) Pantothenic acid
   (*f*) Lipoic acid
   (*g*) Two or more of the above
   (*h*) None of the above

   ( ) Is contained in coenzyme A.
   ( ) Is contained in coenzymes that transport hydrogen.
   ( ) Contains sulfur.
   ( ) Is associated with the enzymatic transfer of acyl groups.
   ( ) Is associated with the enzymatic conversion of succinic acid to fumaric acid.
   ( ) Is a fat-soluble vitamin.
   ( ) Is associated with the enzymatic conversion of oxaloacetic acid to α-ketoglutaric acid.
   ( ) Is associated with transamination.
   ( ) Contains β-alanine in its structure.
   ( ) Contains adenine in its structure.
8. (*a*) How do you explain the fact that when NAD^+ or $NADP^+$ enzyme systems remove two atoms of hydrogen from a substrate, a hydrogen ion is formed?
   (*b*) Is this also true when FAD transports hydrogen? Explain.
9. Write the formula for the thiazole nucleus.
10. What moiety of TPP is responsible for transporting an acyl group?

11. What functional groups are in lipoic acid?

12. What are the products obtained upon the complete hydrolysis of coenzyme A?

13. What type of reactions do biotin enzymes catalyze?

14. Are acyl-SCoA compounds considered high-energy compounds? Explain.

15. What are the sources of energy for nonphotosynthetic and photosynthetic carbon dioxide fixation?

16. Write an equation for the formation of (*a*) oxaloacetic acid from pyruvic acid and (*b*) malonyl-SCoA from acetyl-SCoA.

17. What structure in coenzyme Q is responsible for its ability to transport hydrogen?

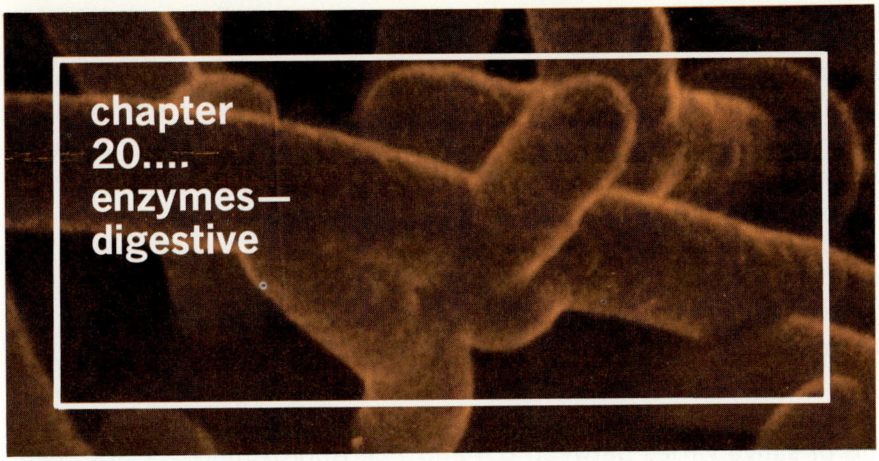

chapter 20.... enzymes—digestive

Digestion is the process whereby large food particles are converted into units small enough to be absorbed. The *only* chemical reaction associated with digestion is *hydrolysis*. Large polymers like starch or protein are eventually hydrolyzed to their respective monosaccharides or amino acids, which are then absorbed. Disaccharides must also be hyrolyzed to their monosaccharides before the body can utilize them.

Digestion is a progressive procedure. It does not all occur in any one part of the digestive tract, and several different enzymes are called upon to effect the hydrolysis of large polymers. Digestive enzymes are secreted in three different sites in the digestive tract, namely, the mouth, the stomach, and the intestines. Digestion of a particular food may be started, for example, in the mouth and be finished in the intestines. For other foods digestion may start in the stomach and be finished in the intestines. Only the simpler molecules like the disaccharides are hydrolyzed in just one place.

We shall discuss digestion from the standpoint of the food itself rather than on the basis of enzymes secreted in a particular digestive fluid. Thus, we shall be able to follow as a unit the complete hydrolysis of a particular food material.

Enzymes which act on polymers do not yield many monomers directly but lower-molecular-weight polymers, which in turn require different enzymes for hydrolysis to monomers. Digestion is the result of the united effect of a number of different enzymes working under different conditions. Enzyme specificity, of course, accounts for this fact.

The chemical reaction associated with digestion is hydrolysis, and the following questions will help the reader recall the hydrolysis reactions of carbohydrates, lipids, and proteins. What type of linkage exists between sugar units of a di-, oligo-, and polysaccharide? Is water eliminated when two monosaccharides combine to form a disaccharide? Is this reaction reversible? Would emulsification of a fat aid in its hydrolysis? Explain. What two groups are involved in the formation of a peptide bond? Is this reaction reversible?

20.1 DIGESTIVE FLUIDS

Saliva is secreted into the mouth from the parotid, submaxillary, and sublingual glands. It averages about 99.4% water and contains various salts, mucin (a glycoprotein), and the enzyme α-amylase (ptyalin). The pH of saliva varies between 6.4 and 7.3. Approximately 1,500 ml of saliva is secreted per day by a normal adult. The control of secretion is entirely under the influence of the nervous system. The thought or sight of food can stimulate secretion. Mucin, which acts as a lubricant, is produced by the sublingual gland, whereas amylase is secreted by the parotid glands. Traces of the enzymes maltase and catalase have been reported in saliva. The composition of saliva is not the same for all species; e.g., the dog, sheep, and cat lack salivary amylase.

Gastric juice, like saliva, is mostly water, 99.4%, and contains mucins and enzymes. A normal adult secretes about 2.5 liters of gastric juice per day. The enzymes, however, differ from those of the saliva and are active at a much lower pH. Enzymes of the gastric juice include proteases (pepsin and rennin) and lipase (gastric lipase).

A distinguishing feature of gastric juice is its high content of hydrochloric acid, which approaches 0.1 *N*. Hydrochloric acid is produced by the parietal cells, and as the acid solution is being secreted its pH is about 0.87. The formation of hydrochloric acid by these cells can be summarized as follows:

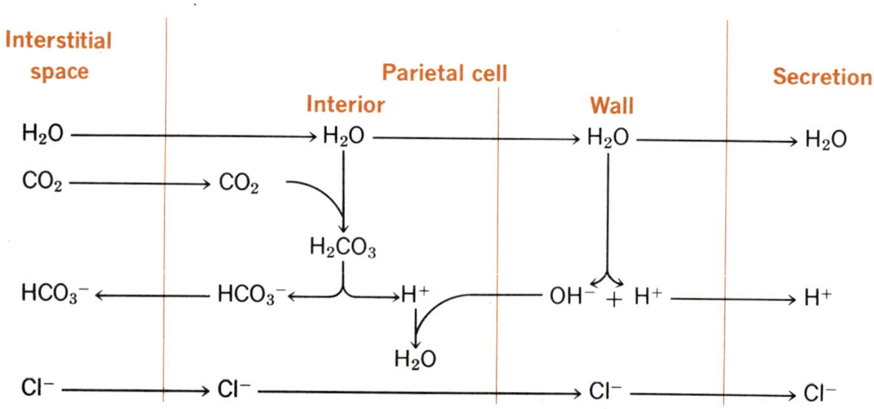

Assume that gastric juice has a pH of 0.1. How does the hydrogen-ion concentration of this juice compare with that of the blood, pH 7.4? The hydrogen-ion concentration in gastric juice is 2.5×10^6 times that in the blood. This high concentration of acid in the stomach prevents fermentation and acts as an antiseptic.

Pancreatic juice, in contrast to gastric juice, has a pH of 7 to 8. Thus, when the food passes from the stomach into the intestines, the hydrochloric acid is neutralized. This entrance of a solution of low pH into the first part of the intestines (duodenum) triggers the release into the blood of the hormone secretin, which stimulates the secretion of pancreatic juice.

About 700 ml of pancreatic juice is secreted per day in a normal adult. The enzymes in this juice are quite varied in their action and include those which hydrolyze proteins, polysaccharides, and lipids.

Intestinal juice is secreted by cells of the small-intestinal mucosa. It is believed that most enzymatic activities of this juice occur in the intestinal mucosa. There is no elaboration of voluminous amounts of enzyme solution as with saliva and gastric, or pancreatic, juice. Among the enzymes of the intestinal juice are those which hydrolyze the smaller molecules, e.g., peptides, oligosaccharides, and lipids. An important enzyme of this secretion is enterokinase, which converts the inactive trypsinogen into the active enzyme trypsin and thereby initiates protein digestion in the intestines.

Bile is produced by the liver and passes into the gall bladder, where it is concentrated and stored until used. It is a viscous brownish-yellow fluid which may contain 10 to 17% solids. The brownish color is due to degradation products of hemoglobin. Hemoglobin from worn-out red blood cells is degraded in the liver and the products secreted in the bile. Bile salts constitute from 6 to 10% of bile and are mainly glycine and taurine peptide derivatives of cholic acid (see Sec. 15.7). These salts play an important role in digestive processes by aiding in the emulsification of lipids and in the absorption of fatty acids and fat-soluble vitamins, especially vitamin K. The bile salts secreted into the intestines are almost completely absorbed with the lipids and then returned to the liver to be secreted again in the bile.

20.2 CARBOHYDRATES

Digestion of carbohydrates occurs in the mouth and the intestines. No enzymes secreted in the stomach hydrolyze carbohydrates. The ultimate product of the digestion of carbohydrates is the simple sugars of which they are composed. Since the simple sugars, or monosaccharides, are not hydrolyzed, it is only the poly- and oligosaccharides that we are concerned with in digestion. Because an enzyme in the saliva hydrolyzes polysaccharides, we shall start our discussion with the digestion of these compounds.

20.3 POLYSACCHARIDES

The two enzymes that hydrolyze amylose, amylopectin, and glycogen are α-amylase and β-amylase. α-Amylase is produced by the salivary glands and secreted into the mouth as salivary amylase. Another α-amylase secreted with the pancreatic juice into the small intestines is pancreatic amylase (Fig. 20.1). Both α-amylases require chloride ion for activity. β-Amylase is a plant enzyme and is found in grain, such as barley. These two amylases attack the starches and glycogen in a different manner although they are both specific for the α-D-glucose-(1—4)-α-D-glucose linkage. Neither hydrolyzes the β-glucosidic linkage found in cellulose and cellobiose.

α-Amylase attacks the (1—4) links in a random manner at any interior point of the polysaccharide chain. It is therefore called an *endoenzyme*. The product of α-amylase activity on amylose is maltose and some glucose since this polysaccharide contains only the α-(1—4) linkage (see Fig. 20.1). However, when amylopectin or glycogen is hydrolyzed by this enzyme, some oligosaccharides containing the α-(1—6) linkage are formed because these polysaccharides contain the α-(1—6) linkage where branching occurs and neither amylase is capable of hydrolyzing this bond. Thus, when the enzyme reaches a 1—6 linkage, its action on that chain stops. The products that remain

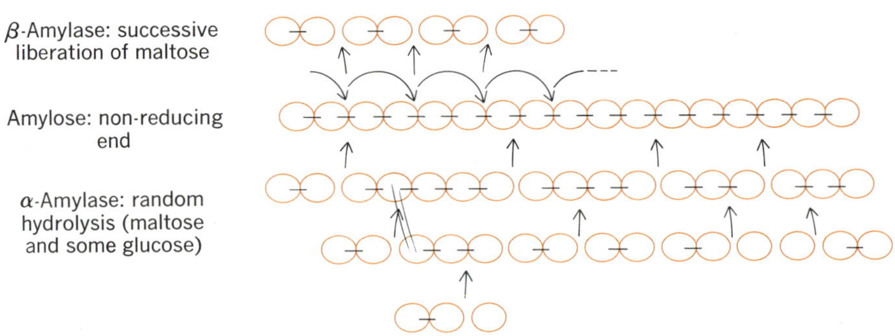

FIGURE 20.1 The action of α- and β-amylase on amylose. ○ = glucose, ⊖⊖ = α-D-glucose-(1—4)-α-D-glucose (maltose).

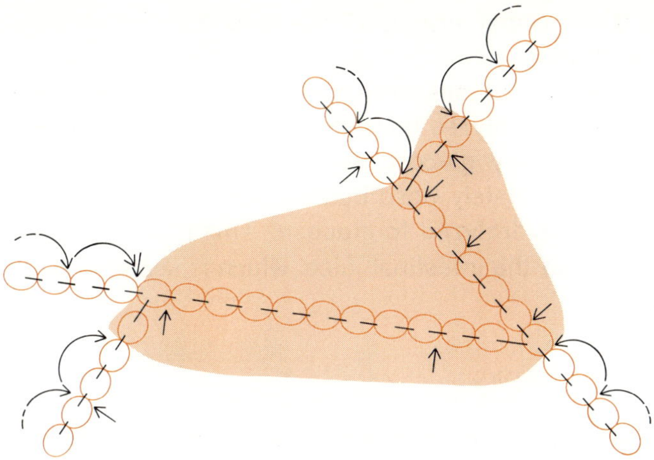

FIGURE 20.2 The action of α- and β-amylase on amylopectin or glycogen. The limit of hydrolysis by β-amylase is indicated by shading; hydrolysis by α-amylase is limited to oligosaccharides, such as ○—○, ○○—○, ○○—○○, ○○○ (with ○ branched from the middle unit), ○—○○. ○ = glucose; ○○ = α-D-glucose-(1—4)-α-D-glucose; ○—○ = α-D-glucose-(1—6)-α-D-glucose; ↷ = β-amylase action; ↑ = α-amylase action.

after digestion are low-molecular-weight oligosaccharides, each containing a α-glucose-(1—6)-glucose unit (Fig. 20.2).

β-Amylase hydrolyzes starch, amylopectin, and glycogen in a more orderly manner than α-amylase. β-Amylase attacks these polysaccharides, beginning at the nonreducing end of the molecule, and successively hydrolyzes off one maltose unit at a time. Therefore, no glucose is produced as it is with α-amylase (see Fig. 20.1). The hydrolysis of amylopectin or glycogen by β-amylase is not as complete as with α-amylase because hydrolysis from the ends of the branched chains occurs until the first 1—6 link is encountered. Enzymatic reaction then ceases on that chain since the enzyme cannot jump over a 1—6 link and attack a 1—4 linkage on the other side. As a result, large *limit dextrins* remain after digestion of these polysaccharides with β-amylase (Fig. 20.2).

The extent of digestion of carbohydrates by saliva is varied. Usually one does not chew food long enough for appreciable digestion to take place in the mouth. However, after the food is swallowed, carbohydrate digestion continues until the acidity of the gastric juice decreases the pH of the stomach contents to a value which inhibits amylase activity. Depending upon the food, this action may continue for perhaps 20 min.

When the food passes from the stomach into the intestine, pancreatic amylase hydrolyzes the polysaccharide left undigested by the salivary amylase. Any glucose which has been formed during this hydrolysis is immediately absorbed as such. The maltose, however, must be further hydrolyzed to glucose. This is accomplished by the enzyme maltase of the intestinal juice, which is secreted by the intestinal mucosa.

20.4 DISACCHARIDES

Disaccharides that have been formed from the reaction of amylases on polysaccharides or taken in with food are hydrolyzed by enzymes in the intestinal juice. In addition to such inorganic salts as bicarbonate, phosphate, chloride, potassium, and sodium, this juice contains the enzymes sucrase, maltase, lactase, and oligo-(1—6)-glucosidase. These enzymes hydrolyze their respective substrates to simple sugars.

$$\text{Maltose} \xrightarrow{\text{Maltase}} 2\ \text{glucose}$$

$$\text{Sucrose} \xrightarrow[\text{Invertase}]{\text{Sucrase}} \text{glucose} + \text{fructose}$$

$$\text{Lactose} \xrightarrow{\text{Lactase}} \text{glucose} + \text{galactose}$$

$$\text{Isomaltose} \xrightarrow[\text{glucosidase}]{\text{Oligo-(1-6)-}} 2\ \text{glucose}$$

20.5 ABSORPTION

The monosaccharides are absorbed through the intestinal wall and enter the blood. Undigested carbohydrate such as cellulose or pentosans is excreted in the feces.

The mechanism of absorption of monosaccharides has not yet been clearly elucidated. Although diffusion through the intestinal lining may play a part in this process, it certainly is not the entire picture. For if it were, the lower-molecular-weight sugars would be absorbed faster than the higher-molecular-weight ones and those of equal molecular weight would be absorbed at the same rate. Experimental evidence indicates that this is not so. The pentoses are absorbed at a slower rate than the hexoses, and not all hexoses are absorbed at the same rate. Based on glucose as 100, the sugars may be arranged in the following order to compare their rate of absorption: galactose, 110; glucose, 100; fructose, 43; mannose, 19; pentoses, 9 to 15. This indicates that there is an active transport of sugars across the intestinal lining. This is further indicated by the fact that inhibitors of respiration, such as cyanide, also inhibit absorption of the sugars. There is a need for energy someplace in the mechanism. The general condition of the animal also influences sugar absorption; e.g., a deficiency of the vitamins thiamin, pantothenic acid, or pyridoxine decreases the efficiency of absorption of sugars.

20.6 LIPIDS

The digestion of triglycerides is not as complete a process as that of carbohydrates. In fact, if fat is well emulsified, some will be absorbed as such without being hydrolyzed. Hydrolysis of lipids proceeds to a varying extent; hence, there is a mixture of diglycerides and monoglycerides, both of which can be absorbed along with the free fatty acids from the partial hydrolysis. Thus, complete digestion of triglycerides is not a prerequisite to their absorption. It is of interest to note that the products of lipid digestion enter mainly into the lymph and occur there as triglycerides. Thus, there is a resynthesis of triglycerides as the hydrolytic products are transported through the lining of the intestinal tract. About 20% of the absorbed lipid finds its way into the portal blood.

20.7 TRIGLYCERIDES

GASTRIC LIPASE. Fats are first exposed to digestive enzymes in the stomach. Gastric lipase is secreted in the gastric juice. The extent of hydrolysis of lipids in the stomach is questionable. The low pH does not favor emulsification of the lipid and hence extensive digestion. Remember that enzymes are water-soluble whereas lipids are not; consequently, the enzyme must react on the surface of the fat particle. It follows that the greater the surface area, the greater the rate of enzymatic action. If fats are already in the emulsified state, as in homogenized milk, when ingested, more digestion occurs in the stomach than of, say, butter on a piece of bread.

PANCREATIC LIPASE. This enzyme is secreted in the intestines in the pancreatic juice and is responsible for most triglyceride hydrolysis. The conditions in the intestines are more favorable to lipid digestion than in the stomach as a result of conditions which favor emulsification of lipids, e.g., the alkaline pH of the intestines and the presence of bile salts, which are emulsifiers. Also, the fatty acids formed by the action of lipase are neutralized, and the resulting salts aid in the emulsification of lipids. The pH of the pancreatic juice is between 7.5 and 8.2.

EXAMPLE

What fraction of fatty acids would exist in the soap form at pH 7.8? Substitution in the Henderson-Hasselbalch equation gives the answer. Assume for easy calculations that the pK of the fatty acid is 4.8 and the pH of the pancreatic juice 7.8. Thus

$$7.8 = 4.8 + \log \frac{[\text{salt}]}{[\text{acid}]}$$

$$3 = \log \frac{[\text{salt}]}{[\text{acid}]}$$

$$\frac{[\text{salt}]}{[\text{acid}]} = \frac{1{,}000}{1}$$

As shown by these calculations, most of the acids released during digestion are converted into soaps.

Lipids that melt appreciably above body temperature are more difficult to emulsify than those that melt at or below body temperature; this is reflected in their ease of digestion. Fat that is liquid at body temperature is more easily digested.

20.8 PROTEIN

Protein digestion starts in the stomach and is completed in the intestines. There is no enzyme in the saliva that hydrolyzes proteins. Since only amino acids are normally absorbed, proteins must eventually be completely hydrolyzed. This is accomplished in steps by several different enzymes.

PEPSIN. Pepsin is synthesized in the cell and secreted in the gastric juice as its zymogen (inactive enzyme), thus preventing the digestion of the producing cells themselves. The zymogen, called *pepsinogen,* is converted to active pepsin by hydrogen ion or by pepsin itself. During this activation several peptides are liberated. One having a molecular weight of about 3,000 is the unit that inhibits pepsin.

$$\text{Pepsinogen} \xrightarrow[\text{Pepsin}]{H^+} \text{pepsin} + \text{peptides}$$

Pepsin is an endopeptidase because it hydrolyzes peptide bonds in the interior of the protein molecule. Although it hydrolyzes peptide bonds in general, it shows a preference for those that include the amino group of the aromatic amino acids tyrosine and phenylalanine. Because of this preferential hydrolysis, peptides of varying lengths are produced; they are further hydrolyzed in the intestines.

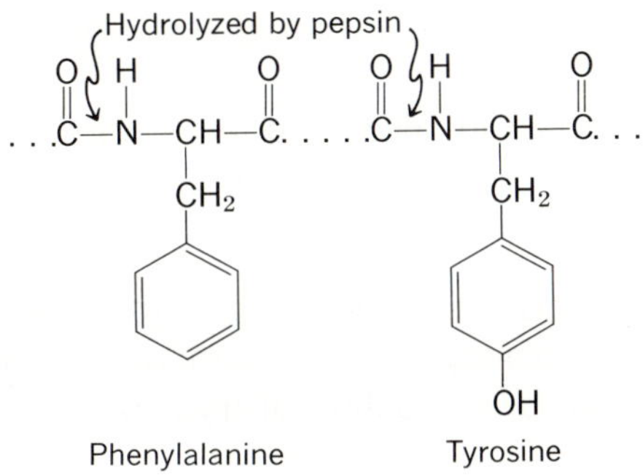

RENNIN. Rennin, like pepsin, is secreted into the stomach as its zymogen, prorennin, and is converted into the active enzyme by the hydrogen ion of the gastric juice. Rennin is especially important in young animals because of its milk-curdling action. It partially hydrolyzes casein to form paracasein, which then reacts with Ca^{++} to form calcium paracaseinate (insoluble curd).

$$\text{Casein} \xrightarrow{\text{Rennin}} \text{paracasein} \xrightarrow{Ca^{++}} \text{calcium paracaseinate}$$

Rennin is absent from the gastric juice of adults. A commercial rennin preparation, obtained from the stomach of young calves, mixed with milk reacts and sets the milk into a jellylike mass of calcium paracaseinate. This is used as a dessert upon addition of flavoring.

TRYPSIN. Trypsin is a proteinase secreted in the pancreatic juice as trypsinogen. Trypsin is an endopeptidase which shows specificity for the peptide bond on the carboxyl side of lysine and arginine.

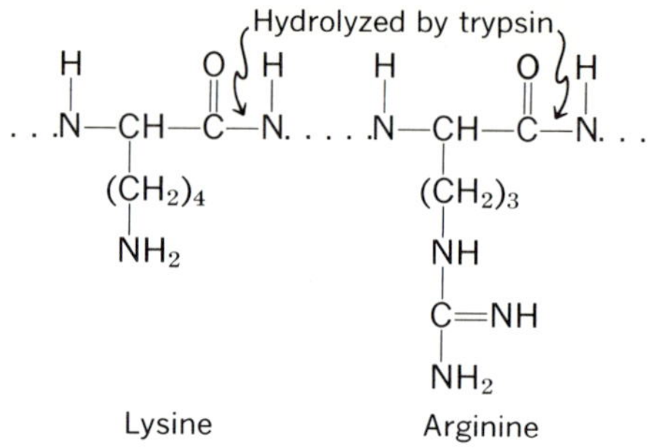

Trypsinogen is converted to trypsin by the proteolytic action of enterokinase secreted in the duodenum. Once trypsin is formed, it acts autocatalytically on trypsinogen to form trypsin.

$$\text{Trypsinogen} \xrightarrow[\text{Trypsin}]{\text{Enterokinase}} \text{trypsin} + \text{valyl—aspartyl}_4\text{—lysine}$$

The activation of trypsin involves the hydrolysis of the hexapeptide, valyl-(aspartyl)$_4$-lysine, from trypsinogen. When attached to trypsin, this hexapeptide masks the active site of the enzyme.

CHYMOTRYPSIN. Chymotrypsin is another proteolytic enzyme secreted as its zymogen, chymotrypsinogen, in the pancreatic juice. It is converted to active chymotrypsin by the action of trypsin.

$$\text{Chymotrypsinogen} \xrightarrow{\text{Trypsin}} \text{chymotrypsin}$$

Like the other proteinases, chymotrypsin is an endoenzyme. It shows preference for peptide bonds in which the carboxyl moiety is supplied by tyrosine or phenylalanine.

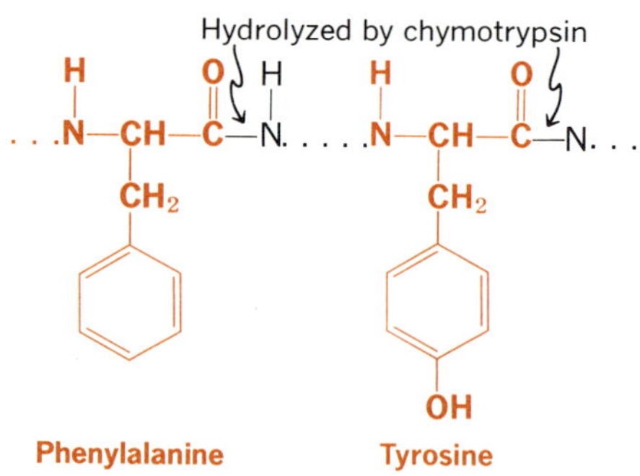

The products of digestion of proteins by proteinases are peptides of varying chain length, frequently called *proteoses* and *peptones*. They must be further hydrolyzed to amino acids.

20.9 PEPTIDES

EXOPEPTIDASE. Exopeptidases hydrolyze off amino acids from the ends of polypeptides (see Fig. 16.3).

CARBOXYPEPTIDASE. Carboxypeptidase is excreted in the pancreatic juice and attacks peptides from the free carboxyl end. Like the proteinases, this enzyme is secreted in an inactive form, which is activated by partial hydrolysis by the action of trypsin. This activation results in the splitting off of about two-thirds of the proenzyme. The action of carboxypeptidase is illustrated as follows:

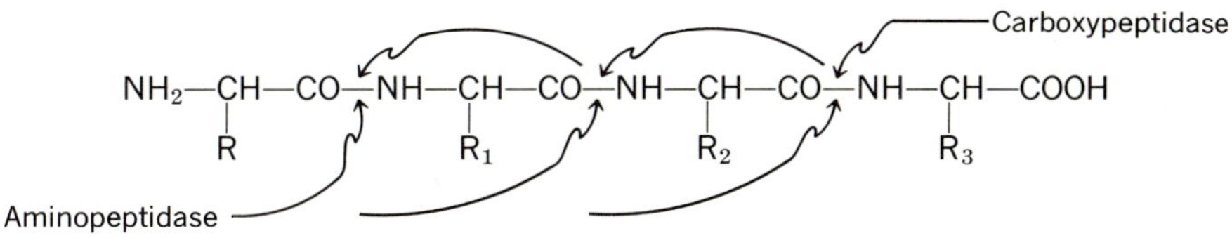

It is of particular interest to note that carboxypeptidase contains zinc, which is essential for enzymatic activity.

AMINOPEPTIDASE. An enzyme that hydrolyzes peptide bonds from the free amino end of a peptide is called an aminopeptidase. It is secreted in the intestinal juice, and its action is illustrated in the equation above. The combined effect of carboxypeptidase and aminopeptidase is to degrade the polypeptides resulting from proteinase activity on proteins. Their action liberates free amino acids, which can then be absorbed.

Several aminopeptidases act only on low-molecular-weight peptides, such as di- and tripeptides. These are also secreted in the intestinal juice and are called di- and tripeptidases, respectively, according to the peptide they hydrolyze most rapidly.

Thus by the action of a number of enzymes, the large protein molecules are eventually hydrolyzed to their respective amino acids. The amino acids are absorbed by an active metabolic process that is not clearly understood and pass into the bloodstream.

QUESTIONS

1. (*a*) Define digestion.
   (*b*) What is the chemical reaction associated with digestion?
2. Name the digestive enzymes in saliva, gastric juice, pancreatic juice, and intestinal juice.
3. (*a*) What acid is contained in gastric juice?
   (*b*) What is the function of this acid?

4. Explain the importance of bile with respect to digestion.

5. (*a*) What is the difference in the action of α- and β-amylase?
(*b*) Which of these enzymes will digest amylopectin most completely? Explain.

6. Complete the following table.

Enzyme	*Substrate*	*Product(s)*
Sucrase	________	________
________	Isomaltose	________
________	________	Glucose and galactose
________	Tripalmitin	________

7. Explain how carboxypeptidase and aminopeptidase can be used to determine the amino acid sequence of a tetrapeptide.

8. What is a zymogen?

9. Explain the difference in the selectivity of pepsin, trypsin, and chymotrypsin.

10. A polypeptide has the following composition. What would be the main products formed upon treating the polypeptide with pepsin? With trypsin?

Ala—Arg—His—Val—Ser—Phe—Lys—Glu—Tyr—Leu

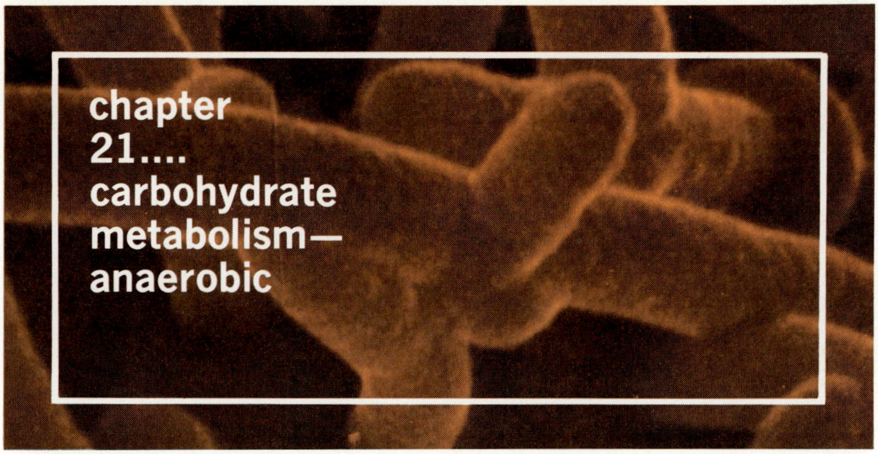

chapter 21.... carbohydrate metabolism—anaerobic

Metabolism, which is considered as any reaction occurring within an organism after metabolites have been absorbed, is a general term including the mechanism both of degrading compounds and of synthesizing new cellular material. The former reactions are collectively termed *catabolism* and the latter *anabolism*.

All cellular reactions are geared to the maintenance and reproduction of the organism and collectively to the survival of the cell. As one can imagine, a large number of reactions must occur to make the many compounds required for cell life. Organisms differ in their capacity to accomplish this end. Plants, for example, are photosynthetic organisms. They require only inorganic material (mineral elements and carbon dioxide) and use light as a source of energy. Some bacteria are also capable of carrying out photosynthetic reactions, whereas other bacteria require various preformed organic compounds. Animals require considerable amounts of nutrients in the organic form, e.g., certain amino acids and vitamins in the diet, as well as a good source of energy since they are not photosynthetic organisms. Animals ultimately rely on the plant kingdom as a dietary source of preformed organic compounds.

Important as the carbohydrates are as a source of energy, they are equally important as a source of carbon chains for synthetic purposes. For example, a number of reactions are involved in the oxidation of glucose to carbon dioxide, water, and energy. Many of the intermediates of these reactions are used for the synthesis of other compounds; e.g., the carbon chain of the nonessential amino acids (those not required in the diet) is made from intermediates of glucose metabolism. The reverse of this is the degradation of amino acids, which involves the entrance of the carbon chain into the scheme of

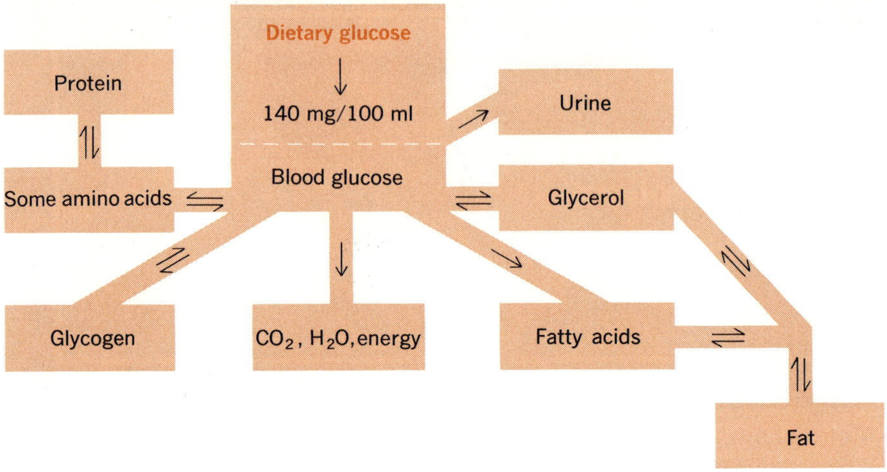

FIGURE 21.1 Distribution of blood glucose.

carbohydrate metabolism. The body also can make fat from carbohydrates. Thus, blood glucose is metabolized to many different products, as briefly diagrammed in Fig. 21.1.

The formation of metabolic intermediates offers a means of controlling metabolic reactions. Products of one reaction may act as a controlling influence on an enzyme associated with another reaction.

When 1 mole of glucose is oxidized, 686,000 cal is released. This is a great deal of energy, and if it were released in large amounts at a time, the cell would not be able to capture it for future use. The cell would die. By oxidizing glucose in many discrete steps, energy can be released in amounts small enough to be used by cellular components without injury.

21.1 THE CELL

Biological processes involve reactions which take place at the cellular level. The various phases of metabolism are associated with definite cellular components. As in a manufacturing plant, not all operations are performed in the same factory room. Unfortunately there is no typical cell that can be used to describe all cells. Cells from different organs of an animal may differ in structure or components. The red blood cells are not identical with liver cells. Bacteria cells differ from animal or plant cells. Photosynthesizing plant cells contain particles not found in nonphotosynthesizing cells. Cells that are not specialized contain much in common, although their organization may vary.

With the advent of the electron microscope, which has a resolution of 5 to 10 Å and a magnification of about 100,000 to 400,000 times, the internal structure of the cell has become better known. A diagram drawn from an electron micrograph of a cell is shown in Fig. 21.2, with various components labeled.

21.2 ANAEROBIC METABOLISM

Glucose is metabolized by several different pathways, three of which are (1) anaerobically (without oxygen), (2) aerobically (with oxygen), and (3) the phosphate shunt, or pentose phosphate pathway. All these pathways have some common intermediates. The anaerobic pathway, which is also known as *glycolysis* or the *Embden-Meyerhof pathway,* will be discussed in this chapter.

In a broad sense, *glycolysis* refers to anaerobic metabolism of carbohydrates. The end product often governs other names which have been given to this process, e.g., alcohol fermentation by yeast, in

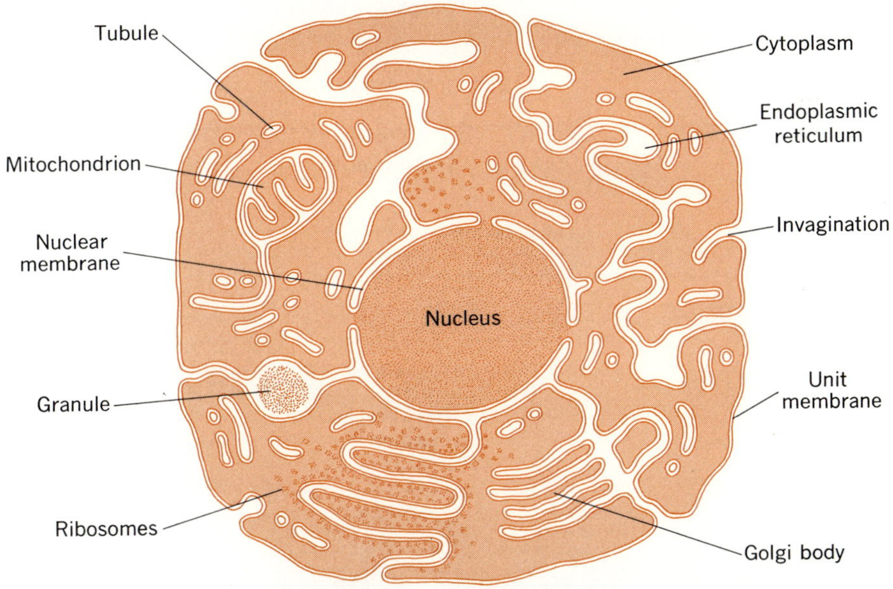

FIGURE 21.2 Schematic cell structure, showing the distribution of the membrane within the cell and some of the many membranous organelles thus far identified. *(Redrawn from J. D. Robertson, The Membrane of the Living Cell, Sci. Amer., April, 1962, p. 72.)*

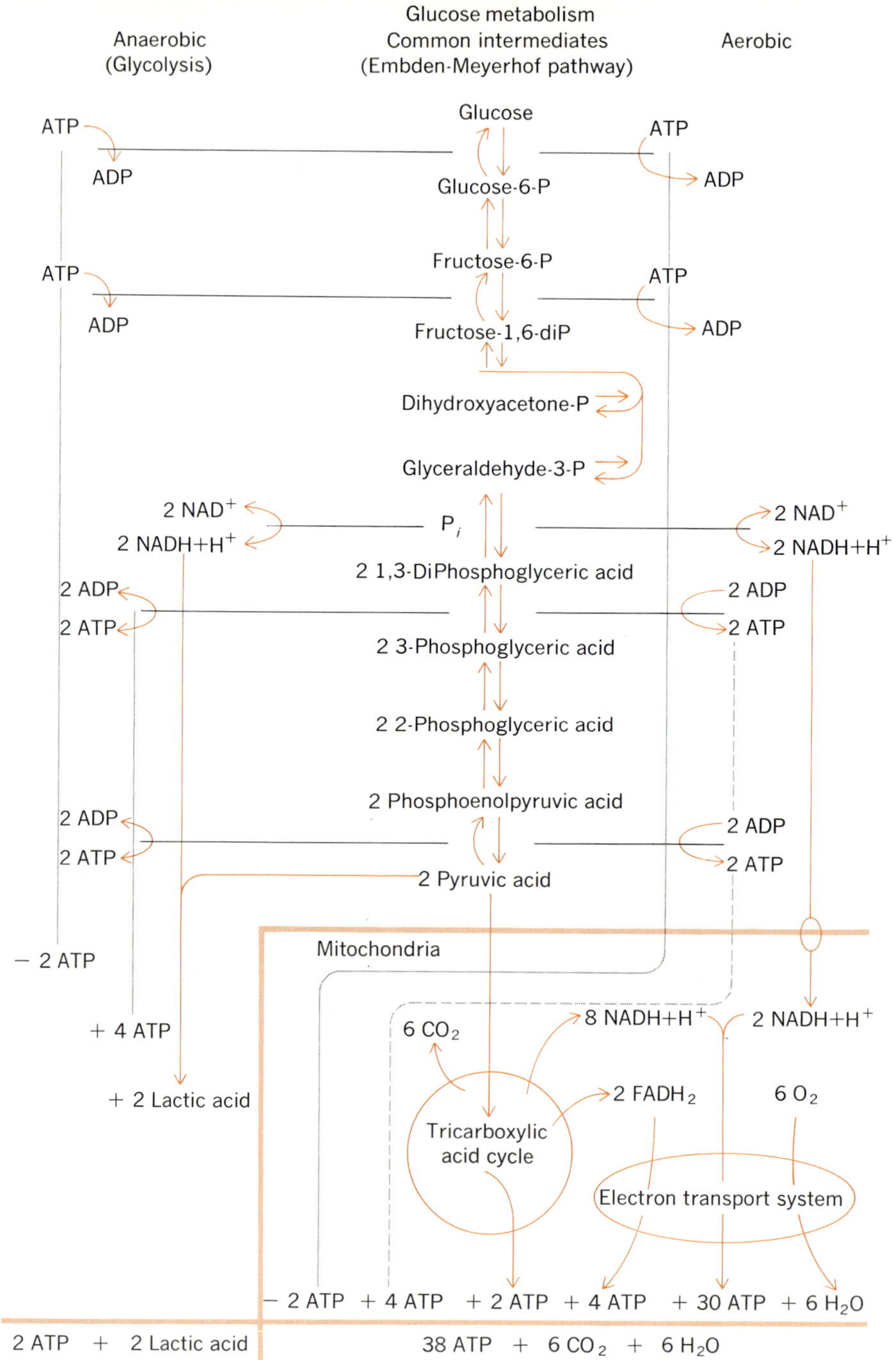

FIGURE 21.3 Summary of glucose oxidation.

which alcohol is the end product, or lactic acid fermentation by microorganisms, in which lactic acid is the end product of glucose metabolism. In muscle one of the end products is also lactic acid.

Starting with glucose, the various steps to be discussed are summarized in Fig. 21.3 and their chemical reactions in Fig. 21.4.

21.3 GLYCOLYSIS (EMBDEN-MEYERHOF PATHWAY)

GLUCOSE ⟶ GLUCOSE-6-PHOSPHATE. The first step in the metabolism of glucose is its conversion to a phosphate ester. Glucose is converted to glucose-6-phosphate by the action of *glucokinase.* Kinase is the general term for enzymes that phosphorylate substrates to form phosphate esters.

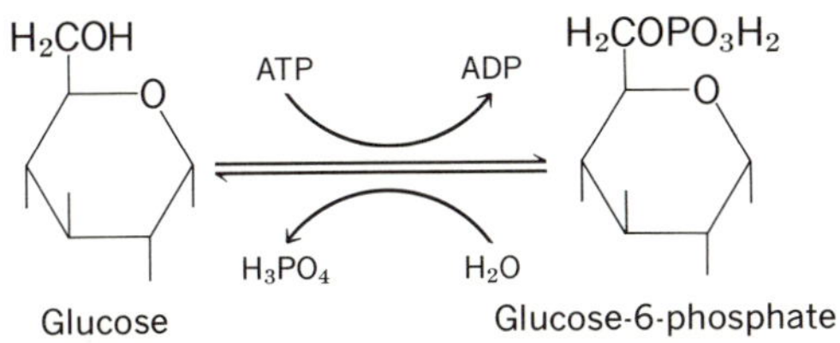

(21.1)

The phosphorylation of an alcohol requires energy; this is derived from ATP. Magnesium is also necessary for phosphorylation reactions.

The hydrolysis of ATP to ADP and H_3PO_4 liberates 8000 cal, whereas the hydrolysis of a mixed phosphate ester liberates only about 3000 cal. Thus, it has cost the cell 5000 cal to form the first metabolic product of glucose metabolism. What has happened to the 5000 cal?

$$\text{ATP} + H_2O \longrightarrow \text{ADP} + H_3PO_4 \qquad \Delta G = 8000 \text{ cal}$$

$$\text{Glucose-6-phosphate} + H_2O \longrightarrow \text{glucose} + H_3PO_4 \qquad \Delta G = 3000 \text{ cal}$$

Neglecting all other thermodynamical considerations, this amount of energy has been converted into heat. This is not the only reaction to be encountered in which a high-energy compound has to be used to form a low-energy substance with a concomitant evolution of heat. The heat thus formed is not available for chemical work; part is used to maintain body temperature, and the remainder is dissipated into the atmosphere.

The use of a high-energy compound to form a low-energy compound makes the reaction *irreversible* under biological conditions. Therefore, to go from glucose-6-phosphate to glucose requires a different enzyme system than glucokinase.

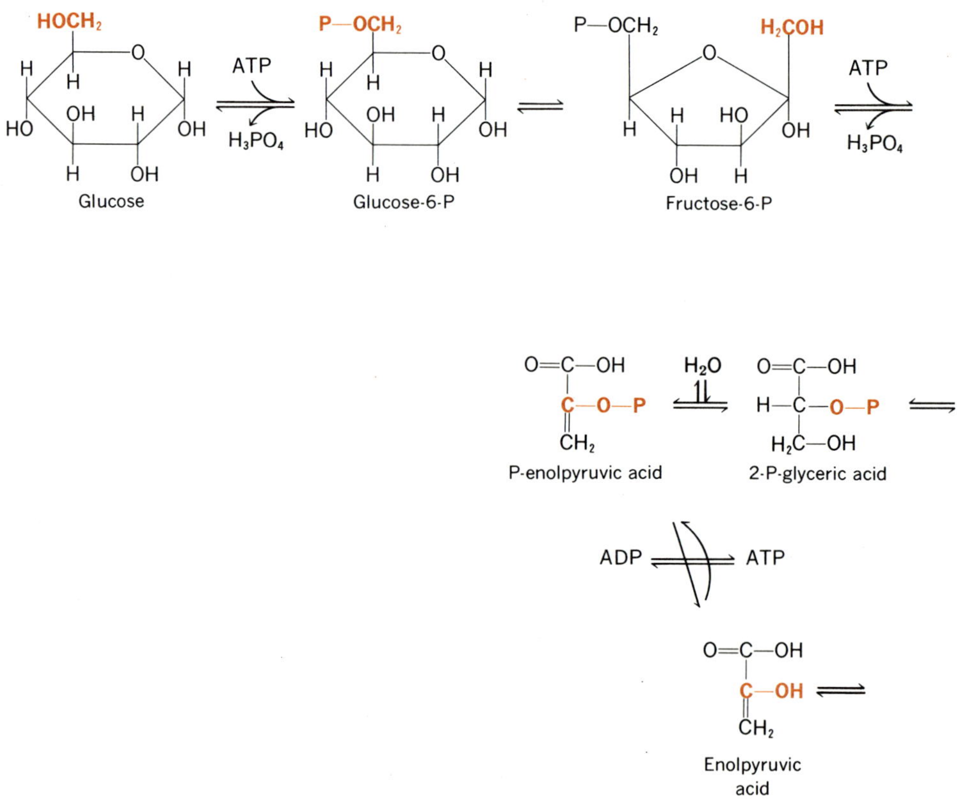

FIGURE 21.4 Anaerobic metabolism of glucose (Embden-Meyerhof pathway, or glycolysis).

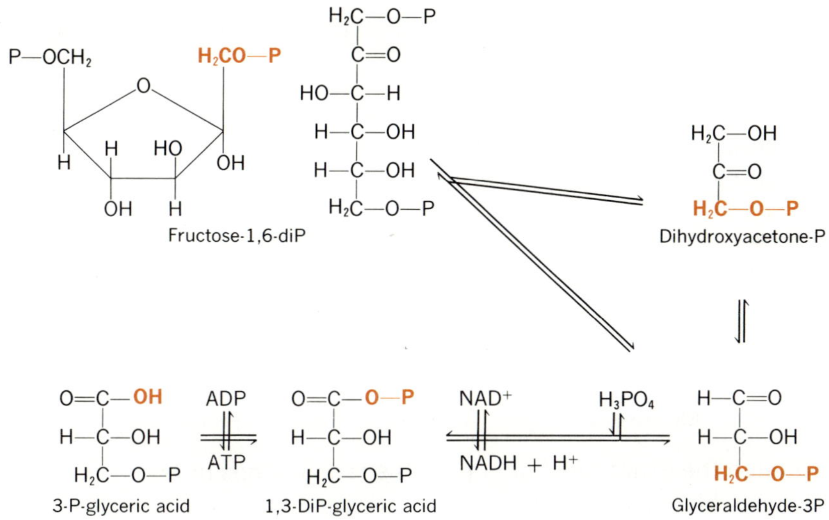
Fructose-1,6-diP
Dihydroxyacetone-P
3-P-glyceric acid
ADP
ATP
1,3-DiP-glyceric acid
NAD+
NADH + H+
H3PO4
Glyceraldehyde-3P

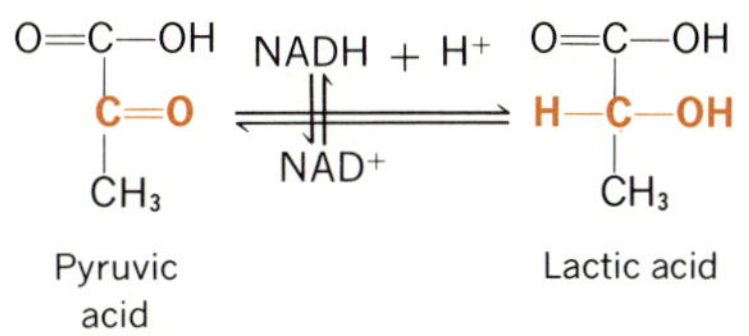
NADH + H+
NAD+
Pyruvic acid
Lactic acid

GLUCOSE-6-PHOSPHATE ⟶ GLUCOSE. Because of the irreversibility of the above reaction, to convert glucose-6-phosphate to glucose another enzyme must be employed. This is indicated in Fig. 21.3 by a curved arrow. Enzymes that catalyze the hydrolysis of a phosphate ester are collectively called *phosphatases. Glucose-6-phosphatase* is the specific phosphatase that hydrolyzes glucose-6-phosphate to glucose and H_3PO_4. This enzyme is found in the liver but not in muscle; therefore, once glucose in muscle has been phosphorylated, it cannot be reconverted into blood glucose.

GLUCOSE-6-PHOSPHATE ⇌ FRUCTOSE-6-PHOSPHATE. The next step is the conversion of glucose into its isomeric sugar fructose. This is accomplished by the enzyme *phosphohexose isomerase.*

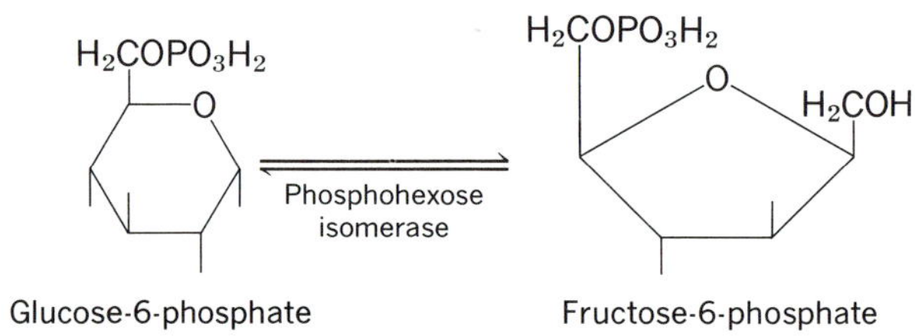

(21.2)

Although the exact mechanism of this reaction has not been elucidated, it is possible that an enediol is an intermediate, as in the alkaline conversion of glucose to fructose and mannose (page 349). This reaction is easily reversible. The equilibrium favors the formation of glucose over fructose in a molar ratio of about 70:30.

FRUCTOSE-6-PHOSPHATE ⟶ FRUCTOSE-1,6-DIPHOSPHATE. Fructose is phosphorylated to the diphosphate derivative by *phosphofructose kinase* before being further metabolized.

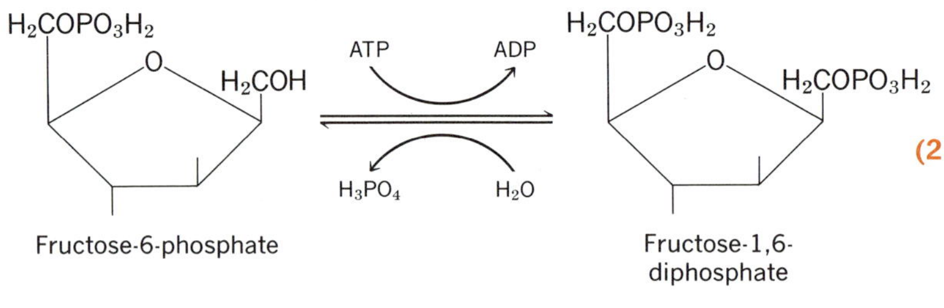

(21.3)

This again is the formation of a low-energy phosphate ester bond and requires the expenditure of energy. One mole of ATP is used in this reaction, and Mg^{++} is required for enzymatic activity. The reaction is not appreciably reversible under biological conditions.

FRUCTOSE-1,6-DIPHOSPHATE ⟶ FRUCTOSE-6-PHOSPHATE. The conversion of fructose diphosphate to the monophosphate derivative is accomplished by the enzyme *fructose-1,6-diphosphatase.* During conditions favorable to reversal of the glycolytic pathway (lactic acid to glucose), this enzyme increases in concentration in the liver.

FRUCTOSE-1,6-DIPHOSPHATE ⇌ DIHYDROXYACETONE PHOSPHATE + GLYCERALDEHYDE-3-PHOSPHATE. Fructose diphosphate is reversibly cleaved into two three-carbon sugars, glyceraldehyde-3-phosphate, an aldotriose, and dihydroxyacetone phosphate, a ketotriose. The equilibrium constant is such that the enzyme favors the formation of fructose diphosphate.

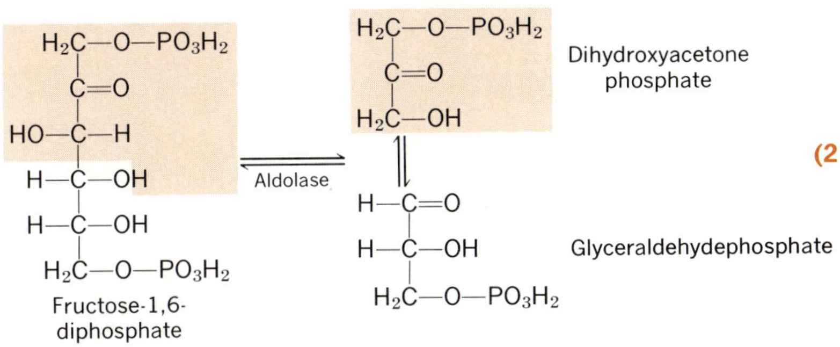

(21.4)

The enzyme responsible for this cleavage reaction is *aldolase,* so named because the type of reaction catalyzed is an aldol condensation of an aldehyde:

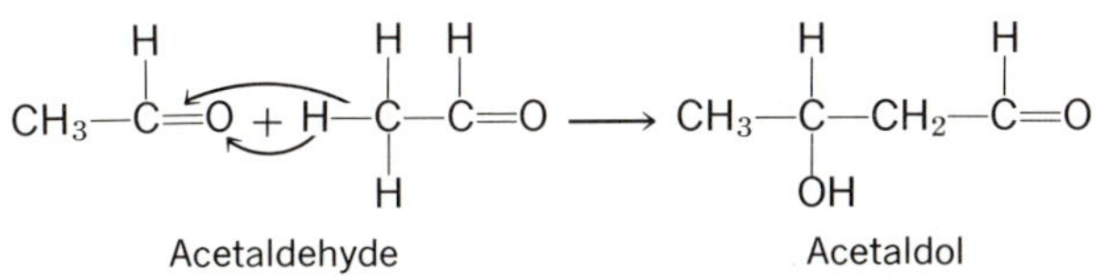

Aldolase has been studied extensively, and yeast aldolase has been crystallized. The yeast enzyme has been found to be activated by mineral elements such as Zn^{++}, Co^{++}, and Fe^{++}, whereas muscle aldolase does not require a metallic activator. Although the enzyme from different sources exhibits these differences, all aldolases have the two characteristics in common: (1) When the aldehyde is condensed with the ketone, the *hydroxy groups* on the carbons joined together are *always* trans to each other. This trans configuration is a prerequisite for cleavage of the carbon bond (note the positions of the OH groups on C-3 and C-4 of fructose). (2) Aldolase shows *complete*

specificity toward the ketone unit; only the ketone dihydroxyacetone phosphate is active. No other ketone will function in this enzymatic reaction. In comparison, several aldehydes enter into the condensation reaction. For example, D-erythrose-4-phosphate can be condensed with dihydroxyacetone phosphate to form sedoheptulose-1,7-diphosphate or, if the aldehyde is acetaldehyde, a methylketotetrose-1-phosphate:

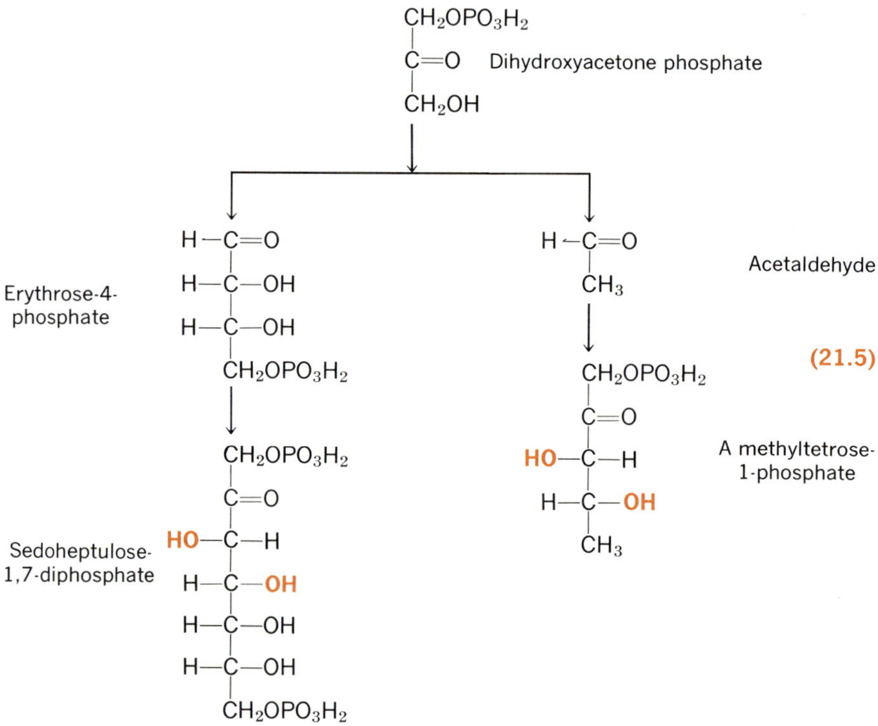

(21.5)

DIHYDROXYACETONE PHOSPHATE ⇌ GLYCERALDEHYDE-3-PHOSPHATE. Further metabolism of carbohydrate involves a transformation of glyceraldehyde-3-phosphate. Hence, to metabolize glucose completely there must be a means of converting dihydroxyacetone phosphate into glyceraldehyde phosphate. This is accomplished with the enzyme *triosephosphate isomerase.*

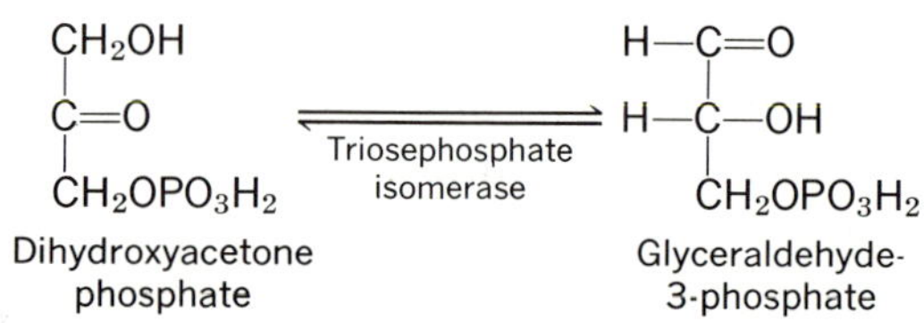

(21.6)

The reaction is analogous to the interconversion of glucose and fructose. Although the equilibrium of this reaction favors the formation of dihydroxyacetone phosphate, the removal of the aldehyde by further metabolism prevents this equilibrium from being attained. Therefore, all the dihydroxyacetone phosphate can be converted into the aldehyde phosphate and metabolized further.

Dihydroxyacetone phosphate is the precursor of glycerol for triglyceride and phospholipid formation. Thus, lipid metabolism is partly integrated with the anaerobic phase of carbohydrate metabolism; see Sec. 25.7.

GLYCERALDEHYDE-3-PHOSPHATE ⇌ 1,3-DIPHOSPHOGLYCERIC ACID. The reactions discussed thus far have required 2 moles of high-energy ATP without yielding any usable energy. The conversion of glyceraldehyde phosphate to diphosphoglyceric acid is the first step in the anaerobic phase which produces a high-energy compound.

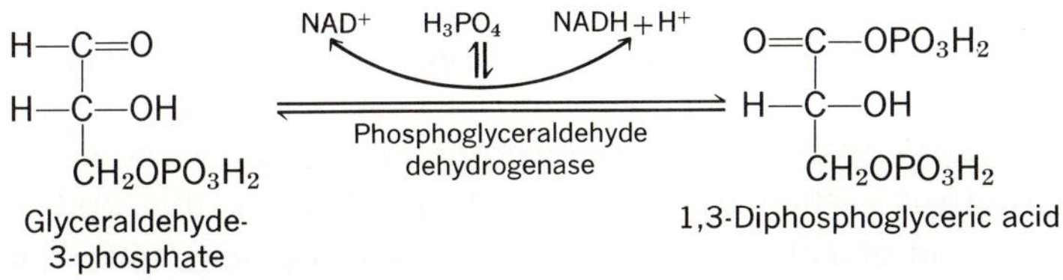

(21.7)

This is an oxidation reaction, but molecular oxygen is not involved, the oxidation being accomplished by the removal of hydrogen. The enzyme for this reaction is appropriately called *phosphoglyceraldehyde dehydrogenase.* The overall reaction is equivalent to removing the hydrogen of the aldehyde group and replacing it with a phosphate group.

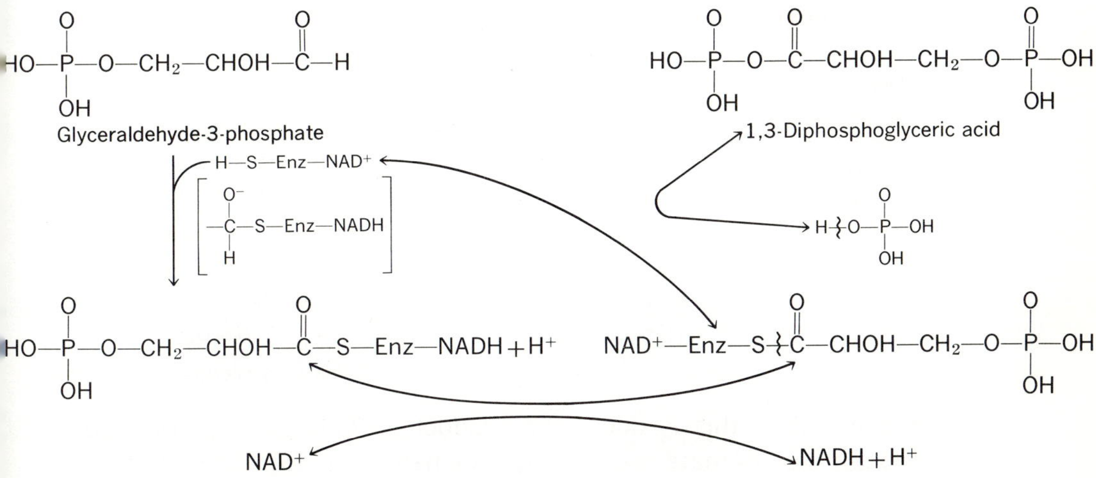

Net reaction:

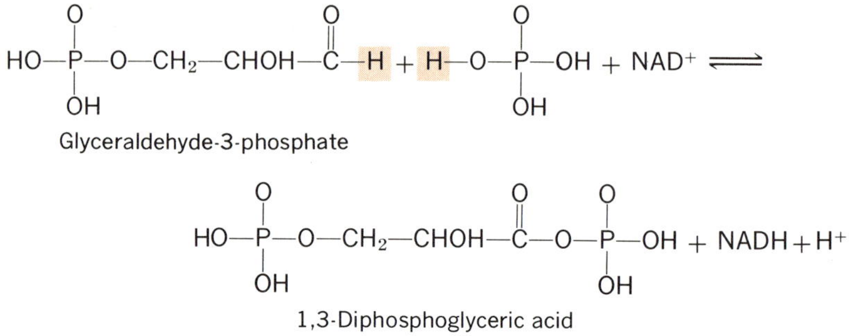

(21.8)

Not only is this reaction of interest from the mechanistic viewpoint, but energetically it is of utmost importance to the cell. This reaction produces a high-energy phosphate anhydride bond, which upon hydrolysis liberates about 14,000 cal/mole.

$$R-\overset{\overset{O}{\|}}{C}-OPO_3H_2 + H_2O \longrightarrow R-\overset{\overset{O}{\|}}{C}-OH + H_3PO_4 \qquad \Delta G = -14{,}000 \text{ cal} \qquad (21.9)$$

Thus the energy from this oxidation reaction is conserved in the anhydride structure rather than lost as heat. This is the first reaction encountered in glycolysis in which a high-energy bond is formed. This reaction is reversible.

Note that inorganic phosphate is utilized in the formation of glyceric acid diphosphate. It has been known for a long time that microorganisms utilize inorganic phosphate from the medium and convert it into organic phosphates. This is an example of such a step.

Phosphate is not a strict requirement for phosphoglyceraldehyde dehydrogenase activity since *arsenate* may substitute for it.

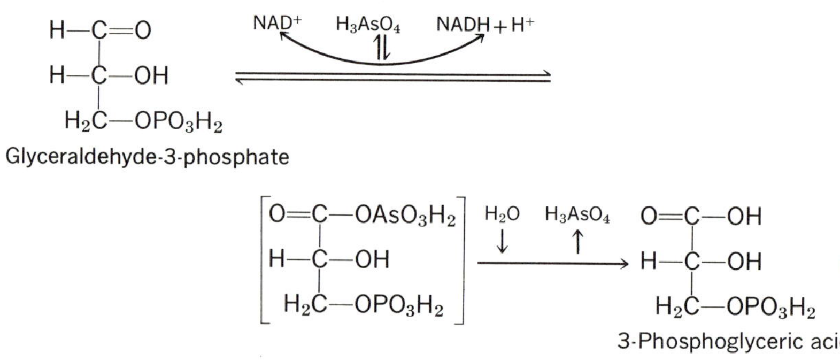

(21.10)

However, when this is done, the product is 3-phosphoglyceric acid. Although the arsenate is originally attached to the first carbon of glyceric acid, the arsenate anhydride bond is so labile that it is hydrolyzed as soon as it is formed. Therefore, it requires only catalytic amounts of arsenate to keep this reaction going since the arsenate is not being tied up in an organic compound but is being regenerated. In the presence of arsenate, the energy produced during the oxidation reaction is not made available for cellular use. It is lost as heat when the arsenate anhydride bond is hydrolyzed.

1,3-DIPHOSPHOGLYCERIC ACID ⇌ 3-PHOSPHOGLYCERIC ACID. This conversion involves the transfer of phosphate from the C-1 position of diphosphoglyceric acid to ADP to form ATP, thus conserving the energy derived from the oxidation step. The enzyme involved in this reaction is *phosphoglyceryl kinase.* The enzyme is named according to the reverse reaction; hence it is classified as a *kinase* which phosphorylates glyceric acid. A divalent metallic element such as Mg^{++} is a requirement for this reaction.

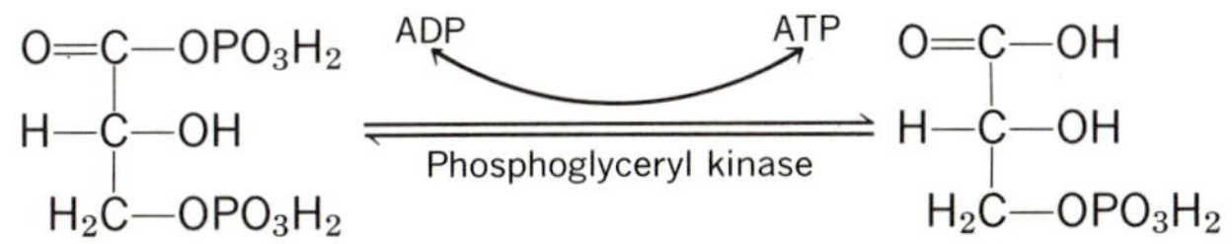

1,3-Diphosphoglyceric acid 3-Phosphoglyceric acid **(21.11)**

Although the equilibrium constant favors the formation of glyceric acid, the reaction is reversible under physiological conditions.

The metabolic scheme presented in Fig. 21.3 shows that 2 moles of ATP are produced in this reaction. This is so when the reactions are balanced for the utilization of 1 mole of glucose; 2 moles of glyceric acid phosphate are produced from 1 mole of glucose. As of now in the metabolic scheme the cell is even from the point of view of energy: 2 moles of ATP have been used, and 2 moles have been formed. Any subsequent energy-yielding reaction will be to the advantage of the cell.

3-PHOSPHOGLYCERIC ACID ⇌ 2-PHOSPHOGLYCERIC ACID. The next step in carbohydrate metabolism is the transfer of the phosphate group from the C-3 to the C-2 of glyceric acid. This is a reversible reaction and is brought about by the enzyme *phosphoglyceromutase,* which requires the cofactors 2,3-diphosphoglyceric acid and Mg^{++}.

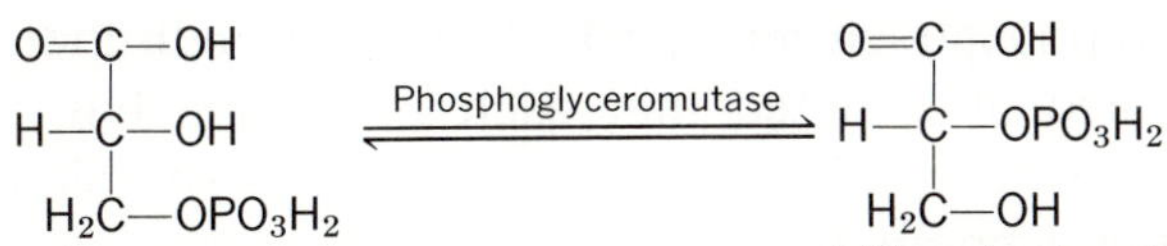

3-Phosphoglyceric acid 2-Phosphoglyceric acid **(21.12)**

The mechanism of this reaction has not been as well established as the one involving the conversion of glucose-1-phosphate to glucose-6-phosphate, which requires glucose-1,6-diphosphate as a cofactor (page 512). Hence, it can only be assumed that the two reactions are similar with respect to the diphosphate cofactor requirement.

The shift of the phosphate group from the third to the second carbon of glyceric acid is a step which leads to the formation of another high-energy compound, as shown below.

2-PHOSPHOGLYCERIC ACID ⇌ PHOSPHOENOLPYRUVIC ACID. Dehydration of 2-phosphoglyceric acid forms enolpyruvate with a phosphate on the enol hydroxy group. The enzyme *enolase* catalyzes this reaction, which is easily reversible and requires Mg^{++} as an activator.

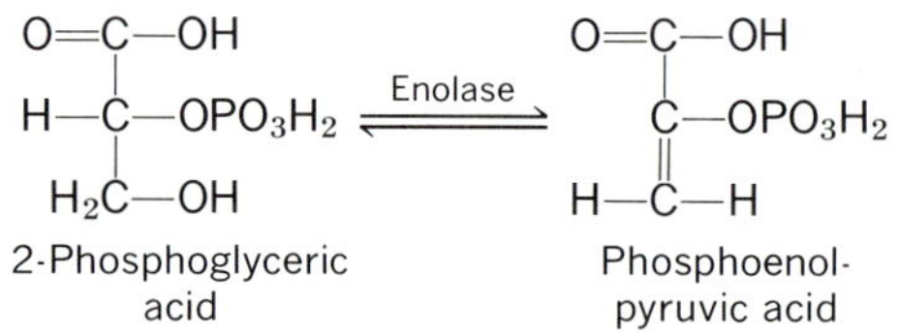

(21.13)

Sodium fluoride at 10^{-3} *M* is a potent inhibitor of carbohydrate metabolism because the fluoride ion combines with phosphate and magnesium to form a magnesium-fluorophosphate complex. The complex combines with the active site of enolase and inactivates the enzyme.

The equilibrium for the enolization of pyruvic acid favors the pyruvic acid form.

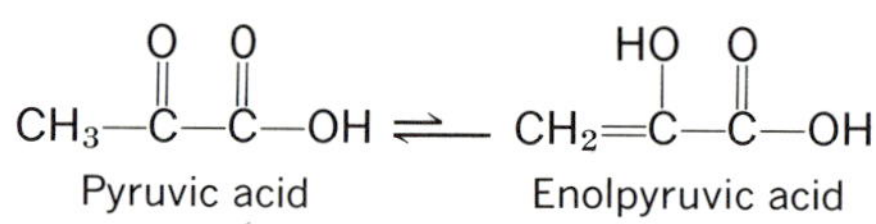

However, when enol pyruvate is phosphorylated, this equilibrium is prevented. Since it requires energy to prevent the conversion of the enol to the keto form, the hydrolysis of phosphoenolpyruvate liberates more energy than the hydrolysis of a phosphate ester. The hydrolysis of the enolphosphate to form the keto acid releases 12,000 cal; therefore, enolpyruvate is considered a high-energy compound. When the formula 1,3-diphosphoglyceric acid is compared with that of phosphoenolpyruvic acid, it is noted that each compound has a phosphate group connected to a carbon with a double bond. This situation makes for a high-energy compound.

An oxidation-reduction has taken place in this conversion. The second carbon of glyceric acid has been oxidized, whereas the third carbon has been reduced. Thus, by the action of enolase an oxidation-reduction has been initiated. In the following reaction the secondary alcohol group of glyceric acid is oxidized to a ketone group.

PHOSPHOENOLPYRUVIC ACID $\rightleftharpoons$ PYRUVIC ACID. As in the conversion of glyceraldehyde to glyceric acid, the energy of oxidation is conserved by the formation of ATP. The energy contained in phosphoenolpyruvate is transferred to ATP by the action of the enzyme *pyruvic kinase*. Mg^{++} is also required.

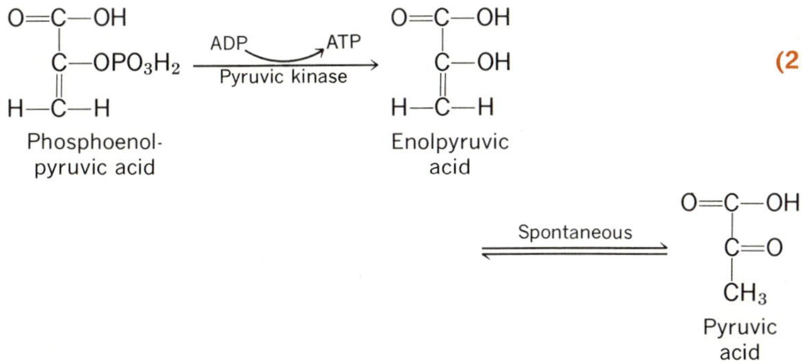

(21.14)

Like phosphoglyceryl kinase, pyruvic kinase is named according to the reverse reaction, the phosphorylation of enolpyruvic acid. This reaction is *not reversible* in liver under biological conditions..Thus, when pyruvate is converted into carbohydrate, a different mechanism must be employed to effect this step; this will be discussed shortly.

Since 2 moles of pyruvic acid are formed from 1 mole of glucose, the conversion of enolpyruvate to pyruvate forms 2 ATP for cellular functions. This makes a total of 4 ATP produced thus far and 2 ATP utilized, or *a net of* 2 ATP *produced per mole of glucose*. This is the last step in the glycolytic scheme where energy is produced. Subsequent anaerobic reactions are concerned with the fate of pyruvic acid and $NADH + H^+$ that have been formed.

21.4 SUMMARY: GLYCOLYSIS

A summary equation for the anaerobic reactions thus far presented is

$$C_6H_{12}O_6 + 2\ ATP + 2\ NAD^+ + 2\ H_3PO_4 + 4\ ADP \longrightarrow 2\ CH_3COCOOH + 4\ ATP + 2\ ADP + 2\ NADH + H^+$$

Net reaction:

$$C_6H_{12}O_6 + 2\ NAD^+ + 2\ ADP + 2\ H_3PO_4 \longrightarrow 2\ CH_3COCOOH + 2\ ATP + 2\ NADH + H^+$$

As the net reaction is written, NAD^+ would have to be present in large amounts (substrate levels) and 2 $NADH + H^+$ would accumulate as an end product for each mole of glucose utilized. This is not so; NAD^+ is a coenzyme occurring in catalytic amounts and consequently *must be reformed* if glucose is to be metabolized to any appreciable extent. A reaction already presented involves both pyruvic acid and $NADH + H^+$, namely, the reduction of pyruvic acid to lactic acid. This is one reaction that reforms NAD^+, which then cycles back to pick up more hydrogen from the conversion of glyceraldehyde to glyceric acid.

PYRUVIC ACID ⇌ LACTIC ACID. Pyruvic acid is easily and reversibly reduced by $NADH + H^+$ and the enzyme *lactic dehydrogenase.*

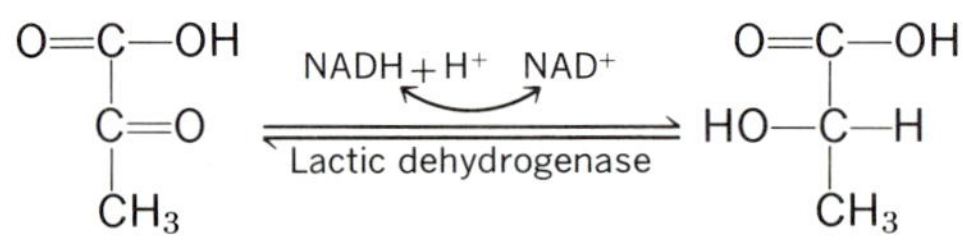

Pyruvic acid Lactic acid **(21.15)**

Lactic acid is the final product of anaerobic glycolysis in muscle and in bacteria which form lactic acid as the only product of fermentation (the homolactic organisms). The conversion of pyruvic to lactic acid produces an asymmetric carbon atom. It is the L(+)-lactic acid that is produced by muscle lactic dehydrogenase; the D(−) form is produced by some bacteria.

Lactic dehydrogenase has been studied in detail, and the active enzyme has been found to be a tetramer (composed of four identical subunits of monomers). This is an illustration of an enzyme where the tetramer is active but the dimers or monomers are completely inactive. The importance of the quaternary structure of this enzyme is easily illustrated by reagents that disassociate the tetramer. Each monomer has an active site, but it is enzymatically inactive unless associated as the tetramer. Each monomer has a molecular weight of 35,000.

SUMMARY: GLUCOSE ⟶ 2 PYRUVIC ACID ⟶ 2 LACTIC ACID. The conversion of 1 mole of glucose into 2 moles of lactic acid uses 2 moles of inorganic phosphate. Also formed are 2 moles of ATP.

$$C_6H_{12}O_6 + 2\ H_3PO_4 + 2\ ADP + 2\ NAD^+ \longrightarrow \begin{cases} 2\ ATP \\ 2\ CH_3COCOOH \\ 2\ NADH + H^+ \end{cases}$$

$$2\ CH_3COCOOH + 2\ NADH + H^+ \longrightarrow 2\ CH_3CHOHCOOH + 2\ NAD^+$$

(The 2 NAD^+ formed returns to the first reaction.)

Although glucose has been split into two C_3 fragments, there has been no degradation to CO_2 and H_2O, which comes later, during the aerobic phase.

Microorganisms which form products such as lactic acid or alcohol by anaerobic fermentation are noted for the large amount of glucose utilized and product formed. This can be explained if one considers the overall equation above. In order to obtain sufficient energy for cellular growth and reproduction, the organism must metabolize large amounts of glucose because only a net of 2 ATP per mole of glucose is biologically available.

The energy theoretically available from the complete oxidation of 1 mole of glucose is 686,000 cal. If the reaction is stopped at lactic acid, $\Delta G = -52{,}000$ cal.

Chemically:

$$C_6H_{12}O_6 \longrightarrow 6\ CO_2 + 6\ H_2O \qquad \Delta G = -686{,}000 \text{ cal}$$

$$C_6H_{12}O_6 \longrightarrow 2\ CH_3CHOHCOOH \qquad \Delta G = -52{,}000 \text{ cal}$$

Metabolically:

$$C_6H_{12}O_6 \longrightarrow 2\ CH_3CHOHCOOH + 2\ ATP \qquad \Delta G = -16{,}000 \text{ cal}$$

Thus the living organism is able to trap 16 of the 52 kcal theoretically obtainable from the formation of lactic acid. This makes the organism about 31% efficient with respect to glycolytic reactions. As noted from the ΔG values, only a small percentage of the total theoretical amount of energy from glucose oxidation has been made available in the anaerobic phase.

The fate of pyruvic acid depends upon the organism and the environmental conditions. In muscle in the resting state, some of the pyruvic acid formed is converted into lactic acid, and some is further oxidized for energy. During strenuous muscular activity, however, there is a period of oxygen shortage, and then lactic acid is formed in greater amounts. It accumulates to some extent in the muscle, but most of it passes into the blood, where it is transported to the liver and eventually converted into glycogen. When sufficient oxygen becomes available, lactic acid does not accumulate in the muscle because the pyruvic acid is being oxidized.

Pyruvic acid is a connecting link between the anaerobic phase of carbohydrate metabolism and amino acid metabolism. Several amino acids can be synthesized from the carbon chain of pyruvic acid. Conversely, these amino acids, when catabolized, enter into the anaerobic phase at pyruvic acid.

The anaerobic phase of carbohydrate metabolism furnishes some intermediates which are converted into other classes of compounds. The conversion of pyruvic acid to amino acids has just been mentioned. The sugar phosphates are the precursors of other carbohydrates, e.g., galactose, lactose, and mannose. Also, the conversion of dihydroxyacetone phosphate to glycerol serves as a link between the anaerobic phase and fat metabolism inasmuch as glycerol is a component of many common lipids.

The reversal of the glycolytic scheme is extremely important; the reactions for going from phosphoenolpyruvate to glucose have already been indicated. Because the conversion of pyruvic acid to phosphoenolpyruvic acid involves an intermediate of the aerobic phase of glucose oxidation, the complete reversal of the anaerobic phase will be discussed following the aerobic phase.

PYRUVIC ACID ⟶ ETHANOL. Pyruvic acid may be converted into ethanol as another end product of anaerobic metabolism. This conversion involves a decarboxylation of pyruvic acid to acetaldehyde and its subsequent reduction to ethanol.

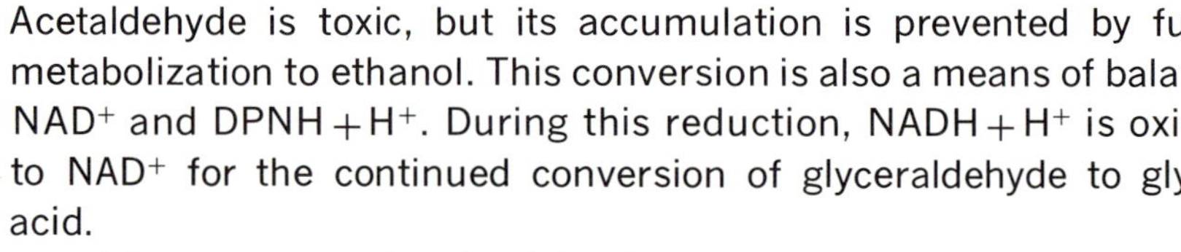

(21.16)

Acetaldehyde is toxic, but its accumulation is prevented by further metabolization to ethanol. This conversion is also a means of balancing NAD^+ and $DPNH + H^+$. During this reduction, $NADH + H^+$ is oxidized to NAD^+ for the continued conversion of glyceraldehyde to glyceric acid.

The coenzyme involved in the decarboxylation of α-keto acids, thiaminpyrophosphate (TPP) (see page 467), functions in conjunction with the enzyme *pyruvic decarboxylase,* which requires Mg^{++} for activation. The intermediate products of this reaction were given in the discussion of coenzyme thiaminpyrophosphate.

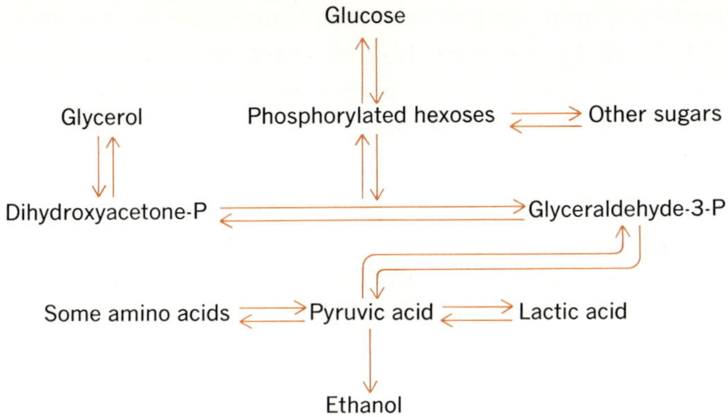

FIGURE 21.5 Some interrelationships in the anaerobic phase of carbohydrate metabolism.

The enzyme *alcohol dehydrogenase* reduces the acetaldehyde to ethanol; $NADH + H^+$ is the cofactor for this reaction.

The net reaction for alcoholic fermentation is

$$C_6H_{12}O_6 + 2\ ADP + 2\ H_3PO_4 \longrightarrow 2\ CH_3CH_2OH + 2\ CO_2 + 2\ ATP$$

A summary of the interrelationships in the anaerobic phase of glucose metabolism is given in Fig. 21.5.

21.5 GLYCOGEN SYNTHESIS

Animals store carbohydrates as glycogen, whereas plants store starch (α- and β-amylose). These polysaccharides are composed of only α-D-glucose units, the difference between them lying in the extent of branching (the number of glucose-1—6-glucose links). In α-amylose

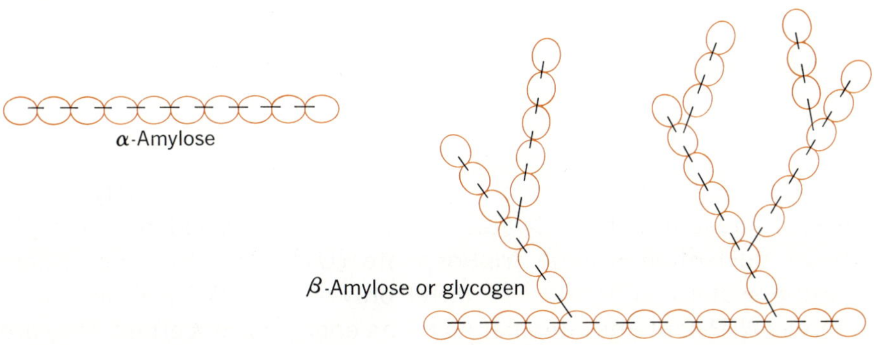

FIGURE 21.6 The schematic structure of α-amylose and β-amylose, or glycogen.

there is no branching, whereas in glycogen there is the most branching (Fig. 21.6). Regardless of the structure of these polysaccharides, they are all made from glucose and can be converted back into glucose. This is accomplished by two different pathways. We shall first discuss the formation of glycogen in the liver.

In the reactions thus far studied, blood glucose enters into metabolic pathways by being phosphorylated to glucose-6-phosphate; hence, we now focus our attention on the steps from glucose-6-phosphate to glycogen.

GLUCOSE-6-PHOSPHATE $\rightleftharpoons$ GLUCOSE-1-PHOSPHATE. Glucose-6-phosphate is converted into glucose-1-phosphate by the enzyme *phosphoglucomutase*. Cofactors for this reaction are glucose-1,6-diphosphate and Mg^{++}. Glucose diphosphate is an intermediate in the conversion of glucose-6-phosphate to glucose-1-phosphate (recall the similarities between this reaction and the conversion of glyceric acid–3-phosphate to glyceric acid–2-phosphate, in which the cofactors were 2,3-diphosphoglyceric acid and Mg^{++}). The following reactions show the mechanism:

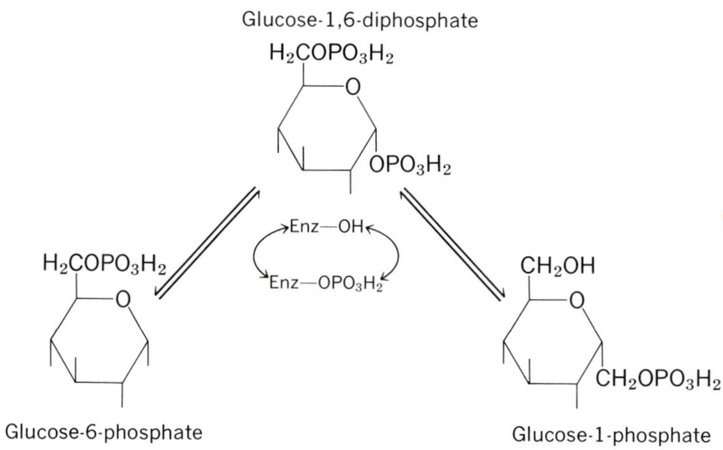

(21.17)

Glucose-6-phosphate $\rightleftharpoons$ glucose-1-phosphate

Before glucose-1-phosphate can be polymerized to polysaccharides, it must first be activated. This is accomplished by the high-energy nucleotide uridine triphosphate (UTP) and the appropriate enzyme system. UTP and ATP differ only in that UTP contains uricil whereas ATP contains adenine. As far as energy is concerned, they are equivalent: ATP + UDP $\rightleftharpoons$ ADP + UTP.

GLUCOSE-1-PHOSPHATE ⟶ GLYCOGEN (STARCH). The activation of glucose and its conversion into the 1—4-glucose polymer are accomplished by the same enzyme, namely, *glycogen synthetase*. The formation of the glucose-1—6-glucose link which occurs in β-amylose and glycogen is catalyzed by another enzyme, called the *branching enzyme*.

Active glucose is formed when glucose-1-phosphate reacts with UTP to form uridine diphosphate glucose (UDPG) and pyrophosphate:

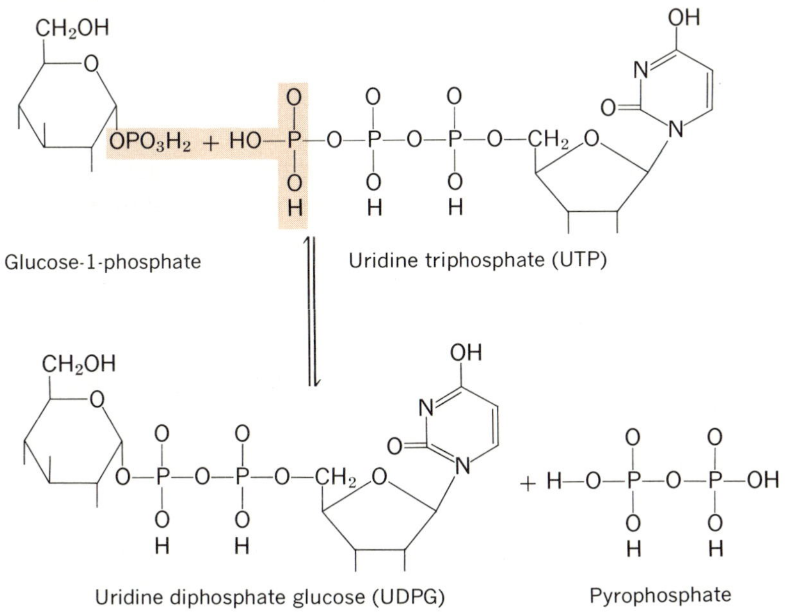

(21.18)

In the presence of a *primer* (a small glucose polymer) the polymerization of glucose units proceeds by adding the glucose moiety of UDPG to the nonreducing end of the polymer primer. Thus a linear polymer of glucose-1—4-glucose is obtained, as in Fig. 21.7.

The branching enzyme transfers small numbers of glucose units from the nonreducing end of a linear polymer to the C-6 of a glucose unit within the chain, thus forming a branched structure. After branching has occurred, the respective chains are lengthened.

By reactions thus far presented, glucose or an intermediary metabolic product in the Embden-Meyerhof pathway or the TCA cycle can be converted to glycogen. Such substances are called glycogenic compounds.

$$\text{Glycogenic compounds} \longrightarrow \text{glycogen}$$

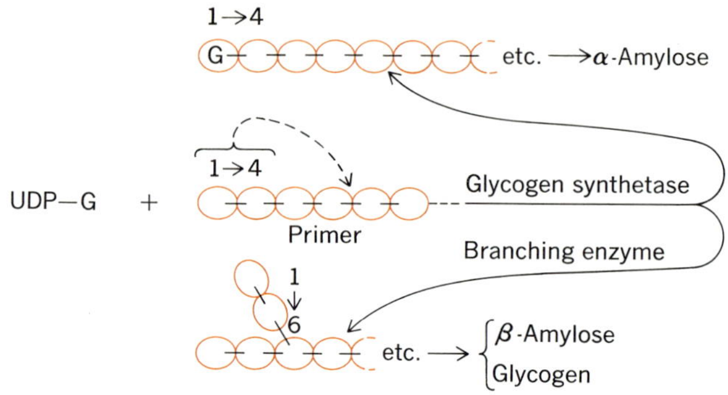

FIGURE 21.7 The polymerization of glucose as facilitated by a primer.

For example, an amino acid that is metabolized to pyruvic acid is called a *glycogenic amino acid.* Likewise, one that forms α-ketoglutaric acid is also known as a glycogenic amino acid. The citric acid of citrus fruits is another glycogenic compound.

The glucose moiety of UDP-glucose can be transferred to a number of sugars in the presence of the appropriate enzyme system to form disaccharides. For example, UDP-glucose and fructose-6-phosphate react to form sucrose.

$$\text{UDP-glucose} + \text{fructose-6-phosphate} \longrightarrow \text{sucrose}$$

GLYCOGEN ⟶ GLUCOSE-1-PHOSPHATE. The term glycogenolysis means the breakdown of glycogen. The first reaction is the formation of glucose-1-phosphate, which is accomplished by the enzyme *phosphorylase.*

$$\text{Glycogen} + H_3PO_4 \longrightarrow \text{glucose-1-phosphate}$$

The reaction is called *phosphorolysis.* Phosphorylase acts on the nonreducing end of a polymer and is specific for the α-(1—4) glucose linkage.

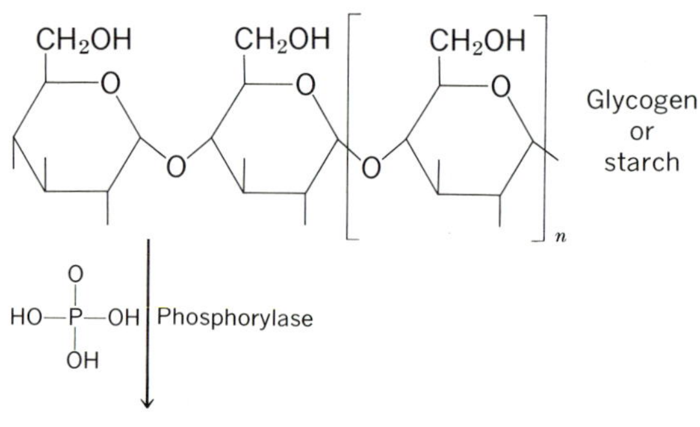

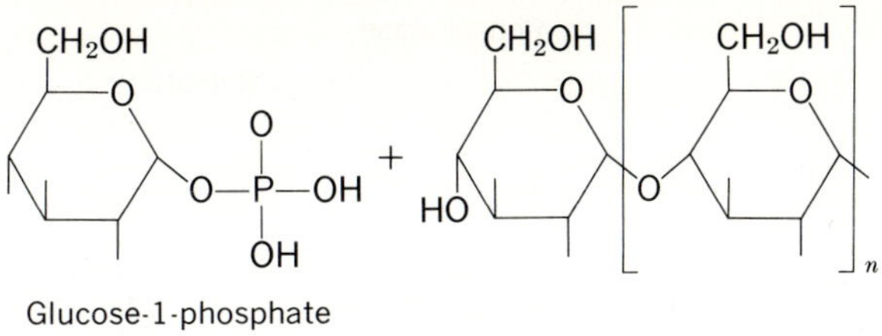

(21.19)

Thus its action stops at an α-(1—6) glucose link. The 1—6 linkage is hydrolyzed by another enzyme in the liver, namely, 1,6-*glucosidase*. It is of interest to note that inorganic phosphate is used in the phosphorolysis of polysaccharides; ATP is not required.

21.6 CONTROL OF GLYCOGEN METABOLISM

Simple as the reactions of glycogen synthetase and phosphorylase may seem, the control of the activity of these enzymes is quite complex. These enzymes are regulated by both hormone and nonhormone products of metabolism. The regulatory influences of these two types of compounds are superimposed upon each other and afford a delicate balancing of glycogenesis versus glycogenolysis. This in turn regulates carbohydrate metabolism in general. For example, if glycogenesis is stimulated and glycogenolysis is inhibited, glycogen synthesis is favored and the forward reactions of the Embden-Meyerhof pathway are suppressed. If the converse is true, glycogenolysis is favored, thus feeding glucose-1-phosphate into the glycolytic pathway.

A schematic representation of factors that influence the activity of glycogen synthetase and phosphorylase is given in Fig. 21.8.

3′,5′-CYCLIC ADENYLIC ACID. A compound common to the control of both glycogenesis and glycogenolysis is the cyclic phosphate 3′,5′-cyclic adenylic acid, which is formed from ATP by the *cyclase enzyme*. This enzyme is inactive until activated by the hormones *adrenaline* (*epinephrine*) or *glucagon*.

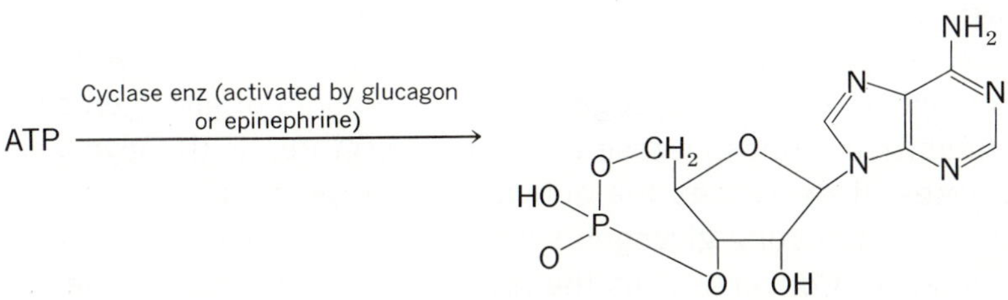

(21.20)

3′,5′-Cyclic adenylic acid

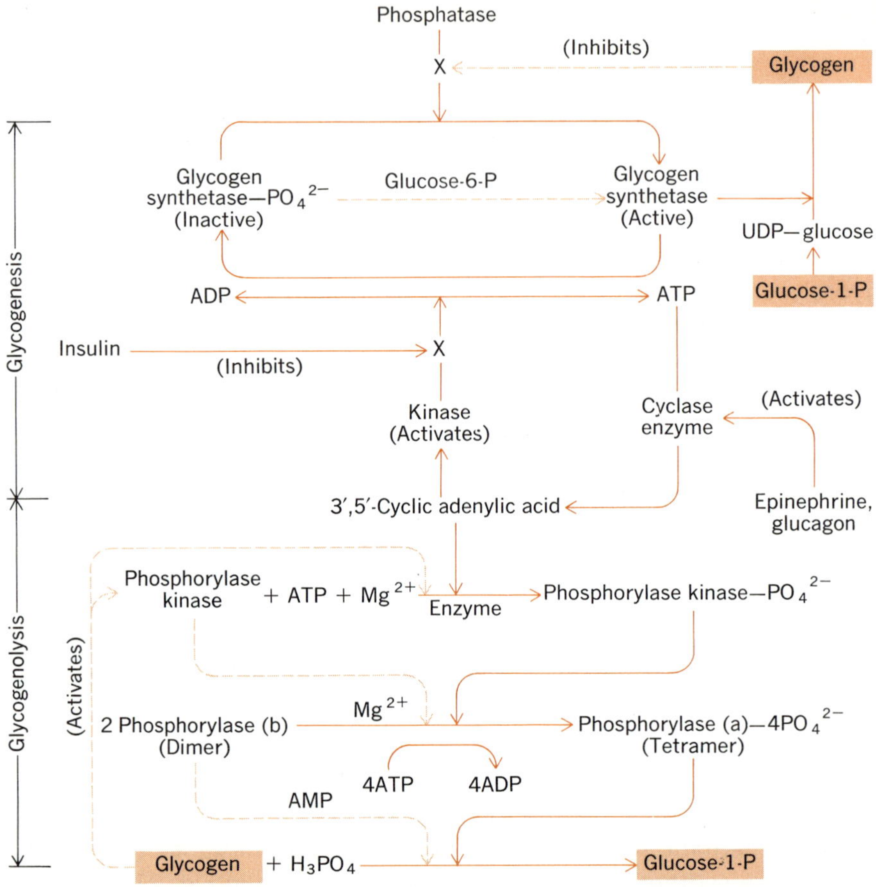

FIGURE 21.8 Glycogen metabolism. Solid arrows = reactions controlled by hormones; broken arrows = reactions controlled by nonhormones. *[Adapted from J. Larner, Hormonal and Nonhormonal Control of Glycogen Metabolism, Trans. N.Y. Acad. Sci.,* **29:***192 (1967).]*

Glycogenesis

Glycogen synthetase exists in two forms, active and inactive. The inactive form is a phosphate derivative of the active form. A specific phosphatase enzyme hydrolyzes off the phosphate of the inactive form to yield the active enzyme. Glycogen, the end product of glycogen synthetase, inhibits the phosphatase enzyme. Therefore, as the glycogen level increases, it decreases the activity of the phosphatase enzyme responsible for activating glycogen synthetase. Thus, glycogen formation is decreased. Conversely, as the glycogen level decreases, the sequence of reactions is as follows. The inhibitory effect of glycogen on phosphatase decreases, and the activation of glycogen synthetase increases, thus increasing the rate of glycogen synthesis. This is a good example of feedback control of enzymatic reactions.

$$\underset{\text{(Active)}}{\text{Glycogen synthetase}} \xrightleftharpoons[\text{Inhibited by glycogen}]{} \underset{\text{(Inactive)}}{\text{glycogen synthetase phosphate}}$$

An auxiliary control of glycogen synthesis is furnished by glucose-6-phosphate, the first product of glucose metabolism. Glucose-6-phosphate activates an enzyme, different from the above phosphatase, that in turn activates glycogen synthetase. Thus, as glucose enters into metabolic reactions, its first derivative stimulates its utilization.

$$\underset{\text{(Inactive)}}{\text{Glycogen synthetase}} \xrightarrow{\text{Glucose-6-phosphate}} \underset{\text{(Active)}}{\text{glycogen synthetase}}$$

The hormonal control of glycogen synthesis acts through 3′,5′-cyclic adenylic acid [see Eq. (21.20)]. This compound activates a kinase that inactivates glycogen synthetase by phosphorylating it to the inactive phosphate derivative. This, of course, decreases glycogen synthesis by removing the active enzyme. The enzyme, *cyclase enzyme,* that converts ATP to the cyclic adenylate is activated by the hormones epinephrine (adrenaline) and glucagon. Thus, by increasing the level of cyclic adenylate, glycogen synthesis is slowed down. Cyclic adenylate plays a dual role; it also stimulates glycogenolysis.

Insulin is a hormone which opposes the effect of epinephrine and glucagon by inhibiting the kinase enzyme that inactivates glycogen synthetase.

$$\underset{\text{(Active)}}{\text{Glycogen synthetase}} \xrightarrow[\text{Insulin} \times]{\text{epinephrine, glucagon}} \underset{\text{(Inactive)}}{\text{glycogen synthetase}}$$

Thus, the effect of insulin is to favor an increase in glycogen synthesis by nullifying the stimulatory influence of epinephrine or glucagon on glycogen synthetase.

In summary, glycogen synthesis is regulated by a balance of the hormones epinephrine, glucagon, and insulin as well as by a balance between the nonhormone compounds glycogen and glucose-6-phosphate. For example, glycogen synthesis is stimulated by a low level of glycogen, epinephrine, and glucagon and by a high level of glucose-6-phosphate and insulin. The converse would have the opposite effect on glycogen synthesis.

Glycogenolysis

The regulation of glycogenolysis is more complex than the regulation of glycogenesis. *Phosphorylase a*–PO_4^{--}, *a tetramer,* is the active enzyme that converts glycogen to glucose-1-phosphate. The inactive form of this enzyme is the *nonphosphorylated dimer phosphorylase b.* The activation of phosphorylase *b* involves its phosphorylation by *phosphorylase kinase*–PO_4^{--}. ATP and Mg^{++} are required for this reaction.

$$\underset{\text{Inactive dimer}}{\text{Phosphorylase } b} \xrightarrow[\text{Kinase}]{\text{ATP, Mg}^{++}} \underset{\text{Active tetramer}}{\text{phosphorylase } a\text{–4 PO}_4^{--}}$$

Phosphorylase kinase–PO_4^{--} also exists in an inactive nonphosphorylated form, *phosphorylase kinase.* The enzyme that phosphorylates the inactive phosphorylase kinase to the active phosphorylase kinase–PO_4^{--} is under the direct control of 3′,5′-cyclic adenylic acid.

$$\underset{\text{Inactive}}{\text{Phosphorylase kinase}} \xrightarrow{\text{Cyclic adenylic acid}} \underset{\text{Active}}{\text{phosphorylase kinase–PO}_4^{--}}$$

As mentioned above, the production of cyclic adenylate is controlled by the hormones epinephrine and glucagon, which exert a double action; they decrease glycogen synthesis and stimulate glycogenolysis simultaneously by regulating the production of the cyclic adenylate. It is of interest to note the requirement of ATP and Mg^{++} in these controlling mechanisms.

Glycogenolysis is also partially nonhormonal-controlled.

$$\text{Glycogen} \xrightarrow{\text{Phosphorylase } b \;\downarrow\; \leftarrow \text{AMP}} \text{glucose-1-phosphate}$$

Phosphorylase *b* itself is somewhat active in the presence of AMP. Furthermore, phosphorylase kinase has a certain amount of influence in directly converting phosphorylase *b* to phosphorylase *a*–4 PO_4^{--}. This is under the control of the level of glycogen. Thus a high level of glycogen has a stimulatory influence on the conversion of glycogen to glucose-1-phosphate. It stimulates phosphorylase kinase to act directly on the conversion of phosphorylase *b* to phosphorylase a–4 PO_4^{--} as well as increasing the activity of the enzymatic conversion of inactive phosphorylase kinase to the active form. The regulation of glycogenolysis seems to be shared by a number of systems. Because of these regulatory systems glycogenolysis is very well controlled.

The controlling reactions of glycogen metabolism are integrated, and the nonhormonal control seems to augment the hormonal control mechanisms, for in the absence of the hormones, the nonhormonal control is not a complete substitute. The hormones epinephrine, glucagon, and insulin are essential for proper balance between the two opposing reactions, glycogenesis and glycogenolysis.

QUESTIONS

1. Define metabolism, anabolism, catabolism.
2. In general, what is the fate of dietary glucose?
3. What do the terms "anaerobic" and "aerobic" signify?
4. How many moles of lactic acid can theoretically be formed from 1 mole of glucose?
5. (*a*) Which carbons of glucose are contained in the carboxyl groups of the two lactic acid molecules produced from a molecule of glucose?
(*b*) Which carbons of glucose have been oxidized during the conversion of glucose to lactic acid? Has any carbon of glucose been reduced?
6. Does the Embden-Meyerhof pathway function under aerobic as well as anaerobic conditions?
7. (*a*) What coenzyme is reduced during glycolysis?
(*b*) By what means is the reduced coenzyme reoxidized?
8. What glycolytic intermediate is the precursor of glycerol?
9. (*a*) Where in the scheme of glycolysis is ATP used, and where is it formed?
(*b*) Write the equations for the reactions in which ATP is formed.
10. Why are 1,3-diphosphoglyceric and phosphoenolpyruvic acids considered high-energy compounds?
11. (*a*) Is ATP utilized in the formation of glucose-1-phosphate from glycogen?
(*b*) Is a high-energy phosphate utilized in going from glucose-1-phosphate to glycogen?
12. What is the dual role of 3′,5′-cyclic adenylic acid with respect to control of glycogen synthesis?
13. What is the influence of insulin and epinephrine (or glucagon) upon the biological activity of 3′,5′-cyclic adenylic acid?
14. What metabolic products function with hormones in the control of glycogen metabolism?

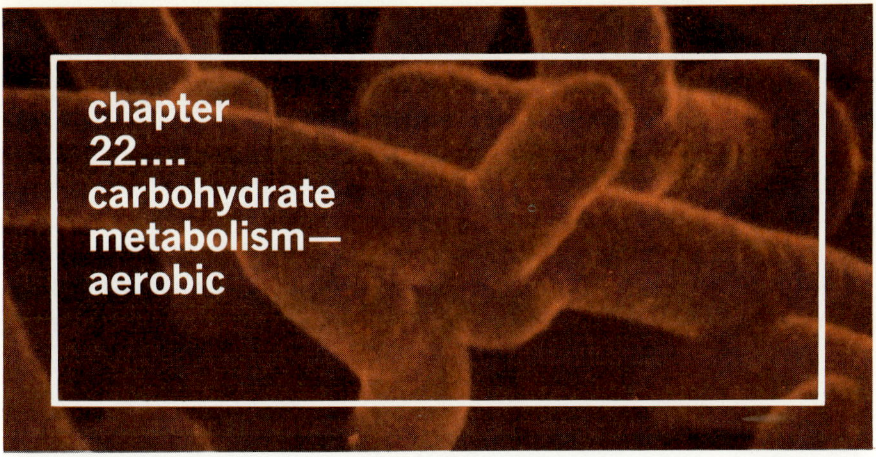

chapter 22.... carbohydrate metabolism—aerobic

To obtain the maximum amount of energy from glucose, pyruvic acid must be completely oxidized to carbon dioxide and water. This is accomplished aerobically partly by a series of reactions known as the Krebs cycle, or the tricarboxylic acid cycle (TCA cycle). In this cycle the carbons of pyruvic acid are oxidized to carbon dioxide, and the hydrogens are trapped by coenzymes. It is noteworthy that the oxygen for carbon dioxide formation comes immediately from water, not molecular oxygen. The TCA cycle, like the anaerobic phase, is not mainly a source of ATP. Only 1 mole of ATP is produced per mole of pyruvate metabolized. Most of the energy from pyruvate is derived from the passage of the hydrogen from the reduced coenzymes to oxygen by way of the electron-transport system, or *oxidative phosphorylation,* as it is frequently called.

Although the Krebs cycle is of great metabolic importance in oxidizing pyruvate carbons to carbon dioxide and in the formation of reduced coenzymes, it is of equal importance as a source of metabolic intermediates that afford a connection between amino acid, fat, and carbohydrate metabolism. Many cellular components are made from these intermediates, and the carbons of many others are metabolized to carbon dioxide via this cycle. This will become more apparent during the discussion of lipid and protein metabolism.

Of particular note is that TCA cycle enzymes and those of the electron-transport system are located in the mitochondria. The glycolytic enzymes, on the other hand, are located in the cytoplasm of the cell. Hence, there must be a transfer of pyruvic acid from the cytoplasm into the mitochondria before the acid can be further oxidized.

The chemistry of the TCA cycle intermediates for the most part has already been presented. It now remains to collect these reactions in a logical order and to discuss some of the enzymes involved. The reactions thus far presented are summarized in Fig. 22.1. Review these reactions.

22.1 THE KREBS, OR TRICARBOXYLIC ACID, CYCLE

Although acetyl-SCoA is the unit that enters into the tricarboxylic acid cycle, the conversion of pyruvic acid into this complex can logically be included under this heading. The various steps of the TCA cycle are

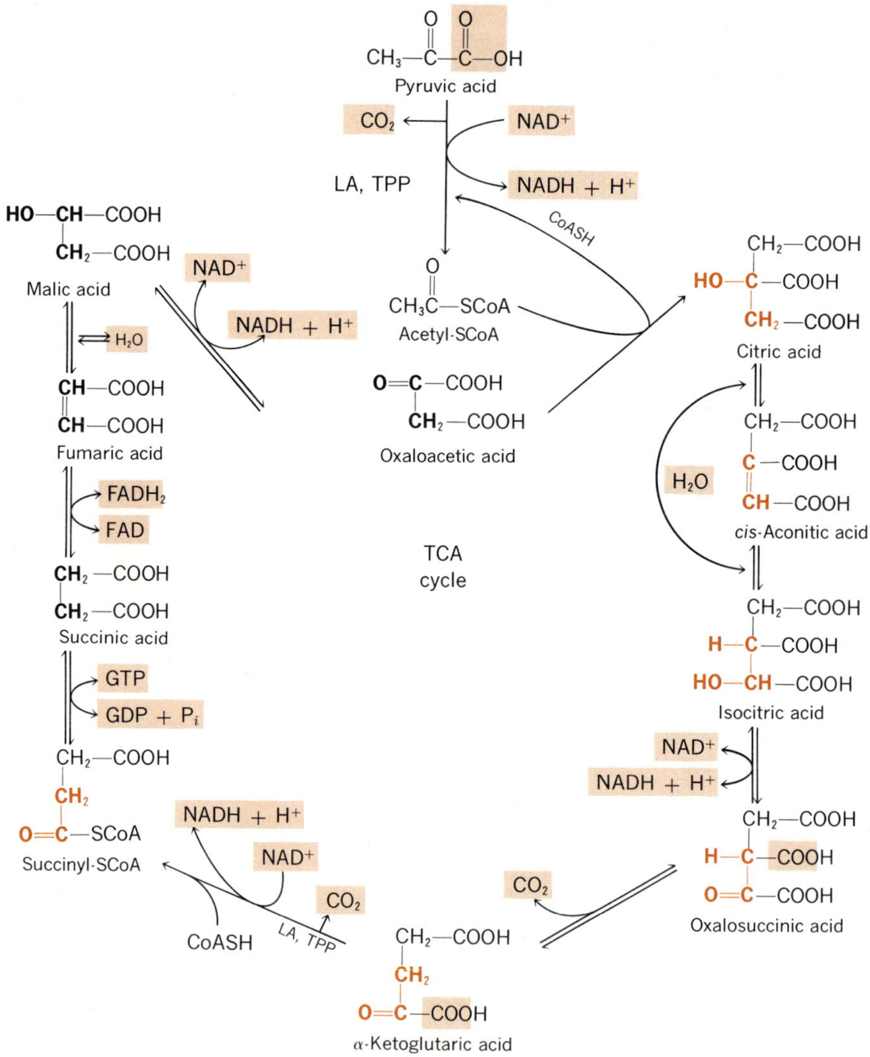

FIGURE 22.1 The reactions of the TCA cycle.

summarized in Fig. 22.1. The net overall reaction of the TCA cycle and the electron-transport system are:

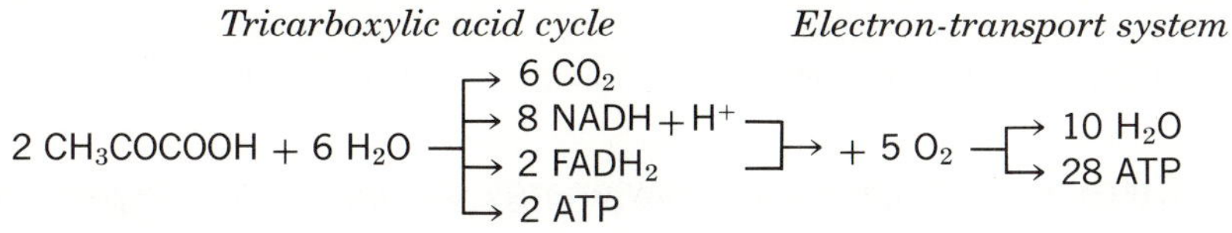

Net reaction: $2\ CH_3COCOOH + 5\ O_2 \longrightarrow 6\ CO_2 + 4\ H_2O + 30\ ATP$

Since these reactions are under aerobic conditions, the reduced coenzymes formed in the TCA cycle are subsequently oxidized by the electron-transport system to yield ATP and water. From the above equation, the oxidation of the reduced coenzymes produces 28 ATP; together with the 2 ATP produced in the TCA cycle this amounts to a total of 30 ATP per 2 moles of pyruvate oxidized.

PYRUVIC ACID $\longrightarrow$ ACETYL-SCoA. The entrance of pyruvic acid into the TCA cycle involves a rather complex series of steps, which start with the decarboxylation of the acid and the condensation of the resulting acetyl group with oxaloacetic acid to form citric acid. This involves several enzymes and the coenzymes thiaminpyrophosphate (TPP), lipoic acid (LA), coenzyme A (CoASH), and nicotinamide dinucleotide (NAD^+). Mg^{++} is also required.

These reactions are brought about by an enzyme complex known as *pyruvic oxidase complex* or *NAD-lipoate-linked pyruvate dehydrogenase system.*

The decarboxylation of pyruvic acid includes the steps discussed on page 468. They can be summarized as

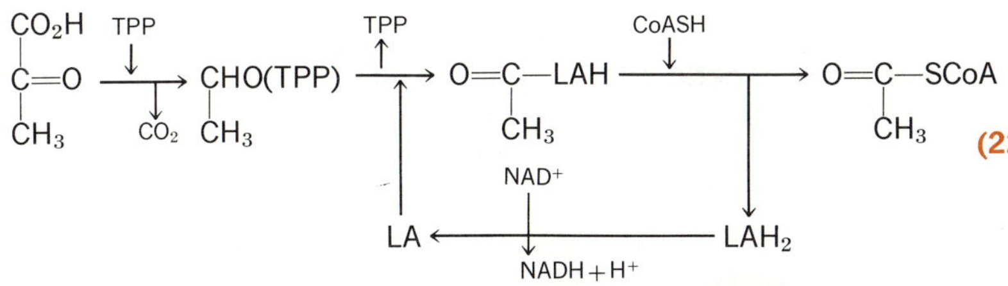

(22.1)

Overall: Pyruvic acid $\xrightarrow[\text{Pyruvate dehydrogenase complex}]{}$ acetyl—SCoA

As the equation shows, TPP is reformed when the C_2 unit is passed from TPP to LA. To keep this reaction functioning, oxidized lipoic acid must be reformed. This is accomplished by an enzyme which catalyzes the transfer of the hydrogens from LAH_2 to coenzyme NAD^+; LA can then recycle to pick up another C_2 unit.

The net equation for this series of reaction is

$$CH_3COCO_2H + HSCoA + NAD^+ \xrightarrow[Mg^{++}]{TPP,\ LA} CO_2 + CH_3COSCoA + NADH + H^+$$

There is a change from a low-energy compound to a high-energy compound in these reactions. When pyruvic acid is decarboxylated to the CH_3CO—TPP complex, a high-energy compound results. Let us consider the final product, $CH_3COSCoA$. If this compound is hydrolyzed, CoASH and CH_3COOH are produced. There are resonance formulas for acetic acid that cannot exist when the acetyl group is attached to CoASH.

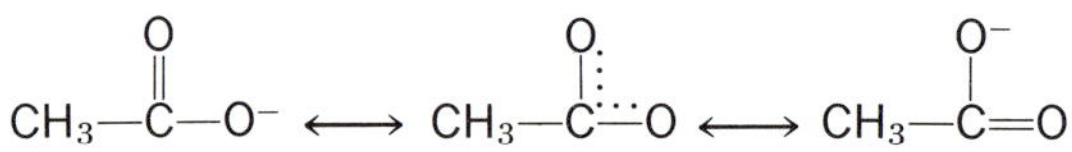

It requires energy to prevent resonance; this extra energy is contained in $CH_3COSCoA$, which is therefore a high-energy compound. Its energy is used to drive the condensation reaction with oxaloacetic acid forward, thus conserving an ATP.

Fate of Acetyl-SCoA

Acetyl-SCoA is an important intermediate. In addition to containing the C_2 carbohydrate unit that enters into the tricarboxylic acid cycle, it is, for example, the C_2 unit from which fatty acids are synthesized. This accounts for the even-numbered carbons in fatty acids. When fatty acids are oxidized, they enter into the TCA cycle by way of acetyl-SCoA. The biochemistry of amino acid metabolism also involves this complex.

ACETYL-SCoA + OXALOACETIC ACID ⟶ CITRIC ACID. The condensation of acetyl-SCoA with oxaloacetic acid is the first reaction of the TCA cycle (see Fig. 22.1).

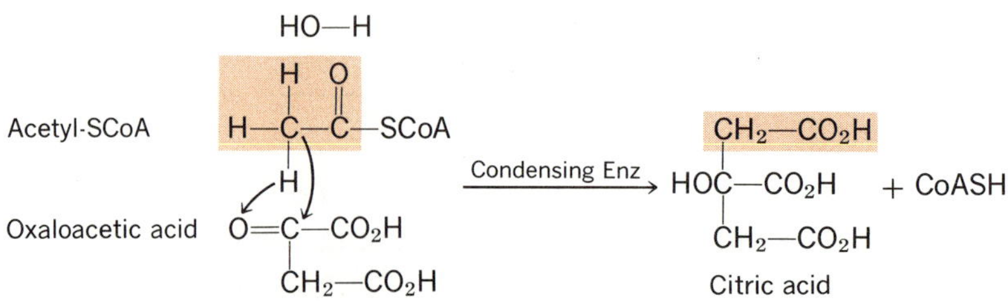

(22.2)

During the discussion of the intermediates of this cycle the carbons of the acetyl group will be shaded to keep them separated from the other carbons and to emphasize that it is not these carbons that appear as carbon dioxide the first time around the cycle.

The so-called *condensing enzyme* is responsible for catalyzing the union between oxaloacetic acid and acetyl-SCoA. The CoASH unit is hydrolyzed off in the reaction, thus forming CoASH, which can then cycle back and pick up another acetyl group. This hydrolytic step also makes the reaction irreversible under biological conditions and furnishes energy for the reaction.

CITRIC ACID $\rightleftharpoons$ *cis*-ACONITIC ACID $\rightleftharpoons$ ISOCITRIC ACID. *Aconitase* is the enzyme that catalyzes the conversion of citric acid to isocitric acid with *cis*-aconitic acid as an intermediate. The metallic element Fe^{++} is required for activation of the enzyme.

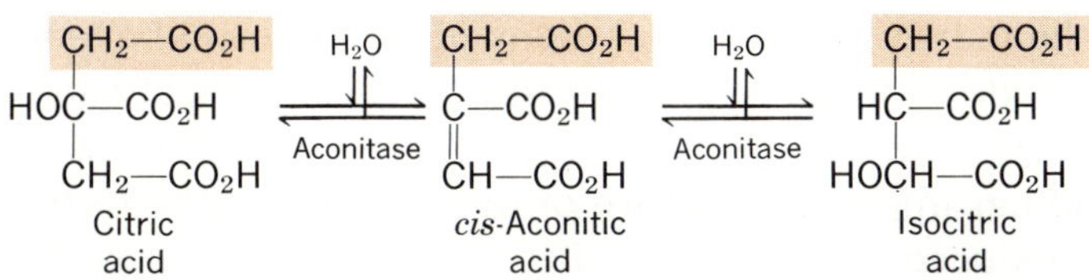

(22.3)

This series of reactions involves the removal of 1 mole of water to form an olefinic bond and then the addition of 1 mole of water in such a way that the OH is on a different carbon than originally. Note that the formation of the double bond did not involve the carbons originating from the acetyl group; isocitric acid contains the CH_2CO_2H group derived from acetyl-SCoA.

ISOCITRIC ACID $\rightleftharpoons$ (OXALOSUCCINIC ACID) $\rightleftharpoons$ α-KETOGLUTARIC ACID. The conversion of isocitric acid to α-ketoglutaric acid includes a dehydrogenation and a decarboxylation step.

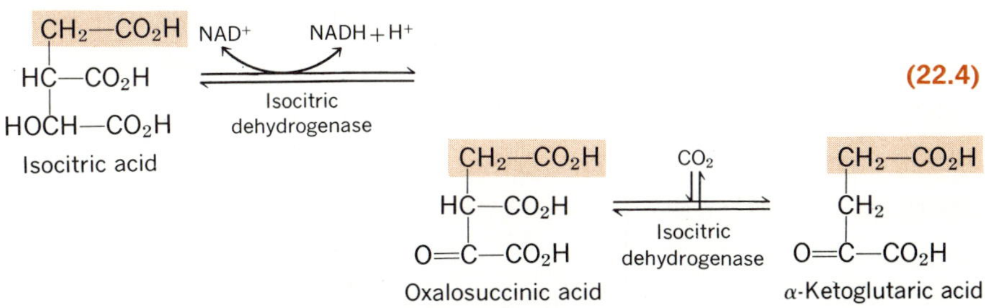

(22.4)

In mitochondria the enzyme *isocitric dehydrogenase* is responsible for this series of reactions. Although oxalosuccinic acid is shown as an intermediate to aid in understanding the steps involved, it is doubtful whether this compound is formed in the mitochondrial system. Apparently the dehydrogenation and decarboxylation reactions occur virtually simultaneously. NAD^+ and Mg^{++} are cofactors for this enzyme system.

Another isocitric dehydrogenase enzyme occurs in the cytoplasm which is specific for the coenzyme $NADP^+$. For it an enzyme-oxalosuccinic acid complex is postulated.

Attention is called to the decarboxylation step, which is a *reductive decarboxylation,* in contrast to that of pyruvic acid, which is an oxidative decarboxylation. In the former instance the carbon holding the CO_2H group is reduced. What is the oxidation number of C-3 of oxalosuccinic acid and the corresponding carbon of α-ketoglutaric acid? Has this carbon been oxidized or reduced?

Beginning with pyruvic acid, this is the second reaction in which CO_2 is eliminated. Thus, two carbons of pyruvic acid have been theoretically oxidized to CO_2—theoretically because the last carbon lost as CO_2 was not from the pyruvic acid but from oxaloacetic acid.

Another important feature of this reaction is the formation of a mole of reduced coenzyme, $NADH + H^+$, for each pyruvate entering the cycle.

α-KETOGLUTARIC ACID $\longrightarrow$ SUCCINYL-SCoA. The conversion of α-ketoglutaric acid to succinyl-SCoA is another example of *oxidative decarboxylation.* The mechanism and cofactors are the same as for the decarboxylation of pyruvic acid to acetyl-SCoA. Upon the decarboxylation of α-ketoglutaric acid the succinyl unit is formed instead of the acetyl unit, which was formed when pyruvic acid was decarboxylated. Thus, it is the succinyl group that is transferred from TPP to lipoic acid to form succinyl lipoate and eventually succinyl coenzyme A.

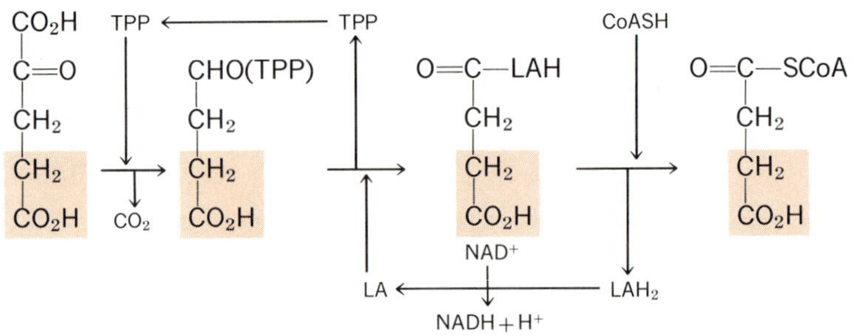

(22.5)

Overall: α-Ketoglutaric acid $\xrightarrow{\alpha\text{-Ketoglutarate dehydrogenase complex}}$ succinyl—SCoA

The enzyme system responsible for this series of reactions is called *α-ketoglutarate dehydrogenase complex.* Although this complex carries out reactions similar to pyruvate dehydrogenase complex, the two enzyme systems are different and each reacts only with its respective substrate.

Besides the liberation of 1 mole of CO_2 and the formation of 1 mole of reduced coenzyme $NADH + H^+$, coenzyme A is attached to the succinyl unit, which results in a high-energy compound similar to acetyl-SCoA. Whereas the energy in acetyl-SCoA was used to drive the condensation reaction forward, the energy content of succinyl-SCoA is conserved by being incorporated into ATP. Succinyl-SCoA does not hold as central a position in biological synthesis as acetyl-SCoA, but it is important in the synthesis of the porphyrin nucleus, which occurs in hemoglobin and the cytochromes. The oxidative decarboxylation of α-ketoglutaric acid, like that of pyruvic acid, is biologically irreversible.

Arsenite inhibits reactions involving reduced lipoic acid and hence prevents the oxidative decarboxylation of both pyruvic and α-ketoglutaric acids. As a result arsenite is lethal because it blocks the main energy-yielding system of most cells.

SUCCINYL-SCoA $\rightleftharpoons$ SUCCINIC ACID. The conversion of succinyl-SCoA to succinic acid is catalyzed by the enzyme *succinic thiokinase.*

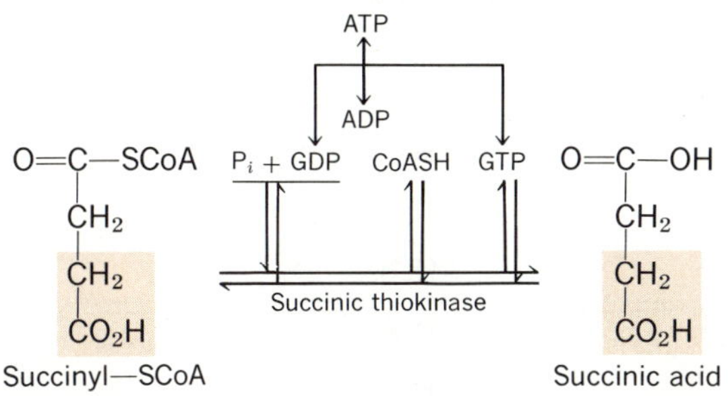

(22.6)

As mentioned above, the energy content of the coenzyme A complex is transferred eventually to ATP. This is not a direct transfer but is mediated by the conversion of GDP (guanosine diphosphate) to GTP (guanosine triphosphate). Like ATP, GTP enters into several energy-requiring reactions and can pass its energy on to ATP [Eq. (22.6)]. The high energy of succinyl-SCoA accounts for one ATP formed per mole of pyruvic acid metabolized. This reaction is the only one in the TCA cycle in which ATP is formed. The net reaction is

$$\text{Succinyl-SCoA} + \text{ADP} + H_3PO_4 \rightleftharpoons \text{succinic acid} + \text{ATP} + \text{CoASH}$$

SUCCINIC ACID ⇌ FUMARIC ACID. *Succinic dehydrogenase* is the enzyme responsible for dehydrogenating succinic acid to *trans*-fumaric acid. This enzyme contains flavin adenine dinucleotide (FAD) bound to the protein moiety, which contains four atoms of iron per molecule of enzyme. This reaction is easily reversible. Fumaric acid is a symmetrical molecule.

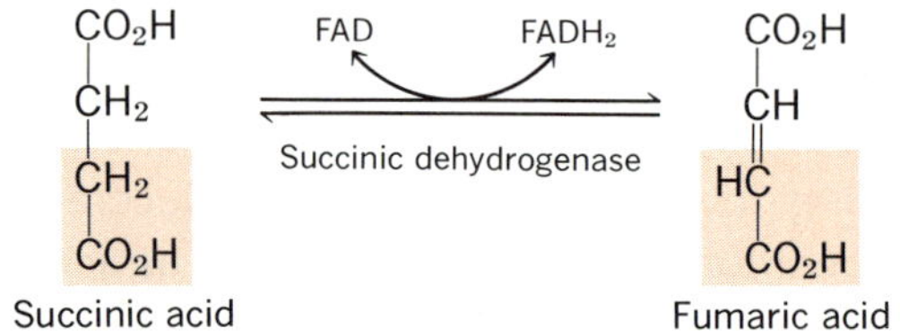

(22.7)

Malonic acid, $HO_2C—CH_2—CO_2H$, is a competitive inhibitor of succinic dehydrogenase. The similarity of this acid to the substrate permits it to be bound to the active site of the enzyme in a loose fashion so that succinic acid can displace it.

FUMARIC ACID ⇌ MALIC ACID. The addition of 1 mole of water to fumaric acid produces the hydroxy acid malic acid. The carbon atom holding the hydroxy group is asymmetric; hence there is a D and an L isomer of malic acid. L-Malic acid is produced in animal tissue.

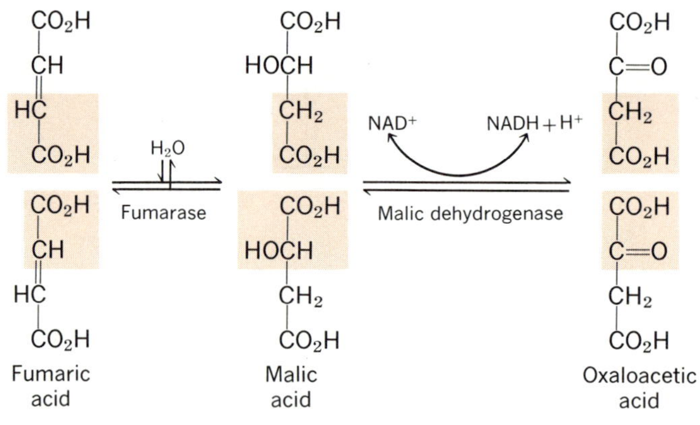

(22.8)

The enzyme *fumarase* catalyzes this reversible reaction. As with other enzymes which add components to a double bond, fumarase causes a trans addition; similarly, the reverse reaction involves a trans elimination of OH and H to form the olefinic bond.

Since the conversion of fumaric acid to malic acid converts a symmetrical molecule into an asymmetrical one, two malic acids are produced with the tagged acetyl carbons. Fumarase has no preference which of the two carbons the OH is added onto; hence, there is a fifty-fifty production of the tagged molecules as indicated in the equation. Follow these tagged carbons through the next step.

MALIC ACID $\rightleftharpoons$ OXALOACETIC ACID. The final step in the TCA cycle is the conversion of malic acid to oxaloacetic acid, which is the compound we started with in this cycle [see Eq. (22.8)]. The enzyme responsible for this reversible conversion is known as *malic dehydrogenase.* In this reaction a secondary alcohol is oxidized to a ketone group by means of a dehydrogenation reaction. Like the succinic dehydrogenase-catalyzed reaction, this reaction yields a reduced coenzyme which can enter the electron-transport cycle to furnish ATP. Whereas the coenzyme for the former reaction was FAD, the coenzyme required for malic dehydrogenase is NAD^+.

The equilibrium for malic dehydrogenase favors the formation of malic acid; however, since the end product of this reaction, oxaloacetic acid, is being removed when it condenses with acetyl-SCoA, the reaction is carried to completion during active metabolism of carbohydrates.

Once around the TCA cycle (starting with acetyl-SCoA) accounts for the formation of 2 moles of carbon dioxide, neither of which was derived from the acetyl group [see the shaded carbons in the two formulas for oxaloacetic acid in Eq. (22.8)].

22.2 SUMMARY

The reactions of the TCA cycle have been given briefly. In the introduction to this section the overall reaction was given, starting with pyruvic acid. It is repeated below, not only for review and emphasis but also for further consideration. As noted by the final equation,

$$CH_3COCO_2H + 3\ H_2O + 4\ NAD^+ + 2\ FAD + ADP + H_3PO_4 \longrightarrow 3\ CO_2 + 4\ NADH{+}H^+ + FADH_2 + ATP$$

the TCA cycle reactions have resulted in the oxidation of pyruvate carbons to CO_2 and the trapping of hydrogens in reduced coenzymes. Since coenzymes are not used in substrate levels, their reduced forms must be oxidized to permit the cycle to continue. This is accomplished by the electron-transport system, which will be discussed shortly.

First, however, some questions about the TCA cycle should be considered. The whole series of reactions in the TCA cycle depends upon an adequate supply of oxaloacetic acid to metabolize acetyl-SCoA. Carbohydrates are not the sole source of acetyl-SCoA; fatty acids and some amino acids are metabolized to this compound, which then enters the TCA cycle for oxidation. There are over a dozen amino acids that feed into the TCA cycle during their catabolism. These amino acids (as well as any other compound except acetyl-SCoA that enters into the TCA cycle) would cause a buildup of oxaloacetic acid. This of course does not happen because the acid is either oxidized or converted to glycogen. How can oxaloacetate be converted to glycogen? Certainly *not* by a reversal of the TCA cycle. Where did oxaloacetic acid come from? There has to be a mechanism for its formation from products of carbohydrate metabolism.

PYRUVIC ACID $\rightleftharpoons$ OXALOACETIC ACID $\rightleftharpoons$ PHOSPHOENOLPYRUVIC ACID. Pyruvic acid is the immediate precursor of oxaloacetic acid. The difference between the molecular formula of pyruvic acid and oxaloacetic acid is one carbon atom and two oxygen atoms, or the equivalent of 1 mole of carbon dioxide. Thus, if pyruvic acid is to be converted into oxaloacetic acid, the equivalent of 1 mole of carbon dioxide must be added to pyruvic acid. This is exactly the reaction that is brought about by enzymes, the ***nonphotosynthetic fixation of carbon dioxide***. Thus, animals under normal conditions are capable of fixing carbon dioxide. The discovery of this fact marked an important advance in understanding the metabolism of carbohydrates, fatty acids, and amino acids.

There are two mechanisms whereby oxaloacetic acid can be synthesized from pyruvate; one forms malic acid, which can be converted into oxaloacetic acid, and the other, the main reaction, forms oxaloacetic acid directly.

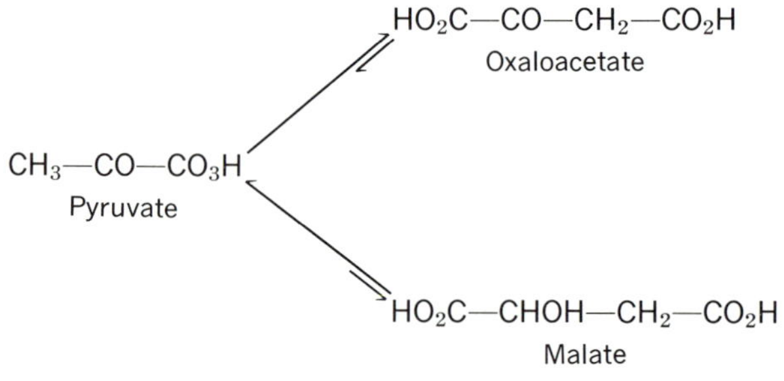

(22.9)

These reactions are doubly significant: they not only guarantee a continued supply of oxaloacetic acid but also offer a means of converting oxaloacetic acid into acetyl-SCoA by way of pyruvic acid.

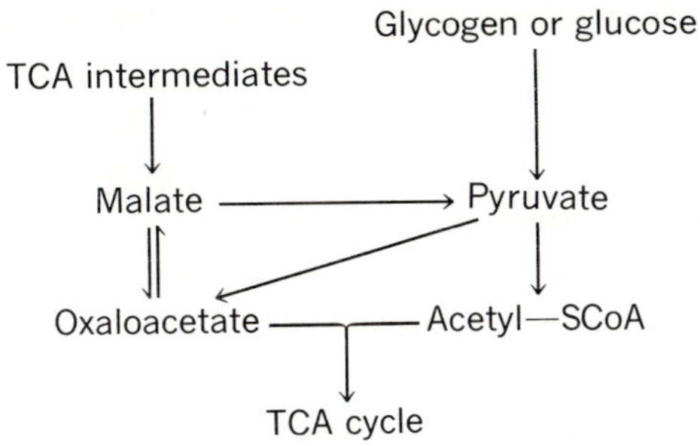

(22.10)

Any true TCA intermediate (citric acid, α-ketoglutaric acid, or succinic acid) can be oxidized completely by way of the malic acid ⟶ pyruvic acid ⟶ acetyl-SCoA pathway.

Acetyl-SCoA is not a source of oxaloacetic acid. The entrance of acetyl-SCoA into the cycle causes no net increase of oxaloacetic acid. The decarboxylation of pyruvic acid is not reversible, and furthermore the C_2 unit is oxidized in going around the cycle.

PYRUVIC ACID ⟶ OXALOACETIC ACID. *Pyruvic carboxylase* contains four moles of biotin per mole of enzyme and catalyzes the formation of oxaloacetic acid by carboxylation of pyruvic acid. The CO_2 forms a complex with biotin

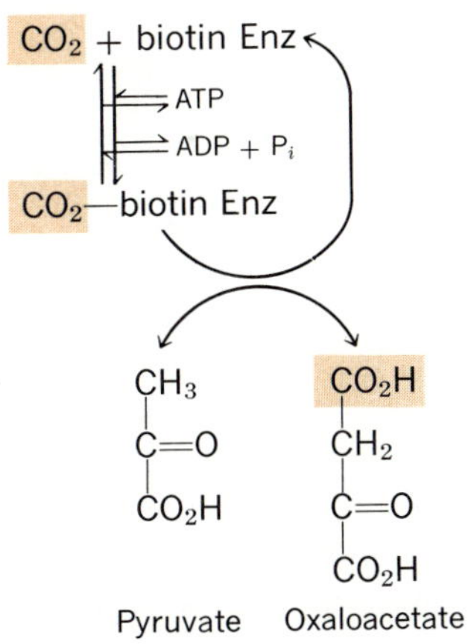

(22-11)

Overall:

$$\text{Pyruvate} \xrightarrow[\text{ATP}]{CO_2} \text{oxaloacetate}$$

(see page 474). ATP is required for this reaction as well as the cofactors Mg^{++} and acetyl-SCoA. The latter is an absolute requirement although it does not enter into the reaction, apparently being needed to hold the enzyme in an active conformation. This undoubtedly plays an important role in regulating carbohydrate metabolism. The oxidation of the acetyl group requires a continued supply of oxaloacetic acid; thus, when acetyl-SCoA is formed, it in turn activates the enzyme that supplies the necessary oxaloacetic acid.

The *equilibrium of this reaction favors the formation of oxaloacetic acid,* thus ensuring an adequate supply of this product for the TCA cycle. It is well to note that the vitamin biotin is associated with carbon dioxide fixation in this reaction.

Pyruvic carboxylase is located in the mitochondria and is quite closely associated with the TCA cycle enzymes.

OXALOACETIC ACID $\rightleftharpoons$ PHOSPHOENOLPYRUVIC ACID. *Phosphoenolpyruvic carboxykinase* is another mitochondrial enzyme associated with the metabolism of oxaloacetic acid. Its main function appears to be associated with the conversion of oxaloacetate to phosphoenolpyruvate, which, when coupled with the enzyme pyruvic carboxylase, furnishes a means of going from pyruvic acid to glucose or glycogen.

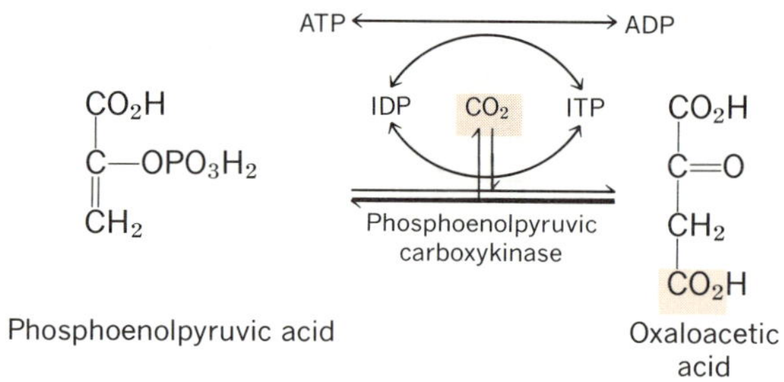

(22.12)

In the Embden-Meyerhof pathway (in liver) the conversion of phosphoenolpyruvic acid to pyruvic acid is not reversible under biological conditions; therefore, once pyruvic acid is formed, it cannot be converted into glycogen unless this irreversible step is circumvented. All the other steps from phosphoenolpyruvic acid to glucose in this pathway can be reversed.

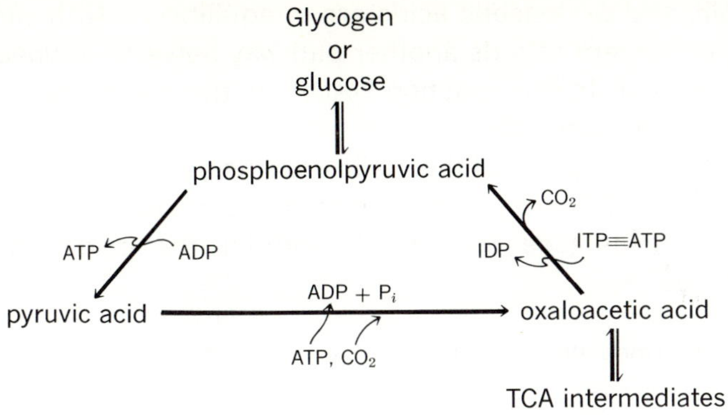

(22.13)

The conversion of oxaloacetic acid to phosphoenolpyruvic acid affords a means of converting pyruvate or lactate to glucose or glycogen for any metabolite that feeds into the TCA cycle and causes a net increase in the level of oxaloacetic acid. Note that the equivalent of an ATP is required to convert oxaloacetate to phosphoenolpyruvate; hence it requires a total of 2 ATP to convert 1 mole of pyruvate to 1 mole of phosphoenolpyruvate.

$$\text{Pyruvic acid} \xrightarrow{\text{2 ATP}} \text{phosphoenolpyruvic acid}$$

Oxaloacetic acid was stressed as an important substance associated with the TCA cycle. In light of the above discussion, it can be seen that oxaloacetic acid is equally important as a connecting link between the TCA and the glycolytic pathway and as an intermediate in the reversal of the glycolytic pathway starting with pyruvic acid. Not to be forgotten in this connection is the role of nonphotosynthetic fixation of carbon dioxide.

MALIC ACID $\rightleftharpoons$ PYRUVIC ACID. *Malic enzyme* is a cytoplasmic enzyme that catalyzes the conversion of malic acid to pyruvic acid and vice versa.

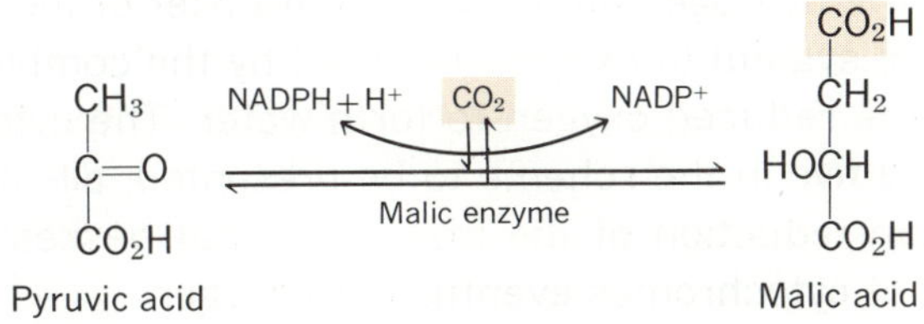

(22.14)

Since malic and oxaloacetic acids are in equilibrium with each other, this enzyme system affords another pathway between oxaloacetic acid and pyruvic acid. In this reaction $NADP^+$ is the coenzyme, and Mn^{++} is required as an activator.

The main functions of the malic enzyme appear to be as a means of furnishing $NADPH + H^+$ for cytoplasmic reactions, e.g., the synthesis of fatty acids (see Sec. 25.5), and for the oxidation of TCA intermediates.

$$\text{TCA intermediates} \longrightarrow \text{malate} \longrightarrow \text{pyruvate} \longrightarrow \text{acetyl-SCoA}$$

22.3 OXIDATIVE PHOSPHORYLATION AND ELECTRON-TRANSPORT SYSTEM

Oxidative phosphorylation is the mechanism whereby hydrogen atoms of reduced coenzymes, $NADH + H^+$, $NADPH + H^+$, and $FADH_2$, are combined with molecular oxygen and the energy of the reaction is used to phosphorylate ADP to ATP. Involved in this overall reaction is the transport of electrons from the hydrogen to oxygen via the electron-transport system.

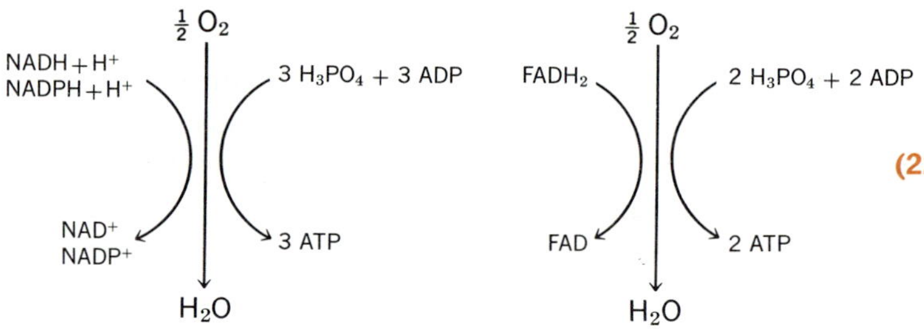

(22.15)

When 1 mole of either $NADH + H^+$ or $NADPH + H^+$ is oxidized, 3 moles of ATP are formed. However, when 1 mole of $FADH_2$ is oxidized, only 2 moles of ATP are formed.

The union of hydrogen atoms with oxygen involves removal of the electron from the hydrogen atoms and the transfer of the electrons via the cytochrome system to oxygen, followed by the combination of the protons with the reduced oxygen to form water. The cytochromes, of which there are five in the scheme to be presented, all contain iron. It is the oxidation-reduction of the iron atom that passes the electron along the line of cytochromes eventually to oxygen.

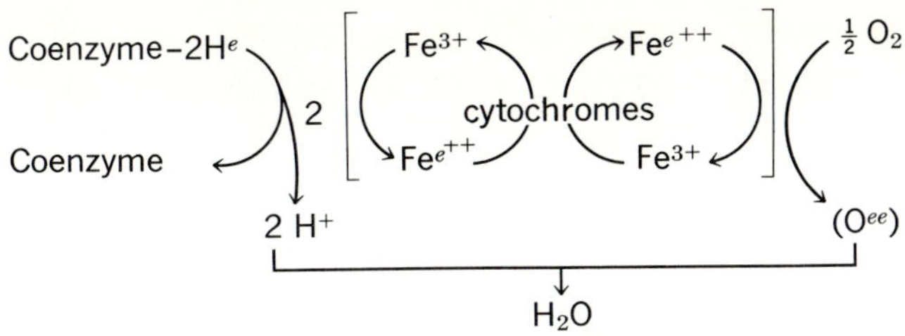

The enzymes for oxidative phosphorylation are closely associated with the inner mitochondrial lining and are therefore readily available to the reduced coenzymes produced in the Krebs cycle (Fig. 22.2.)

The exact mechanism of electron transport along the cytochrome system and factors associated with phosphorylation have not yet been completely elucidated. A general schematic representation of the mechanism is illustrated in Fig. 22.3.

Hydrogens of reduced $NADH+H^+$ or $NADPH+H^+$ are first transferred to a flavoprotein, FP_1 (a non-heme-iron complex), which is closely associated with CoQ. Hydrogens from $FADH_2$ are transferred to a flavoprotein–non-heme-iron complex called FP_2, which, like FP_1, is closely associated with CoQ. CoQ accepts the hydrogen from the reduced FP_1 and FP_2 complexes to form reduced CoQ, $CoQH_2$, and oxidized NAD^+ and $NADP^+$ or FAD, which then are free to pick up more hydrogen from the appropriate substrate. At this point the electrons from the hydrogen atoms enter the cytochrome system via Cyt b. The protons are free in solution. As the electrons combine with Cyt b–Fe^{3+}, the iron is reduced to Fe^{++}. The electron is subsequently passed to Cyt c_1–Fe^{3+}, Cyt c–Fe^{3+}, and Cyt a–Fe^{3+} by the same mechanism of

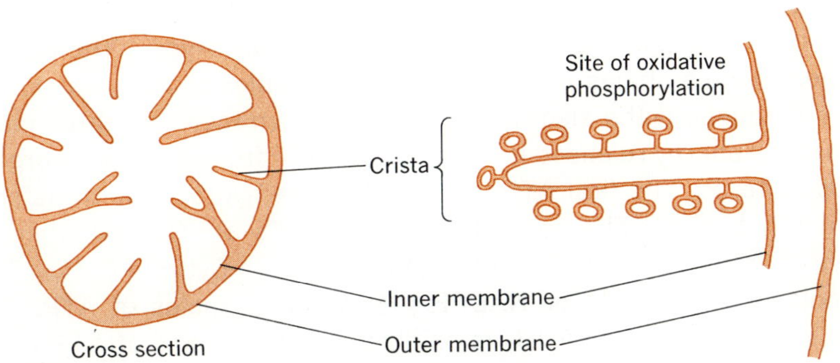

FIGURE 22.2 Cross section of a mitochondrion.

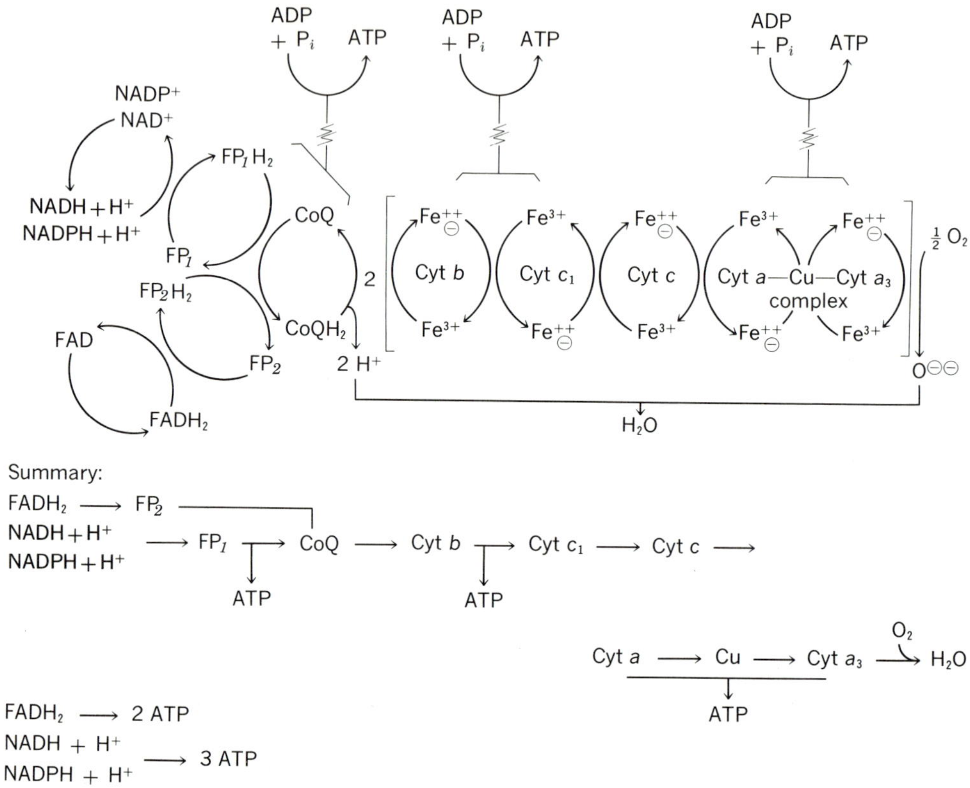

FIGURE 22.3 Electron-transport system and the oxidative phosphorylation system.

oxidation-reduction of the iron in each of these cytochromes. Cyt a is bound in a complex with Cyt a_3; Cu^{++} is an integral component. The electron is passed from Cyt a to Cyt a_3 via oxidation-reduction of the copper. Cyt a_3 is frequently referred to as cytochrome oxidase and is responsible for passing the electron to molecular oxygen. The reduced oxygen then combines with the protons in solution to form water.

Throughout the electron-transport system there are three places where the energy released is used to form ATP from ADP and inorganic phosphate, P_i. The first ATP is formed as the hydrogen is transferred from FP_1 to CoQ. The second and third ATPs are formed when the electron is passed from Cyt b to Cyt c_1 and from Cyt a to Cyt a_3. Thus 1 mole of $NADH+H^+$ or $NADPH+H^+$ yields 3 moles of ATP. Note that when 1 mole of $FADH_2$ is oxidized, only 2 moles of ATP is produced; the first energy-yielding step is bypassed. No energy is released when the hydrogens from FP_2H_2 are passed to CoQ.

Why are so many cytochromes and steps involved in oxidative phosphorylation when the reactions are geared to the production of ATP rather than to the formation of intermediary metabolic products? The transfer of electrons from hydrogen to oxygen yields a considerable quantity of energy. The energy must be released in small enough amounts at a time to be trapped as ATP. Remember that enzyme systems cannot function when too great an energy change is involved.

From Fig. 21.3 we see that from 1 mole of glucose 38 ATP are produced as a result of the oxidation of the reduced coenzymes via the electron-transport system. There were 2 ATP (net) formed in the Embden-Meyerhof pathway, and 2 ATP resulted from the TCA cycle reactions. Thus, out of a total of 38 ATP, 34 ATP were formed by oxidative phosphorylation. The main function of oxidative phosphorylation is to supply energy for biological reactions.

22.4 EXTERNAL MITOCHONDRIAL NADH + H$^+$ OXIDATION

The mitochondrial wall is not particularly permeable to NADH + H$^+$. Therefore, to oxidize cytoplasmic NADH + H$^+$ (formed in the glycolytic pathway) via the mitochondrial electron-transport system a mechanism must be available to transport the reduced coenzyme across the mitochondrial membrane. To accomplish this a so-called shuttle mechanism is employed; i.e., a substrate that can penetrate the mitochondrial membrane is reduced by cytoplasmic NADH + H$^+$; the reduced substrate then diffuses through the mitochondrial membrane and is oxidized by mitochondrial NAD$^+$ or FAD. The oxidized substrate then diffuses out of the mitochondria into the cytoplasm and is available for reduction again:

Cytoplasm			*Mitochondria*	
NADH + H$^+$ ↘	↗ Substrate	⟵	Substrate ↖	↗ NADH + H$^+$ or FADH$_2$
NAD$^+$ ↙	↘ Substrate · H$_2$	⟶	Substrate · H$_2$ ↗	↖ NAD$^+$ or FAD

Two pairs of compounds act in shuttle mechanisms, namely, *α-glycerophosphate* and *dihydroxyacetone phosphate* (most active in muscle tissue) and *β-hydroxybutyrate* and *β-ketobutyrate* (most active in liver tissue).

α-GLYCEROPHOSPHATE CYCLE. Especially in muscle tissue NADH + H$^+$ formed in the Embden-Meyerhof pathway is oxidized by α-glycerophosphate dehydrogenase with the concomitant reduction of dihydroxyacetone

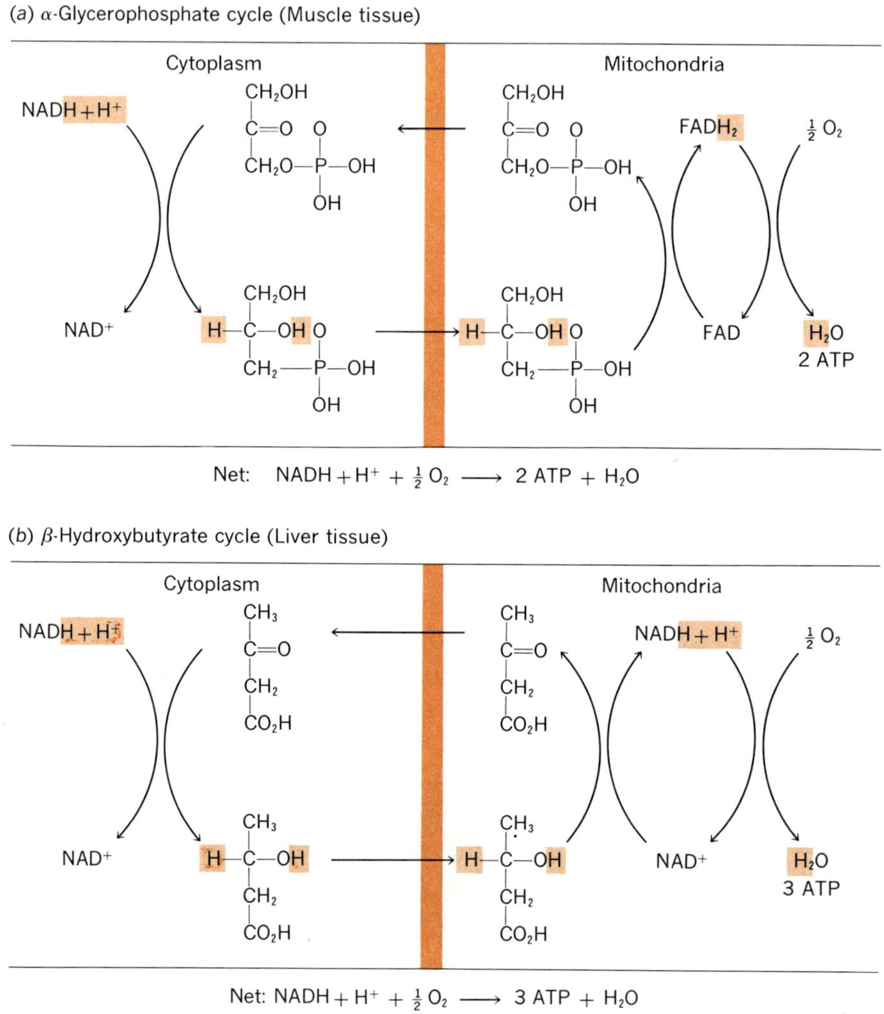

FIGURE 22.4 Oxidation of extramitochondrial NADH+H$^+$.

phosphate to α-glycerophosphate (Fig. 22.4*a*). The glycerophosphate can penetrate the mitochondrial membrane and diffuses into the mitochondria, where it is oxidized to dihydroxyacetone phosphate by a ***FAD-linked enzyme*** (α-glycerophosphate oxidase). The dihydroxyacetone phosphate then diffuses out of the mitochondria and is available for the oxidation of more cytoplasmic NADH+H$^+$. Oxidation of the $FADH_2$ via the cytochrome system produces 2 moles of ATP per mole of reduced coenzyme. By this shuttle mechanism, 1 mole of cytoplasmic NADH+H$^+$ is oxidized to yield 2 moles of ATP.

Cytoplasmic $NADH + H^+$ (muscle) $\longrightarrow$ mitochondrial $FADH_2$ $\longrightarrow$ 2 ATP

Thus the maximum number of moles of ATP obtained from 1 mole of glucose is 36.

β-HYDROXYBUTYRATE CYCLE. Cytoplasmic $NADH + H^+$ can also be oxidized via a shuttle mechanism that involves the oxidation-reduction of β-hydroxybutyrate and β-ketobutyrate (Fig. 22.4*b*). A cytoplasmic *NAD^+-linked enzyme* (β-hydroxybutyric dehydrogenase) reduces β-ketobutyrate to β-hydroxybutyrate, which diffuses through the mitochondrial wall. A mitochondrial NAD^+-linked enzyme oxidizes the hydroxybutyrate to β-ketobutyrate, which then diffuses back out of the mitochondria to perpetuate the cycle. The mitochondrial $NADH + H^+$ formed is oxidized via the electron-transport system to yield 3 moles of ATP per mole of reduced coenzyme oxidized.

Cytoplasmic $NADH + H^+$ (liver) $\longrightarrow$ mitochondrial $NADH + H^+$ $\longrightarrow$ 3 ATP

This shuttle mechanism is more efficient than the previous mechanism in that it yields 3 ATP per mole of reduced $NADH + H^+$ oxidized instead of 2 as in the muscle system.

22.5 OXIDATION OF CYTOPLASMIC $NADPH + H^+$

Under normal conditions cytoplasmic $NADPH + H^+$ is used not for energy purposes but for the synthesis of cellular components such as the fatty acids. However, under abnormal conditions like starvation or diabetes, which interfere with fatty acid synthesis, $NADPH + H^+$ does accumulate, and a mechanism for its oxidation must be available. Cytoplasmic $NADPH + H^+$ is oxidized by an enzyme that reduces acetoacetyl-SCoA (β-ketobutyryl-SCoA) and at the same time removes the —SCoA to form β-hydroxybutyric acid as the end product.

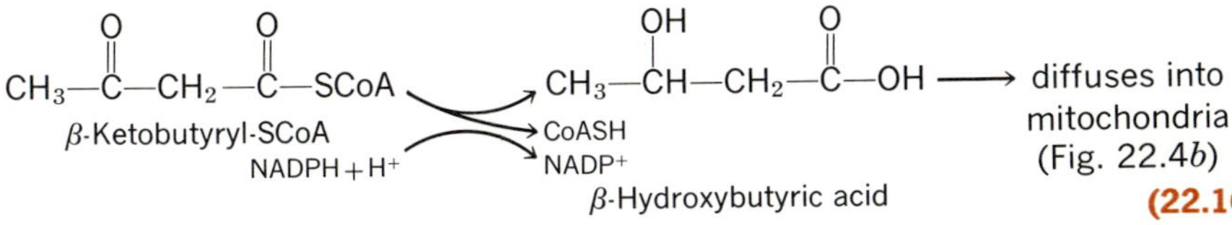

(22.16)

The β-hydroxybutyric acid then diffuses into the mitochondria and is oxidized as above by a NAD^+ enzyme. The net reaction is the oxidation of 1 mole of cytoplasmic $NADPH + H^+$ to yield 3 moles of ATP.

Cytoplasmic $NADPH + H^+$ $\longrightarrow$ mitochondrial $NADH + H^+$ $\longrightarrow$ 3 ATP

QUESTIONS

1. In what unit of the cell are the TCA cycle and the electron-transport systems located?
2. What is the overall significance of the TCA cycle and of oxidative phosphorylation?
3. Which carbons of glucose are eliminated as CO_2 during the conversion of pyruvic acid to acetyl-SCoA?
4. How many moles of ATP are theoretically possible from the metabolism of 1 mole of glucose under aerobic conditions? How many moles of ATP are formed in the TCA cycle?
5. Which phase of glucose metabolism produces more ATP per mole of glucose: glycolytic pathway, TCA cycle, or oxidative phosphorylation?
6. (*a*) What key compound is required for the continuous operation of the TCA cycle?
   (*b*) By what reactions is this key compound produced to assure continued operation of the TCA cycle?
7. Which of the following statements are correct? Carboxylation reactions involving pyruvate and phosphoenolpyruvate are important because:
   (*a*) They provide connections between glycolysis and the TCA cycle.
   (*b*) They allow for the conversion of pyruvate to phosphoenolpyruvate.
   (*c*) The decarboxylation reaction provides a route in animals for the conversion of acetate to glucose.
   (*d*) The carboxylation reaction provides a source of TCA intermediates.
   (*e*) They yield high-energy compounds.
8. Discuss the importance of nonphotosynthetic fixation of carbon dioxide with respect to the TCA cycle.
9. The conversion of phosphoenolpyruvate to pyruvate is not reversible under biological conditions. Explain how the body can utilize pyruvate or oxaloacetate to produce a net synthesis of glycogen.
10. How many moles of ATP are derived from the complete catabolism (oxidation) of 1 mole of α-ketoglutaric acid?
11. Thiamin pyrophosphate is a coenzyme for which of the following enzymes?
    (*a*) α-Ketoglutarate dehydrogenase (*b*) Lactic acid dehydrogenase
    (*c*) Phosphoenolpyruvate carboxylase (*d*) Pyruvic carboxylase

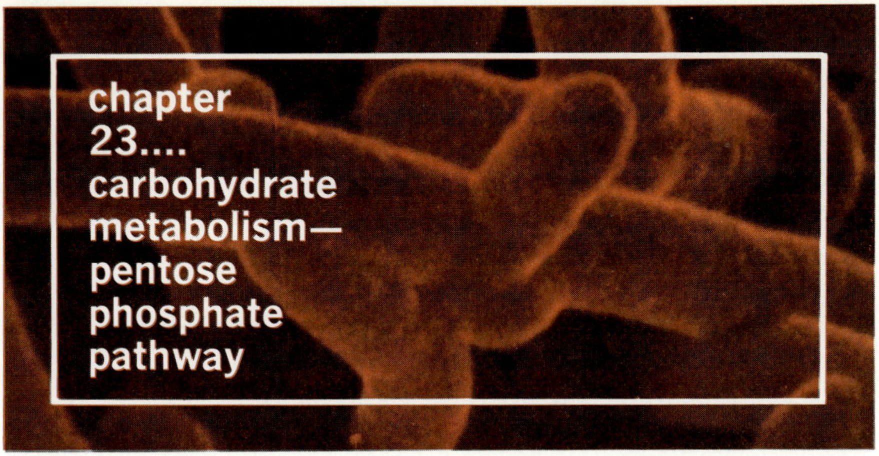

chapter 23.... carbohydrate metabolism—pentose phosphate pathway

The Embden-Meyerhof pathway coupled with the tricarboxylic acid cycle is the major route by which most organisms oxidize glucose. But it is not the only pathway. Another important one, called the pentose phosphate pathway or just the pentose pathway, is the only alternate route we shall discuss.

The pentose phosphate pathway serves not only as an alternate means of oxidizing glucose but also as a source of such compounds as ribose (2-deoxyribose), erythrose, xylose, and ribulose, and it furnishes $NADPH + H^+$ for biological syntheses.

The reactions of the pentose phosphate pathway are summarized in Fig. 23.1. The reactions presented start with glucose-6-phosphate, and the end products are CO_2 and $NADPH + H^+$. There are 12 $NADPH + H^+$ formed per mole of glucose metabolized.

$$\text{Glucose-6-phosphate} + 12\ NADP^+ + 6\ H_2O \longrightarrow 12\ NADPH + H^+ + 6\ CO_2 + H_3PO_4$$

23.1 STEPS IN THE PENTOSE PHOSPHATE PATHWAY

The individual steps in the pentose phosphate pathway will be discussed only briefly. To balance the equations requires 6 moles of glucose-6-phosphate initially.

GLUCOSE-6-PHOSPHATE $\rightleftharpoons$ 6-PHOSPHOGLUCONO-δ-LACTONE. The first step is the dehydrogenation of glucose-6-phosphate to the corresponding lactone.

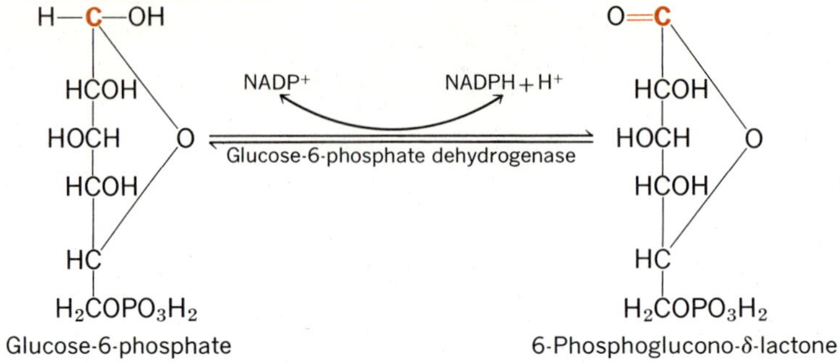

(23.1)

With glucose written in the pyranose ring structure, it is easy to see that the δ-lactone is the expected product. Coenzyme $NADP^+$ picks up the hydrogen. The enzyme catalyzing this reaction, *glucose-6-phosphate dehydrogenase,* is also frequently called *Zwischenferment* and is quite widespread in nature.

6-PHOSPHOGLUCONO-δ-LACTONE ⇌ 6-PHOSPHOGLUCONIC ACID. The hydrolysis of the gluconolactone is catalyzed by the enzyme 6-*phosphoglucono-lactonase.*

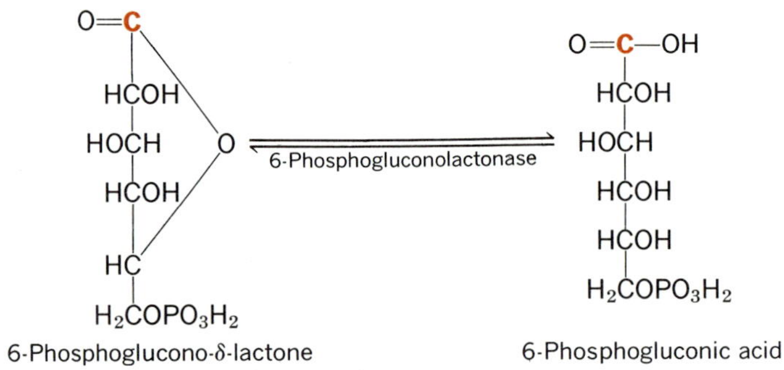

(23.2)

Although this reaction occurs nonenzymatically to an appreciable extent, the presence of the enzyme assures complete conversion of the lactone to the acid. Under biological conditions, this is an irreversible reaction; hence, glucose-6-phosphate is converted quantitatively into 6-phosphogluconic acid.

6-PHOSPHOGLUCONIC ACID ⇌ RIBULOSE-5-PHOSPHATE. The conversion of phosphogluconic acid to ribulose-5-phosphate involves decarboxylation of the carboxyl group and oxidation via dehydrogenation of the third carbon of gluconic acid. *Phosphogluconate dehydrogenase* is the enzyme involved, and $NADP^+$ is the hydrogen carrier. Mg^{++} is the metallic activator for this reaction.

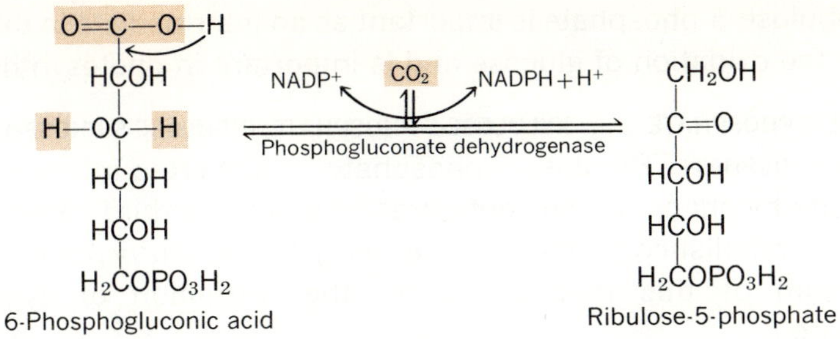

(23.3)

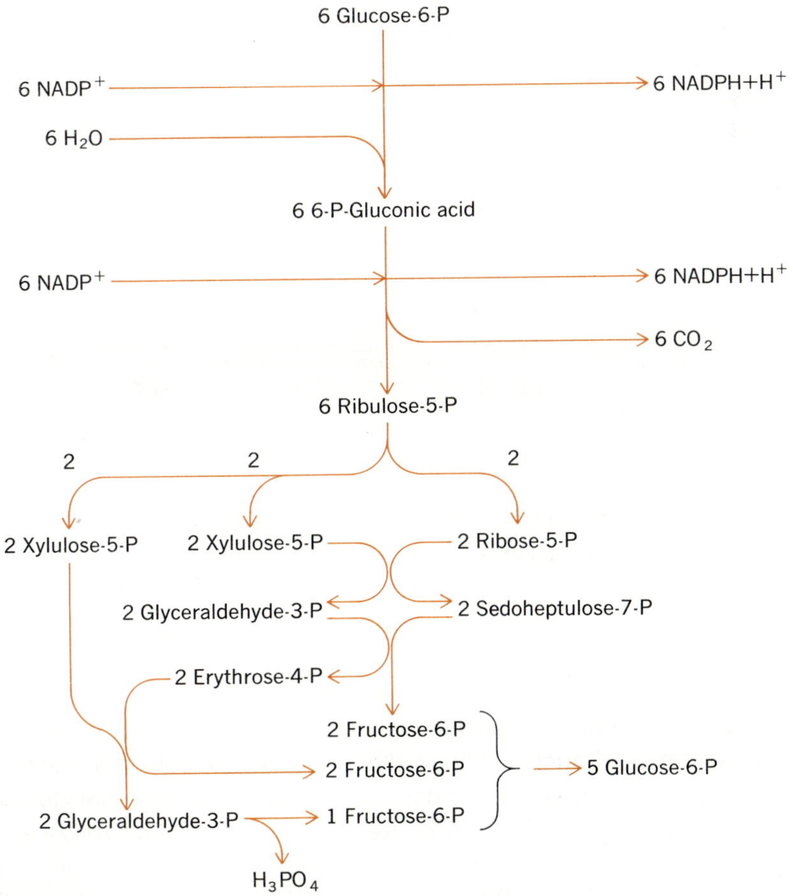

Overall reaction:

$$6\text{ Glucose-6-P} + 12\text{ NADP}^+ + 6\text{ H}_2\text{0} \longrightarrow 5\text{ Glucose-6-P} + 12\text{ NADH+H}^+ + 6\text{ CO}_2 + \text{H}_3\text{PO}_4$$

Net reaction:

$$\text{Glucose-6-P} + 12\text{ NADP}^+ + 6\text{ H}_2\text{0} \longrightarrow 12\text{ NADPH+H}^+ + 6\text{ CO}_2 + \text{H}_3\text{PO}_4$$

FIGURE 23.1 Summary of the pentose phosphate pathway.

Ribulose-5-phosphate is important as an intermediate in this pathway for the oxidation of glucose and is important in photosynthesis.

RIBULOSE-5-PHOSPHATE ⇌ XYLULOSE-5-PHOSPHATE; RIBULOSE-5-PHOSPHATE ⇌ RIBOSE-5-PHOSPHATE. Ribulose-5-phosphate is converted into xylulose-5-phosphate by changing the configuration about the third carbon atom. This is accomplished by the enzyme *phosphoketopentoepimerase*. The equilibrium of this reaction favors the formation of xylulose-5-phosphate.

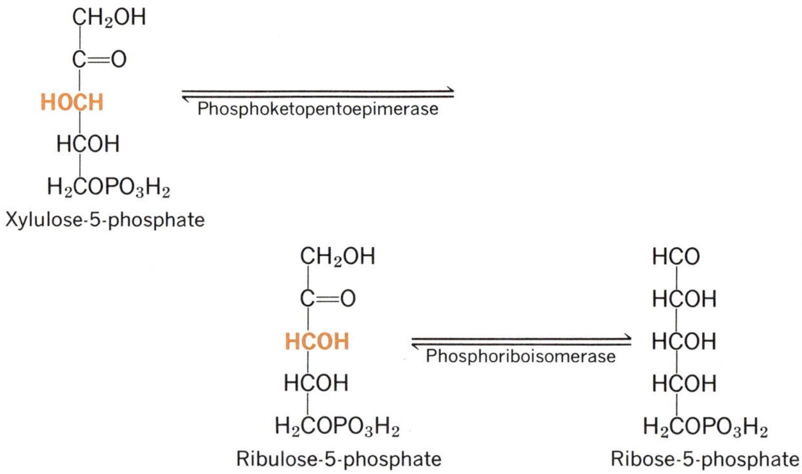

(23.4)

Ribose-5-phosphate is formed from ribulose-5-phosphate under the influence of the enzyme *phosphoriboisomerase*. This reaction reminds one of the conversion of fructose to glucose in the Embden-Meyerhof pathway. The equilibrium favors the formation of ribose-5-phosphate.

The conversion of ribulose-5-phosphate to ribose-5-phosphate is important as a means of furnishing ribose and ultimately deoxyribose for the synthesis of nucleic acids (RNA and DNA). Xylulose-5-phosphate is a precursor of xylose, which occurs in xylans, polysaccharides composed of xylose units. Certain bacteria that are able to utilize xylose do so by converting it to xylulose-5-phosphate, which is a component of the pentose pathway. Ribose is metabolized via the pentose phosphate pathway after being phosphorylated to ribose-5-phosphate.

XYLULOSE-5-PHOSPHATE + RIBOSE-5-PHOSPHATE ⇌ GLYCERALDEHYDE-3-PHOSPHATE + SEDOHEPTULOSE-7-PHOSPHATE. The rest of the reactions of the pentose pathway involve the conversion of xylulose-5-phosphate and ribulose-5-phosphate to fructose-6-phosphate (or glucose-6-phosphate). In the overall scheme in Fig. 23.1, 6 moles of glucose-6-phosphate enter into the reactions; 6 moles each of ribulose-5-phosphate and CO_2 are formed. These 6 moles of pentose account for 30 carbon atoms. By the following reactions they are converted into 5 moles of glucose-6-phosphate, 30 carbons total.

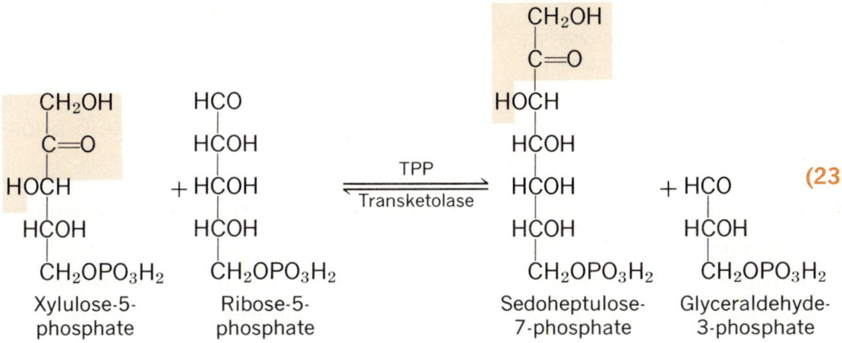

(23.5)

The first step in this series involves the transfer of the ketol (keto-alcohol) group, $CH_2OH—CO$, from xylulose-5-phosphate to the aldehyde sugar ribose-5-phosphate. This forms a C_7 sugar, sedoheptulose-7-phosphate, and glyceraldehyde-3-phosphate.

Transfer of the ketol group is brought about by the enzyme system *transketolase*. Thiamin pyrophosphate plays the same sort of role as it did in transferring the acetyl, CH_3CO, group during pyruvic acid decarboxylation (page 467). The ketol group is complexed with TPP in its transfer from xylulose to ribose. Mg^{++} is needed for activation of the enzyme system.

Several ketoses can donate the ketol group, and several aldehydes can accept the ketol group. However, all ketol donors have the C-3 OH group on the left of the carbon chain. The new ketone sugar produced also has this configuration. Transketolase also catalyzes the reaction between xylulose-5-phosphate and erythrose-4-phosphate (discussed below).

SEDOHEPTULOSE-7-PHOSPHATE + GLYCERALDEHYDE-3-PHOSPHATE $\rightleftharpoons$ ERYTHROSE-4-PHOSPHATE + FRUCTOSE-6-PHOSPHATE. Sedoheptulose phosphate and glyceraldehyde phosphate react under the influence of *transaldolase* to form 1 mole each of fructose-6-phosphate and erythrose-4-phosphate.

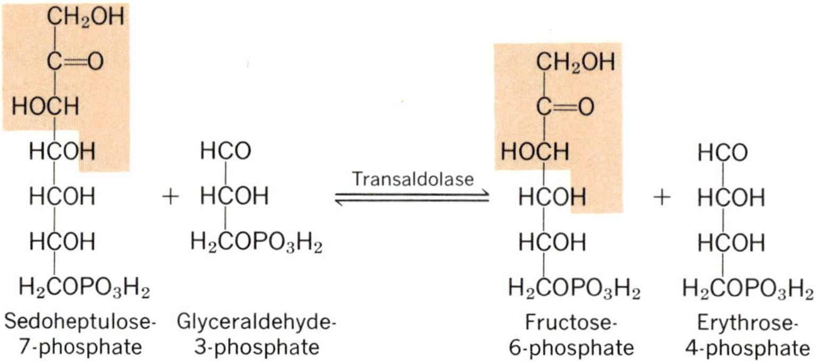

(23.6)

Fructose phosphate can, of course, be converted into glucose-6-phosphate as described in the Embden-Meyerhof pathway.

ERYTHROSE-4-PHOSPHATE + XYLULOSE-5-PHOSPHATE ⇌ FRUCTOSE-6-PHOSPHATE + GLYCERALDEHYDE-3-PHOSPHATE. When 2 moles of xylulose-5-phosphate and 2 moles of erythrose-4-phosphate react, they form 2 moles of fructose-6-phosphate and 2 moles of glyceraldehyde-3-phosphate. This is accomplished by the transfer of the ketol group of xylulose to erythrose.

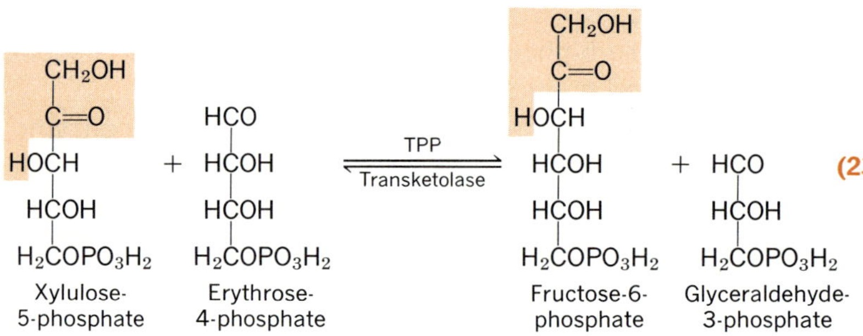

(23.7)

The same cofactors, TPP and Mg^{++}, are needed as for the transketolase-catalyzed reaction (23.5).

The products, glyceraldehyde-3-phosphate and fructose-6-phosphate, are both components of the Embden-Meyerhof pathway. The 2 moles of glyceraldehyde-3-phosphate formed in the balanced equation can be converted into 1 mole of fructose-6-phosphate with the elimination of 1 mole of inorganic phosphate. This completes the pentose phosphate pathway.

Under aerobic conditions the various reactions just presented account for the oxidation of 1 mole of glucose-6-phosphate.

$$\text{Glucose-6-phosphate} + 12\ NADP^+ + 6\ H_2O \longrightarrow 12\ NADPH{+}H^+ + 6\ CO_2 + H_3PO_4$$

$$12\ NADPH{+}H^+ + 6\ O_2 \longrightarrow 12\ H_2O + 12\ NADP^+$$

$$\text{Net:}\quad \text{Glucose-6-phosphate} + 6\ O_2 \longrightarrow 6\ CO_2 + 6\ H_2O + H_3PO_4$$

The energy obtained from 1 mole of glucose-6-phosphate is equivalent to 36 ATP.

23.2 SUMMARY

Although not the major pathway for the oxidation of glucose in most organisms, the pentose phosphate pathway is still significant. It is a means of forming some very important intermediates for cellular synthesis in addition to furnishing ATP.

Attention is called to the net effect of the first three reactions:

$$6\ \text{Glucose-6-phosphate} + 12\ NADP^+ + 6\ H_2O \longrightarrow 6\ \text{ribulose-5-phosphate} + 12\ NADPH{+}H^+ + 6\ CO_2$$

Since the 6 ribulose-5-phosphates are equivalent to 5 glucose-6-phosphates (30 carbons total) the reactions are equivalent to the oxidation of 1 mole of glucose-6-phosphate to CO_2 and $NADPH{+}H^+$; the oxygen is derived from water.

The remainder of the reactions in this pathway are concerned with the conversion of the 6 moles of ribulose-5-phosphate to 5 moles of glucose-6-phosphate.

$$\underset{\text{30 carbons}}{6\ \text{Ribulose-5-phosphate}} \longrightarrow \underset{\text{30 carbons}}{5\ \text{glucose-6-phosphate}}$$

In this process several important intermediates are formed, e.g., xylulose and arabinose for the synthesis of xylans and arabans, ribulose for the photosynthetic process, and ribose for the synthesis of nucleic acids.

In contrast to the Embden-Meyerhof pathway, which yields $NADH{+}H^+$, the pentose pathway forms $NADPH{+}H^+$. Also, the former pathway may operate anaerobically to form lactic acid (or ethanol) from pyruvic acid and thereby reform NAD^+ for perpetuation of the pathway.

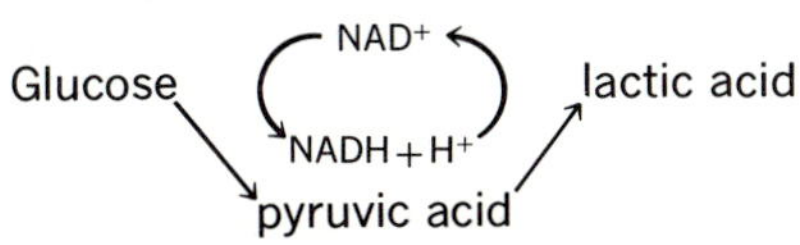

The pentose pathway, however, has no means of oxidizing the NADPH + H^+ formed. Therefore, the extent to which this cycle operates depends upon reactions which utilize the reduced coenzyme. TPNH + H^+ is used in a number of reactions, one of which is the synthesis of fatty acids. Thus fat metabolism is integrated with carbohydrate metabolism. Also, aerobically NADPH + H^+ can be oxidized via the electron-transport system with the concomitant production of ATP.

$$12\ NADPH + H^+ + 6\ O_2 + 36\ ADP + 36\ H_3PO_4 \longrightarrow 36\ ATP + 12\ H_2O + 12\ NADP^+$$

The 12 NADPH + H^+ yields 36 ATP. If we start with free glucose, however, the yield of ATP is 1 less because 1 ATP is required to phosphorylate glucose to glucose-6-phosphate. Thus, the pentose pathway is theoretically capable of yielding 35 ATP per mole of glucose oxidized.

$$\text{Glucose} + 6\ O_2 \xrightarrow[\text{pathway}]{\text{Pentose phosphate}} 6\ CO_2 + 6\ H_2O + 35\ ATP$$

QUESTIONS

1. Is the pentose phosphate pathway for the oxidation of glucose completely independent of the TCA cycle?
2. What are the end products of glucose oxidation via the pentose phosphate pathway?
3. What appear to be the main functions of the pentose phosphate pathway?
4. Assuming all enzyme systems are present and operative, how could you conveniently distinguish whether glucose was being metabolized by the pentose phosphate pathway or by the Embden-Meyerhof pathway?
5. What coenzyme is reduced in the pentose phosphate pathway?
6. How many moles of ATP are theoretically possible from the oxidation of 1 mole of glucose via the pentose phosphate pathway?
7. How is the pentose pathway associated with photosynthesis?
8. Is the coenzyme thiamin pyrophosphate associated with transketolase? With transaldolase?

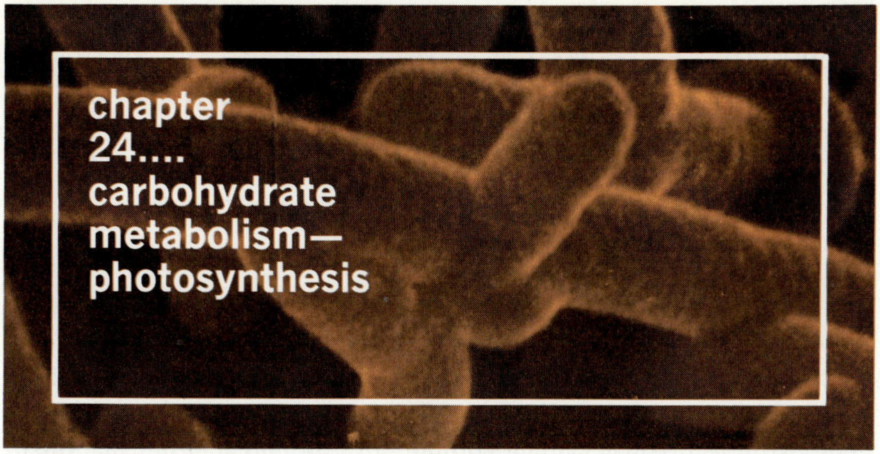

chapter 24.... carbohydrate metabolism—photosynthesis

Reactions that utilize energy derived from light are called *photochemical* reactions. *Photosynthesis* includes reactions associated with reduction of carbon dioxide by the hydrogen of water, or some other electron donor such as H_2S, to carbohydrates with the energy for driving the reactions forward coming from photochemical reactions.

Metabolic reactions discussed thus far have dealt primarily with the oxidation of carbohydrates as a means of forming chemical energy (ATP).

$$\underset{\substack{\text{Carbohydrate}\\ \text{Chemical energy}}}{C(H_2O)} + O_2 \longrightarrow H_2O + CO_2 + \underset{\substack{\text{Chemical}\\ \text{energy}}}{ATP} \tag{24.1}$$

Oxidation: $C(H_2O) \rightarrow CO_2$; Reduction: $O_2 \rightarrow H_2O$

Although many of the intermediate reactions in this process are reversible, the overall reaction is not. The synthesis of carbohydrates from carbon dioxide and water involves photochemical reactions in which energy is derived from light and also a pathway for the fixation of carbon dioxide distinctly different from the carbon dioxide fixation reaction already encountered. The overall reaction for the photosynthesis of carbohydrates can be considered as the reverse of their oxidation,

$$CO_2 + H_2O \xrightarrow{\text{Light energy}} C(H_2O) + O_2 \tag{24.2}$$

Reduction: $CO_2 \rightarrow C(H_2O)$; Oxidation: $H_2O \rightarrow O_2$

Photosynthetic reactions can be divided into two main divisions, the *light reaction* and the *dark reaction*. The light reaction utilizes light energy to produce chemical energy (ATP) and the reducing agent

$NADPH + H^+$, whereas the dark reaction involves the use of ATP to fix carbon dioxide and its subsequent reduction by $NADPH + H^+$ to carbohydrates. Thus in photosynthesis ATP and $NADPH + H^+$ are products of photochemical reactions, whereas in dark reactions ATP and $NADH + H^+$ are products of the oxidation of carbohydrates.

24.1 CHLOROPLASTS

To convert light energy into chemical energy requires specialized cellular components. For example, plants contain the subcellular particles chloroplasts, which are the site of the photochemical reactions of photosynthesis. These cellular components contain chlorophylls, two of which are chlorophyll *a* and chlorophyll *b*, and other pigments, e.g., carotenoids and quinones. The chorophylls give the green color to plants and are mainly responsible for the absorption of light energy, an essential reaction for the first step in photochemical reactions. Chlorophyll contains magnesium and the porphyrin nucleus (see page 303). Chloroplasts also contain the necessary enzymes and cofactors required for the photochemical reactions.

24.2 LIGHT REACTION

Light can be considered as particles of energy called *photons*. The photons of light of a particular wavelength λ contain the same energy E; however, the energy varies inversely with the wavelength of the light, $E \propto 1/\lambda$. Therefore, photons of shorter wavelength contain more energy than photons of the longer wavelength. Often the frequency of light ν, the number of waves passing a given point per second, is used to express the energy of photons. The frequency of light is equal to the velocity of light c, 3×10^{10} cm/sec, divided by the wavelength ($\nu = c/\lambda$).

$$E \propto \frac{1}{\lambda} \qquad E = \frac{hc}{\lambda}$$

but

$$\frac{c}{\lambda} = \nu$$

therefore

$$E = h\nu$$

where λ = wavelength
h = Planck's constant
c = velocity of light = 3×10^{10} cm/sec
ν = frequency of light

Because the energy content of a single photon is very small, light energy is usually expressed in *Einstein units. An Einstein unit is the energy contained in one mole of photons, or* 6.06×10^{23} *photons.* When energy is expressed in calories, the energy of light can be expressed in calories per mole of photons. The energy associated with several wavelengths of light is shown in Table 24.1.

TABLE 24.1 ENERGY CONTENT OF LIGHT

Color	Wavelength, mμ	Einstein, or kcal/mole
Ultraviolet	< 400	
	400	71
Blue	424–490	
	490	57
Yellow	575–585	
	580	49
Red	647–700	
	700	40
Infrared	> 700	
	750	38

When photons of high enough energy collide with an atom, the energy of the photons is transferred to one or more electrons of the atom and produces an excited electron. This excited, or energized, electron has more energy than it did originally in its *ground state* and therefore is pushed into an orbit further away from the positive charged nucleus.

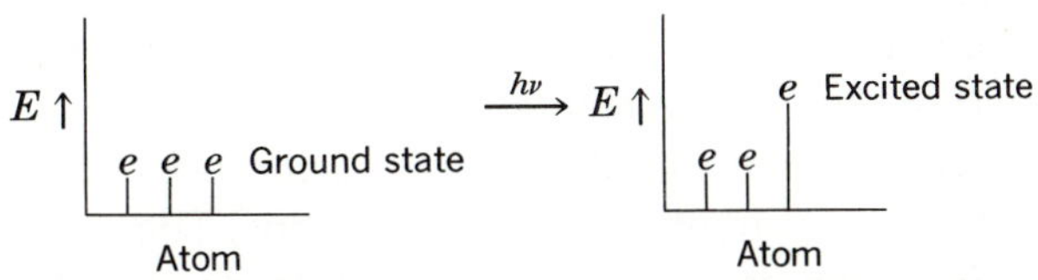

The excited electron may return to its normal ground state either directly or indirectly (see Fig. 24.1). If the electron falls back directly into its original orbit in the atom, the energy released is given off as light, *fluorescense,* and heat (Fig. 24.1a). The wavelength of the fluo-

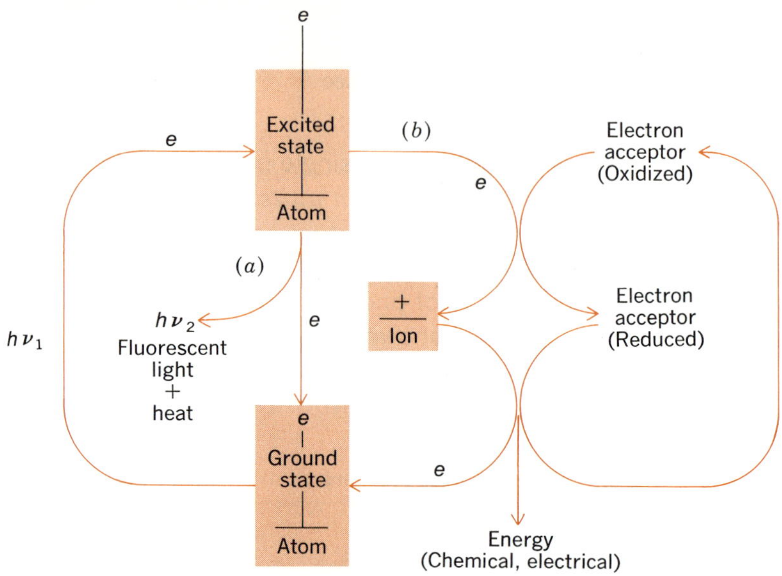

FIGURE 24.1 The possible fate of an excited electron.

rescent light is *always* longer than the wavelength of the incident light that activated the electron. Therefore, there is less energy associated with the fluorescent light than with the incident light

$$h\nu_{\text{inc}} > h\nu_{\text{fluor}}$$

this difference in energy is converted into heat.

The activated electron may return to its ground state by a more circuitous route than above (Fig. 24.1*b*). For example, the excited electron may become so energized that it is easily lost to an electron acceptor and the atom converted into an ion with a positive charge. The electron may be returned to its original position by way of a series of reactions which produce chemical or electric energy.

A photoelectric cell is a simple device which makes use of this phenomenon (Fig. 24.2). The element cesium contains electrons which are easily activated upon the absorption of photons. The stronger the light the greater the number of excited electrons. The excited electrons are removed from the cesium by a very thin layer of platinum or gold, with the resulting formation of positive cesium ions and a negatively charged platinum or gold cathode. In this step light energy has been converted into electric energy. The activated electrons are passed along a copper wire eventually back to the cesium ions. If a sensitive ammeter is placed in series in this circuit (Fig. 24.2*b*), the flow of electrons (current) can be expressed in terms of mechanical energy to indicate the intensity of light striking the photoelectric cell. The electric energy from photoelectric cells can be used to operate appropriate electronic equipment or can be converted into other forms of energy by the use of appropriate devices, e.g., the conversion of electric energy into light energy (Fig. 24.2*c*) or the conversion of electric energy into chemical energy by the electrolysis of water (Fig. 24.2*d*). During electrolysis of water the current is transported between the electrodes by the aqueous medium, and molecular hydrogen and oxygen are produced.

There are several basic similarities between the simple photoelectric cell and the process of photosynthesis. In photosynthesis it is chlorophyll that receives the light energy and becomes activated. The activated electron from chlorophyll is removed by an electron acceptor, with the resulting formation of ionized chlorophyll. The electron is returned at its ground state to the chlorophyll ion by a series of reactions which produce chemical energy (ATP). Water is involved in the transfer of electrons and during this process is decomposed into molecular oxygen, the hydrogen being trapped in $NADPH + H^+$. The electron acceptor and the transport system of photosynthesis are composed of organic compounds, some of which contain the mineral elements iron and copper.

During photosynthesis the activated electron from chlorophyll may be returned to the chlorophyll ion by either of two pathways,

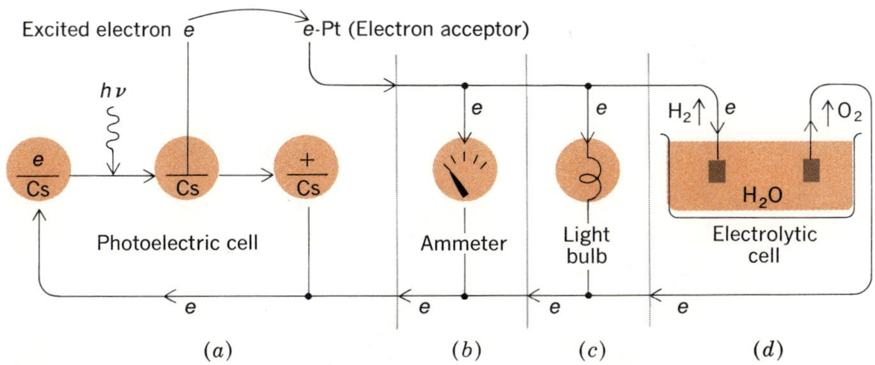

FIGURE 24.2 Conversions of energy from an electron: (*a*) light energy → electric energy; (*b*) electric energy → mechanical energy; (*c*) electric energy → light and heat energy; (*d*) electric energy → chemical energy.

cyclic photophosphorylation or *noncyclic photophosphorylation.* Photophosphorylation is the process whereby light energy is used to produce ATP from ADP and H_3PO_4.

24.3 CYCLIC PHOTOPHOSPHORYLATION

A schematic summary of cyclic photophosphorylation is given in Fig. 24.3, and a more detailed representation is given in Fig. 24.4, system I.

Light is absorbed by chlorophyll (Chl), the chlorophyll becomes activated (Chl*), and the activated electron is transferred to a series of electron carriers. Ionized chlorophyll (Chl^+) is the other product of this photochemical reaction. During the passage of the electron along the electron-transport system back to Chl^+ to reform Chl, the energy is utilized to phosphorylate ADP to ATP. This mechanism does not involve the utilization of substrates such as water but merely the transfer of electrons back to chlorophyll with a concomitant production of ATP.

Cyclic photophosphorylation is briefly summarized in Fig. 24.4, system I. For our discussion only the most probable pathway of electron transport is presented. The actual system is much more complex, and not all the intermediate steps have been elucidated. The light-absorbing material in this system consists chiefly of chlorophyll a and a special chlorophyll a referred to as P700 pigment because it absorbs light of about 700 mμ wavelength. System I is activated by light of wavelength greater than 680 mμ. The activation of electrons and their acceptance by an electron acceptor X (the photochemical step) is illustrated by a heavy line. Because exact nature of substance X is not known (it may be a type of iron-protein complex known as ferrodoxin), for the present, this receptor is indicated merely as X.

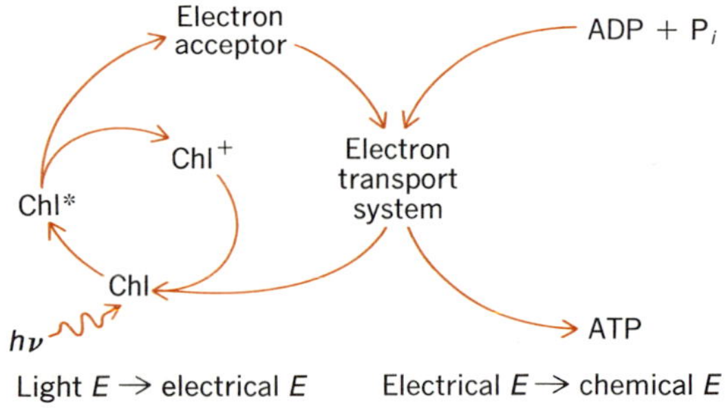

FIGURE 24.3 Schematic summary of cyclic photophosphorylation.

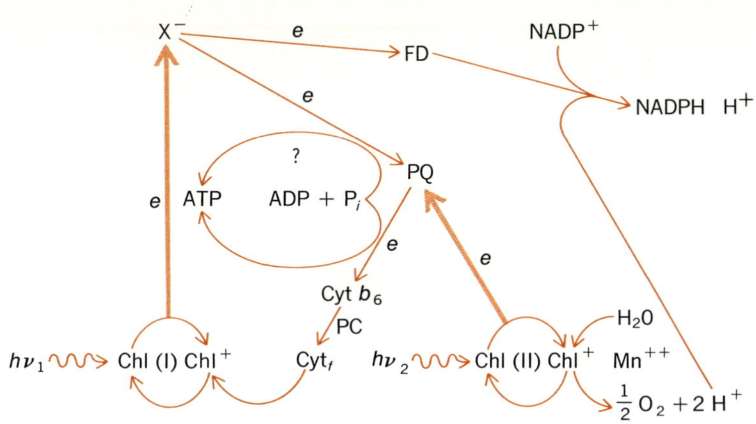

FIGURE 24.4 Photosynthesis. (I) = system I; Chl = chlorophyll a plus "special" chlorophyll a (P700); $h\nu$ = light, wavelength $\lambda > 680$ mμ. (II) = system II; Chl = chlorophyll a and b ($b > a$) plus accessory pigments; $h\nu$ = light, wavelength $\lambda > 680$ mμ; X^- = unknown electron acceptor, may or may not be a ferrodoxin; PQ = plastoquinone, protein-quinone complex; Cyt b_6 and Cyt$_f$ = cytochromes; PC = plastoquinone, copper-containing protein; FD = ferrodoxin, non-heme-iron–containing protein; the heavy lines represent photochemical reactions; an $\xrightarrow{e}$ indicates the flow of electrons. Phosphorylation takes place when electrons are passed from PQ to Chl$^+$(I); phosphorylation may take place in plants as electrons are passed from X^- to PQ.

Acceptor X is reduced (X^-) upon accepting an electron and is subsequently oxidized to X when the electron is passed to a protein-quinone substance called plastoquinone (PQ). During the passage of the electron from X^- to PQ phosphorylation of ADP is thought to occur. While there is some doubt of the importance of this phosphorylation step in green plants, it is a significant energy-conservation step in photosynthesizing bacteria. From PQ the electron is passed along an electron-transport system that involves the oxidation-reduction of cytochrome b_6 (Cyt b_6), plastocyanin (PC), and cytochrome f (Cyt$_f$). From Cyt$_f$ the electron is returned to Chl$^+$, reforming the initial unactivated Chl. Phosphorylation of ADP to ATP also occurs, probably during the transfer of electrons from PQ to Cyt b_6. It is of special interest to note that PC is a copper-protein complex containing two atoms of copper per mole.

24.4 NONCYCLIC PHOTOPHOSPHORYLATION

Noncyclic photophosphorylation is indicated by system II (Fig. 24.4). The pigments of this system consist of chlorophyll a and b, carotenoids, and quinones; the latter two groups are collectively called *accessory*

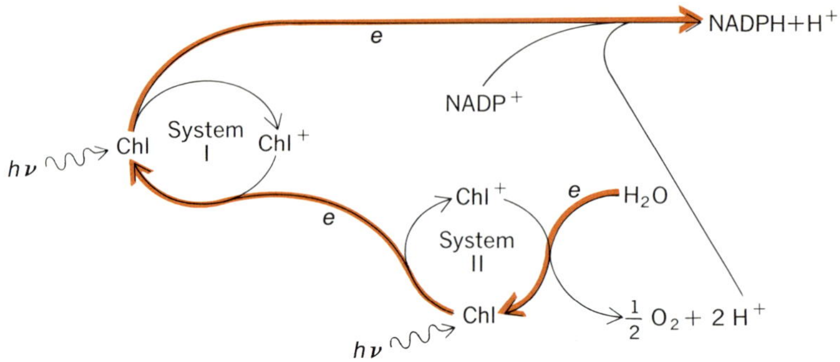

$$NADP^+ + H_2O \xrightarrow{h\nu} NADPH + H^+ + \frac{1}{2}O_2$$

FIGURE 24.5 Summary of electron pathway during photosynthesis.

pigments. System II absorbs light of shorter wavelength ($\lambda < 680$ mμ) than system I. The electron acceptor of system II is PQ, and the electron is returned to Chl$^+$ of system I by way of the electron-transport system mentioned above. Thus, when system II is functioning, the electron from system I Chl must be used to reduce a substance other than PQ, and Chl$^+$ of system II must receive an electron from an outside source in order to be reduced.

First consider the reduction of system II Chl$^+$. The electron donor is oxygen of water.

$$H_2O \xrightarrow{h\nu} 2\,H^+ + \tfrac{1}{2}O_2 + 2e$$

Oxygen is oxidized and the electrons are used to reduce system II Chl$^+$ to Chl. This completes the cycle with respect to system II Chl. The fate of the H$^+$ from water is tied in with the electron-accepting compound of system I.

During photosynthesis the electron produced by system I is passed from X$^-$ to a non-heme-iron–protein compound called ferrodoxin (FD) and thence to NADP$^+$. The reduction of NADP$^+$ to NADP is simultaneous with the union of H$^+$ from water (system II) with NADP to produce NADPH + H$^+$. Thus the electron from system I and the H$^+$ from system II are used to produce NADPH + H$^+$ from NADP$^+$; molecular oxygen is a product of this reaction.

A summary of the pathway of electrons during photosynthesis is shown in Fig. 24.5. The overall reactions for cyclic and noncyclic photophosphorylation are as follows:

Cyclic: $ADP + H_3PO_4 \xrightarrow{h\nu} ATP + H_2O$

Noncyclic: $NADP^+ + H_2O + ADP + H_3PO_4 \xrightarrow{h\nu}$

$$\tfrac{1}{2}\,O_2 + ATP + NADPH + H^+ + H_2O$$

In the cyclic reaction ATP and H_2O result from the reaction of ADP and H_3PO_4. This is also true in the noncyclic reaction, in which $NADPH + H^+$ and O_2 are also formed; however, water is a substrate in this reaction, and the oxygen liberated is all derived from water.

24.5 DARK REACTION

The fixation and reduction of carbon dioxide to form carbohydrate is independent of light, and this mechanism is therefore known as the *dark reaction.* The dark reaction is dependent upon the light reaction for a source of energy (ATP) and for a source of reducing power ($NADPH + H^+$).

24.6 PATH OF CARBON

The path of carbon in photosynthesis involves a cycle that incorporates part of the glycolytic pathway and part of the pentose phosphate pathway. The cyclic nature of carbohydrate formation is briefly illustrated in Fig. 24.6, which also shows the net synthesis accomplished by one turn of the cycle. The cycle begins with a product of the pentose phosphate pathway, ribulose-5-phosphate: 3 moles of ribulose-5-phosphate is converted into 3 moles of ribulose-1,5-diphosphate, which is the acceptor of carbon dioxide. The fixation of 3 moles of carbon dioxide into the 3 ribulose diphosphates results in the formation of 6 moles of 3-phosphoglyceric acid, 5 moles of which can be converted into 3 moles of ribulose-5-phosphate, thus assuring continuation of the cycle. The remaining mole of 3-phosphoglyceric acid (equivalent to 1 mole of glyceraldehyde-3-phosphate or $\frac{1}{2}$ mole of hexose) represents the net synthesis of a triose. A more detailed diagram of the carbon cycle is given in Fig. 24.7, in which the components of the glycolytic and the pentose phosphate pathway are indicated. The conversion of 3-phosphoglyceric acid to glyceraldehyde-3-phosphate represents a partial reversal of glycolysis. It is important to note that in photosynthesis $NADPH + H^+$ and ATP are involved in this conversion, whereas in glycolysis coenzyme $NADH + H^+$ is used.

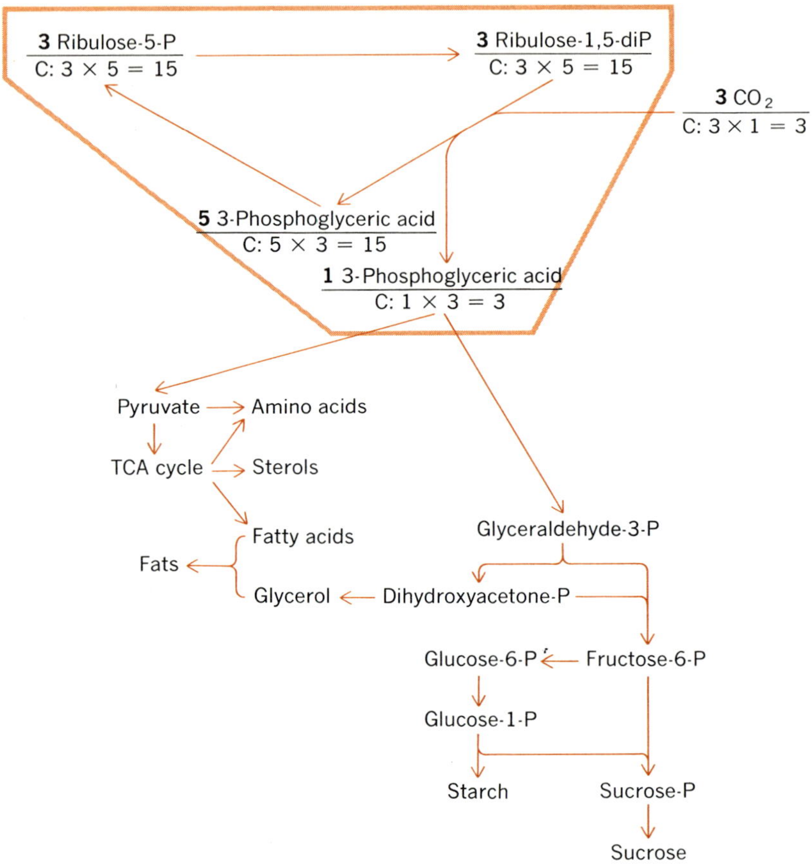

FIGURE 24.6 Abbreviated cyclic nature of carbon reduction.

RIBULOSE-5-PHOSPHATE ⟶ RIBULOSE-1,5-DIPHOSPHATE. Ribulose-5-phosphate is phosphorylated to ribulose-1,5-diphosphate by the enzyme *phosphopentokinase*. The ATP required for the reaction comes from the pho-

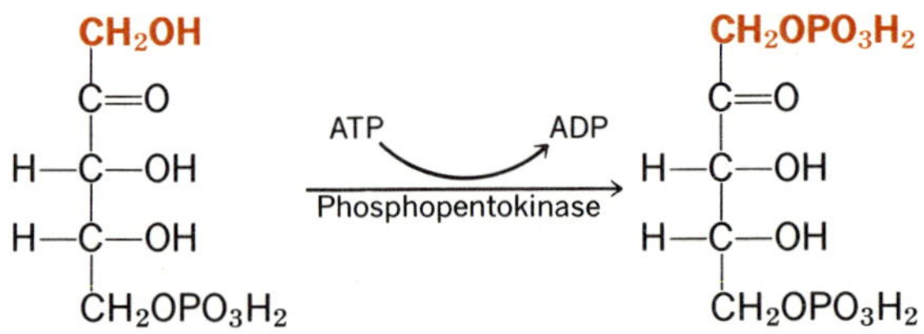

Ribulose-5-phosphate Ribulose-1,5-diphosphate (24.3)

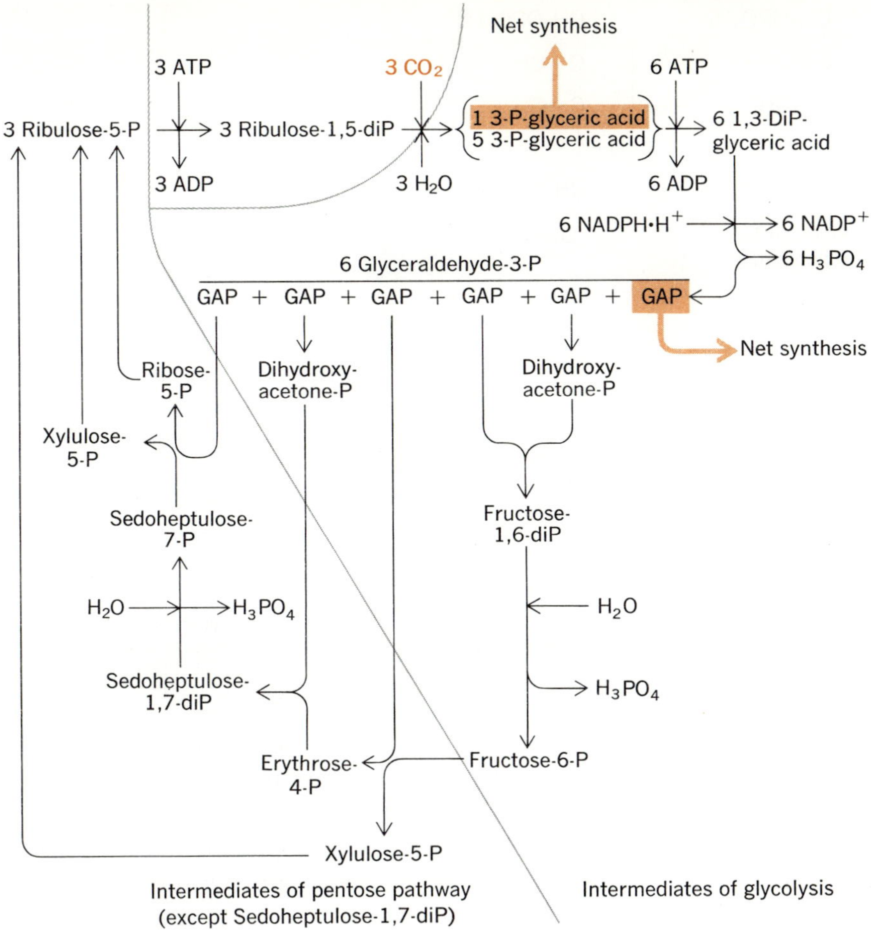

FIGURE 24.7 Reduction of carbon in photosynthesis.

tophosphorylation reactions and the ribulose-5-phosphate from the pentose phosphate pathway (see page 543).

CARBON DIOXIDE FIXATION. Ribulose-1,5-diphosphate is the acceptor of carbon dioxide in photosynthesis. An unstable addition product is initially formed which spontaneously hydrolyzes to yield 2 moles of 3-phosphoglyceric acid. This overall reaction is carried out by the enzyme *carboxydismutase*. 3-Phosphoglyceric acid is an intermediate in the glycolytic pathway and may be metabolized to other products.

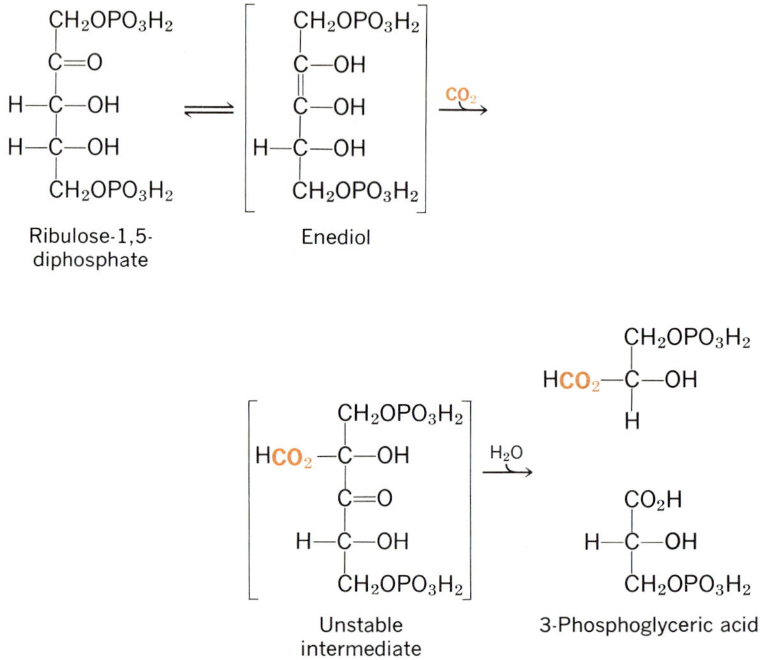

(24.4)

Thus, the incorporation of 1 mole of carbon dioxide into 1 mole of ribulose diphosphate results in the net synthesis of $\frac{1}{3}$ mole of glyceric acid phosphate. It would require 3 moles of ribulose diphosphate for the net synthesis of 1 mole of glyceric acid phosphate; 5 moles of phosphoglyceric acid would also be formed. These 5 moles of phosphoglyceric acid may enter into a series of reactions which reform 3 moles of ribulose-5-phosphate, thus completing and perpetuating the cycle. The remainder of the reactions to be discussed are geared to the conversion of phosphoglyceric acid to ribulose-5-phosphate.

3-PHOSPHOGLYCERIC ACID ⟶ GLYCERALDEHYDE-3-PHOSPHATE. Phosphorylation of 3-phosphoglyceric acid to 1,3-diphosphoglyceric acid is an intermediary step in the conversion of 3-phosphoglyceric acid to glyceraldehyde-3-phosphate by the enzyme system *phosphotriose dehydrogenase.* $NADPH + H^+$ is the coenzyme required for the reduction of the carboxyl group to the aldehyde group. As mentioned above, this reaction is similar to that in glycolysis except that in glycolysis $NADH + H^+$ is used instead of $NADPH + H^+$.

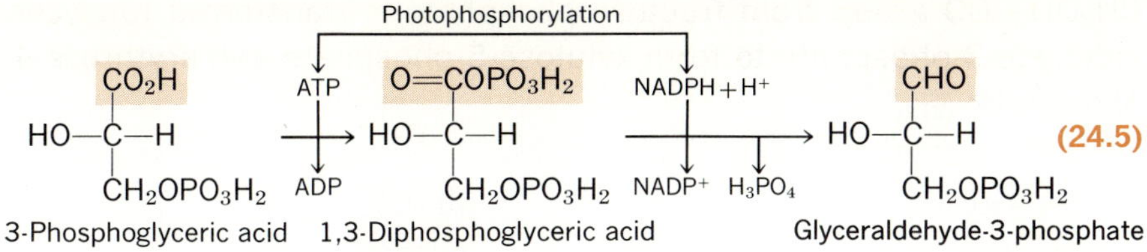

(24.5)

The reduction of 3 moles of carbon dioxide results in the formation of 6 moles of glyceraldehydephosphate, 1 mole of which represents a net synthesis of a triose.

GLYCERALDEHYDE-3-PHOSPHATE ⟶ DIHYDROXYACETONE PHOSPHATE. As shown in Fig. 24.7, part of the glyceraldehydephosphate is converted into dihydroxyacetone phosphate and part to fructose-6-phosphate. These interconversions are identical with the ones encountered in glycolysis:

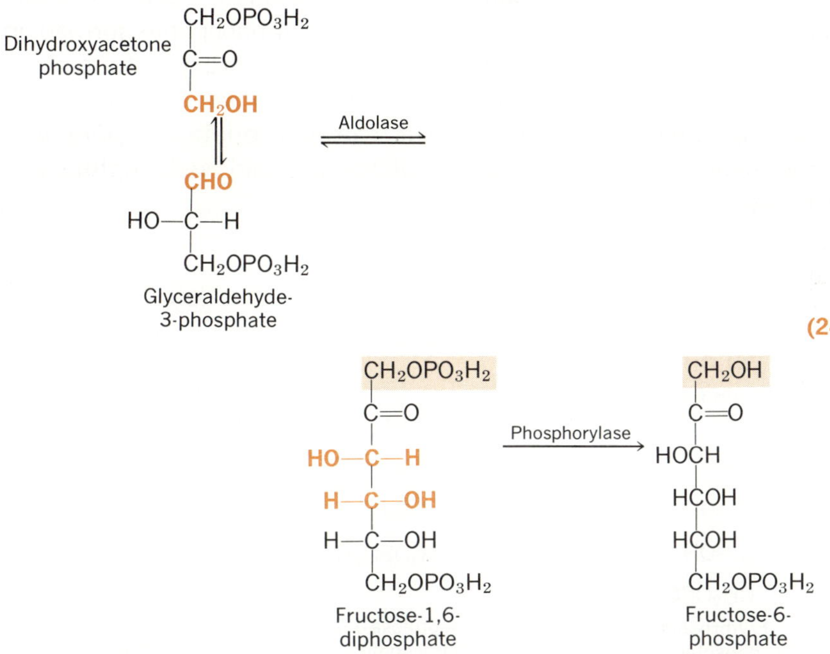

(24.6)

FRUCTOSE-6-PHOSPHATE + GLYCERALDEHYDE-3-PHOSPHATE ⟶ ERYTHROSE-4-PHOSPHATE + XYLULOSE-5-PHOSPHATE. Fructose-6-phosphate and glyceraldehyde-3-phosphate react under the influence of the enzyme *transketolase* to yield erythrose-4-phosphate and xylulose-5-phosphate. These two products are components of the pentose phosphate pathway. The

$CH_2OH—CO$ group from fructose phosphate is transferred to glyceraldehyde-3-phosphate to form xylulose-5-phosphate and erythrose-4-phosphate:

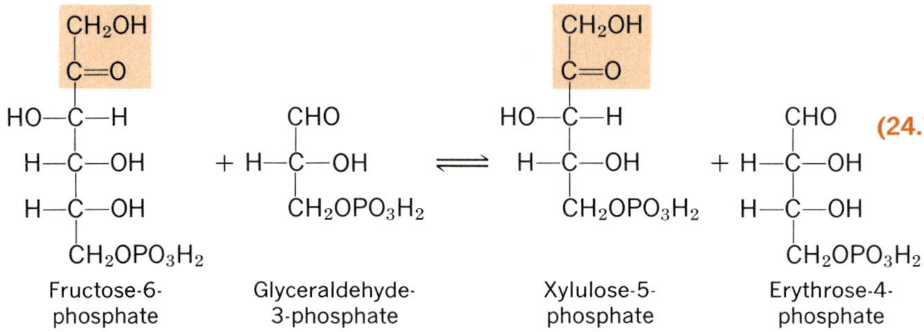

(24.7)

ERYTHROSE-4-PHOSPHATE + DIHYDROXYACETONE PHOSPHATE ⟶ SEDOHEPTULOSE-7-PHOSPHATE. The four-carbon sugar erythrose-4-phosphate formed in the previous reaction is caused to react with dihydroxyacetone phosphate under the influence of the enzyme *transaldolase*. Sedoheptulose-1,7-diphosphate is the product formed. The appropriate *phosphorylase* enzyme hydrolyzes off the 1-phosphate to yield sedoheptulose-7-phosphate:

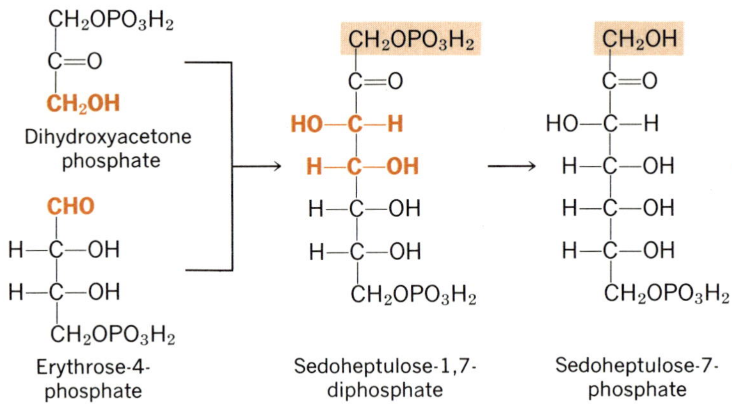

(24.8)

SEDOHEPTULOSE-7-PHOSPHATE ⟶ PENTOSE PHOSPHATES. Sedoheptulose-7-phosphate reacts with glyceraldehyde-3-phosphate to yield 1 mole each of ribose-5-phosphate and xylulose-5-phosphate. The enzyme *transketolase* catalyzes this reaction.

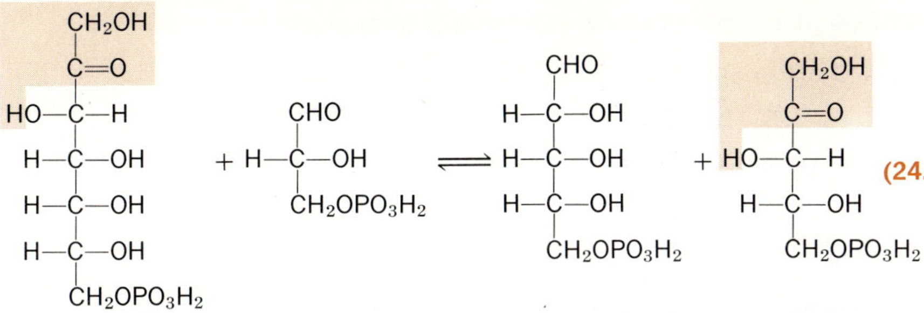

(24.9)

Sedoheptulose-7-phosphate Glyceraldehyde-3-phosphate Ribose-5-phosphate Xylulose-5-phosphate

Ribose-5-phosphate and xylulose-5-phosphate are readily converted to ribulose-5-phosphate by the enzymes *phosphoriboisomerase* and *epimerase,* respectively.

24.7 SUMMARY

In these reactions 5 moles of glyceraldehyde-3-phosphate has been converted into 3 moles of ribulose-5-phosphate, and the carbon reduction cycle is complete with a net synthesis of a triose resulting from the fixation and reduction of 3 moles of carbon dioxide. The source of energy, ATP, and the reducing agent, NADPH+H^+, are obtained from noncyclic photophosphorylation reactions. The net reaction for the

TABLE 24.2 SUMMARY OF CARBON REDUCTION CYCLE (DARK REACTIONS)

Reaction
3 Pentosephosphate + 3 ATP ⟶ 3 ribulose diphosphate + 3 ADP
3 Ribulose diphosphate + 3 CO_3 + 3 H_2O ⟶ 6 phosphoglycerate
6 Phosphoglycerate + 6 ATP ⟶ 6 diphosphoglycerate + 6 ADP
6 Diphosphoglycerate + 6 NADPH+H^+ ⟶ 6 glyceraldehyde-3-phosphate + $NADP^+$ + 6 H_3PO_4
2 Glyceraldehyde-3-phosphate ⟶ 2 dihydroxyacetone phosphate
Dihydroxyacetone phosphate + glyceraldehyde-3-phosphate ⟶ fructose diphosphate
Fructose diphosphate + H_2O ⟶ fructose phosphate + H_3PO_4
Fructose phosphate + glyceraldehyde-3-phosphate ⟶ tetrose phosphate + pentose phosphate
Tetrose phosphate + dihydroxyacetone phosphate ⟶ sedoheptulose diphosphate
Sedoheptulose diphosphate + H_2O ⟶ sedoheptulose phosphate + H_3PO_4
Sedoheptulose phosphate + glyceraldehyde-3-phosphate ⟶ 2 pentose phosphate
Overall: 3 CO_2 + 9 ATP + 5 H_2O + 6 NADPH+H^+ ⟶ glyceraldehyde-3-phosphate + 9 ADP + 8 H_3PO_4 + 6 $NADP^+$

synthesis of 1 mole of glyceraldehyde-3-phosphate is

$$3\ CO_2 + 9\ ATP + 5\ H_2O + 6\ NADPH + H^+ = 1\ \text{glyceraldehyde-3-phosphate} + 9\ ADP + 6\ NADP^+ + 8\ H_3PO_4$$

A summary of the individual reactions is presented in Table 24.2.

QUESTIONS

1. What is the relationship between the energy of light and its frequency?
2. What is meant by an atom in its *ground* state and an atom in its *excited* state?
3. How is fluorescent light produced?
4. What is the relationship of the wavelength of fluorescent light to the wavelength of the incident light?
5. In plants what compounds receive light and become activated?
6. What is meant by the term photophosphorylation?
7. What is the significant difference between cyclic and noncyclic photophosphorylation?
8. Through what common intermediate are cyclic and noncyclic photophosphorylation connected?
9. There are two systems associated with photosynthesis in plants, namely, system I and system II. Each system liberates electrons upon illumination. What is the ultimate acceptor of the electrons from each system?
10. Which metabolic pathway furnishes ribulose phosphate for photosynthesis?
11. What is the overall reaction associated with the dark reaction of photosynthesis?
12. What is the first stable product of photosynthetic carbon dioxide fixation?
13. What is the formula and name of the compound that accepts carbon dioxide during photosynthesis?
14. ATP and $NADPH + H^+$ are products of photophosphorylation. In what reactions are these two products used during photosynthesis?

chapter 25.... lipid metabolism

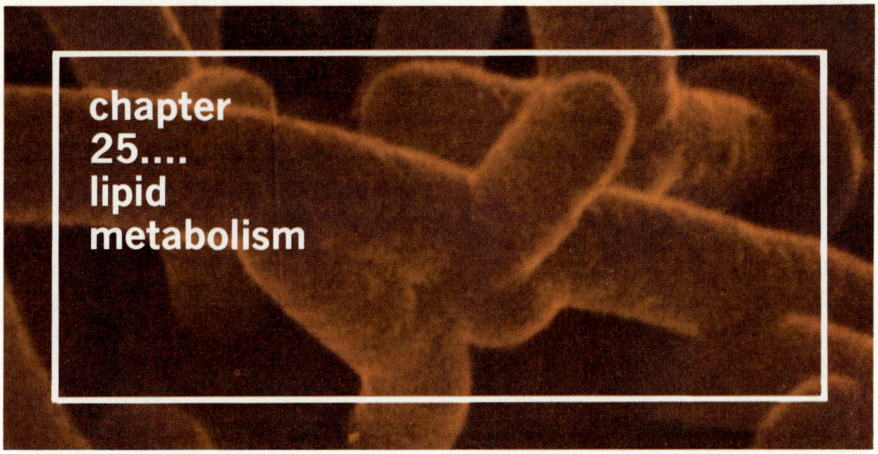

Lipids are a heterogeneous class of compounds defined according to their chemical and physical properties. Lipids are fat-soluble, water-insoluble esters or potential esters that are biologically active (Chap. 15). The last criterion refers to the fact that lipids are synthesized or utilized (or both) by living organisms. Lipids may be simple esters, phosphoesters, or high-molecular-weight compounds like the sterols or carotenes. Some are biologically active as vitamins, vitamins A, D, E, and K being examples of this class of lipid. With respect to lipid metabolism, we shall be mainly concerned with the oxidation and synthesis of triglycerides (neutral fats), which are esters of glycerol and fatty acids.

Triglycerides are stored in body tissue as a source of energy. Glycogen, the immediate energy reserve material, is not stored in large amounts. The liver contains about 5% glycogen and the skeletal muscles about 0.5%. This somewhat limited means of storing carbohydrates necessitates the conversion of excess carbohydrate into another form of stored energy, namely, fat. When reserve energy is needed, it is furnished by glycogen first and then by fat as the glycogen level of tissues decreases. Glycogen is replenished from dietary sugar but not from fatty acids. Fatty acids *cannot* be converted into glycogen in animal tissue.

Upon oxidation, fats furnish more energy on a weight basis than glucose. The body derives 9000 cal/g from fat, whereas carbohydrates and protein each yield 4000 cal/g.

Lipids also serve other functions. For example, the subcutaneous layer of fat serves as an insulation against cold. Such organs as the kidney and spleen are protected against physical shock by a layer of fat. Cellular membranes contain phospholipids as well as neutral lipids.

Lipids are associated with proteins in lipoproteins, which are important, for example, in blood clotting, and with carbohydrates as glycolipids in nerve tissue.

Most of the types of reactions involved in the metabolism of fatty acids have been discussed in previous sections. The following questions serve as a review of these reactions. In the conversion of succinic acid to fumaric acid, what coenzyme is needed? What type of reaction is associated with the conversion of fumaric acid to malonic acid? What coenzyme picks up the hydrogen in the conversion of malonic acid to oxaloacetic acid? Does the conversion of acetic acid to acetyl-SCoA produce a high-energy compound? How many ATPs are produced upon the complete oxidation of one acetyl unit? What nonphotosynthetic carbon dioxide fixation reaction has already been discussed? What two compounds in the TCA cycle contain secondary alcohol groups that are dehydrogenated to ketone groups? In the animal body can acetyl-SCoA be converted into a net increase in glycogen? What products are obtained upon the complete hydrolysis of lecithin? Are acetoacetic acid and β-ketobutyric acid two names for the same compound?

25.1 OXIDATION OF LIPIDS

β oxidation

The liver is the major site of fatty acid metabolism. Elucidation of the β-oxidation mechanism for oxidation of fatty acids evolved from the initial observations of Knoop in 1904. Knoop fed phenyl-substituted low-molecular-weight acids to animals and determined the phenyl-substituted acids excreted in the urine. When the phenyl derivative of the acids with an even number of carbons was fed to animals, the phenyl group appeared in the urine as phenylacetic acid, but, when the phenyl group was attached to the odd-numbered carbon acids, benzoic acid appeared as the end product of metabolism.

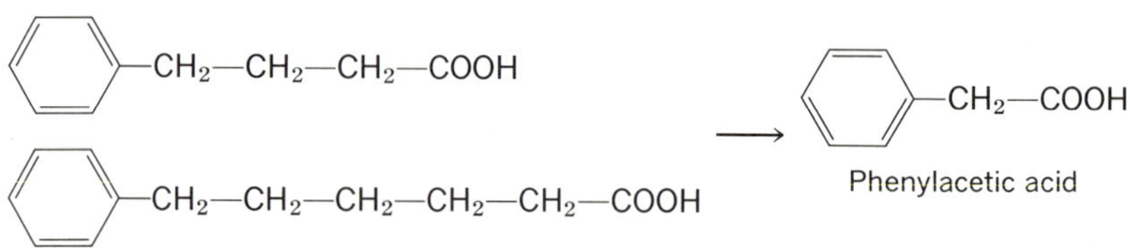

Phenylacetic acid

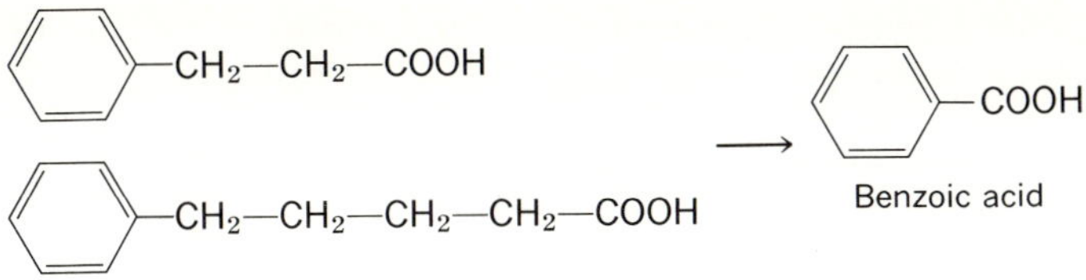

Thus it appeared that fatty acids are metabolized by splitting off two carbons at a time, starting with the COOH end of the acid. This necessitates breaking a carbon-to-carbon bond between the β and γ carbon; hence, the name β oxidation. It was not until 1952 that the enzyme systems for the reactions were isolated. There are five enzymes associated with the oxidation of fatty acids. A summary of the reactions involved is presented in Fig. 25.1.

Activation

The first step in fatty acid metabolism is the conversion of the acid to the CoA complex, acyl-SCoA. This is known as *activating the acid* and *requires energy,* which is derived from ATP. ATP is converted into AMP and pyrophosphate, PP_i. The enzyme *thiokinase,* which requires the mineral ion Mg^{++} for activity, catalyzes this activation reaction.

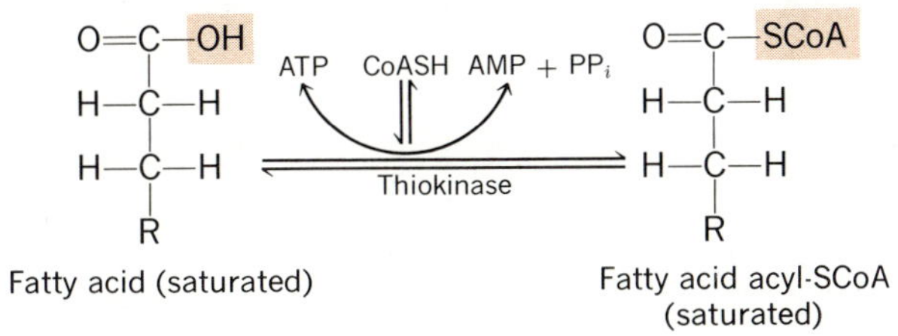

(25.1)

Attention is called to the use of an ATP for the activation of fatty acids. This is the only place in the β-oxidation scheme where a high-energy phosphate is required regardless of the number of carbons in the fatty acid.

Dehydrogenation

The next step is oxidation of the carbon chain via dehydrogenation to form a double bond between the α and β carbons. The enzyme *acyl dehydrogenase* catalyzes this reaction and the flavin coenzyme FAD is the hydrogen acceptor.

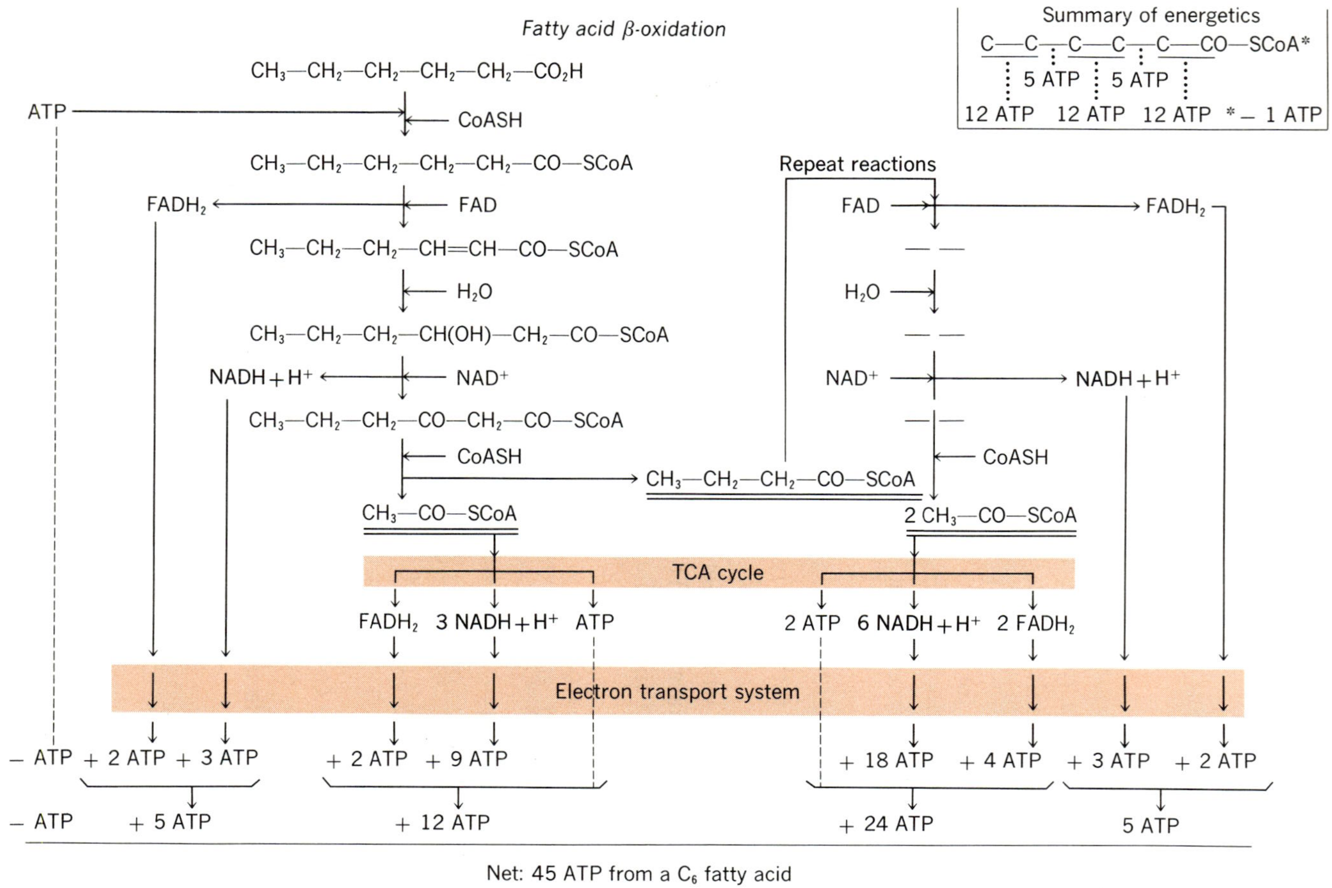

FIGURE 25.1 Fatty acid β oxidation.

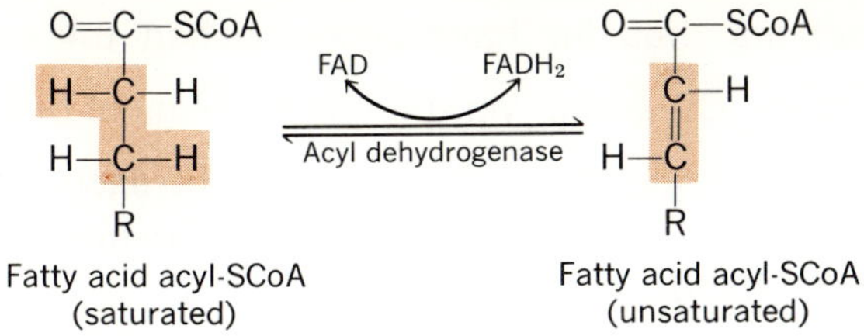

(25.2)

This reaction is similar to the one for converting succinic acid to fumaric acid in the tricarboxylic acid cycle. Also, as in the formation of fumaric acid, the trans configuration exists about the double bond. The next two steps of fatty acid oxidation are also similar to reactions already studied with respect to the tricarboxylic acid cycle, namely, fumaric acid $\longrightarrow$ malic acid $\longrightarrow$ oxaloacetic acid.

Hydration of the double bond

The third step in fatty acid oxidation is the hydration of the double bond to form a secondary alcohol group on the β carbon. *Enol hydrase* is the enzyme responsible for this conversion.

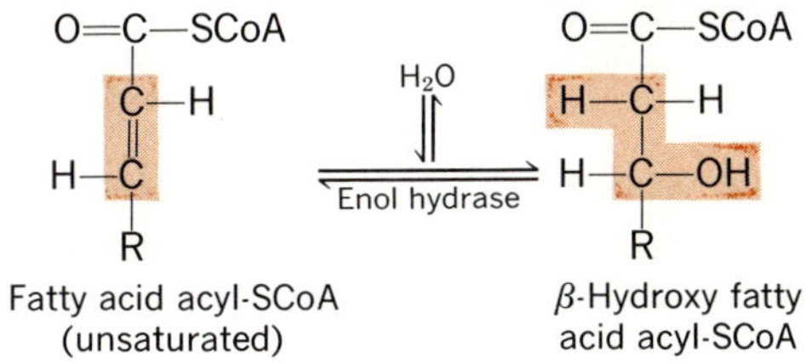

(25.3)

Oxidation of alcohol group

Oxidation of the secondary alcohol group produces a ketone group on the β carbon. This is accomplished by the removal of the hydrogen by the enzyme *β-hydroxylacyl dehydrogenase* and the coenzyme NAD^+.

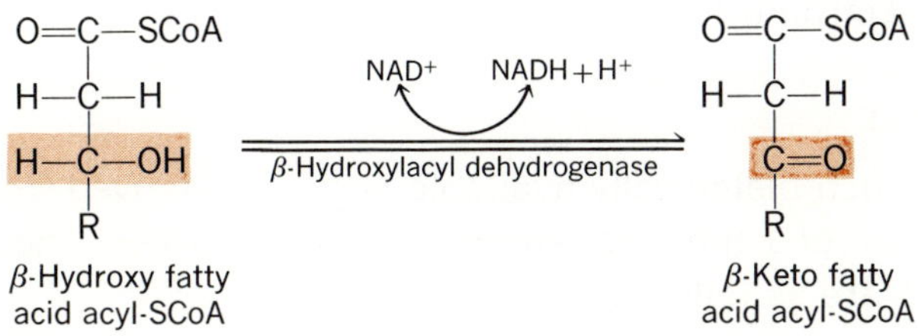

(25.4)

The reactions thus far have accomplished the oxidation of a β CH_2 group to a $>$CO group, via the mechanisms of dehydrogenation and hydration, and the formation of 1 mole each of $FADH_2$ and NADH + H^+. The next reaction in β oxidation involves cleavage of a carbon-to-carbon bond.

Carbon-to-carbon cleavage

The last step is the splitting off of a C_2 unit by cleavage of the bond between the α and β carbons. This is accomplished by the enzyme *β-ketoacyl thiolase* and CoASH. Acetyl-SCoA is one product, and the acyl-SCoA derivative of an acid containing *two carbons less* than the original acid is the other product.

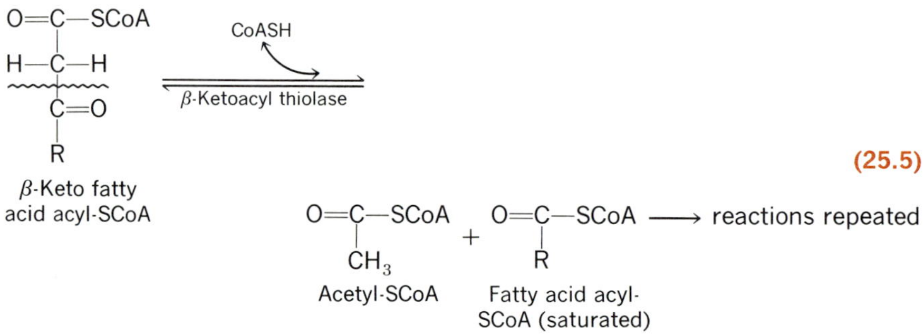

(25.5)

Note that the new acyl-SCoA is already activated and can enter the scheme without requiring an ATP for activation. The acetyl-SCoA formed enters the tricarboxylic acid cycle and is metabolized to carbon dioxide and water.

Energetics

For each carbon bond cleaved 1 mole each of $FADH_2$ and NADH + H^+ is formed. Oxidation of these via the electron-transport system yields 2 and 3 moles of ATP, respectively. Thus for each carbon bond cleaved 5 ATP are formed. Each acetyl-SCoA unit metabolized via the TCA cycle and electron-transport system yields 12 ATP.

EXAMPLE

How much biologically available energy is derived from the oxidation of 1 mole of caproic acid, $C_5H_{11}COOH$? The calculations are as follows:

C—C bonds cleaved = 2:	2 × 5 =	10 ATP
Acetyl-SCoA formed = 3:	3 × 12 =	36 ATP
	Total =	46 ATP
ATP used to activate fatty acid	=	−1 ATP
Net ATP produced/mole caproic acid	=	45 ATP

When 1 mole of glucose is oxidized, a net of 38 ATP is formed. This is 7 ATP less than is produced when a C_6 fatty acid is oxidized.

Calculate the number of moles of ATP produced when stearic acid $C_{17}H_{35}COOH$ is oxidized. Compare this value with the number of ATP produced during the oxidation of 3 moles of glucose (18 carbons total). Stearic acid yields 147 net ATP, whereas 3 moles of glucose yields only 114 ATP.

25.2 ODD-NUMBERED FATTY ACID OXIDATION

There are several odd-numbered fatty acids, e.g., propionic acid and valeric acid. The initial reactions for their oxidations are similar to those of the even-numbered fatty acids, but as the acetyl-SCoA units are progressively split off, eventually a three-carbon acyl unit, propionyl-SCoA, is formed. The difference in the oxidation of the odd-numbered carbon acids lies in the mechanism for the oxidation of the propionyl-SCoA unit. The following discussion also applies to the oxidation of propionic acid.

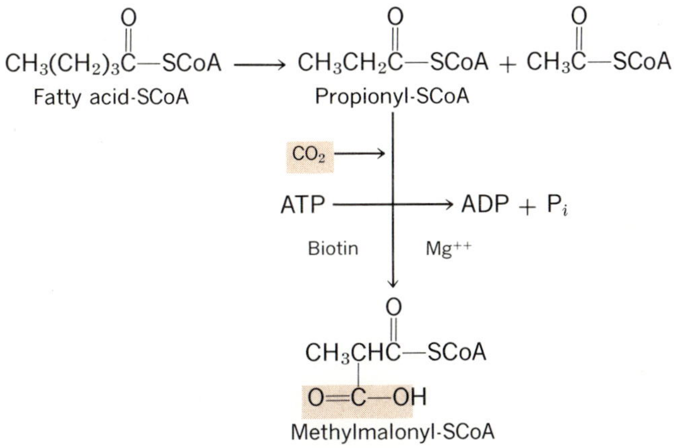

(25.6)

There are several mechanisms for the oxidation of the propionyl group. The only one to be considered here involves the nonphotosynthetic fixation of carbon dioxide into propionyl-SCoA and the ultimate conversion of the resulting product into succinyl-SCoA.

$$\text{Propionyl-SCoA} + CO_2 \xrightarrow[\text{Enz}]{\text{Biotin and } B_{12}} \text{succinyl-SCoA}$$

Carbon dioxide fixation into propionyl-SCoA requires the vitamin biotin as a coenzyme. ATP is required as a source of energy and Mg^{++} as an activator for the reaction. The insertion of the carboxyl group onto the α carbon of propionic acid gives a branched-chain dicarboxylic acid called methylmalonic acid

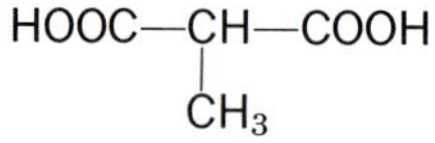

The mechanism for this reaction is similar to the carbon dioxide fixation into acetyl-SCoA encountered in fatty acid biosynthesis.

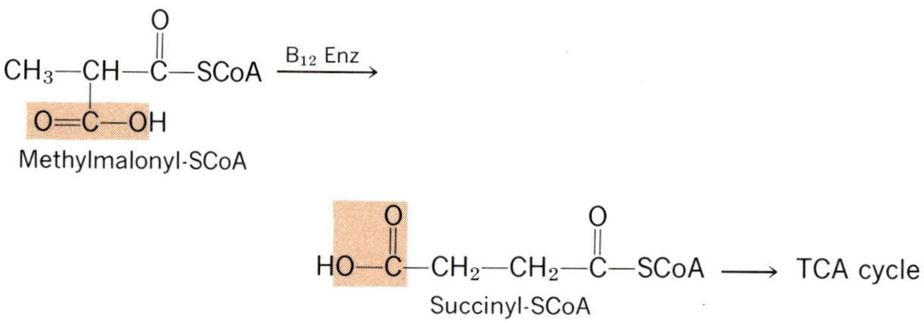

(25.7)

Following carbon dioxide fixation a rearrangement of carbon atoms occurs for which vitamin B_{12} is the coenzyme. The COOH group that resulted from the fixation of carbon dioxide is transferred to the methyl carbon to form succinyl-SCoA which is an intermediate in the TCA cycle and can be converted into oxaloacetic acid.

25.3 GLYCEROL

What is the fate of the glycerol part of the fat? Glycerol is phosphorylated in the liver by *glycerokinase* to form α-glycerophosphate, which is subsequently oxidized with the aid of the enzyme *glycerophosphate dehydrogenase* to dihydroxyacetone phosphate.

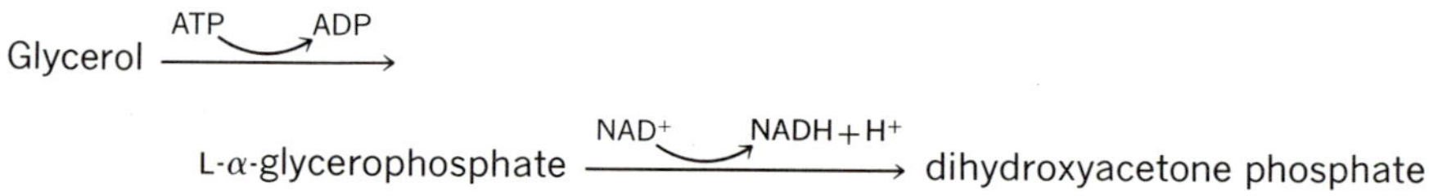

This compound is an intermediate of glycolysis and may be converted into glycogen or oxidized. Adipose tissue does not contain glycerokinase; therefore, glycerol for fat synthesis in this tissue must come from the reduction of dihydroxyacetone phosphate formed during glycolysis.

PROBLEM 1

How many moles of ATP are produced when 1 mole of glycerol is completely oxidized? When 1 mole of glycerol is converted into 1 mole of dihydroxyacetone phosphate, 1 ATP is used and 1 $NADH + H^+$ is formed. This leaves a net of 2 ATP upon oxidation of the reduced coenzyme. In going from dihydroxyacetone phosphate to pyruvic acid 2 ATP and 1 $NADH + H^+$ are formed, an equivalent of 5 ATP. Thus the reaction of glycerol to pyruvic acid produces a net of 7 ATP. The oxidation of 1 mole of pyruvic acid yields 15 ATP, making a total of 22 ATP available from the oxidation of 1 mole of glycerol.

PROBLEM 2

How many moles of ATP are produced when 1 mole of tristearin is oxidized? The answer is

3 moles stearic acid	3×147	= 441 ATP
1 mole glycerol		= 22 ATP
1 mole tristearin		= 463 ATP

Can *fats* be converted into carbohydrates? We have just seen that the glycerol moiety of a fat can be metabolized to yield a net increase in glycogen. How about the fatty acid parts of fats? These are degraded into acetyl-SCoA, which enters into the TCA cycle. To produce a net increase in glycogen, the acetyl-SCoA would have to produce a net increase in oxaloacetic acid. In animal tissues this is impossible, for once around the cycle is equivalent to oxidizing one C_2 unit. There is no bypassing any of the steps. Hence, the fatty acids cannot cause a net increase in glycogen in animal tissue.

25.4 GLYOXYLIC ACID PATHWAY

A different situation exists with respect to oil-bearing seeds. When the seed germinates, it must be able to convert fat into carbohydrates, otherwise it could not grow. These plants have a mechanism whereby

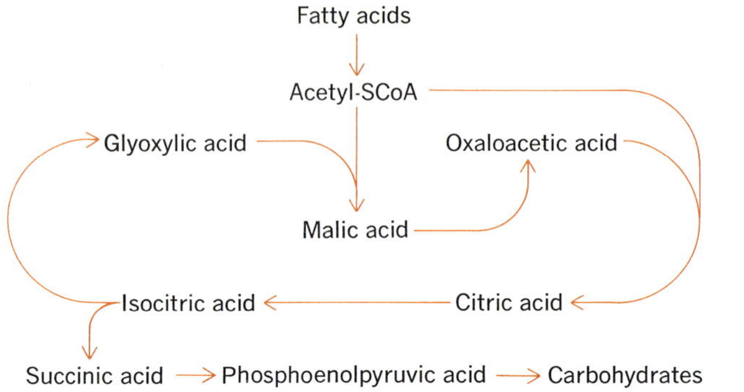

Net effect: 2 Acetyl groups ⟶ Succinic acid ⟶ Carbohydrates

FIGURE 25.2 Glyoxylic acid pathway, the conversion of fatty acids into carbohydrates.

the TCA cycle is partly bypassed in such a way that acetyl-SCoA yields a net increase in oxaloacetic acid that can be converted into phosphoenolpyruvate and into carbohydrates. This pathway is known as the *glyoxylic acid pathway* because glyoxylic acid is an intermediate. This cycle is shown in Fig. 25.2. The two reactions that are new are the cleavage of isocitric acid into succinic acid and glyoxylic acid and the condensation of an acetyl unit with glyoxylic acid to produce malic acid:

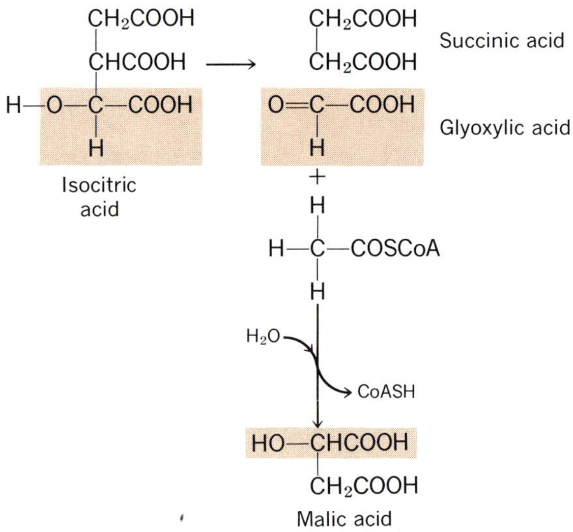

(25.8)

25.5 SATURATED FATTY ACID SYNTHESIS

Following the discovery of the β-oxidation theory, the question naturally arose whether a direct reversal of this pathway might not lead to the synthesis of fatty acids. As the enzymes of the β-oxidative pathway were more closely investigated, it became apparent that fatty acids could not be synthesized by the direct reversal of the oxidative pathway. β oxidation occurs only in the mitochondria, whereas fatty acid synthesis was found to occur to a large extent in the cytoplasm. It was not until 1959 that the role of carbon dioxide and biotin was understood with respect to the synthesis of fatty acids. With the finding that carbon dioxide fixation is essential but the carbon dioxide fixed does not occur in the fatty acids came the discovery that malonyl as well as acetyl groups condense to form fatty acids. Carbon dioxide adds to the acetyl group to form malonyl-SCoA; biotin functions as a coenzyme in this reaction

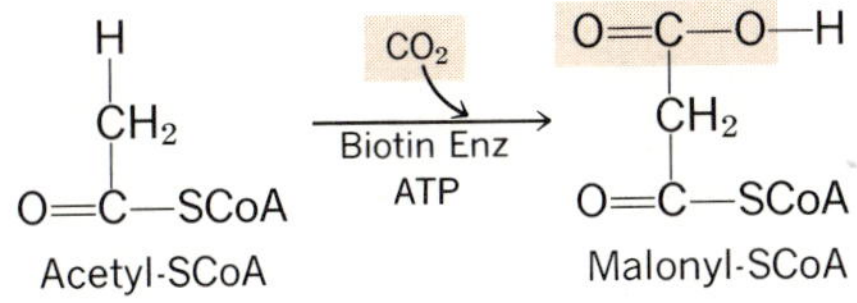

(25.9)

During the condensation of malonyl units, the carbon dioxide that was fixed is eliminated. Hence, the overall reaction is equivalent to the condensation of acetyl units to fatty acids. More recently a so-called acyl-carrier-protein (ACP—SH) has been described which replaces CoASH on the acetyl and malonyl units before condensation takes place.

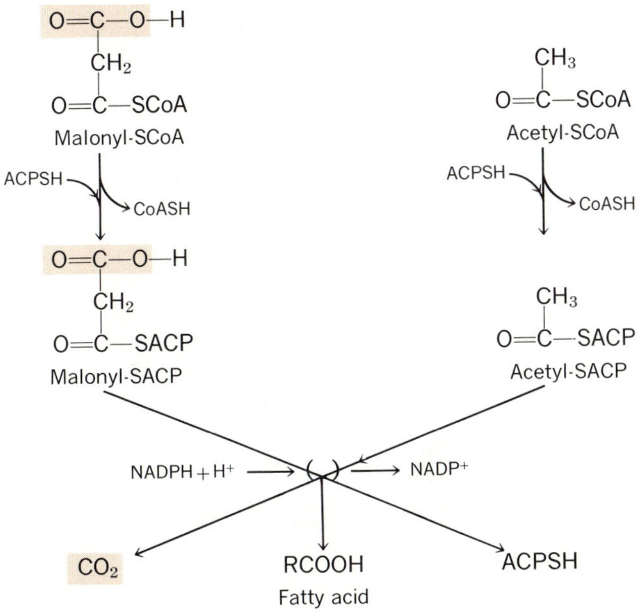

(25.10)

Malonyl-SACP formation

The initial step in the biosynthesis of fatty acids involves the nonphotosynthetic fixation of carbon dioxide for which biotin is required as a cofactor. Also an SH compound other than CoASH is required. These two steps are not involved in the oxidation of fatty acids.

β-Ketobutyryl-SACP formation

The overall reactions of fatty acid biosynthesis are

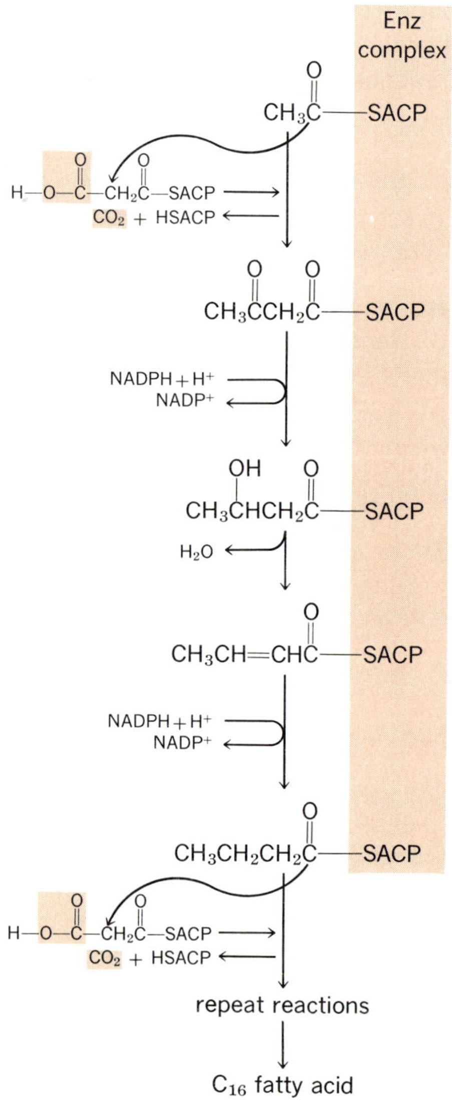

(25.11)

The first units to condense are acetyl-SACP and malonyl-SACP, with the elimination of the carbon dioxide that was fixed into acetyl-SCoA. The result of this condensation is a β-ketobutyryl unit that remains affixed to the enzyme system during its subsequent reduction to a saturated fatty acid chain.

Reduction of β-keto group

The reduction of the β CO to a CH_2 involves steps which are a reversal of the β-oxidation pathway, but there are some differences to be noted. The reduced coenzyme for fatty acid biosynthesis is $NADPH + H^+$ rather than $NADH + H^+$, which was required in fatty acid oxidation. $NADPH + H^+$ is also used to reduce the C=C in biosynthesis, compared to the $FADH_2$ used in the degradative reactions. The enzymes of course are also different for the two pathways.

Dehydration and hydrogenation

Dehydration of the β-hydroxy intermediate forms a C=C which is subsequently hydrogenated to produce a saturated chain. Hydrolysis of the acyl group from the enzyme surface results in the formation of butyric acid, which can be incorporated into fats, e.g., butterfat.

Condensation

Elongation of the C_4 chain occurs by the condensation of malonyl-SACP units while the butyryl unit is still on the enzyme surface. The C_4 unit, formed as above, replaces the carbon dioxide of the malonyl group when condensation occurs. Thus, it is the new C_2 unit that is attached to the enzyme surface. Repetition of the reduction reactions forms a C_6 acyl unit, which may add to another malonyl unit to form a higher-molecular-weight acid or be hydrolyzed off the enzyme to enter into fats.

As noted in the summary, Fig. 25.3, a C_{16} fatty acid is produced in this manner. Only the first two tail-end carbons of this acid derive from the acetyl group directly; the other carbons derive from malonyl condensation.

Palmitic acid is the highest-molecular-weight acid produced in the cytoplasm. Higher-molecular-weight acids are produced in the mitochondria by the condensation of acetyl groups with palmitic acid and the reversal of the β-oxidation mechanism to produce the fatty acid.

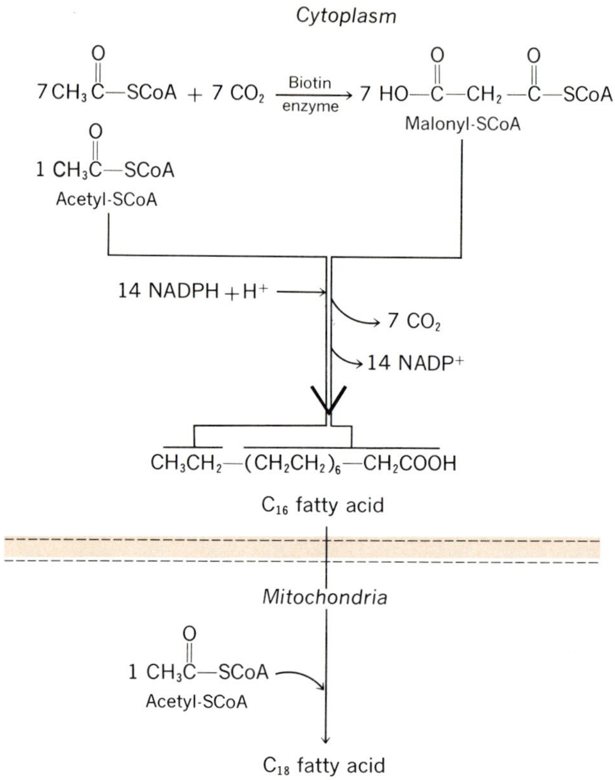

FIGURE 25.3 Summary of fatty acid biosynthesis.

25.6 UNSATURATED FATTY ACID SYNTHESIS

It has been known for many years that animals are capable of limited synthesis of unsaturated fatty acids. For example, they can easily convert palmitic and stearic acid into palmitoleic and oleic acids, respectively, but they cannot synthesize the more highly unsaturated fatty acids linoleic and linolenic from stearic or oleic acids. The body can synthesize linolenic from linoleic and vice versa. Therefore, linoleic and linolenic acids are needed in the diet and are known as *essential fatty acids.*

The mechanism for the synthesis of unsaturated fatty acids is not completely understood. Synthesis is associated with liver microsomes and requires aerobic conditions. The enzyme that converts saturated acids into monounsaturated acids has been identified as a hydroxylase that uses fatty acid–SCoA as the substrate and requires $NADPH + H^+$ and molecular oxygen:

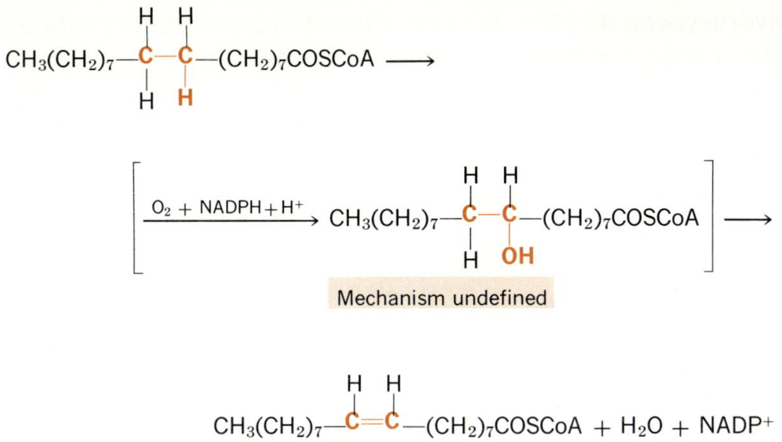

(25.12)

25.7 TRIGLYCERIDE SYNTHESIS

The first step in the oxidation of a triglyceride is its hydrolysis to glycerol and the respective fatty acids. The biosynthesis of triglycerides does not involve the direct reversal of this process but instead starts with α-glycerophosphate and the activated fatty acids. α-Glycerophosphate is derived from dihydroxyacetone phosphate of the glycolytic pathway.

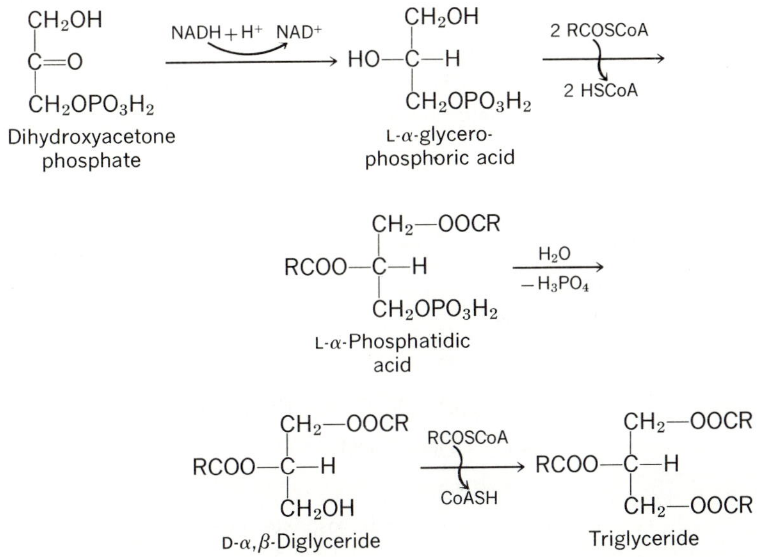

(25.13)

Overall: Dihydroxyacetone phosphate $\xrightarrow[-H_3PO_4]{NADH + H^+,\ RCOSCoA}$ trigylceride

Dihydroxyacetone phosphate is reduced to glycerophosphate by NADH$+$H$^+$ under the influence of the enzyme *α-glycerophosphate dehydrogenase.* The reaction of two fatty acid acyl–SCoA units with glycerophosphate produces L-α-phosphatidic acid. The phosphate group is hydrolyzed off by *phosphatase* to form a D-α,β-diglyceride, which is subsequently esterified with another fatty acid acyl-SCoA to produce a triglyceride.

25.8 PHOSPHATIDE SYNTHESIS

Phosphatides such as lecithin are contained in every tissue and play an important role in membrane structure. Lecithin is also important in the transportation of fat from the liver to other tissue. For example, in the absence of sufficient choline (a component of lecithin) the liver accumulates excessive amounts of fat. Our discussion of phosphatide synthesis will be limited to a brief discussion of the synthesis of lecithin.

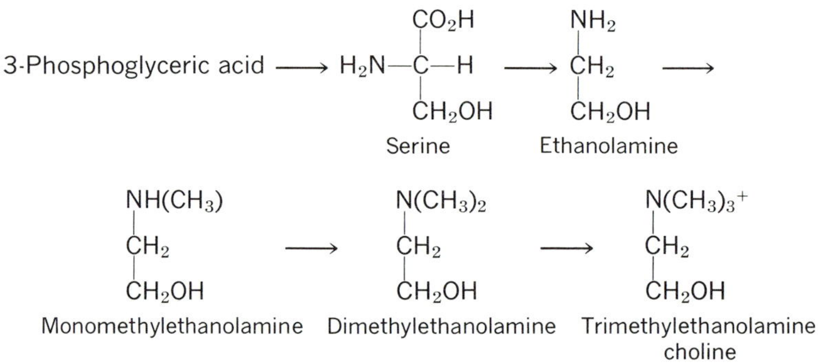

(25.14*a*)

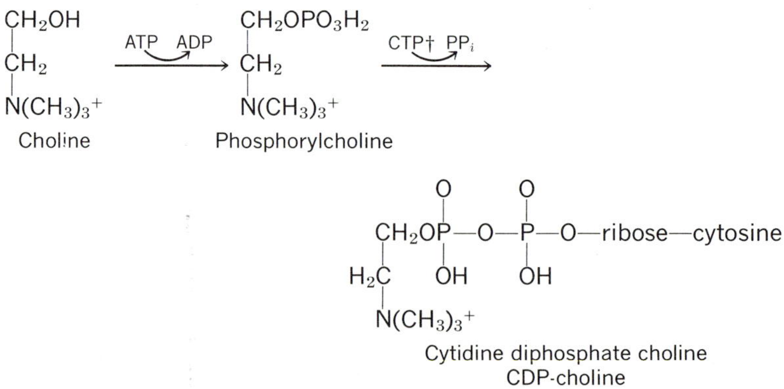

(25.14*b*)

† CTP = cytosine—ribose—phosphate—phosphate—phosphate (cytidine triphosphate), which is similar to ATP

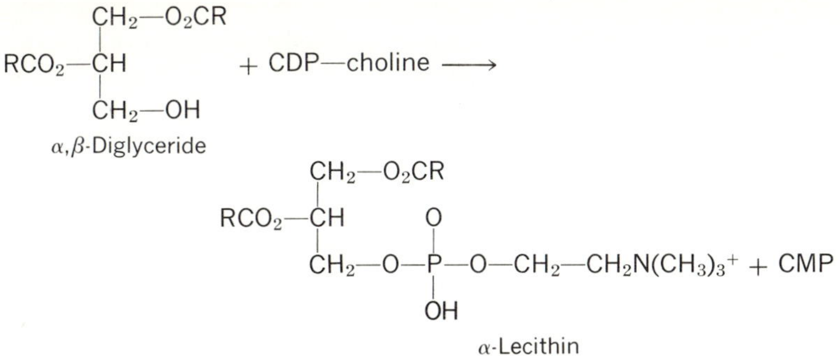

(25.14*c*)

α,β-Diglyceride, an intermediate in the synthesis of triglycerides, reacts with cytidine diphosphate choline (CDP-choline) to form lecithin. CDP-choline is synthesized from choline phosphate, which is in turn obtained from the reaction of choline with ATP. As indicated in Eq. (25.14), choline is obtained from the amino acid serine. Serine is a nonessential amino acid and can be synthesized from 3-phosphoglyceric acid by the appropriate enzyme systems. The decarboxylation of serine yields ethanolamine, which is methylated stepwise to the mono-, di-, and trimethylethanolamine (choline).

25.9 INTERRELATIONSHIP WITH CARBOHYDRATE METABOLISM

An important question remains to be considered, namely, what reactions furnish $NADPH + H^+$, which is in such great demand during fatty acid biosynthesis? This reduced coenzyme is derived from carbohydrate metabolism. Thus fatty acid synthesis cannot proceed normally unless carbohydrates are also being metabolized normally.

There are two sources of $NADPH + H^+$ for synthetic purposes. One is via the pentose phosphate pathway (Fig. 23.1), and the other is as a result of the conversion of malate to pyruvate by the malic enzyme [Eq. (22.14)]. The interrelationship of these two systems to fatty acid synthesis is summarized in Fig. 25.4. The pentose phosphate pathway is normal to the cytoplasm. Malate and acetyl-SCoA have been shown occurring only in the mitochondria. We must therefore consider a mechanism whereby malate and acetyl-SCoA are formed in the cytoplasm. As noted in Fig. 25.4, pyruvate enters the mitochondria and is converted to citrate. The mitochondrial wall is not permeable to acetyl-SCoA, but it is to citrate. A mechanism exists whereby citrate leaves the mitochondria and enters the cytoplasm, where the citrate is enzymatically cleaved to oxaloacetic acid and acetyl-SCoA. The acetyl-

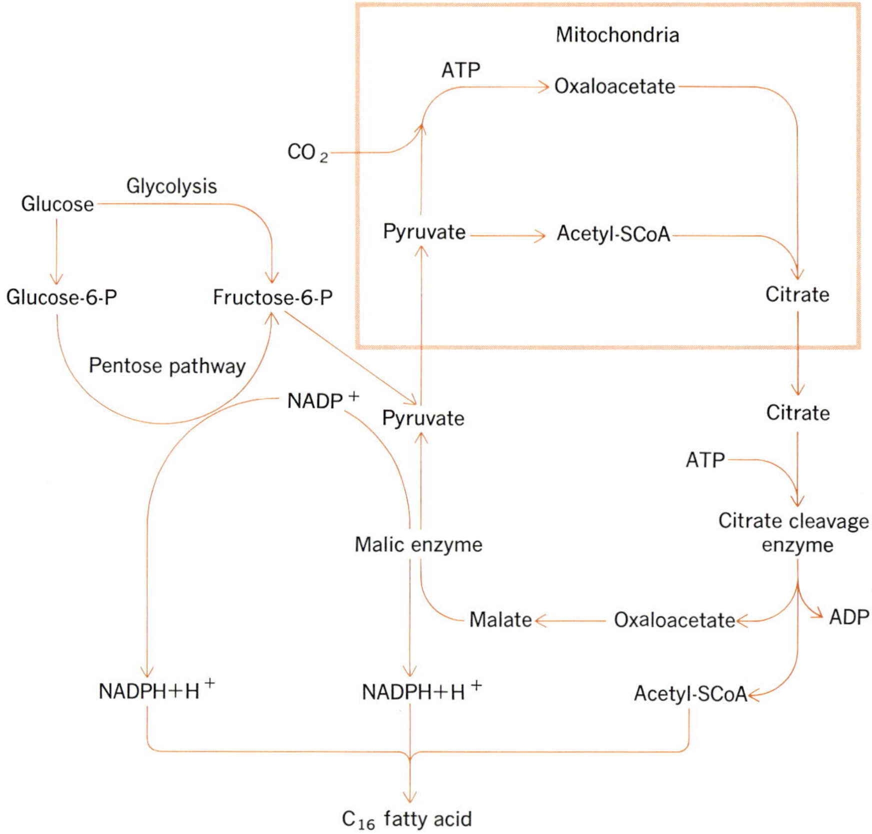

FIGURE 25.4 Interrelationships of carbohydrate metabolism and fatty acid biosynthesis.

SCoA is used for fatty acid synthesis, and the oxaloacetic acid is converted into malate and subsequently into pyruvate with the production of NADPH + H^+, which is used for fatty acid synthesis. Thus part of the NADPH + H^+ for fatty acid synthesis comes from the pentose pathway, and part is dependent upon a product from the TCA cycle. The cytoplasmic acetyl unit is furnished by the TCA cycle intermediate, citric acid.

It is quite apparent that unless both the pentose pathway and the TCA cycle are functioning properly, fatty acid synthesis will be impaired even in the presence of adequate acetyl-SCoA because of an insufficiency of NADPH + H^+. Remember that fatty acid oxidation will be impaired if the TCA cycle is not functioning properly because of an inadequate supply of oxaloacetic acid to oxidize the acetyl units formed from fatty acids.

25.10 KETOSIS

Ketosis is a term used to describe an abnormally high concentration of acetoacetic acid (β-ketobutyric acid), β-hydroxybutyric acid, and acetone in the blood. Under normal conditions these compounds, known as *ketone bodies,* are present in the blood in relatively small amounts. It is the formation of excess amounts that we are now concerned with.

Fatty acid oxidation in the liver is slightly different from the oxidation of these acids elsewhere. In the liver the fatty acids are oxidized via β oxidation to acetoacetyl-SCoA and acetyl-SCoA. Acetyl-SCoA is further oxidized by the TCA cycle, but acetoacetyl-SCoA cannot be converted into acetyl-SCoA in the liver.

The fate of liver acetoacetyl-SCoA is shown in Fig. 25.5. It is converted into acetoacetic acid via the intermediate compound β-hydroxy-β-methylglutaryl-SCoA. The liver does not contain either the enzyme to split acetoacetyl-SCoA into two acetyl-SCoA units or the enzyme which activates acetoacetic acid. Therefore, once acetoacetyl-SCoA is formed, it is not further oxidized but converted into acetoacetic acid. Liver enzymes can reduce acetoacetic acid to β-hydroxybutyric acid. Acetone is formed by the decarboxylation of acetoacetic acid. Thus the normal occurrence of ketone bodies is accounted for. These compounds are transported to other tissues by the blood.

Under normal conditions extrahepatic tissue can oxidize β-hydroxybutyric acid to acetoacetic acid, which it can activate to acetoacetyl-SCoA and metabolize completely via β-oxidation and the TCA cycle.

Under abnormal conditions, such as carbohydrate starvation or diabetes, the TCA cycle is not functioning properly because of a deficiency of oxaloacetic acid; instead, considerable amounts of fat are being oxidized in an attempt to supply the body with adequate energy needs. This situation results in the increased production of acetoacetic acid and an inadequate mechanism for its further complete metabolism. Consequently ketone bodies build up to a level higher than normal, and considerable amounts of them are excreted in the urine.

Acetoacetic and β-hydroxybutyric acids exist as their salts at body pH. Therefore, excessive production of these acids constitutes a continual drain on the buffers of the body as the salts are excreted in the urine and the mineral elements lost to the body. If this process continues long enough, the body becomes unable to neutralize the acids completely. The pH of the blood and cellular fluids decreases,

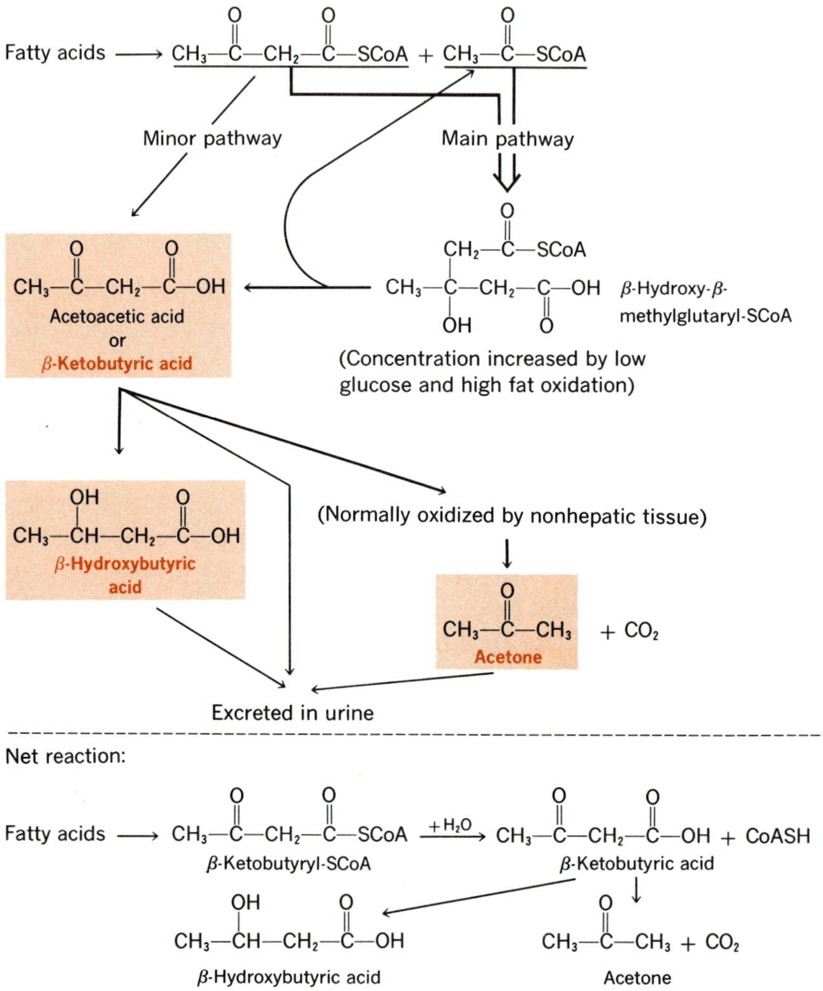

FIGURE 25.5 The formation of ketone bodies in the liver.

and a condition known as *acidosis* results. This, of course, can be quite serious if allowed to continue. Acetone is exhaled in the breath as well as excreted in the urine. The formation of acetone tends to conserve the buffer supply of the body.

Sodium propionate has been used to relieve ketosis in certain cases. Propionate can be activated and converted into succinyl-SCoA and subsequently into oxaloacetic acid [Eq. (25.7)], thus furnishing a supply of oxaloacetic acid to relieve ketosis by permitting the normal metabolism of acetyl-SCoA.

QUESTIONS

1. Define lipids.
2. Write the structural formula for tripalmitin.
3. Write an equation for the hydrolysis of a triglyceride.
4. By what pathways are the glycerol and the fatty acids of fats metabolized?
5. Name the most common way in which fatty acids are oxidized.
6. Classify the various types of reactions associated with β oxidation of fatty acids. Where possible write an equation for an analogous reaction associated with carbohydrate metabolism.
7. Are even- and odd-numbered carbon fatty acids oxidized via the same mechanism? Explain.
8. Explain the following observation: it is found that carbon dioxide is required for the synthesis of fatty acids but the carbon dioxide does not occur in the structure of the fatty acids.
9. The end products of microbial digestion of carbohydrate in the rumen of the cow are butyric, propionic, and acetic acids. From which of these organic acids does the cow form blood glucose? Explain.
10. What is the net number of moles of ATP formed from 1 mole of butyric acid? Show the essential steps involved in arriving at your answer.
11. Citrate in the cytoplasm is associated with fatty acid synthesis. Comment on this statement.
12. When an animal is starved for several days, high amounts of certain normal end products of fatty acid catabolism appear in the blood. Give their formulas.
13. Explain why normal oxidation and synthesis of triglycerides depends upon normal carbohydrate metabolism.
14. Explain the significance of biotin in fatty acid synthesis.
15. Which answers are correct? Fatty acid oxidation:
    (*a*) Takes place in the cytoplasm.
    (*b*) Produces acetyl-SACP as an end product.
    (*c*) Is impaired by abnormally low carbohydrate metabolism.
    (*d*) Requires 1 ATP for activation of the fatty acid regardless of the length of the carbon chain.
    (*e*) Yields ketone bodies as normal end products in the liver.
16. Acetyl-SCoA is formed in the mitochondria from pyruvate. Fatty acid synthesis as catalyzed by fatty acid synthetase complex occurs in the cytoplasm and requires acetyl-SCoA as the ultimate substrate. The mitochondrial membrane is relatively impermeable to acetyl-SCoA. Explain, with the aid of equations, how cytoplasmic acetyl-SCoA is made available for fatty acid synthesis.
17. Acetoacetic acid is a product of fatty acid oxidation. Explain the important role of this acid with respect to aerobic carbohydrate metabolism in the liver.
18. Which answers are correct? β oxidation of fatty acids:
    (*a*) Involves α,β-trans intermediates.
    (*b*) Occurs in the mitochondria of the cell.
    (*c*) Requires FAD enzymes.
    (*d*) Requires NAD enzymes.
    (*e*) Requires malonyl-SCoA.
    (*f*) Requires catalytic amounts of carbon dioxide.

19. Which compounds are *intermediates* in the synthesis of triglyceride from carbohydrate in *adipose* tissue?

(*a*) Lactic acid (*b*) Acetyl-SCoA
(*c*) Glycerol (*d*) Malonyl-SCoA
(*e*) α-Glycerophosphate (*f*) Butyric acid
(*g*) β-Hydroxybutyric acid (*h*) Fatty acid-ADP complex

20. β-Hydroxybutyrate is a normal end product of fatty acid oxidation in the liver. In extrahepatic tissues what reactions must take place before this compound is converted into acetyl-SCoA?

21. Compare the glucogenic nature of fatty acids containing an even and an odd number of carbon atoms.

22. Would tristearin produce some net synthesis of glucose? Explain.

23. Which answers are correct? α-Glycerophosphate:

(*a*) Is produced in adipose tissue by phosphorylation of glycerol.
(*b*) Is an intermediate in the synthesis of triglycerides and phospholipids.
(*c*) Results from the action of lipase on triglycerides.
(*d*) Is produced from glucose in adipose tissue.
(*e*) Is an optically active compound.

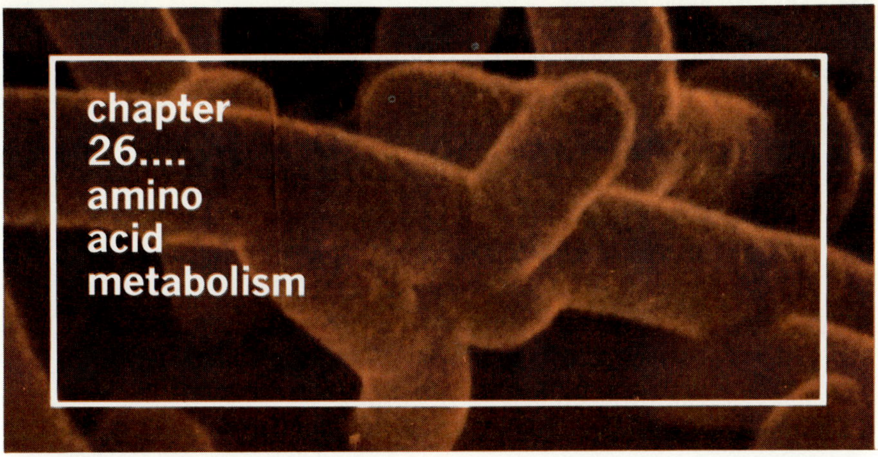

chapter 26.... amino acid metabolism

Unlike carbohydrates and fats, amino acids and proteins are not stored in reserve. The amino acids ingested are used first to meet the body's demands for nitrogenous compounds, e.g., proteins. The carbon chains of any excess amino acids are either metabolized to carbon dioxide and energy or converted into other body constituents, e.g., carbohydrates or fats. The carbon chain of amino acids enters into either the glycolytic pathway or the tricarboxylic acid cycle of carbohydrate metabolism. Entrance into the glycolytic pathway is by way of pyruvic acid, whereas entrance into the TCA cycle is by way of the intermediates α-ketoglutaric, succinic, oxaloacetic, or fumaric acids.

26.1 CLASSIFICATION

Because amino acids that form intermediates in the glycolytic or TCA pathway can be converted into glycogen, they are called *glycogenic amino acids*. Since acetyl-SCoA cannot form glycogen but does form ketone bodies, the amino acids that form acetyl-SCoA are called *ketogenic amino acids*. Some amino acids are metabolized in such a way that one part of their carbon chain is glycogenic and the other ketogenic. A list of some glycogenic and ketogenic amino acids is given in Table 26.1. The fate of the amino group of amino acids depends upon the species of animal; in man the nitrogen of the amino group is excreted as urea or ammonium salts in the urine.

All amino acids can be synthesized by plants and autotropic bacteria. About half of the naturally occurring amino acids can be synthesized by man; the carbon chain is derived from intermediate compounds of carbohydrate metabolism, and the nitrogen comes eventually from ammonia. The carbon chain of the remaining amino acids,

TABLE 26.1 GLYCOGENIC AND KETOGENIC AMINO ACIDS

Amino acid	Product	Type
Alanine Serine Cysteine	Pyruvic	Glycogenic
Glutamic acid Arginine Ornithine Proline Hydroxyproline Histidine	α-Ketoglutaric acid	Glycogenic
Aspartic acid	Oxaloacetic acid	Glycogenic
Tyrosine **Phenylalanine**	**Fumaric acid and acetyl-SCoA**	**Glycogenic and ketogenic**
Isoleucine	**Succinyl-SCoA and acetyl-SCoA**	**Glycogenic and ketogenic**
Leucine	Acetyl-SCoA	Ketogenic

known as essential amino acids, either cannot be synthesized at all or cannot be synthesized in amounts large enough to meet the body demands. The term essential means that the amino acids must be ingested in the diet. *All* amino acids are needed to make the various proteins of living organisms. A list of the essential amino acids is given in Table 26.2, together with the recommended daily requirement.

TABLE 26.2 ESSENTIAL AMINO ACIDS

Amino acid	Recommended human daily intake, g
Essential for all species of higher animals studied:	
Phenylalanine	2.2
Leucine	2.2
Isoleucine	1.4
Valine	1.6
Methionine	2.2
Tryptophan	0.5
Threonine	1.0
Lysine	1.6
Also essential for rat and chicken:	
Histidine	
Arginine	

In this chapter we shall be concerned mainly with the synthesis and degradation of some amino acids and their interrelation with carbohydrate and fat metabolism.

Before discussing chemical reactions associated with the metabolism of amino acids and proteins, the reader should review the following questions. What is meant by transamination? What vitamin is associated with the enzyme transaminase? What do the terms glycogenic and ketogenic mean? Is nonphotosynthetic carbon dioxide fixation an essential step in the conversion of pyruvic acid to glucose or glycogen? Is pyruvic acid involved in the complete oxidation of TCA intermediates, such as α-ketoglutaric acid? Are fatty acids glycogenic or ketogenic compounds? What is meant by activation of fatty acids? What group is associated with the *primary structure* of proteins?

26.2 DEAMINATION

The amino group must be removed before the carbon chain of an amino acid can be metabolized. Deamination means the removal of the NH_2 group as ammonia. This can be accomplished either by oxidative or nonoxidative reactions, which result in the formation of an α-keto acid or a saturated acid, respectively.

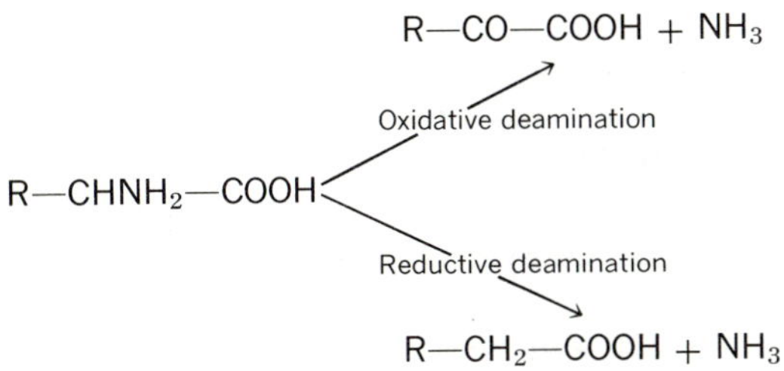

Oxidative

Oxidative deamination is brought about by two different classes of enzyme systems, one requiring NAD^+ and the other one requiring FAD as a coenzyme. The former enzymes are classified as *dehydrogenases* and the latter as *amino acid oxidases.*

GLUTAMIC DEHYDROGENASE. An example of an important amino acid dehydrogenase is glutamic dehydrogenase which converts glutamic acid to α-ketoglutaric acid and NH_3; NAD^+ is the coenzyme for this system.

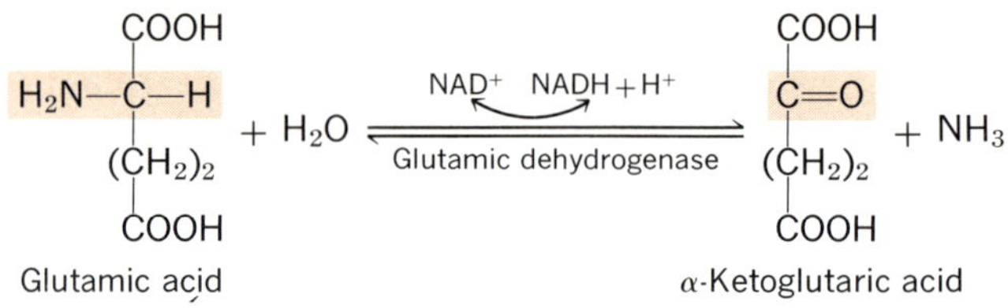

(26.1)

As a readily reversible reaction, it not only affords a means of degrading glutamic acid but also of synthesizing this amino acid from α-ketoglutaric acid and ammonia. Considering only the reaction of glutamic dehydrogenase, would you classify glutamic acid as an essential amino acid?

The entrance of the carbon chain of glutamic acid into the TCA cycle via α-ketoglutaric acid means that glutamic acid is classified as a glycogenic amino acid because of the conversions

α-Ketoglutaric acid $\longrightarrow$ oxaloacetic acid $\longrightarrow$ phosphoenolpyruvic acid $\longrightarrow$ glycogen

ALANINE DEAMINASE. Alanine deaminase is another deaminase enzyme that converts alanine to pyruvic acid and ammonia.

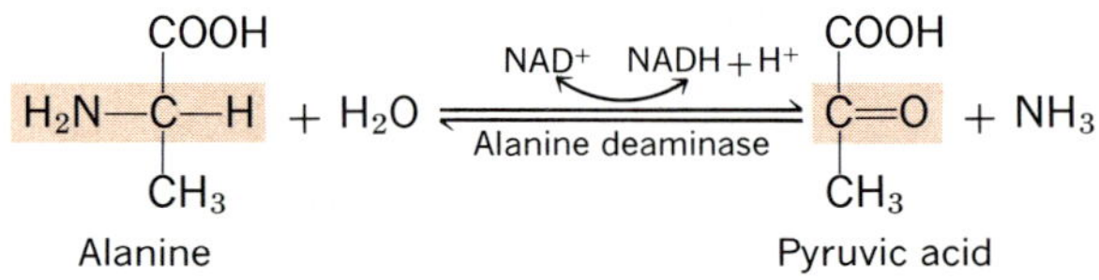

(26.2)

This reaction is easily reversible and requires coenzyme NAD^+. As a result of the action of this enzyme the carbon chain of alanine enters into the glycolytic pathway of carbohydrate metabolism.

Alanine $\longrightarrow$ pyruvic acid $\longrightarrow$ oxaloacetic acid $\longrightarrow$ phosphoenolpyruvic acid $\longrightarrow$ glycogen

AMINO ACID OXIDASE. Amino acid oxidase is the name of a class of enzymes that aerobically deaminate amino acids with the aid of FAD as the coenzyme. This system requires molecular oxygen, and the end products are the corresponding α-keto acid, ammonia, and hydrogen peroxide. The initial reaction involves the removal of two hydrogen atoms from the amino acid to form an imino group.

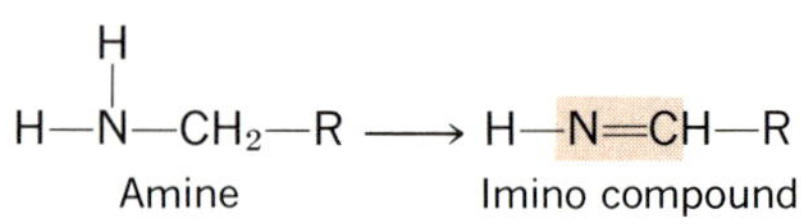

The imino group is quite labile and spontaneously reacts with water to form the corresponding keto compound and ammonia. The hydrogens of the reduced coenzyme, $FADH_2$, are combined with molecular oxygen to form H_2O_2.

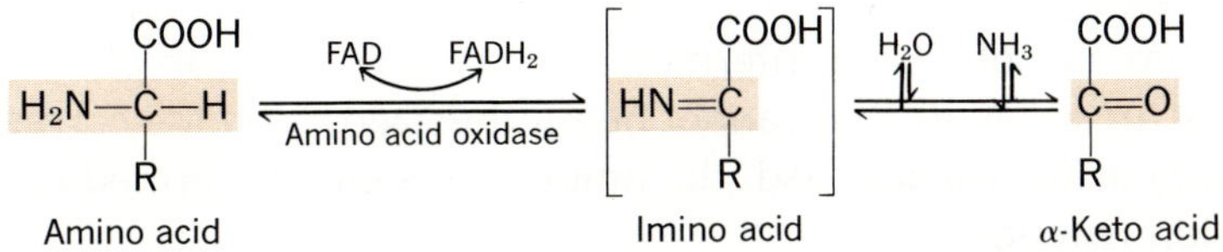

(26.3)

There are two classes of amino acid oxidases, one for the D and one for the L amino acids. D-Amino acid oxidase is a soluble enzyme and is found in the cytoplasm, whereas L-amino acid oxidase is associated with the insoluble mitochondrial particles. In animals the activity of the L-amino acid oxidase is so low that this enzyme is of no apparent importance in the normal breakdown of amino acids. This enzyme system is considerably more active in bacteria and fungi, where it plays an important role in the deamination of amino acids. In animals the D-amino acid oxidase affords a ready means of metabolizing the D amino acids which may be present in the diet. The D form of amino acids is not incorporated into body proteins.

Transamination

Transamination is one of the most important means of removing or adding an amino group to carbon chains. Almost all amino acids enter into this reaction, and it is therefore important in both the catabolism and anabolism of amino acids since the reaction catalyzed by this enzyme is reversible. Transamination reactions constitute a major link between amino acid and carbohydrate intermediates.

The chemistry of transamination was discussed in Chap. 25, under vitamin B_6, which is a cofactor for this enzyme system. Here we shall be concerned with the overall reaction, as shown with glutamic acid and α-ketoglutaric acid:

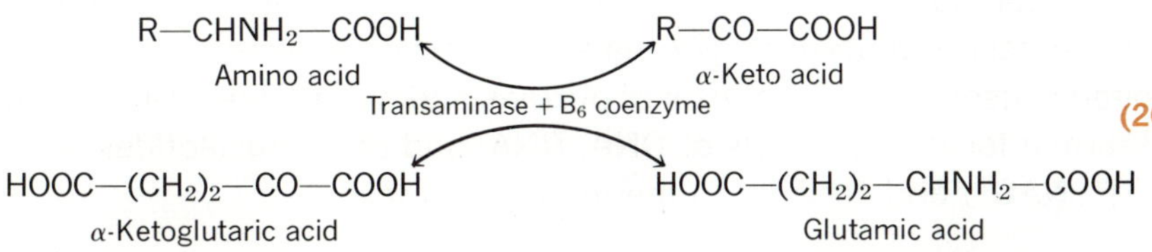

(26.4)

α-Ketoglutaric, pyruvic, and oxaloacetic acids are the keto acids most active in transamination reactions.

Deamination of amino acids by the above reactions is one of the first steps encountered in the breakdown of most amino acids. Metabolism of the carbon chain may or may not be reversible, depending upon the amino acid. The reactions discussed above are reversible and therefore afford a means of synthesizing the nonessential amino acids from carbohydrate intermediates. It is the carbon chain of the essential amino acids that cannot be synthesized, for if the corresponding α-keto acids are supplied, the amino acids can be formed by transamination reactions.

26.3 AMIDE FORMATION

GLUTAMINE SYNTHETASE. Thus far we have been concerned with the formation and removal of the α-amino group of amino acids. *Glutamine synthetase* is the enzyme responsible for the formation of an amide group on the end carbon of glutamic acid. This reaction is equivalent to replacing the OH group of the carboxyl group with an NH_2 group.

Carboxyl group Amide group

Glutamine synthetase requires a source of energy, ATP, and is activated by Mg^{++}.

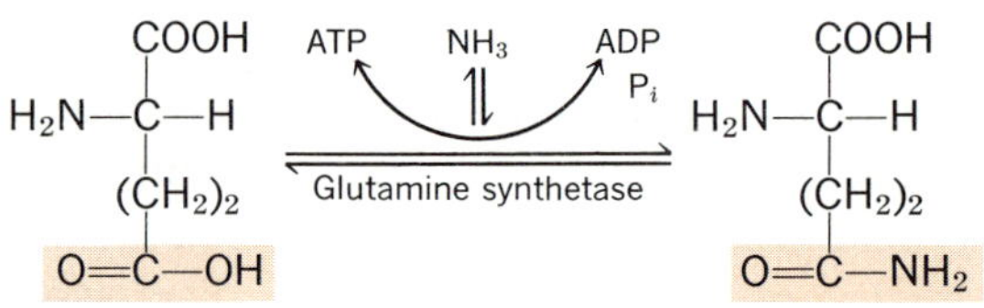

Glutamic acid Glutamine

(26.5)

The formation of glutamine serves as another means of incorporating inorganic ammonia into organic compounds. Glutamine has a high nitrogen content and can serve more or less as nitrogen storage. The synthesis of glutamine is of great importance because its amide nitrogen is used in the synthesis of purines and pyrimidines. Glutamine is essential for the synthesis of DNA, RNA, and other nucleotides such as ATP, NAD^+, and $NADP^+$, to mention only a few.

ASPARAGINE SYNTHETASE. Asparagine synthetase is another enzyme that forms amide groups. It catalyzes the conversion of aspartic acid to asparagine, and, as with glutamine synthetase, ATP is required. However, in contrast to the glutamine synthetase reaction, in which ATP is converted into ADP and P_i, in the arginine synthetase reaction ATP is cleaved to AMP and PP_i (pyrophosphate). The formation of asparagine also constitutes a means of storing available nitrogen.

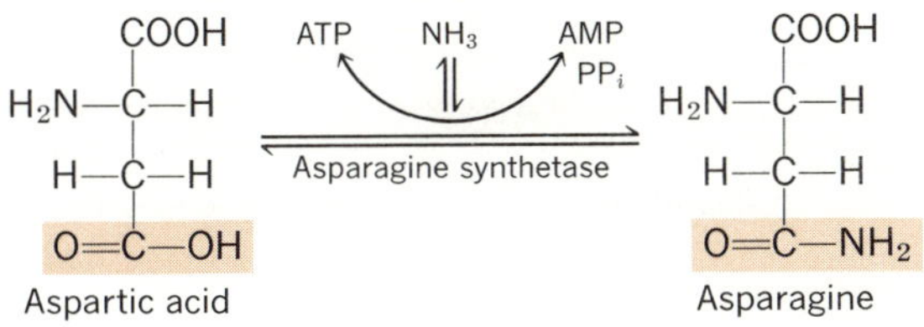

(26.6)

CARBAMYL PHOSPHATE SYNTHETASE. This enzyme synthesizes the fixation of carbon dioxide and ammonia to form carbamyl phosphate, which is an initial product of the biosynthesis of pyrimidines and urea.

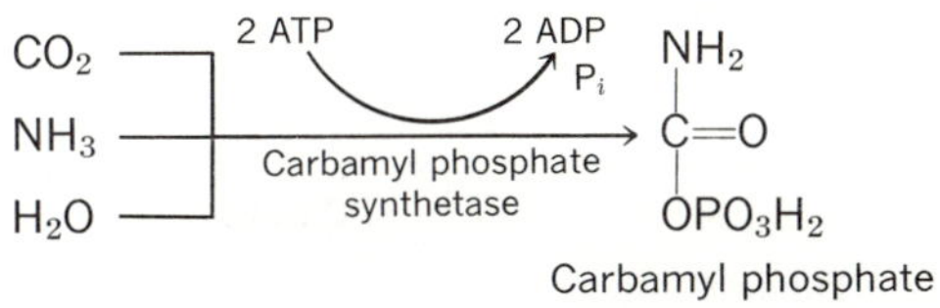

(26.7)

The importance of the pyrimidines in nucleotides is clear from previous discussions, and the importance of urea, which is the end product of nitrogen metabolism in man, will be discussed subsequently. Carbamyl phosphate synthesis is also important in that it represents another reaction in which carbon dioxide is fixed nonphotosynthetically and inorganic nitrogen is converted into organic nitrogen. This synthesis by enzyme systems requires 2 *moles of ATP per mole of carbon dioxide fixed.*

26.4 SUMMARY

A summary flowsheet of the metabolism of the carbon chain of amino acids is given in Fig. 26.1. Note that most amino acids enter into the carbohydrate metabolic scheme at either pyruvate or an intermediate in the TCA cycle. These amino acids are glycogenic. An amino acid such as leucine is metabolized to acetyl-SCoA and acetoacetate and is therefore considered to be a ketogenic amino acid. Several amino acids, tyrosine, phenylalanine, and isoleucine, are both glycogenic and ketogenic. Part of their carbon chain forms a TCA intermediate, and

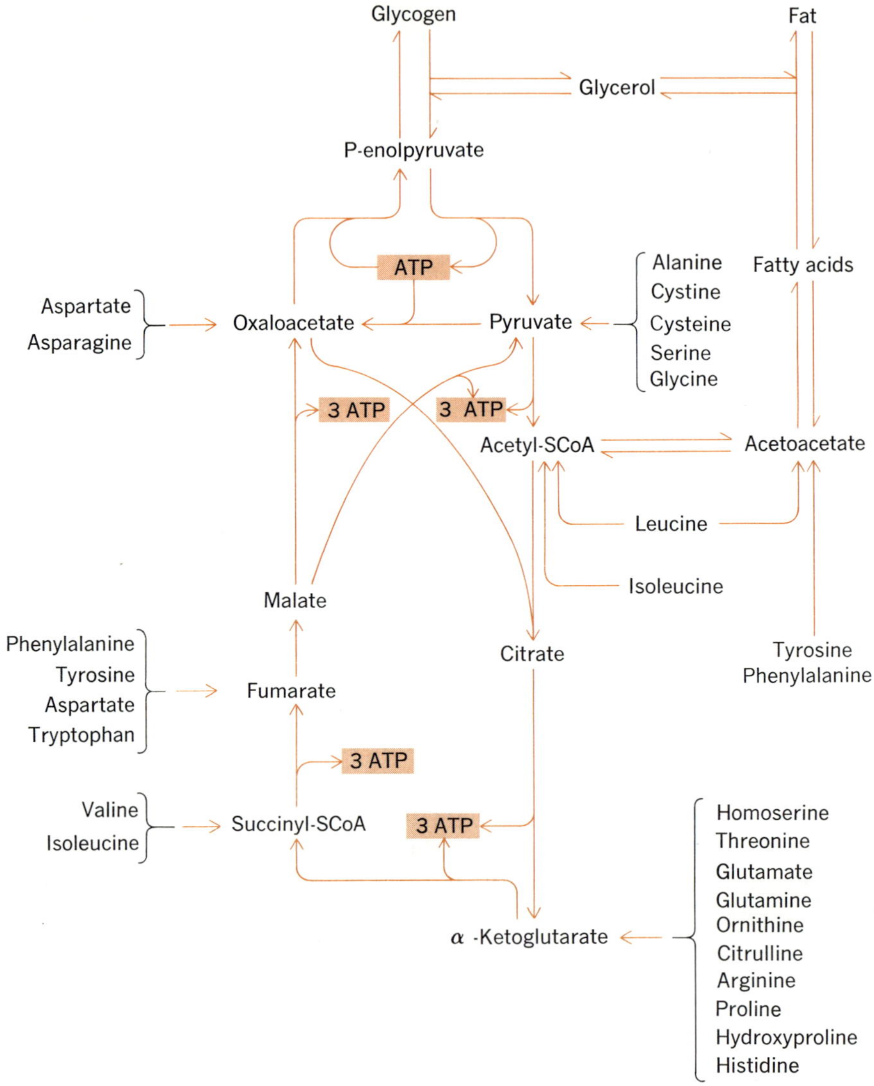

FIGURE 26.1 Interrelationship of amino acid, carbohydrate, and fat metabolism. (Energy utilized by, or liberated from, a reaction is expressed only in terms of ATP.)

part forms acetoacetate or acetyl-SCoA, which are ketogenic compounds. Leucine is a strict ketogenic amino acid and is metabolized to acetoacetate and acetyl-SCoA. Acetyl-SCoA or acetoacetate may be metabolized to fatty acids and incorporated into fats or oxidized via the TCA cycle with the formation of 12 ATP per acetyl-SCoA unit oxidized. These two compounds cannot produce a net increase in glycogen.

Amino acids whose carbon chain enters the glycolytic pathway at pyruvate or the TCA cycle, e.g., at α-ketoglutaric acid, can be converted into any product that can be formed from glucose or can be used for energy purposes. Alanine, for example, would yield pyruvic acid by transamination. The carbon chain could then be metabolized to carbohydrates or fats or be oxidized to yield 15 ATP (see Fig. 26.1). What would be the fate of the carbon chain of an amino acid that formed α-ketoglutarate? α-Ketoglutarate is a member of the TCA cycle and can be converted into a variety of metabolic products. For example, it can enter into the glycolytic pathway via oxaloacetate and phosphoenolpyruvate. From phosphoenolpyruvate the carbon chain can be converted into carbohydrates or fats or oxidized for energy via the TCA cycle.

How many moles of ATP would theoretically be obtained from the carbon chain of 1 mole of glutamic acid if the amino acid is transaminated to α-ketoglutaric acid? In going from α-ketoglutarate to oxaloacetate a maximum of 9 ATP is formed. The conversion of oxaloacetate to pyruvate utilizes 1 ATP and forms 1 ATP; therefore, there is no net formation or utilization of ATP in this conversion. The other pathway, from malate to pyruvate, also yields 9 ATP per mole of α-ketoglutarate metabolized to pyruvate; 1 mole of α-ketoglutarate yields 1 mole of pyruvate, and the oxidation of 1 mole of pyruvate furnishes 15 ATP. Thus, a maximum of 24 ATP would be obtainable from the oxidation of 1 mole of glutamic acid according to the above mechanism. If glutamic acid is converted to α-ketoglutaric acid via glutamic dehydrogenase enzyme, how many moles theoretically of ATP would be produced upon oxidation of one mole of the amino acid? This enzyme forms 1 mole of $NADH + H^+$ during the process of oxidatively deaminating glutamic acid; thus, an additional 3 moles of ATP are formed as a result of oxidizing this reduced coenzyme. There would be a total of 27 moles of ATP produced via this pathway.

26.5 UREA CYCLE

Since ammonia is quite toxic if allowed to accumulate, there must be a mechanism whereby ammonia can be conveniently detoxified and eliminated from the body. The end product of nitrogen metabolism depends upon the animal species. In man some nitrogen is excreted in the urine as simple ammonium salts, some as uric acid, and by far the most as the relatively nontoxic, water-soluble compound urea. The end product of protein nitrogen is urea and of purine nitrogen is uric acid.

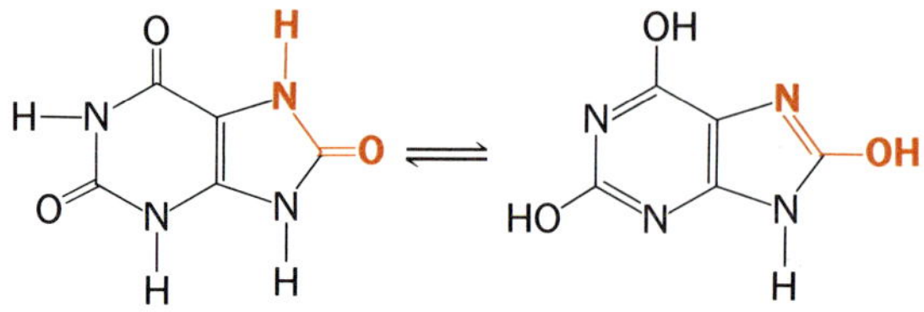

Uric acid (keto) Uric acid (enol)

The formula for urea suggests reacting 2 moles of ammonia with 1 mole of carbon dioxide to form urea and water.

$$CO_2 + 2\,NH_3 \longrightarrow H_2N{-}CO{-}NH_2 + H_2O$$

The overall reaction amounts to just that, but the actual mechanism is quite complicated. The synthesis of urea involves several compounds that are intermediates in the TCA cycle and some amino acids that do not generally occur in proteins. The urea cycle is presented schematically in Fig. 26.2. The individual steps are as follows.

FIXATION OF CARBON DIOXIDE AND AMMONIA. Carbamyl phosphate synthesis is the first step in the urea cycle. This compound is formed by the fixation of carbon dioxide and ammonia with the energy and phosphate coming from ATP [see Eq. (26.7)]. Note that it requires 2 moles of ATP to produce 1 mole of carbamyl phosphate.

CARBAMYL PHOSPHATE + ORNITHINE $\longrightarrow$ CITRULLINE. Ornithine, a C_5 α,ϵ-diamino monocarboxylic acid, reacts with carbamyl phosphate to produce citrulline. In this reaction, the $H_2N{-}CO$ group is transferred from carbamyl phosphate to the ϵ-amino group of ornithine to form citrulline.

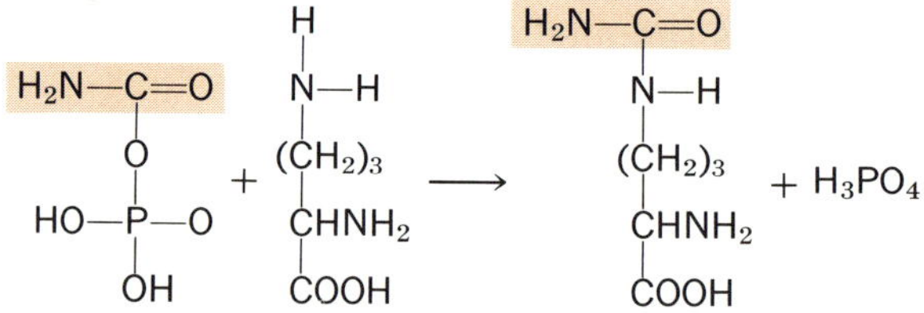

(26.8)

Carbamyl phosphate Ornithine Citrulline

CITRULLINE $\longrightarrow$ ARGININE. The conversion of citrulline to arginine involves the condensation of citrulline with aspartate and decomposition of the resulting intermediate into arginine and fumarate. Asparagine is supplied by transamination of oxaloacetate.

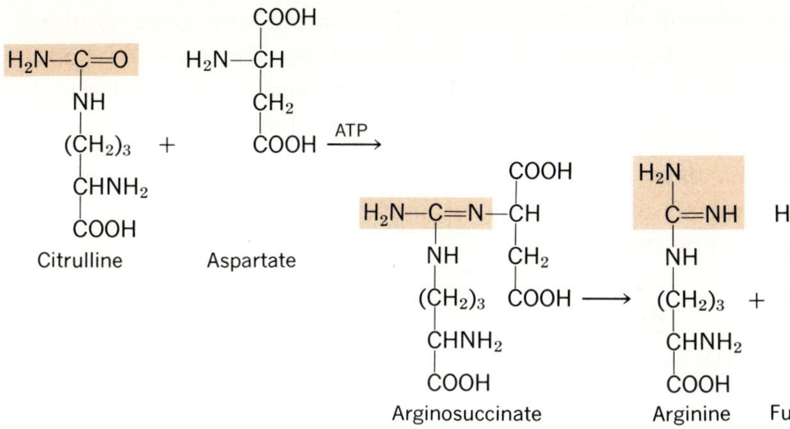

(26.9)

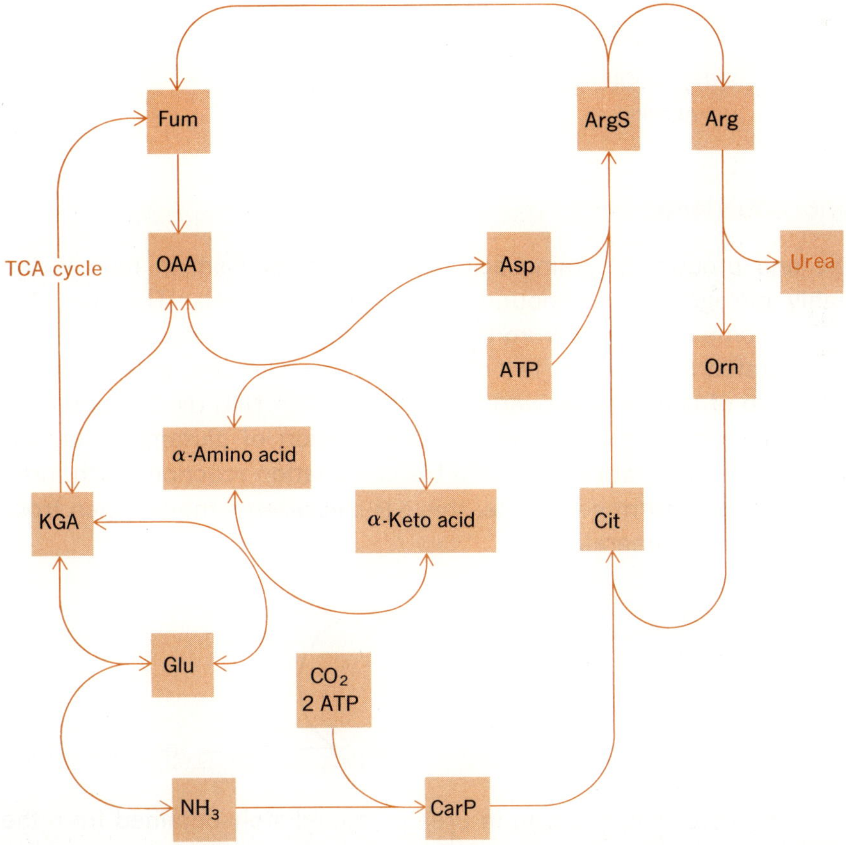

FIGURE 26.2 Interrelationship of the urea and TCA cycles. Arg = arginine, ArgS = arginosuccinic acid, Asp = aspartic acid, CarP = carbamyl phosphate, Cit = citrulline, Fum = fumaric acid, Glu = glutamic acid, KGA = α-ketoglutaric acid, OAA = oxaloacetic acid, Orn = ornithine.

The involvement of oxaloacetate and fumarate with urea synthesis integrates the TCA cycle with the urea cycle. The condensation of citrulline with aspartate requires ATP as a source of energy; this is the third and last mole of ATP required in the synthesis of urea.

ARGININE ⟶ UREA + ORNITHINE. The final step in the formation of urea is the hydrolysis of arginine by arginase to yield urea and ornithine, thus completing the cycle.

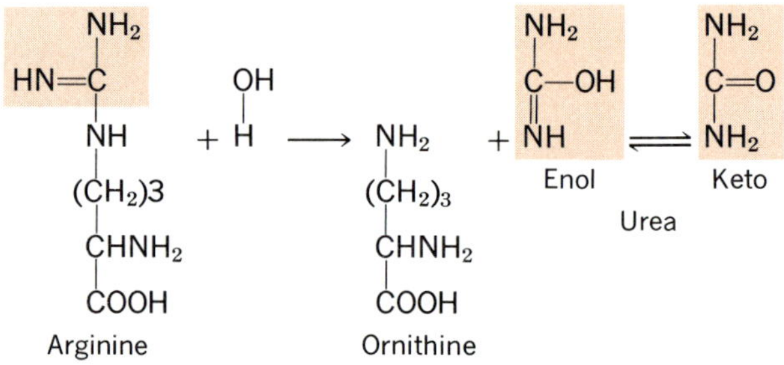

(26.10)

26.6 SUMMARY

The end product of amino acid nitrogen metabolism in man is the highly nitrogenous compound urea, $H_2N—CO—NH_2$. The two NH_2 groups come eventually from the amino groups of amino acids but are incorporated into urea by two different routes. One NH_2 group and the CO group of urea are derived by the fixation of NH_3 (NH_4^+) and CO_2. The NH_3 comes mainly from the deamination of glutamic acid, but since α-ketoglutarate enters into transamination reactions with most amino acids to form glutamate, this NH_3 represents the nitrogen from amino acids in general.

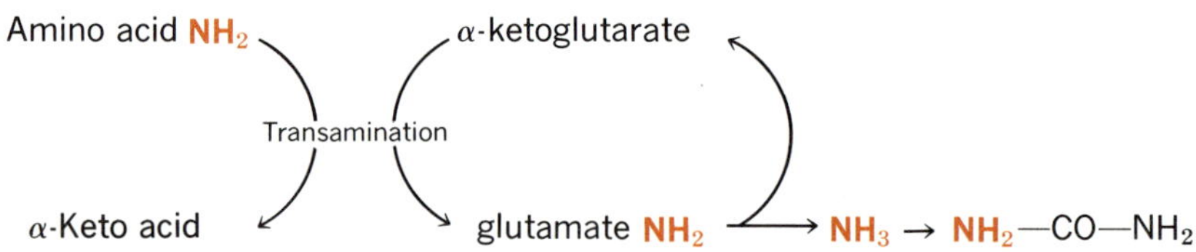

The second NH_2 group in urea is immediately obtained from the NH_2 group of aspartate. However, since aspartate is formed via transamination reactions with oxaloacetate, this NH_2 group can also be considered as ultimately coming from amino acids in general.

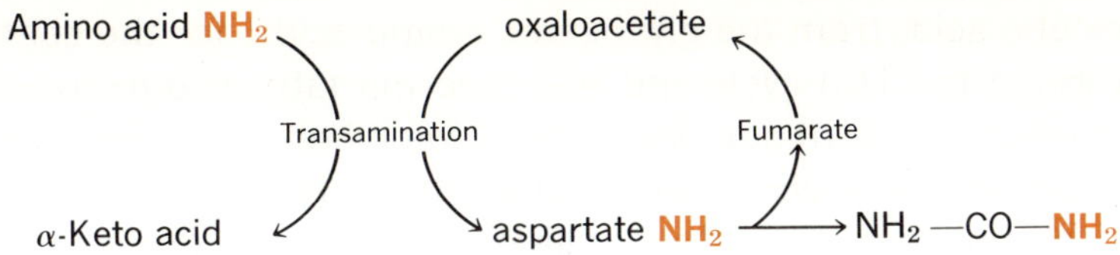

For the synthesis of 1 mole of urea *3 moles of ATP* is required. For the fixation of carbon dioxide and ammonia to form carbamyl phosphate 2 moles of ATP is required. The third mole of ATP furnishes energy for the formation of arginosuccinate from citrulline and aspartate.

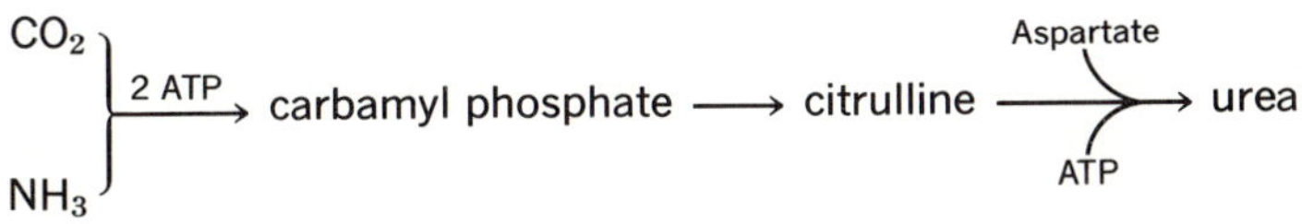

The TCA cycle must be operative for the synthesis of urea. For example, α-ketoglutarate and oxaloacetate are both associated with the first step in nitrogen metabolism, namely, the removal of the amino group from amino acids. This is accomplished by transamination reactions. Further involvement of the TCA cycle is illustrated in the formation of fumarate and the conversion of fumarate to oxaloacetate.

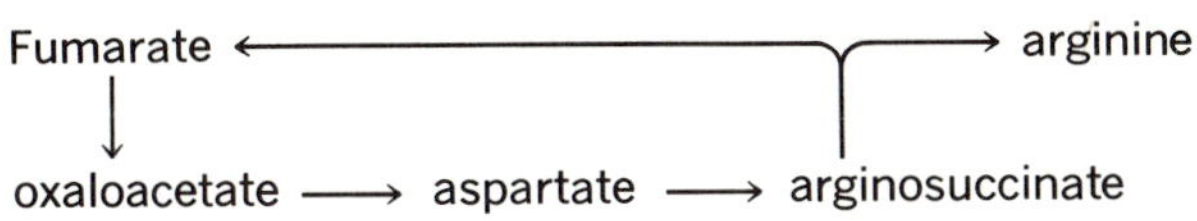

Oxaloacetate is a key compound in the urea cycle as well as in the TCA cycle; therefore, any situation that impairs the functioning of the TCA cycle also impairs the functioning of the urea cycle.

Proper functioning of the electron-transport system (oxidative phosphorylation) is also a requirement for normal urea synthesis inasmuch as the electron-transport system furnishes the ATP required for urea synthesis.

How does starvation influence protein metabolism with respect to urea synthesis? In the previous chapter the influence of a carbohydrate deficiency on fat metabolism was discussed. Now, the metabolism of proteins (amino acids) can be interrelated with this situation. Under the above conditions protein synthesis would become limiting, and catabolism of proteins would increase. This would aid in furnishing oxaloacetic acid, from the glycogenic amino acids, for the continued operation of the TCA cycle and would permit fatty acid oxidation and urea synthesis to continue. Because of the increase in amino acid deamination, urea and NH_4^+ salts in the urine would increase. This condition can continue for only a limited time because proteins are not stored in great excess. Thus, as protein reserves become depleted, the TCA cycle would be impaired, urea synthesis would practically cease, and nitrogen would be excreted as NH_4^+ salts. Also the amount of ketone bodies would increase, as would the level of ammonium salts of acetoacetic and β-hydroxybutyric acids in the urine.

The interrelationships of carbohydrate, fat, and protein metabolism are much more intricate than the simple examples given here, but from the present discussion it can be realized that a good diet includes (1) carbohydrates to furnish energy and intermediates to keep the metabolism of fats and proteins functioning properly; (2) fats to provide the essential fatty acids; and (3) proteins to supply the essential amino acids required by the body.

26.7 PURINES AND PYRIMIDINES

The importance of the chemistry of purines and pyrimidines cannot be over emphasized, but since the reactions for the synthesis and catabolism of these ring structures are beyond the scope of this text, only a schematic representation of the precursors of the atoms in the purine and pyrimidine rings is given.

Purines:

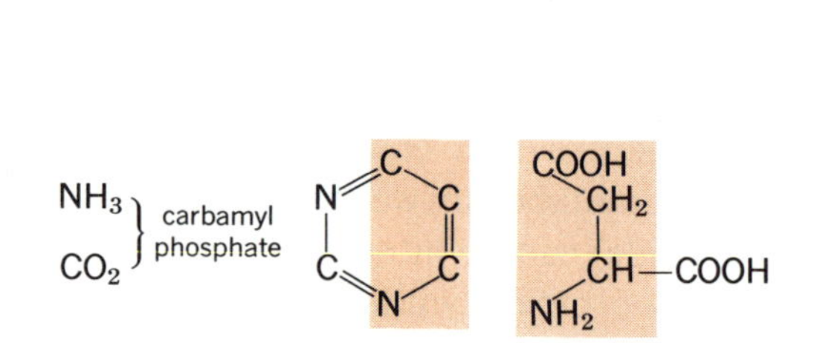

Pyrimidines:

QUESTIONS

1. What is the difference between essential and nonessential amino acids?
2. What is the significance of the terms "glycogenic" and "ketogenic" amino acids?
3. What is the major means of deaminating amino acids?
4. (*a*) What vitamin is associated with transamination reactions?
   (*b*) Name two α-keto acids that enter into transamination reactions in general.
5. Which statements are correct?
   (*a*) Transamination is associated only with essential amino acids.
   (*b*) Vitamin B_6 is contained in the coenzyme needed for transamination reactions.
   (*c*) The same carbon chain results if an amino acid is oxidatively deaminated or transaminated.
6. Which statements are correct?
   (*a*) The carbon chains of amino acids that enter the TCA cycle at compounds other than the acetyl-SCoA are glycogenic.
   (*b*) Components of the TCA cycle are essential for the synthesis of glycogen from amino acids that form pyruvic acid.
   (*c*) Components of the TCA cycle are essential for the proper functioning of the urea cycle.
7. Explain part (*c*) of question 6.
8. (*a*) How many moles of glutamic acid are theoretically needed to produce 1 mole of glucose?
   (*b*) How many moles of ATP are needed or formed to carry out the reactions involved in part (*a*) above?
9. Assuming that carbons of aspartic acid are derived from alanine, how many moles of alanine are required to make 1 mole of aspartic acid? What is the energetics of this conversion?
10. Does nonphotosynthetic carbon dioxide fixation play an important role in the conversion of NH_3 to urea? Explain.
11. What is the structure of carbamyl phosphate? Is this compound important in the synthesis of both urea and purines?
12. Explain how both NH_2 groups of urea can be derived from the NH_2 group of alanine.
13. Explain the effects of a carbohydrate-deficient diet on protein metabolism.

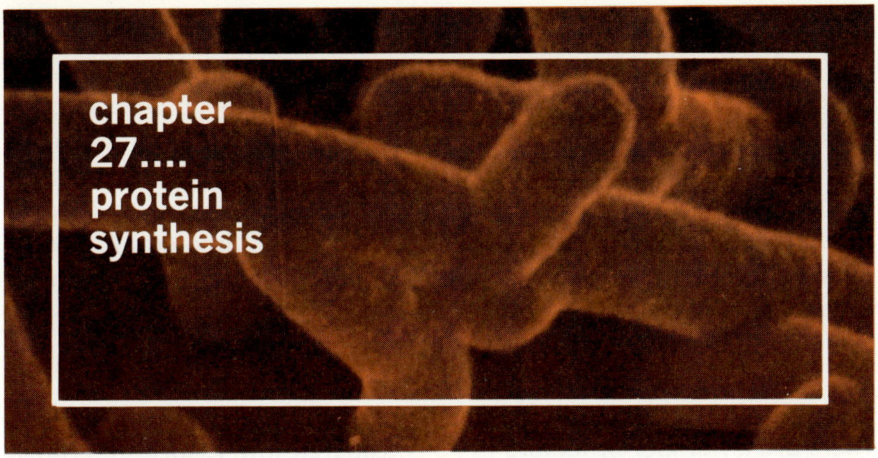

chapter 27.... protein synthesis

Enzymes are proteins responsible for the manifestations of inherited biochemical traits. On the basis of this fact alone protein synthesis and its control are of utmost biological importance.

All inherited traits lead back eventually to molecules of DNA, which contain *all* the necessary information required for the synthesis of every cellular protein. Each cell, therefore, must have a mechanism for *replicating* the DNA and passing it on to the daughter cell as well as a mechanism for *transcribing* the information contained in the DNA and passing this information to the cytoplasm, where the information is *translated* to form proteins of a specific kind, arrangement, and number of amino acids.

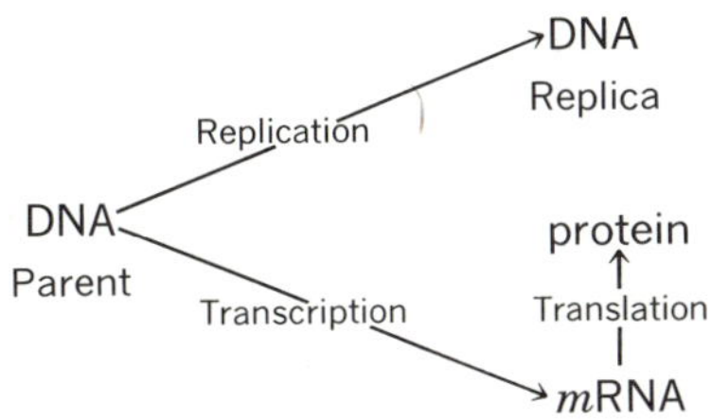

The dynamic state of living organisms necessitates the continuous degradation and synthesis of body components. With respect to the immediate discussion, body proteins are being hydrolyzed to amino acids and resynthesized at a rate dependent upon the organism and tissue in question. For example, the liver is metabolizing proteins at such a rate that the equivalent of half the liver tissue is being replaced with fresh protein every six days. This is known as the half-life of the tissue. Protein of muscle and connective tissue may have a half-life of from 180 to 1,000 days, whereas hair has no half-life because its protein amino acids are not in dynamic equilibrium with free amino acids in the organism (amino acid pool). Enzymes are a good class of compounds to emphasize the need for continuous normal protein synthesis. Of the hundreds of enzymes in each cell, not all are needed all the time or at the same time. Most enzymes are synthesized only when needed; thus the level of enzymes fluctuates with their requirement. A mechanism must exist for starting and stopping protein synthesis in an orderly manner.

27.1 FACTORS IN SYNTHESIS REACTIONS

There are a number of factors to be considered with respect to protein synthesis, e.g., the availability of the amino acids. Since amino acids are not stored appreciably, a continued daily supply of the essential amino acids must be available. For protein synthesis, all the required amino acids must be present at the time the protein is being synthesized; fragments of proteins are not made. Thus, if only one essential amino acid is lacking, no protein containing this amino acid can be made; this would, of course, result in the death of the organism.

Besides the required amino acids, a system must be available for ordering the amino acids to hook up in the correct fashion to form a specific protein. A *template* must be available for each protein, and this template must be inherited. This means that the exact same template must be reproduced for each generation.

The formation of peptide bonds requires energy, ATP, to activate the amino acid and 1 mole of GTP (equivalent to ATP) per peptide bond formed. Each amino acid requires a specific activating enzyme as well as a specific carrier to transport it to the site of protein synthesis. If this specificity did not exist, a definite sequence of amino acids in proteins would not be possible. The definite sequence means that each amino acid carrier must recognize the exact place where it can transfer its amino acid to form a peptide. How is this brought about and controlled?

Considerable progress has been made within the past several years toward elucidating conditions and mechanisms of protein synthesis and its control. This is a very rapidly advancing field of investigation. The 1968 Nobel Prize in medicine was awarded to Holley, Khorana, and Nirenberg for their contributions in this area of research.

Both DNA and RNA are associated with protein synthesis. DNA directs the order of bases in the RNAs, which in turn direct the order of amino acids in proteins. There are three kinds of RNA, namely, messenger RNA (*m*RNA), transfer RNA (*t*RNA), and ribosomal RNA (*r*RNA), to be considered with respect to protein synthesis. Before proceeding further the following review questions should be considered.

Is there a difference between the bases found in DNA and RNA? What base pairing exists between two strands of DNA and between a DNA strand and a RNA strand? What is the sugar found in DNA and in RNA? What is the significance of pApCpGpCpT as an abbreviated manner of writing a polynucleotide? What is the Watson-Crick structure of DNA?

27.2 DNA

Replication

To pass similar biological traits from parent cell to daughter cell, the parent DNA must be duplicated in the *exact same sequence* of bases and passed into the daughter cell. DNA is a double-helix structure; therefore, for reproduction the strands must somehow untwine, com-

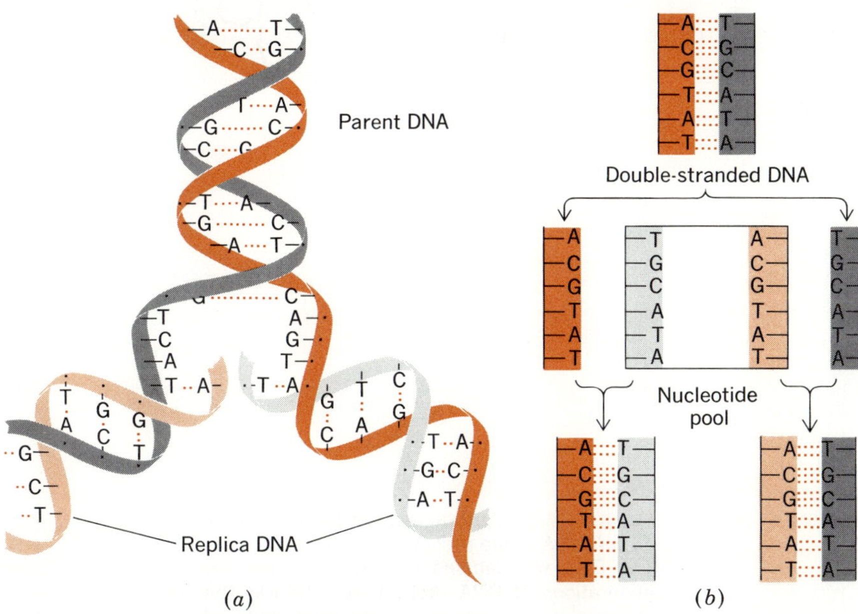

FIGURE 27.1 Replication of DNA: (*a*) overall scheme; (*b*) two pairs of double-stranded DNA identical with the original are formed. One-half of the bases of each new molecule were *not* in the original DNA.

plementary nucleotide phosphates (those which pair by hydrogen bonding) from the surrounding solution must pair with the DNA bases that have just been freed from the double strand, and the enzyme DNA polymerase must catalyze the polymerization of the nucleotides. Thus, in a manner not completely understood, two new strands of DNA of exactly the same base sequence are formed (Fig. 27.1*a*). To emphasize base pairing during replication a section of DNA is shown in Fig. 27.1*b*, where for the sake of simplicity, the segment of DNA is split completely into two single strands and then the nucleotides from the cellular pool are shown combining with the freed bases. Since complementary bases are attached, the end result is the formation of two new double-stranded DNA molecules. Each new molecule now has half of the original bases and half new bases. This has been confirmed

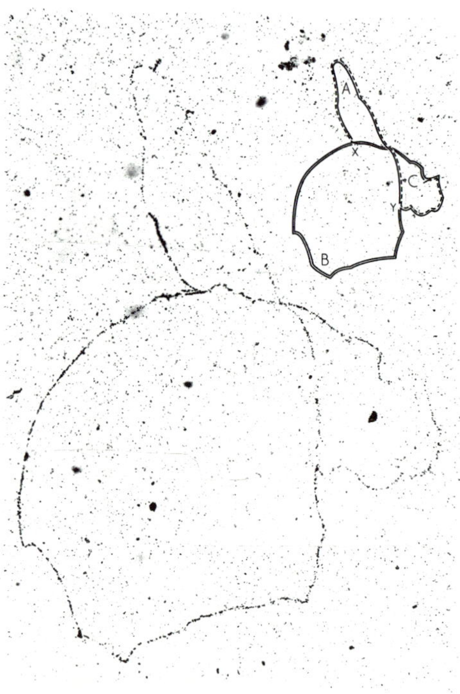

FIGURE 27.2 Autoradiograph showing cyclic *E. coli* DNA in the process of being duplicated. The DNA was labeled with H^3-thymidine for generations of DNA replication. *A* marks the growing point and *B* the finishing point of replication. The scale shows 100 μ. [*Reproduced with permission from J. Cairns, Cold Spring Harbor Symp. Quant. Biol.,* **28:**44 (1963).]

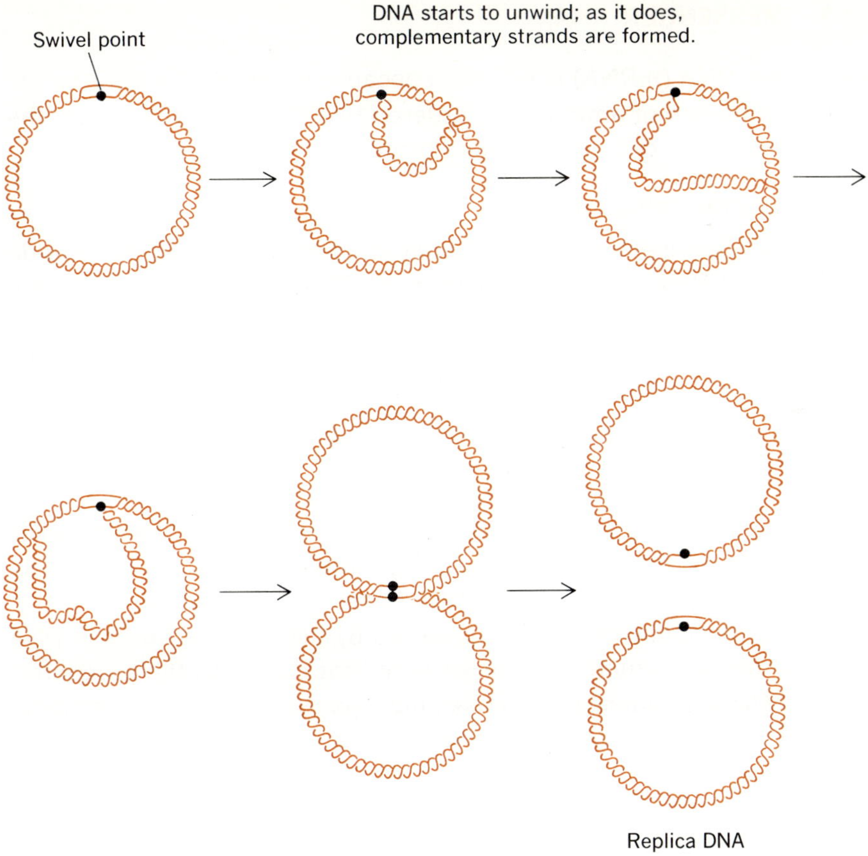

FIGURE 27.3 Duplication of cyclic DNA.

experimentally; however, some questions about how duplication occurs still remain unanswered. A similar replication of these two new DNA molecules would give four molecules of structure identical to the parent DNA, but of these four molecules, only two would have bases originally present in the parent molecule.

Not all DNA molecules are in the open-end chain structure. Some bacterial DNA has a cyclic formation (Fig. 27.2). How can these be duplicated? The double-stranded form has to unwind, which means that at some point in the chain there must be a "swivel" joint to permit the strands to rotate. Then, as a section of the DNA untwines, the DNA bases can be paired with complementary bases to form new strands. A schematic representation of this is shown in Fig. 27.3. When the cycle has been completed, the strands separate and two cyclic two-stranded DNA molecules result.

27.3 MESSENGER RNA

Messenger RNA (*m*RNA) is so called because it carries the information for the sequence of amino acids ordered by DNA of the nucleus to the ribosomes of the cytoplasm, where protein synthesis occurs.

Synthesis (transcription)

What are the bases in DNA and what are the bases in RNA? How do the bases of RNA pair with the bases of DNA? The bases in DNA are adenine, A; guanine, G; cytosine, C; and thymine, T; and the bases in RNA are adenine, A; guanine, G; cytosine, C; and uracil, U, which occurs in place of thymine in DNA. The base pairing of DNA with RNA is

$$\begin{array}{l}\text{DNA}\ \text{—C—G—C—A—T—A—G—}\\ \qquad\ \ \ \vdots\ \ \ \vdots\ \ \ \vdots\ \ \ \vdots\ \ \ \vdots\ \ \ \vdots\ \ \ \vdots\\ \text{RNA}\ \text{—G—C—G—U—A—U—C—}\end{array}$$

Thus, if a section of DNA has the sequence of bases indicated above, the bases in an RNA molecule would be in the order shown; the RNA bases are complementary to the DNA bases, with uracyl replacing thymine. The formation of RNA as governed by the order of bases in DNA is called *transcription*, and the enzyme that catalyzes the reaction is called *DNA-dependent polymerase*. Mg^{++} is required for this reaction. As the nucleotides polymerize, pyrophosphate, PP_i, is formed.

$$\begin{array}{c}\text{GTP}\\+\\\text{CTP}\\+\\\text{UTP}\\+\\\text{ATP}\end{array}\ \xrightarrow[\text{DNA-dependent RNA polymerase, }Mg^{++}]{\text{DNA—3'-G—C—T—A-5'}}\ m\text{RNA—5'-C—G—A—U-3'} + 4\ PP_i$$

Only one strand of a particular section of DNA is transcribed at a time, and transcription occurs on the DNA starting with the 3′ end and proceeds toward the 5′ end. RNA synthesis, however, starts at the 5′ end and proceeds toward the 3′ end of the RNA molecule. The synthesis of *m*RNA is illustrated diagramatically in Fig. 27.4, in which the nucleoside triphosphates are indicated merely as BTP.

Can both strands of DNA be transcribed? Yes, provided there is no appreciable overlapping of the transcription.

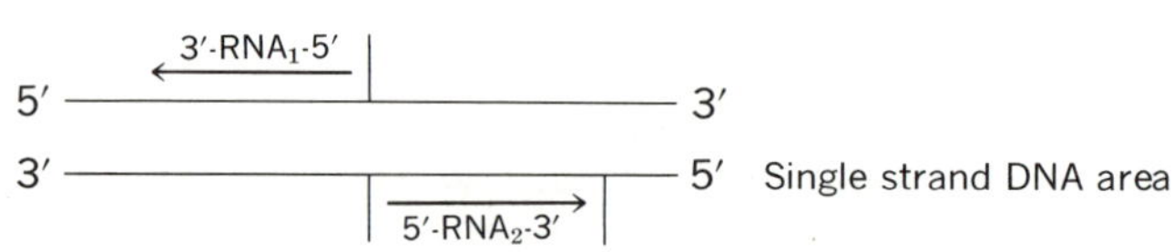

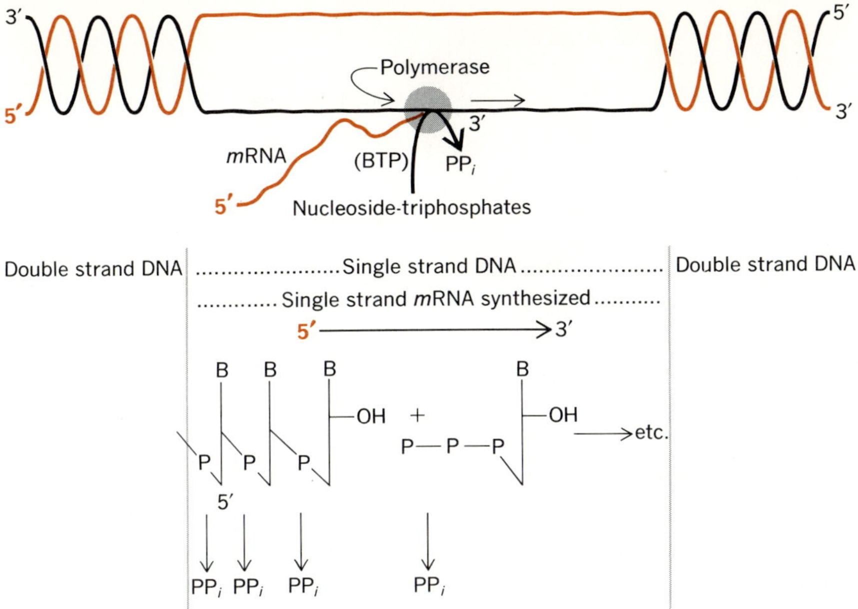

FIGURE 27.4 RNA transcription. (*Redrawn after J. D. Watson, "Molecular Biology of the Gene,"* p. 311, *W. A. Benjamin, Inc., New York,* 1965.)

A specific sequence of bases on the DNA constitutes the signal for the polymerase to start transcribing, and another base sequence signals the enzyme to stop transcribing.

One molecule of *m*RNA may carry the information for the synthesis of more than one molecule of protein. To form this type of *m*RNA the DNA must have an area containing a sequence of structural genes (segment of DNA responsible for the ultimate synthesis of a protein molecule). For example, if enzymes A, B, and C are required in that order for the synthesis of a particular compound or group, the formation of these three enzymes may all be governed by one *m*RNA molecule. The order of synthesis of the proteins would be consecutively from the 5′ to the 3′ end of the *m*RNA molecule.

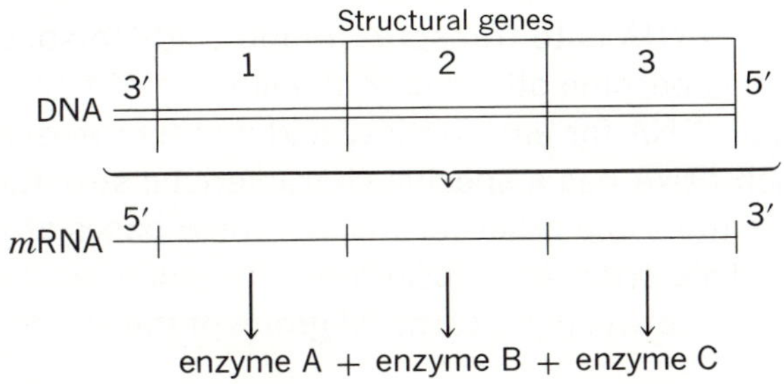

FIGURE 27.5 A complete, active ribosome showing the subunits.

After *m*RNA is synthesized in the nucleus, it migrates into the cytoplasm, where it becomes associated with ribosomes and the appropriate enzyme system for the synthesis of proteins. Neither *m*RNA nor ribosomes can direct the synthesis of proteins until they are associated with each other.

27.4 RIBOSOMES

Ribosomes are composed of about 60% *r*RNA and 40% protein and are approximately 200 Å in diameter. Each ribosome is composed of two subunits, called the 50S and 30S ribosome (the S stands for Svedberg units). A complete active ribosome can be thought of as the association of these two subunits into a 70S ribosome (see Fig. 27.5). Polysomes are merely a group of ribosomes associated with one *m*RNA molecule. Ribosomes bind both *m*RNA and *t*RNA and bring these two units together in such a manner that the amino acids connected to the *t*RNA can interact under the influence of appropriate enzymes to form proteins.

27.5 TRANSFER RNA

The function of *t*RNA is to transport amino acids to specific areas of *m*RNA which has become attached to the surface of a ribosome. There is an individual *t*RNA for each amino acid incorporated into proteins; therefore, each *t*RNA has a specific characteristic structure. However, each *t*RNA has the same order for the last three bases at the 3′ end of the molecule. This order is · · · pCpCpA. The amino acid is bonded to *t*RNA as an ester between the carboxyl group of the amino acid and the

3′ OH group of the ribose of the terminal adenosine moiety. In general the amino acid–*t*RNA complex is called aminoacyl–*t*RNA. There is a specific enzyme for the formation of each aminoacyl–*t*RNA.

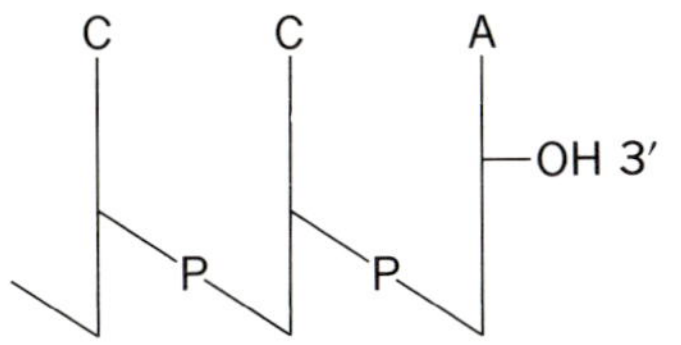

3′ end of *t*RNA

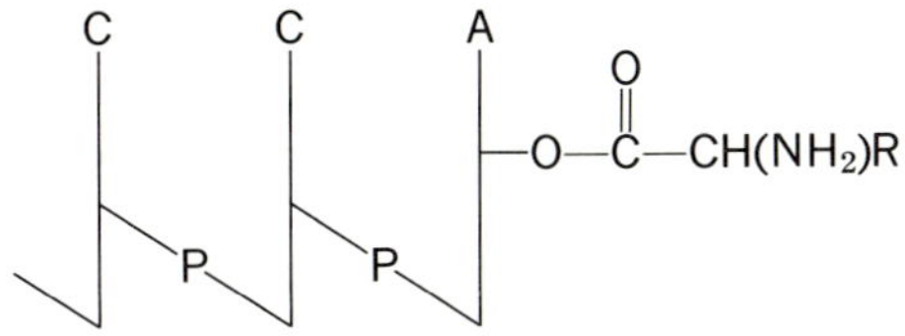

Aminoacyl-*t*RNA

A specific site on each *t*RNA must be associated with recognition of the specific *activating enzyme* which activates the amino acid by forming amino acid–AMP and also esterifies the amino acid with the *t*RNA. For example, alanine, Ala, will be activated and esterified with $tRNA_{Ala}$ only in the presence of the alanine-activating enzyme system. The alanine-activating enzyme will not form aminoacyl-*t*RNA with any other amino acid. There must also be a specific site on each *t*RNA molecule to become attached to the ribosome so that the aminoacyl group can be transferred into protein structure. Furthermore, the *t*RNA has to have another specific site by which to associate with *m*RNA in a predetermined order. As discussed below, this is due to the pairing of three bases of the *m*RNA with three complementary bases of the *t*RNA.

27.6 AMINO ACID ACTIVATION

Amino acids are activated, i.e., are caused to react, with ATP in the presence of Mg^{++} to form the corresponding aminoacyl-AMP complex, which remains attached to the enzyme surface. The enzyme also binds the respective *t*RNA and catalyzes the formation of aminoacyl-*t*RNA, which is subsequently deposited on the ribosome *m*RNA complex. The activation of amino acids and the formation of aminoacyl-*t*RNA is schematically illustrated in Fig. 27.6.

27.7 GENETIC CODE

How does the aminoacyl-*t*RNA complex associate itself with a specific area on the *m*RNA? This is accomplished by what is called the *genetic code*. There is an arrangement of three consecutive bases in *m*RNA for each amino acid. The complement of these three bases is associated with the respective *t*RNA molecule. For example, if a triplet of bases

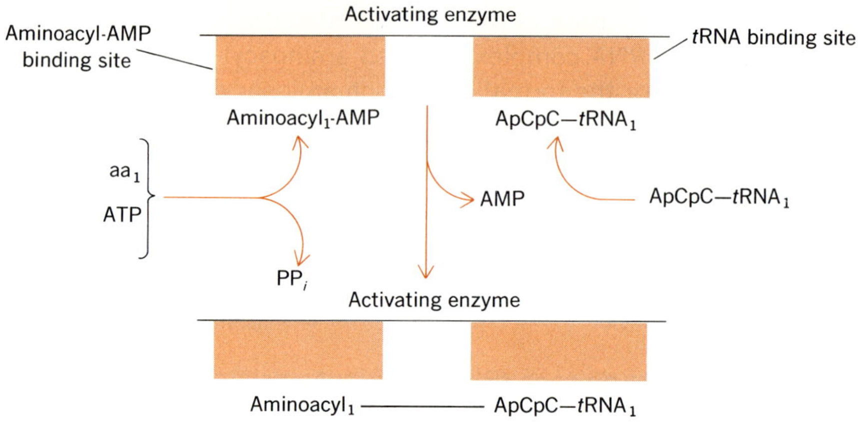

Specific example:

$$\text{Glutamic acid (Glu)} + \text{ATP} + t\text{RNA}_{\text{Glu}} \xrightarrow[\text{enzyme}]{\text{Activating}} \text{Glutamyl-}t\text{RNA}_{\text{Glu}} + \text{PP}_i + \text{AMP}$$

FIGURE 27.6 Schematic diagram of the activation of amino acids and the formation of aminoacyl-*t*RNA.

on *m*RNA is UUU, a *t*RNA having the triplet AAA would pair with the *m*RNA at the particular area where the triplet UUU occurred. Likewise, a triplet of GGA on *m*RNA would bind a *t*RNA possessing the triplet CCU. The triplets on the *m*RNA are called *codons*, and the complementary triplets on the *t*RNA are called *anticodons*. In some anticodons inosine replaces guanine; e.g., the codon for alanine is GCC, and for serine it is UCC; the anticodons are CGI and AGI, respectively.

There are four different bases in *m*RNA as well as in *t*RNA; thus if the code is determined by a sequence of three bases, this accounts for 64 different triplets, a number greater than that theoretically

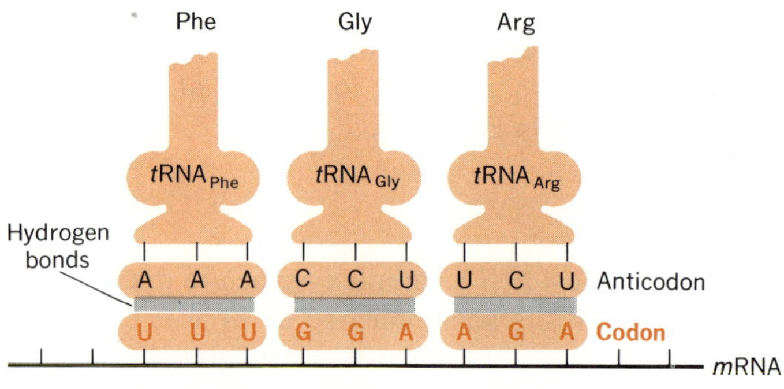

FIGURE 27.7 Hydrogen bonding in anticodon and codon of phenylalamine, glycine, and arginine.

TABLE 27.1 THE GENETIC CODE

Ala	GCA	His	CAC	Ser	AGC
	GCC		CAU		AGU
	GCG				UCA
	GCU	Ile	AUA		UCG
			AUC		UCC
Arg	AGA		AUU		UCU
	AGG				
	CGA	Leu	CUA	Thr	ACA
	CGG		CUC		ACG
	CGC		CUG		ACC
	CGU		CUU		ACU
			UUA		
Asn	AAC		UUG	Try	UGG
	AAU				
		Lys	AAA	Tyr	UAC
Asp	GAC		AAG		UAU
	GAU				
		Met	AUG	Val	GUA
Cys	UGC				GUG
	UGU	Chain initiation: $\{$ Met$_{\text{formyl}}$	AUG		GUC
					GUU
Gln	CAG				
	CAA	Phe	UUU	Chain termination: $\{$	UAA
			UUC		UAG
Glu	GAA				UGA
	GAG	Pro	CCA		
			CCC		
Gly	GGA		CCG		
	GGC		CCU		
	GGG				
	GGU				

needed for the 20 amino acids commonly found in proteins. However, some triplets are not used, whereas in some instances there is more than one triplet for the same amino acid. The genetic code as understood today is presented in Table 27.1. Note that when there is more than one code for a given amino acid, two bases are usually the same but the third base may vary. The triplet that initiates the synthesis of a protein chain has been attributed to AUG, which is the codon for methionine, and the terminator of the protein chain is due to a combination of bases that do not constitute a triplet for an amino acid, UAA, UAG, and UGA. The codons for the amino acids are the same regardless of the origin of the *m*RNA; codons appear to be universal.

The triplet AUG, which codes for methionine, also codes for *N-formyl methionine* (Met$_{\text{formyl}}$).

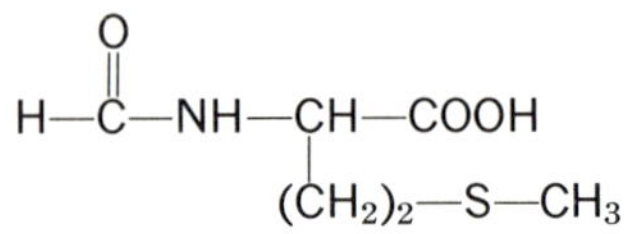

N-Formyl methionine

However, Met and Met_{formyl} are each associated with a specific *t*RNA. *N*-Formyl methionine is the first amino acid to be incorporated into all proteins, but this amino acid derivative does not appear in the final protein. Met_{formyl} is hydrolyzed off the peptide chain before the protein is released from *m*RNA. The formylation of methionine to *N*-formyl methionine requires the cofactor tetrahydrofolic acid; this is an example of another important role this vitamin plays in biological reactions. An exact explanation of the circumstances that control the formylation of methionine is not known, but it apparently occurs after methionine has been bound to *t*RNA. From the structure of Met_{formyl} it seems logical that this compound would bind to the peptidyl-*t*RNA site (see Fig. 27.8) and not to the aminoacyl-*t*RNA binding site and that a peptide bond could not be formed between Met_{formyl} and the carboxyl group of another amino acid since there is no free NH_2 group in Met_{formyl}. Therefore, Met_{formyl} could occur only at the NH_2 terminal end of a protein and not within a protein chain. Conditions that govern the enzymatic hydrolysis of Met_{formyl} off the peptide chain are not known.

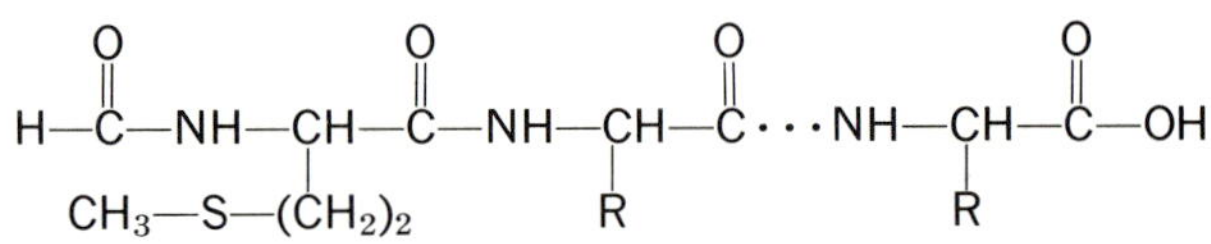

Met_{formyl}

Chain initiation and termination may be illustrated as follows.

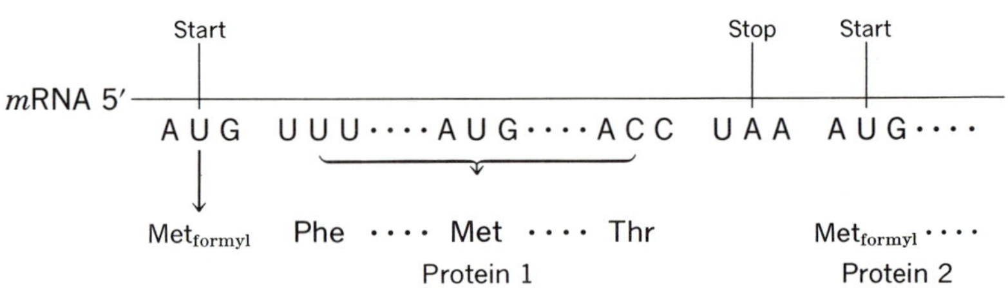

Normally protein synthesis is performed by more than one ribosome (polysome) per *m*RNA molecule. For the average protein it has been estimated that five to eight ribosomes are simultaneously syn-

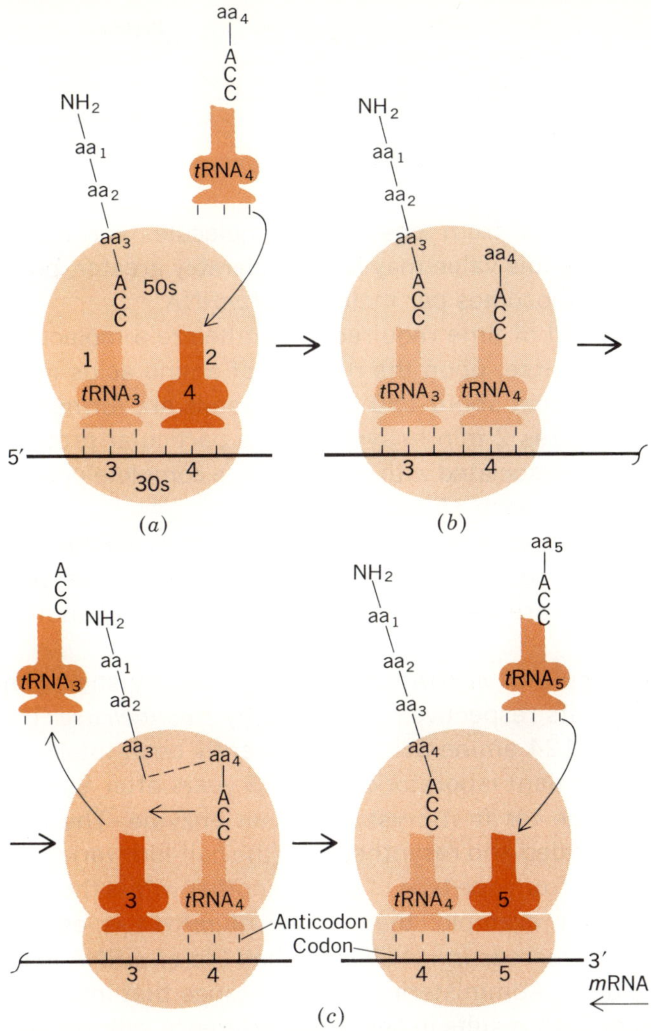

FIGURE 27.8 Binding sites in ribonucleic acids. (*a*) 1-Binding site for peptidyl-*t*RNA; 2-binding site for aminoacyl-*t*RNA (not occupied). Specific aminoacyl-*t*RNA bound is determined by base pairing of *t*RNA anticodon with *m*RNA codon. (*b*) Aminoacyl$_4$-*t*RNA$_4$ hydrogen bonds to codon 4 and is bound at the aminoacyl-*t*RNA binding site. (*c*) As the ribosome moves one codon along the *m*RNA (or the *m*RNA moves), three reactions occur: aa$_3$ is joined to aa$_4$ by a peptide bond, *t*RNA$_3$ is eliminated, and *t*RNA$_4$ with the peptide chain attached becomes associated with the peptidyl-*t*RNA binding site. These reactions are enzymatic, and the enzyme is associated with the ribosome, *t*RNA, and peptidyl unit. The peptidyl unit is not released in the free state from this association until the last amino acid has been added to the peptide chain. (*Redrawn after J. D. Watson, "Molecular Biology of the Gene,"* p. 336, *W. A. Benjamin, Inc., New York,* 1965.)

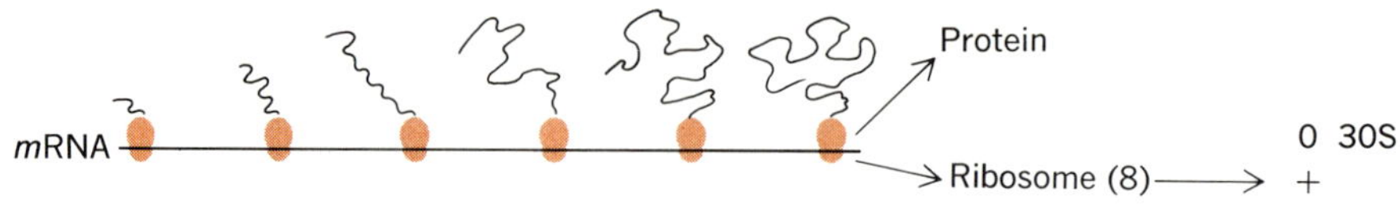

FIGURE 27.9 The "assembly line" for ribosome synthesis.

thesizing proteins from one *m*RNA molecule. For high-molecular-weight proteins this value may be considerably greater, possibly of the order of 100 ribosomes per molecule of *m*RNA.

The period of time required to synthesize a molecule of protein depends, of course, upon the number of amino acids in the protein. It is estimated that about 150 peptide bonds are formed per minute; thus, if it requires 2 min to synthesize a particular protein and eight ribosomes are associated with one *m*RNA, a complete protein molecule would be formed each 15 sec.

After a ribosome has reached the end of an *m*RNA, the ribosome is released and becomes dissociated into its respective subunits (30S and 50S), which are then available for the synthesis of another molecule of protein.

A molecule of *m*RNA exists for a relatively short time. RNA is hydrolyzed to its respective nucleotides by *ribonuclease*. This enzyme, composed of 124 amino acids, has recently been totally synthesized in two independent laboratories only 23 years after Sumner received the Nobel prize for first crystallizing an enzyme. The rate at which *m*RNA is degraded and resynthesized (the half-life) varies from species to species and from one type of cell to another within the same species. In bacteria, for example, the half-life of *m*RNA is estimated to be about 2 min, whereas in rat liver the half-life of *m*RNA is of the order of several days. Administering any substance that interferes with the formation of RNA results in serious interference with protein (enzyme) synthesis, especially in organisms whose *m*RNA has a short half-life. The antibiotic antimycin D is toxic because it inhibits *m*RNA formation by inhibiting RNA polymerase. Antimycin D would be expected to be much more toxic toward bacteria than toward mammals in general because of the relatively short half-life of bacterial *m*RNA.

27.8 CONTROL OF PROTEIN SYNTHESIS

By controlling the rate of *m*RNA synthesis an organism would have control over enzymatic reactions and therefore a control of metabolism in general. For example, if an enzyme is not needed in very large amounts, its rate of synthesis can be decreased, whereas if on occasion the enzyme is needed in high amounts, the rate of synthesis of the enzyme can be speeded up. Specifically, when the bacterium *Escherichia coli* is grown on a medium devoid of galactose, there is no appreciable galactosidase (enzyme that hydrolyzes galactose) in the organism. However, if galactose is a component of the medium, within a relatively short time the galactosidase content of the organism increases appreciably. In this instance the presence of galactose results in the initiation (induction) of the synthesis of the mRNA responsible for the ultimate synthesis of galactosidase. When galactose is withdrawn from the medium, *E. coli* stops the synthesis of the specific mRNA responsible for the formation of galactosidase. Thus, there exists within the cell a mechanism whereby it can start or stop the synthesis of certain mRNAs associated with the synthesis of specific enzymes. Thus, the organism has at its disposal a mechanism for controlling certain metabolic processes. The exact mechanism of initiation and control of mRNA synthesis is still under intensive investigation. Some certain definite facts lead to a working hypothesis, but the final chapter of this interesting and important topic has not been written. Lwoff, Jacob, and Monod developed the first theory on genetic repression and initiation of enzyme synthesis, for which they received the Nobel Prize in 1965.

The control of mRNA synthesis is associated directly with the DNA molecule, in particular with a specific gene that precedes the first structural gene. The DNA molecule must have an area that directs the polymerase to start transcribing from DNA, an area that is responsible for producing a binding site for attachment of the mRNA to ribosomes and also an area that governs the number of polymerase molecules simultaneously transcribing one mRNA, that is, the rate of mRNA formation. For example, the tryptophan operon of *E. coli* is composed of five structural genes responsible for the synthesis of five enzymes associated with tryptophan metabolism. It requires about 9 min for each molecule of polymerase to transcribe the complete operon, but transcription is initiated at intervals of 3 to 4 min. This amounts to approximately 3,000 nucleotides between each molecule of polymerase. Three polymerase molecules simultaneously transcribe one operon at a rate of about 1,200 nucleotides per minute. Thus, an mRNA molecule is formed about every 3 min.

A possible schematic representation of an area on DNA associated with mRNA formation and regulation is shown in Fig. 27.10. A *regulator* gene, which precedes by some distance the operon, dictates the trans-

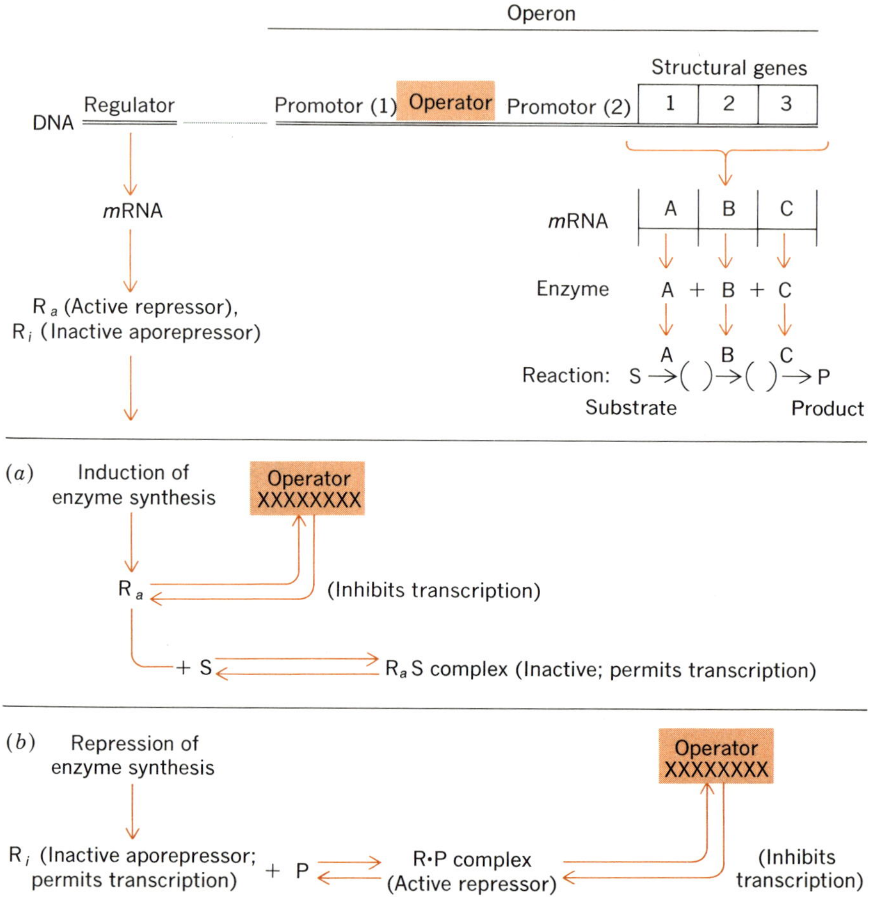

FIGURE 27.10 The control of enzyme synthesis. [*Modified from J. P. Changeux, Sci. Am.,* **36:**212 (1965).]

cription of a particular *m*RNA that is responsible for the synthesis of a protein called a *repressor* or *aporepressor* depending upon the type of reaction to be controlled. Promotors 1 and 2 are sites on DNA which are postulated to govern the binding site for RNA polymerase and ribosomal attachment to *m*RNA. The area of DNA indicated as *operator* governs the starting and stopping of transcription. There is some evidence for the existence of the promotor 1–operator–promotor 2 sequence; however, the relative positions of these areas are not known. A brief discussion of the postulated relationship of regulator to operator follows.

The activation of the operator so that transcription can occur or the inactivation of the operator so that transcription ceases is under the direct or indirect control of the regulator gene. This gene produces a *m*RNA that is responsible for the production of a protein called a *repressor,* which in the active form combines with the operator to prevent transcription. Also certain regulator genes are responsible for the synthesis of an *aporepressor,* an inactive repressor that does not inhibit transcription until activated, e.g., by an end product of a reaction.

Induction of an enzyme system (Fig. 27.10*a*) may be thought of as occurring in the following manner. The regulator gene produces an active repressor, which binds with the operator to prevent transcription. In the presence of a high enough level of substrate, a complex is formed between the repressor and substrate. This complex is inactive and thus permits the operator to initiate transcription. As the level of substrate decreases, the equilibrium is shifted toward more of the active repressor, which stops transcription.

Repression of enzyme synthesis is illustrated in Fig. 27.10*b*. It is assumed that the regulator forms an aporepressor which is inactive and does not complex with the operator to prevent transcription; therefore, enzyme synthesis can take place. However, as the reaction proceeds, the product of the reaction or derivative of the product combines with the aporepressor to change it into an active configuration which combines with the operator and prevents transcription. Thus, the synthesis of enzymes that produce a product is decreased as the product increases in concentration. The stimulation or inhibition of *m*RNA synthesis affords the cell a means of ultimate biological control of enzymatic reactions.

A notable example of the inhibition of a series of reactions via feedback inhibition is the conversion of L-threonine to L-isoleucine. Four enzymes are involved in this conversion. The end product isoleucine inhibits the first enzyme of this series (see Fig. 27.11), thus effecting an immediate control of the whole series of reactions. In addition, a derivative of isoleucine activates the aporepressor, thus inhibiting transcription of the operon for the four enzymes. This represents a slower control of isoleucine synthesis than the direct effect of product on an enzyme. Thus there are two fundamentally different feedback mechanisms for the control of isoleucine synthesis, one in which isoleucine inhibits the first enzyme of a series of enzyme reactions and another in which a derivative of isoleucine inhibits the formation of all of the enzymes associated with the conversion of threonine to isoleucine.

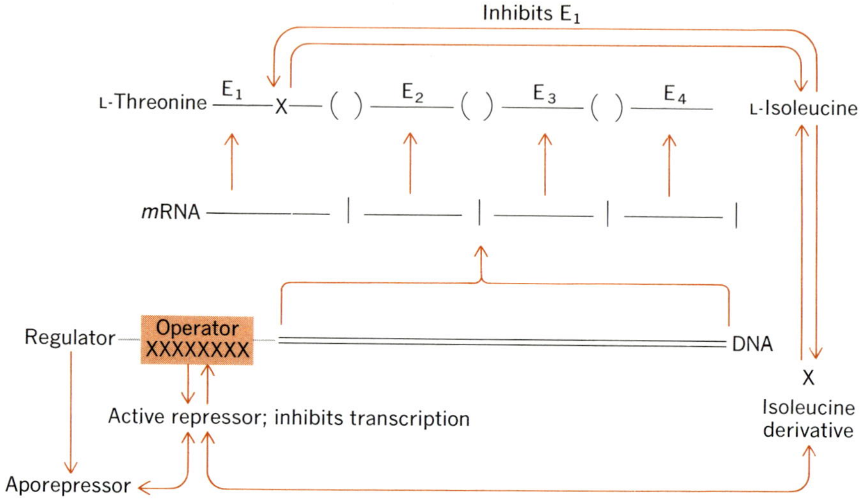

FIGURE 27.11 Schematic representation of the control of L-isoleucine synthesis from L-threonine.

QUESTIONS

1. What ring structure is associated with purines and with pyrimidines?
2. Indicate whether each of the following compounds is a purine or a pyrimidine:
   (*a*) Adenine (*b*) Cytosine (*c*) Uracil
   (*d*) Guanine (*e*) Thymine
3. (*a*) What is the significance of the term "complementary base"?
   (*b*) Explain why DNA has the base relationship A = T and G = C.
4. What is the meaning of the notations *m*RNA, *t*RNA, and *r*RNA?
5. Explain the function of *m*RNA and *t*RNA with respect to protein synthesis.
6. What part do ribosomes play in protein synthesis?
7. Explain briefly the following:
   (*a*) Enzyme induction
   (*b*) Feedback inhibition as affecting *m*RNA synthesis
   (*c*) Operon
   (*d*) Operator
   (*e*) Regulatory gene
   (*f*) Triplet (genetic code)
   (*g*) Codon
   (*h*) Anticodon
8. If a protein contains 200 amino acid residues, what is the minimum number of nucleotides in the *m*RNA to code the amino acid sequence of this protein?
9. A section of DNA has the following sequence of bases:

   5′. . . CCTTGTGGCACACCCGC . . .

   (*a*) What would be the order of bases in the *m*RNA transcribed from this section of DNA?

(*b*) What would be the order of amino acids in the peptide translated from the *m*RNA?

(*c*) What would be the simplest anticodon for each of these amino acids?

10. Schematically indicate the reactions associated with protein synthesis starting with DNA and free amino acids.

11. Assume that a *m*RNA is composed of repeating units of AUU, that is, AUUAUUAUU · · · . What would be the composition of the polypeptides formed in vitro in a system containing all the necessary components for peptide synthesis? (*Hint:* There is no control over the starting point for translation.)

12. List the base sequence in the segment of DNA involved in the following amino acid sequence: Leu—Phe—Gln—Val—Lys—Asp—Ala.

SUGGESTED READINGS FOR PART THREE

Baldwin, Ernest: "The Nature of Biochemistry," 2d ed., Cambridge University Press, London, 1967; "Dynamic Aspects of Biochemistry," 4th ed., Cambridge University Press, London, 1963; "An Introduction to Comparative Biochemistry," 4th ed., Cambridge University Press, London, 1964.

Bassham, J. A., and M. Calvin: "The Path of Carbon in Photosynthesis," Prentice-Hall, Inc., Englewood Cliffs, N.J., 1957.

Bennett, T. Peter, and Earl Frieden: "Modern Topics in Biochemistry," The Macmillan Company, New York, 1966.

———: "Graphic Biochemistry," vols. I and II, The Macmillan Company, New York, 1968.

Block, Konrad E.: "Lipid Metabolism," John Wiley & Sons, Inc., New York, 1960.

Brian, F., C. Clark, and Kjeld A. Marcker: How Proteins Start, *Sci. Am.,* January, 1968, p. 36.

Calvin, Melvin, and J. A. Bassham: "The Photosynthesis of Carbon Compounds," W. A. Benjamin, Inc., New York, 1962.

Cantarow, Abraham, and Bernard Schepartz: "Biochemistry," 3d ed., W. B. Saunders Company, Philadelphia, 1962.

Changeux, J. P.: The Control of Biochemical Reactions, *Sci. Am.,* April, 1965, p. 36.

Conn, Eric E., and P. K. Stumpf: "Outlines of Biochemistry," 2d ed., John Wiley & Sons, Inc., New York, 1966.

Crick, F. H. C.: The Genetic Code, *Sci. Am.,* October, 1962; *ibid.,* October, 1966.

DuPraw, Ernest J.: "Cell and Molecular Biology," Academic Press Inc., New York, 1968.

Fruton, Joseph, and S. Simmonds: "Biochemistry," 2d ed., John Wiley & Sons, Inc., New York, 1958.

Gibson, D. M.: The Biosynthesis of Fatty Acids, *J. Chem. Educ.,* **42:** 236 (1965).

Green, D. E.: The Mitochondrion, *Sci. Am.,* January, 1964, p. 63.

Harrow, Benjamin, and Abraham Mazur: "Textbook of Biochemistry," W. B. Saunders Company, Philadelphia, 1966.

Holley, R. W.: The Nucleotide Sequence of a Nucleic Acid, *Sci. Am.,* February, 1966, p. 30.

Kallenberger, Edward: The Genetic Control of the Shape of a Virus, *Sci. Am.,* December, 1966, p. 32.

Karlson, P.: "Introduction to Modern Biochemistry," 2d ed., Academic Press Inc., New York, 1965.

Lehninger, A. L.: "The Mitochondrion," W. A. Benjamin, Inc., New York, 1964; "Bioenergetics," W. A. Benjamin, Inc., New York, 1965.

Mahler, Henry R., and Eugene H. Cordes: "Biological Chemistry," Harper & Row, Publishers, Incorporated, New York, 1966.

Nirenberg, M. W.: The Genetic Code, II, *Sci. Am.,* March, 1963, p. 80.

Pauling, L., and R. Hayword: "The Architecture of Molecules," W. H. Freeman and Company, San Francisco, 1964.

Pigman, W. W.: "Carbohydrates," Academic Press Inc., New York, 1957.

Racker, E.: The Membrane of the Mitochondrion, *Sci. Am.,* February, 1968, p. 32.

Sinsheimer, Robert L.: Single-stranded DNA, *Sci. Am.,* July, 1962, p. 109.

Watson, James D.: "The Double Helix," Atheneum Publishers, New York, 1968; "Molecular Biology of the Gene," W. A. Benjamin, Inc., New York, 1965.

White, A., P. Handler, and E. Smith: "Principles of Biochemistry," 4th ed., McGraw-Hill Book Company, New York, 1968.

White, Emil H.: "The Chemical Background for the Biological Sciences," Prentice-Hall, Inc., Englewood Cliffs, N.J., 1964.

Yanoksky, Charles: Gene Structure and Protein Structure, *Sci. Am.,* May, 1967, p. 80.

Zubay, G.: Molecular Model for Protein Synthesis, *Science,* **140:** 1092 (1963).

appendix a... mathematical expressions

EXPONENTS

a^n: a = base; n = exponent; a is multiplied by itself n times; $a^{1/n} = \sqrt[n]{a}$; $a^{-n} = 1/n$; $a^0 = 1$

$$10^3 = 10 \times 10 \times 10 = 1{,}000$$
$$4^2 = 4 \times 4 = 16$$
$$4^{1/2} = \sqrt[2]{4} = \sqrt{4} = 2$$
$$16^{1/4} = \sqrt[4]{16} = 2$$
$$\sqrt{10^{-6}} = 10^{-6 \times 1/2} = 10^{-3}$$
$$10^{-3} = \frac{1}{10^3}$$

MULTIPLICATION

$$2^2 \times 4^2 = 4 \times 16 = 64 \qquad \text{or} \qquad (2 \times 4)^2 = 8^2 = 64$$
$$10^3 \times 10^4 = 10^7$$
$$10^{-3} \times 10^4 = 10^{-3+4} = 10^1 = 10$$
$$10^{-4} \times 10^{-5} = 10^{-9}$$

LOGARITHMS

The logarithm of a number is the power to which the base of the logarithm has to be raised to equal the number: logarithm to base 10 of $10^3 = 3$. When no log base is indicated, it means to base 10. $\log 10^x = x$; $\log a = x$; therefore $a = 10^x$.

$$\log 1 = 0 \qquad 10^0 = 1$$
$$\log 10{,}000{,}000 = \log 10^7 = 7 \qquad 10^7 = 10{,}000{,}000$$
$$\log 0.00001 = \log 10^{-5} = -5 \qquad 10^{-5} = 0.00001$$

MULTIPLICATION. Logs are added

$$\log (a \times b \times c) = \log a + \log b + \log c$$

DIVISION. Logs are subtracted

$$\log \frac{a}{b} = \log a - \log b$$

LOGS OF VALUES FROM 1 TO 10. Any number can be expressed in exponential form of a value between 1 and 10.

$$272 = 2.72 \times 10^2 \qquad \log 272 = \log 2.72 + \log 10^2$$

$$0.00343 = 3.43 \times 10^{-3} \qquad \log 0.00343 = \log 3.43 + \log 10^{-3}$$

The log values of numbers between 1 and 10 can be found in logarithmic tables. The log values of whole numbers are

1 = 0.000	6 = 0.778
2 = 0.301	7 = 0.845
3 = 0.477	8 = 0.903
4 = 0.602	9 = 0.954
5 = 0.699	10 = 1.000

To find the log of a number

$$\begin{aligned} \log 5{,}000 &= \log 5 \times 10^3 \\ &= \log 5 + \log 10^3 \\ &= 0.699 + 3 \\ &= 3.699 \\ 5{,}000 &= 10^{3.699} \end{aligned}$$

$$\begin{aligned} \log 0.0004 &= \log 4 \times 10^{-4} \\ &= \log 4 + \log 10^{-4} \\ &= 0.602 - 4 \\ &= -3.398 \\ 0.0004 &= 10^{-3.398} \qquad \text{or} \qquad 10^{0.602} \times 10^{-4} \end{aligned}$$

To find a number if the logarithm is given: $\log n = x$, antilog $x = n$, $n = 10^x$.

$$\begin{aligned} \log n &= 3.699 \\ n &= 10^{3.699} = 10^{0.699+3} = 10^{0.699} \times 10^3 \\ &= 5 \times 10^3 \end{aligned}$$

$$\begin{aligned} \log n &= -5.699 \\ n &= 10^{-5.699} = 10^{0.301-6} = 10^{0.301} \times 10^{-6} \\ &= 2 \times 10^{-6} \end{aligned}$$

appendix b... units of mass, length, and temperature

MASS

STANDARD. Kilogram (kg): a mass of platinum-iridium kept at the International Bureau of Weights and Measures.

$$
\begin{aligned}
1 \text{ kg} &= 2.205 \text{ lb (avoirdupois)}\\
1 \text{ lb (avoirdupois)} &= 453.6 \text{ g}\\
1 \text{ gram (g)} &= 10^{-3} \text{ kg}\\
1 \text{ centigram (cg)} &= 10^{-2} \text{ g}\\
1 \text{ milligram (mg)} &= 10^{-3} \text{ g}\\
1 \text{ microgram } (\mu g) &= 10^{-6} \text{ g}\\
1 \text{ nanogram (ng)} &= 10^{-9} \text{ g}\\
1 \text{ picogram (pg)} = 1 \text{ micromicrogram } (\mu\mu g) &= 10^{-12} \text{ g}
\end{aligned}
$$

LENGTH

STANDARD. Meter (m): 1,650,763.73 wavelengths of reddish-orange wavelength of krypton 86 (adopted 1960).

$$
\begin{aligned}
1 \text{ kilometer (km)} &= 10^{3} \text{ m}\\
1 \text{ decimeter (dm)} &= 10^{-1} \text{ m}\\
1 \text{ centimeter (cm)} &= 10^{-2} \text{ m}\\
1 \text{ millimeter (mm)} &= 10^{-3} \text{ m}\\
1 \text{ micron } (\mu) &= 10^{-6} \text{ m} = 10^{-4} \text{ cm}\\
1 \text{ millimicron } (m\mu) &= 10^{-9} \text{ m} = 10^{-7} \text{ cm}\\
1 \text{ angstrom unit } (\text{Å}) &= 10^{-8} \text{ cm}
\end{aligned}
$$

VOLUME

STANDARD. Liter: volume of 1 kg of pure water at 4°C (temperature at which water has maximum density).

$$
\begin{aligned}
1 \text{ liter} &= 1.000027 \text{ dm}^3\\
&= 1{,}000.027 \text{ cm}^3\\
&= 1.057 \text{ U.S. quarts (liquid)}\\
1 \text{ milliliter (ml)} &= 10^{-3} \text{ liter} \simeq 1 \text{ cm}^3\\
1 \text{ microliter } (\mu l) &= 10^{-6} \text{ liter} \simeq 10^{-3} \text{ ml}
\end{aligned}
$$

MOLAR GAS VOLUME. 22.4 liters, volume occupied by 1 mole of any gas at 0°C and 1 atm pressure.

TEMPERATURE AND HEAT

ABSOLUTE ZERO. Zero degrees Kelvin (°K) = −273.15°C (Celsius).

CELSIUS AND FAHRENHEIT. Standard, 0°C = melting point of pure ice $\simeq$ 32°F; 100°C = boiling point of pure water at 1 atm pressure $\simeq$ 212°F.

$$
\begin{aligned}
0^\circ C &= 32^\circ F \\
100^\circ C &= 212^\circ F \\
^\circ C &= \tfrac{5}{9}(^\circ F - 32) \\
^\circ F &= 32 + \tfrac{9}{5}{}^\circ C
\end{aligned}
$$

CALORIE (CAL). Small calorie or gram calorie, the quantity of heat needed to raise the temperature of one gram of water from 3.5 to 4.5°C.

$$
\begin{aligned}
1 \text{ calorie (cal)} &= 4.184 \times 10^7 \text{ ergs} = 4.184 \text{ joules} \\
1 \text{ kilocalorie (kcal or Cal)} &= 10^3 \text{ cal}
\end{aligned}
$$

appendix c... conversion of $^{238}_{92}U$ to $^{206}_{82}Pb$

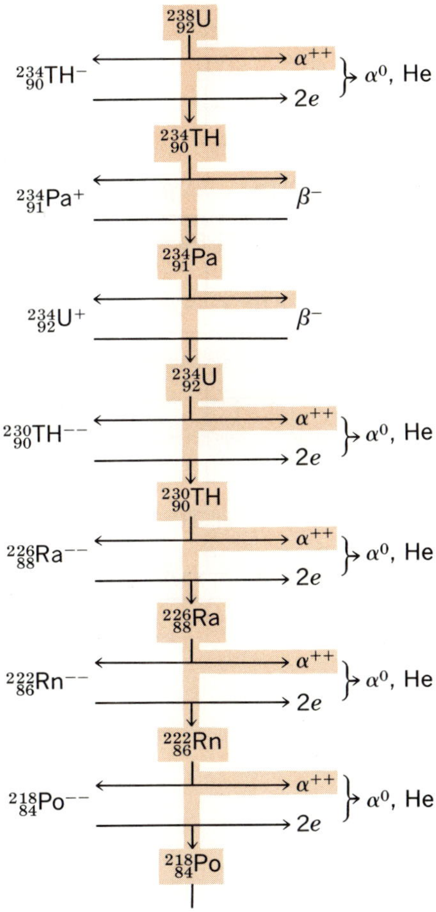

(Continued)

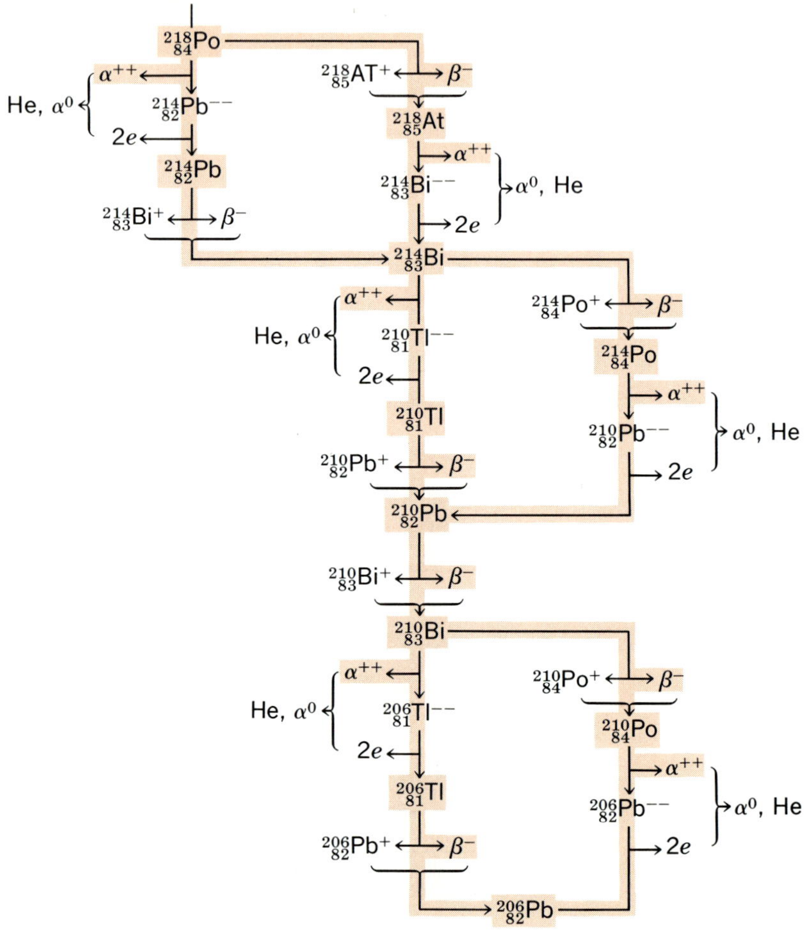

$^{218}_{84}Po$
α^{++}
He, α^0
$^{214}_{82}Pb^{--}$
2e
$^{214}_{82}Pb$
$^{214}_{83}Bi^{+}$
β^-
$^{218}_{85}AT^{+}$
β^-
$^{218}_{85}At$
α^{++}
$^{214}_{83}Bi^{--}$
α^0, He
2e
$^{214}_{83}Bi$
α^{++}
He, α^0
$^{210}_{81}Tl^{--}$
2e
$^{210}_{81}Tl$
$^{210}_{82}Pb^{+}$
β^-
$^{214}_{84}Po^{+}$
β^-
$^{214}_{84}Po$
α^{++}
$^{210}_{82}Pb^{--}$
α^0, He
2e
$^{210}_{82}Pb$
$^{210}_{83}Bi^{+}$
β^-
$^{210}_{83}Bi$
α^{++}
He, α^0
$^{206}_{81}Tl^{--}$
2e
$^{206}_{81}Tl$
$^{206}_{82}Pb^{+}$
β^-
$^{210}_{84}Po^{+}$
β^-
$^{210}_{84}Po$
α^{++}
$^{206}_{82}Pb^{--}$
α^0, He
2e
$^{206}_{82}Pb$

appendix d... the kinetic molecular theory of gases

The kinetic molecular theory of gases explains the behavior of gases when subjected to a change in pressure or temperature or both.

Molecules of gases are in perpetual motion and move in a straight line. For example, the velocity of the hydrogen molecule is normally about 1.1 miles/sec; however, it travels in a straight line only about 1/6,000 mm before it hits another molecule of gas. In reality the molecules are quite far apart; if the hydrogen molecule had a diameter of 1 mile, its nearest neighbor would be some 900 miles away. There is a lot of empty space in a container of gas at 1 atm pressure.

All molecules of a gas are perfectly elastic; i.e. at constant temperature when molecules collide with each other or with the side of the container, there is *no loss of total* (*or average*) *kinetic energy*. All molecules, however, do not have the same kinetic energy. For instance, if one molecule is hit simultaneously by two other molecules, its resultant kinetic energy may be greater or less than originally, depending upon whether the two hit it on the same side or on opposite sides. It is possible for a molecule of gas to be momentarily stationary. However, the average or total kinetic energy of all the molecules is constant at any given temperature.

The gas laws express the interrelationships of temperature, pressure, and volume of a gas.

Boyle's law states that the *volume of a given mass of gas varies inversely* as the pressure at constant temperature, $V \propto 1/P$. The mass of a gas may be expressed in terms of number of moles n. Boyle's law was devised about the middle of the seventeenth century by Robert Boyle, who became interested in studying the atmosphere, which appeared to him to be an elastic fluid. His experiments showed that if a volume of gas is compressed and then the pressure removed, the gas tends to return to its original volume; hence, the term elastic fluid.

The relation of pressure to volume of gas at constant temperature can be visualized as follows. Imagine a cylinder of 1,000 cm^3 capacity filled with a gas at 1 atm pressure. The cylinder is fitted with a weightless, frictionless piston and kept at constant temperature. If the equivalent of 2 atm pressure is placed on the piston, the volume is decreased to 1,000/2, or 500 cm^3; if a pressure of 4 atm is placed on the piston, the volume of the gas is 1,000/4, or 250 cm^3. If the weights on the piston are now removed, the volume of gas will increase until it reaches the original volume of 1,000 cm^3. As the pressure on the gas was increased, the volume became smaller, and, conversely, as the pressure was decreased, the volume became greater. Ideally the product of the pressure and volume is equal to a constant; *this is true at low pressures.* For example

Pressure	×	volume	=	product
P	×	V	=	PV
1 atm	×	1,000 cm^3	=	1,000
2 atm	×	500 cm^3	=	1,000
4 atm	×	250 cm^3	=	1,000

This may be expressed mathematically as

$$PV = k$$

Boyle's law is simply explained from the kinetic-molecular-theory standpoint by assuming that the cylinder contains only *one* molecule of a gas which is moving parallel to the sides of the container so that it strikes only the bottom of the container and the piston (see Fig. D.1). At constant temperature the molecule has a definite constant velocity and bounces off the piston a certain number of times per second, thus

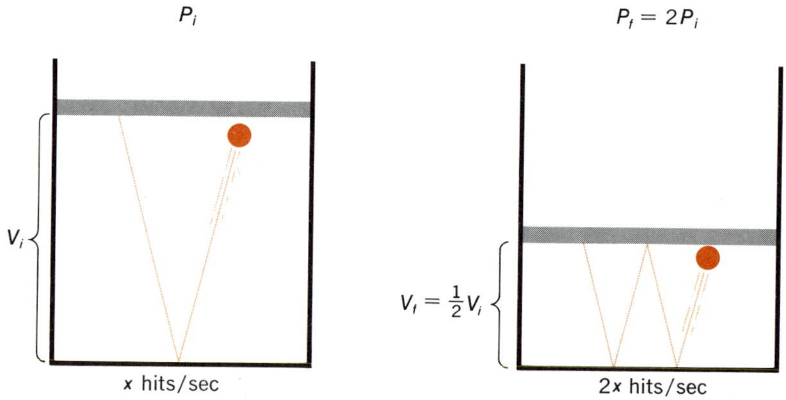

FIGURE D.1 Schematic illustration of volume versus pressure at constant temperature.

exerting a force upon the piston. If the piston is moved to decrease the volume by one-half, the distance the molecule must travel from the piston to the bottom is one-half of what it was previously. Hence, the molecule will hit the piston twice as frequently per unit time. Since the velocity and mass of the molecule have not been changed, the force (pressure) upon the piston must be twice as great as it was initially. Thus, the volume varies inversely with respect to pressure, or

$$V \propto \frac{1}{P} \qquad V = \frac{k}{P} \qquad \text{or} \qquad PV = k$$

In reality many molecules are present in the container; they collide with each other and with all sides of the container at a uniform average rate. Consequently, the pressure exerted by the gas is the same on all sides of the container.

It is beyond the scope of this text to discuss deviations from Boyle's law, but it is important to realize that the molecules of a gas may be crowded so closely together that the molecular attraction becomes of importance. This would cause a deviation from the gas laws. In fact, under certain conditions this attraction is sufficiently great to cause the gas to liquefy. To do this it is necessary to decrease the velocity of the molecules and to crowd them close together; this is accomplished by cooling the gas to a very low temperature under high pressure. The temperature of liquid air is approximately −190°C; liquid helium boils at −268.9°C.

Charles' law expresses the volume-temperature relationship of gases at constant pressure and may be stated as follows. *At constant*

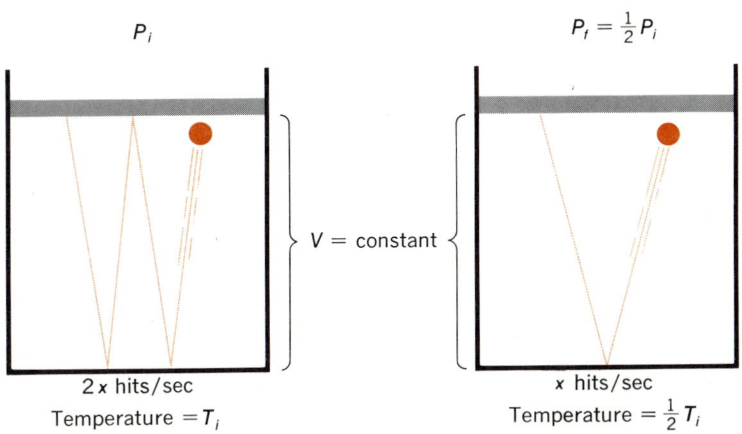

FIGURE D.2 Schematic illustration of pressure versus temperature at constant volume.

pressure and mass (n) the volume of a gas varies directly with the absolute temperature, $V \propto T$, or $V = kT$. This law is true because the kinetic energy of a gas is directly proportional to the temperature; therefore, to maintain a constant pressure while the kinetic energy of the molecules is being decreased, the volume must also be decreased. Resorting again to the cylinder containing only one molecule of gas to simplify this statement (see Fig. D.2), as the velocity of the molecule is decreased, a longer time is required for the molecule to travel from the piston to the bottom of the container and back. Thus, the rate of bombardment of the piston is less; this results in a decrease in pressure within the cylinder. Therefore, to maintain a constant pressure, the piston must be lowered.

It has been observed that for each degree Celsius change in temperature, the volume of a gas changes $\frac{1}{273}$ of its volume (see Fig. D.3). Thus, if 273 cm^3 of a gas is at 0°C and the temperature increased to 1°C, the gas expands sufficiently to occupy a volume of 274 cm^3 if the pressure is kept constant. If the temperature of the gas is increased

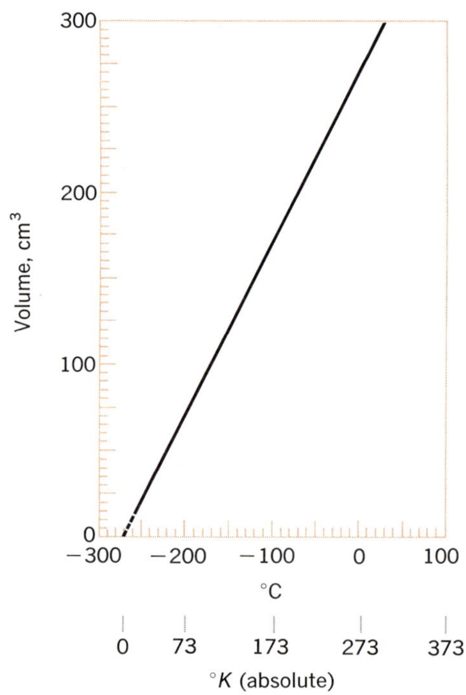

FIGURE D.3 Relationship of temperature to volume of a gas.

from 0 to 10°C, the new volume is 283 cm^3. Likewise, if the original gas is cooled to -10°C, the volume decreases to 263 cm^3.

If the temperature-volume curve (Fig. D.3) is extrapolated to -273°C, or 0°K, the volume occupied by the gas becomes zero; the molecules would be nonexistent! This would be a flagrant violation of the law of conservation of mass, providing Charles' law is valid at this low temperature. It is quite obvious that Charles' law is *not* valid under all conditions; it fails seriously at high pressures and low temperatures as gases approach their liquefaction temperature.

The temperature of -273°C, or 0°K, is called *absolute zero* because it represents the lowest temperature theoretically possible, i.e., when molecules possess no kinetic energy. The lowest temperature experimentally obtained is within a fraction of a degree of absolute zero.

COMBINATION OF BOYLE'S AND CHARLES' LAWS. Boyle's and Charles' laws may be combined to give one equation:

$$V \propto \frac{nT}{P}$$

$$V = \frac{nkT}{P}$$

$$PV = nkT$$

The *molar gas constant* R is substituted for k when $n = 1$; thus

$$PV = nRT$$

STANDARD CONDITIONS. Basic reference points for temperature and pressure must be established. These have been chosen as 273°K (0°C) and 1 atm (760 mm Hg) pressure, called standard temperature and pressure (STP).

MOLECULAR VOLUME OF GASES AT STP. It has been experimentally determined that 1 g mol. wt. (Avogadro's number of molecules) of any gas at STP occupies 22.4 liters. This value is called the *molecular volume of gases.*

CALCULATION OF MOLAR GAS CONSTANT R.

$$PV = nRT$$

$$R = n\frac{VP}{T}$$

When $n = 1$,

$$R = \frac{VP}{T}$$

The units of R are volume-pressure per degree, and its numerical value is determined by the units of standard volume V and pressure P used. The numerical value of R can be calculated as follows.

EXAMPLE 1

$$R = 0.082 \text{ liter-atm/deg}$$

$$V = 22.4 \text{ liters}$$
$$P = 1 \text{ atm}$$
$$T = 273°\text{K}$$
$$n = 1$$

Substitution in

$$R = n\frac{VP}{T}$$

gives

$$R = \frac{22.4}{273}$$
$$= 0.08206 \text{ liter-atm/deg}$$

EXAMPLE 2

$$R = 82.0 \text{ cm}^3\text{-atm/deg}$$
$$22.4 \text{ liters} = 22{,}400 \text{ cm}^3$$
$$R = \frac{22{,}400 \times 1}{273}$$
$$= 82.0 \text{ cm}^3\text{-atm/deg}$$

EXAMPLE 3

$$R = 62{,}370 \text{ cm}^3\text{-mm/deg}$$
$$1 \text{ atm} = 760 \text{ mm Hg}$$
$$P = \frac{22{,}400 \times 760}{273}$$
$$= 62{,}370 \text{ cm}^3\text{-mm/deg}$$

Note: It is more important to know how to calculate R in various units than to rely upon memorizing the isolated values.

PROBLEM 1

What volume in liters would 0.1 mole of H_2 occupy at 50°C and 1.5 atm pressure?

Given	*Known*
$P = 1.5$ atm	$R = \frac{22.4}{273} = 0.082$ liter-atm/deg
$T = 273 + 50 = 323$°K	$PV = nRT$
$n = 0.1$	

$$\text{Unknown} = V \text{ (liters)}$$

$$V = \frac{nRT}{P} = \frac{0.1 \times 0.082 \times 323}{1.5} = 2.65 \text{ liters}$$

PROBLEM 2

How many moles of a gas would be contained in a volume of 250 cm^3 at 25°C and 10 atm pressure?

Given	*Known*
$V = 250$ cm^3	$R = \frac{22{,}400}{273} = 82.0$ cm^3-atm/deg
$T = 273 + 25 = 298$°K	$PV = nRT$
$P = 10$ atm	

$$\text{Unknown} = n \text{ (moles)}$$

$$n = \frac{PV}{RT} = \frac{10 \times 250}{82 \times 298} = 0.102 \text{ mole}$$

PROBLEM 3

To what temperature, in degrees Celsius, must 1×10^{-2} mole of gas contained in 750 cm^3 be heated to have the pressure of the gas equal to 870 mm Hg?

Given | *Known*

$V = 750$ cm^3 $\qquad R = \dfrac{22{,}400 \times 760}{273} = 62{,}370$ cm^3-mm/deg

$P = 870$ mm $\qquad PV = nRT$

$n = 1 \times 10^{-2}$ mole

$$\text{Unknown} = T°\text{K}$$

$$T = \frac{PV}{nR}$$

$$= \frac{870 \times 750}{1 \times 10^{-2} \times 62{,}370}$$

$$= 1047°\text{K}$$

$$= 1047 - 273 = 774°\text{C}$$

These problems can be calculated with other R values provided the appropriate units are used for P and V. For example, Prob. 3 can be calculated as follows:

$$V = 750 \text{ cm}^3 = 0.750 \text{ liter}$$

$$P = 870 \text{ mm} = \tfrac{870}{760} \text{ atm} = 1.14 \text{ atm}$$

$$R = 0.082 \text{ liter-atm/deg}$$

$$T = \frac{1.14 \times 0.75}{1 \times 10^{-2} \times 0.082} = 1043 \times 10^{3}°\text{K}$$

$$= 1043 - 273 = 770°\text{C}$$

The difference between the two answers is due to the rounding off of more numbers.

PROBLEM 4

How many *molecules* of gas are contained in a volume of 850 cm^3 if the pressure is 5 atm and the temperature is 0°C?

Given | *Known*

$V = 850$ cm^3 $\qquad R = \dfrac{22{,}400}{273} = 82.0$ cm^3-atm/deg

$P = 5$ atm $\qquad PV = nRT$

$T = 0°\text{C} = 273°\text{K}$

Unknown $= n$ (moles to be converted into number of molecules)

$$n = \frac{5 \times 850}{82 \times 273}$$

$$= 0.1544 \text{ mole}$$

$$1 \text{ mole} = 6.02 \times 10^{23} \text{ molecules}$$

$$0.1544 \times 6.02 \times 10^{23} = 9.29 \times 10^{22} \text{ molecules}$$

VARIATION OF V, T, AND P AT CONSTANT n. A constant amount n of a gas may be subjected to changes of V, T, and P. The notation V_i, T_i, and P_i is used to indicate the initial conditions and V_f, T_f, and P_f to denote the final conditions of the gas. These may be substituted in the general gas equation as follows:

$$P_iV_i = nRT_i \qquad nR = \frac{P_iV_i}{T_i}$$

$$P_fV_f = nRT_f \qquad nR = \frac{P_fV_f}{T_f}$$

$$\therefore \frac{P_iV_i}{T_i} = \frac{P_fV_f}{T_f}$$

This equation can be solved for any one condition, as illustrated in the following problems.

PROBLEM 5

A gas at 37°C and 800 mm pressure occupies 25 cm^3. What would its pressure be in mm Hg at 105°C and 10 cm^3?

$$\text{Unknown} = P_f$$

$$P_f = P_i\frac{V_iT_f}{V_fT_i}$$

$$= 800 \times \frac{25}{10} \times \frac{273 + 105}{273 + 37}$$

$$= 2{,}438.7 \text{ mm Hg pressure}$$

PROBLEM 6

A balloon contains 1,200 ft^3 of gas at 0.95 atm pressure and 25°C. What is the volume of the gas at 0.5 atm pressure and 12°C?

$$\text{Unknown} = V_f$$

$$V_f = V_i\frac{P_iT_f}{P_fT_i}$$

$$= 1{,}200 \times \frac{0.95}{0.5} \times \frac{273 + 12}{273 + 25}$$

$$= 2{,}180.52 \text{ ft}^3$$

DALTON'S LAW OF PARTIAL PRESSURE. The air we breathe is not just one kind of molecule but a mixture of mostly nitrogen and oxygen, about 79 and 21% by volume, respectively, and varying amounts of carbon dioxide (average about 0.04% by volume) and water vapor. Each kind of gas contributes its share to the total atmospheric pressure.

From the kinetic point of view, if the molecules of a mixture of gases are at the same temperature, the average kinetic energy of the kinds of molecules is the same, each exerting a part of the total pressure proportional to the number of molecules present. Each gas exerts its own pressure independent of other gases, i.e., as though it were the only gas present. If we consider atmospheric pressure to be due only to nitrogen and oxygen, the total atmospheric pressure P equals

$$P = p_{O_2} + p_{N_2}$$

where p represents the partial pressure of the respective gas.

EXAMPLE 4

The atmosphere is about 79% by volume nitrogen and 21% by volume oxygen. If the atmospheric pressure is 790 mm Hg, what is the partial pressure of nitrogen and oxygen? At the same T and P, equal volumes of gases contain the same number of molecules; therefore, the ratio of the percentages by volume is representative of the ratio of the number of molecules. Hence, the ratio of the number of nitrogen molecules and oxygen molecules is $\frac{79}{21}$, or $\frac{3.76}{1}$. Approximately four-fifths of the total pressure is due to nitrogen and about one-fifth is due to oxygen.

$$760 = p_{O_2} + p_{N_2}$$
$$p_{O_2} = 760 \times 0.21 = 159.6 \text{ mm}$$
$$p_{N_2} = 760 \times 0.79 = 600.4 \text{ mm}$$

The partial pressure of gases plays an important role with respect to living organisms. For example, we depend partly upon the difference between the p_{O_2} in the air and the p_{O_2} in the cells to oxygenate hemoglobin of the blood in the lungs and to release this oxygen to the cells. But before hemoglobin can react with oxygen, the oxygen must be absorbed through the lung tissue and dissolved in the blood. The average person breathes in about 15,000 liters/day of air. This amounts to some 3,000 liters of oxygen, of which about 260 liters dissolves in the blood and is transported to the tissues by hemoglobin. This raises the question about the influence of the pressure of a gas on its solubility.

SOLUBILITY OF GASES IN LIQUIDS. The solubility of a gas in a liquid increases with increasing pressure of the gas and as the temperature is lowered. In a mixture of gases, each gas behaves as though it were the only one present. Usually, the presence of one gas does not influence the solubility of another gas in a liquid if no chemical reaction is involved. Pure oxygen at $p_{O_2} = 159.6$ mm will dissolve in water to the same extent as the oxygen from air at a pressure of 760 mm (p_{O_2} in air at 760 mm pressure is 159.6 mm).

The oxygen that dissolves in blood plasma unites with hemoglobin (HHb) to form oxygenated hemoglobin ($HHb \cdot O_2$):

$$HHb + O_2 \rightleftharpoons HHb \cdot O_2$$

This is a reversible reaction; therefore, the amount of $HHb \cdot O_2$ formed is dependent upon the p_{O_2}. Several additional factors affect the oxygenation of hemoglobin. One is the presence of carbon dioxide. Hemoglobin forms a chemical complex with carbon dioxide which causes a shift in the equilibrium toward the left; therefore, in the presence of carbon dioxide the saturation of HHb by O_2 would be a lower value than in the absence of carbon dioxide. In arterial blood p_{O_2} is 100 mm, and p_{CO_2} is 40 mm (approximately), while in venous blood these values are about 35 and 46 mm, respectively. HHb in the lungs is about 90% oxygenated and remains highly oxygenated until p_{O_2} becomes less than about 35 mm. At cellular level, carbon dioxide is eliminated as a metabolic end product, and consequently p_{CO_2} of venous blood is greater (by about 6 mm) than in the arterial blood. This results in a release of O_2 from $HHb \cdot O_2$. Thus the transportation of carbon dioxide and oxygen by the blood is governed by gas-liquid solubility and also by the formation of definite chemical compounds with hemoglobin. All things considered, normally 100 ml of blood releases about 6.4 ml of oxygen to the cellular tissues.

Although nitrogen is biologically inert, the solubility of atmospheric nitrogen in the blood is of vital concern to divers and astronauts. Oxygen is more soluble in water than nitrogen, but p_{N_2} in air is much greater than p_{O_2}; hence, appreciable amounts of nitrogen dissolve in the blood. Great care must be taken to maintain an equilibrium between dissolved blood nitrogen and atmospheric nitrogen; otherwise bubbles of nitrogen gas form in the body fluids. This causes *decompression sickness* when divers who have been working under high pressures and breathing compressed air are brought to the surface too rapidly. This may be likened to removing the stopper from a bottle of pop; the sudden release of pressure causes bubbles of dissolved gas to form in the solution. Astronauts breathe pure oxygen before ascent to decrease the level of blood nitrogen; the pressure in a space cabin is reported to be about 5.1 psi, or about $\frac{1}{3}$ atm.

Helium and oxygen are used in place of air to shorten the decompression time for divers. Helium not only is less soluble in blood than nitrogen but also diffuses through body tissues at a faster rate. Thus, it takes less time for blood helium to reach equilibrium with the inhaled gas. After breathing a mixture of helium and oxygen under 3 atm pressure, the decompression time is about one twenty-third that required when air is breathed under the same conditions.

appendix e... R_f values of some inorganic ions

R_f VALUES OF SOME INORGANIC IONS[a]

Ion	Solvent[b]							
	1	2	3	4	5	6	7	8
Ag	0.10	0.78	0.08	0	0			
Hg^+	0.24	0	0.43					
Pb	0.03	0	0.15		0.27			
Hg^{++}	0.31	0	0.42	0.84	0.81			
Bi	0.02	0	0.63	0.51	0.59			
Cu	0.22	0.76	0.24	0.40	0.20			
Cd	0.05	0.76	0.18	0.83	0.77			
As	0.43	0.65	0.18					
Sn^{++}	0.58	0	0.77					
Sn^{4+}	0.55	0	0.58					
Al	0.03	0	0.03	0.03				
Fe^{3+}	0.95	0	0.10	0.93				
Zn	0.05	0.75	0.08	0.78				
F						0		0
Cl						0.23	0.24	0.10
Br						0.47	0.36	0.16
I						0.71	0.47	0.30
NO_2							0.25	0.20
NO_3							0.4	0.24
CO_3						0.06		
PO_4						0.04		
SO_4						0.07		

[a] Values from R. J. Block, E. L. Durrum, and G. Zweig, "A Manual of Paper Chromatography and Paper Electrophoresis," pp. 306–307, Academic Press Inc., New York, 1955.

[b] Solvents: 1 = 1-butanol-benzoylacetone; 2 = collidine-water; 3 = dioxane-antipyrine; 4 = metals first precipitated with 8-hydroxyquinoline and then developed with butanol–20% HCl; 5 = 1-butanol–3 N HCl; 6 = pyridide–10% H_2O; 7 = 1-butanol–pyridine–1.5 N NH_3 (2:1:2 volume per volume); 8 = 1-butanol–1.5 N NH_4OH.

index

Element	Symbol	Atomic Number	Atomic Weight
Actinium	Ac	89	227
Aluminum	Al	13	26.9815
Americium	Am	95	243
Antimony	Sb	51	121.75
Argon	Ar	18	39.948
Arsenic	As	33	74.9216
Astatine	At	85	210
Barium	Ba	56	137.34
Berkelium	Bk	97	247
Beryllium	Be	4	9.0122
Bismuth	Bi	83	208.980
Boron	B	5	10.811
Bromine	Br	35	79.904
Cadmium	Cd	48	112.40
Calcium	Ca	20	40.08
Californium	Cf	98	251
Carbon	C	6	12.01115
Cerium	Ce	58	140.12
Cesium	Cs	55	132.905
Chlorine	Cl	17	35.453
Chromium	Cr	24	51.996
Cobalt	Co	27	58.9332
Copper	Cu	29	63.546
Curium	Cm	96	247
Dysprosium	Dy	66	162.50
Einsteinium	Es	99	254
Erbium	Er	68	167.26
Europium	Eu	63	151.96
Fermium	Fm	100	253
Fluorine	F	9	18.9984
Francium	Fr	87	223
Gadolinium	Gd	64	157.25
Gallium	Ga	31	69.72
Germanium	Ge	32	72.59
Gold	Au	79	196.967
Hafnium	Hf	72	178.49
Helium	He	2	4.0026
Holmium	Ho	67	164.930
Hydrogen	H	1	1.00797
Indium	In	49	114.82
Iodine	I	53	126.9044
Iridium	Ir	77	192.2
Iron	Fe	26	55.847
Krypton	Kr	36	83.80
Kurchatovium	Ku	104	260
Lanthanum	La	57	138.91
Lawrencium	Lw	103	257
Lead	Pb	82	207.19
Lithium	Li	3	6.939
Lutetium	Lu	71	174.97
Magnesium	Mg	12	24.312
Manganese	Mn	25	54.9380